Quantitative Methods

Quantitative Methods
An Introduction for Business Management

Paolo Brandimarte
Politecnico di Torino
Torino, Italy

A JOHN WILEY & SONS, INC., PUBLICATION

Library of Congress Cataloging-in-Publication Data:

Brandimarte, Paolo.
  Quantitative methods : an introduction for business management / Paolo Brandimarte.
    p. cm.
  Includes bibliographical references and index.
  ISBN 978-0-470-49634-3 (hardback)
1. Management—Mathematical models. I. Title.
  HD30.25.B728 2011
  658.0072—dc22   2010045222

Printed in the United States of America.

10 9 8 7 6 5 4 3 2 1

Contents

Preface

And there I was, waiting for the big door to open, the big door that stood between me and my archnemesis. I found little comfort and protection, if any, sitting in what seemed my thin tin tank, looking around and searching for people in my same dire straits. Then, with a deep rumble, the big steel door of the ship opened, engines were started, and I followed the slow stream of cars. I drove by rather uninterested police officers, and there it was, my archnemesis: the first roundabout in Dover.

For European continental drivers like me, used to drive on the right side of the street (and yes, I do mean *right*), the first driving experience in the Land of Albion has always been a challenge. That difficulty compounded with the lack of roundabouts in Italy at the time, turning the whole thing into sheer nightmare. Yet, after a surprisingly short timespan, maybe thanks to the understanding and discipline of the indigenous drivers, I got so used to driving there, and to roundabouts as well, that after my return to Calais I found driving back in supposedly familiar lanes somewhat confusing.

I had overcome my fear, but I am digressing, am I? Well, this book should indeed be approached like a roundabout: There are multiple entry and exit points, and readers are expected to take their preferred route among the many options, possibly spinning a bit for fun. I should also mention that, however dreadful that driving experience was to me, it was nothing compared with the exam labor of my students of the terrifying quantitative methods course. I hope that this book will help them, and many others, to overcome their fear. By the same token, I believe that the book will be useful to practitioners as well, especially those using data analysis and decision support software packages, possibly in need of a better understanding of those black boxes.

I have a long teaching experience at Politecnico di Torino, in advanced courses involving the application of quantitative methods to production planning, logistics, and finance. A safe spot, indeed, with a fairly homogeneous population of students. Add to this the experience in teaching numerical methods in quantitative finance master's programs, with selected and well-motivated students. So, you may imagine my shock when challenged by more generic and basic courses within a business school (ESCP Europe, Turin Campus), which I started teaching a few years ago. The subject was quite familiar, quantitative methods, with much emphasis on statistics and data analysis.

However, the audience was quite different, as the background of my new students ranged from literature to mathematics/engineering, going through law and economics. When I wondered about how not to leave the whole bunch utterly disappointed, the "mission impossible" theme started ringing in my ears. I must honestly say that the results have been occasionally disappointing, despite my best efforts to make the subject a bit more exciting through the use of business cases, a common mishap for teachers of technical subjects at business schools. Yet, quite often I was delighted to see apparently hopeless students struggle, find their way, and finally pass the exam with quite satisfactory results. Other students, who had a much stronger quantitative background, were nevertheless able to discover some new twists in familiar topics, without getting overly bored. On the whole, I found that experience challenging and rewarding.

On the basis of such disparate teaching experiences, this possibly overambitious book tries to offer to a hopefully wide range of readers whatever they need.

- **Part I** consists of three chapters. Chapter 1 aims at motivating the skeptical ones. Then, I have included two chapters on calculus and linear algebra. Advanced readers will probably skip them, possibly referring back to refresh a few points just when needed, whereas other students will not be left behind. Not all the material provided there is needed; in particular, the second half of Chapter 3 on linear algebra is only necessary to tackle Parts III and IV.

- **Part II** corresponds to the classical core of a standard quantitative methods course. Chapters 4–10 deal with introductory topics in probability and statistics. Readers can tailor their way through this material according to their taste. Especially in later chapters, they can safely skip more technical sections, which are offered to more mathematically inclined readers. Both Chapter 9, on inferential statistics, and Chapter 10, on linear regression, include basic and advanced sections, bridging the gap between cookbook-oriented texts and the much more demanding ones. Also Chapter 11, on time series, consists of two parts. The first half includes classical topics such as exponential smoothing methods; the second half introduces the reader to more challenging models and is included to help readers bridge the gap with the more advanced literature without getting lost or intimidated.

- **Part III** moves on to decision models. Quite often, a course on quantitative methods is declined in such a way that it could be renamed as "business statistics," possibly including just a scent of decision trees. In my opinion, this approach is quite limited. Full-fledged decision models should find their way into the education of business students and professionals. Indeed, statistics and operations research models have too often led separate lives within academia, but they do live under the same roof in the new trend that has been labeled "business analytics." Chapter 12 deals mostly with linear programming, with emphasis on model building; some knowledge on how these problems are actually solved, and which features make them computationally easy or hard, is also provided, but we do not certainly cover solution methods in detail, as quite robust software packages are widely available. This part also relies more heavily on the advanced sections of Chapters 2 and 3. Chapter 13 is quite important, as it merges all previous chapters into the fundamental topic of decision making under risk. Virtually all interesting business management problems are of this nature, and the integration of separate topics is essential from a pedagogical point of view. Chapter 14 concludes Part III with some themes that are unusual in a book at this level. Unlike previous chapters, this is more of an eye-opener, as it outlines a few topics, like game theory and Bayesian statistics, which are quite challenging and can be covered adequately only in dedicated books. The message is that no one should have blind faith in fact-based decisions. A few examples and real-life cases are used to stimulate critical thinking. This is not to say that elementary techniques should be disregarded; on the contrary, they must be mastered in order to fully understand their limitations and to use them consciously in real-life settings. We should always keep in mind that all models are wrong (G.E.P. Box), but some are *useful*, and that nothing is as practical as a *good* theory (J.C. Maxwell).

- **Part IV** completes the picture by introducing selected tools from multivariate statistics. Chapter 15 introduces the readers to the challenges and the richness of this field. Among the many topics, I have chosen those that are more directly related with the previous parts of the book, i.e., advanced regression models in Chapter 16, including multiple linear, logistic, and nonlinear regression, followed in Chapter 17 by data reduction methods, like principal component analysis, factor analysis, and cluster analysis. There is no hope to treat these topics adequately in such a limited space, but I do believe that readers will appreciate the relevance of the basics dealt with in earlier chapters; they will hopefully gain a deeper understanding of these widely available methods, which should not just be used as software black boxes.

Personally, I do not believe too much in books featuring a lot of simple and repetitive exercises, as they tend to induce a false sense of security. On the other hand, there is little point in challenging students and practitioners with overly complicated problems. I have tried to strike a fair compromise, by including a few of them to reinforce important points and to provide readers with some more worked-out examples. The solutions, as well as additional problems, will be posted on the book Webpage.

On the whole, this is a book about fact- and evidence-based decision making. The availability of information-technology-based data infrastructures has made it a practically relevant tool for business management. However, this is not to say that the following simple-minded equation holds:

$$Data = Decisions$$

This would be an overly simplistic view. To begin with, there are settings in which we do not have enough data, because they are hard or costly to collect, or simply because they are not available; think of launching a brand-new and path-breaking product or service. In these cases, knowledge, under the guise of subjective assessments or qualitative insights, comes into play. Yet, some discipline is needed to turn gut feelings into something useful. Even without considering these extremes, it is a fact that knowledge is needed to turn rough data into *information*. Hence, the equation above should be rephrased as

$$Data + Knowledge = Decisions$$

Knowledge includes plenty of things that are not treated here, such as good and sensible intuition or the ability to work in a team, which must be learned on the field. I should also mention that, in my teaching, the discussion of business cases and the practical use of software tools play a pivotal role, but cannot be treated in a book like this. Yet, I believe that an integrated view of quantitative methods, resting on solid but not pedantic foundations, is a fundamental asset for both students and practitioners.

Use of software. In writing this book, a deliberate choice has been not to link it with any software tool, even though the application of quantitative methods does require such a support in practice.[1] One the one hand, whenever you select a specific tool, you lose a share of readers. On the other hand, there is no single software environment adequately covering the wide array of methods discussed in the book. Microsoft Excel is definitely a nice environment for introducing quantitative modeling, but when it comes, e.g., to complex optimization models, its bidimensional nature is a limitation; furthermore, only dedicated products are able to cope with large-scale, real-life models. For the reader's convenience, we offer a nonexhaustive list of useful tools:

- MATLAB (http://www.mathworks.com/) is a numerical computing environment, including statistics and optimization toolboxes.[2] Indeed, many diagrams in the book have been produced using MATLAB (and a few using Excel).

---

[1]The software environments that are mentioned here are copyrights and/or trademarks of their owners. Please refer to the listed Websites.

[2]The virtues of MATLAB are well illustrated in my other book: P. Brandimarte, *Numerical Methods in Finance and Economics: A MATLAB-Based Introduction*, 2nd. ed., Wiley, New York, 2006.

- Stata (`http://www.stata.com/`) and SAS (`http://www.sas.com/`) are examples of rich software environments for statistical data analysis and business intelligence.

- Gurobi (`http://www.gurobi.com/`) is an example of a state-the-art optimization solver, which is necessary when you have to tackle a large-scale, possibly mixed-integer, optimization model.

- AMPL (`http://www.ampl.com/`) is a high-level algebraic modeling language for expressing optimization models in a quite natural way. A tool like AMPL provides us with an interface to optimization solvers, such as Gurobi and many others. Using this interface, we can easily write and maintain a complex optimization model, without bothering about low-level data structures. We should also mention that a free student version is available on the AMPL Website.

- COIN-OR (`http://www.coin-or.org/`) is a project aimed at offering a host of free software tools for Operations Research. Given the cost of commercial licenses, this can be a welcome resource for students.

- By a similar token, the R project (`http://www.r-project.org/`) offers a free software tool for statistics, which is continuously enriched by free libraries aimed at specific groups of statistical methods (time series, Bayesian statistics, etc.).

Depending on readers' feedback, I will include illustrative examples, using some of the aforementioned software packages, on the book Website. Incidentally, unlike other textbooks, this one does not include old-style statistical tables, which do not make much sense nowadays, given the wide availability of statistical software. Nevertheless, tables will also be provided on the book Website.

Acknowledgments. Much to my chagrin, I have to admit that this book would not have been the same without the contribution of my former coauthor Giulio Zotteri. Despite his being an utterly annoying specimen of the human race, our joint teaching work at Politecnico di Torino has definitely been an influence. Arianna Alfieri helped me revise the whole manuscript; Alessandro Agnetis, Luigi Buzzacchi, and Giulio Zotteri checked part of it and provided useful feedback. Needless to say, any remaining error is their responsibility. I should also thank a couple of guys at ESCP Europe (formerly ESCP-EAP), namely, Davide Sola (London Campus) and Francesco Rattalino (Turin Campus); as I mentioned, this book is in large part an outgrowth of my lectures there. I gladly express my gratitude to the authors of the many books that I have used, when I had to learn quantitative methods myself; all of these books are included in the end-of-chapter references, together with other textbooks that helped me in preparing my courses. Some illuminating examples from these sources have been included here, possibly with some adaptation. I have provided the original reference for (hopefully) all of them, but it might be the case that I omitted some due reference because, after so many years of teaching, I could not trace all of the original sources; if so, I apologize with the authors, and I will be happy to include the reference in the list of errata. Last but not least, the suffering of quite a few cohorts of students at both Politecnico di Torino and ESCP Europe, as well as their reactions and feedback, contributed to shape this work (and improved my mood considerably).

Supplements. A solution manual for the problems in the book, along with additional ones and computational supplements (Microsoft Excel workbooks, MATLAB scripts, and AMPL models), will be posted on a Webpage. My current URL is:

- http://staff.polito.it/paolo.brandimarte

A hopefully short list of errata will be posted there as well. One of the many corollaries of Murphy's law says that my URL is going to change shortly after publication of the book. An up-to-date link will be maintained on the Wiley Webpage:

- http://www.wiley.com/

For comments, suggestions, and criticisms, my e-mail address is

- paolo.brandimarte@polito.it

PAOLO BRANDIMARTE
Turin, February 2011

Part I

Motivations and Foundations

1

Quantitative Methods: Should We Bother?

If you are reading this, chances are that you are on your way to becoming a manager. Or, maybe, you are striving to become a better one. It may also be the case that the very word *manager* sounds dreadful to you and conjures up images of unjustified bonuses; yet, you might be interested in how *good* management decisions should be made or supported, in both the private and public sectors. Whatever your personal plan and taste, what makes a good manager or a good management decision? The requirements for a career in management make a quite long list, including interpersonal communication skills, intuition, human resource management, accounting, finance, operations management, and whatnot. Maybe, if you look down the list of courses offered within master's programs in the sector, you will find quantitative methods (QMs). Often, students consider this a rather boring, definitely hard, maybe moderately useful subject. I am sure that a few of my past students would agree that the greatest pleasure they got from such a course was just passing the exam and forgetting about it. More enlightened students, or just less radical ones, would probably agree that there is something useful here, but you may just pay someone else to carry out the dirty job. Indeed, they do have a point, as there are plenty of commercially available software packages implementing both standard and quite sophisticated statistical procedures. You just load data gathered somewhere and push a couple of buttons, so why should one bother learning too much about the intricacies of QMs? Not surprisingly, a fair share of business schools have followed that school of thought, as the role of QMs and management science in their curricula has been reduced,[1] if they have not been eliminated altogether.

Even more surprisingly however, there is another bright side of the coin. The number of software packages for data analysis and decision support is increasing, and they are more and more pervasive in diverse application fields such as supply chain management, marketing, and finance. Their role is so important that even books aimed at non specialists try to illustrate the relevance of quantitative methods and analytics to a wide public; the key concept of books like *Analytics at Work* and *The Numerati* is that these tools make an excellent *competitive weapon*.[2] Indeed, if someone pays good money for expensive software tools, there must be a reason. How can we explain such a blatant contradiction in opinions about QMs? The mathematics has been there for a while, but arguably the main breakthrough has been the massive availability of data thanks to Web-based information systems. Add to that the availability of cheap computing power and better software architectures, as well as smart user interfaces. These are relatively recent developments, and it will take time to overcome the inertia, but the road is clear.

Still, one of the objections above still holds: I can just pay a specialist or, maybe, learn a few pages of a software manual, without bothering with the insides of the underlying methods. However, relying on a tool without a reasonable knowledge of its traps and hidden assumptions can be quite dangerous. The role of quantitative strategies in many financial debacles has been the subject of heated debate. Actually, the unpleasing outcome of bad surgery executed by an incompetent person with distorted incentives can hardly be blamed on the scalpel, but it is true that quantitative analysis can give a false sense of security in an uncertain world. This is why anyone involved in management needs a decent knowledge of analytics. If you are a top manager, you will not be directly involved in the work of the specialists, but you should share a common language with them and you should be knowledgeable enough to appreciate the upsides and the downsides of their work. At a lower level, if you get an esoteric error message when running a software application, you should not be utterly helpless; by the same token, if there are alternative methods to solve the same problem, you should figure out what is the best one in your case. Last but not least, a few other students of mine accepted the intellectual challenge and discovered that studying QMs can be rewarding, interesting, and professionally relevant, after all.[3]

---

[1]The actual term that is often used is "dumbed down," but this could sound not too politically correct to some.

[2]See Refs. [1, 7], as well as T.H. Davenport, Competing on Analytics, *Harvard Business Review*, Jan. 2006, pp. 1–9.

[3]I am quite proud to say that one of my best QMs master's students had a degree in classics, from Oxford University. I am also proud to say that a few of them changed their mind about statistics: "You know, prof, last year I got a pretty good grade in statistics, but I couldn't figure out why..."

I will spend quite a few pages trying to convince you that a good working knowledge of QMs is a useful asset for your career.

- When information is available, decisions should be based on data. True, a good manager should also rely on intuition, gut feelings, and the ability to relate to people. However, there are notable examples of managers who were considered geniuses after a lucky decision, and eventually destroyed their reputation, endangered their business, and went to jail in some remarkable cases. Without going to such extremes, even the best manager may make a wrong decision, because something absolutely unpredictable can happen. A good decision should be somewhat robust, but when things go really awry, being able to justify your move on a formal analysis of data may save your neck.

- QMs can make you a sort of universal blood donor. The mathematics behind is general enough to be applied in different settings, such as supply chain management, finance, and marketing. QMs can open many doors for you. Indeed, throughout the book I will insist on this point by alternating examples from quite different areas.

- Even if you are not a specialist, you should be able to work with consultants who have specialized quantitatively. You should be able to interact constructively with them, which means neither refusing good ideas merely because they seem complicated, nor taking for granted that sophistication always works. At the very least, you should be aware of what they are doing.

I have met some people whose idea of applying QMs is collecting data and coming up with a few summary measures, maybe some fancy plots to spice up a presentation, and that's it. In fact, QMs are much more than collecting basic descriptive statistics:

1. If QMs are to be of any utility to a manager, they should help her in *making decisions*. Unfortunately, modeling to make decisions is a rather hard topic.

2. By the same token, basic probability and statistics are not enough to meet the challenge of a complex reality. Multivariate analysis tools have been applied, but there is a gap between books covering the standard procedures and those at an advanced level.

We will try to bridge that gap, which is somewhat hard to do by just walking through a lengthy and dry list of theorems and proofs. In this chapter I will illustrate a few toy examples, that will hopefully provide you with enough motivation to proceed.

We have emphasized the role of data to make decisions. If we knew all of the relevant data in advance, then our task would be considerably simplified. Nevertheless, we show in Section 1.1 that even in such an ideal situation some quantitative analysis may be needed. More often than not, uncertainty makes our life harder (or more interesting). In Section 1.2 we deal with different examples in which we have to make a decision under uncertainty. The standard tools that help us in such an endeavor are provided by probability and statistics, which constitute a substantial part of the book. Nevertheless, we will show that some concepts, such as probability, can be somewhat dependent on the context. Indeed, many features of real life may make a straightforward application of simple methods difficult, and we will see a few examples in Section 1.3. Finally, in Section 1.4 we will discuss how, when, and why QMs can be useful, while pointing out their limitations.

1.1   A DECISION PROBLEM WITHOUT UNCERTAINTY: PRODUCT MIX

Product mix decisions are essentially resource allocation problems. We have limited resources, such as machines, labor, and raw materials, and the problem calls for their optimal use in order to maximize profit, which is earned by producing and selling a set of items. The decision problem consists of finding the right amounts to produce for each item over a certain timespan. Profit depends on the cost of producing each item and the price at which they can be sold. Produced quantities should comply with several constraints, such as production capacity and market limitations, since we should not produce what we are not going to sell anyway.

One of the fundamental pieces of information we need is demand. The time period we work with can be a day, a week, or a month. In practice, demand varies over time and can be quite uncertain. Here we consider an idealized problem in which demand is known and constant over time. Furthermore, demand is not completely exogenous in real life, as we might influence it by pricing decisions. Price can be more or less under direct control, depending on the level of competition and the type of market we deal with; in a product mix problem we typically assume that we are price takers.

In the first example below, products are similar in the sense that they consume similar amounts of resources. In the second one, we will complicate resource consumption a bit.

1.1.1   The case of similar products

A firm[4] produces red and blue pens, whose unit production cost is 15 cents (including labor and raw material). The firm incurs a daily fixed cost, amounting to €1000, to run the plant, which can produce at most 8000 pens per day in total (i.e., including both types). Note that we are expressing the capacity

---

[4]This example is based on Chapter 2 of Ref. [5].

constraint in terms of the total number of pens produced, which makes sense if resource requirements are the same for both products; in the case of radically different products (say, needles and air carriers), this makes no sense, as we shall see in the next section. We are not considering changeover times to switch production between the two different items, so the above information is all we need to know from the technological perspective.

From the market perspective, we need some information about what the firm might sell and at which price. The blue pens sell for 25 cents, whereas things are a tad more complicated for the red ones. On a daily basis, the first 5000 red pens can be sold for 30 cents each, but additional ones can be sold for only 20 cents. This may sound quite odd at first, but it makes sense if we think that the same product can be sold in different markets, where competition may be different, as well as general economic conditions. Such a price discrimination can be maintained if markets are separated, i.e., if one cannot buy on the cheaper market and resell on the higher-priced market.[5] In general, there may be a complex relationship between price and demand, and in later chapters we will consider QMs to estimate and take advantage of this relationship.

The problem consists of finding how many red and how many blue pens we should produce each day. Note that we are assuming constant demand; hence, the product mix is just repeated each day. In the case of time-varying demand and changeover costs, there could be an incentive to build some inventory, which would make the problem dynamic rather than static.

1. The production manager, an ugly guy with little business background, decides to produce 5000 red and 3000 blue pens, yielding a daily profit of €50 (please, check this result). This may not sound too exciting, but at least we are in the black.

2. A brilliant consultant (who has just completed a renowned master, including accounting classes) argues that this plan does not consider how the fixed cost should be allocated between the two product types. Given the produced quantities, he maintains that €625 ($\frac{5}{8}$ of the fixed cost) should be allocated to red pens, and €375 to blue pens. Subtracting this fraction of the fixed cost from the profit contribution by blue pens, he shows that blue pens are not profitable at all, as their production implies a loss of €75 per day! Hence, the consultant concludes that the firm should just produce red pens.

What do you think about the consultant's idea? Please, *do* try finding an answer before reading further!

---

[5]International students may be familiar with book editions that are marked as "Not for sale in the USA."

A straightforward calculation shows that the second solution, however reasonable it might sound, implies a daily loss:

$$1000 - 5000 \times (0.30 - 0.15) - 3000 \times (0.20 - 0.15) = €100$$

It is also fairly easy to see that the simple recipe of the production manager is just based on the idea of giving priority to the item that earns the largest profit margin. Apart from that, we should realize that the fixed cost is not really affected by the decisions we are considering at this level. If the factory is kept open, the fixed cost must be paid, whatever product mix is selected. However, this does not mean that the fixed cost is irrelevant altogether. At a more strategic decision echelon, the firm could consider shutting the plant down because it is not profitable. The point is that any cost is variable, at some hierarchical level and with a suitably long time horizon.

From a formal point of view, what we have been trying to solve is a problem such as

$$\begin{aligned} \max \quad & \pi(x_r, x_b) \\ \text{s.t.} \quad & x_r + x_b \leq 8000 \\ & x_r, x_b \geq 0 \end{aligned}$$

In this mathematical statement of the problem we distinguish the following:

- Two *decision variables*, x_r and x_b, which are the amounts of red and blue pens that we produce, respectively.

- An *objective function*, $\pi(x_r, x_b)$, representing the profit we earn, depending on the selected mix, i.e., on the value assigned to the two decision variables. Our task is maximizing profit with respect to decision variables.

- A set of *constraints* on the decision variables. We should maximize profit with respect to the decision variables, *subject to* (s.t. in the model formulation) this set of constraints. The first constraint here is an inequality corresponding to the capacity limitation. Further, we have included nonnegativity requirements on sold amounts. Granted, unless you are pretty bad with marketing, you are not going to sell negative amounts, which would reduce profit. Yet, from a mathematical perspective, manufacturing negative amounts of an item could be an ingenious way to create capacity for another item, which makes little sense and must be forbidden. Constraints pinpoint a *feasible region*, i.e., a set of solutions that are acceptable, among which we should find the best one, according to our criterion.

The feasible region in our case is just the shaded triangle depicted in Fig. 1.1. If you have trouble understanding how to get that figure, you might wish to refer to Section 2.3; yet, we may recall from high school

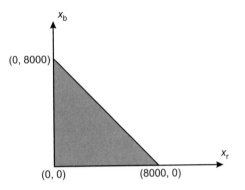

Fig. 1.1   The feasible set for the problem of red and blue pens.

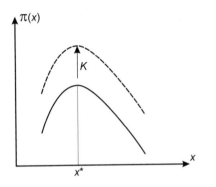

Fig. 1.2   Shifting a function up and down does not change the optimal solution.

mathematics that an equation like $ax_1 + bx_2 = c$ is the equation of a
line in the plane; an inequality like $ax_1 + bx_2 \leq c$ represents one of the
two half-planes separated by that line. To see which one, the easy way
is checking if the origin of the plane, i.e., the point of coordinates $(0, 0)$
satisfies the inequality, in which case it belongs to the half-plane, or not.

Intuitively, since the firm makes money by selling whatever pen it produces,
the capacity constraint should be binding at the optimal solution, which means
that we should look for solutions on the line segment joining points of coordinates $(0, 8000)$ and $(8000, 0)$. In Chapter 2 we will see how one can maximize
a profit function (or minimize a cost function) in simple cases; a more thorough treatment will be given in Chapter 12. For now, we may immediately
see why the fixed cost should be ignored in finding the optimal mix. Assume,
for the sake of simplicity, that we have just one decision variable and consider
the objective function $\pi(x)$ in Fig. 1.2. Let us denote the optimal solution
of the maximization problem, $\max \pi(x)$, by x^*. We see that if the function
is shifted up (or down) by a given amount K, i.e., if we solve $\max \pi(x) + K$,
the optimal solution does not change. Yet, the optimal value does, and this

may make the difference between a profitable business and an unprofitable one. Whether this matters or not depends on the specific problem we are addressing.

Takeaways Even from a simple problem like this, there are some relevant lessons that deserve being pointed out:

- A simple decision problem consists of decision variables, constraints on them, and some performance measure that we want to optimize, such as minimizing cost or maximizing profit.

- Not all costs are always relevant; this may depend on the level at which we are framing the problem.

- The relationship between price and demand can be complex. In real life, data analysis can be used to quantify their link, as well as the uncertainty involved.

1.1.2   The case of heterogeneous products

We solved the previous example by a simple rule: Let us pick the most profitable item and try producing as much as we can; if we hit a market limitation, consider the next most profitable item, and go on until we run out of resource availability. However, there must be something more to it. To begin with, we had just one resource; what if there are many? Well, maybe one of them will prove to be the bottleneck and will limit overall production. But there is another issue, as we expressed the capacity constraint as the number of overall items that we could produce each day. What if each item consumes a different amount of each resource? In order to see that things may be a tad more complicated, let us consider another toy example.[6]

We are given

- Two item types (P1 and P2) that we are supposed to produce and sell

- Four resource types (machine groups A, B, C, and D) that we use to produce our end items

Note that *all* of the above resources are needed to produce an item of either type; they are not alternatives, and each part type must visit all of the machine groups in some sequence. The information we have to gather from a production engineer is the time that each piece must spend being processed on each machining center. This information is given in Table 1.1, where columns labeled $T_A, \ldots, T_D$ are the processing times (say, minutes) for each part type on each machine type. At this level, we are not really interested in the exact

---

[6]This example is based on Chapter 16 of Ref. [10].

Table 1.1 Data for the optimal mix problem.

Item	T_A	T_B	T_C	T_D	Cost	Price	Demand
P1	15	15	15	25	45	90	100
P2	10	35	5	15	40	100	50

sequence of machine visits; probably, some technological reason will force a sequence of operations, but we want to determine how many pieces we produce during each period. To make this point clearer, let us say that we want to find a weekly production mix. Someone else will have the task to specify what has to be processed on each machine, on each hour of each day during the week. In most problem settings there is a decision hierarchy, whereby we first specify an aggregate plan, that becomes progressively more detailed while going down the hierarchy.

From Table 1.1 we immediately see that end items differ in their resource requirements. Hence, it makes no sense to express a capacity constraint in terms of the total number of items that we can produce each week. What we need to know is how many minutes of resource availability we have each week. This depends on the work schedule, labor and machines available, etc. Each machine group may consist of many similar or identical machines; hence, we are interested in the aggregate capacity, rather than the time that each single physical machine is available. To consider a simple case, let us assume that machine availability is the same for all of the four groups: 2400 minutes. Note that this is the availability, or capacity, for *each* machine group.

Another limitation on production stems from market size. If demand is limited, there is no point in making something we can't sell (remember that, according to our assumptions, both capacity and demand are constant over time, so there is no point in building and carrying any inventory). Furthermore, we should consider the cost of producing an item and the price at which we may sell it. These market and economical data are given in the last three columns of Table 1.1. The cost given in the third column from the right refers to each single item and it may also include raw material, labor, etc. Further to that, let us say that we also incur a fixed cost of €5000 per week. We have already pointed out that this will not influence the optimal mix, but it makes the difference between being in the black or in the red. In the two last columns we see the price at which we sell each unit, which we assumed constant and independent from the number of items produced, and the weekly demand for each part type, which places an upper bound on sales.

Our task is to find the optimal production mix, i.e., a production plan maximizing profit. The task is not that difficult, as we just need two numbers. Let us denote by x_1 and x_2 the amounts of item P1 and P2 that we produce, respectively. Yet, we must be careful to meet all of the capacity and market size constraints.

A trial-and-error approach   One thing we may try is to apply the same principle of the red and blue pens: P2 looks more profitable, since its profit margin is $100 - 40 = $€60, which is larger than the $90 - 45 = $€45 of item P1. So, let us try to maximize production of item P2. From the technological data, we see immediately that the bottleneck machine group, on which P2 spends the most time, is machining center B. An upper bound on x_2 is obtained by assuming that we use all of the capacity of group B to manufacture P2:

$$35x_2 \leq 2400 \Rightarrow x_2 \leq 68.57$$

One could object that the true bound is 68, as we cannot manufacture fractional amounts of an item. Anyway, we cannot sell more than 50 pieces, so we set $x_2 = 50$, and then we maximize production of P1 using residual capacity. We should figure out which of the four capacity constraints will turn out to be binding. We can write the following set of inequalities, one per machine group, and check which one actually limits production:

$$15x_1 + 10 \cdot 50 \leq 2400 \quad \Rightarrow \quad x_1 \leq 126.67$$
$$15x_1 + 35 \cdot 50 \leq 2400 \quad \Rightarrow \quad x_1 \leq 43.33$$
$$15x_1 + 5 \cdot 50 \leq 2400 \quad \Rightarrow \quad x_1 \leq 143.33$$
$$25x_1 + 15 \cdot 50 \leq 2400 \quad \Rightarrow \quad x_1 \leq 66$$

which yields $x_1 = 43.33$. For the sake of simplicity, let us assume that we are indeed able to make fractional amounts of items. This is somewhat true when we deal with things such as paint, and it is a sensible approximation for large numbers; rounding 1,000,000.489 up or down induces a small error. We will see in Chapter 12 why forcing integrality of decision variables may complicate things, and we should do it only when really needed. The production plan $x_1 = 43.33$, $x_2 = 50$ is feasible; unfortunately, total profit is negative:

$$45 \times 43.33 + 60 \times 50 - 5000 = -50$$

What went wrong? Maybe this is the best we can do, and we should just shut the business down, or try reducing cost, or try increasing price without reducing demand too much. Or maybe we missed something. With red and blue pens, resource consumption was the same for both items, but in our case P2 features the larger resource consumption on machine B. Maybe we should somehow consider a tradeoff between profit and resource consumption; maybe we should come up with a ratio between profit contribution and resource consumption. It is not quite clear how we should do this, since it is not true that P2 requires more time than P1 on *all* of the four resources. Nevertheless, it could well be the case that, carrying out this analysis, P1 would turn out to be more profitable. So, let us see what we get if we maximize production of P1 first. In this case, machine group D is the bottleneck, and the same reasoning as above yields

$$25x_1 \leq 2400 \Rightarrow x_1 \leq 96$$

Now we do not reach the market bound, which is 100 for P1, but then, since we use all of the capacity of group D for item P1, we must set $x_2 = 0$. Fair enough, but profit is even worse than before: $45 \cdot 96 - 5000 = -680$.

Hopefully, the reader is starting to see that even for a toy problem such as this one, the art of quick calculations based on plausible and intuitive reasoning may fall short of our expectations. But before giving up, let us try to see if there is a way to make the problem simpler. After all, the difficulty comes mainly form capacity constraints and differentiated resource consumption. If we look a bit more carefully at Table 1.1, we see something interesting. Consider resources A and B: Are they equally important? Note that a plan that is feasible for group B must be feasible for A as well: P1 requires the same amount of time on both groups, whereas P2 has a larger requirement on B. We may conclude that group A will never be a binding resource. If we compare resource requirements for groups B and C, we immediately reach a similar conclusion. In fact, only resources B and D need to be considered.[7]

Now the perspective looks definitely better: We just need to find a solution which uses all of the resources B and D, as this will maximize production. Unless we hit a market constraint, there is no point in leaving critical resources unused. We should find two values for our two decision variables, x_1 and x_2, such that both machine groups B and D are fully utilized. This results in a system of two equations:

$$\begin{cases} 15x_1 + 35x_2 = 2400 \\ 25x_1 + 15x_2 = 2400 \end{cases} \qquad (1.1)$$

We will see a bit more about solving such a system of linear equations in Chapter 3. For now, let us just say that solving this system yields the production mix $x_1 = 73.84$ and $x_2 = 36.92$, rounding numbers down to the second decimal digit; this results in a total profit of €538.46, which is positive! Intuition worked pretty well for the red and blue pens problem, but this solution is a bit harder to get by sheer intuition.

If this seems too hard, please have a reality check. We had to solve just a toy problem, ignoring all of the complications that make real life so fun:

- We had to deal with just two end items (they may easily be thousands).

- Demand was known with certainty (you wish).

---

[7]A whole managerial philosophy, called the *theory of constraints*, has been born, based on the principle that you may simplify a problem and better focus your effort by concentrating on bottlenecks, i.e., the factors that really limit performance.

- All of the relevant data were constant over time (same as above).

- We did not consider interactions between demands for different end items (if a customer wants both items P1 and P2, and we have not enough of one of them, we might well lose the whole order).

- We did not consider availability of raw materials (one of the most amusing moments you might experience in life is when you cannot finish the assembly of a $100,000 item because you miss a little screw worth a few cents).

- We did not consider changeover times between different item types (on very old press lines in the automotive industry, setting up production for another model required 11 hours).

- We did not consider detailed execution and timing.

- We did not consider substitution between raw materials; in some blending processes (food and oil), there are some degrees of freedom making the choice even more complicated (we cover blending problems in Section 12.2.3).

- We did not include integrality constraints on the decision variables, which would probably make our approach unsuitable (we will see how to cope with this complication in Section 12.6.2).

If we realize the true complexity of a real-life problem, it is no surprise that sometimes even getting a *feasible* solution (let alone an optimal one) may be difficult without some quantitative support. Hence, we need a more systematic approach.

A model-based approach   In the case of red and blue pens, we hinted at the possibility of building a mathematical representation of a decision problem. Maybe, this can be helpful in a complex setting. To begin with, we want to maximize profit. Formally, this means that we want to maximize a function such as

$$45x_1 + 60x_2$$

We have already remarked that fixed costs do not change where the optimal solution is, so subtracting €5,000 is inconsequential. From the work we have carried out before, we see that capacity constraints can be represented as a set of inequalities:

$$15x_1 + 10x_2 \leq 2400$$
$$15x_1 + 35x_2 \leq 2400$$
$$15x_1 + 5x_2 \leq 2400$$
$$25x_1 + 15x_2 \leq 2400$$

If we also include nonnegativity of decision variables and market bounds, we end up with the following mathematical problem:

$$\max \quad 45x_1 + 60x_2 \tag{1.2}$$
$$\text{s.t.} \quad 15x_1 + 10x_2 \leq 2400$$
$$15x_1 + 35x_2 \leq 2400$$
$$15x_1 + 5x_2 \leq 2400$$
$$25x_1 + 15x_2 \leq 2400$$
$$0 \leq x_1 \leq 100$$
$$0 \leq x_2 \leq 50$$

This is an example of a linear programming problem, where *linear* is due to the fact that decision variables occur linearly: you do not see products such as $x_1 \cdot x_2$, powers such as x_1^2, or other weird functions such as $\sin x_2$. Real-life problems may involve thousands of decision variables, but they can be solved by many computer packages implementing a solution strategy called the *simplex method*, and (guess what?) using this magic you get the optimal solution above. By the way, good software will also spot and get rid of irrelevant constraints to speed up the solution process.

More on this in Chapter 12, but in this simple case we may visualize things graphically in order to better understand why the first simple-minded approach failed.

A graphical solution   As with red and blue pens, we are dealing here with a bidimensional problem. Each (linear) inequality corresponds to a half-plane. Since we must satisfy a set of such constraints, the set of feasible solutions is the intersection of half-planes, and is illustrated in Fig. 1.3. The shaded figure is a polyhedron, resulting from the intersection of the relevant constraints: these are the capacity constraints for groups B and D, and the market bound for item P2.

The parallel lines shown in the figure are the level curves of the profit function. For instance, to visualize all of the product mixes yielding a profit contribution of €2000 (neglecting the fixed cost), we should draw the line corresponding to the linear equation

$$45x_1 + 60x_2 = 2000$$

Changing the desired value of profit contribution, we draw a set of parallel lines; three of them are displayed in Fig. 1.3. It is also easy to see that profit increases by moving in the northeast direction, i.e., by increasing production of both part types.

There is an infinite set of feasible mixes (barring integrality requirements on decision variables), but we see that a only a very few of them are relevant: those corresponding to the vertices (or extreme points) of the polyhedron, i.e., points $M_0, M_1, \ldots, M_4$. Point M_0, the origin of the axes, corresponds to

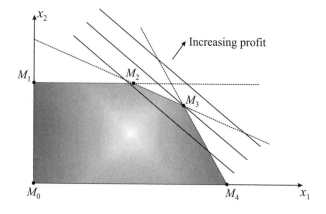

Fig. 1.3   Graphical solution of the optimal mix problem.

making nothing and is not quite interesting. Point M_1 corresponds to making 50 items of P2, and none of P1; in fact, during our first attempt, we moved to that point first, and then to point M_2, with coordinates (43.33, 50), which was our first mix. Point M_4, with coordinates (96, 0), represents the second tentative mix we came up with. We see that the second solution was worse than the first one by checking the level curves of profit. Since level curves are parallel lines, and we should move along the direction of increasing profit, we see that the optimal solution must be a feasible point that "touches" the level curve with the highest profit. This happens at point M_3, which in fact corresponds to the optimal mix.

The slope of the level curves depends on the profit margin of each item. For instance, if we increase the profit margin of P1, the lines rotate clockwise; if profit margin of P1 is increased enough, the optimal mix turns out to be point M_4. In general, changing the economics of the problem will result in different optimal mixes, as expected, but they will always be extreme points of the feasible set, and there are not so many of them. Whatever profit margins are, only points M_2, M_3, and M_4 can be candidate optimal solutions. If level curves happen to be parallel to an edge of the feasible set, we have an infinite number of optimal solutions, but we may just consider one corresponding to a vertex. In fact, the standard approach to solving a linear programming model, via the simplex method, exploits this property to find an optimal solution with stunning efficiency even for large-scale problems involving thousands of variables and constraints.

Incidentally, if we insist on producing integer amounts, we should only consider points with integer coordinates within the polyhedron. We may draw this feasible set as a grid of discrete points. Doing so, the optimal mix turns out to be $x_1 = 73$, $x_2 = 37$, with total profit 505. Understandably, profit is reduced by adding a further constraint on production volume. It is tempting to conclude that we may easily get this solution by solving the previous problem and then rounding the solution to the closest integer point on the grid. Unfortunately, this is not always the case, and quite sophisticated methods are needed to solve problems with integer decision variables efficiently.

Takeaways

- Intuition may fail when tackling problems with many constrained decision variables.

- Mathematics may yield an optimal solution for the *model*. Because modeling calls for simplification, this need not be the best solution of our *problem*, but it may be a good starting point.

- Sophisticated software packages are available to tackle mathematical model formulations. Hence, we need to concentrate on modeling rather than on complicated solution procedures. Indeed, in Chapter 12 we focus on models for decision making, while just giving a glimpse of the computational solution procedures. This is extended in Chapter 13 to cope with uncertainty.

- Nevertheless, a suitable background in calculus and algebra is needed to gain a proper understanding of the involved approaches; this is the subject of Chapters 2 and 3.

1.2 THE ROLE OF UNCERTAINTY

We often have to make decisions here and now, without complete knowledge about problem data or the occurrence of future events. In distribution logistics, significantly uncertain demand must be faced; in finance, several sources of risks affect the return of an investment portfolio. In all of these settings, the future effect of actions is not known for sure. Uncertainty can take several forms. In the simplest case, we may be able to gather past information and use that to generate a set of plausible future scenarios. This is where the standard tools of probability and statistics come into play. They will be the subject of Part II of the book, and are typically considered the core of any course on QMs. To get gradually acquainted with them, let us consider a few toy problems.

1.2.1 A problem in supply chain management

In the product mix problem, we assumed perfect knowledge of future demand, but, unfortunately, exact demand forecasts are a bit of scarce commodity in the true world. Indeed, the standard trouble in supply chain management is purchasing an item for which demand information is quite uncertain. If we order too much, one or more of the following scenarios might occur:

- Finance will suffer, as money is tied up in inventories.

- Items may become obsolete because of fads or product innovation, and money will be lost in inventory writeoffs.

- Perishable items may run out of their shelf life before being sold, and money will be lost again.

On the other hand, if we do not order enough items, we may not be able to meet customer demand and revenue will suffer (as well as our career; life is hard, isn't it?).

To take our first baby steps, let us consider a relatively simple version of the problem. We are in charge of purchasing an item with a very limited shelf life. Both purchased quantities and demand are given as small integer numbers, which makes sense for a niche product. Items are purchased for delivery at the beginning of each week, and any unsold item is scrapped at the end of the same week; hence, each time we face a brand-new problem, in the sense that nothing is left in inventory from the previous time periods. Demand for the next week is not known, but we do have some information about past demand. The following list shows demand for the past 20 weeks:

$$3, \ 1, \ 3, \ 2, \ 3, \ 2, \ 2, \ 4, \ 5, \ 2$$
$$1, \ 2, \ 2, \ 3, \ 5, \ 2, \ 3, \ 2, \ 4, \ 1 \tag{1.3}$$

The big question is: How many items should we order right now?

When asked this question, most students suggest considering the *average* demand, which is easily calculated as

$$\overline{D} = \frac{3 + 1 + 3 + 2 + \cdots + 4 + 1}{20} = \frac{52}{20} = 2.6$$

Not too difficult, even though this result may leave us a bit uncertain, as we cannot really order fractional amounts of items. Yet, it seems that a reasonable choice could be between 2 and 3.

Other students suggest that we should stock the *most likely* value of demand. To see what this means exactly, it would be nice to see some more structure in the demand history, maybe by counting the frequency at which each value has occurred in the past. If we sort demand data, we get the following picture:

$$\underbrace{1, \ 1, \ 1,}_{3 \text{ times}} \ \underbrace{2, \ 2, \ 2, \ 2, \ 2, \ 2, \ 2, \ 2,}_{8 \text{ times}} \ \underbrace{3, \ 3, \ 3, \ 3, \ 3,}_{5 \text{ times}} \ \underbrace{4, \ 4,}_{2 \text{ times}} \ \underbrace{5, \ 5}_{2 \text{ times}}$$

These numbers provide us with the frequencies at which each value occurred in the observed timespan. If we divide each frequency by the number of observations, we get *relative frequencies*. For instance, the relative frequency

Table 1.2 Frequencies (F), relative frequencies (F_{rel}), and cumulated (relative) frequencies (F_{cum}) for demand data.

Value	F	F_{rel}	F_{cum}
1	3	0.15	0.15
2	8	0.40	0.55
3	5	0.25	0.80
4	2	0.10	0.90
5	2	0.10	1.00

of the value 2 is $\frac{8}{20} = 0.4$ or, in percentage terms, 40%. We may also calculate average demand by using relative frequencies:

$$
\begin{aligned}
\overline{D} &= \frac{3 \times 1 + 8 \times 2 + 5 \times 3 + 2 \times 4 + 2 \times 5}{20} \\
&= \frac{3}{20} \times 1 + \frac{8}{20} \times 2 + \frac{5}{20} \times 3 + \frac{2}{20} \times 4 + \frac{2}{20} \times 5 \\
&= 0.15 \times 1 + 0.40 \times 2 + 0.25 \times 3 + 0.10 \times 4 + 0.10 \times 5 \\
&= 2.6
\end{aligned}
$$

Not surprisingly, we get the same average as above. We see that average demand is a weighted average of observed values, where weights correspond to relative frequencies. If we believe that the future will reflect the past, relative frequencies provide us with useful information about the likelihood of each demand value in the future.

Frequencies and relative frequencies are tabulated in columns 2 and 3 of Table 1.2. Be sure to note that relative frequencies cannot be negative and add up to 1, or 100%. Frequencies and relative frequencies may also be visualized using a histogram, as shown in Fig. 1.4. The observed values are reported on the horizontal axis (abscissa); the vertical axis (ordinate) may represent frequencies (a) of relative frequencies (b). The two plots are qualitatively the same, as relative frequencies are just obtained by normalizing frequencies with respect to the number of observations. After a quick glance at the graphical representation of relative frequencies, the intuitive idea of a "likelihood measure" of each demand value comes to mind rather naturally. Indeed, it is possible to interpret relative frequencies as *probabilities*. However, some caution should be exercised and we will see in Chapters 5 and 14 that probability is not such a trivial concept, as there are alternative interpretations. Still, this intuitive interpretation may be useful in many practical cases.

Looking at Table 1.2, we see that the most likely value (or the most frequent value in the past, to be honest with ourselves) is 2, which is not too different from the average value. In descriptive statistics, the most likely value

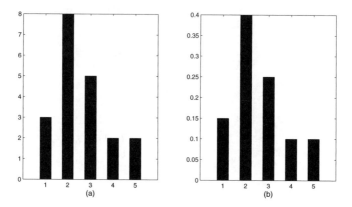

Fig. 1.4   Histograms visualizing frequencies and relative frequencies for demand data.

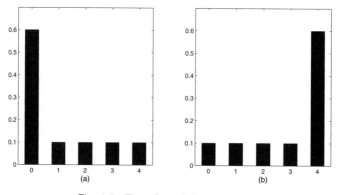

Fig. 1.5   Two skewed distributions.

is called *mode*. Since we get similar solutions by considering either mean or mode, we could be tricked into believing that we will always make a sensible choice by relying on them. Before we get so overconfident, let us consider the histograms of relative frequencies in Fig. 1.5. In histogram (a), we see that the most likely value is zero, but would we really stock nothing? Probably not. The two histograms in Fig. 1.5 are two examples of asymmetric cases. They are "skewed" into opposite directions, and we probably need a way to characterize skewness. We will deal with this and other summary measures in Chapter 4, but it is already clear that mean and mode do not tell the whole story and they are not always sufficient to come up with a solution for a decision problem. Lack of symmetry is likely to affect our stocking decisions, but there is still another essential point that we are missing: dispersion. Consider the two histograms in Fig. 1.6. Histogram (a) looks more concentrated, which arguably suggests less uncertainty about future demand with respect to histogram (b). We need some ways to measure dispersion as well, and to

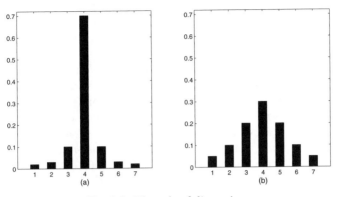

Fig. 1.6 The role of dispersion.

figure out how it can affect our choice. Indeed, we need some ways to characterize uncertainty, and this motivates the study of descriptive statistics (to be carried out in Chapter 4). This is fine, but it is utterly useless, unless we find a way to use that information to come up with a decision. It is important to realize how many points we are missing, if we just consider relative frequencies.

The role of economics. If we have a stockout, i.e., we run out of stock and do not meet the whole customer demand, how much money do we lose? And what if we have an overage, i.e., we stock too much and have to scrap perished or obsolete items? To see the point, consider the following problem. We have to decide how many T-shirts to make (or buy) for an upcoming major sport event. Producing and distributing a T-shirt costs €5; each T-shirt sells for €20, but unsold items at the end of the event must be sold at a markdown price, resulting in a loss.[8] Let us assume that the discount on sales after the event is 80%, so that the markdown price is €4. A credible forecast, based on similar events, suggests that the expected value of sales is 12,000 pieces. We will clarify what we mean by *expected value* exactly, but you may think of it as the "best forecast" given our knowledge. However, demand is quite uncertain. A consultant, considering demand uncertainty and the risk of unsold items, suggests to keep on the safe side and produce just 10,000 pieces. Is this a good idea?

Please! Wait and think about the question before going on.

---

[8]The example may not sound too significant, but this kind of situation is typical of fashion items and, given the pace of technological innovation, presents some features shared by huge markets such as consumer electronics.

When we sell a T-shirt, our profit is €15; if we have to mark down, we lose only €1. Given that, most people would probably suggest a more aggressive strategy and buy a bit more than the expected value. Indeed, most fashion stores mark prices down at some time, which means that they tend to overstock. Would you change your idea if profit margin were €2 and the cost of an unsold item were €5? Economics must play a role here, as well as dispersion. Without any information about uncertainty, we cannot specify how much above or below the expected value of demand we should place our order. A plain point forecast, i.e., a single number, is not enough for robust decision making, a point that we will stress again in Chapters 10 and 11, when dealing with regression and time series models for forecasting.

Predictable vs. unpredictable variability. Consider once again the demand data in (1.3), but this time imagine that the time series, in chronological order, is

$$1, 1, 1, 2, 2, 2, 2, 2, 2, 2, 2, 3, 3, 3, 3, 3, 4, 4, 5, 5$$

Mean, mode, etc., are not affected by this reshuffling of data, but should we neglect the clear pattern that we see? There is a trend in demand, which is not captured by simple summary measures. And what should we do with a demand pattern such as the following one?

$$1, 2, 3, 4, 3, 2, 1, 2, 3, 4, 3, 2, 1, 2, \ldots$$

In this case, we notice a seasonal pattern, with regular up- and downswings in demand. Trend and seasonality contribute to demand variability, but we should set predictable and unpredictable components of variability apart. In chapter 11 we describe some simple methods for doing so.

The role of time and intertemporal dependence. The previous point shows that time does play a role, when we can identify partially predictable patterns such as trend and seasonality. Time may also play a role when our assumptions about ordering and shelf life are less restrictive. Assume that the shelf life is longer than the time between the orders we issue to suppliers. In making our decision, we should also consider the inventory level, and this would make the problem dynamic rather than static. A safe guess is that this is no simplification.

An even subtler point must be considered in order to properly represent unpredictable variability. I will illustrate it with a real-life story. A few years ago in Turin, where I live, there was a period of intense rain followed by an impressive flood. A weird thing with such an event is that there is way too much water in the streets, but you do not get any from your water tap at home. In that case, the high level of the main river in the city prevented the pumping stations from working. This problem, as I recall, was solved quickly, but the immediate consequence was a race to buy any bottle of mineral water around (with plenty of amicable exchange of ideas between tactful customers at retail stores). Now, if you were the demand manager for a company selling mineral water, would you interpret that spike in demand as a signal of an increasing market share? Well, not really, I guess. On the contrary, you could expect a period of low demand, when households deplete their unusually high inventories. More generally, if consumption of an item is relatively steady over time, a spike in demand (maybe due to a trade promotion) is likely to be followed by a period of low demand.[9] In order to take such issues into account, we need statistical tools to investigate correlation, which is the subject of Chapter 8. Correlation is useful in many settings where we ask questions about random variables rather than random events, such as

- If demand for an item has been larger than usual today, can we say something about demand tomorrow? Will it be larger or smaller than usual?

- If the return from a financial investment has been good, does this tell us something about the return of the same investment in the future, or maybe about the return from other investments?

The role of alternative items and competitors. We have just considered one item, disregarding possible interactions with other ones. In practice, items may interact in many ways:

- *Shelf space.* Limited shelf space at a retail store must be allocated to different items; then, stocking decisions are not independent.

- *Substitute products.* A stockout on one item may be almost irrelevant, if the customer switches easily to a substitute item that we sell.

- *Complementary products.* Stocking out on one item may have a detrimental effect on other items, too; imagine a customer order consisting of several lines, related to different items; the customer could cancel the order if some order line is not satisfied.

- *New products.* If the assortment is changed by the introduction of a new item, sales of older items are likely to be affected by cannibalization.

---

[9]Incidentally, in such a case, a trade promotion might have the only effect of making the life of logistic managers a misery, without contributing much to the bottom line. In fact, some retail stores adopt a policy of *everyday low prices*.

By the same token, we should not disregard the role of competitors. If a competitor is about to launch a new product, we should plan in advance a suitable reaction. In such a case, just looking at past sales will be as safe and smart as driving a car by just looking into the rear mirror. Furthermore, pricing is likely to affect sales along multiple dimensions: the price of the item, the price of related items, and the price asked by competitors. We should investigate, among other things, the relationship between price and demand. One way of doing that involves regression models, which are the subject of Chapters 10 and 16.

The role of sampling uncertainty. When we evaluate summary measures such as the mean of a variable, we look at a limited set of past data. But how reliable is that information? Intuitively, the more data we have, the better. However, looking too far into the past is dangerous, as we might take into account information that is hardly relevant for new market conditions. Would you use information about stock returns before World War II to manage a pension fund? In Chapter 9 we deal with inferential statistics, which may help us in assessing the degree of confidence we can have in our estimates.

Observables vs. unobservables. Finally, available data may not be what is actually relevant and needed. We have taken for granted that demand data were at our disposal. Unfortunately, in many practical settings we do not really observe demand, but *sales*. If there is a stockout on an item at our retail store, will the customer inform anyone that her demand was unmet? Not necessarily; maybe she will just go and buy at another retail store. If we are lucky, she will just settle for a product substitute. But even in this case, what we gather is sales data by scanning bar codes at the cash desk or point of sale. Clearly, this can result in an underestimation of actual demand. In other cases, data are available, but they are not gathered because of a wrong business process. Imagine a business to business setting, where a potential customer calls about immediate product availability. When the desired items are not in stock, she tries with another supplier. If the business process is such that only agreed-on orders are entered into the information system, disregarding lost sales, we are underestimating demand again. Often we have to settle for a proxy of what we cannot observe directly, and this might affect decisions.

After this long list of complicating factors,[10] you may feel a little overwhelmed, but please don't: There is a long array of powerful quantitative methods that we may integrate with good old common sense in order to tackle challenging

---

[10]If you hate supply chain management, and you are fond of finance, do not fret. The good news is that you will see plenty of examples related to finance in what follows. The bad news is that the list of complicating factors is overwhelming in that setting as well.

problems. Anyway, if you want to take a short route, you could always try to do just a little better than your competitor. Say that he is able to meet demand completely in 80% of the weeks. The probability of not having a stockout is one of the many ways in which one can measure the *service level*. By choosing a stocking level S, you are setting the probability of satisfying all customers. For instance, with the data of Table 1.2, if you choose $S = 1$, demand will be met only when it is 1, which happens with probability 0.15. If $S = 2$, you meet demand when it is 1 or 2; summing the respective relative frequencies, we see that this happens with probability $0.15 + 0.40 = 0.55$. The pattern is clear: The service level for increasing stocking levels is obtained by summing probabilities. This leads us to the concept of *cumulative relative frequencies*, which are displayed in the last column of Table 1.2. If you want to do a little better than the competitor, setting a 85% service level, you should stock four items, implying a 90% service level.

This last observation leads us to concepts such as cumulative distributions and quantiles. They are fundamental in many areas, such as inferential statistics and risk management, and will be among the most important topics in Chapters 6 and 7.

1.2.2 Squeezing information out of known facts

As we have remarked in the previous section, we often have to cope with the impossibility of gathering in advance all the information we need to make a decision, partly because of uncertainty about future events, and partly because some variable cannot be observed directly. However, this is no good reason not to do our best to exploit *partial* information. In customer relationship management, we are not able to read a customer's mind; nevertheless, we may observe her behavioral patterns and try to infer if she will be loyal and hopefully profitable, or not. The analytical tools of probability and statistics are commonly used for this kind of analysis.

A typical and quite familiar domain in which we have to settle for an observable proxy of an unobservable variable is medicine. Hence, to get a feeling for the involved issues, let us consider a medical problem. Quite often we undergo an exam to determine whether we are suffering from a certain illness. Thankfully, it is a rare occurrence that our bodies must be ripped open in order to check the state of the matter. Clinical tests are an indirect way to draw conclusions about something that cannot, or should not, be inspected too directly. Now, suppose that we have the following information:

- A kind of illness affects 0.2% of a population.

- A test can reveal the illness, *if a person is ill*, with a probability of 99.9%; i.e., if you are ill, there is a small probability (0.1%) that the exam will fail to detect your true state.

- The probability of a false positive is 1%; in other words, if one is perfectly healthy, the test may wrongly report the illness with probability 1%.

The issues of how this information is obtained and how reliable it is pertain to the domain of inferential statistics; for now, we will take someone's word for it. We see that there are two kinds of potential errors:

- We may fail to detect illness in an ill subject (missed positive).

- We may see a problem where there is none (false positive).

We would like to know whether a person is ill, but the only thing we can observe is the result of the test. If the test is positive, we would like to draw the conclusion that the patient is ill in order to start an adequate treatment. However, we cannot be that sure. Hence, the question is: If the test is positive for a person, what is the probability that he is actually ill?

When I ask students this question, I make it somewhat easier and just ask them to tell in which of the following intervals the required probability lies: $(0, 25)\%$, $(25, 50)\%$, $(50, 75)\%$, or $(75, 100)\%$?

Please! Pause for a while before reading further, and try giving your answer.

If you have selected an interval with high probability, you are in good company. From my experience, the most voted interval is $(50, 75)\%$. I guess the reason is that the professor maliciously insists on pointing out that the test is reliable, in the sense that if you are ill, the test will tell you that with high probability. Some students take me seriously, and vote $(75, 100)\%$. Most students probably temper that with a bit of skepticism and choose $(50, 75)\%$, based on some *good old intuition.* Very few students choose the correct answer, i.e., $(0, 25)\%$.

Let us try to get the answer by a somewhat crude line of reasoning.

1. Say that our population consists of 10,000 persons. How many should we expect to be ill? Taking 0.2% of 10,000 yields 20 persons; 99.9% of them will be reported ill, which is 19.98 (almost 20).

2. How many false positives do we expect to get? There are (on the average) 9980 healthy persons out of the total of 10,000; 1% of them will be incorrectly reported ill, i.e., 99.80 (which is almost 100, i.e., 1% of the whole population).

3. Then, on the average we will get

$$19.98 + 99.8 = 119.78$$

positives, but only 19.98 of them are in fact ill. So, the correct answer to our question is

$$\frac{19.98}{119.78} = 16.68\%$$

This is certainly much less than 50%. The problem is that the fraction of ill people is (luckily) low, in fact much lower than the number of false positives. Indeed, we could get close to the right answer simply by noting that the fraction of false positives is about 1% and the fraction of correct positives is about 0.2%; if we take the ratio

$$\frac{0.2}{0.2 + 1} = \frac{1}{6} = 0.1667 \approx 16.68\%$$

we see that it was *bad* intuition leading us into the trap, as a closer look into the data is enough to show that false positives have a large impact here.

Now, on the basis of this result, you should sit a while and broaden your view by asking a few questions:

1. If this is a cheap test, should we use a more expensive but more reliable one?

2. Should we use this test for mass screening?

3. What happens in the case of a false positive?

4. What are the social and psychological costs of a false positive?

Of course, there are no general answers, as they depend on the impact of this illness, its mortality rate, etc. Moreover, spending resources for this illness may subtract funds from programs aimed at preventing and curing other types of illnesses. The last question, in particular, shows that in a decision problem there are issues that are quite hard to quantify.[11] So, quantitative methods will definitely *not* provide us with all of the answers. Still, they may be quite useful in providing us with information that is fundamental for a correct analysis, information that we are going to miss if we rely on *bad* intuition.

Generalizing a bit, we note that a priori, if you pick a person at random the probability that he is ill is 0.2%. If you know that the test has been positive, that probability is updated to a larger value. The new probability of the event of interest (he is ill) is *conditional* on the occurrence of another event (the test is positive). We will learn about conditional probabilities in Chapter 5, where we develop a tool to tackle such problems more systematically, namely, Bayes' theorem. Bayes' theorem is the foundation of Bayesian statistics, which is increasingly relevant in practical settings in which either most past data are not relevant or you have none, and you have to rely on your subjective assessment (see Chapter 14).

---

[11] I recall the story of that guy who was diagnosed in a terminal state a while ago. He started spending all of his money to live what precious time was left to him in the most entertaining way. The good news was that the doctors were wrong; you imagine the bad news.

1.2.3   Variations on coin flipping

In this section, we consider the mother of all random experiments: coin flipping. This is a good way to get acquainted with probabilities, which are not necessarily relative frequencies, as well as with investments under uncertainty. We illustrate plain coin flipping first, and then some more interesting variations on the theme.

Plain coin flipping   Assuming that we flip a fair coin, what is the probability of its landing head? Probably, you will not flip the coin one million times to conclude that this probability is 0.5. The reasoning you are following is, in fact, based on some form of symmetry that is exploited whenever you think about gambling by simple mechanisms such as throwing dice, picking cards at random, or spinning a roulette. Hopefully, you start to see that there may be different concepts of probability, a topic that we will discuss a bit in Section 5.1. The very nature of probabilities has been the subject of heated debate, but let us leave such philosophical issues aside and assume that you play a simple lottery. A fair coin is flipped: You win $10 when it lands head, and you lose $5 when it lands tail. If the game is repeated a large number of times, how much money should you win with each flip, on the average? Looking back at how we computed average demand in Section 1.2.1, you might suggest the idea of multiplying each possible payoff outcome by its probability:

$$\mathrm{E}[P] = 0.5 \times 10 + 0.5 \times (-5) = 2.5$$

We are using a new notation here, as $\mathrm{E}[\cdot]$ denotes the *expected value* of a random variable. Indeed, we are not taking averages based on observed data, looking backward; rather, we are stating something about the future, looking forward.

It would be nice to play that game a large number of times without having to pay for the privilege of flipping the coin. Still, a basic law of economics says that there is no free lunch; hence, someone will ask you for some money to play the game. How much money would you be willing to pay exactly? Suppose that someone says that, in order to play a fair game, the price is $2.5; note that, if you pay that amount, the expected overall profit is

$$-2.5 + 0.5 \times 10 + 0.5 \times (-5) = 0$$

which does sound fair. Would you accept? Maybe yes, as the worst that can happen is losing $7.5, which will not change your life. But let's scale the game up a million times: You may win $10,000,000, you may lose $5,000,000. Would you still be willing to pay $2,500,000 for playing the game? I guess that the answer is no; probably, you would not play the game even for free.

What you have seen is an example of how risk aversion affects our behavior. This is a fundamental ingredient of any decision under uncertainty, including investments in financial assets or new products. Decision making under uncertainty and risk aversion are the subject of Chapter 13. For now, let us consider a simple investment decision that looks quite similar to coin flipping.

Fancy coin flipping Suppose that you are the lucky manager of a movie company, which has just signed a contract with a new and promising director.[12] The contract provides that you may produce zero, one, or two movies with this director, during the next 2 years. To clarify, we could produce a movie now, investing some money immediately and hopefully collecting some payoff in, say, one year. At the end of the first year, we could produce a new movie,[13] collecting revenues at the end of the second year. One decision we have to make is whether we should produce those movies or not. In pondering the decision, note that any advance money you have paid to the good director is gone; in other words, it is a *sunk* cost. If, on second thought, you feel that producing a movie with him will turn out a disaster, you should not suffer from the sunk cost syndrome, i.e., the tendency to go on at any cost since you have already spent money on your endeavor.[14] On the contrary, you should only consider the money you are about to spend now to produce the movie, and the prospects for future revenue.

To spice things up, let us say that there is another dimension, as we may select one out of two production and marketing plans, a conservative and an aggressive one. We may invest more money in well-known actors and/or special effects, as well as in marketing the movie around the world.

1. Production/marketing plan A costs 2500 whatever monetary unit you like. This is the money you have to spend for the privilege of flipping the coin. We have some uncertainty about the success of our pet group. If the movie is successful, revenue after one year will be 4400 the same whatever; if it's a flop, you make nothing. To consider a case as close as possible to coin flipping, and to simplify calculations, let us assume that the probability of a success is 50%. Of course, coin flipping is a rather crude model of uncertainty for such a situation, and you should really ponder how to come up with a set of sensible scenarios along with their probabilities. By the way, the astute reader will suspect that here we should work with yet another, more subjective concept of probability. The symmetry of coin flipping or dice throwing does not apply and if this is an unknown director, looking at the past relative frequencies case might just make no sense. This happens whenever you deal with a brand new product.[15]

---

[12]This section is based on an example described in HBS business case [8].

[13]Well, maybe it will not really be a *new* movie; sequels are all the rage, aren't they?

[14]While we are talking about movies, imagine that you have paid a ticket for a movie that you find disgusting and/or insufferably boring after 20 minutes. Should you stay seated there for two more painful hours, just because you paid for the ticket?

[15]As a case in point, when the first IBM computer, a piece of hardware as large as a big room, was working (and heating whoever was around it), an IBM boss said that it was

2. Production/marketing plan B is more aggressive and expensive, as it costs 4000, but it increases revenue by 50% in the case of a hit, without changing probabilities (hence, we have two possible outcomes: 6600 or 0, with 50% probability). Arguably, a different marketing plan could also change success/flop probabilities, but for the sake of simplicity we will neglect this.

Both plans look like a coin flipping experiment, but in this case we should consider time or, to be more precise, the time value of money. Let us digress a bit in order to clarify this issue.

A little digression: time value of money   Would you prefer to receive $100 right now or $100 in one year? Well, easy answer, and you surely concur that $100 in one year is not the same as $100 today. But what if you have to choose between $100 right now and $106 in one year? In order to make the two amounts comparable, the standard approach is to take money in the future and discount it back to now. To do so, we apply a *discount rate*, which is related to an interest rate.

Say that you may invest money at an annual risk-free interest rate of 5%; if you invest $100 now at that rate, in one year you will own

$$\$100 \times (1 + 0.05) = \$105$$

The general rule is that to evaluate money in one year, you multiply money now by $(1+r)$, where r is the prevailing interest rate. If you leave that money invested for 2 years, and interest on interest is earned, the money in 2 years will be

$$\$100 \times (1 + 0.05) \times (1 + 0.05) = \$100 \times (1 + 0.05)^2 = \$110.25$$

Note the effect of compounding; without that, in 2 years you would get only $110, i.e., the initial capital plus twice the interest payment. Now, if you see things the other way around, how much are $106 in one year worth now? A little thought should suggest the following rule: Discount money in the future using the interest rate, and compare it with money now:

$$\frac{\$106}{1.05} = \$100.95 > \$100$$

which suggests the opportunity of waiting one year to grab almost one extra dollar.

This is what cash flow discounting is all about. We will see a bit more about financial mathematics later.[16]   In practice, the real issue is estimating the

---

a very nice piece of equipment and that probably they were going to sell 10 in the entire world. Hardly a good prophecy, and the same difficulty in forecasting sales applied when the first cell phones were being produced.

[16] See Examples 2.7 and 2.33.

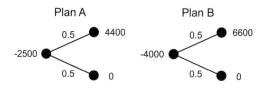

Fig. 1.7 Graphical representation of alternative plans for producing and marketing a movie.

uncertain cash flows and choosing a suitable discount rate. This may be difficult, as the discount rate should take investment risk into account. For now, let us assume that we discount cash flows with an annual discount rate of 10%.

Back to fancy coin flipping   One sensible starting point in deciding whether we should trust the good director with our money is to find out how much we expect to make using each alternative marketing plan to produce one movie. Let us consider plan A. We pay 2500 for sure now, and there is a 50–50 chance of making 4400 or 0 in one year. Taking discounting into account, the money we expect to make is

$$-2500 + \frac{0.5 \times 4400 + 0.5 \times 0}{1.10} = -2500 + \frac{2200}{1.10} = -2500 + 2000 = -500 \quad (1.4)$$

This −500 is the (expected) *net present value* (NPV) of our investment. A basic rule of investment analysis says that we should invest in projects with positive NPV. Unfortunately, the idea of adopting marketing plan A to produce a movie now does not seem quite promising. What about using the bolder plan B? Well, repeating the analysis yields the following NPV:

$$-4000 + \frac{0.5 \times 6600 + 0.5 \times 0}{1.10} = -1000$$

an even bleaker perspective. A useful way of visualizing lotteries is illustrated in Fig. 1.7, which depicts a *scenario tree*. The leftmost node corresponds to "now," i.e., the current state of the world, where we have to decide whether we should spend 2500 or 4400 to start one of the marketing plans. The two nodes on the right correspond to possible states of the world in the future. In our case, we have just two possible scenarios, success or flop, which are associated to given payoffs and probabilities. Clearly, this a two-state scenario tree is a very crude representation of uncertainty; in practice, scenario trees may be definitely richer.

It seems there is no hope, but wait: We just considered one movie! What if we produce two over the next 2 years? If one game of coin flipping is a bad lottery, can we make it any better by just repeating it? To illustrate the point with plain coin flipping, what is the probability of getting two heads in a row, or any other possible outcome? The possible head–tail sequences when you

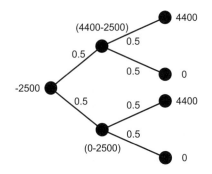

Fig. 1.8 Repeating the fancy coin flipping experiment twice.

flip a coin twice are

$$HH, \qquad HT, \qquad TH, \qquad HH$$

The usual symmetry argument suggests that none of these four outcomes is more likely than the others; hence, the probability of each sequence should just be $\frac{1}{4} = 0.25$ (note once again that probabilities add up to 1). A deeper reasoning should reveal that, unless the coin has some memory, the result of the second flip has nothing to do with the result of the first one; the two events are *independent*. We will learn in Chapter 5 that the joint probability of two independent events is just the product of their probabilities. For instance, we obtain the following equation, which confirms the previous result:

$$P\{HH\} = P\{H\} \times P\{H\} = \frac{1}{2} \times \frac{1}{2} = \frac{1}{4}$$

Now, what should we conclude if we consider the movie production problem as no more than a fancy variation on coin flipping? The situation is depicted in Fig. 1.8, where we produce two movies using plan A (incidentally, we ignore inflation and assume that future production and marketing costs will be the same as today). We have four equally likely scenarios: success–success, success–flop, flop–success, flop–flop. Each scenario has probability 0.25. We should discount cash flows occurring in 1 and 2 years; we should also consider the *net* cash flow in one year, resulting from collecting revenues (if any) and investing in the new movie. For the success–success scenario, the NPV is as follows:

$$-2500 + \frac{4400 - 2500}{1.10} + \frac{4400}{(1.10)^2} \approx 2863.64$$

Please carry out the calculation for the other three scenarios, and verify that the expected NPV is

$$0.25 \times 2863.64 + 0.25 \times (-772.73) + 0.25 \times (-1136.36)$$
$$+0.25 \times (-4772.73) = -954.55$$

The result is negative again, but this is hardly a surprise; if flipping a coin once has a negative expected profit, repeating that gamble twice will not produce a positive result, because we are replicating the same experiment. In fact, a much smarter way to obtain the result is by noting that we are just carrying out the experiment with plan A twice in a row, and we may use the result of Eq. (1.4). We have just to discount the expected NPV for the second movie back from the end of year 1 to now:

$$-50 + \frac{(-50)}{1.10} = -954.55$$

By the same token, if we try mixing plans A and B in some sequence, we will hardly get any better, *if* our problem is just a fancy variation of coin flipping.

But is the movie production case really like coin flipping? In Section 1.2.2 we learned about *conditional* probabilities: If the test is positive, this does change your probability of being ill. If the first movie is a success (or a flop), can we say that producing the next one will just be another coin flipping experiment? Probably not, as the result of the first trial tells us something about market reaction to our product. In fact, we need an assessment of *conditional* probabilities, i.e., the probability that the second movie is a success (or a flop), conditional on the fact that the first one has been a success (or a flop). Reasonably, if the first movie is a success, the probability that the second one is a success as well is greater than 0.5. By the same token, if we have a flop in the first trial, we raise our expectation for another flop with the second movie. Coins are supposed to have no memory, but the tendency to stack movie sequels one after another suggests that moviegoers are not memoryless coins. Estimating those conditional probabilities may be quite hard, but let us consider a very extreme and overly simplified case, just to illustrate the point. Assume that if the first attempt is a success, the second one will certainly be as well (i.e., the conditional probability of a second hit after the first one is 1); on the other hand, assume that after a flop, we will have another flop for sure. In such a case, it is easy to see that if the first movie is a success, we should certainly produce the second one; on the other hand, if the first movie is a flop, we will just cut our losses and forget about it. This is what managerial flexibility is all about; we do not plan everything in advance, but we adapt our decisions when we gather additional information and revise our beliefs.

One sensible idea is to be conservative at the first trial and using plan A for the first movie; if we are lucky, and discover that the director can deliver a successful movie, we will be more aggressive and take plan B next year. This strategy is depicted in Fig. 1.9. In the figure, we do not clearly distinguish nodes at which we make a decision and nodes at which we observe results, but we will see a clearer representation of a dynamic decision-making process under uncertainty in Chapter 13, where we deal with decision trees. If we

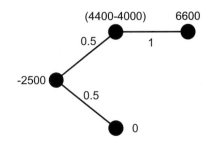

Fig. 1.9 Fancy coin flipping with memory.

calculate the expected NPV, we obtain

$$-2500 + \frac{0.5 \times (4400 - 4000) + 0.5 \times 0}{1.10} + \frac{0.5 \times 1 \times 6600}{(1.10)^2} = 409.09$$

Please beware of a common mistake. The last cash flow in the expression, 6600, occurs with probability 0.5, not 0.25. In order to evaluate the probability of each node in the scenario tree we must take the product of the conditional probabilities along the path leading from the root of the scenario tree (the "now" node) to that node. In our case, $0.5 \times 1 = 0.5$. This positive result shows that maybe our investment in the director was not so bad. Clearly, we should assess how much this conclusion depends on the probabilities that we have assumed. This process of checking the influence of uncertain data on the decision suggested by a model is called *sensitivity analysis*.

If you want to generalize a bit, a situation like this is common in risky research and development (R&D) projects. You could try a risky venture that, if successful, will pave the way to huge revenues by expanding the business. When analyzing investments, one should account for managerial flexibility and the possibility of learning over time by gathering new information. This approach leads to so-called real options;[17] in the real options literature, our example is known as a *growth* option.

Coin flipping in the short or long run   Before leaving coin flipping aside, it may be instructive to consider in more detail the idea of repeating bets. Suppose that you can play the plain coin-flipping game, where you may either win 10 million or lose 5 million of whatever monetary unit you like. The coin is fair and memoryless, and you can play the game for free, so that the expected payoff is 2.5 million.

1. Would you play the game once?

2. Would you play the same game 1000 times in a row?

---

[17]They are called "real," as opposed to financial options based on financial assets such as stock shares. We will consider options in a few examples in the following chapters.

The answer to the first question might be no, as the chance of being broke is far from negligible, even though the expected payoff is rich. But if you may repeat the game several times, you have the possibility of recovering your losses. In the long run, the bet is quite profitable indeed, but this is no solace if you can get out of the business after a losing streak of bad outcomes, without the time to recover.[18] Again, risk aversion does play an important role and considering long-run, or expected, profits may not be always appropriate. In other words, the plain expected value of the payoff fails to take into account risk in the short term. In the movie production case, we have taken for granted that a proper discount rate can take risk aversion into account, but the issue is far from trivial and is the subject of quite some controversy in corporate finance. We investigate the role of risk in decision making under uncertainty in Chapter 13.

1.3 ENDOGENOUS VS. EXOGENOUS UNCERTAINTY: ARE WE ALONE?

A colleague of mine draws the line between engineers (like myself) and economists as follows:

- An engineer believes that fellow human beings are a little dumb, but otherwise they are generally good chaps. All you have to do is to find a clever solution for them, and they will live happily ever after.

- An economist believes that fellow human beings are not dumb at all. The problem is that you cannot turn your back on them, unless you have a taste for an instant stab.

In fact, the need for analyzing the interactions between noncooperative decision makers was the driving force for the development of game theory. We will consider related issues in Section 14.3, where we investigate the role of misaligned incentives when multiple stakeholders are involved in a decision-making process. The interaction of multiple actors may also change the nature of the uncertainty involved in decision making. Simple probability models assume that uncertainty is known and fixed; if we have enough information, we can make a good decision. However, it might be the case that the very uncertainty is influenced by decisions.

Example 1.1 To illustrate, let us go once again back to the inventory management problem of Section 1.2.1. There, we used historical data to characterize demand uncertainty and planned on using that information to come up with a decision. One tough question is: Does our decision affect the demand

---

[18]As the economist John Maynard Keynes aptly put it, in the long run we are all dead.

distribution? To get the point, imagine that you stock very few items. In the long run, you will end up losing customers, not just orders. What's even worse, maybe this will also affect the demand for other items.   □

In practice, there are many cases in which demand is affected by stocking decisions. At retail stores, the shelf space allocated to a product may have a remarkable effect on demand. More generally, uncertainty need not be purely *exogenous*, i.e., given by outside factors that are not going to change (at least in the short term); uncertainty may be partially *endogenous*, i.e., influenced by our actions. Representing the dependence between our decisions and uncertainty may be very difficult, but at least one should be aware of the issue.

More generally, we should consider that management decisions are made in a social context, where other people could

- React to our decisions

- Use the same information we have

- Observe our behavior to gather information

Example 1.2 To see another example in a supply chain management context, suppose that we are selling a perishable item. We start selling it in the morning, and we know that at the end of the day we will scrap what's left. One could consider a dynamic pricing strategy, whereby price is marked down at the end of the day; after all, it seems better to recover some money even if this implies selling below cost. That's fine, but we are not considering the possibility of *strategic* behavior on the part of customers. Knowing that price will be reduced, they could wait and buy at dusk, just before closing time.[19] The point is that there is no separation between the set of customers who buy during the first part of the day and customers who buy later. We may apply price discrimination, i.e., asking different prices for the same product on different markets, provided they are well separated, possibly in space. The above revenue management policy is a sort of failed price discrimination strategy in time, rather than in space.

There could also be other reasons not to mark down some products. Some fresh and quickly perishable produce is not really sold for profit at retail stores. The real intent is to promote a positive image of "freshness" that has a positive impact on sales of other, more profitable items. The message would be lost by selling almost perished items in the evening.   □

Social issues are even more fundamental in finance. A whole discipline, behavioral finance, was born to investigate the interplay between psychology and

---

[19]This is what happened at the fish counter at a big retail store in front of my home. Very few people bought fish all day long, but queues materialized after 6 p.m., not to mention a terrible smell. As you may imagine, this brilliant revenue management policy disappeared rather quickly. Much to the chagrin of my nostrils, the smell didn't follow.

finance. When you buy a good, you know its selling price and you may plan accordingly. However, financial markets are based on a competitive auction mechanism, so that prices depend on investors' behavior, on their beliefs (or lack of information), and on their beliefs about other investors' beliefs. The resulting pattern may be quite complex, and indeed it leads to bubbles and crashes. Everything is made even more complicated by deliberate trading strategies.

Example 1.3 (Short selling) A *short sale* is a trading strategy whereby you sell an asset, say, a stock share, that you do not own. To do so, you have to borrow that asset from someone else. It is a strategy that makes sense if you expect the asset price to drop. To see this, assume that the current price is $S_0 = \$50$ at time $t = 0$. If you sell the asset short, you borrow the asset and sell it for $50. Clearly, you will have to give the asset back to its legitimate owner at some later time $t = T$, which means that you will have to buy the asset at a price S_T in the future. If you are right and the asset price falls, e.g., to $S_T = \$40$, your reward will be $10 per share.

At times of financial turmoil, wealthy speculators could do a lot of short selling, which may itself depress prices, resulting in a self-fulfilling prophecy. Selling an asset short in large volumes can lead to lower prices, even if the economic and business fundamentals of the firm are good. Uninformed traders may just follow the lead, just because they see a drop in price, even if they do not know why this is happening. A perverse feedback mechanism can be the result, and indeed short selling is sometimes prohibited, when markets are under severe pressure. The role of short sellers (the "shorts") is actually debated, but the point here is that the interaction between actors can result in weird patterns.

A similar speculative practice is *predatory trading*. Suppose that you own a large amount of an asset and you know that Mr. X has to sell a large amount of the same asset, because he is running short of liquidity. You could sell the asset before Mr. X, which will result in a significant price drop, as the large trading volume will affect prices. Mr. X will have to sell, anyway, which will depress prices further. Then, you may buy the asset back at a lower price, getting back to your previous portfolio, plus some extra cash. □

Empirical research is carried out to verify if certain speculative practices such as predatory trading are actually carried out at a significant level. These very empirical research lines do use a lot of statistical methods, and have a practical impact as market regulation should be devised in such a way to avoid excessive distortions. Empirical finance is beyond the scope of this book, but in the next section we analyze in more detail a somewhat stylized, yet very significant, example to see the interplay between organized markets and uncertainty.

1.3.1   The effect of organized markets: pricing a forward contract

An asset (e.g., a stock share) is sold now at a spot price $S_0 = \$50$, where $t = 0$ is current time. The spot price is the prevailing price at which the asset is exchanged at any moment. The spot price in one year is, of course, uncertain, but say that the expected price in one year is $E[S_1] = \$60$. A type of contract that is commonly traded on many assets is a forward contract.[20] This contract is signed between two parties who agree on exchanging the asset at some future time (say, one year) for a price that is agreed on *now*. This fixed price, that we denote by F_0, is the *forward price* at time $t = 0$. In one year, one of the two parties will sell the item to the other one at price F_0, no matter what spot price S_1 will prevail in one year. The party agreeing to buy is said to hold the *long position*; the party agreeing to sell is said to hold the *short position*. Depending on the relationship between the agreed-on forward price F_0 and the future spot price S_1, one of the two parties will turn out to be happy, at the expense of the counterpart. If it turns out that $S_1 > F_0$, the long position will benefit since she can buy at a price lower than the current spot price. The short position will benefit if $S_1 < F_0$, since she can sell at a higher price. A priori, there are different reasons for trading such contracts. One reason is hedging risks away: By fixing the price for the future, long and short parties eliminate risk ex ante. Ex post, one of them will regret the hedging decision, but risk management strategies should be evaluated *before* uncertainty is resolved. Another reason could be speculation of traders having firm beliefs about future spot prices and willing to bet on their forecasts.

Now the problem is: What should F_0 be? Actually, it is the balance between demand and offer that determines prices; still, there are many reasons why one should be interested in finding a sensible estimate of a fair price. One possible guess is that a fair price should be $E[S_1] = \$60$. Note that money and the underlying asset are exchanged in the future, so we do not need to discount cash flows, as the forward price F_0 will be paid in one year. We have seen before that using expected values to value assets or lotteries ignores risk aversion. Taking this into account, one could argue that the price should be a bit less than \$60. But how much less? There are many actors in the markets; whose risk aversion matters most? It seems that there is no hope to find a sensible forward price! By the way, each actor might have a different view of the future, as well as different information, but to keep it simple, let us assume that everyone agrees on the above expectation about S_1.

To get some clue, let us assume that the forward price is set as the expected future (spot) price, $F_0 = E[S_1] = \$60$. Also assume that we may lend or borrow money at an annual risk-free rate of 10%. In this setting, you could adopt the following trading strategy: Borrow \$50 now, buy the asset, and

---

[20] In practice, futures contracts are more commonly traded. They differ from forward contracts with respect to some institutional features that enhance their liquidity, such as standardization of contracts and measures to avoid defaults by one of the two parties.

enter a short position in the forward (i.e., you agree to sell the asset at the forward price in the future, no matter what the spot price will be). This trade is costless at time $t = 0$, since cash flows cancel each other and (in principle) no money is needed to enter a forward contract. In one year, you will pay back $55 to the guy who loaned you that money but, whatever the spot price, you will be able to sell the asset for $60, earning $5 *for sure*. If this were the case, we would have found the perfect money-making machine. Please note that you make money *without taking chances* and *without any need for initial cash*, as the initial net cash flow is zero. If you enter a long position at $F_0 = \$60$ and the spot price turns out to be $S_1 = \$100$, you will make $40 by using the forward contract to buy the asset at $60 and selling it immediately on the spot market. This is much better than the $5 you could make by the riskless strategy, but in doing so, you are taking chances, as the spot price could be well below the forward price and you would lose money. A riskless trading strategy that does *not* require any money is an example of an *arbitrage* opportunity.

Could we scale the investment up and make an arbitrary amount of money out of nothing?

Now, remember that you are not alone. If such an arbitrage opportunity were available, many other investors would follow your pattern and buy the underlying asset; this would increase the spot price (not to mention the fact that many people would borrow money, which would affect the interest rate). What could happen is not quite clear, as it depends on complex market dynamics, but we can conclude that those three numbers (spot price $50, forward price $60, risk-free interest rate 10%) are not consistent; they cannot be equilibrium prices and rates.

So, we know that $60 is too high a forward price. It is easy to see that any price higher than $55 leads to a similar arbitrage opportunity. On the other hand, assume that F_0 is smaller than $55, say, $52. In this case, you could reverse the trade by selling the asset short (see Example 1.3 on short selling). You borrow the asset, sell it at the current spot price $S_0 = 50$, invest the proceeds at the risk-free rate, and enter a long position in the forward contract. This long position comes in handy, as you will have to give the asset back to the lender in one year. You will do so at the given forward price, $52, after collecting the money you lent plus interest, which is now $55; then, you made $3 for free. A similar strategy can be applied for any forward price less than $55.

The conclusion is that the only arbitrage free forward price is $55. Generalizing a bit, what we found is that, in order to rule out arbitrage opportunities, the forward price should be

$$F_0 = S_0(1 + R_f)$$

which is a somewhat stunning conclusion: uncertainty does not play any role, as it is the risk-free interest rate R_f that determines the forward price. This conclusion should be taken with care, as it is at odds, e.g., with what we read on newspapers about speculation with futures on oil prices. In fact, the reasoning assumes somewhat idealized financial markets, and it can be more or less justified, depending on the actual asset we are dealing with. While a financial asset can be practically stored at no cost, oil cannot; by the same token, selling oil short is certainly not that easy. Pricing forward and futures contracts is affected by many issues such as transaction costs, differences between lending and borrowing rates, etc. Still, there is an important lesson here:

Uncertainty is important in decision making, but collective behavior may have some surprising effects on it.

1.4   QUANTITATIVE MODELS AND METHODS

Hopefully, the examples in the previous sections have shown the relevance, as well as the limitations, of quantitative analysis for real-life business decisions. In the following chapters we cover a wide variety of tools, which could be somewhat confusing for the reader. Hence, it is a good idea to set a conceptual framework to classify and interpret the different approaches, which should not be regarded as a disordered array of technicalities. Applying a quantitative analysis typically requires two steps:

- Building a quantitative *model* of a system, or part of it

- Solving it by some suitable *method*

To see the difference, consider the linear programming model to find the optimal mix in Section 1.1.2. After you have written the model down, a numerical solution procedure should be applied to come up with the answer we need. Lightning fast software packages to solve linear programming are widely available; hence, emphasis should be placed on model building, rather than model solving. In fact, data collection, a correct assessment of objectives and constraints, and the ability to interpret the solution are definitely more critical to success than plain number crunching.

Indeed, the title of the book is arguably somewhat misleading, even though quantitative methods is the standard name of courses covering these subjects. Nevertheless, it is often essential to have at least a rough idea of the inner working of solution methods, because

1. The way the model is formulated may have an impact on available methods, as well as their efficiency.

2. The solution method may be valid only if some assumptions about the data hold; ignoring these assumptions may result in gross mistakes if a method is applied to the wrong problem.

Another important healthy principle that we should keep in mind is[21]

All models are wrong, but some are useful.

Indeed, any model is a *simplified* representation of reality, and it need not be quantitative. Qualitative models are very useful in business process reengineering (BPR), and are often based on some graphical formalism to clarify relationships between actors, chain of events by activity diagrams, etc. Quantitative models, on the contrary, are based on *numerical* information.

1.4.1   Descriptive vs. prescriptive models

A quantitative model can be

- *Descriptive*, if its purpose is to shed some light on the relationships between two (or more) variables of interest (e.g., do sales depend significantly on advertisement expenditure?) or to predict system performance as a function of some design variable (e.g., the average waiting time in the queue, as a function of the number of tellers in a bank).

- *Prescriptive*, when the aim is (more ambitiously) to find a solution, subject to economical or technological constraints, so that costs are minimized or profits are maximized (see, e.g., the optimal mix problem).

Typical examples of descriptive models that we cover in the book are

- Simulation models for performance evaluation (Section 9.7)

- Linear regression models (Chapters 10 and 16)

- Time series models for forecasting (Chapter 11)

All of these models are used to generate information that helps in coming up with a decision, but they are not aimed at generating the decision directly. Examples of prescriptive models whose output is the decision itself are

- The economic order quantity model (Section 2.1)

- Linear programming models (Chapter 12)

It should be clear that prescriptive models are more ambitious and, in principle, they could even be used to automate the decision process. This applies to strictly technical problems, especially when the time to make a decision is quite limited. In most business settings, however, prescriptive models should be regarded as decision *supports*, and we must be aware of their limitations:

---

[21] The quote is generally attributed to the statistician George E.P. Box, 1979.

- The output is affected by data uncertainty (which is not only of a statistical nature; demand is uncertain because we cannot forecast the future exactly, whereas some cost are hard to quantify–how much is the inventory holding cost for an item?). Sensitivity analysis should be an integral part of a quantitative analysis.

- Some tradeoffs between conflicting objectives are difficult to assess quantitatively, possibly requiring the interaction between multiple stakeholders.

- More often than not, we have to approximate parts of a problem to make it tractable, and expert judgment is needed to evaluate the impact on these simplifications on the viability of the model and of the solution that we obtain from solving it. This process is called *model validation.*

Because of these limitations, it is sometimes argued that quantitative analysis should be confined to academia, and that common sense and a lot of practical experience are what is really needed. It is certainly true that human knowledge is an extremely valuable asset; what is not true is that it is incompatible with quantitative analysis. Furthermore, there is another more and more important factor: time. The rate of change in business conditions is faster and faster. For instance, when you change a product line, you have to redesign the whole supply chain to support its production. There is simply no time to do that manually, and some automatic support is needed. Intuition and experience are essential, but they are not enough.

1.5   QUANTITATIVE ANALYSIS AND PROBLEM SOLVING

Even if the problem is too complex to rely on the decision proposed by the solution of a model, we should not underestimate the value of model building per se. The model building process itself is a valuable activity as it requires the following ingredients:

- *Gathering data.* Quite often, complex organizations do not pursue a disciplined approach to data management. Important data are missed, some are duplicated with possible inconsistencies, some are not shared between different offices, and errors are not discovered because no one really uses that information. The need to collect data to build a model may force an improvement of the related processes. Furthermore, model building can help in transforming a huge amount of useless data, into useful and shared *information.*

- *Structuring the problem.* This requires sharing information and understanding the multiple dimensions of a problem, as well as the points of view of other stakeholders. This is important in large organizations, where conflicting views are the rule rather than the exception, and entities within a firm pursue hidden agendas without agreeing on a shared understanding. It is always important to keep in mind that any quantitative analysis is doomed to failure if some relevant actor is not involved and motivated.

If model building and model solving result in a solution to the business problem, the solution itself must be implemented, monitored, and adapted when required by new circumstances. Fostering the culture and the discipline needed to monitor a solution and to assess the improvement in key performance indices is again valuable per se.

We close this section by remarking that quantitative analysis can play an important role in the following circumstances:

- When an objective support for decisions is needed. Rationalizing the analysis may ease potential conflicts between stakeholders.

- When the decision process leading to a recommendation must be explicitly documented. Because of uncertainty, even the best decision may result in a bad outcome. If you can back your decision with serious analysis, chances are that you will be able to save your neck.

- When the relationship between variables is too complex to be analyzed intuitively; in such a case, intuition can lead to wrong decisions because we are not able to fully grasp the impact of decisions.

- When the number of decision variables is too large to be managed even by the best human experts.

- When there are many difficult constraints and even finding a *feasible* (let alone optimal) solution is very hard. One example is train timetabling, which is a daunting task because of the number of shared resources (trains, crews, rails) and the constraints on their use, such as rules constraining how personnel shifts are scheduled.

- When tradeoffs between conflicting criteria must be assessed objectively (e.g., customer service vs. inventory holding cost).

Problems

1.1 Consider again the growth option problem of Section 1.2.2. We want to check the impact of less extreme assumptions about the conditional probabilities for the second movie, but we are unsure which values we should use. So, we assume that the conditional probability of a second success, after a first success, is $0.5 + \alpha$, for some unknown value of α; this is also the probability of a second flop, after a first flop. The analysis in Section 1.2.2 shows that if $\alpha = 0.5$, we should produce the first movie. What is the limit value of α below which we should change our mind?

1.2   Consider the optimal mix model of Section 1.1.2. How could we extend the model to cope with

- Third-party suppliers offering items at given cost?

- The possibility of overtime work at some resource centers?

For further reading

- There are many books covering quantitative methods for business applications. Most of them deal only with one side of the coin, either probability/statistics or decision models. A book illustrating many different kinds of applications is Ref. [6], where many kinds of problems and methods are presented, even though skipping many details and applicability conditions. Another book aimed at a relatively broad treatment of quantitative models and methods is Ref. [2], which covers less material than the previous reference, but at a deeper level, and is also rich in interesting cases.

- The practical impact of quantitative analysis on business is emphasized in Refs. [1, 7].

- Last but not least, depending on your personal taste, you might be interested in books covering the application of quantitative analysis to specific application domains:

  Logistics. A wide range of applications related to supply chain management and logistics are covered in Ref. [4], ranging from strategic network design to operational routing of vehicles.

  Operations management. Both descriptive and prescriptive models for production planning and control are covered in Ref. [10]; in Ref. [3] emphasis is given to prescriptive models.

  Pricing and revenue management. Dynamic pricing strategies and revenue management are an essential tool for certain kinds of business application; a relatively informal treatment is given in Ref. [13], whereas Ref. [15] is more challenging and illustrates the application of sophisticated quantitative models.

  Marketing. Interesting applications of statistics to marketing are illustrated in Ref. [12].

  Finance. This is a domain where the role of quantitative analysis has boomed in the last two decades, with a considerable trail of controversy; introductory books that can help you appreciate the role of quantitative models in financial markets are, e.g., Refs. [11] and [14]; a good reference for corporate finance is Ref. [9]; you may also appreciate the connection between the two fields in Ref. [8], which discusses option valuation for investment analysis.

REFERENCES

1. S. Baker, *The Numerati*, Houghton-Mifflin, Boston, 2008.

2. D. Bertsimas and R.M. Freund, *Data, Models and Decisions: The Fundamentals of Management Science*, Dynamic Ideas, Belmont, MA, 2004.

3. P. Brandimarte and A. Villa, *Advanced Models for Manufacturing Systems Management*, CRC Press, Boca Raton, FL, 1995.

4. P. Brandimarte and G. Zotteri, *Introduction to Distribution Logistics*, Wiley, New York, 2007.

5. L.M.B. Cabral, *Introduction to Industrial Organization*, MIT Press, Cambridge, MA, 2000.

6. J. Curwin and R. Slater, *Quantitative Methods for Business Decisions,* 6th ed., Cengage Learning EMEA, London, 2008.

7. T.H. Davenport, J.G. Harris, and R. Morison, *Analytics at Work: Smarter Decisions, Better Results*, Harvard Business Press, Boston, 2010.

8. M.E. Edleson, *Real Options: Valuing Managerial Flexibility*, Harvard Business Publishing, 2002, (Teaching Note 9294109).

9. D. Hillier, S.A. Ross, R.W. Westerfield, J. Jaffe, and B.D. Jordan, *Corporate Finance*, McGraw-Hill, New York, 2010.

10. W. Hopp and M. Spearman, *Factory Physics,* 3rd ed., McGraw-Hill, New York, 2008.

11. D.G. Luenberger, *Investment Science*, Oxford University Press, New York, 1998.

12. J.H. Myers and G.M. Mullet, *Managerial Applications of Multivariate Analysis in Marketing*, South-Western, Cincinnati, OH, 2003.

13. R.L. Phillips, *Pricing and Revenue Optimization*, Stanford University Press, Stanford, CA, 2005.

14. S.M. Ross, *An Introduction to Mathematical Finance: Options and Other Topics*, Cambridge University Press, Cambridge, UK, 1999.

15. K.T. Talluri and G.J. Van Ryzin, *The Theory and Practice of Revenue Management*, Springer, New York, 2005.

2

Calculus

Calculus is a classical branch of mathematics, dealing with the study of functions. A function is essentially a rule for association of one or more input variables with an output value. For instance, we might be interested in relating a decision, say, how much to produce, to the business outcome, say, profit. In Chapter 1 we have seen that this is needed, for instance, to figure out the best mix of products. Building a relationship linking managerial levers and the resulting outcome is essential in tackling any decision problem. In other cases, our aim is somewhat more instrumental. In statistical model building, we want to find a mathematical representation that yields the best fit between the empirically observed data and the predictions of the model. To accomplish this task, first we need to choose a functional form depending on some unknown parameters, and then we must choose another function expressing the lack of fit, which in turn is related to prediction errors. Empirical model building calls for the minimization of such lack of fit. In real life, we deal with functions of many variables, but in this chapter we just deal with functions of one variable, i.e., rules mapping one input value x to an output value $y = f(x)$. We may introduce all of the required concepts in this simplified setting. Later, precisely at the end of the next chapter, we will generalize to functions of multiple variables.

Given a function, the first task that comes to our mind is plotting it, in order to get an intuitive feeling for the relationship between x and y, and to analyze the sensitivity of the output to variations in the input. Typical questions are as follows:

- How will a slight variation in interest rates affect an investment decision?

- How will a change in unit cost of some raw material affect total profit from manufacturing a product?

To deal with these important issues, we introduce a fundamental concept: the derivative of a function. The same concept is also essential in optimizing a function, i.e., in finding decisions maximizing profit or minimizing cost. Studying the related issues, such as convexity and concavity of functions, paves the way for later chapters dealing with optimization models. We will also hint at some other important topics such as sequences, series, and integrals, even though our treatment will be quite brief, and limited to what is really essential to understand a few topics in later chapters.

The treatment below is reasonably self-contained, as we just take for granted a basic familiarity with sets and real/integer numbers. Many readers are probably familiar with the topics covered in this chapter. If so, the best way to proceed could be just skimming through the chapter and return here, if needed, to clarify a concept used in more application-oriented chapters. In general, our treatment will leave much to be desired in terms of mathematical rigor.[1] A more solid foundation would certainly be necessary to prove some useful results, and to characterize and deal with some pathological cases. However, we just strive to get an intuitive understanding of a few fundamental concepts of calculus, which is more than adequate from our perspective; the interested reader is referred to references for a deeper treatment.

We start in Section 2.1 with a motivating example from inventory control theory, the economic order quantity (EOQ) model. Then in Section 2.2 we provide readers with a little background on numbers, intervals, and permutations. Functions, the core business of calculus, are introduced in Section 2.3, followed by Sections 2.4 on continuous functions, 2.5 on building functions by composition, and 2.6 on inverting functions. As we mentioned, one of the main tools of calculus is the derivative of a function, which is introduced in Section 2.7; practical rules for finding the derivative of a function are dealt with in Section 2.8. The first application of derivatives is in graphing functions, as shown in Section 2.9; then, in Section 2.10, we illustrate their role in sensitivity analysis and in approximating complicated functions by Taylor's expansions. Section 2.11 lays down the foundations of optimization methods, including the essential concepts of convexity and concavity. We close the chapter by outlining very briefly two other useful tools: series in Section 2.12, and definite integrals in Section 2.13. Later, in Section 3.9, we will briefly deal with derivatives and integrals of functions depending on multiple variables.

---

[1] For instance, we will never use the epsilon–delta approach to properly define limits and continuous functions. We believe that, for our purposes, an intuitive understanding of these concepts is sufficient.

2.1 A MOTIVATING EXAMPLE: ECONOMIC ORDER QUANTITY

Before getting into formal details of calculus, it is essential to arm ourselves with some motivation for doing so. In management science, we often want to relate decisions to cost or profit. This is necessary in order to find an "optimal" decision yielding the best performance in some well-specified sense. We should quote "optimal," because our decision is just optimal with respect to our cost or profit *model*. It is often said that all models are wrong but some are useful. Indeed, a model can be a quite rough representation of a possibly complicated and uncertain reality. Still, it can be useful to sharpen our understanding and to find a good decision. We should never forget that a quantitative model is just a decision aid and not a magical oracle. A case in point is the economic order quantity (EOQ) model. This is a somewhat old-style, archetypal model for inventory management, relying on a lot of debatable assumptions; yet, it can have some value in practice and is a good way to show mathematical modeling and calculus in action.

Imagine that you are in charge of managing the inventory of an item at a retail shop. Inventory is depleted at a rate depending on demand, and every now and then you have to replenish inventory by issuing a purchase order to a supplier, or maybe a production order to your own manufacturing facility. To keep it simple, let us assume that demand is perfectly constant in time and let d denote the demand rate. The demand rate is the demand per unit time, such as 10 items per day or 75 items per week. The specific time unit is not important, provided we are consistent in specifying all of the relevant data.

Which kind of decisions are you supposed to make? Actually, there are two decisions involved here:

1. When to order

2. How much to order

The first decision, in an ideal world, is actually trivial. By "ideal" world we mean a world in which there is no uncertainty involved:

- Demand is perfectly predictable (even constant, under our assumption); we also assume that demand is "fluid," in the sense that we approximate the discrete flow of items out of inventory, one piece at a time, as a continuous flow.

- The time it takes to get a shipment from the supplier is fixed and known; this time is known as *lead time* or *time to delivery*.

In this setting, you should issue a replenishment order when inventory reaches a level corresponding to demand during lead time. For instance, if time to delivery is 2 days and the demand rate is 10 items per day, it is easy to see that

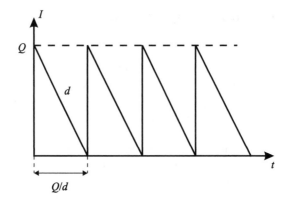

Fig. 2.1 Time evolution of inventory levels in the EOQ model.

you should issue an order when your inventory level decreases to 20 items.[2] Doing so, you will receive the new stuff exactly when you run out of stock! Clearly, such magic does not happen in real life, but for our little example we may ignore issues related to variability in time to delivery and timing of orders.

As to the second decision, since demand is constant, it is reasonable to assume that whenever you replenish, you always order the same amount, i.e., your preferred order size. Let Q be the order size we choose and, without loss of generality, assume that we start with Q units on hand. The resulting inventory pattern will look like Fig. 2.1, where I denotes the inventory level and t denotes time. We see that the inventory pattern is periodic, as we issue an order every Q/d time units. For instance, if the demand rate were 100 items per week, an order size $Q = 300$ would result in a replenishment order every 3 weeks. In practice, the inventory plot would be a stepwise function, as we sell one item at a time, and not fractions of items; yet, if demand volume is large enough, a continuous pattern like the one in the figure is a fair approximation. The case of sporadic demand for a niche product should be addressed by a different modeling approach.

So, under our quite strong assumptions, we have just one decision to make: Finding the right amount Q to order, i.e., the economic order quantity (EOQ). Indeed, this is a matter of compromise between conflicting needs:

---

[2]In real life, ordering decisions are never just based on on-hand (physically available) inventory, since we should take into account the possibility of replenishment orders that have already been issued to suppliers, but have not been received yet. Furthermore, when dealing with uncertainty, you should take into account the possibility of backlogged customer orders that you were not able to meet immediately, because you ran out of stock. For a more in-depth analysis, please refer, e.g., to Chapter 5 of Ref. [4].

- Imagine that, whenever you issue an order, a fixed charge[3] is incurred. For instance, if your supplier ships a container, quite reasonably the shipping charge will include a fixed component that does not depend on what and how much is transported. Hence, from this perspective you would like to order a rather large amount, whenever you replenish, in order to spread the fixed ordering charge on as many items as possible. Put another way, the less orders are issued per year, the better.

- On the other hand, you would like to order just a few items quite often. Doing so is important when items, such as fresh produce, are perishable and their limited shelf life precludes stocking large amounts. Perishability should not be confused with a similar issue, obsolescence, which is typical of fashion products or items characterized by a fast pace of technological innovation, such as consumer electronics.[4] Even if you do not consider such issues, there is a good financial reason to keep inventories low. When you receive a shipment from your supplier, you pay her, and this means that you have tied up some liquid cash to your inventory. When you sell, you will get your money back, plus a profit margin, but having too much liquidity immobilized there may be a bad idea. Apart from this opportunity cost of capital, large on-hand inventories also imply high insurance costs. Even worse, you may end up with a huge inventory of a slow-moving item, locking liquidity and precluding the purchase of items that are in demand, unless you borrow money from a bank, incurring possibly high interest rates.

Note that, in our setting, we should not consider profit, since we are assuming that we will satisfy all of the customers anyway; hence, our problem is just satisfying demand at minimum cost, and this in turn requires spotting the right compromise between the two conflicting requirements above. How can we do that?

There are three tasks that we should accomplish in sequence:

1. We should make the relationship between the decision Q and the total cost explicit. To this aim, we may write down a function $C_{tot}(Q)$ mapping order size Q into total cost; to be more precise, we will consider the average total cost per unit time. To fix ideas, we assume that our time unit is exactly one year.

2. We should sketch a graph of the total cost function to see what impact our decisions have on average total cost per year.

---

[3]There may be some ambiguity between fixed *charge* and fixed *cost*. We use the latter term to refer to a cost that is paid, whether we execute an activity or not; the only way to avoid the fixed cost is shutting our business down. By *fixed charge* we mean a cost that we pay whenever we execute an activity, like ordering a batch of items. If we do not order, the fixed charge is not paid.

[4]In a market with fast obsolescence, the problem is best tackled by using variants of the newsvendor model; see Example 6.9 and Section 7.4.4.

3. Finally, we should find the best decision, i.e., the optimal order size Q^*, such that the cost function attains its minimum.

2.1.1 Task 1: representing the total cost function

In order to express average total cost per unit time as a function of the order size Q, we should consider all of the factors contributing to the overall cost. The first one that comes to mind is purchase cost. If the unit item cost is c, measured in money per item, we have to pay cQ whenever we replenish; this variable cost per order must be translated in terms of cost per unit time. If we issue an order every Q/d years (our time unit), then we will issue d/Q orders per year, on the average. For instance, if demand rate is 1200 items per year and the order size is $Q = 100$, we will issue 12 orders per year. Hence, the contribution of purchasing to total cost is just the product of cost per order times average number of orders:

$$C_{\text{pu}} = cQ \cdot \frac{d}{Q} = cd$$

On second thought, this is obvious: In order to satisfy total demand during one year, we have to order d items, anyway, whether in small or large batches. We notice immediately that this term does not depend on Q, as we are disregarding possible discounts based on ordered quantity, which would make the unit cost a function $c(Q)$ of the order size.

Now let us consider the contribution to total cost from the fixed-charge component. Whenever we order, we have to pay A (euros, dollars, or whatever), and we have already found that the average number of orders we issue per year is d/Q, on the average. Then, the contribution from the fixed ordering charge is

$$C_{\text{or}} = A \cdot \frac{d}{Q}$$

Finally, we need the contribution due to inventory holding. This is a bit trickier, and we should clarify which kind of cost we are dealing with. If we keep one item in inventory for one year, we incur a cost. Let us denote this unit inventory holding cost by h. It is important to realize that the dimensions of this unit inventory holding cost are money per item, per unit time; in a sense, this is a "twice" unit cost, with respect to volume and with respect to time. There are a few ways to figure out the contribution from holding cost and all of them require a careful look at Fig. 2.1. There, we see that the inventory level ranges between 0 and Q and changes according to a smooth and constant rate. The pattern is repeated every Q/d time units. Consider an order for Q items. A good question is: How much time does an item of that shipment spend sitting in our warehouse? Actually, the first item that gets issued from inventory waits no time at all, in our abstract model; it is placed into the warehouse when we receive the shipment, but it is immediately sold.

The less lucky item is the last one, as it will wait Q/d time units. On average, the waiting time for an item will just be the average between the two limit values 0 and Q/d, i.e., $Q/(2d)$. Since there are d items going through the inventory each year, the contribution to total cost from inventory holding is

$$C_{\text{in}} = h \cdot d \cdot \frac{Q}{2d} = h\frac{Q}{2}$$

As an alternative way to obtain the same result, we may consider the average inventory level. Since the inventory level ranges between Q to 0 and changes uniformly in time, its average is $Q/2$. Multiplying average inventory level by h yields the result above. This alternative view has the advantage of being generalized to any inventory pattern, not necessarily a constant and uniform one. Now we are ready to put all of it together. The average total cost per year, as a function of Q, is

$$C_{\text{tot}}(Q) = C_{\text{in}} + C_{\text{or}} + C_{\text{pu}} = h\frac{Q}{2} + \frac{Ad}{Q} + cd$$

One thing is immediately clear: The purchase cost component is constant and does not play any role in determining the optimal order size. If you have any difficulty in seeing this, please have another look at Fig. 1.2.

As we see, a function $f(\cdot)$ is essentially a rule mapping an *independent* variable, Q in our case, into a value $y = f(Q)$ of a *dependent* variable. Notations like $f(\cdot)$ or f are used to emphasize the difference between the function itself (a rule to map variables to values) and the output value $f(Q)$ taken for an input value Q (a specific numerical value).

2.1.2 Task 2: plotting the total cost function

Having figured out a relationship between the order size and the average total cost per year, it would be useful to plot the function in order to see the effect of Q and to figure out a good decision. There are plenty of powerful software packages that, given a range of the independent variable Q, compute the corresponding values $f(Q)$ and display the result graphically. This is certainly quite useful, but we should also be able to figure out some basic properties of the function by just looking at its expression. There are a few reasons justifying the need for this ability:

- It is helpful when evaluating the function numerically is difficult or prone to numerical errors.

- It is often the case that a function cannot be evaluated for some values of the independent variable, but a piece of software will not help you in understanding what is going wrong.

- In more complicated settings, we may not just rely on brute computational force; to see this, imagine plotting a function depending on more than two independent variables.

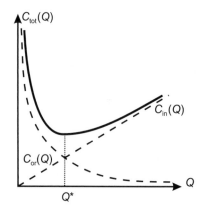

Fig. 2.2   Inventory holding and fixed ordering charge components in the EOQ model.

- Last but not least, this exercise helps in honing skills and improving understanding.

In general, plotting a function can be difficult, but it is fairly easy in the case of the EOQ cost model. Disregarding the constant term, the essential form of the total cost function is

$$C_{\text{tot}}(Q) = aQ + \frac{b}{Q}$$

where $a = h/2$ and $b = Ad$. From our high school background, we should immediately see that this function is just the sum of a straight line aQ, going through the origin, and a hyperbola b/Q (see Fig. 2.2). To get a rough picture of the function, let us consider the two extreme cases:

- When Q is very small, the linear component aQ is small, too, but the nonlinear component b/Q gets larger and larger; in fact, for a very small order size, the inventory holding cost component goes to zero, but the fixed-charge component increases without limits, because we are spreading the fixed ordering charge over a tiny number of items. We cannot evaluate $C_{\text{tot}}(Q)$ for $Q = 0$, since we cannot divide by zero, but we can write

$$\lim_{Q \to 0^+} C_{\text{tot}}(Q) = +\infty$$

This notation has a precise mathematical meaning, but the informal interpretation is that the *limit* of the total cost goes to (plus) infinity, when Q *tends* to zero. Notations like

$$\lim_{Q \to 0^+} \qquad \text{or} \qquad \lim_{Q \downarrow 0}$$

are used to indicate that Q goes to zero while staying on the positive side. (A negative order size makes no sense!) We say that Q goes to zero "from the right" or "from above." So, in the range of small order sizes, the nonlinear component prevails and the function looks like a hyperbola.

- When Q is very large, the linear component aQ is large, too, as large holding costs are incurred. On the other side of the coin, the nonlinear component b/Q goes to zero, since the fixed charge component is spread on a huge amount of items. In this range of large order sizes it is the linear component that prevails, and the function looks like a straight line. In this case too, the limit of the cost function goes to infinity:

$$\lim_{Q \to +\infty} C_{\text{tot}}(Q) = +\infty$$

The result of these observations is a sketch like that in Fig. 2.2. Indeed, we see that there are very bad decisions, corresponding to very large or very small order sizes, and some good ones, for which cost is much lower. Now we have to find the optimal compromise.

2.1.3 Task 3: finding the best decision

In plotting the function, we have ignored the purchase cost component cd, which is constant and would just push the graph up a bit. This is not relevant to us, since what we are interested in is finding an order size Q^* minimizing total cost. Indeed, since the function goes to infinity for very small and very large order sizes, we would expect that somewhere in between there is an optimal order size, the EOQ. One possible way of finding the EOQ is by trial and error, i.e., by calculating the cost for several input values and spotting the best choice. Unfortunately, this brute-force approach is time-consuming, not quite informative, and not feasible in more complicated cases. In this chapter we will learn a more straightforward way to spot a minimum-cost or maximum-profit solution, based on the concept of function derivative. For now, we can observe that the minimum cost solution is a point where the tangent line to the function graph is horizontal. This is illustrated in Fig. 2.3(a): Three tangent lines are shown and, indeed, the minimum-cost solution is where the tangent line is horizontal. It seems that, if the tangent line corresponding to an order size is not horizontal, we may find an improvement by moving to the left or to the right. Actually, this reasoning is not 100% correct, and this intuition needs some refinement. Still, using concepts explained later along these lines, it will be easy to see that the optimal order size, according to the EOQ model, is

$$Q^* = \sqrt{\frac{2Ad}{h}} \tag{2.1}$$

We see that the EOQ size is increasing with respect to fixed charge A and decreasing with respect to inventory holding cost h, which is quite reasonable.

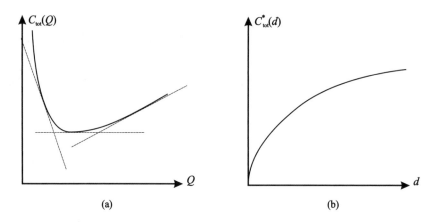

Fig. 2.3 (a) Tangent lines to the total cost function, and (b) the optimal cost in the EOQ model for varying demand rate d.

If we plug the optimal order size into the total cost function, we get the average cost per year for the optimal solution as a function of the demand rate d:

$$C_{\text{tot}}^*(d) = \sqrt{2Ahd} + cd \qquad (2.2)$$

This function is plotted in 2.3(b), and it is interesting to note its shape. As expected, it is an increasing function of the demand rate. This is no surprise, after all: The larger the demand that must be satisfied, the larger the incurred cost. What is not that obvious is that the rate at which the function increases is decreasing. To see this, imagine drawing tangent lines at different points on that graph. The slopes of the tangent lines are decreasing with respect to demand rate d. This is a typical behavior of cost functions exhibiting economies of scale. We will learn later that a function like that is a *concave* function.

2.2   A LITTLE BACKGROUND

As we have already pointed out, the reader is assumed to be equipped with a basic mathematical background about sets as well as integer and real numbers. In this section we briefly recall a few basic concepts for convenience.

2.2.1   Real vs. integer numbers

If we order cars from a car manufacturer, we cannot order 10.56986 cars; we may order either 10 or 11 cars, but any value in between makes no sense. It should be intuitively clear what we mean by an *integer number*; integer numbers are used to measure variables that have a intrinsically discrete nature. A *real number* is a number with some, possibly infinite, decimal part. Real

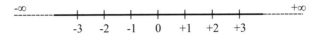

Fig. 2.4 The real line.

numbers are not that obvious to characterize formally, but intuitively a real
variable is used to measure something that has an intrinsically continuous (or
"fluid") nature. In the following, we denote the set of real numbers by $\mathbb{R}$
and the set of integer numbers by $\mathbb{Z}$. If we are just interested in nonnegative
numbers, we use notations $\mathbb{R}_+$ and $\mathbb{Z}_+$.

Real numbers can be represented on a real line, as illustrated in Fig. 2.4,
ranging from $-\infty$ (minus infinity) to $+\infty$ (plus infinity). Given any pair of
real numbers on the real line, you can always find a real number between
them. For instance, given

$$x_1 = 2.3982, \qquad x_2 = 2.3983$$

we can actually find infinitely many real numbers between them; one example
is $x = 2.3982133$. The set of real numbers is "dense," whereas the set of
integer numbers is not: There is no integer number between 10 and 11.

2.2.2 Intervals on the real line

Inequalities like $a \leq x \leq b$, where a and b are arbitrary real numbers such
that $a < b$, define intervals on the real line. The inequality above defines an
interval that includes its extreme points. In such a case, we use the notation
$[a, b]$ to denote the interval, and we speak of a *closed interval*. On the contrary,
inequalities $a < x < b$ define the *open interval* (a, b). For instance, the
point $x = 10$ belongs to the closed interval $[0, 10]$ but does not belong to
open interval $(0, 10)$. We may also consider open–closed intervals like $[a, b)$,
corresponding to inequalities $a \leq x < b$, or $(a, b]$, corresponding to inequalities
$a < x \leq b$.

If both a and b are finite, we have a bounded interval. We may also consider
unbounded intervals by extending the real line to include $\pm\infty$. Examples of
unbounded intervals are

- $(-\infty, a]$, corresponding to $x \leq a$

- $[a, +\infty)$, corresponding to $x \geq a$

- $(-\infty, +\infty)$, which is the whole real line

Note that there are infinite integer numbers in $\mathbb{Z}$, as well as infinite real
numbers in $\mathbb{R}$. However, we cannot say that the two sets have the same
"order of infinity." To see this informally, note that there are infinite real
numbers even in a bounded interval like $[0, 1]$. To get an infinite set of integer
numbers, we have to consider an unbounded interval.

Whenever we have an infinite collection of items that can be counted, i.e., can be placed in correspondence with the set of integer numbers, we speak of a *countably infinite* set. For instance, a sequence of arbitrary real numbers x_k, for $k = 1, 2, 3, \ldots$, is an infinite sequence, but it includes less points than the bounded interval $[0, 1]$. More generally, we speak of *denumerable* or *countable* sets, including also finite collections of elements.

2.2.3 The sum notation

Consider an expression like

$$x_1 + x_2 + x_3 + x_4$$

We will meet similar expressions quite often in the book, and a nice shorthand notation for this expression is

$$\sum_{i=1}^{4} x_i$$

which should be read as the sum of "x subscript i," for i ranging from 1 to 4. Sometimes, the sum limits can be symbolic, as in

$$\sum_{i=1}^{n} x_i = x_1 + x_2 + x_3 + \cdots + x_{n-1} + x_n$$

We may even consider an infinite sum like

$$\sum_{i=1}^{+\infty} x_i = x_1 + x_2 + x_3 + \cdots$$

In this case, we should wonder whether the expression makes sense, since summing an infinite number of terms may result in a sum going to infinity. A thorough study of the involved issues requires the theory of mathematical series; in Section 2.12 we deal with a few examples that are most relevant in applications. In some cases, we might wish to skip values corresponding to some subscript:

$$\sum_{\substack{i=1 \\ i \neq 2}}^{5} x_i = x_1 + x_3 + x_4 + x_5$$

Finally, when variables have two subscripts, we may consider double sums like

$$\begin{aligned}
\sum_{i=1}^{3}\sum_{j=1}^{4} x_{ij} &= \sum_{j=1}^{4} x_{1j} + \sum_{j=1}^{4} x_{2j} + \sum_{j=1}^{4} x_{3j} \\
&= (x_{11} + x_{12} + x_{13} + x_{14}) + (x_{21} + x_{22} + x_{23} + x_{24}) \\
&\quad + (x_{31} + x_{32} + x_{33} + x_{34})
\end{aligned}$$

In this example we also see that the order of the two sums is irrelevant and they may be swapped:

$$\sum_{i=1}^{3}\sum_{j=1}^{4} x_{ij} = \sum_{j=1}^{4}\sum_{i=1}^{3} x_{ij}$$

We will always take for granted that this is the case, even though some care should be taken with infinite sums. When sum limits are irrelevant, we may use streamlined notations like

$$\sum_{i,j} x_{ij} \qquad \text{or} \qquad \sum_{i\neq j} x_{ij}$$

The latter notation comes in handy when we want to exclude terms of the form x_{11}, x_{22}, x_{33}, etc.

2.2.4 Permutations and combinations

Many practical problems involve permutations and combinations of objects. A first question is: Given a collection of n objects, in how many ways can we permute them? For instance, let us consider the set $\{a, b, c\}$. Since the set is quite small, we can enumerate all of the possible permutations systematically. First we consider permutations beginning with a; we can form two such permutations, (a, b, c) and (a, c, b). Then, considering permutations starting with b, we have (b, a, c) and (b, c, a), and by the same token we have (c, a, b) and (c, b, a), beginning with c. Hence, there are six possible permutations of three objects, but what can we say in general for n objects? Again, let us be systematic in our enumeration of permutations:

- When we choose the first item in the sequence, we have n possibilities.

- When we choose the second item, we have $n - 1$ possibilities, since one object has already been used to fill the first slot.

- When we choose the third item, we have $n - 2$ possibilities, since two objects have already been used to fill the first and second slots.

- When we have only two items to go, we can choose among two.

- When we have one item left, there is just one possible choice.

Hence, the total number of permutations of n objects is

$$n \times (n - 1) \times (n - 2) \times (n - 3) \times \cdots \times 2 \times 1$$

The kind of product above is so common that it has been given a name.

DEFINITION 2.1 (Factorial) *Given a positive integer number n, the factorial of n, denoted by n! is defined as*

$$n! \equiv n \times (n-1) \times (n-2) \times (n-3) \times \cdots \times 2 \times 1$$

The factorial can also be defined recursively as

$$n! = n \times (n-1)!$$

and, by convention, we set $0! = 1$.

A noteworthy feature of the factorial function is that it grows very quickly with n:

$$3! = 6, \quad 5! = 120, \quad 10! = 3{,}628{,}800$$

This process is known as *combinatorial explosion* and has practical implications.

Example 2.1 Imagine that you are a traveling salesperson. You live in a city, and your task is to visit customers living in other cities and then come back home. Let n be the number of cities you have to visit, not including yours; all of them must be visited exactly once. Arguably, you would like to follow the shortest route in your tour. Geographic information systems can provide you with distances d_{ij} between any pair of cities (i, j), where $i \neq j$ and the indices range over the set $\{0, 1, 2, \ldots, n\}$ (say that 0 is your city). How can we find the optimal tour?

One obvious idea is that, since there must be a finite number of tours, we could just enumerate all of them and pick up the shortest one. Using a fast computer, this should be no big deal. But how many tours must be evaluated? We have $n + 1$ cities to visit. There are $(n + 1)!$ permutations of them, but actually only $n!$ are really different. One way of seeing this is observing that the first city is your home and is fixed. An alternative view is that we do have $(n+1)!$ different tours, since there are $n+1$ possible starting points, but the total length is not really influenced by the starting city as the tour is a closed cycle; hence, the number of different tours in term of total length is $(n+1)!/(n+1) = n!$. The problem is further simplified if we assume symmetric distances, i.e., $d_{ij} = d_{ji}$. In fact, there is little difference in terms of mileage between traveling from Boston to New York or from New York to Boston. This may not apply on a smaller scale: Within a city, one-way streets make distances asymmetric. If we assume symmetry, we have two equivalent ways of traveling along each tour; you may visualize them as a clockwise and a counterclockwise tour. Hence, the total number of different tours is $n!/2$.

Now, say that you have to visit 25 customers. Using any pocket calculator, we obtain

$$25! \approx 1.55112 \times 10^{25}$$

Some difficult number to read! It means that there are about 15.5112 millions of billions of billions of possible permutations of 25 cities. Luckily, we must only consider half of those, and we also have a very fast computer. Say that we are able to generate and evaluate one billion permutations, i.e., 10^9, per second. Since there are 3600 seconds per hour, 24 hours per day, and 365 days per year, we will get the optimal tour after

$$\frac{1.5511 \times 10^{25}}{2 \times 10^9 \times 3600 \times 24 \times 365} = 245{,}928{,}621.95 \quad \text{years}$$

Patience is indeed a virtue, but *you* need a solution, not your (many times grand)-nephews; luckily there are much smarter ways to find the optimal solution, using some good mathematics. □

Now let us consider a slightly different combinatorial problem. We have n items, and we want to select a *combination* of r of them. To see a practical setting in which a combination of items must be selected, imagine a batch of n manufactured parts, from which a random sample of r items is drawn for a quality check. How many combinations of r items out of n can we form? One way to look at the task is the following:

- We take any permutation of n items, and we know that there are $n!$ possible permutations.

- Then we select the first r items and we include them into our combination, leaving the remaining $n - r$ items outside. However, we are not really interested in the order of the r items; all of their $r!$ permutations correspond to the same combination, and the same applies for the $(n - r)!$ permutations of the objects that we do not include in the combination.

Hence, there are

$$\frac{n!}{r!(n-r)!}$$

combinations of r items out of n. As this quantity is quite common, too, it has earned a specific name.

DEFINITION 2.2 (Binomial coefficient) *Given positive integer numbers n and r, $r < n$, the binomial coefficient is defined as*

$$\binom{n}{r} \equiv \frac{n!}{r!(n-r)!}$$

Example 2.2 Consider $n = 4$ objects. How many combinations of two items can we form? Using binomial coefficients, we find

$$\binom{4}{2} = \frac{4!}{2!(4-2)!} = \frac{4 \times 3 \times 2}{2 \times 2} = 6$$

Indeed, in soccer championships groups of four teams are formed and six matches between all of them played. Note that in this case the order of the two teams in defining a match is inconsequential, as there are no return matches. ▯

To see where the name "binomial" comes from, we recall the *binomial expansion formula*:

$$(a+b)^n = \sum_{k=0}^{n} \binom{n}{n-k} a^{n-k} b^k \qquad (2.3)$$

Example 2.3 We all remember the following formulae from high school math:

$$
\begin{aligned}
(a+b)^2 &= a^2 + 2ab + b^2 \\
(a+b)^3 &= a^3 + 3a^2b + 3ab^2 + b^3
\end{aligned}
$$

Let us work them out using the binomial expansion formula. For the first case we have

$$
\begin{aligned}
(a+b)^2 &= \sum_{k=0}^{2} \binom{n}{n-k} a^{n-k} b^k \\
&= \binom{2}{2} a^2 b^0 + \binom{2}{1} a^1 b^1 + \binom{2}{0} a^0 b^2 \\
&= \frac{2!}{2! \times 0!} a^2 + \frac{2!}{1! \times 1!} ab + \frac{2!}{0! \times 2!} b^2 \\
&= a^2 + 2ab + b^2
\end{aligned}
$$

where we see the usefulness of setting $0! = 1$ in Definition 2.1. The second case is also easy to check, by observing that

$$\binom{3}{2} = \frac{3!}{2! \times 1!} = \frac{6}{2} = 3$$

▯

In practice, when computing a binomial coefficient, care must be taken as large numbers can result in overflow errors on a computer.[5] A good way to simplify the calculation is to note that

$$\frac{n!}{r!(n-r)!} = \frac{n \times (n-1) \times (n-2) \times \cdots \times (n-r+1)}{r \times (r-1) \times (r-2) \times \cdots \times 2 \times 1}$$

---

[5]An overflow error occurs when you try to compute a number so large, that it cannot be represented on the limited size registers of the central processing unit (CPU) of your machine.

2.3 FUNCTIONS

Functions are rules that map input values to output values in a well determined way. They come in many guises, depending on what is mapped on what. Generally, a function is specified as

$$f : \mathcal{D} \to \mathcal{I}$$

where $\mathcal{D}$ is the *domain* of the function, i.e., the set of possible input values on which the function is defined, and $\mathcal{I}$ is the *image* or *range* of the function, i.e., the set of values that we obtain by applying the function to each x in the domain $\mathcal{D}$.

In this book, we are essentially interested in numerical functions like

$$f(x) = x + 2$$

This is an example of a single-variable function mapping real numbers into real numbers; formally, this is denoted by $f : \mathbb{R} \to \mathbb{R}$. Sometimes, we have to restrict the domain to integer values. Consider, for instance, a function mapping the number of electric motors we buy to the purchasing cost. Obviously, we can only buy a nonnegative and integer number of motors; hence, we should specify a function $g : \mathbb{Z}_+ \to \mathbb{R}_+$ where, as we said, $\mathbb{Z}_+$ and $\mathbb{R}_+$ denote the set of nonnegative integer and nonnegative real numbers, respectively.[6] If the number of items we buy is large, it may be convenient to express a function $g : \mathbb{R}_+ \to \mathbb{R}_+$, which is a sensible approximation and may be easier to work with from a computational perspective, as we shall see.

In other cases, the domain $\mathcal{D}$ is restricted because of conditions that ensure the proper definition of the function.

Example 2.4 Consider the function $f(x) = \sqrt{1 - x^2}$. The square root function is defined on nonnegative real numbers;[7] hence, the domain of f is restricted by the condition:

$$1 - x^2 \geq 0 \quad \Rightarrow \quad -1 \leq x \leq +1$$

The domain restriction is reflected by denoting the function as $f : [-1, 1] \to \mathbb{R}_+$. In the case of $g(x) = 1/x$, we have $g : \mathbb{R} \backslash \{0\} \to \mathbb{R} \backslash \{0\}$, where we use $A \backslash B$ to denote set difference, i.e., the set consisting of elements of A that are *not* elements of B. In fact, function g is not defined for $x = 0$, and there is no solution to the equation $1/x = 0$, as the function *tends* to zero for $x \to \pm\infty$, but it is never exactly zero. Hence, both domain and range of g are the set of real numbers minus the singleton $\{0\}$. The domain of g could also be denoted by $\{x \in \mathbb{R} \mid x \neq 0\}$. $\square$

---

[6]Well, prices are usually quoted with at most two decimals, but we need not be that picky.
[7]Unless one wishes to consider complex numbers, which can be introduced by defining an imaginary unit $i \equiv \sqrt{-1}$. We do not consider complex numbers in this book.

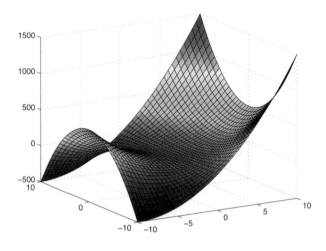

Fig. 2.5   Surface plot for the function of Example 2.5.

We also use functions with multiple input values, such as function $f(x, y)$. Here, the domain is the set of pairs of real numbers, (x, y), denoted as the Cartesian product $\mathbb{R} \times \mathbb{R} \equiv \mathbb{R}^2$. We defer a study of multivariable functions to Section 3.9 at the end of next chapter. Even the very basic task of plotting a multivariable function has no obvious solution, as shown in the example below.

Example 2.5 (Plotting functions of multiple variables) Drawing the graph of a single-variable function is, at least conceptually, a straightforward task. The case of multiple variables is not that easy and there are different ways of representing them. Since we are just able to draw in three dimensions, we can only draw a surface corresponding to a function of two variables. For instance, the *surface plot* corresponding to function

$$f(x, y) = 5x^2 + xy^2$$

is illustrated in Fig. 2.5. The surface plot is obtained by drawing points in three dimensions, where the vertical coordinate z is associated with function values, $z = f(x, y)$. Another useful plotting tool is the *contour plot*; Fig. 2.6 shows the contour plot for the function above. This is a two-dimensional plot displaying a set of *level curves*. A level curve consists of points on the plane (x, y), such that the function value is constant on the curve; there is a level curve for each value of the function. More formally, to draw a set of level curves we fix a set of function values $z_1, z_2, z_3, \ldots$, and plot curves defined by equations

$$f(x, y) = z_i, \qquad i = 1, 2, 3, \ldots$$

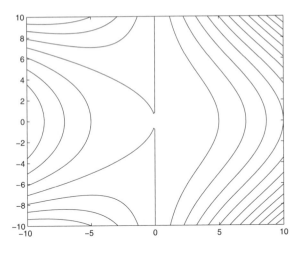

Fig. 2.6 Contour plot for the function of Example 2.5.

Sometimes, level curves may provide us with a clearer view of how a function behaves. ⬚

From the example, we can immediately appreciate the limitations in our ability to "see" a function of multiple variables. This is why we need to develop tools that are useful in characterizing functions of many variables, even though we are not able to visualize them. Those tools are also the foundations of numerical methods to solve complicated equations and to find optimal sets of decisions when many of them are involved (see Chapter 12). Luckily, we may build the intuition and the essential concepts that we need by dealing with the single-variable case. In the remainder of this section we illustrate the most common function classes, limiting our attention to those that are more relevant in a business setting.[8]

2.3.1 Linear functions

A linear affine function has the following general form:

$$f(x) = mx + q \tag{2.4}$$

Figure 2.7 shows a few linear functions. Strictly speaking, only the first function is linear. A function f is linear if the following condition holds:

$$f(\alpha_1 x_1 + \alpha_2 x_2) = \alpha_1 f(x_1) + \alpha_2 f(x_2)$$

---

[8]A most notable omission, in our introductory and business-oriented treatment, is the set of trigonometric functions like $\sin x$ and $\cos x$.

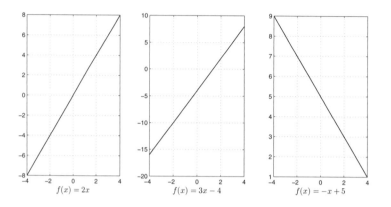

Fig. 2.7 Graphs of linear (affine) functions.

for arbitrary numbers α_i and x_i, $i = 1, 2$. However, this holds only when the coefficient q in (2.4) is zero. To see this, note that, for a generic linear affine function,

$$f(\alpha_1 x_1 + \alpha_2 x_2) = m(\alpha_1 x_1 + \alpha_2 x_2) + q$$

whereas

$$\alpha_1 f(x_1) + \alpha_2 f(x_2) = \alpha_1(m x_1 + q) + \alpha_2(m x_2 + q)$$

Equating the two expressions leads to $q = 0$. Nevertheless, we will often use the term "linear" in a loose sense, referring to linear affine functions. Since coefficient q is the value of f when $x = 0$, it represents the value of coordinate y at which the graph crosses the vertical axis. In fact, coefficient q is called the *intercept*, whereas coefficient m represents the *slope* of the linear function. The concept of slope is linked to the increment ratio of a function.

DEFINITION 2.3 (Increment ratio) *Given a function f and two points x_a and x_b, such that $x_a < x_b$, the increment ratio of f over those two points is defined as*

$$\frac{f(x_b) - f(x_a)}{x_b - x_a}$$

The increment ratio is the average *rate at which the function increases (or decreases, if the ratio is negative) over the interval $[x_a, x_b]$.*

For a linear function, the increment ratio is

$$\frac{(m x_b + q) - (m x_a + q)}{x_b - x_a} = \frac{m x_b - m x_a}{x_b - x_a} = m$$

This is always the same for any pair of point; indeed a linear function is characterized by the fact that its slope is constant and tells us by how much the function changes for a unit change in the independent variable (the reader is

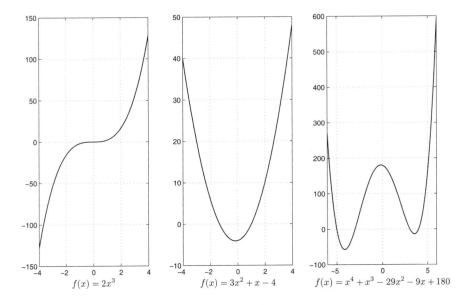

$$f(x) = 2x^3 \qquad f(x) = 3x^2 + x - 4 \qquad f(x) = x^4 + x^3 - 29x^2 - 9x + 180$$

Fig. 2.8 Graphs of polynomial functions.

invited to check the slopes of linear functions in Fig. 2.7). We may specify a particular line by giving its slope m and its intercept q, i.e., the vertical coordinate of the point $(0, q)$ at which it crosses the vertical axis. Alternatively, we can specify a line by giving its slope and one point (x_0, y_0) through which the function passes. This line is represented by the equation

$$y - y_0 = m(x - x_0)$$

or

$$y = y_0 + m(x - x_0) \tag{2.5}$$

2.3.2 Polynomial functions

The next step is to consider powers of the independent variable x. A term of the form ax^m is called a *monomial* of degree m. Summing monomials, we get a *polynomial* function:

$$f(x) = a_0 + a_1 x + a_2 x^2 + \cdots + a_n x^n$$

Here n is the *degree* of the polynomial. A few polynomial functions are shown in Fig. 2.8. A quick glance at the three plots suggests a few observations:

- The first polynomial function is always increasing and does not have a point minimizing or maximizing its value (the maximum goes to $+\infty$ and the minimum goes to $-\infty$).

- The second one has a point that is both a local and a global minimum; that point is where the function stops decreasing and starts increasing.

- The third one has two local minima, one of which is also the global one, as well as a local maximum.

We define concepts like local or global minimum later, but the intuition should be clear from the figure. Informally, a local minimum is a point such that we see an increase in the function while moving both to its right and to its left, but it need not be the true globally minimum of the function, as the function could take a lower value somewhere else.

While a linear function is either increasing or decreasing over the whole real line, a polynomial function may oscillate. In general, the larger the degree of a polynomial, the larger the possibility for up/downswings. As a consequence, a polynomial may have several roots, i.e., solutions of the equation $f(x) = 0$. In Fig. 2.8, the graph of the last polynomial crosses the horizontal axis a few times, while a linear function does so exactly once (unless it is a constant function). The number of roots is impossible to determine for a generic function, but a polynomial of degree n may have up to n (real) roots.

2.3.3 Rational functions

If $P(x)$ and $Q(x)$ are polynomial functions, the function

$$f(x) = \frac{P(x)}{Q(x)}$$

is a *rational* function. In other words, a rational function is just a ratio of two polynomials. Unlike linear and polynomial functions, the domain of a rational function need not be the whole real line. We are in trouble when the denominator polynomial is zero, i.e., when $Q(x) = 0$. Loosely speaking, a rational function "goes to infinity" near the roots of the denominator polynomial.

Example 2.6 Consider the rational function

$$f(x) = \frac{2x^2 - 3x + 10}{x^3 - 4x^2 - 7x + 10}$$

The numerator polynomial has no real root; in fact, the function graph shown in Fig. 2.9 does not cross the horizontal axis anywhere. The denominator polynomial has roots -2, 1, and 5, and the function behavior is critical near these roots, where the function goes to $\pm\infty$.   ▯

Example 2.7 (Internal rate of return) Let us consider a more interesting example from finance, involving polynomial and rational functions. In analyzing an investment, we often deal with a sequence of periodic cash flows C_t,

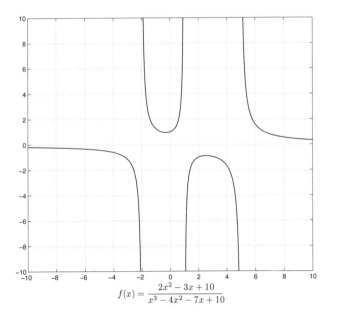

$$f(x) = \frac{2x^2 - 3x + 10}{x^3 - 4x^2 - 7x + 10}$$

Fig. 2.9 Graph of a rational function.

$t = 0, 1, \ldots, T$. A positive cash flow at time t means that the investor receives some money, whereas a negative cash flow is, from her viewpoint, a payment. We deal with integer-valued time instants t, which are actually integer multiples of a basic time period, which could be a month or a year. To fix ideas, say that the basic time period is one year. Then, C_0 is an immediate cash flow, C_1 is a cash flow occurring in one year, etc. Quite often, C_0 is the initial capital outlay to invest in a project, whereas C_t, $t = 1, \ldots, T$, are the net cash inflows from the investment (they are not necessarily positive, though). A typical question is whether the project is worth financing. Arguably, if the straightforward sum of cash flows is negative, we are about to loose money. However, we already know from Section 1.2.3 that the time value of money should also be taken into account: Cash flows should be discounted using a discount rate r. The sum of discounted cash flows is the *net present value* (NPV) of the investment and is given by the following function of the discount rate:

$$\text{NPV}(r) = C_0 + \frac{C_1}{1+r} + \frac{C_2}{(1+r)^2} + \cdots + \frac{C_T}{(1+r)^T}$$

This function is actually a rational function of r, as by a straightforward manipulation we could recast it as the ratio of two polynomials:

$$\text{NPV}(r) = \frac{C_0(1+r)^T + C_1(1+r)^{T-1} + C_2(1+r)^{T-2} + \cdots + C_T}{(1+r)^T}$$

In practice, this manipulation is of little use, as we shall see immediately.

According to financial theory, provided that we are able to estimate cash flows and to select a suitable discount rate accounting for risk, we should select investments with a positive NPV. As a numerical example, consider a project that requires $100 to be started, and will pay $20, $40, and $60 at the end of the first, second, and third years, respectively. Note that the sum of the three cash inflows is $120, which is larger than the initial cash outflow, but is the project really a good deal? That depends on the discount rate r. If we choose $r = 10\%$, then NPV is

$$-100 + \frac{20}{1.1} + \frac{40}{1.1^2} + \frac{60}{1.1^3} = -100 + 18.1818 + 33.0579 + 45.0789 = -3.6814$$

which suggests giving up the project. The decision depends critically on the chosen discount rate. The theory of corporate finance gives us some clue about this choice, but it is always wise to carry out a *sensitivity analysis* to investigate the impact on NPV of changes in the discount rate as well as in the predicted cash flows.

It is natural to wonder for which critical value of the discount rate the NPV of a cash flow sequence turns out to be zero, since this value draws the line between acceptance or rejection of an investment proposal. The critical discount rate is called *internal rate of return* (IRR) and is found by solving the following equation:

$$\text{NPV}(r) = 0$$

The IRR is sometimes used as an alternative tool to analyze investments. If we rely on IRR, then we should select an investment such that the IRR is larger than some required rate of return. This benchmark rate of return could be associated with an alternative project, or it could be a rate of return high enough that we are willing to take the risk of investing in the project. But how can we find the IRR? Although the NPV is a rational function of r, it is much better to transform it into a more manageable form. By the change of variable

$$y = \frac{1}{1+r}$$

we may transform the equation, which involves a rational function, into a polynomial form:

$$C_0 + C_1 y + C_2 y^2 + \cdots + C_T y^T = 0$$

Luckily, very efficient numerical procedures to find roots a polynomial are included in many commercially available software packages. All we have to do is solve for y and transform the solution back to discount rates using $r = (1 - y)/y$. The IRR of the above cash flow stream is 8.21%. This also means that the NPV is negative for $r > 8.21\%$, but positive for $r < 8.21\%$. The larger the required rate of return, the larger the chance that the project is rejected.

However, we should see a potential trouble: A polynomial equation might have several positive real roots. In such a case, which is the correct IRR? Indeed, this is why the NPV criterion is typically considered a better one. One fortunate case is when only the first cash flow is negative, i.e., $C_0 < 0$ and $C_t > 0$ for $t > 0$; for such a cash flow sequence, it can be shown that there is a unique real and positive IRR. In complex projects, with multiple capital outlays in time, there is no guarantee that only the first cash flow is negative. Nevertheless, there are quite relevant kinds of investment in which the condition applies. One such example are bonds. A bond is a financial instrument that is used by governments and corporations to finance their activities. When an investor buys a bond, she is basically lending money to the bond issuer. The *face value* F of the bond is the amount of money that is loaned to the bond issuer and will be repaid to the investor at a future time instant known as *bond maturity*, plus some interest. Typically, the bond issuer also promises periodic payments during the life of the bond; these payments are called *coupons*.[9] For instance, if the face value is $F = \$1000$ and the coupon rate is 6% per year, the coupons would be \$60 each year, and the last cash flow would be \$1060. Often, coupon payments are semiannual (every 6 months). In principle, bonds can be purchased from the issuer on primary markets, but they are typically traded on secondary markets at prices depending on many factors including prevailing interest rates. In bond investing, C_0 is the price of the bond, the cash flows for $t = 1, \ldots, T-1$ correspond to coupon payments, $C_t = C$, and the last cash flow includes the face value of the bond, $C_T = C + F$. In bond valuation and management, the IRR is referred to as *yield to maturity*. ▯

2.3.4   Exponential functions

Polynomial functions involve powers like x^k, where the exponent k is an integer number. We recall some fundamental rules that are quite handy when dealing with powers and should be familiar from high school mathematics:

$$
\begin{aligned}
x^r x^s &= x^{r+s} \\
x^{-r} &= \frac{1}{x^r} \\
\frac{x^r}{x^s} &= x^{r-s} \\
(x^r)^s &= x^{rs} \\
x^0 &= 1
\end{aligned}
$$

In a monomial function $f(x) = \alpha x^k$, the basis x is the independent variable and the exponent k is a fixed parameter. In exponential functions we reverse

---

[9]In the past, a physical paper coupon was detached from the bond and was used to claim each periodic payment.

their roles and deal with expressions such as

$$f(x) = a^x$$

One problem is that while it is quite clear what this expression means when x is an integer number, the same cannot be said when x is a real number. Actually, we may easily find a meaning when x is a rational number, i.e., a ratio m/n of integer numbers. We recall that

$$\sqrt{x} = x^{1/2}, \qquad \frac{1}{\sqrt{x}} = x^{-1/2}$$

It is easy to see that defining the square root this way agrees with the general rules above. We may also define a cubic root

$$a = \sqrt[3]{x} = x^{1/3}$$

i.e., a number such that $x = a^3$. In general

$$x^{m/n} = \sqrt[n]{x^m}$$

which is again consistent with the rule for powers. However, if the exponent x is a generic real number, it is not clear how a^x can be computed, and it is not clear why it should be useful at all. The best way to answer the second question is by a practical example.

Example 2.8 (Continuous compounding in finance) If we invest an amount B at an annual interest rate r, we will end up with a capital $B(1+r)$ after 1 year. If interest is paid after the first year and we reinvest capital plus accrued interest, thus earning interest on interest, we will own $B(1+r)^2$ after 2 years. Now imagine that interests are paid semiannually, i.e., every 6 months (let us ignore complications due to the fact that months do not consist of the same number of days). In this case, the annual rate r is used just for quotation purposes, but in practice we earn a rate $r/2$ every 6 months. Hence, after 1 year, we own $B(1+r/2)^2$. For instance, if we invest $100 at $r = 10\%$ with annual compounding, after 1 year our wealth will be

$$100 \times (1 + 0.10) = \$110$$

whereas semiannual compounding yields a slightly greater wealth:

$$100 \times \left(1 + \frac{0.10}{2}\right)^2 = \$110.25$$

As you may imagine, the smaller the time interval at which interest is compounded, the faster our capital grows. You may see this by evaluating

$$100 \times \left(1 + \frac{0.10}{k}\right)^k$$

Table 2.1   Discovering Euler's number e.

k	1	2	3	4	5
$g(k)$	2.0000	2.2500	2.3704	2.4414	2.4883

k	10	50	100	1000	5000
$g(k)$	2.5937	2.6916	2.7048	2.7169	2.7180

What happens if $k \rightarrow +\infty$, i.e., if compounding occurs so often that we have a *continuously compounded* interest? To provide a clue, Table 2.1 shows values of function

$$g(k) = \left(1 + \frac{1}{k}\right)^k$$

for increasing values of k. It seems that this sequence does converge to a number. This number is so important that it has been named *Euler's number* and is denoted as

$$e \equiv \lim_{k \to +\infty} \left(1 + \frac{1}{k}\right)^k \approx 2.718281828459046$$

Whatever base we use, we know that a power like e^n has a clear meaning for an integer exponent n. Rather surprisingly, it turns out that

$$e^x = 1 + x + \frac{x^2}{2!} + \frac{x^3}{3!} + \frac{x^4}{4!} + \cdots \tag{2.6}$$

where x is a generic real number. A proof of this equality requires some advanced concepts that we introduce later, so we have to defer this to Example 2.31. The equality (2.6) provides us with one clear and well-defined procedure for computing exponentials with base e. This need not be the best one, but we need not worry about that, since the exponential function is implemented in many software tools, including spreadsheets. We show a plot of the exponential function e^x and the *negative* exponential function $e^{-x} = 1/e^x$ in Fig. 2.10. The exponential e^x grows quite rapidly even for moderately large values of x; this is where the term *exponential growth* comes from.

Let us go back to our financial example and consider the growth of one dollar invested over one year at a nominal annual rate r compounded k times per year, when k goes to infinity. We rewrite the expression a bit, introducing a new variable $y = k/r$, which goes to $+\infty$ when k does so:

$$\left(1 + \frac{r}{k}\right)^k = \left(1 + \frac{1}{k/r}\right)^{r \cdot k/r} = \left[\left(1 + \frac{1}{y}\right)^y\right]^r \longrightarrow e^r \tag{2.7}$$

Hence, continuous compounding results in an exponential growth of capital. An initial capital B_0, earning a continuously compounded interest rate r,

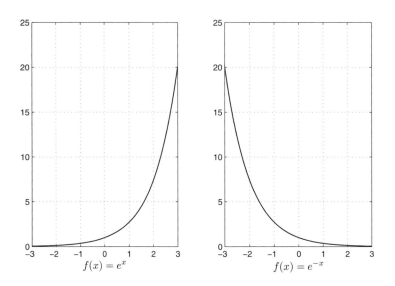

Fig. 2.10 Graphs of exponential and negative exponential functions.

grows in time according to the following exponential function:

$$B(t) = B_0 e^{rt}$$

The negative exponential function comes into play when going the other way around, i.e., when discounting cash flows. If the discount factor r is compounded semiannually, a cash flow C occurring in one year should be discounted as

$$\frac{C}{(1 + r/2)^2}$$

If there are k periods per year, we use

$$\frac{C}{(1 + r/k)^k}$$

With continuous compounding, the present value of a cash flow C occurring at time t is

$$Ce^{-rt}$$

Note that with this concept we may easily discount cash flows occurring at arbitrary time instants. Indeed, continuous compounding does a great job at simplifying calculations in financial mathematics. ☐

Now we have made a little step forward, since we know how to compute an exponential function with base e; still, we do not know how to compute something like a^x for an arbitrary value a. To do this, we need to introduce the logarithm as an inverse of the exponential function (see Section 2.6).

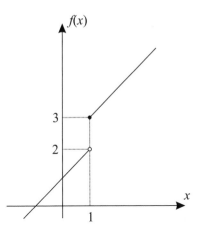

Fig. 2.11 A discontinuous function.

2.4 CONTINUOUS FUNCTIONS

Before we proceed in our treatment of functions, we should pause a little and discuss a fundamental feature of functions: continuity, or lack thereof. Compare the graphs of polynomial functions in Fig. 2.8 against the graph of the rational function in Fig. 2.9. There is a striking qualitative difference between the two figures; in the first case, if we imagine drawing the plot with a pencil, we never "detach" the pencil from paper. In the second case, the function "jumps" from $-\infty$ to $+\infty$ and vice versa. Quite informally, it might be argued that if we can draw a graph without such interruptions, then the function is continuous.[10] Lack of continuity may be the result of problems in the domain of the function, like the rational function above, which goes to infinity when we approach points where it is not defined, but this need not be the only case.

Example 2.9 Consider the following function:

$$f(x) = \begin{cases} x+1 & x < 1 \\ x+2 & x \geq 1 \end{cases}$$

A qualitative plot of the function is depicted in Fig. 2.11. Please note the standard convention used in the figure: The filled bullet corresponds to the true value taken by the function for $x = 1$; the empty circle below corresponds to a value that is approached only when x gets close to 1 *from the left*, before the function jumps. If we get closer and closer to this critical point *from the*

---

[10]This is not quite correct, as we are disregarding the possibility of pathological oscillations, but such weird cases do not play any role in the book.

right, the value of the function tends to 3, which is indeed the value $f(1)$ (please note the exact way in which the function is defined). We can express this by writing

$$\lim_{x \to 1^+} f(x) = 3 = f(1)$$

where $x \to 1^+$ denotes the fact that we approach $x = 1$ from the right, i.e., considering values like $1 + \epsilon$, where $\epsilon > 0$ gets smaller and smaller. An alternative notation is $\lim_{x \downarrow 1}$. On the other hand, if we approach $x = 1$ from the left, we have a different limit:

$$\lim_{x \to 1^-} f(x) = 2 \neq f(1)$$

An alternative notation for this limit is $\lim_{x \uparrow 1}$. We say that the function is *continuous from the right*, but *discontinuous from the left*.[11]   ☐

The kind of discontinuity featured in the example is only one among the possible occurrences. In other situations, the trouble might stem from wild oscillations in the function, but we will not dwell too much in such pathological examples.[12]

DEFINITION 2.4 (Continuous function) *A function f is said to be continuous at point x_0 if*

$$\lim_{x \to x_0} f(x) = f(x_0)$$

If this condition applies to all points within an interval or domain, we say that the function is continuous on that interval or domain.

We will not define the mathematical concept of limit too formally, as intuition suffices for our purposes.[13] Still, you should note that convergence to a limit must be the same from both sides, left and right.

2.5 COMPOSITE FUNCTIONS

So far, we have considered linear, polynomial, rational, and exponential functions. From our high school math, we might recall something about trigonometric functions; since we will not use them in the following, we leave them

---

[11]You might even stumble on exoteric jargon like a *càdlàg function*! This just a French acronym for "continue à droite, limitée à gauche," since the function is continuous from the right, and is limited (or bounded; i.e., it does not go to infinity) from the left.

[12]One such case is the function $\sin(1/x)$.

[13]From a historical perspective, it is interesting to note that a proper definition of limit and continuity was given by Augustin Cauchy and Karl Weierstrass, by the epsilon–delta approach. They did so in the 1800s, about a couple of centuries after Isaac Newton and Gottfried Leibniz, who are considered the founding fathers of calculus.

aside. A natural way to build quite complicated, but hopefully useful, functions is function composition. Given functions g and h, we may build the composite function:

$$f(x) = g(h(x))$$

The idea is that, given x, we compute $z = h(x)$, and then $g(z)$. Strictly speaking, the notation we have used is a bit sloppy as it refers to values taken by the function. The proper notation for denoting the composition of g and h, when we want to refer to the function itself, should be

$$f = g \circ h$$

This notation makes it clear that h maps an input argument into a result, which is in turn mapped by g into another result; the composite mapping is function f.

Some care is needed in checking the domain on which a composite function makes sense. As we have already noted, the function

$$\sqrt{1 - x^2}$$

is defined only for $-1 \leq x \leq 1$.

Example 2.10 (Gaussian function) By composing the functions

$$h(x) = \frac{x^2}{2}, \qquad g(z) = e^{-z}$$

we obtain function

$$f(x) = g(h(x)) = e^{-x^2/2} \tag{2.8}$$

Figure 2.12 illustrates the two building blocks and the resulting function. This function has a classical bell shape and plays a prominent role in probability and statistics, which make quite some use of normal, or Gaussian, probability distributions.   ▯

There are two simple function compositions that are quite common and have a natural interpretation. Given a function $f(x)$, let us consider functions

$$g(x) = f(x - \alpha), \qquad h(x) = f\left(\frac{x}{\beta}\right)$$

for $\beta > 0$. Function g is actually just function f shifted to the right by an amount α, if $\alpha > 0$; if $\alpha < 0$, then the function is shifted to the left. Dividing the independent variable x by $\beta > 0$ has the effect of changing of scale, i.e., stretching or shrinking the function graph horizontally, depending on whether $\beta > 1$ or $\beta < 1$.

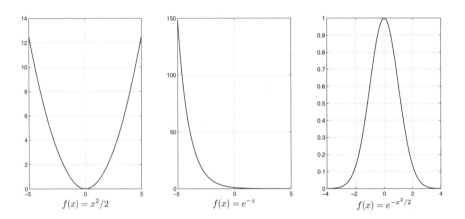

$f(x) = x^2/2$ $f(x) = e^{-z}$ $f(x) = e^{-x^2/2}$

Fig. 2.12   Composition of two functions.

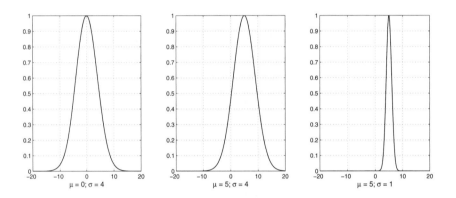

$\mu = 0; \sigma = 4$ $\mu = 5; \sigma = 4$ $\mu = 5; \sigma = 1$

Fig. 2.13   Scaling and shifting the bell-shaped function of Eq. (2.9).

Example 2.11 (Shifting and scaling) Consider again the bell-shaped function (2.8) and apply the following transformation

$$f(x) = \exp\left\{ -\frac{1}{2}\left(\frac{x - \mu}{\sigma}\right)^2 \right\} \tag{2.9}$$

where $\exp(\cdot)$ is just an alternative notation for the exponential function. The parameter μ governs the amount (and direction) of shifting, whereas σ changes the scale, making the graph more or less dispersed. This is illustrated in Fig. 2.13. The first plot illustrates the function for $\mu = 0$ and $\sigma = 4$. For that value of μ, the graph is symmetric with respect to the origin. If we set $\mu = 5$ the graph is shifted by five units to the right. If we further set $\sigma = 1$, the effect is compressing the horizontal scale by a factor 4, thus reducing dispersion.   □

2.6   INVERSE FUNCTIONS

A function maps an input value x into an output value $y = f(x)$. There are cases in which we want to go the other way around; i.e., given y, we would like to find a value x such that $y = f(x)$. Actually, this is what we do whenever we want to solve an equation. For instance, given a function that evaluates the NPV of an investment depending on the discount rate r, finding its IRR calls for the solution of the equation $\text{NPV}(r) = 0$. Solving an equation like that requires inversion of the mapping associated with the function for one *specific* value of NPV, zero in this case. Now imagine that we want to do something more. We would like to find a trick that allows us to solve an entire range of equations $f(x) = y$ for *many* values of y. This idea leads to the definition of the *inverse function* of f, mapping y to the corresponding x. As you may imagine, this cannot be done for every function f, as equations may have multiple solutions or none at all, whereas an inverse function should map one input value into one output value.

DEFINITION 2.5 (Inverse function) *Given a function $f(x) : E_1 \to \mathbb{R}$, where E_1 is the domain of f, the inverse function of f is a function $g(x) : E_2 \to \mathbb{R}$ such that*

1. *$g(f(x)) = x$, for all x in the domain E_1 of f*

2. *$f(g(z)) = z$, for all z in the domain E_2 of g*

Typically, the notation f^{-1} is used to denote the inverse of function x. This should not be confused with the composite function $g(x) = 1/f(x)$. Some care in notation may avoid the confusion: $f^{-1}(x)$ refers to the inverse function, whereas $[f(x)]^{-1} = 1/f(x)$ refers to the composite function.

Example 2.12 Consider function $f(x) = (x-1)/(x+1)$. To find its inverse, we set up and solve the following equation:

$$y = \frac{x-1}{x+1} \quad \to \quad yx + y = x - 1 \quad \to \quad 1 + y = x(1 - y) \quad \to \quad x = \frac{1+y}{1-y}$$

Hence, the inverse function of f is

$$g(y) = \frac{1+y}{1-y}$$

We see that the inverse function is not defined for $y = 1$. Indeed

$$\frac{x-1}{x+1} = 1 \quad \to \quad x - 1 = x + 1 \quad \to \quad -1 = 1$$

which is absurd. The image of function f does not include the value 1, and the inverse function is not defined there.   ☐

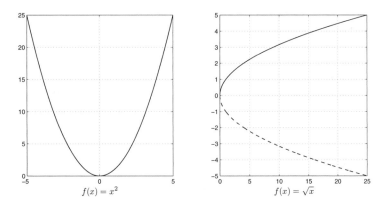

Fig. 2.14 Square root as an inverse function on a restricted domain.

Example 2.13 Consider function $f(x) = x^2$. It may be tempting to say that its inverse is the square-root function $g(x) = \sqrt{x}$, but a quick look at plots in Fig. 2.14 points out a difficulty. An equation like $x^2 = 4$ has two roots, $x = \pm 2$; which one should we use when inverting the function? As shown in the figure, inverting the function essentially means swapping the two coordinate axes, but in doing so we do not always define a true function, which should map each value of the independent variable into one value. In this case, the domain of $f(x) = x^2$ must be restricted in order to define its inverse function. The customary choice is restricting the domain to positive numbers, so we just consider the positive root. In other words, we cancel the dashed lower part of the rotated parabola in the second plot of Fig. 2.14.

□

The last example shows that not every function can be inverted over its whole domain. In order to be invertible, f should not assign the same value to two different arguments:

$$x_1 \neq x_2 \quad \Rightarrow \quad f(x_1) \neq f(x_2)$$

A function whose graph goes up and down does not have this property. Another unpleasing feature that may prevent inversion is lack of continuity. If a function jumps, then we may fail to invert it. A condition ensuring invertibility of a function is that it is continuous and strictly increasing:

$$x_1 < x_2 \quad \Rightarrow \quad f(x_1) < f(x_2)$$

If this condition is met on an interval, then the function is invertible on that interval. It is also easy to see this also applies to a continuous strictly decreasing function.

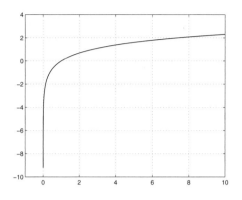

Fig. 2.15 The natural logarithm.

2.6.1 The logarithm

The logarithm arises as the inverse of an exponential function. To further motivate this, let us consider again continuous compounding of interest rates. As we have pointed out, continuous compounding leads to an exponential function that streamlines financial calculations considerably. However, in practice, interest rates are not quoted like this. Typically, interest rates are quoted on an annual basis. Yet, given an annually compounded interest rate r, we could find the equivalent continuously compounded rate r_c and perform calculations based on this rate. Doing so requires solving the equation

$$e^{r_c} = 1 + r$$

and this in turn requires inverting the exponential function. If we look back at Fig. 2.10 we see that the exponential function $f(x) = e^x$ is a very nice continuous and monotonically increasing function, and therefore it can be inverted over the whole real line. This leads to the definition of the *natural logarithm function*:

$$y = e^x \quad \Leftrightarrow \quad x = \ln y$$

This definition implies that

$$e^{\ln x} = x, \qquad \ln e^x = x$$

Plotting the logarithmic function $\ln x$ just requires plotting the exponential and swapping the axes, which results in the graph of Fig. 2.15. Natural logarithm is defined only for strictly positive x; it is positive for $x > 1$, negative for $x < 1$, and $\ln 1 = 0$. Many years ago tables were provided to carry out calculations with logarithms. Now, logarithms are easily calculated using many software tools, including spreadsheets.

Example 2.14 (A financial example) Say that the annually compounded interest rate is $r = 5\%$. Which continuously compounded rate corresponds to this rate? We have

$$e^{r_c} = 1 + r = 1.05, \quad \rightarrow \quad r_c = \ln 1.05 \approx 4.88\%$$

Please note that r_c is smaller than r, because of the effect of continuous compounding on capital growth.   □

PROPERTY 2.6 *Here we summarize a few useful properties of natural logarithm, which are a direct consequence of the properties of exponential function:*

- $\ln(xy) = \ln x + \ln y$

- $\ln(1/x) = -\ln x$

- $\ln(x/y) = \ln x - \ln y$

- $\ln x^y = y \ln x$

- $\ln 1 = 0$

Example 2.15 Using logarithms and the properties above, we may finally answer a question we left open: How can we evaluate a^x, when x is an arbitrary real number? This can be accomplished using properties of the exponential:

$$a^x = e^{\ln a^x} = e^{x \ln a}$$

This can actually be calculated, e.g., using expansion (2.6).   □

Natural logarithms enjoy a privileged status as they are the inverse of the exponential function using base e. If we consider an exponential function with an arbitrary base a, its inverse is base a logarithm:

$$y = \log_a x \quad \Leftrightarrow \quad a^y = x.$$

A common case is $a = 10$, leading to decimal logarithms. The properties that we listed above apply to any logarithm in any base. Incidentally, the notation $\log_a$ can be used with any base a. Clearly, $\ln \equiv \log_e$, and normally $\log \equiv \log_{10}$; however, notation is not always standard, as log is often used to denote natural logarithms.

2.7 DERIVATIVES

We have seen that a linear (affine) function $f(x) = mx + q$ has a well-defined slope. Whatever value of the independent variable we consider, the slope of

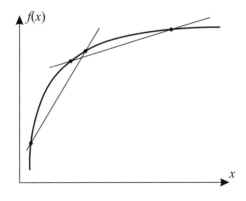

Fig. 2.16 A nonlinear function does not have constant increment ratios.

the function is always the same. If we are at point x and we move to point $x + h$, by any displacement h, the increment ratio[14] is:

$$\frac{f(x + h) - f(x)}{(x + h) - x} = \frac{[m(x + h) + q] - [mx + q]}{h} = m$$

which is constant and does not depend either on x or h. The increment ratio, i.e., the rate of change in the function, is constant everywhere for a linear function and corresponds to the slope of the line. This does not apply to a nonlinear function like the exponential or the logarithm. The issue is illustrated in Fig. 2.16. Still, there are many reasons why we should be interested in investigating the "slope" of a nonlinear function:

- It may help in checking if the function is increasing or decreasing at some specific point x.

- It may help in evaluating the rate at which the function increases or decreases.

- It may help in finding a point at which the function is maximized or minimized.

This leads to the definition of a concept that is arguably the most important one in calculus: the derivative of a function.

2.7.1   Definition of the derivative

Consider a point x_0 and the increment ratio of function f at that point:

$$\frac{f(x_0 + h) - f(x_0)}{h}$$

[14]See Definition 2.3.

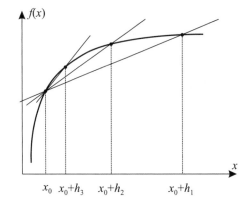

Fig. 2.17   The derivative is the limit of an increment ratio.

For a nonlinear function, keeping x_0 fixed, this ratio is a function of h. Now consider smaller and smaller steps h, as illustrated in Fig. 2.17. If we let $h \to 0$, we get the "tangent" line to the graph of f at point x_0. The slope of this line is called the *derivative* of f at point x_0:

$$\frac{df}{dx}(x_0) \equiv \lim_{h \to 0} \frac{f(x_0 + h) - f(x_0)}{h}$$

The notation reminds us that the derivative is the limit of the increment ratio $\Delta f / \Delta x$, when the step Δx becomes infinitesimal. We should also note that we may evaluate the derivative of a given function at several points. Indeed, taking the derivative of function $f(x)$, we define another function. The notation

$$f'(x)$$

is also often used to indicate the derivative of $f(x)$. Given the derivative, we may easily find the equation of the tangent line at any point on the graph of $f(x)$. We recall that the equation of a line going through point (x_0, y_0), with slope m, is given by

$$y = y_0 + m(x - x_0)$$

Hence, given the graph of $f(x)$, we can express the tangent line at point $(x_0, f(x_0))$ as follows:

$$y = f(x_0) + f'(x_0) \cdot (x - x_0) \tag{2.10}$$

Example 2.16 In some easy cases, it is possible to find the derivative of a function by direct application of the definition. For instance, consider the

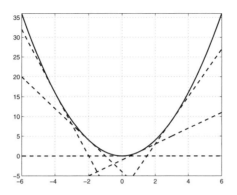

Fig. 2.18 The x^2 function and its tangent lines at points $x_0 = -4, -2, 0, 1, 3$.

quadratic function $f(x) = x^2$. Applying the definition, we have

$$
\begin{aligned}
f'(x) &\equiv \lim_{h \to 0} \frac{f(x+h) - f(x)}{h} = \lim_{h \to 0} \frac{(x+h)^2 - x^2}{h} \\
&= \lim_{h \to 0} \frac{x^2 + 2hx + h^2 - x^2}{h} = \lim_{h \to 0} \frac{2hx + h^2}{h} \\
&= \lim_{h \to 0} \frac{2hx}{h} = 2x
\end{aligned}
$$

The key point here is that the expression $2hx + h^2$ can be simplified for small values of h: A term like h^2 is negligible with respect to h, as it goes to zero at a faster rate. It is very useful to interpret the result by visualizing the plot of function x^2 and checking its derivative, as well as its tangent lines, as shown in Fig. 2.18. Analytically, the equation of the tangent lines, with slopes $2x_0$ are given by straightforward application of Eq. (2.10):

$$
y = x_0^2 + 2x_0(x - x_0) = 2x_0 x - x_0^2
$$

- For $x_0 < 0$, the derivative is negative. Indeed, the function is decreasing on the negative part of the real line, and the slope of the tangent line is negative. This slope goes to $-\infty$ when $x_0 \to -\infty$ and the rate at which the function decreases diminishes when x approaches zero.

- The derivative is zero for $x_0 = 0$; in fact, the tangent line there is horizontal.

- For positive values of x_0, the function is increasing; the rate of increase is itself increasing and it goes to infinity when $x_0 \to +\infty$.

The example shows that the derivative is quite useful in figuring out the behavior of a function. More generally, it can be shown that, for a positive integer n

$$(x^n)' = nx^{n-1} \tag{2.11}$$

This is a consequence of the binomial expansion formula (2.3). Consider the binomial $(x + h)^n$ for a small value of h

$$(x + h)^n = \sum_{k=0}^{n} \binom{n}{n-k} x^n h^{n-k} \approx x^n + nx^{n-1}h$$

where the approximation is justified by the fact that higher powers of h are negligible when $h \to 0$. Then, applying the definition of derivative, we get

$$
\begin{aligned}
(x^n)' &\equiv \lim_{h \to 0} \frac{(x + h)^n - x^n}{h} \\
&= \lim_{h \to 0} \frac{x^n + nx^{n-1}h - x^n}{h} = nx^{n-1}
\end{aligned}
$$

Later, we will see that Eq. (2.11) generalizes to an exponential function a^x, where x is a real number. Applying the result for $n = 0$, we see that the derivative of the constant function $f(x) = 1$ is zero. Alternatively, using the definition, it is easy to see that the derivative of any constant function is zero, since the increment of a constant function is identically zero.

More often than not, we do not apply the definition to find a derivative directly. Rather, we use a set of rules that are illustrated in Section 2.8, together with results about the derivative of basic functions, such as the exponential. But before doing that, it is important to realize that the derivative need not always exist.

2.7.2   Continuity and differentiability

If the derivative of function f at point x_0 exists, then we say that the function is *differentiable* at point x_0; if this holds for all points on an interval or domain, the function is differentiable on that interval or domain. If the derivative $f'(x)$ exists at all points x on an interval and the derivative is a continuous function, we say that the function is *continuously differentiable* on that interval.

In Section 2.4 we have encountered discontinuous functions. If a function f is discontinuous at x_0, then

$$\lim_{x \to x_0} f(x) \neq f(x_0)$$

We have also noted that a limit should not depend on the way x approaches x_0; the limit, if it exists, must be the same for $x \to x_0^+$ (from the right) and $x \to x_0^-$ (from the left). Since the derivative is defined as a limit, we may have similar concerns. To see a couple of standard cases in which the derivative does not exist, have a look at Fig. 2.19. Graph (a) shows a typical "kinky" function. This function is actually a piecewise linear function, and for most

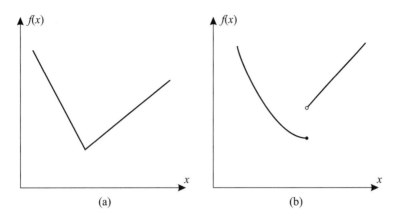

Fig. 2.19   The derivative may fail to exist.

points the derivative is just the slope of the corresponding linear piece. But where the two lines join, we have a point of nondifferentiability: the left and the right limit are different, and all we have are a left derivative and a right derivative.

Example 2.17 The absolute value is a function defined as

$$|x| = \begin{cases} -x & x < 0 \\ x & x \geq 0 \end{cases}$$

The derivative of this function is

$$(|x|)' = \begin{cases} -1 & x < 0 \\ +1 & x > 0 \end{cases}$$

At point $x = 0$ we do not have a derivative since

$$\lim_{h \to 0^+} \frac{|0 + h| - |0|}{h} = \lim_{h \to 0^+} \frac{h}{h} = +1$$

whereas

$$\lim_{h \to 0^-} \frac{|0 + h| - |0|}{h} = \lim_{h \to 0^+} \frac{-h}{h} = -1$$

The two limits above are different and do not define a derivative, but only a right derivative and a left derivative, respectively. We may also observe that there is no well-defined tangent line at the kinky point. □

Figure 2.19(b) shows a typical discontinuous function. It is also clear that we do not have a well-defined derivative where a function jumps.

Putting the two graphs together, we notice an interesting fact. If the function jumps, i.e., is discontinuous, we fail to find the derivative. However, even if the function is continuous, we may fail to find the derivative. Indeed, the kinky function in Fig. 2.19(a) is a continuous function. This suggests that continuity is a necessary condition for differentiability, but not a sufficient one. In other words, if the function is differentiable, then it must be continuous, but not vice versa. To see this, consider this equivalent definition of derivative

$$\lim_{x \to x_0} \frac{f(x) - f(x_0)}{x - x_0}$$

If this limit exists, it must be the case that, for $x \to x_0$, the numerator gets closer and closer to zero:

$$\lim_{x \to x_0} f(x) = f(x_0) = f\left(\lim_{x \to x_0} x \right) \tag{2.12}$$

Note that, for a continuous function, the limit of the function is the same as the function of the limit.

It is tempting to think that, after all, such weird functions are the result of mathematical phantasy, nowhere to be found in the real world. The following example shows that this is not the case.

Example 2.18 (All-unit and incremental discounts) Quite often, when purchasing a product, we are offered price discounts, provided we buy a number of items exceeding a given breakpoint. For instance, the unit price could be €10 per piece, but if we buy at least 100, the unit price drops to €9. Actually, there are two different cases:

1. In *all-unit* discounts, the discounted price applies to all of the units we buy.

2. In *incremental* discounts, the discounted price applies only to items beyond the breakpoint.

Let us write down the total cost function $C(x)$ depending on the purchased amount x, assuming for simplicity that x is a real number, which is sensible for items bought in units of weight or volume and is a fair approximation for discrete items bought in large quantities. The cost for all-unit discount is

$$C_a(x) = \begin{cases} 10x & x < 100 \\ 9x & x \geq 100 \end{cases}$$

whereas cost for incremental discount is

$$C_i(x) = \begin{cases} 10x & x < 100 \\ 1000 + 9(x - 100) & x \geq 100 \end{cases}$$

Qualitative sketches of the two functions are illustrated in Fig. 2.20. Graph (a) shows a weird feature of all-unit discount: Close to the price breakpoint,

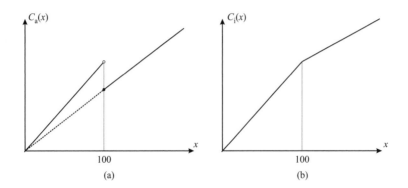

Fig. 2.20 Nondifferentiable functions arising from discount opportunities in purchasing decisions.

we pay less if we buy more. Disregarding this issue, we see how discontinuous or nondifferentiable functions may actually arise in practice. As another example, think of the different tax brackets as a function of income. The plot of the tax amount we have to pay would be similar to the incremental discount case, but when dealing with taxes, the slopes of the linear pieces are increasing rather than decreasing.   □

2.8 RULES FOR CALCULATING DERIVATIVES

The direct application of the definition to find the derivative of a function is typically a rather difficult and cumbersome procedure, possibly requiring some intuition.

Example 2.19 (Derivative of logarithm and exponential function)
One of the most useful results concerning derivatives is that the derivative of the exponential is just the exponential itself:

$$(e^x)' = e^x$$

As a first step to prove this deceptively simple result, it is better to find the derivative of the logarithm:

$$(\ln x)' = \lim_{h \to 0} \frac{\ln(x + h) - \ln x}{h}$$

Both numerator and denominator of the increment ratio go to zero, so some manipulation is needed in order to figure out what really happens. To begin with, we may use properties of the logarithms to transform the increment ratio a bit:

$$\frac{\ln(x + h) - \ln x}{h} = \frac{1}{h} \ln \left(\frac{x + h}{x} \right) = \ln \left(1 + \frac{h}{x} \right)^{1/h} = \ln \left(1 + \frac{1/x}{1/h} \right)^{1/h}$$

If $h \to 0$, then $1/h \to \infty$.[15] Furthermore, the logarithm is a continuous function. Then, Eq. (2.12) applies and the limit of a logarithm is the logarithm of the limit. So

$$\lim_{h \to 0} \frac{\ln(x+h) - \ln x}{h} = \ln \left[\lim_{h \to 0} \left(1 + \frac{1/x}{1/h} \right)^{1/h} \right] = \ln \left[\lim_{z \to \infty} \left(1 + \frac{1/x}{z} \right)^{z} \right]$$

where $z = 1/x$. But from the basic results about Euler's number[16] we know that

$$\lim_{z \to \infty} \left(1 + \frac{1/x}{z} \right)^{z} = e^{1/x}$$

which in turn implies that

$$(\ln x)' = \ln \left(e^{1/x} \right) = \frac{1}{x}$$

This result is convincing if we look at the plot of the logarithm function in Fig. 2.15. The function is always increasing, but the rate of increase is very large for a small value of x; actually, it goes to infinity for $x \to 0+$. The function flattens when x increases, as the rate of increase diminishes and goes to zero when $x \to +\infty$.

Now we would also like to find the derivative of the exponential function. To this aim, we need a result about the derivative of inverse functions, which is outlined below.   ⬜

The procedure that we have just illustrated, although a bit informal and shaky, should convince you that we really need some handy way to find derivatives. In practice, we do the following:

1. We take advantage of basic results about the derivative of a few fundamental functions. For instance, we already know the derivative of building blocks such as the monomial x^n and the exponential e^x.

2. Then, we apply rules that allow to decompose the task of differentiating a complicated function into more manageable subtasks.

In this section we describe rules to find the derivative in the following cases:

• Functions obtained by summing, multiplying, or dividing other functions

• Functions obtained by composition

• Functions obtained by inversion

---

[15]To be precise, $1/h$ may tend to $-\infty$ or $+\infty$, depending on the sign of h going to zero. However, it turns out that this does not affect our little exercise. Still, this is a weak point in the argument and if you would like to see a thorough and more rigorous treatment, please refer, e.g., to Chapter 8 of Ref. [6].
[16]See Eq. (2.7).

2.8.1 Derivative of functions obtained by sum, multiplication, and division

Given two functions f and g, there are a few easy ways to build other functions by ordinary arithmetic operations such as sum, multiplication, and division.

- We define the *sum* of functions as follows:

$$(f + g)(x) \equiv f(x) + g(x)$$

  By the same token, we define the *difference* of two functions. To be precise, this makes sense on the intersection of the two domains, but we will not be bothered by such details.

- Given a number α and a function f, we can apply *multiplication by a constant*:

$$(\alpha f)(x) \equiv \alpha f(x)$$

- Given two functions f and g, we define the *product of functions*:

$$(fg)(x) \equiv f(x)g(x)$$

- Finally, we may use *division of functions*:

$$\left(\frac{f}{g}\right)(x) \equiv \frac{f(x)}{g(x)}$$

  Again, this makes sense on a common domain, where both functions f and g are defined, provided that $g(x) \neq 0$.

If we are able to find the derivative of f and g, the following theorem shows how to find the derivative of functions defined by the mechanisms above.

THEOREM 2.7 *Let f and g be functions and α be a real number. If f and g are defined and differentiable at x_0, then*

$$
\begin{aligned}
(f \pm g)'(x_0) &= f'(x_0) \pm g'(x_0) \\
(\alpha f)'(x_0) &= \alpha f'(x_0) \\
(fg)'(x_0) &= f'(x_0)g(x_0) + f(x_0)g'(x_0) \\
\left(\frac{f}{g}\right)'(x_0) &= \frac{f'(x_0)g(x_0) - f(x_0)g'(x_0)}{g^2(x_0)}
\end{aligned}
$$

The last result also requires $g(x_0) \neq 0$.

The following examples illustrate the application of the theorem.

Example 2.20 A polynomial function is basically a sum of monomials, obtained by multiplying a number and an integer power of x. Then, to find the derivative of a polynomial we can use the first two results as follows. Consider

$$f(x) = 3x^3 - 2x^2 + 5x - 10$$

The derivative of the first term is

$$\left(3x^3\right)' = 3\left(x^3\right)' = 3 \times 3x^{3-1} = 9x^2$$

The same approach yields the derivatives of the second and third terms. The last term is just a constant, and its derivative is zero. Putting everything together, we have

$$f'(x) = 9x^2 - 4x + 5$$

⬜

Example 2.21 Let us illustrate the product of functions. Consider the product of a polynomial and an exponential:

$$f(x) = \left(2x^2 + 1\right)e^x$$

We may easily take the derivative of each factor of the product:

$$\left(2x^2 + 1\right)' = 4x, \qquad \left(e^x\right)' = e^x$$

Applying the result for the product of functions, we obtain

$$f'(x) = 4xe^x + (2x^2 + 1)e^x = \left(2x^2 + 4x + 1\right)e^x$$

⬜

Example 2.22 Finally, let us illustrate the case of a rational function, obtained by dividing two polynomials:

$$f(x) = \frac{x}{x^2 + 1}$$

We should break down the overall task into smaller pieces. First we compute the derivatives of numerator and denominator:

$$(x)' = 1, \qquad \left(x^2 + 1\right)' = 2x$$

Then we put everything together:

$$f'(x) = \frac{1 \times (x^2 + 1) - x \times (2x)}{\left(1 + x^2\right)^2} = \frac{1 - x^2}{\left(1 + x^2\right)^2}$$

⬜

2.8.2 Derivative of composite functions

Given two functions g and h, we may build a new function by composition, namely, $g \circ h$. It would be nice to have a way of finding the derivative of the composite function by decomposing the task and exploiting knowledge about the derivatives of g and h.

THEOREM 2.8 (Chain rule) *Given functions g and h, we obtain the derivative of their composition as*

$$\frac{d(g \circ h)}{dx}(x) = g'(h(x)) \cdot h'(x)$$

provided that all of the involved derivatives exist.

A good way to remember this rule is by rephrasing it in terms of increment ratios:

$$\frac{d(g \circ h)}{dx} = \frac{dg}{dh}\frac{dh}{dx}$$

The idea is that we should take the derivative of function $g(z)$ and evaluate it for $z = h(x)$; then, this is multiplied by the derivative of $h(x)$.

Example 2.23 As a first example, consider the exponential function $f(x) = e^{4x}$. This is best viewed as the composition of functions $g(z) = e^z$ and $h(x) = 4x$. We know that

$$\frac{dg}{dz} = e^z, \qquad \frac{dh}{dx} = 4$$

Hence, we apply the chain rule, where we set $z = 4x$ in the first derivative:

$$\frac{df}{dx} = 4e^{4x}$$

More generally, $(e^{\alpha x})' = \alpha e^{\alpha x}$. This also applies to negative values of α. The case $\alpha = -1$ can also be tackled by different route, considering the composition of $g(z) = 1/z$ and $h(x) = e^x$:

$$\frac{dg}{dz} = -\frac{1}{z^2}, \qquad \frac{dh}{dx} = e^x$$

which yields

$$\frac{df}{dx} = -\frac{1}{(e^x)^2}e^x = -e^{-x}$$

We invite the reader to check that

$$f(x) = e^{-x^2} \quad \Rightarrow \quad f'(x) = -2xe^{-x^2}$$

☐

Example 2.24 Consider function $f(x) = (x^2 + 1)^3$. To find its derivative, we could use binomial expansion to transform it into a polynomial, but it is much easier to apply the chain rule:

$$f'(x) = 3(x^2 + 1)^2 \times 2x = 6x(x^2 + 1)^2$$

More generally, when we have a power of a function, like $[f(x)]^n$, we obtain

$$\frac{d[f(x)]^n}{dx} = n[f(x)]^{n-1}\frac{df}{dx}$$

☐

2.8.3   Derivative of inverse functions

The rules of previous section do not help us in finding derivatives of functions like the square root or, given the derivative of logarithm, in finding the derivative of the exponential. We need a rule to deal with the derivative of an inverse function.

THEOREM 2.9 (Derivative of an inverse function) *Let* $x = g(y)$ *be the inverse function of* $y = f(x)$. *Then, subject to some technical conditions, the derivative of* g *is*

$$g'(y) = \frac{1}{f'(g(y))}$$

We will not insist on the conditions ensuring that the theorem holds. Of course, f must be invertible, which means that it must be monotonic on some interval; we also expect that some continuity is required in order to ensure differentiability, and the derivative f' should not be zero. Leaving all of this aside, the best way to remember the result is by regarding it as

$$\frac{dx}{dy} = \frac{1}{dy/dx}$$

In practice, we should

1. Find the derivative of f, $f'(x)$

2. Evaluate it for $x = g(y)$

3. Take the reciprocal of this value

Example 2.25 In Example 2.19 we saw that the derivative of $f(x) = \ln x$ is $f'(x) = 1/x$. On the basis of this finding, we may use Theorem 2.9 to find the derivative of its inverse, the exponential function $g(y) = e^y$. Incidentally, the logarithm is continuous and increasing on its domain, so there is no problem in inverting it. The rule is applied as follows:

$$f'(x) = \frac{1}{x} \quad \rightarrow \quad \frac{1}{f'(x)} = x \quad \rightarrow \quad \frac{1}{f'(e^y)} = e^y$$

Thus, we see that the derivative of the exponential function, with base e, is the exponential itself. ▯

Example 2.26 Now we are also able to find the derivative of the square root $g(y) = \sqrt{y}$ as the inverse of $f(x) = x^2$. Following the inverse function drill, we have

$$f'(x) = 2x \quad \rightarrow \quad \frac{1}{f'(x)} = \frac{1}{2x} \quad \rightarrow \quad \frac{1}{f'(\sqrt{y})} = \frac{1}{2\sqrt{y}}$$

The behavior of the rate of increase for the square root function is similar to the natural logarithm.

One further thing that is worth noting is that we could have found the same result by *formally* applying the rule $(x^n)' = nx^{n-1}$, for $n = \frac{1}{2}$:

$$\left(\sqrt{x}\right)' = \left(x^{1/2}\right)' = \frac{1}{2}x^{(1/2)-1} = \frac{1}{2}x^{-1/2} = \frac{1}{2\sqrt{x}}$$

This formula can generalized as follows. Using the derivative of the inverse function, we might prove that the rule above can be applied to any power $x^{m/n}$ for integer numbers m and n; this means that the result applies to rational exponents; then, by limit arguments, it can be shown that it applies to any power x^α, when α is a real number: $(x^\alpha)' = \alpha x^{\alpha-1}$. ⬚

Clearly, all of the rules that we have encountered can be applied together to deal with complicated functions, as Example 2.27 illustrates.

Example 2.27 Consider

$$f(x) = \ln\left(x^3 + 1\right)$$

Using the derivative of natural logarithm and the derivative of composite functions, we get

$$f'(x) = \frac{3x^2}{x^3 + 1}$$

Needless to say, this makes sense for the domain of $f(x)$, i.e., $x^3 + 1 > 0$, or $x > -1$. By the same token, if

$$f(x) = \sqrt{x^3 + 3x - 18}$$

then

$$f'(x) = \frac{3x^2 + 3}{2\sqrt{x^3 + 3x - 18}}$$
⬚

2.9   USING DERIVATIVES FOR GRAPHING FUNCTIONS

The derivative is the slope of the tangent line to the graph of a function. Hence, the sign of the derivative at a point tells us whether the function is increasing or decreasing there and how rapidly. We can use this to figure out essential features of a function and to sketch its graph.

Example 2.28 Let us try to figure out the behavior of the rational function

$$f(x) = \frac{x - 1}{x + 1}$$

To begin with, a rational function is in trouble when the denominator is zero. Hence, the domain of this function does not include the point $x = -1$. Since the numerator is negative for $x = -1$, we can see that

$$\lim_{x \to -1^-} \frac{x-1}{x+1} = -\infty, \qquad \lim_{x \to -1^+} \frac{x-1}{x+1} = +\infty$$

The function is not continuous at $x = -1$. It is also easy to see that the function is zero when $x = 1$; there, the function crosses the horizontal axis. Another feature which is worth noting is the behavior for very large values of x, where x dominates the constants -1 and $+1$ at numerator and denominator, respectively:

$$\lim_{x \to \pm\infty} \frac{x-1}{x+1} = \lim_{x \to \pm\infty} \frac{x}{x} = 1$$

To see the rate of increase/decrease of the function, we need its derivative. Applying the rule for the ratio of functions, we obtain

$$f'(x) = \frac{1 \times (x+1) - (x-1) \times 1}{(x+1)^2} = \frac{2}{(x+1)^2}$$

We see that the derivative is always positive, and it goes to $+\infty$ when x tends to -1, both from the left and from the right.

Putting everything together, we see that this function:

- Tends to 1 (from above) for $x \to -\infty$

- Increases on the interval $(-\infty, -1)$ and goes to $+\infty$ when x approaches -1 from the left

- Goes to $-\infty$ when x approaches -1 from the right

- Increases on the interval $(1, \infty)$

- Has value 0 for $x = 1$

- Tends to 1 (from below) for $x \to +\infty$

On the basis of these features, we could sketch a plot like Fig. 2.21.   ▯

As a more useful example, we are now able to figure out the behavior of the total cost function in the EOQ model of Section 2.1.

Example 2.29 The average total cost per unit time in the EOQ model is given by function

$$C_{\text{tot}}(Q) = h\frac{Q}{2} + \frac{Ad}{Q} + cd$$

The last term does not depend on the order quantity Q, and its effect is to shift the graph up a bit. The linear component $hQ/2$ goes to zero for small order sizes and goes to $+\infty$ for large order sizes. The nonlinear component

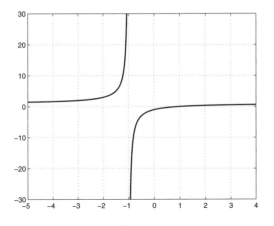

Fig. 2.21 Graph of the function of Example 2.28.

Ad/Q goes to zero for large order sizes and goes to $+\infty$ for small order sizes. Hence, we obtain

$$\lim_{Q\to 0^+} C_{\text{tot}}(Q) = +\infty, \qquad \lim_{Q\to +\infty} C_{\text{tot}}(Q) = +\infty$$

The domain of interest is the interval of strictly positive order quantities, $(0, +\infty)$. Unlike the rational function of the previous example, there is no point at which the function goes into trouble (is discontinuous) within this domain. Hence, it must be the case that the function is decreasing first, and then starts to increase.[17] Indeed, the derivative is

$$C'_{\text{tot}}(Q) = \frac{h}{2} - \frac{Ad}{Q^2}$$

Note that the constant term cd disappears in the derivative. Solving for Q, we see that the derivative is zero for

$$Q^* = \sqrt{\frac{2Ad}{h}}$$

For $Q < Q^*$, the derivative is negative and the function is decreasing. At point $Q = Q^*$, the slope of the tangent line is horizontal. For $Q > Q^*$, the

---

[17]Well, this is not quite precise. We can claim that the function must be decreasing for x close to $-\infty$ and increasing for x close to $+\infty$. In the middle, however, it could oscillate, resulting in possibly many local minima and maxima. We will see later that the function has the right curvature, as its second-order derivative is positive and the function is convex. (See Example 2.36.)

derivative is positive and the function is increasing. This results in the plot of Fig. 2.2. □

Since the total cost function is decreasing on interval $(0, Q^*)$ and increasing on interval $(Q^*, +\infty)$, it must be the case that Q^* is the optimal order size. This point is associated with a horizontal tangent line and we have good reasons to guess that such stationarity points are essential when optimizing a function.

DEFINITION 2.10 (Stationarity point) *Given a function $f(x)$, a point x^* such that $f'(x^*) = 0$ is called a* stationarity point *for the function. We also say that f is stationary at x^*.*

The EOQ example suggests that a function can be optimized by finding a stationarity point. In fact, life is not that easy, as many things can go wrong.

1. To begin with, the function that we want to optimize may fail to be differentiable; for instance, the nondifferentiable function illustrated in Fig. 2.19(a) is minimized just where the derivative is not defined.

2. Moreover, even if we assume continuity and differentiability, a stationarity point may be a minimum, a maximum, or neither. We are in trouble when functions oscillate; to see such examples, please take a look back at Fig. 2.8.

3. Last but not least, a stationarity point might just be a local rather than a global optimum.

In order to better characterize points at which a function is maximized or minimized, we need some information that is not really provided by the derivative; we need information about the *curvature* of the function. This information is provided by second-order derivatives.

2.10 HIGHER-ORDER DERIVATIVES AND TAYLOR EXPANSIONS

The derivative tells us something about the rate at which a function f increases or decreases at some point x. This rate is the slope of the tangent line to the graph of f at x. So, the derivative tells us something about the "linear" behavior of a function. However, this does not tell us anything about its curvature. To visualize the issue, compare the behavior patterns of functions $f(x) = x^2$ and $g(x) = -x^2$ for $x = 0$. Since $x^2 \geq 0$, it is obvious that $x = 0$ is a minimizer of f and a maximizer of g. This is a stationarity point, as it is easily checked by finding the derivatives of the two functions:

$$f'(x) = 2x, \qquad g'(x) = -2x$$

The slope of f is negative for $x < 0$ and positive for $x > 0$. This implies that the function is decreasing for $x < 0$ and increasing for $x > 0$, but this in turn implies that $x = 0$ is a minimum. The pattern for g is just the opposite one. In the case of f, the slope itself is increasing, whereas it is decreasing (turning from positive to negative) for g. We can see this by taking the derivatives of f' and g':

$$f''(x) = 2, \qquad g''(x) = -2$$

Here we have used $f''(x)$ to denote the derivative of the derivative. This is referred to as the *second-order* derivative, whereas the derivative that we have seen so far is actually called *first-order* derivative.

DEFINITION 2.11 (Higher-order derivatives) *Given a continuously differentiable function f, its second-order derivative at point x is the derivative of the first-order derivative. This is denoted as $f''(x)$ or*

$$\frac{d^2 f}{dx^2}(x) = \left[\frac{d}{dx} \left(\frac{df}{dx} \right) \right](x)$$

Taking the derivative of the second-order derivative, we get the third-order derivative:

$$f'''(x) \qquad or \qquad \frac{d^3 f}{dx^3}(x)$$

More generally, we define the k-th order derivative of f as

$$f^{(k)}(x) \qquad or \qquad \frac{d^k f}{dx^k}(x)$$

In the definition, we use the term *continuously differentiable*. We recall that this simply means that the function is differentiable and its derivative is a continuous function; otherwise, we could not take the derivative of the derivative.

Example 2.30 Given the polynomial function $f(x) = 3x^3 - 2x^2 + 5x - 10$, we have

$$
\begin{aligned}
f'(x) &= 9x^2 - 4x + 5 \\
f''(x) &= 18x - 4 \\
f'''(x) &= 18 \\
f^{(k)}(x) &= 0, \qquad k = 4, 5, 6, \ldots
\end{aligned}
$$

We see that a polynomial of degree n has zero derivative from order $n + 1$ on.

$\square$

It is worth noting that for a linear function $f(x) = a + bx$, we have $f''(x) = 0$; in fact, a linear function has no curvature. For nonlinear functions, knowing the kind of curvature helps us in plotting them more accurately, and this information can be exploited to find their maxima and minima. In fact, first- and second-order derivatives can be used to characterize the local behavior of a function near some point x_0. It is natural to wonder if we could exploit knowledge about higher-order derivatives, provided they exist, to shed some more light on the function. Indeed, it can be shown that by using the derivatives of a function near some point x_0, we can find an arbitrarily good approximation of the function by a polynomial, subject to some assumptions, including continuity. We will not state the theorem exactly, as it would require a few technicalities; nevertheless, the result is quite useful, as polynomial functions are relatively easy to deal with.

DEFINITION 2.12 (Taylor's expansion) Taylor's expansion *of order* n *for function* f *around point* x_0 *is given by the following approximation:*

$$f(x_0 + h) \approx f(x_0) + f'(x_0)h + \frac{1}{2}f''(x_0)h^2 + \cdots + \frac{1}{n!}f^{(n)}(x_0)h^n \qquad (2.13)$$

A few comments are in order:

- The formula relies on the existence of all the derivatives involved.

- The formula above tells us how we can approximate function f by a polynomial, for a small displacement h around x_0.

- The quality of the approximation improves if we increase the order of the involved derivatives.

- On the contrary, the quality of the approximation worsens if we increase the displacement h.

Actually, the formula of Taylor's expansion relies on a theorem that states that the function can be expressed by the formula above *plus* a remainder term. This remainder depends on derivative of order $n + 1$ evaluated at some point in the neighborhood of x_0. In practice, this remainder is negligible if we do not get "too far" from x_0.

It is quite useful to consider in more detail Taylor's expansion if we stop with the first- or second-order derivative. If we stop with first-order derivative, we get the *first-order Taylor expansion*, which can be equivalently rewritten by setting $h = x - x_0$:

$$f(x) \approx f(x_0) + f'(x_0)(x - x_0) \qquad (2.14)$$

We immediately see that the first-order approximation is equivalent to a finite-difference approximation of the first-order derivative:

$$f'(x_0) \approx \frac{f(x) - f(x_0)}{(x - x_0)}$$

Formally, this finite-difference approximation can also be expressed as

$$\frac{df}{dx} \approx \frac{\Delta f}{\Delta x}$$

Table 2.2 Accuracy of first- and second-order approximations of the exponential function.

x	e^x	$f_1(x)$	$f_2(x)$
0.1	1.1051710	1.1000000	1.1050000
0.2	1.2214028	1.2000000	1.2200000
0.5	1.6487212	1.5000000	1.6250000

to remind us that we are substituting infinitesimal increments with finite ones. Furthermore, we see that (2.14) approximates the function by its tangent line;[18] hence, it is a *linear* or first-order approximation. If we want to keep curvature information in the approximation, we can include a quadratic term from Taylor's expansion:

$$f(x) \approx f(x_0) + f'(x_0)(x - x_0) + \tfrac{1}{2}f''(x_0)(x - x_0)^2 \qquad (2.15)$$

This is a quadratic approximation around x_0; the second-order Taylor expansion is just a parabola.

Example 2.31 To illustrate Taylor's expansions, we can prove Eq. (2.6), which gives a concrete way to evaluate an exponential function with base e. The exponential function e^x is a peculiar one, as $f^{(n)}(x) = e^x$ for any n. Its Taylor's expansion around x_0 is then

$$e^x \approx e^{x_0} + e^{x_0} \cdot (x - x_0) + \frac{1}{2} e^{x_0} \cdot (x - x_0)^2 + \cdots + \frac{1}{n!} e^{x_0} \cdot (x - x_0)^n$$

Now, if we set $x_0 = 0$ and let $n \to +\infty$, we get

$$e^x = 1 + x + \frac{x^2}{2!} + \frac{x^3}{3!} + \cdots = \sum_{k=0}^{+\infty} \frac{x^k}{k!}$$

To check the accuracy of this expansion, let us compare the true value of e^x against the first- and second-order approximations $f_1(x) = 1 + x$, and $f_2(x) = 1 + x + x^2/2$. The numerical results are displayed in Table 2.2; in Fig. 2.22 we may also visually compare the exponential function with its first- and second-order approximations around $x_0 = 0$. $\qquad \square$

As expected, low-order Taylor's approximations deteriorate rather quickly when we depart from the point at which they are developed. Indeed, practical numerical approximations often rely on more sophisticated approaches. Nevertheless, Taylor's expansions are both conceptually and practically relevant. One fundamental application is sensitivity analysis.

---

[18]See Eq. (2.10).

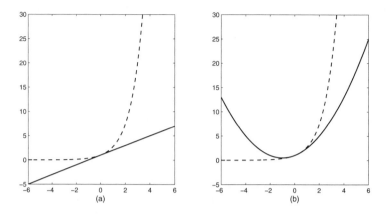

Fig. 2.22 The exponential function (dashed line) against its first- and second-order Taylor expansions.

2.10.1 Sensitivity analysis

Reading on, you will notice that a large part of the book deals with uncertainty. Uncertainty comes in many forms:

- We might be unsure about the value of input data we use in making decisions.

- We might be unsure about the future values of some relevant exogenous variable.

One way to deal with uncertainty is to rely on the tools of probability theory and statistics. The main limitation of these tools is that they may require a lot of past data to characterize uncertainty, assuming that past data do tell us something useful about the future. Alternatively, we may analyze how changes in the input data affect our decision, in order to check its robustness. This may be more appropriate for the first situation above, which is more linked to ignorance than to genuine randomness. This fundamental process is called *sensitivity analysis* and we illustrate the concept with a few examples.

Example 2.32 (Sensitivity analysis of EOQ model) In the EOQ model we rely on a few pieces of information that in practice must be estimated: the fixed ordering charge A, the holding cost h, and the demand rate d. If we knew these quantities exactly, we would be sure that the optimal order quantity and the ensuing average cost per unit time are

$$Q^* = \sqrt{\frac{2Ad}{h}}, \qquad C_{\text{tot}}(Q^*) = \sqrt{2Adh}$$

respectively (if the assumptions behind the model are satisfied). In the total cost function we have neglected the total purchasing cost which does not really depend on the order size and just contributes a constant term. But if we make some mistake in estimating the input data, we will end up with some other order quantity Q, which will be suboptimal: $C_{\text{tot}}(Q) > C_{\text{tot}}(Q^*)$. What is the effect of our mistake in economic terms? We can measure this by taking the ratio of the two costs:

$$\frac{C_{\text{tot}}(Q)}{C_{\text{tot}}(Q^*)} = \frac{\frac{1}{2}Qh + (Ad/Q)}{\sqrt{2Adh}} = \frac{1}{2}Q\sqrt{\frac{h}{2Ad}} + \frac{1}{2Q}\sqrt{\frac{2Ad}{h}} = \frac{1}{2}\left(\frac{Q}{Q^*} + \frac{Q^*}{Q}\right)$$

where the last step relies on the expression of Q^*. To see the value of this relationship, let us assume that we make a rather gross mistake and apply an order quantity that is twice the right one, i.e., $Q = 2Q^*$. Plugging this into the ratio above yields

$$\frac{C_{\text{tot}}(Q)}{C_{\text{tot}}(Q^*)} = \frac{1}{2}\left(2 + \frac{1}{2}\right) = 1.25$$

In practice, if we make a 100% mistake in setting the order quantity, the total cost will increase by just 25%. This illustrates a robustness property of the EOQ formula. Of course, this tells us something useful only as far as the assumptions behind the model hold, at least approximately. □

The example above is, in a sense, an example of sensitivity analysis "in the large." We have not actually examined the effect of a change in a single factor, such as h, A, or d; rather, we have put everything together, analyzing the overall effect of making a mistake in setting the order quantity. However, one could ask what the effect of a small change in, say, the ordering cost A, is on both cost and the resulting EOQ. When we analyze the effect of a small change in a single factor, Taylor's expansions come into play. We illustrate with an example from finance.

Example 2.33 (Bond duration) We got acquainted with bonds in Example 2.7. The price of a bond is essentially the present value of the cash flows up to bond maturity, using yield to maturity as the discount rate. Most long-term bonds offer the periodic payment of a coupon, in addition to repayment of the face value, but there are bonds that do not. A *zero-coupon bond* is a bond that promises only repayment of the face value F at maturity.

If we hold a zero-coupon bond maturing in 5 years, are we safe? When asked this question, students typically start mentioning a few risks that may be associated with such a bond:

- One obvious risk is default. A default occurs when bond issuers do not repay their debt. Default risk is rather high for bonds issued by distressed corporations or governments of unstable countries.

- Another possible risk is inflation. Indeed, with a long-term bond, even if we are repaid, the real value of the money we get back might have been significantly eroded.

What most students fail to mention is the fact that the general level of interest rates may change. This happens as central banks adjust interest rates in order to control inflation or to help economy against recession. Another important point is: Why are we holding that bond in our portfolio? If we want to keep the bond until maturity, and there is neither default nor inflation risk, we are perfectly safe. But if we want to sell it before, because we need immediate cash, an unfavorable change in the interest rate might affect bond value.

Actually, all of the risk factors that we mentioned are somehow captured by the required bond yield y.[19] This is typically larger than the current level of risk-free rates, as it reflects different kinds of risk premia. Reasonably, when holding a risky security, an investor expects to be compensated for the risk she is bearing, and this implies a rate of return higher than that of a risk-free asset. The price of a zero-coupon bond with face value F, maturing in T years, is given by

$$P(y) = \frac{F}{(1+y)^T} \tag{2.16}$$

If, for whatever reason, the required yield y goes up, the bond price will go down. As an illustration, consider the price of a zero-coupon bond, with face value $1000, maturing in 10 years, when yield to maturity is $y = 5\%$:

$$P(y) = \frac{F}{(1+y)^T} = \frac{1000}{(1.05)^{10}} = \$613.91$$

If the required yield increases by 100 basis points,[20] i.e., the new yield is $y = 0.05 + 0.01 = 0.06$, the new price turns out to be $P(0.05 + 0.01) = 558.39$. The resulting loss is

$$\frac{558.39 - 613.91}{613.91} = -9.04\%$$

It would be very nice to have a sensitivity measure telling us the effect of a small variation δy of the required yield, possibly applying to coupon-bearing bonds as well.

Before doing so for a general bond, let us check the quality of a Taylor expansion of the yield–price relationship for the zero-coupon bond. This requires taking first- and second-order derivatives with respect to yield in (2.16):

$$P(y + \delta y) \approx P(y) - \frac{TF}{(1+y)^{T+1}} \cdot \delta y + \frac{T(T+1)F}{(1+y)^{T+2}} \cdot (\delta y)^2$$

The first-order approximation gives $555.45 as an estimate of the new price, whereas the second-order approximation gives $558.51. We see that the approximations are rather good. Indeed, the change in the required yield cannot

---

[19] We recall from Example 2.7 that the bond yield is the internal rate of return of the investment, i.e., the discount rate such that its NPV is zero.
[20] A basis point is one-hundredth of 1%, i.e., 0.0001.

be a huge number, since we are talking about interest rates. Now, let us explore the concept further for a general coupon-bearing bond. The price–yield relationship is

$$P(y) = \sum_{t=1}^{T-1} \frac{C}{(1+y)^t} + \frac{C+F}{(1+y)^T} \tag{2.17}$$

where C is the coupon paid every period; note again that the last cash flow at bond maturity $t = T$ includes both payment of last coupon and refund of face value F. For the sake of simplicity, we assume that one coupon is paid per year, so time subscripts $t = 1, \ldots, T$ correspond to years. The sensitivity of price with respect to yield is associated with the first-order derivative of Eq. (2.17):

$$
\begin{aligned}
P'(y) &= -\sum_{t=1}^{T-1} \frac{tC}{(1+y)^{t+1}} - \frac{T(C+F)}{(1+y)^{T+1}} \\
&= -\frac{1}{1+y} \left[\sum_{t=1}^{T-1} t \frac{C}{(1+y)^t} + T \frac{C+F}{(1+y)^T} \right] \\
&= -\frac{1}{1+y} \sum_{t=1}^{T} t d_t
\end{aligned}
$$

where d_t is the discounted cash flow for time period t:

$$
d_t = \begin{cases}
\dfrac{C}{(1+y)^t}, & t = 1, 2, \ldots, T-1 \\[2mm]
\dfrac{C+F}{(1+y)^T}, & t = T
\end{cases}
$$

Not surprisingly, the derivative is negative, as an increase in yield implies a drop in bond price. We can rewrite the relationship in a more informative way by noting that the bond price is just the sum of the discounted cash flows d_t:

$$P'(y) = \left(-\frac{1}{1+y} \sum_{t=1}^{T} t d_t \right) \times \frac{P(y)}{P(y)} = -\frac{P(y)}{1+y} \times \frac{\sum_{t=1}^{T} t d_t}{\sum_{t=1}^{T} d_t}$$

The ratio of the two sums can be regarded as a weighted average of the time instants t at which cash flows occur, where weights are just the discounted cash flows d_t. The ratio

$$D \equiv \frac{\sum_{t=1}^{T} t d_t}{\sum_{t=1}^{T} d_t} \tag{2.18}$$

is called *Macaulay duration*. In order to get rid of the annoying $1+y$ factor,[21] we introduce *modified duration*

$$D_M \equiv \frac{D}{1+y} \qquad (2.19)$$

which allows us to write the following relationships:

$$\delta P \approx -P \times D_M \times \delta y \qquad \Rightarrow \qquad \frac{\delta P}{P} \approx -D_M\,\delta y \qquad (2.20)$$

This is just a rewriting of Eq. (2.18) based on the finite-difference approximation[22] $P'(y) \approx \delta P/\delta y$. The second relationship states that the percentage change in bond price can be approximated by the product of modified duration times the change in yield. The minus sign accounts for the fact that bond prices and yields are inversely related. Equation (2.20) shows that duration is a risk measure for bonds. The larger the duration, the larger the sensitivity of bond prices to unexpected changes in the required yield.

Other things being equal, the longer the bond maturity, the larger the bond duration. Hence, to stay safe, one should invest in short-term bonds. By the way, not surprisingly, bond duration for a zero-coupon bond is just bond maturity. However, other factors may affect bond duration. Since this is a weighted average of the times at which cash flows occur, we may also see that a large coupon C reduces duration. This happens because the weights d_t for periods $t = 1, 2, \ldots, T - 1$ are relatively larger when C is significant with respect to face value F.

Bond duration is a useful concept for bond portfolio management, but it is rich in shortcomings. For instance, it assumes that there is only one catch-all risk factor y, whereas in practice one should account, e.g., for relative variations in the short- versus the long-term interest rates. Moreover, it is associated with a first-order approximation that is likely to give a poor approximation unless the change δy in required yield is small. To ease the last difficulty, one can resort to a second-order Taylor expansion

$$P(y + \delta y) \approx P(y) - D_M \cdot P(y) \cdot \delta y + \tfrac{1}{2} \cdot C \cdot (\delta y)^2$$

The term C, which is just the second order derivative $P''(y)$ is called *bond convexity*. ⬜

---

[21] We leave as an exercise for the reader to show that if cash flows are discounted by a continuously compounded yield y_c, using negative exponentials as discount factors, this annoying factor disappears. Indeed, given the properties of the exponential function, financial mathematics is much nicer when compounding in continuous rather than discrete time.

[22] When denoting a finite-difference approximation of a derivative, we may use either δf or Δf. Since in finance Δ may refer to a sensitivity measure for options, here we prefer the former notation.

2.11 CONVEXITY AND OPTIMIZATION

What we have learned so far about function derivatives suggests that in order to optimize a function, assuming that it is differentiable, a good starting point is to set its first-order derivative to zero. However, we know that this first-order, stationarity condition may not be enough, as it does not even discriminate between a maximum and a minimum. In practice, the matter is further complicated by two features of optimization models:

- They involve a possibly large number of decision variables, whereas we are dealing here with functions of a single variable.

- They involve constraints, as we have seen in the optimal mix example of Section 1.1.2. Generally speaking, our decisions are constrained to stay within a set S called *feasible region* or *feasible set*.

We will learn a lot more about these issues in Chapter 12. However, it would be nice to find a feature that makes an optimization problem relatively easy. This feature is termed *convexity*. We start by defining convexity for sets; then, we generalize the concept to functions;[23] finally, we illustrate what role convexity plays in economic modeling and in optimization. Before doing so, though, it may be useful to clarify what we mean by local and global optimality formally.

2.11.1 Local and global optimality

Earlier in the chapter we plotted the polynomial function

$$f(x) = x^4 + x^3 - 29x^2 - 9x + 180$$

whose graph is reported again for convenience in Fig. 2.23. The stationarity points can be found by setting its derivative to zero:

$$f'(x) = 4x^3 + 3x^2 - 58x - 9 = 0$$

Using numerical methods, we find the following roots of $f'(x)$:

$$x_1 = -4.1294, \quad x_2 = -0.1542, \quad x_3 = 3.5336$$

which are indeed the points at which f is stationary. Observing the graph, we see that x_1 is the global minimum, x_2 is a local maximum, and x_3 is a local

---

[23] Although this chapter deals with functions of a single variable, we will state convexity of functions in more general terms, which apply to functions of more variables as well. Here, it is more of a formality than anything else; we will appreciate its importance in Chapter 12; see also the last sections of Chapter 3, where we outline a few concepts of calculus for functions of multiple variables.

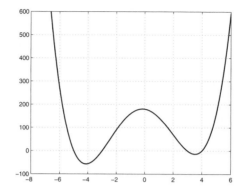

Fig. 2.23   Graph of $f(x) = x^4 + x^3 - 29x^2 - 9x + 180$.

minimum. In order to formalize these intuitive concepts, we need to define the *neighborhood* of a point on the real line.

DEFINITION 2.13 (Neighborhood of a point) *Given point $x_0 \in \mathbb{R}$, we define its (open) neighborhood of radius ϵ as the interval $(x_0 - \epsilon, x_0 + \epsilon)$, which is also denoted by $N_\epsilon(x_0)$.*

In practice, for a small radius ϵ the neighborhood $N_\epsilon(x_0)$ consists of the points close to x_0, i.e., points within a distance from x_0 bounded by ϵ. The concept can be generalized to multiple dimensions easily.

DEFINITION 2.14 (Global optimality) *Given a function f defined over a domain $\mathcal{D}$, we say that $x^*_{\min} \in \mathcal{D}$ is a global minimizer of function f if $f(x^*_{\min}) \leq f(x)$, for any other $x \in \mathcal{D}$. We speak of a strict global minimizer if $f(x) \neq f(x^*_{\min})$ for $x \neq x^*_{\min}$. By the very same token we define a global maximizer $x^*_{\max}$ by requiring $f(x^*_{\max}) \geq f(x)$, for any other $x \in \mathcal{D}$. The definition of strict global maximizer is obvious.*

In this definition we should not confuse, e.g., the minimizer $x^*_{\min}$ with the minimum $f^* = f(x^*_{\min})$. In a minimum-cost problem, the minimizer is a decision, whereas the minimum is the cost of that decision. The domain $\mathcal{D}$ could be restricted by the conditions under which the function is defined or by constraints on the decisions. Finally, a global minimizer need not be unique, as the minimum value f^* could be attained by more than one point, unless it is a strict minimum.

DEFINITION 2.15 (Local optimality) *Given a function f defined over a domain $\mathcal{D}$, we say that $x^*_{\min} \in \mathcal{D}$ is a local minimizer of function f if there exists a neighborhood $N_\epsilon(x^*)$ such that $f(x^*_{\min}) \leq f(x)$, for any other $x \in \mathcal{D} \bigcap N_\epsilon(x^*)$. The definition for a local maximizer is obtained similarly.*

In local optimality we consider the intersection of the domain $\mathcal{D}$ and the neighborhood $N_\epsilon(x^*)$. In plain words, a local minimum is a point such that we cannot find anything better in its neighborhood, even though there might be a better solution if we look far enough. When dealing with a function of a single variable, a local minimizer is a point such that if we look both to the left and to the right, we see an increase in the function. Clearly, this local view does not imply that, outside this neighborhood, there is no point associated with a smaller function value. It is rather intuitive to generalize the concept to a function of two variables (we move on a plane), or a function of three variables (we move within a three-dimensional space), and so on.

2.11.2   Convex sets

Convexity can be introduced as a fairly intuitive concept that applies to n-dimensional subsets of $\mathbb{R}^n$. Spaces with multiple dimensions will be the subject of next chapter, but we can visualize things on a plane, which is just the set $\mathbb{R}^2$ of points with two coordinates. We use boldface characters when referring to a point $\mathbf{x} \in \mathbb{R}^2 \equiv \mathbb{R} \times \mathbb{R}$, with coordinates (x_1, x_2). Subscripts refer to coordinates; for instance, point $\mathbf{y}$ has coordinates (y_1, y_2). Generally speaking, in the next chapter we consider tuples $\mathbf{x} = (x_1, x_2, \ldots, x_n)$ in $\mathbb{R}^n$, but two-dimensional visualization is what we need to grasp the concepts. Intuitively, a set S is convex if all of the points on the line segment joining two points $\mathbf{x}$ and $\mathbf{y}$ in it belong to S itself, for any choice of $\mathbf{x}$ and $\mathbf{y}$. Figure 2.24 illustrates the idea:

1. Set S_1 is a polyhedron, and is convex.

2. Set S_2 is a standard example of a nonconvex set.

3. The last set S_3 is a bit different, but nonconvex nevertheless. It is a discrete set obtained by considering only points with integer coordinates.[24]

Here is a formal definition of convexity that applies to subsets of $\mathbb{R}^n$.

DEFINITION 2.16 (Convex sets) *A set $S \subset \mathbb{R}^n$ is convex if, given any pair of points $\mathbf{x}$ and $\mathbf{y}$ in S, for any real number λ in the interval $[0, 1]$ we have*

$$\lambda \mathbf{x} + (1 - \lambda)\mathbf{y} \in S$$

For an arbitrary value λ, $\lambda \mathbf{x} + (1 - \lambda)\mathbf{y}$ is just a way to describe the line passing through that pair of points (we know from elementary geometry in the plane that there is exactly one such line; this also applies to lines in spaces with multiple dimensions). If we restrict λ to the interval $[0, 1]$, we are just considering the line segment between them. Indeed, if we set $\lambda = 0$, we get $\mathbf{y}$; if we set $\lambda = 1$, we get $\mathbf{x}$.

---

[24] In fact, as we will learn in Chapter 12, lack of convexity is what makes optimization with discrete decision variables difficult.

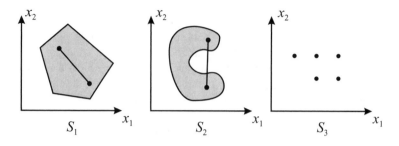

Fig. 2.24   Convex and nonconvex sets.

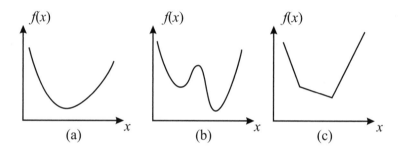

Fig. 2.25   Convex and nonconvex functions.

2.11.3   Convex functions

Convexity can be easily generalized to functions by applying the idea of convexity for sets to the *epigraph* of the function. For functions of a single variable, which can be plotted on a plane, the epigraph of the function is just the set of points lying above the function graph. The idea generalizes to an arbitrary number of dimensions. Figure 2.25 illustrates the concept for functions of one variable. Function (a) is obviously convex, whereas function (b) is not. We see immediately that function (a) is easy to minimize, since there is one point where the first-order derivative is zero, and it is a global minimum Function (b) is not that easy, as it features local maxima and minima. Function (c) may look a bit weird, as it is not differentiable; nevertheless, it is a convex function. We also see that the epigraph of function (c) is a polyhedron; this is why a function like that is called a *polyhedral convex function*. In Chapter 12 we will see some tricks that allow us to minimize a polyhedral convex function quite easily, despite lack of differentiability.

There is an alternative way to regard convexity, which is better suited to emphasize the essential features of a convex function. If a function is convex and we take two arbitrary points on its graph, the line segment joining them will always stay *above* the function graph. This allows us to define convex functions as follows.

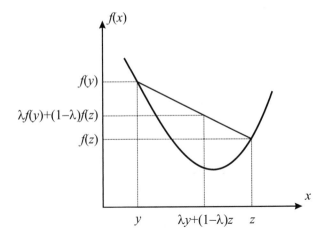

Fig. 2.26 A schematic illustration of the definition of a convex function.

DEFINITION 2.17 (Convex and concave functions) *A function f, de-fined on a domain S, is convex on S if, given any pair of points $\mathbf{y}$ and $\mathbf{z}$ in S, the following condition holds for any real number λ in the interval $[0,1]$:*

$$f\left[\lambda\mathbf{y} + (1-\lambda)\mathbf{z}\right] \le \lambda f(\mathbf{y}) + (1-\lambda) \cdot f(\mathbf{z})$$

A function f is concave if $(-f)$ is convex.

This definition is illustrated in Fig. 2.26. In this case the set S is a portion of the real line, within which we consider an interval $[y, z]$; if we take any point $x = \lambda y + (1-\lambda)z$ within that interval, the true function value at x is less than the linearly interpolated value $\lambda f(y) + (1-\lambda)f(z)$. A concave function is just a convex function flipped upside down. If the inequality is strict for $\mathbf{y} \ne \mathbf{z}$ and λ in the open interval $(0, 1)$, we have a *strictly convex function*. Figure 2.27 illustrates the difference between a convex and a strictly convex function.

Intuition suggests that for a convex function we do not have trouble with local minima, whereas for a concave function we do not have trouble with local maxima. If convexity is not strict, we may have multiple but equivalent optima. It is also important to realize that the definition above does not rely on differentiability of function f and applies to functions defined on any n-dimensional space. If we restrict our attention to differentiable functions of one variable, then the following properties hold for a function of one variable.[25]

PROPERTY 2.18 *If f is a convex differentiable function of one variable, on a domain $S \subseteq \mathbb{R}$, then the following condition holds, for any x_0 and x in*

---

[25] These properties do generalize to functions of multiple variables, as we will see in Section 12.1.3.

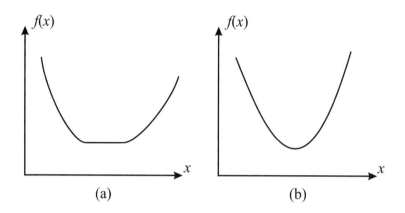

Fig. 2.27 (a) Convex vs. (b) strictly convex functions.

S:

$$f(x) \geq f(x_0) + f'(x_0)(x - x_0)$$

The inequality is reversed for a concave function.

PROPERTY 2.19 *If f is a convex differentiable function of one variable, on a domain $S \subseteq \mathbb{R}$, then the following condition holds, for any x in S:*

$$f''(x) \geq 0$$

The inequality is reversed for a concave function.

Property 2.18 essentially says that if we draw the tangent line $y = f(x_0) + f'(x_0)(x - x_0)$ at any point with coordinates $[x_0, f(x_0)]$ on the graph of a convex (differentiable) function, then the linear approximation (first-order Taylor expansion) will consistently *underestimate* the true value of the function (see Fig. 2.28). Property 2.19 states that the slope of tangent lines to f, drawn at different points x, is always increasing with respect to x, since the second-order derivative is always positive. Again, this property assumes suitable differentiability conditions, and is illustrated in Fig. 2.29. The following example shows the connection between convex functions and sets.

Example 2.34 We show that the region S described by the inequality $g(\mathbf{x}) \leq 0$ is a convex set if g is a convex function. If $\mathbf{x} \in S$, then $g(\mathbf{x}) \leq 0$; by the same token, if $\mathbf{y} \in S$, then $g(\mathbf{y}) \leq 0$. What we want to prove is that $\lambda \mathbf{x} + (1 - \lambda)\mathbf{y} \in S$, for all $\lambda \in [0, 1]$, i.e., that $g(\lambda \mathbf{x} + (1 - \lambda)\mathbf{y}) \leq 0$. But, since g is convex, we have

$$g(\lambda \mathbf{x} + (1 - \lambda)\mathbf{y}) \leq \lambda g(\mathbf{x}) + (1 - \lambda)g(\mathbf{y}) \leq 0$$

where the last inequality depends on the fact that we are summing nonpositive terms, which are obtained by multiplying a nonpositive quantity, $g()$,

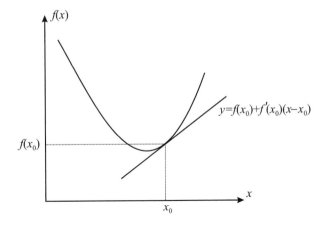

Fig. 2.28   Tangent lines always underestimate a convex function.

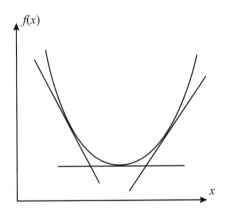

Fig. 2.29   Slopes of tangent lines are increasing for a convex function.

by nonnegative coefficients λ and $(1 - \lambda)$. This proves that $\lambda \mathbf{x} + (1 - \lambda)\mathbf{y}$ is in S. From a practical point of view, this property plays an important role when inequalities are used to define the feasible set of an optimization problem.   ▯

2.11.4   The role of convexity

Convexity and concavity play a major role in optimization. Consider a one-dimensional optimization problem, $\min_{x \in \mathbb{R}} f(x)$; this problem is *unconstrained*, since x can be any point on the real line. Furthermore, assume that f is convex on the whole real line $\mathbb{R}$ and that x^* is a stationarity point. Property 2.18 applies to x^*:

$$f(x) \geq f(x^*) + f'(x^*)(x - x^*) = f(x^*)$$

for any $x \in \mathbb{R}$, but this implies that x^* is a global minimizer. We have proved the following theorem.

THEOREM 2.20 *If function f is convex and differentiable on $\mathbb{R}$, then stationarity is a necessary and sufficient condition for global optimality in an unconstrained minimization problem. If f is concave, the result applies to an unconstrained maximization problem.*

Example 2.35 The theorem above applies to an unconstrained optimization problem. If there are constraints, stationarity need not be associated with optimality. To see this, consider the minimization problem

$$\min_{x \geq 2} x^2$$

The function $f(x) = x^2$ is convex, but it is easy to see that the minimizer is $x^* = 2$, which is not a stationarity point. We need some more theory to cope with constrained problems. However, it may be the case that a constraint is nonbinding where the function is stationary and we find the minimizer by setting first-order derivative to zero. This happens, for instance, for the problem

$$\min_{x \geq -2} x^2$$

Here the feasible region is $[-2, +\infty)$ but the constraint is nonbinding at stationarity point $x^* = 0$, which is indeed the global minimizer. We speak in this case of an *interior* solution, as the minimizer lies in the interior of the feasible region. ▯

We may also note that, since the second-order derivative of a convex function is always positive by property 2.19, then we do not run the risk of mistaking a maximum for a minimum. This, per se, does not imply global optimality, however. Still, property 2.19 is important because it provides us with an easy criterion to check convexity.

Example 2.36 Consider function

$$f(x) = ax + \frac{b}{x}$$

Its second-order derivative is $f''(x) = 2b/x^3$, which is positive on the domain $x > 0$, if $b > 0$. Indeed, this is the case for the total cost function of the EOQ model, where $b = Ad$ is the product of demand rate and fixed ordering charge, and we are interested in strictly positive order quantities. Hence, stationarity is in fact a necessary and sufficient condition for a globally optimal solution of the EOQ problem, which turns out to be an interior solution. ▯

Apart from optimization, convexity and concavity are useful for characterizing functions that model essential features of a problem from an economic

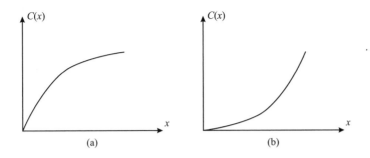

Fig. 2.30   Schematics of economies and diseconomies of scale.

perspective. Consider a cost function $c(x)$, quantifying the cost of some activity that we carry out at level x; this could be the cost of purchasing an item, but also the cost of producing an amount x of it. Two such functions are illustrated in Fig. 2.30. Both functions are increasing, as expected, but they are qualitatively different. In case (a) we see that the first-order derivative of the cost function is decreasing, whereas it is increasing in case (b). In economics, the first-order derivative of a cost function, evaluated at some point, is called the *marginal cost*. The marginal cost tells us the rate at which a cost is increasing when we raise the level x of the activity we are engaged in. For concave cost function (a), the marginal cost is decreasing; this implies that there is an *economy of scale*. On the contrary, convex function (b) models a *diseconomy of scale*, which may occur if the production process is less and less efficient when production volume is increased.

In later chapters we will see other uses of convexity and concavity, which are just worth mentioning here:

- In Chapter 12 we illustrate constrained optimization problems such as

$$\min_{\mathbf{x} \in S} f(\mathbf{x})$$

  As it turns out, a problem like this is relatively easy to solve if the feasible set S and the objective function f are both convex. If we are dealing with a maximization problem, it is easy if the function f is concave and the set S is convex.

- In Chapter 13 we deal with decision under uncertainty. In that setting, risk aversion of the decision maker plays a prominent role. We will see that one way of capturing risk aversion is by modeling preferences of the decision maker by a concave utility function.

2.11.5   An application to economics: optimal pricing

One of the most fruitful application fields of quantitative methods is revenue management. Revenue management is actually a group of techniques that can be applied in quite diverse settings, such as pricing of aircraft seats or perishable products. In this section we consider an idealized case in which a manufacturer has to find an optimal price for the items she produces and sells. We have to clarify the context in which she operates, however, as one cannot really set prices at will if there is stiff competition. To keep it as simple as possible, let us assume that there is no competition at all; the manufacturer is a *monopolist*. The fact that there is no competition does not imply that one can set arbitrarily high prices, since doing so will just kill demand. Indeed, the monopolist, barred other "political" constraints, must find the best tradeoff between the conflicting needs of selling as much as possible and setting as high a price as possible. This is a classical problem in microeconomics.

The first step to cope with our task is to find a suitable formalization. We want to maximize profit, which is revenue minus cost. The total cost depends on the manufactured amount q. Hence, it is a function $\mathrm{TC}(q)$. As an example, we might consider a cost function

$$\mathrm{TC}(q) = F + cq^2$$

where F is a fixed cost and the variable cost component involves a squared term; in such a model, cost increases rather rapidly for large values of q, reflecting possible inefficiencies in large-scale production.

We also need to find a relationship between price and demand. Usually, one would expect that an increase in price implies a drop in demand. One possible model of that is a demand function such as

$$\mathrm{D}(p) = \alpha - \beta p$$

expressing demand as a function of price p, with $\beta > 0$. This linear demand function is just a simplistic model used for illustration; it is clearly valid on only a limited range, as for suitably large prices it would imply a negative demand. In practice, one would use statistical tools described in later chapters to analyze data and identify such a relationship on an empirical basis. One problem we have is that cost is naturally expressed as a function of q, whereas demand is expressed as a function of price. Since cost depends on the produced amount, it is convenient to express everything as a function of q, which calls for inversion of the demand function:

$$q = \alpha - \beta p \quad \Rightarrow \quad p = \frac{\alpha - q}{\beta}$$

Hence, we define the *inverse demand function*, expressing price as a function of quantity

$$\mathrm{P}(q) = a - bq$$

where $a \equiv \alpha/\beta$ and $b \equiv 1/\beta$. Now we are finally able to express our problem. Profit $\Pi(q)$ is the difference between revenue $\mathrm{R}(q)$ and cost $\mathrm{TC}(q)$. Since revenue in turn is just price times demand, the profit maximization problem is as follows:

$$\max_{q \geq 0} \Pi(q) = R(q) - TC(q) = P(q)q - TC(q)$$

We will disregard the nonnegativity constraint on the decision variable q; we assume an interior solution, which we may check after solving the problem. Revenue as a function of quantity is

$$R(q) = (a - bq)q = aq - bq^2$$

We could just apply the stationarity condition $\Pi'(q) = 0$ to the profit function, but it is more instructive to take an alternative but equivalent route that yields a better understanding of the economics involved. The first-order optimality condition can be written as

$$R'(q) - TC'(q) = MR(q) - MC(q) = 0$$

where $MR(q)$ and $MC(q)$ are called the marginal revenue and marginal cost, respectively. *Marginal revenue*, which is just the derivative of revenue with respect to quantity, tells us the rate at which revenue increases if we increase quantity; by the same token, *marginal cost* is the rate at which total cost increases when increasing production. The optimality condition tells us that marginal revenue should equal marginal cost, and it is quite instructive to interpret this condition:

- If marginal revenue is larger than marginal cost, then we should increase production, since revenue increases faster than cost.

- If marginal revenue is smaller than marginal cost, then we should decrease production, since cost decreases faster than revenue.

- The optimal production corresponds to an amount q^* such that there is no reason to either increase or decrease production.

Using our simplistic model, we have

$$MR(q) = a - 2bq, \qquad MC(q) = 2cq$$

and the first-order optimality condition reads as follows:

$$a - 2bq = 2cq \quad \Rightarrow \quad q^* = \frac{a}{2(b + c)}$$

Note that, indeed, we get an interior solution $q^* > 0$; hence, the nonnegativity constraint is not binding and we verify ex post that it can be disregarded. Then, the optimal price is

$$p^* = a - bq^* = \frac{a(b + 2c)}{2(b + c)}$$

Since all of the involved parameters are nonnegative, optimal price is nonnegative as well. It is also easy to check that

$$\Pi''(q) = -2(b + c) < 0$$

so we really found a minimizer.

The fixed cost F did not play any role in the analysis and this is consistent with what we pointed out in Section 1.1.1. A constant term, which is not affected by q, does not influence our quest for q^*. However, this does *not* imply that the fixed cost is irrelevant if we need to decide whether our business should be shut down for good! Indeed, we should also check whether overall profit

$$\Pi(q^*) = \frac{a^2(b + 2c)}{4(b + c)^2} - F - \frac{ca^2}{4(b + c)^2} = \frac{a^2}{4(b + c)} - F$$

is positive. We see that we should produce only if

$$F < \frac{a^2}{4(b + c)}$$

2.12   SEQUENCES AND SERIES

Series are another important topic in classical calculus. They have limited use in the remainder of the book, so we will offer a very limited treatment, covering what is strictly necessary. To motivate the study of series, let us consider once again the price of a fixed-coupon bond, with coupon C and face value F, maturing at time T. If we discount cash flows with rate r, we know that its price is

$$P = \sum_{t=1}^{T} \frac{C}{(1 + r)^t} + \frac{F}{(1 + r)^T}$$

If we have to compute the price of a long-term bond, the formula above results in a quite tedious calculation. Can we find a more compact expression for the bond price? Furthermore, suppose that maturity goes to infinity. This may sound weird and unrealistic, but years ago bonds were issued in the UK with infinite maturity; a bond like this was called a *console*. How can we calculate the price of a console, since this requires us to discount and add an infinite number of cash flows?

Before proceeding further, let us define a **series**. Consider an infinite sequence of numbers:

$$a_1, a_2, a_3, a_4, \ldots$$

For a finite n, the *partial sum* of the first n terms in the sequence is

$$s_n = a_1 + a_2 + a_3 + \cdots + a_n = \sum_{i=1}^{n} a_i \qquad (2.21)$$

Apparently, summing an infinite number of terms in the sequence makes no sense. It is a reasonable bet that the result will be either $+\infty$ or $-\infty$. However, if a_n goes to zero fast enough, when $n \to +\infty$, the partial sum s_n can converge to a finite limit. In such a case, the value of the infinite series is defined as the limit of the sum.

DEFINITION 2.21 *Given an infinite sequence* a_n, $n = 1, 2, 3, \ldots$, *the corresponding series is defined as the limit of the partial sums (2.21):*

$$\sum_{i=1}^{\infty} a_i \equiv \lim_{n \to +\infty} s_n$$

provided that the limit exists.

A very useful case is the so-called geometric series.

Example 2.37 (Geometric series) Let α be a real number and consider the series

$$\sum_{i=0}^{\infty} \alpha^i$$

This kind of series is called *geometric*. If we choose $\alpha = 2$, there is no doubt that the series goes to infinity, as each element of the sequence α^i is increasing and we sum an infinite number of larger and larger terms. If we choose $\alpha = 0.5$, the elements of the sequence tend to zero; if they do so fast enough, maybe, the series will converge to a finite limit. If α is negative, there is the additional complication of oscillations of terms α^i, which are positive for even powers and negative for odd powers.

To keep it simple, let us just consider $0 < \alpha < 1$ and rewrite the series as follows:

$$S = \sum_{i=0}^{\infty} \alpha^i = 1 + \sum_{i=1}^{\infty} \alpha^i = 1 + \alpha \sum_{i=0}^{\infty} \alpha^i = 1 + \alpha S$$

Solving for S yields

$$S = \frac{1}{1 - \alpha} \tag{2.22}$$

which makes sense if $\alpha < 1$; if $\alpha > 1$ the formula above yields a negative number, which is not reasonable as the series is a sum of positive terms. Sometimes, the starting element of the series we are interested in is α, rather than $\alpha^0 = 1$. Then, it is easy to adapt formula (2.22):

$$\sum_{i=1}^{\infty} \alpha^i = \sum_{i=0}^{\infty} \alpha^i - 1 = \frac{\alpha}{1 - \alpha} \tag{2.23}$$

Alternatively, we may regard the series starting from $i = 1$ as αS to obtain the same result. □

Example 2.38 (Pricing a console bond) A *console bond* is a bond with infinite maturity. Hence, its price is obtained by discounting an infinite sequence of cash flows:

$$P = \sum_{t=1}^{\infty} \frac{C}{(1+r)^t}$$

All we have to do is using the result for the geometric series by plugging $\alpha = 1/(1+r)$ into Eq. (2.23):

$$P = C \cdot \frac{1/(1+r)}{1 - 1/(1+r)} = \frac{C}{r}$$

In order to check whether this result makes financial sense, consider the value of a console bond paying \$1 per year, when $r = 10\%$. The formula yields $P = 1/0.1 = \$10$. Indeed, if we invest \$10 at an interest rate $r = 10\%$, each year we will get the dollar we need to pay the coupon, while maintaining the capital. ▯

The basic result from the geometric series can be applied to find a compact expression for a finite sum. A finite sum can be expressed as the difference of two infinite series:

$$\sum_{i=1}^{n} \alpha^i = \sum_{i=1}^{\infty} \alpha^i - \sum_{i=n+1}^{\infty} \alpha^i = \sum_{i=1}^{\infty} \alpha^i - \alpha^{n+1} \sum_{i=0}^{\infty} \alpha^i$$

$$= \frac{\alpha}{1-\alpha} - \frac{\alpha^{n+1}}{1-\alpha} = \frac{\alpha(1-\alpha^n)}{1-\alpha} \qquad (2.24)$$

Example 2.39 (Pricing a fixed-coupon bond) To find the price of a bond paying a coupon C per year, we plug $\alpha = 1/(1+r)$ into Eq. (2.24):

$$\sum_{t=1}^{T} \frac{C}{(1+r)^t} = C \times \frac{\frac{1}{1+r}\left[1 - \frac{1}{(1+r)^T}\right]}{1 - \frac{1}{1+r}} = \frac{C}{r}\left[1 - \frac{1}{(1+r)^T}\right] \qquad (2.25)$$

In order to get the bond price, we should also include the discounted value of the face value refunded at maturity:

$$P = \frac{C}{r}\left[1 - \frac{1}{(1+r)^T}\right] + \frac{F}{(1+r)^T}$$

If we let $T \to +\infty$, we recover the pricing formula for the console. ▯

In this book, we will just use variations of the geometric series. Hence, we will not proceed further with series theory. We just mention a useful result stating that, under suitable conditions, a series can be differentiated term by term. We clarify what we mean by a useful example.

Example 2.40 (Derivative of the geometric series) We know that the geometric series converges to $S(\alpha) = 1/(1-\alpha)$. Let us take its derivative with respect to α:

$$S'(\alpha) = \frac{1}{(1-\alpha)^2}$$

Now, let us consider a slightly different series

$$\sum_{i=1}^{\infty} i\alpha^i$$

With a little intuition, we see that this series is related to the derivative of the geometric series. In fact, using term-by-term differentiation, we get

$$\sum_{i=1}^{\infty} i\alpha^i = \alpha \sum_{i=1}^{\infty} i\alpha^{i-1} = \alpha \frac{d}{d\alpha}\left(\sum_{i=0}^{\infty} \alpha^i\right) = \alpha \frac{d}{d\alpha}\left(\frac{1}{1-\alpha}\right) = \frac{\alpha}{(1-\alpha)^2}$$

$\square$

2.13 DEFINITE INTEGRALS

The last section of this chapter deals with definite integrals. The concept of integral plays a fundamental role in calculus and applied mathematics and, as we shall see, it is in a sense the opposite operation with respect to taking derivatives. In the book, we use definite integrals essentially to deal with continuous random variables in probability theory.[26] In that context, it is sufficient to understand definite integrals as a way to compute an area, which has the meaning of a probability. Hence, we introduce the definite integral as the area below the graph of a function. We also try to get some business motivation for introducing integrals in models where time is continuous. What we will not do is lay down rigorous foundations for the concept, which would be outside the scope of the book, or developing sophisticated technical skills and tricks to find integrals in intricate cases.

2.13.1 Motivation: definite integrals as an area

Consider a function f on interval $[a, b]$. If the function assumes nonnegative values on that interval, it will define a region below its graph; this is illustrated as the shaded region in Fig. 2.31. Now imagine that we are interested in the area of that region. If the function were constant or linear, we would get the result by recalling a few concepts from elementary geometry. In the case of a constant function, we would just compute the area of a rectangle, and the area of a trapeze is needed for a linear function; but if the function is arbitrary,

---

[26]See Chapter 7.

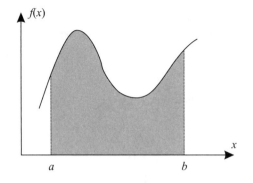

Fig. 2.31   The area below a function graph.

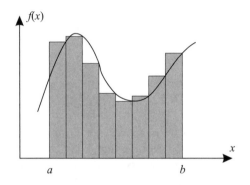

Fig. 2.32   Approximating area by piecewise constant functions.

things are not that simple. Nevertheless, we could try to find a suitable approximation of the area. Since we know pretty well how to cope with a constant function, we could try to approximate f by a piecewise constant function. The idea is illustrated in Fig. 2.32. More precisely, we can partition the interval $[a, b]$ into n "slices" of width

$$\Delta = \frac{b - a}{n}$$

Each slice is a subinterval with extreme points of the form

$$x_i = a + i\Delta, \qquad i = 0, 1, 2, \ldots, n$$

Note that $x_0 \equiv a$ and $x_n \equiv b$. The first subinterval is $[x_0, x_1] \equiv [a, a + \Delta]$, and in general we have subintervals like $[x_{i-1}, x_i]$, for $i = 1, \ldots, n$. Then, we should decide how to assign the height of each rectangle in Fig. 2.32. The choice in the figure is to consider $f(x_i)$, i.e., the value of the function on the right endpoint of each subinterval $[x_{i-1}, x_i]$, but different choices could make sense as well. With our approximation, the area is

$$S(\Delta) = \sum_{i=1}^{n} f(x_i)\Delta$$

where the notation highlights the dependence of sum S on the selected subinterval width Δ. Reasonably, the smaller this width, the better our approximation, and (maybe) taking a very fine partition with a very small Δ, we would really find the area we need. This results in the following (informal) *definition of the definite integral* of function f on interval $[a, b]$:

$$\int_{a}^{b} f(x)\,dx \equiv \lim_{\Delta \to 0} S(\Delta)$$

The notation $\int$ looks a bit like a slanted S; indeed, this reminds us that the integral is basically the limit of a sum. By the same token, the notation dx refers to an infinitesimal increment in the x variable; the notation Δx or δx is used when referring to small but finite increments.

Our treatment leaves a lot to be desired, to say the least, as we should clarify under which conditions the sum will converge to a limit. Furthermore, one could wonder whether our choice of rectangles is the best one or if it really makes any sense. Indeed, a rigorous treatment of integral should follow a slightly different route. Nevertheless, this is the simplest way to grasp the essential concept of an integral. Incidentally, if the function takes negative values on some subintervals, we can just consider the associated area as a negative one, which can be summed algebraically to positive areas.

2.13.2 Calculating definite integrals

Using the definition above to compute an integral is cumbersome, to say the least. It may work in some simple cases, but we certainly need something more handy. Luckily, the following theorem, which really deserves the name *fundamental*,[27] provides us with a practical way to compute definite integrals.

THEOREM 2.22 (Fundamental theorem of calculus) *Let $F(x)$ be a function such that $F'(x) = f(x)$, and assume that f is continuous on the interval $[a, b]$. Then*

$$\int_{a}^{b} f(x)\,dx = F(b) - F(a)$$

Function F is called the *antiderivative* of f. Actually, if F is the antiderivative of f, this also holds for any function $F(x) + C$ obtained by adding an arbitrary

---

[27] To be precise, the claim of the theorem below is a *consequence* of the fundamental theorem of calculus, but we will allow a little shortcut.

constant C. However, since in the theorem we take the difference $F(b) - F(a)$, the constant is irrelevant. The following shorthand notation is used:

$$F(x) \Big|_a^b \equiv F(b) - F(a)$$

Example 2.41 To illustrate the theorem, recall that $(x^2)' = 2x$. Then, if we consider $f(x) = x$, its antiderivative is $F(x) = x^2/2$ (plus an irrelevant constant):

$$\int_0^1 x \, dx = \frac{x^2}{2} \Big|_0^1 = \frac{1}{2}$$

which is indeed the area of the triangle associated with the function $f(x) = x$ on interval $[0, 1]$.

Armed with the fundamental theorem, we are certainly able to compute integrals of relatively simple functions formed using powers and exponentials. Sometimes, quite ingenious tricks are needed to find the antiderivatives of intricate functions, but we will not need that for what follows. The next property of definite integrals is quite helpful in dealing with sums of functions.

PROPERTY 2.23 *If α and β are arbitrary real numbers and all of the integrals below exist, then*

$$\int_a^b [\alpha f(x) + \beta g(x)] \, dx = \alpha \int_a^b f(x) \, dx + \beta \int_a^b g(x) \, dx$$

As you may imagine, this property is linked to the similar properties of derivatives.[28]

Example 2.42 Consider function $f(x) = x^2 + 1 - 3e^{-2x}$, to be integrated on interval $[0, 4]$. It is easy to see that the antiderivative of $x^2 + 1$ is $x^3/3 + x$. The exponential is a bit trickier, but we may recall that

$$(e^{\alpha x})' = \alpha e^{\alpha x}$$

So, the antiderivative of $-3e^{-2x}$ is $3e^{-2x}/2$, and the integral is

$$\int_1^4 (x^2 + 1 - 3e^{-2x}) \, dx = \left(\frac{x^3}{3} + x + \frac{3}{2}e^{-2x} \right) \Big|_0^4 = \frac{64}{3} + 4\frac{3}{2}e^{-8} - \frac{3}{2} = 23.8338$$

---

[28] In the next chapter, we get acquainted with the concept of *linear combination*. The property shows that the integral, just like the derivative, is a linear operator in the sense that the integral (derivative) of a linear combination of functions is just the linear combination of the integrals (derivatives) of each function.

2.13.3   Improper integrals

So far, we have considered the integral of a continuous function on a bounded interval. The idea can be generalized to unbounded intervals and to functions featuring certain types of discontinuity. In fact, the integral might not exist, because the function has pathological behavior; in other cases, it could go to infinity, which may well be the case if we consider an infinite area. Still, in some fortunate cases we do find an integral that should be regarded as the limit of an integral of a continuous function on a bounded interval. We illustrate the concept with a couple of examples, without providing adequate theoretical background.

Example 2.43 Consider the exponential function $f(x) = \lambda e^{-\lambda x}$, where $\lambda > 0$, and the improper integral

$$\int_0^{+\infty} \lambda e^{-\lambda x}\, dx$$

on the unbounded interval $[0, +\infty)$. This should be regarded as the following limit:

$$\lim_{z \to +\infty} \int_0^z \lambda e^{-\lambda x}\, dx$$

Since the antiderivative of f is $F(x) = -e^{-\lambda x}$, all boils down to the following calculation:

$$-e^{-\lambda x}\,\Big|_0^{+\infty} = -0 + 1 = 1$$

In this case, the integral exists because the negative exponential function goes to zero fast enough, when $x \to +\infty$, so that the area on the unbounded interval is still finite. Clearly, the improper integral on the interval $(-\infty, 0]$ cannot exist, since the positive function goes to infinity there. □

Example 2.44 Consider the definite integral

$$\int_0^1 \frac{1}{x}\, dx$$

Unlike the previous case, the integration interval is finite here, but the function is not continuous for $x = 0$ and it goes to $+\infty$. We could still wonder whether the area is finite or, more precisely, whether the limit

$$\lim_{\epsilon \to 0^+} \int_\epsilon^1 \frac{1}{x}\, dx$$

exists. The antiderivative we need is $F(x) = \ln x$. Unfortunately, if we try using it, we fail, since

$$\ln x\,\Big|_0^1 = \ln 1 - \ln 0 = 0 - (-\infty) = +\infty$$

The area is unbounded, as $1/x$ goes to infinity too quickly. The trick, however, works for $g(x) = 1/\sqrt{x}$. In this case, the antiderivative is $G(x) = 2\sqrt{x}$ and the integral is

$$2\sqrt{x} \Big|_0^1 = 2$$

The rate of increase of function g, when $x \to 0^+$, is slow enough to result in a finite area.   ☐

2.13.4 A business view of definite integral

In this section we try to further motivate the use of definite integral, at least conceptually, for business management problems. To do so, we use the EOQ model of Section 2.1 once again. There, we have claimed that the contribution of inventory holding cost to average total cost per unit time is

$$C_{\text{in}} = h\frac{Q}{2}$$

In the reasoning, a key role is played by the fact that demand rate is constant, and given by D items per unit time.[29] But what can we do if demand rate is an arbitrary function $D(t)$ of time? The holding cost, in such a case, depends on the ordering policy we follow, as well as on the initial inventory level.[30] Whatever we do, it will result in a time-varying on-hand inventory level that we may represent by a function $I(t)$. Then, the holding cost of our ordering policy, over time interval $[0, T]$ is just

$$\int_0^T hI(t)\, dt$$

To see this, think of an inventory level $I(t)$ that varies in time; since time is continuous, we should take the instantaneous inventory level $I(t)$ along the time interval, multiply it by the unit inventory holding cost h, and "sum" the results. In this case, the sum is actually an integral, which informally is the sum of infinite tiny, infinitesimal terms.

To see that this works for the EOQ model, consider the first triangle in Fig. 2.1. This corresponds to the first ordering cycle. We start with $I(0) = Q$, and after an interval corresponding to the time between two consecutive orders, $T = Q/D$, we have $I(T) = 0$. On that interval, inventory as a function of time is given by

$$I(t) = Q - Dt$$

[29] Here we denote demand rate by D, rather than d, in order to avoid confusion with the dt notation in the integrals below.

[30] The initial inventory level is irrelevant in the EOQ model, since the ordering policy is periodic, but this need not be true in general.

The inventory cost on this cycle is

$$\int_0^{Q/D} h(Q - Dt)\, dt = h \left(Qt - \frac{Dt^2}{2} \right) \Big|_0^{Q/D} = h \left(\frac{Q^2}{D} - \frac{DQ^2}{2D^2} \right) = \frac{hQ^2}{2D}$$

This expression is just h times the area of a triangle with base Q/D and height Q. Since there are D/Q such cycles per unit time, on average, the average holding cost is:

$$\frac{hQ^2}{2D} \times \frac{D}{Q} = \frac{hQ}{2}$$

as we claimed before.

Thus, integrals can be used when we have to model complicated business problems where variables of interest change in continuous time. In general, building and/or solving such a model is difficult, and one resorts to *time discretization*. The idea is to partition time interval $[0, T]$ into n subintervals of width Δt, as we did in Fig. 2.32. Then, we have n time buckets of the form $[(i - 1)\, \Delta t, i\, \Delta t]$, for $i = 1, \dots, n$. If we denote the inventory level *at the end* of ith time bucket by $I_i = I(i\, \Delta t)$, we can approximate the total holding cost as

$$h \int_0^T I(t)\, dt \approx h' \sum_{i=1}^n I_i \tag{2.26}$$

where $h' \equiv h\, \Delta t$ is the unit holding cost for a time period of length Δt. This expression can be definitely more manageable when building possibly large-scale decision models.[31]

A possible objection is that we should take the *average* inventory level rather than the inventory level at the end of a time bucket. Doing so results in the following expression:

$$h' \sum_{i=1}^n \frac{I_{i-1} + I_i}{2}$$

This may look quite different from (2.26), indeed. However, in practice the two expressions are equivalent. To see this, let us expand the sum:

$$\sum_{i=1}^n \frac{I_{i-1} + I_i}{2} = \frac{I_0 + I_1}{2} + \frac{I_1 + I_2}{2} + \cdots + \frac{I_{n-2} + I_{n-1}}{2} + \frac{I_{n-1} + I_n}{2}$$

$$= \frac{I_0}{2} + I_1 + I_2 + \cdots + I_{n-1} + \frac{I_n}{2}$$

Comparing this against (2.26), we see a difference in the first and last terms. However, the first term is a constant, as the initial inventory level is given and

---

[31] This is the subject of Chapter 12. In particular, we will use time discretization in the dynamic production planning model of Section 12.2.2.

it is not a decision variable. The last term, if it is not fixed to a positive value, will be set to zero by the optimization algorithm; in fact, the holding cost is minimized, and since the model does not see any demand beyond the last time bucket, there is no reason to have any item left in inventory. To avoid running out of stock at the end of the planning horizon, we should assign a value to the terminal inventory level; but then, the last term is a constant just like the first one. So, we see that considering the inventory level at the end of each time bucket is equivalent to considering the average inventory level. Coincidentally, this is consistent with the finite sums we have considered in Section 2.13.1.[32] The choice between a continuous-time and a discrete-time model is largely a matter of convenience, depending on the business problem at hand.

Problems

2.1   Find the domain of functions

$$f(x) = \frac{1}{\sqrt{1-x^2}-1}, \qquad g(x) = \frac{1}{\sqrt{x^2+1}-x}$$

2.2   Find the equation of a line

- With slope -3 and intercept 10

- With slope 5 and passing through point (-2,4)

- Passing through points (1,3) and (3,-5)

2.3   Find the first-order derivative of the following functions:

$$f_1(x) = \frac{3x}{x^2+1}, \quad f_2(x) = e^{x^3-x^2+5x-3}, \quad f_3(x) = \sqrt{\exp\left(\frac{x+2}{x-1}\right)}$$

2.4   Consider functions $f(x) = x^3 - x$ and $g(x) = x^3 + x$. Use derivatives to sketch the function graphs, and look for maxima and minima.

2.5   Consider the following functions defined on piecewise domains:

$$f_1(x) = \begin{cases} -x, & x < 0 \\ 2x^2, & x \geq 0 \end{cases}, \quad f_2(x) = \begin{cases} x^2+1, & x < 0 \\ 3x, & x \geq 0 \end{cases}, \quad f_3(x) = \begin{cases} x^3, & x < 1 \\ x, & x \geq 1 \end{cases}$$

---

[32] In other cases, one should define an integral by considering the value of the function *at the beginning* of each time bucket. A notable case occurs in financial engineering models, where the so-called Ito stochastic integral is extensively used. In fact, there are many kinds of integrals, but this is way beyond the scope of this book. See, e.g., Ref. [2] for an elementary introduction to such models.

For each function, indicate whether it is continuous and differentiable at the point of transition between each portion of the domain.

2.6 Consider function

$$f(x) = \exp\left(-\frac{1}{1+x^2}\right)$$

and find linear (first-order) and quadratic (second-order) approximations around points $x_0 = 0$ and $x_0 = 10$. Check the quality of approximations around these points.

2.7 In Example 2.33 we assume discrete-time compounding of interest. This results in the need for introducing modified duration, in Eq. (2.19), to get rid of an annoying factor $1 + y$, where y is yield to maturity. Compute bond duration when assuming continuous-time compounding, at yield y_c, and show that this correction is not necessary in this case.

2.8 Consider the following functions and tell if they are convex, concave, or neither:

$$f_1(x) = e^{x^2+1}, \quad f_2(x) = \ln(x+1), \quad f_3(x) = x^3 - x^2 + 2$$

2.9 Prove that the intersection of two convex sets S_1 and S_2 is a convex set.

2.10 Consider a stream of constant cash flows $F_t = C$, for $t = 1, \ldots, T$. From Example 2.39, we know how to find its present value at time $t = 0$. Now imagine that these cash flows are the amount you invest to build your pension wealth; the current amount of wealth is invested at an interest rate r, which we assume fixed and constant over time. How can you compute your *future* pension wealth at time T?

2.11 Assume that you work from age 25 to age 65. At the end of each year, you contribute C to your pension fund, which is invested at a yearly rate $r = 5\%$. At age 65 you retire and plan to consume \$20,000 each year, until you die. Assuming that you die at age 85, and that interest rates remain constant, how much should you save each year? How does the result change if, due to your reckless lifestyle, you assume that departure will occur at age 75? (*Hint:* Use Problem 2.10.)

2.12 Consider the following extension of the EOQ model, often labeled economic manufacturing quantity (EMQ). Most of the assumptions of the EOQ and EMQ models are the same, but in the latter, rather than assuming instantaneous replenishment of inventory, inventory is replenished at production rate p, which gives the number of parts produced per unit time. Find the order quantity minimizing the sum of inventory and ordering cost, averaged over time. (*Hint:* Draw the inventory level during an ordering cycle.)

For further reading

- In this chapter, we have been often rather informal. There are many rigorous treatments of calculus, but a suitable starting point is Ref. [6], which is a good and quite readable compromise. If you prefer a classical treatment, Ref. [7] is recommended.

- Another nice book that, among many other things, illustrates calculus with an emphasis on economic applications is Ref. [9].

- We have motivated part of the chapter by the need to solve optimization models, which are the subject of Chapter 12. There, too, we will pursue a rather informal approach. Readers interested in a full-fledged and extensive introduction to the mathematics of optimization are referred to Refs. [1] or [10].

- Furthermore, we have only considered the optimization of differentiable functions. Readers interested in an elementary introduction to the optimization of nondifferentiable functions might consult Appendix 1 of Ref. [3].

- We have used the EOQ model as a motivation. Readers wishing a more in-depth treatment of EOQ model and its generalizations may have a look at Ref. [4].

- Calculus also plays a role in microeconomic theory. Nice, application-oriented examples are proposed in Refs. [5] and [8].

- We have also hinted at revenue management, which indeed is a rich application field for quantitative business models; an extensive reference is the text by Talluri and Van Ryzin [11].

REFERENCES

1. M.S. Bazaraa, H.D. Sherali, and C.M. Shetty, *Nonlinear Programming. Theory and Algorithms,* 2nd ed., Wiley, Chichester, West Sussex, UK, 1993.

2. P. Brandimarte, *Numerical Methods in Finance and Economics: A Matlab-Based Introduction,* 2nd ed., Wiley, New York, 2006.

3. P. Brandimarte and A. Villa, *Advanced Models for Manufacturing Systems Management,* CRC Press, Boca Raton, FL, 1995.

4. P. Brandimarte and G. Zotteri, *Introduction to Distribution Logistics,* Wiley, New York, 2007.

5. L.M.B. Cabral, *Introduction to Industrial Organization*, MIT Press, Cambridge, MA, 2000.

6. S. Lang, *A First Course in Calculus,* 5th ed., Springer, New York, 1986.

7. W. Rudin, *Principles of Mathematical Analysis,* 3rd ed., McGraw-Hill, New York, 1976.

8. O. Shy, *Industrial Organization: Theory and Applications*, MIT Press, Cambridge, MA, 1995.

9. C.P. Simon and L. Blume, *Mathematics for Economists*, W.W. Norton & Company, New York, 1994.

10. R.K. Sundaram, *A First Course in Optimization Theory*, Cambridge University Press, Cambridge, UK, 1996.

11. K.T. Talluri and G.J. Van Ryzin, *The Theory and Practice of Revenue Management*, Springer, New York, 2005.

3

Linear Algebra

This chapter contains some very basic and some relatively advanced material, which is needed only in a few later chapters. Hence, it can be read in quite different ways, depending on readers' interest:

- Readers interested in elementary probability and statistics (Chapters 4–11) may just have a look at Sections 3.1 and 3.2, which motivate and illustrate basic approaches to solve systems of linear equations; this is an important topic that any reader should be familiar with.

- The remaining sections of this chapter are more advanced, but they are mostly needed by those interested in the multivariate statistics techniques described in Chapters 15–17. One possibility is reading this whole chapter now; alternatively, one might prefer just having a cursory look at the material, and returning here when tackling those chapters.

- Finally, readers who are interested in optimization modeling can do the same when grappling with Chapter 12. Some concepts related to quadratic forms (Section 3.8) and multivariable calculus (Section 3.9) are useful to obtain a better grasp of some technical aspects of optimization. Apparently, these are not necessary if you are just interested in building an optimization model, but not in the algorithms needed to solve it; nevertheless, an appreciation of what a convex (i.e., easy-to-solve) optimization model is better gained if you are armed with the background we build here.

In Section 3.1 we propose a simple motivating example related to pricing a financial derivative. A simple binomial valuation model is discussed, which leads to the formulation of a system of linear equations. Solving such a system is the topic of Section 3.2, where we disregard a few complicating issues and cut some corners in order to illustrate standard solution approaches. In Sections 3.3 and 3.4 we move on to more technical material: vector and matrix algebra. Matrices might be new to some readers, but they are a ubiquitous concept in applied mathematics. On the contrary, you are likely to have some familiarity with vectors from geometry and physics. We will abstract a little in order to pave the way for Section 3.5, dealing with linear spaces and linear algebra, which can be considered as a generalization of the algebra of vectors. Another possibly familiar concept from elementary mathematics is the determinant, which is typically taught as a tool to solve systems of linear equations. Actually, the concept has many more facets, and a few of them are dealt with in Section 3.6. The last concepts from matrix theory that we cover are eigenvalues and eigenvectors, to which Section 3.7 is devoted. They are fundamental, e.g., in data reduction methods for multivariate statistical analysis, like principal component analysis and factor analysis.[1] The two last sections are, in a sense, halfway between linear algebra and calculus. In Section 3.8 we consider quadratic forms, which play a significant role in both statistics and optimization. They are, essentially, the simplest examples of a nonlinear function of multiple variables, and we will consider some of their properties, illustrating the link between eigenvalues of a matrix and convexity of a function. Finally, in Section 3.9 we hint at some multivariable generalizations of what we have seen in the previous chapter on calculus for functions of a single variable.

3.1   A MOTIVATING EXAMPLE: BINOMIAL OPTION PRICING

Options are financial derivatives that have gained importance, as well as a bad reputation, over the years. In Section 1.3.1 we considered forward contracts, another type of derivative. With a forward contract, two parties agree on exchanging an asset or a commodity, called the *underlying asset*, at a prescribed time in the future, for a fixed price determined now. In a call option, one party (the option holder, i.e., the party buying the option) has the right but not the obligation to buy the asset from the other party (the option writer, i.e., the party selling the option) at a price established now. This is called the *strike price*, which we denote by K. Let us denote the current price of the underlying asset by S_0 (known when the option is written) and the future price at a time $t = T$ by S_T. Note that S_T, as seen at time $t = 0$, is a random variable; the time $t = T$ corresponds to the option maturity, which is when

---

[1]See Chapter 17.

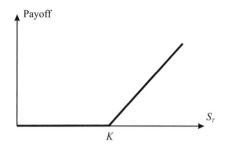

Fig. 3.1 Payoff of a call option.

the option can be exercised.[2]   Clearly, the option will be exercised only if $S_T \geq K$, since there would be no point in using the option to buy at a price that is larger than the prevailing market price at time T . Then the payoff of the option, for the holder, is

$$\max\{S_T - K, 0\}$$

To interpret this, consider the possibility of exercising the option to buy at K an asset that can be immediately sold at S_T. Market structure is not really that simple, but some options are settled in cash, so the payoff really has a monetary nature; this occurs for options written on a nontraded market index or on interest rates. This payoff is a piecewise linear function and is illustrated in Fig. 3.1. When the future asset price S_T will be realized, at time $t = T$, the payoff will be clearly determined. What is not that clear is the fair price for the contract at time $t = 0$: The payoff for the option holder cannot be negative, but the option writer can suffer a huge loss; hence, the latter requires an option premium, paid at time $t = 0$, to enter into this risky contract. Here we consider the most elementary model for option pricing. Whatever model we may think of, it should account for uncertainty in the price of the underlying asset, and the simplest model of uncertainty is a binomial model.[3]   As shown in Fig. 3.2, a binomial model for the price of the underlying asset assumes that its price in the future can take one of two values. It is quite common in financial modeling to represent uncertainty using *multiplicative shocks*, i.e., random variables by which the current stock price is multiplied to obtain the new price. In a binomial model, this means that the current price S_0, after a time period of length T, is multiplied by a shock that may take values d or u, with probabilities p_u and p_d, respectively. It is natural to think that d is the shock if the price goes down, whereas u is the shock when the price goes up; so, we assume $d < u$. The asset price may

---

[2]We assume that the option can only be exercised at maturity. Such an option is called *European-style*. On the contrary, *American-style options* can be exercised at any time before their expiration date.
[3]The binomial model is basically related to coin flipping; see Section 1.2.3.

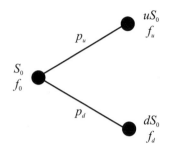

Fig. 3.2 One-step binomial model for option pricing.

be either $S_0 u$ or $S_0 d$, and the option will provide its holder a payoff f_u and f_d corresponding to the two outcomes. These values are also illustrated in Fig. 3.2. For instance, in the case of a call option with strike price K, we have

$$f_d = \max\{S_0 d - K, 0\}, \qquad f_u = \max\{S_0 u - K, 0\}$$

The pricing problem consists of finding a fair price f_0 for the option, i.e., its value at time $t = 0$. Intuitively, one would guess that the two probabilities p_u and p_d play a major role in determining f_0. Furthermore, individual risk aversion could also play a role. Different people may assign a different value to a lottery, possibly making the task of finding a single fair price a hopeless endeavor, unless we have an idea of the aggregate risk aversion of the market as a whole.

However, a simple principle can be exploited in order to simplify our task. The option is a single, somewhat complicated, asset; now suppose that we are able to set up a portfolio of simpler assets, in such a way that the portfolio yields the same payoff as the option, in any possible future state. The value of this *replicating portfolio* at $t = 0$ is known, and it cannot differ from the option value f_0. Otherwise, there would be two essentially identical assets traded at different prices. Such a violation of the law of one price would open up arbitrage opportunities.[4]

The replicating portfolio should certainly include the underlying asset, but we need an additional one. Financial markets provide us with such an asset, in the form of a risk-free investment. We may consider a bank account yielding a risk-free interest rate r; we assume here that this rate is continuously compounded,[5] so that if we invest \$1 now, we will own \$$e^{rt}$ at a generic time instant t. Equivalently, we may think of a riskless zero-coupon bond, with initial price $B_0 = 1$ and future price $B_1 = e^{rT}$, when the option matures at time $t = T$.

---

[4]An arbitrage opportunity is essentially a riskless money-making machine exploiting price inconsistencies. We have considered arbitrage opportunities when discussing a forward contract in Section 1.3.1.

[5]See Section 2.3.4 for an introduction to continuous compounding.

What we have to find is the number of stock shares and riskless bonds that we should include in the portfolio, in order to replicate the option payoff. Let us denote the number of stock shares by Δ and the number of bonds by Ψ. The initial value of this portfolio is

$$\Pi_0 = \Delta S_0 + \Psi \tag{3.1}$$

and its future value, depending on the realized state, will be either

$$\Pi_u = \Delta S_0 u + \Psi e^{rT} \qquad \text{or} \qquad \Pi_d = \Delta S_0 d + \Psi e^{rT}$$

Note that future value of the riskless bond does not depend on the realized state. If this portfolio has to replicate the option payoff, we should enforce the following two conditions:

$$\begin{aligned} \Delta S_0 u + \Psi e^{rT} &= f_u \\ \Delta S_0 d + \Psi e^{rT} &= f_d \end{aligned} \tag{3.2}$$

This is a system of two linear equations in two unknown variables Δ and Ψ. The equations are linear as variables occur linearly. There is no power like x^3, or cross-product like $x_1 \cdot x_2$, or any weird function involved. All we have to do is solve this system of linear equations for Δ and Ψ, and plug the resulting values into Eq. (3.1) to find the option price f_0. A simple numerical example can illustrate the idea.

Example 3.1 Let us assume that option maturity is $T = 1$ year, the risk-free interest rate is 10%, the current price of the stock share is \$10, and the strike price is \$11. Finally, say that the two possible returns of the stock share in one year are either 20% or -10%, which imply $u = 1.2$ and $d = 0.9$. The corresponding payoffs are

$$f_u = \max\{10 \times 1.2 - 11, 0\} = 1, \qquad \max\{10 \times 0.9 - 11, 0\} = 0$$

The system of linear equations is

$$\begin{aligned} 12\Delta + \Psi e^{0.1} &= 1 \\ 9\Delta + \Psi e^{0.1} &= 0 \end{aligned}$$

If we subtract the second equation from the first one, we get

$$(12 - 9)\Delta = 1 \quad \Rightarrow \quad \Delta = \tfrac{1}{3} \approx 0.3333$$

Plugging this value back into the first equation yields

$$\Psi = \frac{1 - \frac{12}{3}}{e^{0.1}} \approx -2.7145$$

and putting everything together, we obtain

$$f_0 = \Delta S_0 + \Psi = \tfrac{10}{3} - 2.7145 \approx \$0.6188$$

Note that a negative value of Ψ means that we borrow some money, at the risk-free rate.

To get a financial understanding of what is going on, you should take the point of view of the option writer, i.e., the guy selling the option. Say that he takes a "naked" position, i.e., he collects the option price and waits until maturity. If the stock price goes up, he will have to buy one share at $12 just to hand it over to the option holder for $11, losing $1. If he does not like taking this chance, he could consider the opposite position: Buy one stock share, just in case the option gets exercised by its holder. This "covered" position does not really solve the problem, as the writer loses $1 if the price goes down, the option is not exercised, and he has to sell for $9 the share that was purchased for $10.

The replicating strategy suggests that, to hedge the risk away, the writer should actually buy $\Delta = \frac{1}{3}$ shares. To this aim, he should use the option premium, plus some borrowed money ($2.7145, as given by Ψ). If the stock price goes down to $9, the option writer will just sell that share for $3, i.e., what he needs to repay the debt, which at the risk-free rate has risen to $3. Hence, the option writer breaks even in the "down" scenario. The reader is encouraged to work out the details for the other case, which requires purchasing two additional thirds of a stock share at the price of $12, in order to sell a whole share to the option holder, and to verify that the option writer breaks even again (in this case, the proceeds from the sale of the share at the strike price are used to repay the outstanding debt).   ▯

This numerical example could leave the reader a bit puzzled: Why should anyone write an option just to break even? No one would, of course, and the fair option price is increased somewhat to compensate the option writer. Furthermore, we have not considered issues such as transaction costs incurred when trading stock shares. Most importantly, the binomial model of uncertainty is oversimplified, yet it does offer a surprising insight: The probabilities p_u and p_d of the up/down shocks do not play any role, and this implies that the expected price of the stock share at maturity is irrelevant. This is not surprising after all, as we have matched the option payoff in any possible state, irrespective of its probability. Moreover, the careful reader will recall that we reached a similar conclusion when dealing with forward contracts in Section 1.3.1. There, by no-arbitrage arguments that are actually equivalent to what we used here, it was found that the fair forward price for delivery in one year is $F_0 = S_0(1 + r)$, where S_0 is the price of the underlying asset at time $t = 0$ and r is the risk-free rate with annual compounding. In the case of continuous compounding and generic maturity T, we get $F_0 = S_0 e^{rT}$.

Let us try to see if there is some link between forward and option pricing. To begin with, let us solve the system of linear equations (3.2) in closed form. Using any technique described in the following section,[6] we get the following

---

[6]The result is easily obtained by applying Cramer's rule of Section 3.2.3.

composition of the replicating portfolio:

$$\Delta = \frac{f_u - f_d}{S_0(u - d)}$$

$$\Psi = e^{-r \cdot T} \frac{u f_d - d f_u}{u - d}$$

Hence, by the law of one price, the option value now is

$$
\begin{aligned}
f_0 &= \Delta S_0 + \Psi \\
&= \frac{f_u - f_d}{u - d} + e^{-rT} \frac{u f_d - d f_u}{u - d} \\
&= e^{-rT} \left\{ \frac{e^{rT} - d}{u - d} f_u + \frac{u - e^{rT}}{u - d} f_d \right\}
\end{aligned}
\tag{3.3}
$$

As we already noticed, this relationship does *not* depend on the objective probabilities p_u and p_d. However, let us consider the quantities

$$\pi_u = \frac{e^{rT} - d}{u - d}, \qquad \pi_d = \frac{u - e^{rT}}{u - d}$$

which happen to multiply the option payoffs in Eq. (3.3). We may notice two interesting features of π_u and π_d:

1. They add up to one, i.e., $\pi_u + \pi_d = 1$.

2. They are nonnegative, i.e., $\pi_u, \pi_d \geq 0$, provided that $d \leq e^{rT} \leq u$.

The last condition does make economic sense. Suppose, for instance, that $e^{rT} < d < u$. This means that, whatever the realized scenario, the risky stock share will perform better than the risk-free bond. But this cannot hold in equilibrium, as it would pave the way for an arbitrage: Borrow money at the risk-free rate, buy the stock share, and you will certainly be able to repay the debt at maturity, cashing in the difference between the matured debt and the stock price. On the contrary, if the risk-free rate is always larger than the stock price in the future, one could make easy money by selling the stock share short.[7]

We already know that probabilities may be related to relative frequencies and that the latter do add up to one and are nonnegative. Then we could interpret π_u and π_d as probabilities and Eq. (3.3) can be interpreted in turn as the discounted expected value of the option payoff

$$f_0 = e^{-rT} (\pi_u f_u + \pi_d f_d)$$

provided that we do not use the "objective" probabilities p_u and p_d, but π_u and π_d. These are called *risk-neutral probabilities*, a weird name that we can

---

[7]See Example 1.3 for an illustration of short selling.

easily justify. Indeed, if we use the probabilities π_u and π_d, the expected value of the underlying asset at maturity is

$$\mathrm{E}^{\mathbb{Q}}\left[S_T\right] = \pi_u S_0 u + \pi_d S_0 d = S_0 e^{rT}$$

where the notation $\mathrm{E}^{\mathbb{Q}}$ emphasizes that change in probability measure.[8] This suggests that, under these unusual probabilities, the return of the risky asset is just the risk-free return. Furthermore, this is just the forward price of the underlying asset. In other words, the forward price is in fact the expected value of the underlying asset, but under modified probabilities. But how is it possible that the expected return of a risky asset is just the risk-free rate? This might happen only in a world where investors do not care about risk, i.e., in a risk-neutral world, which justifies the name associated with the probabilities we use in option pricing. Indeed, what we have just illustrated is the tip of a powerful iceberg, the risk-neutral pricing approach, which plays a pivotal role in financial engineering.

Of course, the binomial model we have just played with is too simple to be of practical use, but it is the building block of powerful pricing methods. We refer the interested reader to references at the end of this chapter, but now it is time to summarize what we have learned from a general perspective:

- Building and solving systems of linear equations may be of significant practical relevance in many domains, including but not limited to financial engineering. We have already stumbled on systems of linear equations when solving the optimal production mix problem in Section 1.1.2.

- We have also seen an interplay between algebra and probability theory; quantitative, fact-based business management requires the ability of seeing these and other connections.

- We have considered a simple binomial model of uncertainty. We had two states, and we used two assets to replicate the option. What if we want a better model with three or more states? How many equations do we need to find a solution? More generally, under which conditions are we able to find a unique solution to a system of linear equations?

We will not dig deeper into option pricing, which is beyond the scope of the book, but these questions do motivate further study, which is carried out in the rest of this chapter. But first, we should learn the basic methods for solving systems of linear equations.

---

[8]The concept of expected value $\mathrm{E}[X]$ of a random variable will be fully treated later. For now, you may safely assume that it is equivalent to the intuitive concept of mean, using probabilities of future events rather than relative frequencies resulting from the observation of past data.

3.2 SOLVING SYSTEMS OF LINEAR EQUATIONS

The theory, as well as the computational practice, of solving systems of linear equations is relevant in a huge list of real-life settings. In this section we just outline the basic solution methods, without paying due attention to bothering issues such as the existence and uniqueness of a solution, or numerical precision.

A linear equation is something like

$$a_1 x_1 + a_2 x_2 + \cdots + a_n x_n = b$$

i.e., it relates a linear function of variables $x_1, \ldots, x_n$ to a right-hand side b. A system of linear equations involves m such equations:

$$
\begin{aligned}
a_{11} x_1 + a_{12} x_2 + \cdots + a_{1n} x_n &= b_1 \\
a_{21} x_1 + a_{22} x_2 + \cdots + a_{2n} x_n &= b_2 \\
\vdots &= \vdots \\
a_{m1} x_1 + a_{m2} x_2 + \cdots + a_{mn} x_n &= b_m
\end{aligned}
$$

where a_{ij} is the coefficient multiplying variable x_j, $j = 1, \ldots, n$, in equation i, $i = 1, \ldots, m$, whose right-hand side is b_i.

A preliminary question is: How many equations should we have to pinpoint exactly one solution? An intuitive reasoning about degrees of freedom suggests that we should have as many equations as unknown variables. More precisely, it seems reasonable that

- If we have more variables than equations ($n > m$), then we may not find one solution, but many; in other words, we have an underconstrained problem, and the residual degrees of freedom can be exploited to find a "best" solution, possibly optimizing some cost or profit function, much along the lines of Section 1.1.2; we will see how to take advantage of such cases in Chapter 12 on optimization modeling.

- If we have more equations than variables ($n < m$), then we may fail to find any solution, as the problem is overconstrained; still, we might try to find an assignment of variables such that we get as close as possible to solving the problem; this leads to linear regression models and the least-squares method (see Chapter 10).

- Finally, if $n = m$, we may hope to find a unique solution.

Actually, this straightforward intuition will work in most cases, but not always. This is easy to visualize in a two-dimensional space, as illustrated in Fig. 3.3. In such a setting, a linear equation like $a_1 x_1 + a_2 x_2 = b$ corresponds to a line (see Section 2.3). In case (a), we have one equation for two unknown variables; there are infinite solutions, lying on the line pinpointed by the equation. If we have two equations, as in case (b), the solution is the unique point

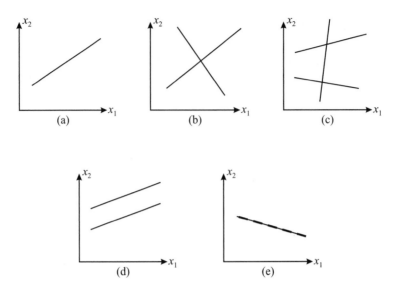

Fig. 3.3   Schematic representation of issues in existence and uniqueness of the solution of systems of linear equations.

where the two lines intersect. However, if we have three lines as in case (c), we may fail to find a solution, because three lines have no common intersection, in general. These cases correspond to our intuition above. Nevertheless, things can go wrong. In case (d), we have two lines, but they are parallel and there is no solution. To see an example, consider the system

$$2x_1 + 3x_2 = 10$$
$$2x_1 + 3x_2 = 5$$

If we subtract the second equation from the first one, we get $0 = 5$, which is, of course, not quite true. By a similar token, in case (e) we have two equations, but the solution is not unique. Such a case may arise when two apparently different equations boil down to the same one, as in the following system:

$$2x_1 + 3x_2 = 10$$
$$4x_1 + 6x_2 = 20$$

It is easy to see that the two equations are not really independent, since the second one is just the first one multiplied by 2.

Indeed, to address issues related to existence and uniqueness of solutions of systems of linear equations, we need the tools of linear algebra, which we cover in the advanced sections of this chapter. The reader interested in a basic treatment might just skip these sections. Here, we illustrate the basic approaches to solve systems of linear equations, assuming that the solution is unique, and that we have as many equations as variables.

3.2.1 Substitution of variables

A basic (highschool) approach to solving a system of linear equations is substitution of variables. The idea is best illustrated by a simple example. Consider the following system:

$$x_1 - 2x_2 = 8$$
$$3x_1 + x_2 = 3$$
(3.4)

Rearranging the first equation, we may express the first variable as a function of the second one:

$$x_1 = 2x_2 + 8$$

and plug this expression into the second equation:

$$3 \times (2x_2 + 8) + x_2 = 3 \quad \rightarrow \quad 7x_2 = -21 \quad \rightarrow \quad x_2 = -3$$

Then we may plug this value back into the expression of x_1, which yields $x_1 = 2$.

Despite its conceptual simplicity, this approach is somewhat cumbersome, as well as difficult to implement efficiently in computer software. A related idea, which is more systematic, is *elimination of variables*. In the case above, this is accomplished as follows:

1. Multiply the first equation by -3, to obtain $-3x_1 + 6x_2 = -24$

2. Add the new equation to the second one, which yields $7x_2 = -21$ and implies $x_2 = -3$

3. Substitute back into the first equation, which yields $x_1 = 2$

We see that elimination of variables, from a conceptual perspective, is not that different from substitution. One can see the difference when dealing with large systems of equations, where carrying out calculations without due care can result in significant numerical errors. Gaussian elimination is a way to make elimination of variables systematic.

3.2.2 Gaussian elimination

Gaussian elimination, with some improvements, is the basis of most numerical routines to solve systems of linear equations. Its rationale is that the following system is easy to solve:

$$
\begin{aligned}
a_{11}x_1 + a_{12}x_2 + \cdots + a_{1n}x_n &= b_1 \\
a_{22}x_2 + \cdots + a_{2n}x_n &= b_2 \\
\vdots &= \vdots \\
a_{nn}x_n &= b_n
\end{aligned}
$$

Such a system is said to be in *upper triangular form*, as nonzero coefficients form a triangle in the upper part of the layout. A system in upper triangular form is easy to solve by a process called *backsubstitution*. From the last equation, we immediately obtain

$$x_n = \frac{b_n}{a_{nn}}$$

Then, we plug this value into the second-to-last (penultimate) equation:

$$a_{n-1,n}x_{n-1} + a_{n-1,n}x_n = b_{n-1} \quad \rightarrow \quad x_{n-1} = \frac{b_{n-1} - a_{nn}x_n}{a_{n-1,n}}$$

In general, we proceed backward as follows:

$$x_k = \frac{1}{a_{kk}} \left(b_k - \sum_{j=k+1}^{n} a_{kj}x_j \right), \qquad k = n-1, n-2, \ldots, 1 \tag{3.5}$$

Now, how can we transform a generic system into this nice triangular form? To get a system in triangular form, we form combinations of equations in order to eliminate some coefficients from some equations, according to a systematic pattern. Starting from the system in its original form

$$
\begin{aligned}
(E_1) && a_{11}x_1 + a_{12}x_2 + \cdots + a_{1n}x_n &= b_1 \\
(E_2) && a_{21}x_1 + a_{22}x_2 + \cdots + a_{2n}x_n &= b_2 \\
&& \vdots \quad &= \quad \vdots \\
(E_n) && a_{n1}x_1 + a_{n2}x_2 + \cdots + a_{nn}x_n &= b_n
\end{aligned}
$$

where (E_k) denotes the kth equation, $k = 1, \ldots, n$, we would like to get an equivalent system featuring a column of zeros under the coefficient a_{11}. By "equivalent" system we mean a new system that has the same solution as the first one. If we multiply equations by some number, we do not change their solution; by the same token, if we add equations, we do not change the solution of the overall system.

For each equation (E_k) $(k = 2, \ldots, n)$, we must apply the transformation

$$(E_k) \leftarrow (E_k) - \frac{a_{k1}}{a_{11}}(E_1)$$

which leads to the equivalent system:

$$
\begin{aligned}
a_{11}x_1 + a_{12}x_2 + \cdots + a_{1n}x_n &= b_1 \\
a_{22}^{(1)}x_2 + \cdots + a_{2n}^{(1)}x_n &= b_2^{(1)} \\
\vdots \quad &= \quad \vdots \\
a_{n2}^{(1)}x_2 + \cdots + a_{nn}^{(1)}x_n &= b_n^{(1)}
\end{aligned}
$$

Now we may repeat the procedure to obtain a column of zeros under the coefficient $a_{22}^{(1)}$, and so on, until the desired form is obtained, allowing for backsubstitution. The idea is best illustrated by a simple example.

Example 3.2 Consider the following system:

$$x_1 + 2x_2 + x_3 = 0$$
$$2x_1 + 2x_2 + 3x_3 = 3$$
$$-x_1 - 3x_2 = 2$$

It is convenient to represent the operations of Gaussian elimination in a tabular form, which includes both the coefficients in each equation and the related right-hand side:

$$\left[\begin{array}{ccc|c} 1 & 2 & 1 & 0 \\ 2 & 2 & 3 & 3 \\ -1 & -3 & 0 & 2 \end{array}\right] \Rightarrow \left[\begin{array}{ccc|c} 1 & 2 & 1 & 0 \\ 0 & -2 & 1 & 3 \\ 0 & -1 & 1 & 2 \end{array}\right] \Rightarrow \left[\begin{array}{ccc|c} 1 & 2 & 1 & 0 \\ 0 & -2 & 1 & 3 \\ 0 & 0 & \frac{1}{2} & \frac{1}{2} \end{array}\right]$$

It is easy to apply backsubstitution to this system in triangular form, and we find $x_3 = 1$, $x_2 = -1$, and $x_1 = 1$.   ▯

Gaussian elimination is a fairly straightforward approach, but it involves a few subtle issues:

1. To begin with, is there anything that could go wrong? The careful reader, we guess, has already spotted a weak point in Eq. (3.5): What if $a_{kk} = 0$? This issue is easily solved, provided that the system admits a solution. We may just swap two equations or two variables. If you prefer, we might just swap two rows or two columns in the tabular form above.

2. What about numerical precision if we have a large-scale system? Is it possible that small errors due to limited precision in the calculation creep into the solution process and ultimately lead to a very inaccurate solution? This is a really thorny issue, which is way outside the scope of the book. Let us just say that state-of-the-art commercial software is widely available to tackle such issues in the best way, which means avoiding or minimizing the difficulty if possible, or warning the user otherwise.

3.2.3   Cramer's rule

As a last approach, we consider Cramer's rule, which is a handy way to solve systems of two or three equations. The theory behind it requires more advanced concepts, such as matrices and their determinants, which are introduced below. We anticipate here a few concepts so that readers not interested in advanced multivariate statistics can skip the rest of this chapter without any loss of continuity.

In the previous section, we have seen that Gaussian elimination is best carried out on a tabular representation of the system of equations. Consider a system of two equations in two variables:

$$a_{11}x_1 + a_{12}x_2 = b_1$$
$$a_{21}x_1 + a_{22}x_2 = b_2$$

We can group the coefficients and the right-hand sides into the two tables below:

$$\mathbf{A} = \begin{bmatrix} a_{11} & a_{12} \\ a_{21} & a_{22} \end{bmatrix}, \qquad \mathbf{b} = \begin{bmatrix} b_1 \\ b_2 \end{bmatrix}$$

These "tables" are two examples of a *matrix*; $\mathbf{A}$ is a two-row, two-column matrix, whereas $\mathbf{b}$ has just one column. Usually, we refer to $\mathbf{b}$ as a *column vector* (see next section).

The *determinant* is a function mapping a square matrix, i.e., a matrix with the same number of rows and columns, to a number. The determinant of matrix $\mathbf{A}$ is denoted by $\det(\mathbf{A})$ or $|\mathbf{A}|$. The concept is not quite trivial, but computing the determinant for a 2×2 matrix is easy. For the matrix above, we have

$$\det(\mathbf{A}) = \begin{vmatrix} a_{11} & a_{12} \\ a_{21} & a_{22} \end{vmatrix} = a_{11}a_{22} - a_{21}a_{12}$$

In practice, we multiply numbers on the main diagonal of the matrix and we subtract the product of numbers on the other diagonal. Let us denote by $\mathbf{B}_i$ the matrix obtained by substituting column i of matrix $\mathbf{A}$ with the vector $\mathbf{b}$ of the right-hand sides. Cramer's rule says that the solution of a system of linear equations (if it exists) is obtained by computing

$$x_i = \frac{\det(\mathbf{B}_i)}{\det(\mathbf{A})}, \qquad i = 1, \ldots, n \tag{3.6}$$

Example 3.3 To illustrate, let us consider system (3.4) again. We need the following determinants:

$$\det(\mathbf{A}) = \begin{vmatrix} 1 & -2 \\ 3 & 1 \end{vmatrix} = 1 \times 1 - 3 \times (-2) = 7$$

$$\det(\mathbf{B}_1) = \begin{vmatrix} 8 & -2 \\ 3 & 1 \end{vmatrix} = 8 \times 1 - 3 \times (-2) = 14$$

$$\det(\mathbf{B}_2) = \begin{vmatrix} 1 & 8 \\ 3 & 3 \end{vmatrix} = 1 \times 3 - 3 \times 8 = -21$$

Applying Cramer's rule yields

$$x_1 = \frac{14}{7} = 2, \qquad x_2 = \frac{-21}{7} = -3$$

which is, of course, what we obtained by substitution of variables.   ▯

The case of a 3×3 matrix is dealt with in a similar way, by computing the determinant as follows:

$$\begin{vmatrix} a_{11} & a_{12} & a_{13} \\ a_{21} & a_{22} & a_{23} \\ a_{31} & a_{32} & a_{33} \end{vmatrix} = a_{11} \begin{vmatrix} a_{22} & a_{23} \\ a_{32} & a_{33} \end{vmatrix} - a_{12} \begin{vmatrix} a_{21} & a_{23} \\ a_{31} & a_{33} \end{vmatrix} + a_{13} \begin{vmatrix} a_{21} & a_{22} \\ a_{31} & a_{32} \end{vmatrix}$$

In general, the calculation of an $n \times n$ determinant is recursively boiled down to the calculation of smaller determinants.

Example 3.4 To illustrate the calculation of the determinant in a three-dimensional case, let us apply the formula to the matrix of coefficients in Example 3.2 above:

$$\begin{vmatrix} 1 & 2 & 1 \\ 2 & 2 & 3 \\ -1 & -3 & 0 \end{vmatrix} = 1 \times \begin{vmatrix} 2 & 3 \\ -3 & 0 \end{vmatrix} - 2 \times \begin{vmatrix} 2 & 3 \\ -1 & 0 \end{vmatrix} + 1 \times \begin{vmatrix} 2 & 2 \\ -1 & -3 \end{vmatrix}$$

$$= 1 \times (2 \times 0 + 3 \times 3) - 2 \times (2 \times 0 + 1 \times 3) +$$
$$1 \times (-2 \times 3 + 1 \times 2)$$

$$= -1$$

The reader is encouraged to compute all of the determinants, such as

$$\det(\mathbf{B}_1) = \begin{vmatrix} 0 & 2 & 1 \\ 3 & 2 & 3 \\ 2 & -3 & 0 \end{vmatrix} = -1$$

that we require to apply Cramer's rule and check that we obtain the same solution as in Example 3.2.[9]   □

We will generalize this rule and say much more about the determinant in Section 3.6. We will also see that, although computing the determinant is feasible for any square matrix, this quickly becomes a cumbersome approach. Indeed, determinants are a fundamental tool from a conceptual perspective, but they are not that handy computationally. Still, it will be clear from Section 3.6 how important they are in linear algebra.

For now, we should just wonder if anything may go wrong with the idea. Indeed, Cramer's rule may crumble down if $\det(\mathbf{A}) = 0$. We will see that in such a case the matrix $\mathbf{A}$ suffers from some fundamental issues, preventing us from being able to solve the system. In general, as we have already pointed out, we should not take for granted that there is a unique solution to a system of linear equations. There could be none, or there could be many (actually, infinite). To investigate these issues, we need to introduce powerful and far-reaching mathematical concepts related to vectors, matrices, and linear algebra.

---

[9]The careful reader should wonder if we can take advantage of the element $a_{33} = 0$ in the matrix by developing the determinant along the last row, rather than along the first one; indeed, this can be done by the rules illustrated in Section 3.6.

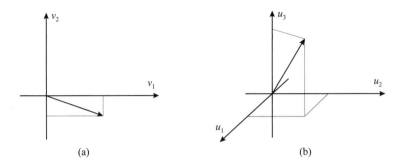

Fig. 3.4   Vectors in two- and three-dimensional spaces.

3.3   VECTOR ALGEBRA

Vectors are an intuitive concept that we get acquainted with in highschool mathematics. In ordinary two- and three-dimensional geometry, we deal with points on the plane or in the space. Such points are associated with coordinates in a Cartesian reference system. Coordinates may be depicted as vectors, as shown in Fig. 3.4; in physics, vectors are associated with a direction (e.g., of motion) or an intensity (e.g., of a force). In this book, we are interested in vectors as tuples of numbers. The vectors of Fig. 3.4 can be represented numerically as a pair and a triple of numbers, respectively:

$$\mathbf{v} = \begin{bmatrix} 2 \\ -1 \end{bmatrix}, \qquad \mathbf{u} = \begin{bmatrix} 2 \\ 3 \\ 6 \end{bmatrix}$$

.

More generally, we define a vector in an n-dimensional space as a tuple of numbers:

$$\mathbf{v} = \begin{bmatrix} v_1 \\ v_2 \\ \vdots \\ v_n \end{bmatrix}$$

As a notational aid, we will use boldface lowercase letters (e.g., $\mathbf{a}$, $\mathbf{b}$, $\mathbf{x}$) to refer to vectors; components (or elements) of a vector will be referred to by lowercase letters with a subscript, such as y_1 or v_i. In this book, components are assumed to be real numbers; hence, an n-dimensional vector $\mathbf{v}$ belongs to the set $\mathbb{R}^n$ of tuples of n real numbers. If $n = 1$, i.e., the vector boils down to a single number, we speak of a *scalar* element.

Note that usually we think of vectors as columns. Nothing forbids the use of a row vector, such as

$$[2, \ -3, \ 0, \ 9]$$

In order to stick to a clear convention, we will always assume that vectors are column vectors; whenever we insist on using a row vector, we will use the transposition operator, denoted by superscript T, which essentially swaps rows and columns:

$$\mathbf{v}^T = [v_1, \; v_2, \; \ldots, \; v_n]$$

Sometimes, transposition is also denoted by $\mathbf{v}'$.

In business applications, vectors can be used to represent several things, such as

- Groups of decision variables in a decision problem, such as produced amounts x_i for items $i = 1, \ldots, n$, in a production mix problem

- Successive observations of a single random variable X, as in a random sample $(X_1, X_2, \ldots, X_m)$

- Attributes of an object, measured according to several criteria, in multivariate statistics

3.3.1   Operations on vectors

We are quite used to elementary operations on numbers, such as addition, multiplication, and division. Not all of them can be sensibly extended to the vector case. Still, we will find some of them quite useful, namely:

- Vector addition

- Multiplication by a scalar

- Inner product

Vector addition   Addition is defined for pairs of vectors having the same dimension. If $\mathbf{v}, \mathbf{u} \in \mathbb{R}^n$, we define:

$$\mathbf{v} + \mathbf{u} \equiv \begin{bmatrix} u_1 + v_1 \\ u_2 + v_2 \\ \vdots \\ u_n + v_n \end{bmatrix}$$

For instance

$$\begin{bmatrix} 10 \\ -4 \\ 6 \end{bmatrix} + \begin{bmatrix} -3 \\ -1 \\ 10 \end{bmatrix} = \begin{bmatrix} 7 \\ -5 \\ 16 \end{bmatrix}$$

Since vector addition boils down to elementwise addition, it enjoys commutativity and associativity:

$$\mathbf{v} + \mathbf{u} = \mathbf{u} + \mathbf{v}; \qquad (\mathbf{v} + \mathbf{u}) + \mathbf{w} = \mathbf{v} + (\mathbf{u} + \mathbf{w})$$

and we may add an arbitrary number of vectors by just summing up their elements.

We will denote by **0** (note the boldface character) a vector whose elements are all 0; sometimes, we might make notation a bit clearer by making the number of elements explicit, as in $\mathbf{0}_n$. Of course, $\mathbf{u} + (-\mathbf{u}) = \mathbf{0}$.

Multiplication by a scalar Given a scalar $\alpha \in \mathbb{R}$ and a vector $\mathbf{v} \in \mathbb{R}^n$, we define the product of a vector and a scalar:

$$\alpha \mathbf{v} \equiv \begin{bmatrix} \alpha v_1 \\ \alpha v_2 \\ \vdots \\ \alpha v_n \end{bmatrix}$$

For instance

$$\mathbf{v} = \begin{bmatrix} 10 \\ -4 \\ 6 \end{bmatrix}, \quad \alpha = 3 \quad \Rightarrow \quad \alpha \mathbf{v} = \begin{bmatrix} 30 \\ -12 \\ 18 \end{bmatrix}$$

Again, familiar properties of multiplication carry over to multiplication by a scalar.

Now, what about defining a product between vectors, provided that they are of the same dimension? It is tempting to think of componentwise vector multiplication. Unfortunately, this idea does not lead to an operation with properties similar to ordinary products between scalars. For instance, if we multiply two nonzero numbers, we get a nonzero number. This is not the case for vectors. To see why, consider vectors

$$\mathbf{v} = \begin{bmatrix} 1 \\ 0 \end{bmatrix}, \quad \mathbf{v} = \begin{bmatrix} 0 \\ 1 \end{bmatrix}$$

If we multiply them componentwise, we get the two-dimensional zero vector $\mathbf{0}_2$. This, by the way, does not allow to define division in a sensible way. Nevertheless, we may define a useful concept of product between vectors, the inner product.[10]

Inner product The *inner product* is defined by multiplying vectors componentwise and summing the resulting terms:

$$\mathbf{u} \cdot \mathbf{v} = \sum_{i=1}^{n} u_i v_i$$

Since the inner product is denoted by a dot, it is also known as the *dot product*; in other books, you might also find a notation like $\langle \mathbf{u}, \mathbf{v} \rangle$. The inner product

---

[10]In physics, there is another definition of vector product, which is not useful in our setting.

yields a scalar.[11] Clearly, an inner product is defined only for vectors of the same dimension. For instance

$$\begin{bmatrix} 3 \\ -1 \\ 4 \end{bmatrix} \cdot \begin{bmatrix} 2 \\ 6 \\ -1 \end{bmatrix} = 3 \times 2 + (-1) \times 6 + 4 \times (-1) = -4$$

One could wonder why such a product should be useful at all. In fact, the inner-product concept is one of the more pervasive concepts in mathematics and it has far-reaching consequences, quite beyond the scope of this book. Nevertheless, a few geometric examples suggest why the inner product is so useful.

Example 3.5 (Euclidean length, norm, and distance) The inner-product concept is related to the concept of vector length. We know from elementary geometry that the length of a vector is the square root of the sum of squared elements. Consider the two-dimensional vector $\mathbf{v} = [2, -1]^T$, depicted in Fig. 3.4. Its length (in the Euclidean sense) is just $\sqrt{2^2 + (-1)^2} = \sqrt{5}$. We may generalize the idea to n dimensions by defining the Euclidean norm of a vector:

$$\| \mathbf{v} \| = \sqrt{\sum_{i=1}^{n} v_i^2} = \sqrt{\mathbf{u} \cdot \mathbf{u}}$$

The Euclidean norm $\| \cdot \|$ is a function mapping vectors to nonnegative real numbers, built on the inner product. Alternative definitions of norm can be proposed, provided that they preserve the intuitive properties associated with vector length. By the same token, we may consider the distance between two points. Referring again to a two-dimensional case, if we are given two points in plane

$$\mathbf{u} = \begin{bmatrix} u_1 \\ u_2 \end{bmatrix}, \qquad \mathbf{v} = \begin{bmatrix} v_1 \\ v_2 \end{bmatrix}$$

their Euclidean distance is

$$d(\mathbf{u}, \mathbf{v}) \equiv \sqrt{(u_1 - v_1)^2 + (u_2 - v_2)^2}$$

This can be related to the norm and the inner product:

$$d(\mathbf{u}, \mathbf{v}) = \| \mathbf{u} - \mathbf{v} \| = \sqrt{(\mathbf{u} - \mathbf{v}) \cdot (\mathbf{u} - \mathbf{v})}$$

In such a case, we are measuring distance by assigning the same importance to each dimension (component of the difference vector). Later, we will generalize

---

[11] Indeed, the inner product is sometimes referred to as the "scalar product." We will avoid this term, as it may be confused with multiplication by a scalar.

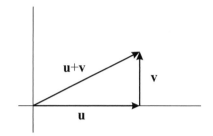

Fig. 3.5   Orthogonal vectors illustrating the Pythagorean theorem.

distance by allowing for a vector of weights $\mathbf{w}$, such as in

$$d_{\mathbf{w}}(\mathbf{u}, \mathbf{v}) = \sqrt{\sum_{i=1}^{n} w_i (u_i - v_i)^2}$$

A vector $\mathbf{e}$ such that $\| \mathbf{e} \| = 1$ is called a *unit vector*. Given a vector $\mathbf{v}$, we may obtain a unit vector parallel to $\mathbf{v}$ by considering vector $\mathbf{v} / \| \mathbf{v} \|$.   ▯

Example 3.6 (Orthogonal vectors) Consider the following vectors:

$$\mathbf{e}_1 = \begin{bmatrix} 1 \\ 0 \end{bmatrix}, \qquad \mathbf{e}_2 = \begin{bmatrix} 0 \\ 1 \end{bmatrix}$$

They are unit-length vectors parallel to the two coordinate axes, and they are orthogonal. We immediately see that

$$\mathbf{e}_1 \cdot \mathbf{e}_2 = 1 \times 0 + 0 \times 1 = 0$$

The same thing happens, for instance, with $\mathbf{v}_1 = [1, 1]^T$ and $\mathbf{v}_2 = [-1, 1]^T$. These vectors are orthogonal as well (please draw them and check).

Orthogonality is not an intuitive concept in an n-dimensional space, but in fact we may define orthogonality in terms of the inner product: We say that two vectors are *orthogonal* if their inner product is zero.   ▯

Example 3.7 (Orthogonal projection) From elementary geometry, we know that if two vectors $\mathbf{u}$ and $\mathbf{v}$ are orthogonal, then we must have

$$\| \mathbf{u} + \mathbf{v} \|^2 = \| \mathbf{u} \|^2 + \| \mathbf{v} \|^2$$

This is simply the Pythagorean theorem in disguise, as illustrated in Fig. 3.5. Now consider two vectors $\mathbf{u}$ and $\mathbf{v}$ that are not orthogonal, as illustrated in Fig. 3.6. We may decompose $\mathbf{u}$ in the sum of two vectors, say, $\mathbf{u}_1$ and $\mathbf{u}_2$, such that $\mathbf{u}_1$ is parallel to $\mathbf{v}$ and $\mathbf{u}_2$ is orthogonal to $\mathbf{v}$. The vector $\mathbf{u}_1$ is the *orthogonal projection* of $\mathbf{u}$ on $\mathbf{v}$. Basically, we are decomposing the

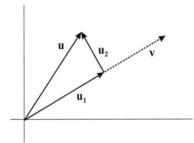

Fig. 3.6 Orthogonal projection of vector **u** on vector **v**.

"information" in **u** into the sum of two components. One contains the same information as **v**, whereas the other one is completely "independent."

Since $\mathbf{u}_1$ is parallel to **v**, for some scalar α we have:

$$\mathbf{u}_1 = \alpha\mathbf{v}, \qquad \mathbf{u}_2 = \mathbf{u} - \alpha\mathbf{v}$$

The first equality derives from the fact that two parallel vectors must be related by a proportionality constant; the second one states that **u** is the sum of two component vectors. What is the right value of α that makes them orthogonal? Let us apply the Pythagorean theorem:

$$
\begin{aligned}
\|\mathbf{u}\|^2 &= \|\alpha\mathbf{v}\|^2 + \|\mathbf{u} - \alpha\mathbf{v}\|^2 \\
&= \alpha^2\|\mathbf{v}\|^2 + (\mathbf{u} - \alpha\mathbf{v})\cdot(\mathbf{u} - \alpha\mathbf{v}) \\
&= \alpha^2\|\mathbf{v}\|^2 + \mathbf{u}\cdot\mathbf{u} - 2\mathbf{u}\cdot(\alpha\mathbf{v}) + (\alpha\mathbf{v})\cdot(\alpha\mathbf{v}) \\
&= \alpha^2\|\mathbf{v}\|^2 + \|\mathbf{u}\|^2 - 2\alpha(\mathbf{u}\cdot\mathbf{v}) + \alpha^2\|\mathbf{v}\|^2
\end{aligned}
$$

This, in turn, implies

$$\alpha = \frac{\mathbf{u}\cdot\mathbf{v}}{\|\mathbf{v}\|^2}$$

If **v** is a unit vector, we see that the inner product gives the length of the projection of **u** on **v**. We also see that if **u** and **v** are orthogonal, then this projection is null; in some sense, the "information" in **u** is independent from what is provided by **v**. ☐

3.3.2 Inner products and norms

The *inner product* is an intuitive geometric concept that is easily introduced for vectors, and it can be used to define a vector norm. A *vector norm* is a function mapping a vector **x** into a nonnegative number $\|\mathbf{x}\|$ that can be interpreted as vector length. We have see that we may use the dot product to define the usual Euclidean norm:

$$\|\mathbf{x}\| = \sqrt{\mathbf{x}\cdot\mathbf{x}} = \sqrt{x_1^2 + x_2^2 + \cdot + x_n^2}$$

It turns out that both concepts can be made a bit more abstract and general, and this has a role in generalizing some intuitive concepts such as orthogonality and Pythagorean theorem to quite different fields.[12]

Generally speaking, the inner product on a space is a function mapping two elements $\mathbf{x}$ and $\mathbf{y}$ of that space into a real number $\langle \mathbf{x}, \mathbf{y} \rangle$. There is quite some freedom in defining inner products, provided our operation meets the following conditions:

- $\langle \mathbf{x}, \mathbf{x} \rangle$ is a real number such that $\langle \mathbf{x}, \mathbf{x} \rangle \geq 0$ and $\langle \mathbf{x}, \mathbf{x} \rangle = 0$ only when $\mathbf{x} = \mathbf{0}$.

- $\langle \mathbf{x}, \mathbf{y} \rangle = \langle \mathbf{y}, \mathbf{x} \rangle$.

- $\langle \mathbf{x}, \alpha \mathbf{y} \rangle = \alpha \langle \mathbf{x}, \mathbf{y} \rangle$ for any scalar α.

- $\langle \mathbf{x}, \mathbf{y} + \mathbf{z} \rangle = \langle \mathbf{x}, \mathbf{y} \rangle + \langle \mathbf{x}, \mathbf{z} \rangle$.

It is easy to check that our definition of the inner product for vectors in $\mathbb{R}^n$ meets these properties. If we define alternative inner products, provided they meet the requirements described above, we end up with different concepts of orthogonality and useful generalizations of Pythagorean theorem. A straightforward generalization of the dot product is obtained by considering a vector $\mathbf{w}$ of positive weights and defining:

$$\langle \mathbf{x}, \mathbf{y} \rangle_{\mathbf{w}} = \sum_{i=1}^{n} w_i x_i y_i$$

This makes sense when we have a problem with multiple dimensions, but some of them are more important than others.

By a similar token, we may generalize the concept of norm, by requiring the few properties that make sense when defining a vector length:

- $\| \mathbf{x} \| \geq 0$, with $\| \mathbf{x} \| = 0$ if and only if $\mathbf{x} = \mathbf{0}$; this states that length cannot be negative and it is zero only for a null vector.

- $\| \alpha \mathbf{x} \| = | \alpha | \, \| \mathbf{x} \|$ for any scalar α; this states that if we multiply all of the components of a vector by a number, we change vector length accordingly, whether the scalar is negative or positive.

- $\| \mathbf{x} + \mathbf{y} \| \leq \| \mathbf{x} \| + \| \mathbf{y} \|$; this property is known as *triangle inequality* and can be interpreted intuitively by looking at Fig. 3.7 and interpreting vectors as displacement in the plane, and their length as distances. We can move from point $\mathbf{P}_0$ to point $\mathbf{P}_1$ by the displacement corresponding to vector $\mathbf{x}$, and from point $\mathbf{P}_1$ to $\mathbf{P}_2$ by displacement $\mathbf{y}$; if we move directly from $\mathbf{P}_0$ to $\mathbf{P}_2$, the length of the resulting displacement, $\| \mathbf{x} + \mathbf{y} \|$, cannot be larger than the sum of the two distances $\| \mathbf{x} \|$ and $\| \mathbf{y} \|$.

---

[12]Later, we will learn that two random variables are "orthogonal" if they are uncorrelated.

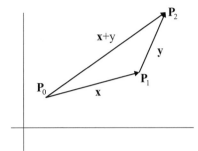

Fig. 3.7 Orthogonal projection of vector **u** on vector **v**.

Using the standard dot product, we find the standard Euclidean norm, which can be denoted by $\|\cdot\|_2$. However, we might just add the absolute value of each coordinate to define a length or a distance:

$$\|\mathbf{x}\|_1 = \sum_{i=1}^{n} |x_i| \tag{3.7}$$

The notation is due to the fact that the two norms above are a particular case of the general norm

$$\|\mathbf{x}\|_p = \left(\sum_{i=1}^{n} |x_i|^p \right)^{1/p}$$

Letting $p \to +\infty$, we find the following norm:

$$\|\mathbf{x}\|_\infty = \max_{i=1,\ldots,n} |x_i| \tag{3.8}$$

All of these norms make sense for different applications, and it is also possible to introduce weights to assign different degrees of importance to multiple problem dimensions.[13]

Norms need not be defined on the basis of an inner product, but the nice thing of the inner product is that if we use an inner product to define a norm, we can be sure that we come up with a legitimate norm.

3.3.3 Linear combinations

The two basic operations on vectors, addition and multiplication by a scalar, can be combined at wish, resulting in a vector; this is called a *linear combination* of vectors. The linear combination of vectors $\mathbf{v}_j$ with coefficients α_j,

---

[13] For instance, Euclidean norm is used in standard linear regression by least squares; see Section 10.1.1 for alternative approaches.

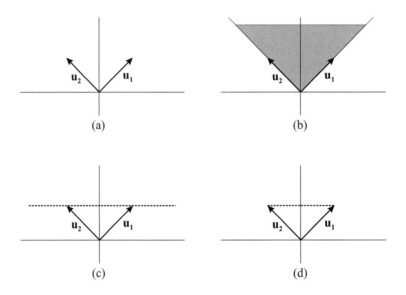

Fig. 3.8 Illustrating linear combinations of vectors.

$j = 1, \ldots, m$ is

$$\sum_{j=1}^{m} \alpha_j \mathbf{v}_j$$

If we denote each component i, $i = 1, \ldots, n$, of vector j by v_{ij}, the component i of the linear combination above is simply

$$\sum_{j=1}^{m} \alpha_j v_{ij}$$

It is useful to gain some geometric intuition to shed some light on this concept. Consider vectors $\mathbf{u}_1 = [1, 1]^T$ and $\mathbf{u}_2 = [-1, 1]^T$, depicted in Fig. 3.8(a), and imagine taking linear combinations of them with coefficients α_1 and α_2:

$$\mathbf{v} = \alpha_1 \mathbf{u}_1 + \alpha_2 \mathbf{u}_2$$

Which set of vectors do we generate when the two coefficients are varied? Here are a few examples:

$$
\begin{array}{lll}
\alpha_1 = 2, \ \alpha_2 = -3 & \Rightarrow & \mathbf{v} = [5, -1]^T \\
\alpha_1 = 2, \ \alpha_2 = 3 & \Rightarrow & \mathbf{v} = [-1, 5]^T \\
\alpha_1 = 2, \ \alpha_2 = -1 & \Rightarrow & \mathbf{v} = [3, 1]^T \\
\alpha_1 = 0.5, \ \alpha_2 = 0.5 & \Rightarrow & \mathbf{v} = [0, 1]^T
\end{array}
$$

The reader is strongly urged to draw these vectors to get a feeling for the geometry of linear combinations. On the basis of this intuition, we may try to generalize a bit as follows:

- It is easy to see that if we let α_1 and α_2 take any possible value in $\mathbb{R}$, we generate any vector in the plane. Hence, we span all of the two-dimensional space $\mathbb{R}^2$ by expressing any vector as a linear combination of $\mathbf{u}_1$ and $\mathbf{u}_2$.

- If we enforce the additional restriction $\alpha_1, \alpha_2 \geq 0$, we take *positive* linear combinations. Geometrically, we generate only vectors in the shaded region of Fig. 3.8(b), which is a *cone*.

- If we require $\alpha_1 + \alpha_2 = 1$, we generate vectors on the horizontal dotted line in Fig. 3.8(c). Such a combination is called *affine combination*. Note that in such a case we may express the linear combination using one coefficient α as $\alpha\mathbf{u}_1 + (1 - \alpha)\mathbf{u}_2$. We get vector $\mathbf{u}_1$ when setting $\alpha = 1$ and vector $\mathbf{u}_2$ when setting $\alpha = 0$.

- If we enforce both of the additional conditions, i.e., if we require that weights be nonnegative and add up to one, we generate only the line segment between $\mathbf{u}_1$ and $\mathbf{u}_2$. Such a linear combination is said a *convex combination* and is illustrated in Fig. 3.8(d).[14]

3.4 MATRIX ALGEBRA

We began this chapter by considering the solution of systems of linear equations. Many issues related to systems of linear equations can be addressed by introducing a new concept, the matrix. Matrix theory plays a fundamental role in quite a few mathematical and statistical methods that are relevant for management.

We have introduced vectors as one-dimensional arrangement of numbers. A *matrix* is, in a sense, a generalization of vectors to two dimensions. Here are two examples of matrices:

$$
\mathbf{A} = \begin{bmatrix} 1 & 3 & -1 \\ 0 & 4 & -2 \\ 3 & 5 & 0 \end{bmatrix}; \quad
\mathbf{B} = \begin{bmatrix} 1 & 3 & -1 & 6 \\ 0 & 4 & -2 & 5 \end{bmatrix}
$$

In the following, we will denote matrices by boldface, uppercase letters. A matrix is characterized by the number of rows and the number of columns. Matrix $\mathbf{A}$ consists of three rows and three columns, whereas matrix $\mathbf{B}$ consists of two rows and four columns. Generally speaking, a matrix with m rows and n columns belongs to a space of matrices denoted by $\mathbb{R}^{m,n}$. In the two examples

---

[14]The reader might appreciate the definition of convex combination a little better by referring back to Section 2.11 on convexity. We will also see that coefficients with those properties might be interpreted as probabilities, as they are nonnegative and add up to one; hence, a convex combination bears some resemblance to the probabilistic concept of expected value.

above, $\mathbf{A} \in \mathbb{R}^{3,3}$ and $\mathbf{B} \in \mathbb{R}^{2,4}$. When $m = n$, as in the case of matrix $\mathbf{A}$, we speak of a *square matrix*.

It is easy to see that vectors are just a special case of a matrix. A column vector belongs to space $\mathbb{R}^{n,1}$ and a row vector is an element of $\mathbb{R}^{1,n}$.

3.4.1   Operations on matrices

Operations on matrices are defined much along the lines used for vectors. In the following, we will denote a generic element of a matrix by a lowercase letter with two subscripts; the first one refers to the row, and the second one refers to the column. So, element a_{ij} is the element in row i, column j.

Addition   Just like vectors, matrices can be added elementwise, provided they are of the same size. If $\mathbf{A} \in \mathbb{R}^{m,n}$ and $\mathbf{B} \in \mathbb{R}^{m,n}$, then

$$
\mathbf{A} + \mathbf{B} = \begin{bmatrix} a_{11} & \cdots & a_{1n} \\ \vdots & a_{ij} & \vdots \\ a_{m1} & \cdots & a_{mn} \end{bmatrix} + \begin{bmatrix} b_{11} & \cdots & b_{1n} \\ \vdots & b_{ij} & \vdots \\ b_{m1} & \cdots & b_{mn} \end{bmatrix}
$$

$$
= \begin{bmatrix} a_{11} + b_{11} & \cdots & a_{1n} + b_{1n} \\ \vdots & a_{ij} + b_{ij} & \vdots \\ a_{m1} + b_{m1} & \cdots & a_{mn} + b_{mn} \end{bmatrix}
$$

For example

$$
\begin{bmatrix} 1 & -1 & 0 \\ -2 & 4 & 6 \\ 7 & -3 & -2 \end{bmatrix} + \begin{bmatrix} 6 & -3 & 1 \\ 0 & 10 & 3 \\ 4 & 5 & 1 \end{bmatrix} = \begin{bmatrix} 7 & -4 & 1 \\ -2 & 14 & 9 \\ 11 & 2 & -1 \end{bmatrix}
$$

Multiplication by a scalar   Given a scalar $\alpha \in \mathbb{R}$, we define its product with a matrix $\mathbf{A} \in \mathbb{R}^{m,n}$ as follows:

$$
\alpha \mathbf{A} = \begin{bmatrix} \alpha a_{11} & \cdots & \alpha a_{1n} \\ \vdots & \alpha a_{ij} & \vdots \\ \alpha a_{m1} & \cdots & \alpha a_{mn} \end{bmatrix}
$$

For example

$$
3 \begin{bmatrix} 2 & -1 & 0 \\ -3 & 5 & 1 \\ 7 & -4 & 2 \end{bmatrix} = \begin{bmatrix} 6 & -3 & 0 \\ -9 & 15 & 3 \\ 21 & -12 & 6 \end{bmatrix}
$$

Matrix multiplication   We have already seen that elementwise multiplication with vectors does not lead us to an operation with good and interesting properties, whereas the inner product is a rich concept. Matrix multiplication is even trickier, and it is not defined elementwise, either. The matrix product $\mathbf{AB}$ is defined only when the number of columns of $\mathbf{A}$ and the number of rows of $\mathbf{B}$ are the same, i.e., $\mathbf{A} \in \mathbb{R}^{m,k}$ and $\mathbf{B} \in \mathbb{R}^{k,n}$. The result is a matrix, say, $\mathbf{C}$, with m rows and n columns, whose element c_{ij} is the inner product of the ith row of A and the jth column of $\mathbf{B}$:

$$c_{ij} = \sum_{h=1}^{k} a_{ih} b_{hj}$$

For instance

$$\begin{bmatrix} 1 & 3 \\ -2 & 7 \\ 9 & 0 \end{bmatrix} \begin{bmatrix} 4 & 2 \\ 1 & -1 \end{bmatrix} = \begin{bmatrix} (1 \times 4 + 3 \times 1) & (1 \times 2 - 3 \times 1) \\ (-2 \times 4 + 7 \times 1) & (-2 \times 2 - 7 \times 1) \\ (9 \times 4 + 0 \times 1) & (9 \times 2 - 0 \times 1) \end{bmatrix}$$

$$= \begin{bmatrix} 7 & -1 \\ -1 & -11 \\ 36 & 18 \end{bmatrix} \tag{3.9}$$

If two matrices are of compatible dimension, in the sense that they can be multiplied using the rules above, they are said to be *conformable*. The definition of matrix multiplication does look somewhat odd at first. Still, we may start getting a better felling for it, if we notice how we use it to press a system of linear equations in a very compact form. Consider the system

$$a_{11}x_1 + a_{12}x_2 + \cdots a_{in}x_n = b_1$$
$$a_{21}x_1 + a_{22}x_2 + \cdots a_{2n}x_n = b_2$$
$$\vdots$$
$$a_{n1}x_1 + a_{n2}x_2 + \cdots a_{nn}x_n = b_n$$

and group coefficients a_{ij} into matrix $\mathbf{A} \in \mathbb{R}^{n,n}$, and right-hand sides b_i into column vector $\mathbf{b} \in \mathbb{R}^n \equiv \mathbb{R}^{n,1}$. Using the definition of matrix multiplication, we may rewrite the system as follows:

$$\mathbf{Ax} = \mathbf{b}$$

By repeated application of matrix product, we also define the *power* of a matrix. Squaring a matrix $\mathbf{A}$ does *not* mean squaring each element, but rather multiplying the matrix by itself:

$$\mathbf{A}^2 \equiv \mathbf{AA}$$

It is easy to see that this definition makes sense only if the number of rows is the same as the number of columns, i.e., if matrix $\mathbf{A}$ is a square matrix. By the same token, we may define a generic power:

$$\mathbf{A}^m \equiv \underbrace{\mathbf{AA} \cdots \mathbf{A}}_{m \text{ times}}$$

We may even define the square root of a matrix as a matrix $\mathbf{A}^{1/2}$ such that:

$$\mathbf{A} = \mathbf{A}^{1/2} \mathbf{A}^{1/2}$$

The existence and uniqueness of the square root matrix are not to be taken for granted. We will discuss this further when dealing with multivariate statistics.

A last important observation concerns commutativity. In the scalar case, we know that inverting the order of factors does not change the result of a multiplication: $ab = ba$. This does not apply to matrices in general. To begin with, when multiplying two rectangular matrices, the number of rows and column may match in such a way that $\mathbf{AB}$ is defined, but $\mathbf{BA}$ is not. But even in the case of two square matrices, commutativity is not ensured, as the following counterexample shows:

$$\begin{bmatrix} 2 & 1 \\ 1 & 1 \end{bmatrix} \begin{bmatrix} 1 & -1 \\ 0 & 2 \end{bmatrix} = \begin{bmatrix} 2 & 0 \\ 1 & 1 \end{bmatrix}$$

$$\begin{bmatrix} 1 & -1 \\ 0 & 2 \end{bmatrix} \begin{bmatrix} 2 & 1 \\ 1 & 1 \end{bmatrix} = \begin{bmatrix} 1 & 0 \\ 2 & 2 \end{bmatrix}$$

Matrix transposition   We have already met transposition as a way of transforming a column vector into a row vector (and vice versa). More generally, matrix transposition entails interchanging rows and columns: Let $\mathbf{B} = \mathbf{A}^T$ be the transposed of $\mathbf{A}$; then $b_{ij} = a_{ji}$. For instance

$$\mathbf{A} = \begin{bmatrix} 6 & 0 \\ -1 & 5 \\ 3 & 8 \end{bmatrix} \quad \Rightarrow \quad \mathbf{A}^T = \begin{bmatrix} 6 & -1 & 3 \\ 0 & 5 & 8 \end{bmatrix}$$

So, if $\mathbf{A} \in \mathbb{R}^{m,n}$, then $\mathbf{A}^T \in \mathbb{R}^{n,m}$.

If $\mathbf{A}$ is a square matrix, it may happen that $\mathbf{A} = \mathbf{A}^T$; in such a case we say that $\mathbf{A}$ is a *symmetric* matrix. The following is an example of a symmetric matrix:

$$\mathbf{A} = \begin{bmatrix} 2 & 0 & -3 \\ 0 & 3 & 5 \\ -3 & 5 & -1 \end{bmatrix}$$

Symmetry should not be regarded as a peculiar accident of rather odd matrices; in applications, we often encounter matrices that are symmetric by nature.[15] Finally, it may be worth noting that transposition offers another way to denote the inner product of two vectors:

$$\mathbf{x} \cdot \mathbf{y} = \mathbf{x}^T \mathbf{y}$$

By the same token

$$\|\mathbf{x}\|^2 = \mathbf{x} \cdot \mathbf{x} = \mathbf{x}^T \mathbf{x}$$

The following is an important property concerning transposition and product of matrices.

---

[15] One such common example that we will meet later is the matrix collecting all of the covariances between a vector of random variables.

PROPERTY 3.1 (Transposition of a matrix product) *If* $\mathbf{A}$ *and* $\mathbf{B}$ *are conformable matrices, then*

$$(\mathbf{AB})^T = \mathbf{B}^T \mathbf{A}^T$$

We encourage the reader to check the result for the matrices of Eq. (3.9).

Example 3.8 An immediate consequence of the last property is that, for any matrix $\mathbf{A}$, the matrices $\mathbf{A}^T\mathbf{A}$ and $\mathbf{A}\mathbf{A}^T$ are symmetric:

$$\left(\mathbf{A}^T\mathbf{A}\right)^T = \mathbf{A}^T(\mathbf{A}^T)^T = \mathbf{A}^T\mathbf{A} \qquad \text{and} \qquad \left(\mathbf{A}\mathbf{A}^T\right)^T = (\mathbf{A}^T)^T\mathbf{A}^T = \mathbf{A}\mathbf{A}^T$$

Note that, in general, the matrices $\mathbf{A}^T\mathbf{A}$ and $\mathbf{A}\mathbf{A}^T$ have different dimensions.

□

3.4.2 The identity matrix and matrix inversion

Matrix inversion is an operation that has no counterpart in the vector case, and it deserves its own section. In the scalar case, when we consider standard multiplication, we observe that there is a "neutral" element for that operation, the number 1. This is a neutral element in the sense that for any $x \in \mathbb{R}$, we have $x \cdot 1 = 1 \cdot x = x$.

Can we define a neutral element in the case of matrix multiplication? The answer is yes, provided that we confine ourselves to square matrices. If so, it is easy to see that the following *identity matrix* $\mathbf{I} \in \mathbb{R}^{n,n}$ will do the trick:

$$\mathbf{I} = \begin{bmatrix} 1 & 0 & \cdots & 0 \\ 0 & 1 & \cdots & 0 \\ \vdots & \vdots & \ddots & \vdots \\ 0 & 0 & \cdots & 1 \end{bmatrix}$$

The identity matrix consists of a diagonal of "ones," and it is easy to see that, for any matrix $\mathbf{A} \in \mathbb{R}^{n,n}$, we have $\mathbf{AI} = \mathbf{IA} = \mathbf{A}$. By the same token, for any vector $\mathbf{v} \in \mathbb{R}^n$, we have $\mathbf{Iv} = \mathbf{v}$ (but in this case we cannot commute the two factors in the multiplication).

In the scalar case, given a number $x \neq 0$, we define its inverse x^{-1}, such that $x \cdot x^{-1} = x^{-1} \cdot x = 1$. Can we do the same with (square) matrices? Indeed, we can sometimes find the *inverse of a square matrix* $\mathbf{A}$, denoted by $\mathbf{A}^{-1}$, which is a matrix such that

$$\mathbf{A}^{-1}\mathbf{A} = \mathbf{I}, \qquad \mathbf{A}\mathbf{A}^{-1} = \mathbf{I}$$

However, the existence of the inverse matrix is not guaranteed. We will learn in the following text that matrix inversion is strongly related to the possibility of solving a system of linear equations. Indeed, if a matrix $\mathbf{A}$ is invertible, solving a system of linear equations like $\mathbf{A}\mathbf{x} = \mathbf{b}$ is easy. To see this, just premultiply the system by the inverse of $\mathbf{A}$:

$$\mathbf{A}\mathbf{x} = \mathbf{b} \quad \Rightarrow \quad \mathbf{A}^{-1}\mathbf{A}\mathbf{x} = \mathbf{A}^{-1}\mathbf{b} \quad \Rightarrow \quad \mathbf{x} = \mathbf{A}^{-1}\mathbf{b}$$

Note the analogy with the solution of the linear equation $ax = b$. We know that $x = b/a$, but we are in trouble if $a = 0$. By a similar token, a matrix might not be invertible, which implies that we fail to find a solution to the system of equations. To really understand the issues involved, we need a few theoretical concepts from linear algebra, which we introduce in Section 3.5.

For now, it is useful to reinterpret the solution of a system of linear equations under a different perspective. Imagine "slicing" a matrix $\mathbf{A} \in \mathbb{R}^{n,n}$, i.e., think of it as made of its column vectors $\mathbf{a}_j$:

$$\mathbf{A} = \begin{bmatrix} \vdots & \vdots & & \vdots \\ \mathbf{a}_1 & \mathbf{a}_2 & \cdots & \mathbf{a}_n \\ \vdots & \vdots & & \vdots \end{bmatrix}, \quad \text{where} \quad \mathbf{a}_j = \begin{bmatrix} a_{1j} \\ a_{2j} \\ \vdots \\ a_{nj} \end{bmatrix}, \, j = 1, \ldots, n$$

Then, we may see that solving a system of linear equations amounts to expressing the right-hand side $\mathbf{b}$ as a linear combination of the columns of $\mathbf{A}$. To illustrate

$$\begin{cases} x_1 + 2x_2 + x_3 = 0 \\ 2x_1 + 2x_2 + 3x_3 = 3 \\ -x_1 - 3x_2 = 2 \end{cases} \iff \begin{bmatrix} 1 \\ 2 \\ -1 \end{bmatrix} x_1 + \begin{bmatrix} 2 \\ 2 \\ -3 \end{bmatrix} x_2 + \begin{bmatrix} 1 \\ 3 \\ 0 \end{bmatrix} x_3 = \begin{bmatrix} 0 \\ 3 \\ 2 \end{bmatrix}$$

More generally

$$\mathbf{a}_1 x_1 + \mathbf{a}_2 x_2 + \cdots \mathbf{a}_n x_n = \mathbf{b}$$

There is no guarantee that we can always express $\mathbf{b}$ as a linear combination of columns $\mathbf{a}_j$, $j = 1, \ldots, n$, with coefficients x_j. More so, if we consider a rectangular matrix. For instance, if we have a system of linear equations associated with a matrix in $\mathbb{R}^{m,n}$, with $m > n$, then we have many equations and just a few unknown variables. It stands to reason that in such a circumstance we may fail to solve the system, which means that we may well fail to express a vector with many components, using just a few vectors as building blocks. This kind of interpretation will prove quite helpful later.

3.4.3 Matrices as mappings on vector spaces

Consider a matrix $\mathbf{A} \in \mathbb{R}^{m,n}$. When we multiply a vector $\mathbf{u} \in \mathbb{R}^n$ by this matrix, we get a vector $\mathbf{v} \in \mathbb{R}^m$. This suggests that a matrix is more than just an arrangement of numbers, but it can be regarded as an operator mapping $\mathbb{R}^n$ to $\mathbb{R}^m$:

$$\mathbf{v} = f(\mathbf{u}) = \mathbf{A}\mathbf{u}$$

Given the rules of matrix algebra, it is easy to see that this mapping is linear, in the sense that:

$$f(\alpha_1\mathbf{u}_1 + \alpha_2\mathbf{u}_2) = \mathbf{A}(\alpha_1\mathbf{u}_1 + \alpha_2\mathbf{u}_2) = \alpha_1\mathbf{A}\mathbf{u}_1 + \alpha_2\mathbf{A}\mathbf{u}_2$$
$$= \alpha_1 f(\mathbf{u}_1) + \alpha_2 f(\mathbf{u}_2) \qquad (3.10)$$

This means that the mapping of a linear combination is just a linear combination of the mappings. Many transformations of real-world entities are (at least approximately) linear.

If we restrict our attention to a square matrix $\mathbf{A} \in \mathbb{R}^{n,n}$, we see that this matrix corresponds to some mapping from $\mathbb{R}^n$ to itself; in other words, it is a way to transform a vector. What is the effect of an operator transforming a vector with n components into another vector in the same space? There are two possible effects:

1. The length (norm) of the vector is changed.

2. The vector is rotated.

In general, a transformation will have both effects, but there may be more specific cases.

If, for some vector $\mathbf{v}$, the matrix $\mathbf{A}$ has only the first effect, thus means that, for some scalar $\lambda \in \mathbb{R}$, we have

$$\mathbf{A}\mathbf{v} = \lambda\mathbf{v}$$

A trivial case in which this happens is when $\mathbf{A} = \lambda\mathbf{I}$, i.e., the matrix is a diagonal of numbers equal to λ:

$$\mathbf{A} = \lambda\mathbf{I} = \begin{bmatrix} \lambda & 0 & \cdots & 0 \\ 0 & \lambda & \cdots & 0 \\ \vdots & \vdots & \ddots & \vdots \\ 0 & 0 & \cdots & \lambda \end{bmatrix}$$

Actually, this is not that interesting, but we will see in Section 3.7 that the condition above may apply for specific scalars (called *eigenvalues*) and specific vectors (called *eigenvectors*).

If a matrix has the only effect of rotating a vector, then it does not change the norm of the vector:

$$\| \mathbf{A}\mathbf{v} \| = \| \mathbf{v} \|$$

This happens if the matrix $\mathbf{A}$ has an important property.

DEFINITION 3.2 (Orthogonal matrix) *A square matrix* $\mathbf{P} \in \mathbb{R}^{n,n}$ *is called an* **orthogonal matrix** *if* $\mathbf{P}^T\mathbf{P} = \mathbf{I}$. *Note that this property also implies that* $\mathbf{P}^{-1} = \mathbf{P}^T$.

To understand the definition, consider each column vector $\mathbf{p}_j$ of matrix $\mathbf{P}$. The element in row i, column j of $\mathbf{P}^T\mathbf{P}$ is just the inner product of $\mathbf{p}_i$ and $\mathbf{p}_j$. Hence, the definition above is equivalent to the following requirement:

$$\mathbf{p}_i \cdot \mathbf{p}_j = \mathbf{p}_i^T\mathbf{p}_j = \begin{cases} 1 & \text{if } i = j \\ 0 & \text{otherwise} \end{cases}$$

In other words, the columns of $\mathbf{P}$ is a set of orthogonal vectors. To be more precise, we should say *orthonormal*, as the inner product of a column with itself is 1, but we will not be that rigorous. Now it is not difficult to see that an orthogonal matrix is a rotation matrix. To see why, let us check the norm of the transformed vector $\mathbf{y} = \mathbf{P}\mathbf{v}$:

$$\|\mathbf{y}\|^2 = \mathbf{y}^T\mathbf{y} = (\mathbf{P}\mathbf{v})^T \mathbf{P}\mathbf{v} = \mathbf{v}^T\mathbf{P}^T\mathbf{P}\mathbf{v} = \mathbf{v}^T\mathbf{I}\mathbf{v} = \mathbf{v}^T\mathbf{v} = \|\mathbf{v}\|$$

where we have used Property 3.1 for transposition of the product of matrices. Rotation matrices are important in multivariate statistical techniques such as principal component analysis and factor analysis (see Chapter 17).

3.4.4 Laws of matrix algebra

In this section, we summarize a few useful properties of the matrix operations we have introduced. Some have been pointed out along the way; some are trivial to check, and some would require a technical proof that we prefer to avoid.

A few properties of matrix addition and multiplication that are formally identical to properties of addition and multiplication of scalars:

- $(\mathbf{A} + \mathbf{B}) + \mathbf{C} = \mathbf{A} + (\mathbf{B} + \mathbf{C})$
- $\mathbf{A} + \mathbf{B} = \mathbf{B} + \mathbf{A}$
- $(\mathbf{A}\mathbf{B})\mathbf{C} = \mathbf{A}(\mathbf{B}\mathbf{C})$
- $\mathbf{A}(\mathbf{B} + \mathbf{C}) = \mathbf{A}\mathbf{B} + \mathbf{A}\mathbf{C}$

Matrix multiplication is a bit peculiar since, in general, $\mathbf{A}\mathbf{B} \neq \mathbf{B}\mathbf{A}$, as we have already pointed out. Other properties involve matrix transposition:

- $(\mathbf{A}^T)^T = \mathbf{A}$
- $(r\mathbf{A})^T = r\mathbf{A}^T$
- $(\mathbf{A} \pm \mathbf{B})^T = \mathbf{A}^T \pm \mathbf{B}^T$
- $(\mathbf{A}\mathbf{B})^T = \mathbf{B}^T\mathbf{A}^T$

They are all fairly trivial except for the last one, which we have already pointed out. A few more properties involve inversion. If $\mathbf{A}$ and $\mathbf{B}$ are square and invertible, as well as of the same dimension, then:

- $\left(\mathbf{A}^{-1}\right)^{-1} = \mathbf{A}$

- $\left(\mathbf{A}^{T}\right)^{-1} = \left(\mathbf{A}^{-1}\right)^{T}$

- $\mathbf{AB}$ is invertible and $\left(\mathbf{AB}\right)^{-1} = \mathbf{B}^{-1}\mathbf{A}^{-1}$

3.5 LINEAR SPACES

In the previous sections, we introduced vectors and matrices and defined an algebra to work on them. Now we try to gain a deeper understanding by taking a more abstract view, introducing linear spaces. To prepare for that, let us emphasize a few relevant concepts:

- *Linear combinations.* We have defined linear combination of vectors, but in fact we are not constrained to take linear combinations of finite-dimensional objects like tuples of numbers. For instance, we might work with infinite-dimensional spaces of functions. To see a concrete example, consider monomial functions like $m_k(x) \equiv x^k$. If we take a linear combination of monomials up to degree n, we get a polynomial:

$$p(x) = a_0 + a_1 x + a_2 x^2 + \cdots + a_n x^n = \sum_{j=0}^{n} a_j x^j$$

We might represent the polynomial of degree n by the vector of its coefficients:

$$\mathbf{p} = [a_0, a_1, a_2, \ldots, a_n]^T \in \mathbb{R}^{n+1}$$

Note that we are basically expressing the polynomial using "coordinates" in a reference system represented by monomials. However, we could express the very same polynomial using different building blocks. For instance, how can we express the polynomial $p(x) = 2 - 3x + 5x^2$ as a linear combination of $e_1(x) = 1$, $e_2(x) = 1 - x$, and $e_3(x) = 1 + x + x^2$? We can ask a similar question with vectors. Given the vector $\mathbf{v} = [1, 2]^T$, how can we express it as a linear combination of vectors $\mathbf{u}_1 = [1, 0]^T$ and $\mathbf{u}_2 = [1, 1]^T$? More generally, we have seen that solving a system of linear equations entails expressing the right-hand side as a linear combination of the columns of the matrix coefficients. But when are our building blocks enough to represent anything we want? This leads us to consider concepts such as linear independence and the basis of a linear space. We will do so for simple, finite-dimensional spaces, but the concepts that we introduce are much more general and far-reaching. Indeed, if we consider polynomials of degree only up to n, we are dealing with a finite-dimensional space, but this is not the case when dealing with general functions.[16]

- *Linear mappings.* We have seen that when we multiply a vector $\mathbf{u} \in \mathbb{R}^n$ by a square matrix $\mathbf{A} \in \mathbb{R}^{n,n}$, we get another vector $\mathbf{v} \in \mathbb{R}^n$, and that matrix multiplication can be considered as a mapping f from the space of n-dimensional vectors to itself: $\mathbf{v} = f(\mathbf{u}) = \mathbf{A}\mathbf{u}$. Moreover, this mapping is linear in the sense of Eq. (3.10). Viewing matrix multiplication in this sense helps in viewing the solution of a system of linear equations as a problem of function inversion. If the image of the linear mapping represented by $\mathbf{A}$ is the whole space $\mathbb{R}^n$, then there must be (at least) one vector $\mathbf{x}$ such that $\mathbf{b} = \mathbf{A}\mathbf{x}$. Put another way, the set of columns of matrix $\mathbf{A}$ should be rich enough to express $\mathbf{b}$ as a linear combination of them.

- *Linear spaces.* Consider the space $\mathbb{R}^n$ of n-dimensional vectors. On that space, we have defined elementary operations such as addition and multiplication by a scalar, which allow us to take linear combinations. If we take an arbitrary combination of vectors in that space, we get another vector in $\mathbb{R}^n$. By the same token, we see that a linear combination of polynomials of degree not larger than n yields another such polynomial. A linear space is a set equipped with the operations above, which is closed under linear combinations.[17]

Linear algebra is the study of linear mappings between linear spaces. Before embarking into a study of linear mappings and linear spaces, we consider a motivating example that generalizes the option pricing model of Section 3.1.

3.5.1 Spanning sets and market completeness

Consider a stylized economy with three possible future states of the world, as illustrated in Fig. 3.9. Say that three securities are available and traded on financial markets, with the following state-contingent payoffs:

$$\boldsymbol{\pi}_1 = \begin{bmatrix} 1 \\ 3 \\ 0 \end{bmatrix}, \quad \boldsymbol{\pi}_2 = \begin{bmatrix} 3 \\ 1 \\ 0 \end{bmatrix}, \quad \boldsymbol{\pi}_3 = \begin{bmatrix} 1.2 \\ 1.2 \\ 1.2 \end{bmatrix}$$

These vectors indicate, e.g., that asset 1 has a payoff 1 if state 1 occurs, a payoff 2 if state 2 occurs, and a payoff 0 in the unfortunate case of state 3. We may notice that the third security is risk-free, whereas state three is bad

---

[16]To see this, consider the Taylor expansion of Section 2.10; to approximate a continuous function to an arbitrary accuracy, we should let the degree of the polynomial go to infinity.
[17]A rigorous definition involves a much longer listing of properties (see, e.g., Ref. [4]), but we will just settle for the basic idea.

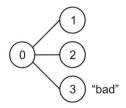

Fig. 3.9 A three-state economy: insuring the "bad" state.

for the holder of the first and the second security. Let us also assume that the three securities have the following prices now: $p_1 = p_2 = p_3 = 1$. What should be the price of an insurance against the occurrence of the bad state, i.e., a security paying off 1 if state 3 occurs, and nothing otherwise?

Following the ideas that we have already applied in binomial option pricing, we can try to replicate the payoff of the insurance by a portfolio consisting of an amount α_i of each security, for $i = 1, 2, 3$. Finding the right portfolio requires finding weights such that the payoff vector of the insurance is a linear combination of payoffs π_i:

$$\alpha_1 \begin{bmatrix} 1 \\ 3 \\ 0 \end{bmatrix} + \alpha_2 \begin{bmatrix} 3 \\ 1 \\ 0 \end{bmatrix} + \alpha_3 \begin{bmatrix} 1.2 \\ 1.2 \\ 1.2 \end{bmatrix} = \begin{bmatrix} 0 \\ 0 \\ 1 \end{bmatrix}$$

This is just a system of three equations in three unknown variables and yields:

$$\alpha_1 = -\tfrac{1}{4}, \quad \alpha_2 = -\tfrac{1}{4}, \quad \alpha_3 = \tfrac{5}{6}$$

Note that we should sell the two risky securities short. But then, invoking the law of one price, in order to avoid arbitrage opportunities, we may find the fair price of the insurance:

$$p_f = \alpha_1 \cdot 1 + \alpha_2 \cdot 1 + \alpha_3 \cdot 1 = \tfrac{1}{3} \approx 0.33333$$

In this case, we are able to price a contingent claim because there is a unique solution to a system of linear equations. In fact, given those three priced assets, we can price by replication any contingent claim, defined on the three future states of the world. In financial economics, it is said that the market is complete.

However, if we enlarge the model of uncertainty to four states, we should not expect to price any contingent claim as we did, using just three "basic" securities. Furthermore, assume that we have three traded securities with the following payoffs:

$$\beta_1 = \begin{bmatrix} 1 \\ 1 \\ 0 \end{bmatrix}, \quad \beta_2 = \begin{bmatrix} 0 \\ 0 \\ 1 \end{bmatrix}, \quad \beta_3 = \begin{bmatrix} 1 \\ 1 \\ 1 \end{bmatrix}$$

We immediately see that $\beta_3 = \beta_1 + \beta_2$. The third security is just a linear combination of the other two and is, in a sense that we will make precise shortly, dependent on them. If we take linear combinations of these three securities, we will not really generate the whole space $\mathbb{R}^3$:

$$\alpha_1 \begin{bmatrix} 1 \\ 1 \\ 0 \end{bmatrix} + \alpha_2 \begin{bmatrix} 0 \\ 0 \\ 1 \end{bmatrix} + \alpha_3 \begin{bmatrix} 1 \\ 1 \\ 1 \end{bmatrix} = \begin{bmatrix} \alpha_1 + \alpha_3 \\ \alpha_1 + \alpha_3 \\ \alpha_2 + \alpha_3 \end{bmatrix}$$

We see that, on the basis of these three assets, we cannot generate a security with two different payoffs in states 1 and 2.

Generalizing a bit, when we take a linear combination of vectors and we set weights in any way that we can, we will *span* a set of vectors; however, the spanned set need not be the whole space $\mathbb{R}^n$.

3.5.2   Linear independence, dimension, and basis of a linear space

The possibility of expressing a vector as a linear combination of other vectors, or lack thereof, plays a role in many settings. In order to do so, we must ensure that the set of vectors that we want to use as a building blocks is "rich enough." If we are given a set of vectors $\mathbf{v}_1, \mathbf{v}_2, \ldots, \mathbf{v}_m \in \mathbb{R}^n$, and we take linear combinations of them, we generate a linear subspace of vectors. We speak of a subspace, because it is in general a subset of $\mathbb{R}^n$; it is linear in the sense that any linear combination of vectors included in the subspace belongs to the subspace as well. We say that the set of vectors is a spanning set for the subspace. However, we would like to have a spanning set that is "minimal," in the sense that all of the vectors are really needed and there is no redundant vector. It should be clear that if one of the vectors in the spanning set is a linear combination of the others, we can get rid of it without changing the spanning set.

Example 3.9 Consider the unit vectors in $\mathbb{R}^2$:

$$\mathbf{e}_1 = \begin{bmatrix} 1 \\ 0 \end{bmatrix}, \qquad \mathbf{e}_2 = \begin{bmatrix} 0 \\ 1 \end{bmatrix}$$

It is easy to see that we may span the whole space $\mathbb{R}^2$ by taking linear combinations of $\mathbf{e}_1$ and $\mathbf{e}_2$. If we also consider vector $\mathbf{e}_3 = [1, 1]^T$, we get another spanning set, but we do not change the spanned set, as the new vector is redundant:

$$\mathbf{e}_3 = \mathbf{e}_1 + \mathbf{e}_2$$

This can be rewritten as follows:

$$\mathbf{e}_1 + \mathbf{e}_2 - \mathbf{e}_3 = \mathbf{0}$$

□

The previous example motivates the following definition.

DEFINITION 3.3 (Linear dependence) *Vectors* $\mathbf{v}_1, \mathbf{v}_2, \ldots, \mathbf{v}_k \in \mathbb{R}^n$ *are linearly dependent if and only if there exist scalars* $\alpha_1, \alpha_2, \ldots, \alpha_k$, *not all zero, such that*

$$\alpha_1 \mathbf{v}_1 + \alpha_2 \mathbf{v}_2 + \cdots + \alpha_k \mathbf{v}_k = \mathbf{0}$$

Indeed, vectors $\mathbf{e}_1$, $\mathbf{e}_2$, and $\mathbf{e}_3$ in the example above are linearly dependent. However, vectors $\mathbf{e}_1$ and $\mathbf{e}_2$ are clearly not redundant, as there is no way to express one of them as a linear combination of the other one.

DEFINITION 3.4 (Linear independence) *Vectors* $\mathbf{v}_1, \mathbf{v}_2, \ldots, \mathbf{v}_k \in \mathbb{R}^n$ *are linearly independent if and only if*

$$\alpha_1 \mathbf{v}_1 + \alpha_2 \mathbf{v}_2 + \cdots + \alpha_k \mathbf{v}_k = \mathbf{0}$$

implies $\alpha_1 = \alpha_2 = \cdots = \alpha_k = 0.$

Consider now the space $\mathbb{R}^n$. How many vectors should we include in a spanning set for $\mathbb{R}^n$? Intuition suggests that

- The spanning set should have at least n vectors.

- If we include more than n vectors in the spanning set, some of them will be redundant.

This motivates the following definition.

DEFINITION 3.5 (Basis of a linear space) *Let* $\mathbf{v}_1, \mathbf{v}_2, \ldots, \mathbf{v}_k$ *be a collection of vectors in a linear space* V. *These vectors form a basis for* V *if*

1. $\mathbf{v}_1, \mathbf{v}_2, \ldots, \mathbf{v}_k$ *span* V.

2. $\mathbf{v}_1, \mathbf{v}_2, \ldots, \mathbf{v}_k$ *are linearly independent.*

Example 3.10 Referring again to Example 3.9, the set $\{\mathbf{e}_1, \mathbf{e}_2, \mathbf{e}_3\}$ is a spanning set for $\mathbb{R}^2$, but it is not a basis. To get a basis, we should get rid of one of those vectors. Note that the basis is not unique, as any of these sets does span $\mathbb{R}^2$: $\{\mathbf{e}_1, \mathbf{e}_2\}$, $\{\mathbf{e}_1, \mathbf{e}_3\}$, $\{\mathbf{e}_2, \mathbf{e}_3\}$. These sets have all dimension 2.

We cannot span $\mathbb{R}^2$ with a smaller set, consisting of just one vector. However, let us consider vectors on the line $y = x$. This line consists of the set of vectors $\mathbf{w}$ of the form

$$\mathbf{w} = \begin{bmatrix} x \\ x \end{bmatrix}$$

This set W is a linear subspace of $\mathbb{R}^2$ in the sense that

$$\mathbf{w} \in W \quad \Rightarrow \quad \alpha \mathbf{w} \in W, \qquad \forall \alpha \in \mathbb{R}$$

Vector $\mathbf{e}_3$ is one possible basis for subspace W. ☐

The last example suggests that there are many possible bases for a linear space, but they have something in common: They consist of the same number of vectors. In fact, this number is the *dimension* of the linear space. We also see that a space like $\mathbb{R}^n$ may include a smaller linear subspace, with a basis smaller than n.

As a further example, we may consider the plane xy as a subspace of $\mathbb{R}^3$. This plane consists of vectors of the form

$$\begin{bmatrix} x \\ y \\ 0 \end{bmatrix}$$

and the most natural basis for it consists of vectors

$$\mathbf{e}_1 = \begin{bmatrix} 1 \\ 0 \\ 0 \end{bmatrix}, \qquad \mathbf{e}_2 = \begin{bmatrix} 0 \\ 1 \\ 0 \end{bmatrix}$$

Such unit vectors yield natural bases for many subspaces, but they need not be the only (or the best) choice.

We close this section by observing that if we have a basis, and a vector is represented using that basis, the representation is unique.

3.5.3   Matrix rank

In this section we explore the link between a basis of a linear space and the possibility of finding a unique solution of a system of linear equations $\mathbf{Ax} = \mathbf{b}$, where $\mathbf{A} \in \mathbb{R}^{m,n}$, $\mathbf{x} \in \mathbb{R}^n$, and $\mathbf{b} \in \mathbb{R}^m$. Here, n is the number of variables and m is the number of equations; in most cases, we have $m = n$, but we may try to generalize a bit.

Recall that our system of equations can be rephrased as

$$\mathbf{a}_1 x_1 + \mathbf{a}_2 x_2 + \cdots + \mathbf{a}_n x_n = \mathbf{b}$$

where $\mathbf{a}_j$, $j = 1, \ldots, n$, are the columns of $\mathbf{A}$. We can find a solution only if $\mathbf{b}$ is in the linear subspace spanned by the columns of $\mathbf{A}$. Please note that, in order to find a solution, we are not necessarily requiring that the columns of $\mathbf{A}$ be linearly independent. This cannot happen if, for instance, we have $m = 3$ equations and $n = 5$ variables, because in a space of dimension 3 we cannot find five linearly independent vectors; indeed, the solution need not be unique in such a case. On the other hand, if we have $m = 5$ equations and $n = 3$ variables, we will not find a solution in general, unless we are lucky and vector $\mathbf{b}$ lies in the subspace (of dimension ≤ 3) spanned by those three columns.

If there is a solution, when is it unique? By now, the reader should not be surprised to learn that this happens when the columns of the matrix are linearly independent, as this imply that there is at most one way to express **b** using them.

If we consider a square system, i.e., $m = n$, we are sure to find a unique solution of the system if the columns of the matrix are linearly independent. In such a case, the n columns form a basis for $\mathbb{R}^n$ and we are able to solve the system for any right-hand side **b**.

Given this discussion, it is no surprise that the number of linearly independent columns of a matrix is an important feature.

DEFINITION 3.6 (Column–row rank of a matrix) *Given a matrix* $\mathbf{A} \in \mathbb{R}^{m,n}$, *its column and row ranks are the number of linearly independent columns and the number of linearly independent rows of* **A**, *respectively.*

A somewhat surprising fact is that the row and the column rank of a matrix are the same. Hence, we may simply speak of the rank of a matrix $\mathbf{A} \in \mathbb{R}^{m,n}$, which is denote by rank(A). It is also obvious that rank(A) $\leq \min\{m, n\}$. When we have rank(A) = $\min\{m, n\}$, we say that the matrix **A** is full-rank. In the case of a square matrix $\mathbf{A} \in \mathbb{R}^{n,n}$, the matrix has full rank if there are n linearly independent columns (and rows as well).

Example 3.11 The matrix

$$\mathbf{A} = \begin{bmatrix} 1 & 1 & 5 \\ 2 & 0 & 6 \\ 1 & 2 & 7 \end{bmatrix}$$

is not full rank, as it is easy to check that the third column is a linear combination of the first two columns: $\mathbf{a}_3 = 3\mathbf{a}_1 + 2\mathbf{a}_2$. Since there are only two linearly independent columns, its rank is 2. The ranks is exactly the same if we look at rows. In fact, the third row can be obtained by adding twice the first row and minus half the second row.

The matrix

$$\mathbf{B} = \begin{bmatrix} 1 & 1 & 4 \\ 1 & 0 & 3 \end{bmatrix}$$

has rank 2, since it is easy to see that the rows are linearly independent. We say that this matrix has full row rank, as its rank is the maximum that it could achieve. ▯

We have said that a square system of linear equations $\mathbf{Ax} = \mathbf{b}$ essentially calls for the expression of the right-hand side vector **b** as a linear combination of the columns of **A**. If these n columns are linearly independent, then they are a basis for $\mathbb{R}^n$ and the system can be solved for any vector **b**. We have also noted that the solution can be expressed in terms of the inverse matrix, $\mathbf{x} = \mathbf{A}^{-1}\mathbf{b}$, if it exists. Indeed, the following theorem links invertibility and rank of a square matrix.

THEOREM 3.7 *A square matrix* $\mathbf{A} \in \mathbb{R}^{n,n}$ *is* **invertible** *(nonsingular) if and only if* $\text{rank}(\mathbf{A}) = n$. *If the matrix is not full-rank, it is not invertible and is called* **singular**.

The rank of a matrix is a powerful concept, but so far we have not really found a systematic way of evaluating it. Actually, Gaussian elimination would do the trick, but what we really need is a handy way to check whether a square matrix has full rank. The determinant provides us with the answers we need.

3.6   DETERMINANT

The determinant of a square matrix is a function mapping square matrices into real numbers, and it is an important theoretical tool in linear algebra. Actually, it was investigated before the introduction of the matrix concept. In Section 3.2.3 we have seen that determinants can be used to solve systems of linear equations by Cramer's rule. Another use of the determinant is to check whether a matrix is invertible. However, the use of determinants quickly becomes cumbersome as their calculation involves a number of operations that increases very rapidly with the dimension of the matrix. Hence, the determinant is mainly a conceptual tool.

We may define the determinant inductively as follows, starting from a two-dimensional square matrix:

$$\det(\mathbf{A}) = \begin{vmatrix} a_{11} & a_{12} \\ a_{21} & a_{22} \end{vmatrix} = a_{11}a_{22} - a_{21}a_{12}$$

We already know that for a three-dimensional matrix, we may define the determinant by selecting a row and multiplying its elements by the determinant of the submatrix obtained by eliminating the row and the column containing that element:

$$\begin{vmatrix} a_{11} & a_{12} & a_{13} \\ a_{21} & a_{22} & a_{23} \\ a_{31} & a_{32} & a_{33} \end{vmatrix} = a_{11} \begin{vmatrix} a_{22} & a_{23} \\ a_{32} & a_{33} \end{vmatrix} - a_{12} \begin{vmatrix} a_{21} & a_{23} \\ a_{31} & a_{33} \end{vmatrix}$$

$$+ a_{13} \begin{vmatrix} a_{21} & a_{22} \\ a_{31} & a_{32} \end{vmatrix} \tag{3.11}$$

The idea can be generalized to an arbitrary square matrix $\mathbf{A} \in \mathbb{R}^n$ as follows:

1. Denote by $\mathbf{A}_{ij}$ the matrix obtained by deleting row i and column j of $\mathbf{A}$.

2. The determinant of this matrix, denoted by $M_{ij} \equiv \det(\mathbf{A}_{ij})$, is called the (i,j)th *minor* of matrix $\mathbf{A}$.

3. We also define the (i,j)th *cofactor* of matrix $\mathbf{A}$ as $C_{ij} \equiv (-1)^{i+j} M_{ij}$. We immediately see that a cofactor is just a minor with a sign. The sign is positive if $i + j$ is even; it is negative if $i + j$ is odd.

Armed with these definitions, we have two ways to compute a determinant:

- *Expansion by row.* Pick any row k, $k = 1, \ldots, n$, and compute

$$\det(\mathbf{A}) = \sum_{j=1}^{n} a_{kj} C_{kj}$$

- *Expansion by column.* Pick any column k, $k = 1, \ldots, n$, and compute

$$\det(\mathbf{A}) = \sum_{i=1}^{n} a_{ik} C_{ik}$$

The process can be executed recursively, and it results ultimately in the calculation of many 2×2 determinants. Indeed, for a large matrix, the calculation is quite tedious. The reader is invited to check that Eq. (3.11) is just an expansion along row 1.

3.6.1 Determinant and matrix inversion

From a formal perspective, we may use matrix inversion to solve a system of linear equations:

$$\mathbf{A}\mathbf{x} = \mathbf{b} \qquad \Rightarrow \qquad \mathbf{x} = \mathbf{A}^{-1}\mathbf{b}$$

From a practical viewpoint, this is hardly advisable, as Gaussian elimination entails much less work. To see why, observe that one can find each column $\mathbf{a}_j^*$ of the inverse matrix by solving the following system of linear equations:

$$\mathbf{A}\mathbf{a}_j^* = \mathbf{e}_j$$

Here, vector $\mathbf{e}_j$ is a vector whose elements are all zero, with a 1 in position j: $\mathbf{e}_j = [0, 0, \ldots 0, 1, 0, \ldots, 0]^T$. In other words, we should apply Gaussian elimination n times to find the inverse matrix.

There is another way to compute the inverse of a matrix, based on the cofactors we have defined above.

THEOREM 3.8 *Let $\tilde{\mathbf{A}}$ be a matrix whose element (i, j) is the (j, i)th cofactor C_{ji} of an invertible matrix $\mathbf{A}$: $\tilde{a}_{ij} = C_{ji}$. Then*

$$\mathbf{A}^{-1} = \frac{1}{\det(\mathbf{A})} \cdot \tilde{\mathbf{A}}$$

The matrix $\tilde{\mathbf{A}}$ is called the *adjoint* of $\mathbf{A}$. In fact, Cramer's rule (3.6) is a consequence of this theorem. Still, computing the inverse of a matrix is a painful process, and we may see why inverse matrices are not computed explictly quite often. Nevertheless, the inverse matrix is conceptually relevant, and we should wonder if we may characterize invertible matrices in some useful manner.

THEOREM 3.9 *A square matrix is invertible (nonsingular) if and only if* $\det(\mathbf{A}) \neq 0$.

This should not come as a surprise, considering that when computing an inverse matrix by the adjoint matrix or when solving a system of linear equations by Cramer's rule, we divide by $\det(\mathbf{A})$.

Example 3.12 Using Theorem 3.8, we may prove the validity of a handy rule to invert a 2×2 matrix:

$$\mathbf{A} = \begin{bmatrix} a & b \\ c & d \end{bmatrix} \quad \Rightarrow \quad \mathbf{A}^{-1} = \frac{1}{ad - bc} \begin{bmatrix} d & -b \\ -c & a \end{bmatrix}$$

In plain terms, to invert a bidimensional matrix, we must

1. Swap the two elements on the diagonal.

2. Change the sign of the other two elements.

3. Divide by the determinant.

⬚

3.6.2   Properties of the determinant

The determinant enjoys a lot of useful properties that we list here without any proof.

- For any square matrix $\mathbf{A}$, $\det(\mathbf{A}^T) = \det(\mathbf{A})$.

- If two rows (or columns) of $\mathbf{A}$ are equal, then $\det(\mathbf{A}) = 0$.

- If matrix $\mathbf{A}$ has an all-zero row (or column), then $\det(\mathbf{A}) = 0$.

- If we multiply the entries of a row (or column) in matrix $\mathbf{A}$ by a scalar α to obtain matrix $\mathbf{B}$, then $\det(\mathbf{B}) = \alpha \det(\mathbf{A})$.

- The determinant of a lower triangular, upper triangular, or diagonal matrix is the product of entries on the diagonal.

- The determinant of the product of two conformable matrices $\mathbf{A}$ and $\mathbf{B}$ is the product of the determinants: $\det(\mathbf{AB}) = \det(\mathbf{A}) \det(\mathbf{B})$.

- If $\mathbf{A}$ is invertible, then $\det(\mathbf{A}^{-1}) = 1/\det(\mathbf{A})$.

3.7   EIGENVALUES AND EIGENVECTORS

In Section 3.4.3 we observed that a square matrix $\mathbf{A} \in \mathbb{R}^{n,n}$ is a way to represent a linear mapping from the space of n-dimensional vectors to itself.

Such a transformation, in general, entails both a rotation and a change of vector length. If the matrix is orthogonal, then the mapping is just a rotation. It may happen, for a specific vector $\mathbf{v}$ and a scalar λ, that

$$\mathbf{A}\mathbf{v} = \lambda\mathbf{v}$$

Then, if $\lambda > 0$ we have just a change in the length of $\mathbf{v}$; if $\lambda < 0$ we also have a reflection of the vector. In such a case we say

1. That λ is an *eigenvalue* of the matrix $\mathbf{A}$

2. That $\mathbf{v}$ is an *eigenvector* of the matrix $\mathbf{A}$

It is easy to see that if $\mathbf{v}$ is an eigenvector, then any vector obtained by multiplication by a scalar $\alpha \in \mathbb{R}$, i.e., any vector of form $\alpha\mathbf{v}$, is an eigenvector, too. If we divide a generic eigenvector by its norm, we get a *unit eigenvector*:

$$\mathbf{u} = \frac{\mathbf{v}}{\|\mathbf{v}\|}$$

Example 3.13 Consider

$$\mathbf{A} = \begin{bmatrix} -1 & 3 \\ 2 & 0 \end{bmatrix}, \qquad \mathbf{v} = \begin{bmatrix} 1 \\ 1 \end{bmatrix} \tag{3.12}$$

Then we have

$$\mathbf{A}\mathbf{v} = \begin{bmatrix} 2 \\ 2 \end{bmatrix} = 2\mathbf{v}$$

Hence, $\lambda = 2$ is an eigenvalue of $\mathbf{A}$, and $\mathbf{v}$ is a corresponding eigenvector. Actually, any vector of the form $[\alpha, \alpha]^T$ is an eigenvector of $\mathbf{A}$ corresponding to the eigenvalue $\lambda = 2$. The eigenvector

$$\mathbf{u} = \frac{\mathbf{v}}{\|\mathbf{v}\|} = \frac{1}{\sqrt{1^2 + 1^2}}\begin{bmatrix} 1 \\ 1 \end{bmatrix} = \begin{bmatrix} \frac{1}{\sqrt{2}} \\ \frac{1}{\sqrt{2}} \end{bmatrix}$$

is a unit eigenvector of $\mathbf{A}$. □

Matrix eigenvalues are an extremely useful tool with applications in mathematics, statistics, and physics that go well beyond the scope of this book. As far as we are concerned, the following points will be illustrated in this and later chapters:

- Eigenvalues may be used to investigate convexity and concavity of a function.

- They are relevant in optimization applications.

- They are useful in multivariate statistical methods, like principal component analysis, which have a lot of applications, e.g., in marketing and quantitative finance.

Now we should wonder whether there is a way to compute eigenvalues systematically. In practice, there are powerful numerical methods to do so, but we will stick to the most natural idea, which illustrates a lot of points about eigenvalues. Note that if λ is an eigenvalue and $\mathbf{v}$ an eigenvector, we have

$$\mathbf{A}\mathbf{v} = \lambda\mathbf{v} \quad \rightarrow \quad (\mathbf{A} - \lambda\mathbf{I})\mathbf{v} = \mathbf{0} \tag{3.13}$$

This means that we may express the zero vector by taking a linear combination of the columns of matrix $\mathbf{A} - \lambda\mathbf{I}$. A trivial solution of this system is $\mathbf{v} = \mathbf{0}$. But a nontrivial solution can be found if and only if the columns of that matrix are not linearly independent. This, in turn, is equivalent to saying that the determinant is zero. Hence, to find the eigenvalues of a matrix, we should solve the following equation:

$$\det(\mathbf{A} - \lambda\mathbf{I}) = 0 \tag{3.14}$$

This equation is called *characteristic equation* of the matrix, as eigenvalues capture the essential nature of a matrix.

Example 3.14 Let us apply Eq. (3.14) to the matrix $\mathbf{A}$ in Eq. (3.12):

$$\det(\mathbf{A} - \lambda\mathbf{I}) = \begin{vmatrix} -1 - \lambda & 3 \\ 2 & -\lambda \end{vmatrix} = \lambda^2 + \lambda - 6 = 0$$

This second-order equation has solutions $\lambda_1 = 2$ and $\lambda_2 = -3$. These numbers are the eigenvalues of matrix $\mathbf{A}$. To find the eigenvectors corresponding to an eigenvalue, just plug an eigenvalue (say, 2) into the equation $(\mathbf{A} - \lambda\mathbf{I})\mathbf{v} = \mathbf{0}$:

$$(\mathbf{A} - 2\mathbf{I})\mathbf{v} = \begin{bmatrix} -3 & 3 \\ 2 & -2 \end{bmatrix} \begin{bmatrix} v_1 \\ v_2 \end{bmatrix} = \begin{bmatrix} 0 \\ 0 \end{bmatrix}$$

These two equations are redundant (the matrix is singular), and by taking either one, we can show that any vector $\mathbf{v}$ such that $v_1 = v_2$ is a solution of the system, i.e., an eigenvector associated with the eigenvalue $\lambda = 2$.   ☐

Can we say something about the number of eigenvalues of a matrix? This is not an easy question, but we can say that a $n \times n$ matrix can have up to n distinct eigenvalues. To see why, observe that the characteristic equation involves a polynomial of degree n, which may have up to n distinct roots. In general, we can state what follows:

- Eigenvalues may be complex conjugates, rather than real numbers; for instance, consider the following matrix:

$$\mathbf{B} = \begin{bmatrix} 0 & -1 \\ 1 & 0 \end{bmatrix}$$

The characteristic polynomial is

$$\det(\mathbf{B} - \lambda\mathbf{I}) = \lambda^2 + 1$$

Hence, the two eigenvalues are $\lambda = \pm i$, where i is the unit imaginary number defined by $i^2 = -1$. We do not find real eigenvalues, but this is not surprising as matrix $\mathbf{B}$ is a matrix rotating a vector by $\pi/2$, i.e., $90°$ on the plane. (Please check this graphically!)

- Eigenvalues may be multiple roots of the characteristic polynomial, as in the case $\lambda^2 + 2\lambda + 1 = (\lambda + 1)^2 = 0$. When there are multiple eigenvalues, finding eigenvalues may be tricky, but we can do without these technicalities.

The following properties of the eigenvalues of a matrix $\mathbf{A}$ are worth mentioning here:

- The determinant of the matrix is the product of the eigenvalues: $\det(\mathbf{A}) = \Pi_{k=1}^n \lambda_k$.

- The trace of the matrix, which is just the sum of its entries on the diagonal, is the sum of the eigenvalues: $\operatorname{tr}(\mathbf{A}) \equiv \sum_{i=1}^n a_{ii} = \sum_{k=1}^n \lambda_k$.

The first property has an interesting consequence: The matrix $\mathbf{A}$ is singular (hence, not invertible) when one of its eigenvalues is zero.

3.7.1   Eigenvalues and eigenvectors of a symmetric matrix

In applications, it is often the case that the matrix $\mathbf{A}$ is symmetric.[18]   A symmetric matrix has an important property.

THEOREM 3.10 *Let $\mathbf{A} \in \mathbb{R}^n$ be a symmetric matrix. Then*

1. *The matrix has only real eigenvalues.*

2. *The eigenvectors are mutually orthogonal.*

The second point in the theorem implies that if we form a matrix $\mathbf{P}$ using normalized (unit) eigenvectors

$$\mathbf{P} = \begin{bmatrix} \mathbf{v}_1 & \mathbf{v}_2 & \cdots & \mathbf{v}_n \end{bmatrix}$$

---

[18]In probability and statistics, we use covariance matrices, which are symmetric; see Section 8.3. In optimization, symmetric matrices are used to define a quadratic form; see Section 3.9 on multidimensional calculus.

this matrix is orthogonal, i.e., $\mathbf{P}^T\mathbf{P} = \mathbf{I}$ and $\mathbf{P}^{-1} = \mathbf{P}^T$. This allows us to build a very useful factorization of a symmetric matrix $\mathbf{A}$:

$$\begin{aligned} \mathbf{A} &= \mathbf{A}\mathbf{P}\mathbf{P}^T = \mathbf{A} \begin{bmatrix} \mathbf{v}_1 & \mathbf{v}_2 & \cdots & \mathbf{v}_n \end{bmatrix} \mathbf{P}^T \\ &= \begin{bmatrix} \mathbf{A}\mathbf{v}_1 & \mathbf{A}\mathbf{v}_2 & \cdots & \mathbf{A}\mathbf{v}_n \end{bmatrix} \mathbf{P}^T \\ &= \begin{bmatrix} \lambda_1\mathbf{v}_1 & \lambda_2\mathbf{v}_2 & \cdots & \lambda_n\mathbf{v}_n \end{bmatrix} \mathbf{P}^T \\ &= \mathbf{P} \begin{bmatrix} \lambda_1 & 0 & \cdots & 0 \\ 0 & \lambda_2 & \cdots & 0 \\ \vdots & \vdots & \ddots & \vdots \\ 0 & 0 & \cdots & \lambda_n \end{bmatrix} \mathbf{P}^T \end{aligned}$$

If we denote by $\mathbf{D}$ the diagonal matrix consisting of the n eigenvalues, we see that we may factor matrix $\mathbf{A}$ as follows:

$$\mathbf{A} = \mathbf{P}\mathbf{D}\mathbf{P}^T \tag{3.15}$$

Going the other way around, we may "diagonalize" matrix $\mathbf{A}$ as follows:

$$\mathbf{P}^T\mathbf{A}\mathbf{P} = \mathbf{D} \tag{3.16}$$

Example 3.15 Consider the following matrix:

$$\mathbf{A} = \begin{bmatrix} 2 & 1 & 0 \\ 1 & 2 & 0 \\ 0 & 0 & 3 \end{bmatrix}$$

Its characteristic polynomial can be obtained by developing the determinant along the last row:

$$\det(\mathbf{A} - \lambda\mathbf{I}) = \begin{vmatrix} 2-\lambda & 1 & 0 \\ 1 & 2-\lambda & 0 \\ 0 & 0 & 3-\lambda \end{vmatrix} = (3-\lambda)\left[(2-\lambda)^2 - 1\right]$$

We immediately see that its roots are

$$\lambda_1 = 3, \qquad \lambda_2 = 3, \qquad \lambda_3 = 1$$

They are three real eigenvalues, as expected, but one is a multiple eigenvalue. To find the eigenvectors, let us plug $\lambda = 3$ into Eq. (3.13):

$$\begin{bmatrix} -1 & 1 & 0 \\ 1 & -1 & 0 \\ 0 & 0 & 0 \end{bmatrix} \begin{bmatrix} v_1 \\ v_2 \\ v_3 \end{bmatrix} = \begin{bmatrix} 0 \\ 0 \\ 0 \end{bmatrix}$$

From the last row of the matrix, we see that v_3 can be chosen freely. Furthermore, we see that the first and the second equation are not linearly independent. In fact, the rank of matrix $\mathbf{A}$ is 1, and we may find two linearly independent eigenvectors of the form

$$\mathbf{v}_1 = [\alpha, \alpha, 0]^T, \qquad \mathbf{v}_2 = [0, 0, \beta]^T$$

Let us choose

$$\mathbf{v}_1 = \left[\frac{1}{\sqrt{2}}, \frac{1}{\sqrt{2}}, 0 \right]^T, \qquad \mathbf{v}_2 = [0, 0, 1]^T$$

in order to have two orthogonal and unit eigenvectors. We leave as an exercise for the reader to verify that, if we plug the eigenvalue $\lambda = 1$, we obtain eigenvectors of the form

$$\mathbf{v}_3 = [\alpha, -\alpha, 0]^T$$

Now, we may check that

$$\mathbf{A} = \mathbf{PDP}^T$$

where

$$\mathbf{P} = \begin{bmatrix} \dfrac{1}{\sqrt{2}} & \dfrac{1}{\sqrt{2}} & 0 \\[2mm] -\dfrac{1}{\sqrt{2}} & \dfrac{1}{\sqrt{2}} & 0 \\[2mm] 0 & 0 & 1 \end{bmatrix}, \qquad \mathbf{D} = \begin{bmatrix} 1 & 0 & 0 \\ 0 & 3 & 0 \\ 0 & 0 & 3 \end{bmatrix}$$

□

Luckily, there are efficient and robust numerical methods for finding eigenvalues and eigenvectors, as well as for diagonalizing a matrix. These methods are implemented by widely available software tools. What is more relevant to us is the remarkable number of ways in which we may take advantage of the results above. As an example, the computation of powers of a symmetric matrix, $\mathbf{A}^k$, is considerably simplified:

$$\mathbf{A}^k = \left(\mathbf{PDP}^T \right) \left(\mathbf{PDP}^T \right) \cdots \left(\mathbf{PDP}^T \right) = \mathbf{PDD} \cdots \mathbf{DP}^T = \mathbf{PD}^k \mathbf{P}^T$$

where

$$\mathbf{D}^k = \begin{bmatrix} \lambda_1^k & 0 & \cdots & 0 \\ 0 & \lambda_2^k & \cdots & 0 \\ \vdots & \vdots & \ddots & \vdots \\ 0 & 0 & \cdots & \lambda_n^k \end{bmatrix}$$

By the same token, we may compute the square root of a matrix, i.e., a matrix $\mathbf{A}^{1/2}$ such that $\mathbf{A} = \mathbf{A}^{1/2} \mathbf{A}^{1/2}$

$$\mathbf{A}^{1/2} = \mathbf{PD}^{1/2} \mathbf{P}^T$$

where

$$\mathbf{D}^{\frac{1}{2}} = \begin{bmatrix} \sqrt{\lambda_1} & 0 & \cdots & 0 \\ 0 & \sqrt{\lambda_2} & \cdots & 0 \\ \vdots & \vdots & \ddots & \vdots \\ 0 & 0 & \cdots & \sqrt{\lambda_n} \end{bmatrix}$$

Matrix inversion is easy, too:

$$\mathbf{A}^{-1} = \mathbf{PD}^{-1} \mathbf{P}^T$$

where $\mathbf{D}^{-1} = \mathrm{diag}(\lambda_1^{-1}, \lambda_2^{-1}, \ldots, \lambda_n^{-1},)$. Here we have used a shorthand notation to denote a diagonal matrix by just giving its diagonal elements.

3.8 QUADRATIC FORMS

In the last two sections of this chapter, we explore the connections between linear algebra and calculus. This is necessary in order to generalize calculus concepts to functions of several variables; since any interesting management problem involves multiple dimensions, this is a worthy task. The simplest nonlinear function of multiple variables is arguably a *quadratic form*:

$$Q(x_1, x_2, \ldots, x_n) = \sum_{i \leq j} a_{ij} x_i x_j \tag{3.17}$$

Denoting the double sum as $\sum_{i \leq j}$ is typically preferred to $\sum_{i=1}^n \sum_{j=1}^n$, as the latter is a bit ambiguous. To see the issue, consider a term such as $4x_1 x_2$; we could rewrite this as $3x_1 x_2 + x_2 x_1$ or $2x_1 x_2 + 2x_2 x_1$. The notation above results in a unique way of expressing a quadratic form. Concrete examples of quadratic forms are

$$f(x_1, x_2) = x_1^2 + x_2^2, \qquad g(x_1, x_2) = x_1^2 - 2x_1 x_2 - x_2^2 \tag{3.18}$$

Quadratic forms are important for several reasons:

1. They play a role in the approximation of multivariable functions using Taylor expansions; in Section 3.9 we will see how we can generalize the Taylor expansion of Section 2.10.

2. Quadratic forms are also important in nonlinear optimization (see Section 12.5).

3. In probability and statistics, the variance of a linear combination of random variables is linked to a quadratic form (Section 8.3); this is fundamental, among many other things, in financial portfolio management.

Quadratic forms are strongly linked to linear algebra as they can be most conveniently associated with a symmetric matrix. In fact, we may also express any quadratic form as

$$q(\mathbf{x}) = \mathbf{x}^T \mathbf{A} \mathbf{x} \tag{3.19}$$

where matrix $\mathbf{A}$ is symmetric. For instance, we can express the quadratic forms in Eq. (3.18) as follows:

$$f(x_1, x_2) = [x_1 \; x_2] \begin{bmatrix} 1 & 0 \\ 0 & 1 \end{bmatrix} \begin{bmatrix} x_1 \\ x_2 \end{bmatrix} = x_1^2 + x_2^2$$

$$g(x_1, x_2) = [x_1 \; x_2] \begin{bmatrix} 1 & -1 \\ -1 & -1 \end{bmatrix} \begin{bmatrix} x_1 \\ x_2 \end{bmatrix} = x_1^2 - 2x_1 x_2 - x_2^2$$

Note that cross-product terms correspond to off-diagonal entries in the matrix; in fact, the a_{ij} coefficients in expression (3.17), for $i \neq j$, occur divided by 2 in the matrix. These two quadratic forms are plotted in Figs. 3.10 and

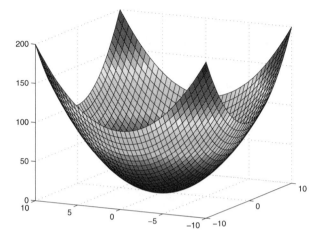

Fig. 3.10   A convex quadratic form.

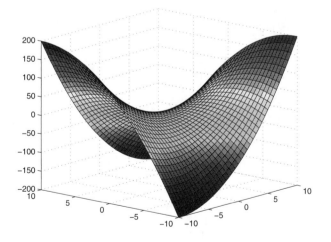

Fig. 3.11   An undefined quadratic form.

3.11, respectively. From what we know about convexity (Section 2.11), we immediately see that f is convex, whereas g is neither convex nor concave, as it features a "saddle point." Convexity and concavity of a quadratic form are linked to properties of the corresponding matrix.

DEFINITION 3.11 (Definiteness of quadratic forms) *We say that a quadratic form* $\mathbf{x}^T \mathbf{A} \mathbf{x}$

- *Is positive definite if* $\mathbf{x}^T \mathbf{A} \mathbf{x} > 0$ *for all* $\mathbf{x} \neq \mathbf{0}$

- *Is positive semidefinite if* $\mathbf{x}^T \mathbf{A} \mathbf{x} \geq 0$ *for all* $\mathbf{x}$

- *Is negative definite if* $\mathbf{x}^T \mathbf{A} \mathbf{x} < 0$ *for all* $\mathbf{x} \neq \mathbf{0}$

- *Is negative semidefinite if* $\mathbf{x}^T \mathbf{A} \mathbf{x} \leq 0$ *for all* $\mathbf{x}$

Otherwise, we say that the quadratic form is indefinite.

The definiteness of a quadratic form is strictly related to convexity by the following properties:

- A positive definite quadratic form is a strictly convex function.

- A positive semidefinite quadratic form is a convex function.

- A negative definite quadratic form is a strictly concave function.

- A negative semidefinite quadratic form is a concave function.

Incidentally, we may observe that for a positive definite quadratic form, the origin $\mathbf{x}^* = \mathbf{0}$ is a global minimizer, as it is the only point at which the function is zero, whereas it is strictly positive for any $\mathbf{x} \neq \mathbf{0}$. By the same token, $\mathbf{x}^* = \mathbf{0}$ is a global maximizer for a negative definite quadratic form. For semidefinite forms, there may be alternative optima, i.e., several vectors $\mathbf{x}$ for which the function is zero.

Now the main problem is to find a way to check the definiteness of a quadratic form. A good starting point is to observe that this task is easy to accomplish for a diagonal matrix:

$$
\mathbf{D} = \begin{bmatrix} \lambda_1 & 0 & \cdots & 0 \\ 0 & \lambda_2 & \cdots & 0 \\ \vdots & \vdots & \ddots & \vdots \\ 0 & 0 & \cdots & \lambda_n \end{bmatrix}
$$

In such a case, the quadratic form does not include cross-products:

$$
\mathbf{x}^T \mathbf{D} \mathbf{x} = \sum_{i=1}^{n} \lambda_i x_i^2
$$

It is easy to see that definiteness of this quadratic form depends on the signs of the diagonal terms λ_i. If they are all strictly positive (negative), the form is positive (negative) definite. If they are nonnegative (nonpositive), the form is positive (negative) semidefinite. But for a diagonal matrix, the entries on the diagonal are just its eigenvalues. Since the matrix associated with a quadratic form is symmetric, we also know that all of its eigenvalues are real numbers. Can we generalize and guess that definiteness depends on the sign of the eigenvalues of the corresponding symmetric matrix? Indeed, this is easily verified by factorizing the quadratic form using Eq. (3.15)

$$\mathbf{x}^T \mathbf{A} \mathbf{x} = \mathbf{x}^T \mathbf{P} \mathbf{D} \mathbf{P}^T \mathbf{x} = \mathbf{y}^T \mathbf{D} \mathbf{y}$$

where $\mathbf{y} = \mathbf{P}^T \mathbf{x}$. This is just another quadratic form, which has been diagonalized by a proper change of variables. We may immediately conclude that a quadratic form

- Is positive definite if and only if the eigenvalues of the corresponding matrix $\mathbf{A}$ are all strictly positive (> 0)

- Is positive semidefinite if and only if the eigenvalues of the corresponding matrix $\mathbf{A}$ are all nonnegative (≥ 0)

- Is negative definite if and only if the eigenvalues of the corresponding matrix $\mathbf{A}$ are all strictly negative (< 0)

- Is negative semidefinite, if and only if the eigenvalues of the corresponding matrix $\mathbf{A}$ are all nonpositive (≤ 0)

3.9   CALCULUS IN MULTIPLE DIMENSIONS

In this section we extend some concepts that we introduced in the previous chapter, concerning calculus for functions of one variable. What we really need for what follows is to get an intuitive idea of how some basic concepts are generalized when we consider a function of multiple variables, i.e., a function $f(x_1, x_2, \ldots, x_n) = f(\mathbf{x})$ mapping a vector in $\mathbb{R}^n$ to a real number. In particular, we would like to see

1. How we can extend the concept of derivative

2. How we can extend the Taylor's expansion

3. What is an integral in multiple dimensions

The first two issues are relevant from an optimization perspective, and they are strictly linked to linear algebra. Multiple integrals play a more limited role in the book, as they will be used only to deal with probability distributions of multiple random variables.

3.9.1   Partial derivatives: gradient and Hessian matrix

In Section 2.7 we defined the derivative of a function of a single variable as the limit of an increment ratio:

$$\frac{df}{dx}(x_0) = \lim_{h \to 0} \frac{f(x_0 + h) - f(x_0)}{h}$$

If we have a function of several variables, we may readily extend the concept above by considering a point $\mathbf{x}^0 = [x_1^0, \ldots, x_n^0]^T$ and perturbing one variable at a time. We obtain the concept of a *partial derivative* with respect to a single variable x_i:

$$\frac{\partial f}{\partial x_i}(\mathbf{x}^0) = \lim_{h \to 0} \frac{f(x_1^0, \ldots x_i^0 + h, \ldots, \ldots x_n^0) - f(x_1^0, \ldots x_i^0, \ldots, \ldots x_n^0)}{h}$$

As in the single-variable case, we should not take for granted that the limit above exists, but we will not consider too many technicalities. In practice, the limit above is readily computed by applying the usual rules for derivatives, considering one variable at a time and keeping the other variables fixed. For a function of n variables, we have n (first-order) partial derivatives at a point $\mathbf{x}^0$. They can be grouped into a column vector in $\mathbb{R}^n$, which is called the *gradient* of f at point $\mathbf{x}^0$:

$$\nabla f(\mathbf{x}_0) = \begin{bmatrix} \dfrac{\partial f}{\partial x_1}(\mathbf{x}^0) \\[2mm] \dfrac{\partial f}{\partial x_2}(\mathbf{x}^0) \\[2mm] \vdots \\[2mm] \dfrac{\partial f}{\partial x_n}(\mathbf{x}^0) \end{bmatrix}$$

Example 3.16 Consider the quadratic form $f(x_1, x_2) = x_1^2 + x_2^2$. We may compute the following partial derivatives:

$$\frac{\partial f}{\partial x_1} = 2x_1, \qquad \frac{\partial f}{\partial x_2} = 2x_2$$

When computing the partial derivative with respect to x_1, we consider x_2 as a constant, and this explains why x_2 does not contribute to the first partial derivative. The two partial derivatives can be grouped into the gradient

$$\nabla f(\mathbf{x}) = [2x_1, \ 2x_2]^T$$

To see a more contrived example, consider the following function:

$$f(x_1, x_2, x_3) = x_1 x_2^2 + x_3(1 - x_1 x_2)$$

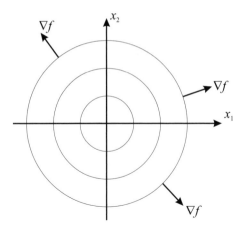

Fig. 3.12   The gradient gives the direction of maximum ascent at each point.

We invite the reader to verify that its gradient is

$$\nabla f(\mathbf{x}) = \left[\begin{array}{c} x_2^2 - x_2 x_3 \\ 2x_1 x_2 - x_1 x_3 \\ 1 - x_1 x_2 \end{array} \right]$$

□

How can we interpret the gradient? We know from single variable calculus that a stationary point is a natural candidate to be a minimum or a maximum of a function; at that point, the tangent line is horizontal. In the multivariable case, we can say something similar by referring to the gradient. A *stationary point* (or point of stationarity) is a point $\mathbf{x}^*$ such that the gradient is the zero vector: $\nabla f(\mathbf{x}^*) = \mathbf{0}$. For instance, the origin is a stationary point of the quadratic form in the example above. Geometric interpretation is difficult for several variables, but in the case of two variables, the function is graphed in three dimensions and a stationary point is characterized by a horizontal tangent *plane*. In general, it can be shown that the gradient is a vector pointing toward the direction of maximum ascent of the function.

Example 3.17 Consider again the quadratic form $x_1^2 + x_2^2$, and its gradient. The function is just the squared distance of each point from the origin and its level curves are concentric circles, as shown in Fig. 3.12. The figure also shows the gradient vector for a few points on a level curves. We see that the gradient $\nabla f(x_1, x_2)$ at each point is a vector moving away from the origin toward infinity, and along that direction the function has the steepest ascent. If we change the sign of the gradient, we get a vector pointing toward the origin, and spotting the path of steepest descent.

It is easy to understand that this feature of the gradient is relevant to function maximization and minimization. □

The gradient vector collects the first-order partial derivatives of a function with respect to all of the independent variables. We may also consider second-order derivatives by repeated application of partial differentiation. If we take a derivative of function f twice with respect to the same variable x_i, we have the second-order partial derivative, which is denoted as

$$\frac{\partial^2 f}{\partial x_i^2}$$

We may also take the derivative with respect to two different variables, which yields a mixed derivative; if we take partial derivative first with respect to x_i, and then with respect to x_j, we obtain

$$\frac{\partial^2 f}{\partial x_j \, \partial x_i}.$$

An immediate question is if the order with which derivatives are taken is relevant or not. The answer is that, when suitable continuity conditions are met, the order in which we take derivatives is inconsequential:[19]

$$\frac{\partial^2 f}{\partial x_i \, \partial x_j} = \frac{\partial^2 f}{\partial x_j \, \partial x_i}.$$

If we group second-order partial derivatives into a matrix, we obtain the *Hessian matrix*:

$$\mathbf{H}(\mathbf{x}_0) = \begin{bmatrix} \dfrac{\partial^2 f}{\partial x_1^2}(\mathbf{x}_0) & \dfrac{\partial^2 f}{\partial x_1 x_2}(\mathbf{x}_0) & \cdots & \dfrac{\partial^2 f}{\partial x_1 x_n}(\mathbf{x}_0) \\[2mm] \dfrac{\partial^2 f}{\partial x_2 x_1}(\mathbf{x}_0) & \dfrac{\partial^2 f}{\partial x_2^2}(\mathbf{x}_0) & \cdots & \dfrac{\partial^2 f}{\partial x_2 x_n}(\mathbf{x}_0) \\[2mm] \vdots & \vdots & \ddots & \vdots \\[2mm] \dfrac{\partial^2 f}{\partial x_n x_1}(\mathbf{x}_0) & \dfrac{\partial^2 f}{\partial x_n x_1}(\mathbf{x}_0) & \cdots & \dfrac{\partial^2 f}{\partial x_n^2}(\mathbf{x}_0) \end{bmatrix}$$

Since the order of variables in mixed terms is not relevant, the Hessian matrix is symmetric.

Example 3.18 Let us find the Hessian matrix for the quadratic forms:

$$f(x_1, x_2) = x_1^2 + 3x_2^2, \qquad g(x_1, x_2) = x_1^2 - 2x_1 x_2 - x_2^2$$

---

[19] This result is known as *Young's theorem*.

First, we calculate all of the relevant partial derivatives for f:

$$\frac{\partial f}{\partial x_1} = 2x_1, \quad \frac{\partial f}{\partial x_2} = 6x_2,$$

$$\frac{\partial^2 f}{\partial x_1^2} = 2, \quad \frac{\partial^2 f}{\partial x_2^2} = 6, \quad \frac{\partial^2 f}{\partial x_1 \, \partial x_2} = 0$$

The Hessian matrix of f is diagonal:

$$\mathbf{H} = \begin{bmatrix} 2 & 0 \\ 0 & 6 \end{bmatrix}$$

The reader is invited to verify that, in the case of g, we obtain

$$\mathbf{H} = \begin{bmatrix} 2 & -2 \\ -2 & -2 \end{bmatrix}$$

From the example, we immediately notice that this Hessian matrix is just twice the matrix associated with the quadratic form. Indeed, if we write a quadratic form as

$$q(\mathbf{x}) = \tfrac{1}{2}\mathbf{x}^T \mathbf{A}\mathbf{x} \qquad (3.20)$$

we see that matrix $\mathbf{A}$ is its Hessian. Another useful result concerning quadratic forms written in the form of Eq. (3.20) is

$$\nabla q(\mathbf{x}) = \mathbf{A}\mathbf{x} \qquad (3.21)$$

The result is easy to check directly.

3.9.2   Taylor's expansion for multivariable functions

Using the gradient and Hessian matrix, we may generalize Taylor's expansion to functions of multiple variables. The second-order expansion around point $\mathbf{x}_0$ is as follows:

$$f(\mathbf{x}) \approx f(\mathbf{x}_0) + [\nabla f(\mathbf{x}_0)]^T (\mathbf{x} - \mathbf{x}_0) + \tfrac{1}{2}(\mathbf{x} - \mathbf{x}_0)^T \mathbf{H}(\mathbf{x}_0)(\mathbf{x} - \mathbf{x}_0)$$

We see that this approximation boils down to Eq. (2.13) when considering a function of a single variable. If we stop the expansion to first-order terms, we get a linear approximation, which is just a plane when working with functions of two variables. Such a tangent plane is illustrated in Fig. 3.13. The inclusion of the second-order terms implies that the approximation involves a quadratic form. The approximation can be convex, concave, or neither, depending on the definiteness of the quadratic form corresponding to the Hessian matrix. Hence, the eigenvalues of the Hessian matrix are useful in analyzing convexity/concavity issues and in checking optimality conditions.

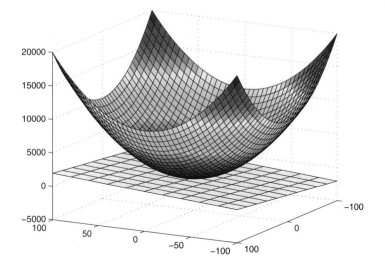

Fig. 3.13 First-order Taylor expansion yields a tangent plane.

Example 3.19 Let us compute Taylor's expansion of function

$$f(x_1, x_2) = \sqrt{x_1 x_2} = x_1^{1/2} x_2^{1/2}$$

in a neighborhood of point $(1, 1)$. The partial derivatives are

$$\frac{\partial f}{\partial x_1} = \tfrac{1}{2} x_1^{-1/2} x_2^{1/2}, \qquad \frac{\partial f}{\partial x_2} = \tfrac{1}{2} x_1^{1/2} x_2^{-1/2}$$

$$\frac{\partial^2 f}{\partial x_1^2} = -\tfrac{1}{4} x_1^{-3/2} x_2^{1/2}, \qquad \frac{\partial^2 f}{\partial x_2^2} = -\tfrac{1}{4} x_1^{1/2} x_2^{-3/2}$$

$$\frac{\partial^2 f}{\partial x_1 \, \partial x_2} = \tfrac{1}{4} x_1^{-1/2} x_2^{-1/2}$$

Evaluating gradient and Hessian at $(1, 1)$ yields

$$\nabla f(1, 1) = \begin{bmatrix} \tfrac{1}{2} \\ \tfrac{1}{2} \end{bmatrix}, \qquad \mathbf{H}(1, 1) = \begin{bmatrix} -\tfrac{1}{4} & \tfrac{1}{4} \\ \tfrac{1}{4} & -\tfrac{1}{4} \end{bmatrix}$$

Note that, unlike a quadratic form, Taylor's expansion of f depend on the point at which it is taken. Considering small displacements δ_1 and δ_2 around

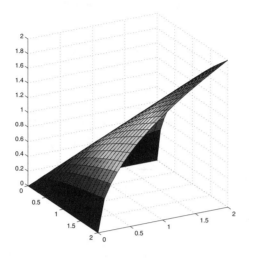

Fig. 3.14 Surface plot of function $\sqrt{x_1 x_2}$.

$x_1 = 1$ and x_2, we get

$$f(1 + \delta_1, 1 + \delta_2)$$

$$\approx f(1, 1) + \begin{bmatrix} \frac{1}{2} & \frac{1}{2} \end{bmatrix} \begin{bmatrix} \delta_1 \\ \delta_2 \end{bmatrix} + \begin{bmatrix} \delta_1 & \delta_2 \end{bmatrix} \begin{bmatrix} -\frac{1}{4} & \frac{1}{4} \\ \frac{1}{4} & -\frac{1}{4} \end{bmatrix} \begin{bmatrix} \delta_1 \\ \delta_2 \end{bmatrix}$$

$$= 1 + \frac{\delta_1}{2} + \frac{\delta_2}{2} - \frac{\delta_1^2}{4} - \frac{\delta_2^2}{4} + \frac{\delta_1 \delta_2}{2}$$

The eigenvalues of the Hessian matrix are $\lambda_1 = -0.5$ and $\lambda_2 = 0$. One of the eigenvalues is zero, and indeed it is easy to see that the Hessian matrix is singular. Moreover, the eigenvalues are both nonpositive, suggesting that the function f is locally concave, but not strictly. A surface plot of the function is illustrated in Fig. 3.14, for positive values of x_1 and x_2; from the figure, we see the concavity of the function in this region. A closer look at the surface explains why the function is not *strictly* concave there. For $x_1 = x_2$, we have $f(x_1, x_2) = \sqrt{x_1^2} = |x_1|$. Hence, the function is linear along that direction.

□

3.9.3 Integrals in multiple dimensions

Definite integrals have been introduced in Section 2.13 as a way to compute the area below the curve corresponding to the graph of a function of one variable. If we consider a function (x, y) of two variables, there is no reason why

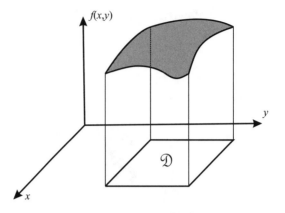

Fig. 3.15 Double integral as a volume below a surface graph of a function.

we should not consider its surface plot and the *volume* below the surface, corresponding to a region $\mathcal{D}$ on the (x, y) plane. This double integral is denoted as follows:

$$\iint\limits_{(x,y)\in\mathcal{D}} f(x, y)\, dx\, dy$$

and the idea is illustrated in Fig. 3.15. Here, the domain on which the function is integrated is a rectangle, but more general shapes are allowed. Nevertheless, rectangular tiles are easy to deal with, and indeed they are the basis for a rigorous definition of the double integral. In the following chapters, we will encounter double integrals only when characterizing the joint distribution of two random variables, in Section 8.1, and we just need an intuitive understanding.

To compute a double integral, a convenient way is to regard double integrals as iterated integrals, over a rectangular domain $[a, b] \times [c, d]$:

$$\int_a^b \int_c^d f(x, y)\, dy\, dx$$

Please note the order of differentials dy and dx: Here we want to point out the ordering of variables with which the integration is carried out. If we want to be precise, we could write

$$\int_a^b \left[\int_c^d f(x, y)\, dy \right] dx$$

The idea behind iterated integration is straightforward: We should first integrate with respect to y, treating x as a constant, obtaining a function of x that is then integrated with respect to x.

Example 3.20 Consider function $f(x, y) = x^2 y$ and the rectangular domain $[1, 2] \times [-3, 4]$ obtained by taking the Cartesian product of intervals $[1, 2]$ on the x-axis and $[-3, 4]$ on the y-axis. We want to find the following integral:

$$\int_1^2 \int_{-3}^4 x^2 y \, dy \, dx$$

In the inner integral, x can be regarded as a constant and, in this case, it can be just taken outside:

$$\int_{-3}^4 x^2 y \, dy = x^2 \int_{-3}^4 y \, dy = x^2 \left. \frac{y^2}{2} \right|_{-3}^4 = \frac{7x^2}{2}$$

Then, we proceed with the outer integral:

$$\int_1^2 \frac{7x^2}{2} \, dx = \frac{7}{2} \left. \frac{x^3}{3} \right|_1^2 = \frac{49}{6}$$

☐

The conditions under which a double integral can be tackled as an iterated integral are stated by Fubini's theorem, and the idea can be generalized to multiple dimensions.

Problems

3.1   Solve the system of linear equations:

$$\begin{cases} x_1 + 2x_2 - x_3 = -3 \\ x_1 + 4x_3 = 9 \\ 2x_2 + x_3 = 0 \end{cases}$$

using both Gaussian elimination and Cramer's rule.

3.2   Express the derivative of polynomials as a linear mapping using a matrix.

3.3   Prove that the representation of a vector using a basis is unique.

3.4   Let $\mathbf{A} \in \mathbb{R}^{m,n}$, and let $\mathbf{D}$ be a diagonal matrix in $\mathbb{R}^{n,n}$. Prove that the product $\mathbf{AD}$ is obtained by multiplying each element in a row of $\mathbf{A}$ by the corresponding element in the diagonal of $\mathbf{D}$. Check with

$$\mathbf{A} = \begin{bmatrix} 1 & 3 & 5 \\ 2 & 6 & 4 \end{bmatrix}, \qquad \mathbf{D} = \begin{bmatrix} 2 & 0 & 0 \\ 0 & -3 & 0 \\ 0 & 0 & 7 \end{bmatrix}$$

3.5   Unlike usual algebra, in matrix algebra we may have $\mathbf{AX} = \mathbf{BX}$, even though $\mathbf{A} \neq \mathbf{B}$ and $\mathbf{X} \neq \mathbf{0}$. Check with

$$\mathbf{A} = \begin{bmatrix} 1 & 0 & 2 \\ 0 & 1 & 1 \\ 2 & 0 & 2 \end{bmatrix}, \qquad \mathbf{B} = \begin{bmatrix} 1 & 3 & 0 \\ 0 & 4 & -1 \\ 2 & 3 & 0 \end{bmatrix}, \qquad \mathbf{X} = \begin{bmatrix} 6 & 5 & 7 \\ 2 & 2 & 4 \\ 3 & 3 & 6 \end{bmatrix}$$

3.6   Consider the matrix $\mathbf{H} = \mathbf{I} - \mathbf{h}\mathbf{h}^T$, where $\mathbf{h}$ is a column vector in $\mathbb{R}^n$ and $\mathbf{I}$ is the properly sized identity matrix. Prove that $\mathbf{H}$ is orthogonal, provided that $\mathbf{h}^T\mathbf{h} = 1$. This matrix is known as the *Householder matrix*.

3.7   Consider the matrix $\mathbf{C} = \mathbf{I}_n - \frac{1}{n}\mathbf{J}_n$, where $\mathbf{I}_n \in \mathbb{R}^{n,n}$ is the identity matrix and $\mathbf{J}_n \in \mathbb{R}^{n,n}$ is a matrix consisting of 1. This matrix is called a centering matrix, since $\mathbf{x}^T\mathbf{C} = \{x_i - \bar{x}\}$, where $\mathbf{x} = [x_1, x_2, \ldots, x_n]$ is a vector of observations. Prove this fact. Also prove that

$$\sum_{i=1}^{n}(x_i - \bar{x})^2 = \mathbf{x}^T\mathbf{C}\mathbf{x}$$

3.8   Check that the determinant of diagonal and triangular matrices is the product of elements on the diagonal.

3.9   Find the inverse of each of the following matrices

$$\mathbf{A}_1 = \begin{bmatrix} 6 & 0 & 0 \\ 0 & 2 & 0 \\ 0 & 0 & -5 \end{bmatrix}, \quad \mathbf{A}_2 = \begin{bmatrix} 0 & 0 & 5 \\ 0 & 2 & 0 \\ 3 & 0 & 0 \end{bmatrix}, \quad \mathbf{A}_3 = \begin{bmatrix} 1 & 1 & 0 \\ 0 & 1 & 1 \\ 1 & 0 & 1 \end{bmatrix}$$

3.10   For a square matrix $\mathbf{A}$, suppose that there is a vector $\mathbf{x} \neq \mathbf{0}$ such that $\mathbf{A}\mathbf{x} = \mathbf{0}$. Prove that $\mathbf{A}$ is singular.

3.11   Prove that $\mathbf{h}\mathbf{h}^T - \mathbf{h}^T\mathbf{h}\mathbf{I}$ is singular.

3.12   Prove that two orthogonal vectors are linearly independent.

3.13   Show that if λ is an eigenvalue of $\mathbf{A}$, then $1/(1 + \lambda)$ is an eigenvalue of $(\mathbf{I} + \mathbf{A})^{-1}$.

3.14   Show that, if the eigenvalues of $\mathbf{A}$ are positive, those of $\mathbf{A} + \mathbf{A}^{-1}$ are not less than 2.

3.15   Prove that, for a symmetric matrix $\mathbf{A}$, we have

$$\sum_{i=1}^{n}\sum_{j=1}^{n}a_{ij}^2 = \sum_{k=1}^{n}\lambda_k^2$$

where λ_k, $k = 1, \ldots, n$, are the eigenvalues of $\mathbf{A}$.

For further reading

- A short and readable introduction to linear algebra can be found, e.g., in the text by Lang [4].

- A more advanced treatment, paying some more attention to matrix analysis, is offered by Meyer [6]

- Since the actual reason why we are interested in matrix algebra is its role in multivariate statistics, the reader might be interested in references which are specifically aimed at this kind of application, such as Refs. [2] or [7]; a few exercises have been taken from the latter reference.

- A good reference for multivariable calculus is Ref. [5]. Since this excellent reference takes a quite general perspective, including some concepts that are more relevant in physics, the reader might wish to consult a more economically oriented reference: Ref. [8] covers single- and multi-variable calculus, as well as linear algebra and matrix analysis.

- We have opened this chapter mentioning option pricing. The standard reference for financial derivatives is the book by Hull [3]; a text that is more focused on computational methods in finance is Ref. [1].

REFERENCES

1. P. Brandimarte, *Numerical Methods in Finance and Economics: A Matlab-Based Introduction,* 2nd ed., Wiley, New York, 2006.

2. D.H. Harville, *Matrix Algebra from a Statistician's Perspective*, Springer, New York, 1997.

3. J.C. Hull, *Options, Futures, and Other Derivatives,* 5th ed., Prentice Hall, Upper Saddle River, NJ, 2003.

4. S. Lang, *Introduction to Linear Algebra,* 2nd ed., Springer, New York, 1986.

5. S. Lang, *Calculus of Several Variables,* 3rd ed., Springer, New York, 1987.

6. C.D. Meyer, *Matrix Analysis and Applied Linear Algebra*, Society for Industrial and Applied Mathematics, Philadelphia, 2000.

7. S.R. Searle, *Matrix Algebra Useful for Statistics*, Wiley, New York, 1982.

8. C.P. Simon and L. Blume, *Mathematics for Economists*, W.W. Norton, New York, 1994.

Part II

*Elementary Probability
and Statistics*

4

Descriptive Statistics: On the Way to Elementary Probability

Some fundamental concepts of descriptive statistics, like frequencies, relative frequencies, and histograms, have been introduced informally in Chapter 1. Here we want to illustrate and expand those concepts in a slightly more systematic way. Our treatment will be rather brief since, within the framework of this book, descriptive statistics is essentially a tool for building some intuition paving the way for later chapters on probability theory and inferential statistics.

We introduce basic statistical concepts in Section 4.1, drawing the line between descriptive and inferential statistics, and illustrating the difference between sample and population, as well as between qualitative and quantitative variables. Descriptive statistics provides us with several tools for organizing and displaying data, some of which are outlined in Section 4.2. While displaying data graphically is useful to get some feeling for their distribution, we typically need a few numbers summarizing their essential features; quite natural summary measures such as mean and variance are dealt with in Section 4.3. Then, in Section 4.4 we consider measures of relative standing such as percentiles, which are a less obvious but quite important tool used to analyze data. We should mention that basic descriptive statistics does not require overly sophisticated concepts, and it is rather easy to understand. However, sometimes concepts are a bit ambiguous, and a few subtleties can be better appreciated when armed with a little more formal background. Percentiles are a good case in point, as there is no standard definition and software packages may compute them in different ways; yet, they are a good way to get some intuitive feeling for probabilistic concepts, like quantiles, that are relevant in many applications in logistics and finance. Finally, in Section 4.5 we move from data in a single dimension to data in multiple dimensions. We limit the discussion to two dimensions, but the discussion here is a good way to understand the need for the data reduction methods discussed in Chapter 17.

4.1   WHAT IS STATISTICS?

A rather general answer to this question is that statistics is a group of methods to collect, analyze, present, and interpret data (and possibly to make decisions). We often consider statistics as a branch of mathematics, but this is the result of a more recent tendency. From a historical perspective, the term "statistics" stems from the word "state." Originally, the driving force behind the discipline was the need to collect data about population and economy, something that was felt necessary in the city states of Venice and Florence during Renaissance. Many governments did the same in the following centuries. Then, statistics got a more quantitative twist, mainly under the impulse of French mathematicians. As a consequence, statistics got more intertwined with the theory of probability, a tendency that was not free from controversy.

Over time, many statistical tools have been introduced and they are often looked at as a bunch of cookbook recipes, which may result in quite some confusion. In order to bring some order, a good starting point is drawing the line between two related subbranches:

- *Descriptive Statistics* consists of methods for organizing, displaying, and describing data by using tables, graphs, and summary measures.

- *Inferential Statistics* consists of methods that use *sampling* to help make decisions or predictions about a *population*.

To better understand the role of sampling, we should introduce the following concepts.

DEFINITION 4.1 (Population vs. sample) *A* **population** *consists of all elements (individuals, items, etc.) whose characteristics are being studied. A* **sample** *is a portion of the population, which is selected for study.*

To get the point, it suffices to reflect a bit on the cost and the time required to carry out a census of the whole population of a state, e.g., to figure out average household income. A much more common occurrence is a sample survey. For the study to be effective, the sample must be *representative* of the whole population. If you sample people in front of a big investment bank, you are likely to get a misleading picture, as the sample is probably biased toward a very specific type of individual.

Example 4.1 One of the best-known examples of bad sample selection is the 1936 presidential election poll by the *Literary Digest*. According to this poll, the Republican governor of Kansas, Alf Landon, would beat former president Franklin Delano Roosevelt by 57–43%. The sample size was not tiny at all, as the *Digest* mailed over 10 million questionnaires and over 2.3 million people responded. The real outcome was quite different, as Roosevelt won with 62%. One of the reasons commonly put forward to explain such a blunder is that many respondents were selected from lists of automobile and telephone owners. Arguably, a selection process like that would be correct nowadays, but in the past the sample was biased towards relatively wealthy people, which in turn resulted in a bias towards republican voters.   ☐

A sample drawn in such a way that each element in the target population has a chance of being selected is called a *random sample*. If the chance of being selected is the same for each element, we speak of a *simple random sample*.[1]

Household income is an example of a *variable*. A variable is a characteristic of each member of the population, and below we discuss different types of variable we might be interested in. Income is a quantitative variable, and we may want some information about average income of the population. The average income of the population is an example of a *parameter*. Typically, we do not know the parameters characterizing a whole population, and we have to resort to some form of estimate. If we use sampling, we have to settle for the average income of the sample, which is a *statistic*. The statistic can be used to estimate the unknown parameter.

If sampling is random, whenever we repeat the experiment, we get different results, i.e., different values of the resulting statistic. If the results show wide swings, any conclusion that we get from the study cannot be trusted. Intuition suggests that the larger the sample, the more reliable the conclusions. Furthermore, if the individuals in the population are not too different from one another, the sample can be small. In the limit, if all of the individuals were identical, any one of them would make a perfect sample. But if there is much variability within the population, a large sample must be taken. In practice, we need some theoretical background to properly address issues related to the size of the sample and the reliability of the conclusions we get from sampling, especially if such conclusions are the basis of decision making. In Chapter 9, on inferential statistics, we will consider such issues in detail. On the contrary, basic descriptive statistics does not strictly rely on quite sophisticated concepts. However, probability theory is best understood by using descriptive statistics as a motivation. Descriptive statistics is quite useful when conducting an *exploratory* study, i.e., if we want to analyze data to see if an interesting pattern emerges, suggesting some hypothesis or line

---

[1]In practice, to obtain acceptable results with a small sample, we resort to stratification, i.e., we build a sample that reflects the essential features of overall population. In this book we will only deal with simple random samples.

Table 4.1 Illustrating types of variable.

ID number	Weight (kg)	Height (m)	Married	n. children
1	68.3	1.77	Yes	0
2	81.7	1.85	Yes	2
3	94.5	1.88	No	0
4	55.5	1.60	Yes	4
5	61.8	1.68	No	1

of action. However, when a *confirmatory* analysis is carried out, to check a hypothesis, inferential statistics comes into play.

4.1.1 Types of variable

If we are sampling a population to figure out average household income, we are considering income as the variable of interest.

DEFINITION 4.2 (Variables and observations) *A* **variable** *is a characteristic under study, which assumes different values for different elements of a population or a sample. The value of a variable for an element is called an* **observation** *or* **measurement**.

This definition is illustrated in Table 4.1, where hypothetical data are shown. An anonymous person is characterized by weight, height, marital status, and number of children. Variables are arranged on columns, and each observation corresponds to a row. We immediately see differences between those variables. A variable can be

- *Quantitative*, if it can be measured numerically

- *Qualitative* or **categorical**, otherwise

Clearly, weight and number of children are quantitative variables, whereas marital status is not. Other examples of categorical variables are gender, hair color, or make of a computer.

If we look more carefully at quantitative variables in the table, we see another difference. You cannot have 2.1567 children; this variable is restricted to a set of discrete values, in this case integer numbers. On the contrary, weight and height can take, in principle, any value. In practice, we truncate those numbers to a suitable number of significant digits, but they can be considered as real numbers. Hence, quantitative variables should be further classified as

- *Discrete*, if the values it can take are countable (number of cars, number of accidents occurred, etc.)

- *Continuous*, if the variable can assume any value within an interval (length, weight, time, etc.)

In this book we will generally associate discrete variables with integer numbers, and continuous variables with real numbers, as this is by far the most common occurrence. However, this is not actually a rule. For instance, we could consider a discrete variable that can take two real values such as $\ln(18)$ or 2π. We should also avoid the strict identification of "a variable that can take an infinite number of values" with a continuous variable. It is true that a continuous variable restricted to a bounded interval, e.g., [2,10], can assume an infinite number of values, but a discrete variable can take an infinite number of integer values as well.[2] For instance, if we consider the number of accidents that occurred on a highway in one month, there is no natural upper bound on them, and this should be regarded as a variable taking integer values $i = 1, 2, 3, \ldots$, even though very large values are (hopefully) quite unlikely.

The classification looks pretty natural, but the following examples show that sometimes a little care is needed.

Example 4.2 (Dummy and nominal variables) Marital status is clearly a qualitative variable. However, in linear regression models (Chapters 10 and 16) it is quite common to associate them with binary values 1 and 0, which typically correspond to yes/no or true/false. In statistics, such a variable is often called "dummy." The interpretation of these numerical values is actually arbitrary and depends on modeler's choice. It is often the case that numerical values are attached to categorical variables for convenience, but we should consider these as *nominal* variables. A common example are the Standard Industrial Classification (SIC) codes.[3] You might be excited to discover that SIC code 1090 corresponds to "Miscellaneous Metal Ores" and 1220 to "Bituminous Coal & Lignite Mining." No disrespect intended to industries in this sector and, most importantly, no one should think that the second SIC code is really larger than the first one, thereby implying some ranking among them. ☐

The example above points out a fundamental feature of truly numerical variables: They are ordered, whereas categorical variables cannot be really ordered, even though they may be associated with numerical (nominal but *not* ordinal) values. We cannot double a qualitative or a nominal variable, can we? But even doubling a quantitative variable is trickier than we may think.

Example 4.3 Imagine that, between 6 and 11 a.m., temperature on a day rises from 10°C to 20°C. Can we say that temperature has doubled? It is

---

[2]Intervals and real vs. integer numbers are introduced in Section 2.2. We will avoid considering pathological cases such as the sequence $x_k = 1 - 1/k$, $k = 1, 2, 3, \ldots$, where a countably infinite number of values is contained in a bounded interval.

[3]See http://www.sec.gov/info/edgar/siccodes.htm

tempting to say yes, but imagine that you measure temperature by Fahrenheit, rather than Celsius degrees. In this case, the two temperatures are 50°F and 68°F, respectively,[4] and their ratio is certainly not 2.   ☐

What is wrong with the last example is that the origin of the temperature scale is actually arbitrary. On the contrary, the origin of a scale measuring the number of children in a family is not arbitrary. We conclude this section with an example showing again that the same variable may be used in different ways, associated with different types of variable.

Example 4.4 (What is time?) Time is a variable that plays a fundamental role in many models. Which type of variable should we use to represent time?

Time as a continuous variable. From a "philosophical" point of view, time is continuous. If you consider two time instants, you can always find a time instant between them. Indeed, time in physics is usually represented by a real number. Many useful models in finance are also based on a continuous representation of time, as this may result in handy formulas.[5]

Time as a discrete variable. Say that we are in charge of managing the inventory of some item, that is ordered at the end of each week. Granted, time is continuous, but from our perspective what really matters is demand during each week. We could model demand by a variable like d_t, where subscript t refers to weeks $1, 2, 3, \ldots$. In this model, time is discretized because of the structure of the decision making process. We are not interested in demand second by second. In the EOQ model (see Section 2.1) we treated time as a continuous variable, because demand rate was constant. In real life, demand is unlikely to be constant, and time must be discretized in order to build a manageable model. Indeed, quite often time is discretized to come up with a suitable computational procedure to support decisions.[6]

Time as a categorical variable. Consider daily sales at a retail store. Typically, demand on Mondays is lower than the average, maybe because the store is closed in the morning. Demand on Fridays is greater, and it explodes on Saturdays, probably because most people are free from their work on weekends. We observe similar seasonal patterns in ratings of TV programs and in consumption of electrical energy.[7] We could try to analyze the statistical properties of d_{mon}, d_{fri}, etc. In this case, we see

---

[4]The formula used to convert Celsius degrees to Fahrenheit is $F = (9/5) \times C + 32$.
[5]Actually, too handy a formula may be dangerous if not properly understood and misused. A notable example is the celebrated Black–Scholes–Merton formula for pricing options.
[6]That applies to physics, too; computational procedures to simulate physical systems are always based on some form of discretization, including discretization of time.
[7]Time series, including seasonal ones, are dealt with in Chapter 11.

Table 4.2 Raw data may be hard to interpret.

2	4	4	2	1	5
3	4	5	4	3	4
2	3	4	6	1	2
5	1	2	3	4	3
5	2	2	2	3	2
3	2	1	1	4	6
4	4	4	4	3	

that "Monday" and "Friday" subscripts do not correspond to ordered time instants, as there are different weeks, each one with a Monday and a Friday. A time subscript in this case is actually related to categorical variables.

Time can be modeled in different ways, and the choice among them may depend on the purpose of the model or on computational conveniency. ▯

4.2 ORGANIZING AND REPRESENTING RAW DATA

We have introduced the basic concepts of frequencies and histograms in Section 1.2.1. Here we treat the same concepts in a slightly more systematic way, illustrating a few potential difficulties that may occur even with these very simple ideas.

Imagine a car insurance agent who has collected the weekly number of accidents occurred during the last 41 weeks, as shown in Table 4.2. This raw representation of data is somewhat confusing even for such a small dataset. Hence, we need a more systematic way to organize and present data. A starting point would be sorting the data in order to see the *frequency* with which each of the values above occurs. For instance, we see that in 5 weeks we have observed one accident, whereas two accidents have been observed in 12 cases. Doing so for all of the observed values, from 1 to 6, we get the second column of Table 4.3, which shows frequencies for the raw data above.

An even clearer picture may be obtained by considering *relative frequencies*, which are obtained by taking the ratio of observed frequencies and the total number of observations:

$$\text{Relative frequency of a category} = \frac{\text{frequency of that category}}{\text{number of observations}}$$

For instance, two accidents have been observed in 10 cases out of 41; hence, the relative frequency of the value 2 is $\frac{10}{41} = 0.2439 = 24.39\%$. Relative frequencies are also displayed in Table 4.3. Sometimes, they are reported in

Table 4.3 Organizing raw data using frequencies and relative frequencies.

Accidents per week	Frequency	Relative frequency	Percentage (%)
1	5	0.1219	12.19
2	10	0.2439	24.39
3	8	0.1951	19.51
4	12	0.2927	29.27
5	4	0.0976	9.76
6	2	0.0488	4.88

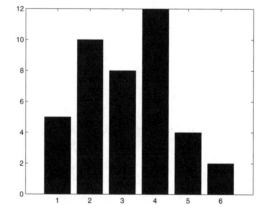

Fig. 4.1 A bar chart of frequencies for the data in Table 4.3.

percentage terms. What is really essential to notice is that relative frequencies should add up to 1, or 100%; when numbers are rounded, a small discrepancy can occur.

Frequencies and relative frequencies may be represented graphically in a few ways. The most common graphical display is a *bar chart*, like the one illustrated in Fig. 4.1. The same bar chart, with a different vertical axis would represent relative frequencies.[8] A bar chart can also be used to illustrate frequencies of categorical data. In such a case, the ordering of bars would have no meaning, whereas for quantitative variables we have a natural ordering of bars. A bar chart for quantitative variables is usually called a *histogram*. For qualitative variables, we may also use an alternative representation like a pie chart. Figure 4.2 shows a pie chart for answers to a hypothetical question

---

[8]See also Fig. 1.4.

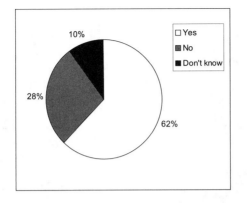

Fig. 4.2   A pie chart for categorical data.

Table 4.4   Frequency and relative frequencies for grouped data.

Average travel time to work (minutes)	Frequency	Percentage (%)
< 18	7	14
18–21	7	14
21–24	23	46
24–27	9	18
27–30	3	6
≥ 30	1	2

(answers can be "yes," "no," and "don't know"). Clearly, a pie chart does not show any ordering between categories.

A histogram is naturally suited to display discrete variables, but what about continuous variables, or discrete ones when there are many observed values? In such a case, it is customary to group data into intervals corresponding to *classes*. As a concrete example, consider the time it takes to get to workplace using a car. Time is continuous in this case, but there is little point in discriminating too much using fractions of seconds. We may consider "bins" characterized by a width of three minutes, as illustrated in Table 4.4. To formalize the concept, each "bin" corresponds to an interval. The common convention is to use closed-open intervals.[9] This means, for instance, that the class 18–21 in Table 4.4 includes all observations ≥ 18 and < 21 or, in other words, it corresponds to interval [18, 21). To generalize, we use bins of the

---

[9]See Section 2.2.2 for the definition of open and closed intervals.

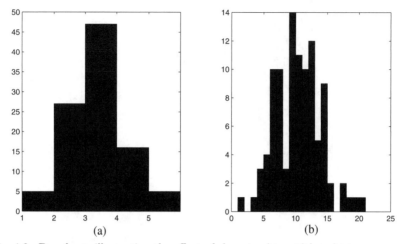

Fig. 4.3   Bar charts illustrating the effect of changing bin width in histograms.

following form:

$$B_j(x_0, h) = [x_0 + (j-1)h, \ x_0 + jh), \quad j = 1, 2, 3, \ldots$$

where x_0 is the origin of this set of bins and should not be confused with an observation, and h is the bin width. Actually, h need not be the same for all of the bins, but this may be a natural choice. The first bin, $B_1(x_0, h)$, corresponds to interval $[x_0, x_0 + h)$; the second bin, $B_2(x_0, h)$, corresponds to interval $[x_0 + h, x_0 + 2h)$, and so on. For widely dispersed data it might be convenient to introduce two side bins, i.e., an unbounded interval $(-\infty, x_l)$ collecting the observations below a lower bound x_l, and an unbounded interval $[x_u, +\infty)$ for the observations above an upper bound x_u.

Histograms are a seemingly trivial concept, that can be used to figure out basic properties of a dataset in terms of symmetry vs. skewness (see Fig. 1.5), as well as in terms of dispersion (see Fig. 1.6). In practice, they may not be as easy to use as one could imagine. A first choice we have to make concerns the bin width h and, correspondingly, the number of bins. The same dataset may look differently if we change the number of bins, as illustrated in Fig. 4.3. Using too few bins does not discriminate data enough and any underlying structure is lost in the blur [see histogram (a) in Fig. 4.3]; using too many may result in a confusing jaggedness, that should be smoothed in order to see the underlying pattern [see histogram (b) in Fig. 4.3]. A common rule of thumb is that one should not use less than 5 bins, and no more than 20. A probably less obvious issue is related to the choice of origin x_0, as illustrated by the next example.

Example 4.5   Consider the dataset in Table 4.5, reporting observed values along with their frequency. Now let us choose $h = 0.2$, just to group data a

Table 4.5 Data for Example 4.5.

Values	55.1	55.2	55.3	55.4	55.5	55.6	55.7	55.8	55.9	56.0
Frequency	2	4	3	7	5	2	3	8	9	1

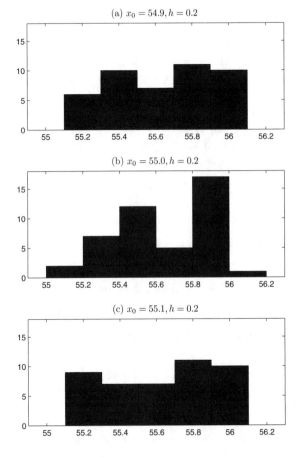

Fig. 4.4 Effect of shifting the origin of histograms for the data in Table 4.5.

little bit. Figure 4.4 shows three histograms obtained by setting $x_0 = 54.9$, $x_0 = 55.0$, and $x_0 = 55.1$, respectively. At first sight, the change in histogram shape due to an innocent shift in the origin of bins is quite surprising. If we look more carefully into the data, the source of the trouble is evident. Let us check which interval corresponds to the first bin in the three cases. When

$x_0 = 54.9$, the first bin is $[54.9, 55.1)$ and is empty. When $x_0 = 55.0$, the first bin is changed to $[55.0, 55.2)$; now two observations fall into this bin. Finally, when $x_0 = 55.1$, the first bin is changed to $[55.1, 55.3)$, and $2 + 4 = 6$ observations fall into this bin. The point is that, by shifting bins, we change abruptly the number of observations falling into each bin, and a wild variation in the overall shape of the histogram is the result.[10]   □

The example above, although somewhat pathological, shows that histograms are an innocent-looking graphical tool, but they may actually be dangerous if used without care, especially with small datasets. Still, they are useful to get an intuitive picture of the distribution of data. In the overall economy of the book, we should just see that the histogram of relative frequencies is a first intuitive clue leading to the idea of a probability distribution.

We close this section by defining two concepts that will be useful in the following.

DEFINITION 4.3 (Order statistics) *Let $X_1, X_2, \ldots, X_n$ be a sample of observed values. If we sort these data in increasing order, we obtain order statistics, which are denoted by $X_{(1)}, X_{(2)}, \ldots, X_{(n)}$. The smallest observed value if $X_{(1)}$ and the largest observed value is $X_{(n)}$.*

DEFINITION 4.4 (Outliers) *An outlier is an observation that looks quite apart from the other ones. This may be a very unlikely value, or an observation that actually comes from a different population.*

Example 4.6 Consider observations

$$X_1 = 18, \quad X_2 = 5, \quad X_3 = 189, \quad X_4 = 21, \quad X_5 = 13$$

Ordering these values yields the following order statistics:

$$X_{(1)} = 5, \quad X_{(2)} = 13, \quad X_{(3)} = 18, \quad X_{(4)} = 21, \quad X_{(5)} = 189$$

The last value is quite apart from the remaining ones and is a candidate outlier.   □

Spotting an outlier is a difficult task, and the very concept looks quite arbitrary. There are statistical procedures to classify an outlier in a sensible and objective manner, but in practice we need to dig a bit deeper into data to figure out why a value looks so different. It may be the result of a data entry error or a wrong observation, in which case the observation should be eliminated. It could be just an unlikely observation, in which case eliminating the observation may result in a dangerous underestimation of the actual uncertainty. In other cases, we may be mixing observations from what are actually different populations. If we take observations of a variable in small towns and then we throw New York into the pool, an outlier is likely to result.

----

[10]To avoid these abrupt changes we can smooth data using so-called kernel density functions; this is beyond the scope of this book, and we refer the reader, e.g., to Chapter 1 of Ref. [2].

4.3 SUMMARY MEASURES

A look at a frequency histogram tells us many things about the distribution of values of a variable of interest within a population or a sample. However, it would be quite useful to have a set of numbers capturing some essential features quantitatively; this is certainly necessary if we have to compare two histograms, since visual perception can be misleading. More precisely, we need a few *summary measures* characterizing, e.g., the following properties:

- Location, i.e., the central tendency of the data

- Dispersion

- Skewness, i.e., lack of symmetry

4.3.1 Location measures: mean, median, and mode

We are all familiar with the idea of taking averages. Indeed, the most natural location measure is the mean.

DEFINITION 4.5 (Mean for a sample and a population) *The mean for a population of size n is defined as*

$$\mu = \frac{1}{n} \sum_{i=1}^{n} x_i$$

The mean for a sample of size n is

$$\bar{X} = \frac{1}{n} \sum_{i=1}^{n} X_i$$

The two definitions above may seem somewhat puzzling, since they look identical. However, there is an essential difference between the two concepts. The mean of the population is a well-defined number, which we often denote by μ. If collecting information about the whole population is not feasible, we take a sample resulting in a mean $\bar{X}$. But if we take two different samples, possibly random ones, we will get *different* values for the mean. In later chapters, we will discover that the population mean is related to the concept of expected value in probability theory, whereas the sample mean is used in inferential statistics as a way to estimate the (unknown) expected value. The careful reader might also have noticed that we have used a lowercase letter x_i when defining the mean of a population and an uppercase letter X_i for the mean of a sample. Again, this is to reinforce the conceptual difference between them; in later chapters we will use lowercase letters to denote numbers and uppercase letters to denote random variables. Observations in a random sample are, indeed, random variables.

Example 4.7 We want to estimate the mean number of cars entering a parking lot every 10 minutes. The following 10 observations have been gathered, over 10 nonoverlapping time periods of 10 minutes: 10, 22, 31, 9, 24, 27, 29, 9, 23, 12. The sample mean is

$$\bar{X} = \frac{10 + 22 + 31 + 9 + 24 + 27 + 29 + 9 + 23 + 12}{10} = 19.6 \text{ cars}$$

Note that the mean of integer numbers can be a fractional number. Also note that a single small observation can affect the sample mean considerably. If, for some odd reason, the first observation is 1000; then

$$\bar{X} = \frac{1000 + 22 + 31 + 9 + 24 + 27 + 29 + 9 + 23 + 12}{10} = 118.6 \text{ cars}$$

⬚

The previous example illustrates the definition of mean, but when we have many data it might be convenient to use frequencies or relative frequencies. If we are given n observations, grouped into C classes with frequencies f_i, the sample mean is

$$\bar{X} = \frac{1}{n} \sum_{k=1}^{C} f_k y_k \tag{4.1}$$

Here, y_k is a value representative of the class. Note that y_k need not be an observed value. In fact, when dealing with continuous variables, y_k might be the midpoint of each bin; clearly, in such a case grouping data results in a loss of information and should be avoided. When variables are integer, one single value can be associated with a class, and no difficulty arises.

Example 4.8 Consider the data in Table 4.6, which contains days of unjustified absence per year of a group of employees. Then: $C = 6$, $n = \sum_{i=1}^{C} f_i = 410 + 430 + 290 + 180 + 110 + 20 = 1440$, and

$$\bar{X} = \frac{0 \times 410 + 1 \times 430 + 2 \times 290 + 3 \times 180 + 4 \times 110 + 5 \times 20}{1440} = 1.451 \text{ days}$$

⬚

If relative frequencies $p_k = f_k/n$ are given, the mean is calculated as

$$\bar{X} = \sum_{k=1}^{C} p_k y_k$$

where again y_k is the value associated with class k. It is easy to see that this is equivalent to Eq. (4.1). In this case, we are computing a weighted average of values, where weights are nonnegative and add up to one.

The *median*, sometimes denoted by m, is another measure of central tendency. Informally, it is the value of the middle term in a dataset that has been ranked in increasing order.

Table 4.6 Data for Example 4.8.

Days of absence	Frequency
0	410
1	430
2	290
3	180
4	110
5	20

Example 4.9 Consider the dataset: 10, 5, 19, 8, 3. Ranking the dataset (3,5,8,10,19), we see that the median is 8.   □

More generally, with a dataset of size n the median should be the order statistic

$$X_{\left((n+1)/2\right)}$$

An obvious question is: What happens if we have an even number of elements? In such a case, we take the average of the two middle terms, i.e., the elements in positions $n/2$ and $n/2 + 1$.

Example 4.10 Considered the ordered observations

$$74.1, 76.4, 79.4, 79.9, 80.2, 82.1, 86.8, 89.3, 98, 103.5, 109.7, 121.2 \qquad (4.2)$$

We have $n = 12$ observations; since $(n + 1)/2 = 6.5$, we take the average of the sixth and seventh observations:

$$\frac{82.1 + 86.8}{2} = 84.45$$

□

The median is less sensitive than the mean to extreme data (possibly outliers). To see this, consider the dataset (4.2) and imagine substituting the smallest observation, $X_{(1)} = 74.1$, with a very small number. The mean is likely to be affected significantly, as the sample size is very small, but the median does not change. The same happens if we change $X_{(12)}$, i.e., the largest observation in the sample. This may be useful when the sample is small and chances are that an outlier enters the dataset. Generally speaking, there are statistics that may be *more robust* than other ones, and they should be considered when we have a small dataset that is sensitive to outliers.

The median can also be used to measure *skewness*. Observing the histograms in Fig. 4.5, we may notice that:

- For a perfectly symmetric distribution, mean and median are the same.

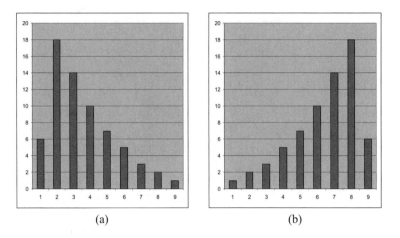

Fig. 4.5   Bar charts illustrating right- and left-skewed distributions.

- For a right-skewed distribution [see histogram (a) in Fig. 4.5], the mean is larger than the median (and we speak of positively skewed distributions); this happens because we have rather unlikely, but very high values that bias the mean to the right with respect to the median.

- By the same token, for a left-skewed distribution [see histogram (b) in Fig. 4.5], the mean is smaller than the median (and we speak of negatively skewed distributions).

In descriptive statistics there is no standard definition of skewness, but one possible definition, suggested by K. Pearson, is

$$\frac{3(\overline{X} - m)}{\sigma}$$

where m is the median and σ is the standard deviation, a measure of dispersion defined in the next section. This definition indeed shows how the difference between mean and median can be used to quantify skewness.[11]

Finally, another summary measure is the *mode*, which corresponds to the most frequent value. In the histograms of Fig. 4.5 the mode corresponds to the highest bar in the plot. In some cases, mean, mode, and median are the same. This happens in histogram (a) of Fig. 4.6. It might be tempting to generalize and say that the three measures are the same for a symmetric distribution, but a quick glance at Fig. 4.6(b) shows that this need not be the case.

---

[11]There are alternative definitions of skewness in descriptive statistics. Later, we will see that there is a standard definition of skewness in probability theory; see Section 7.5.

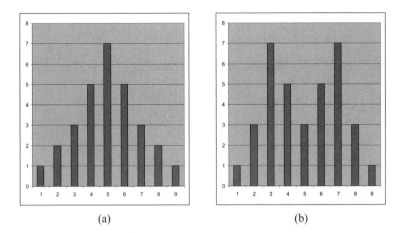

Fig. 4.6 Single and bimodal distributions.

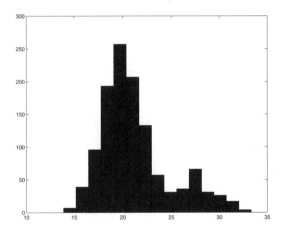

Fig. 4.7 A bimodal distribution.

Example 4.11 The histogram in Fig. 4.6(b) is somewhat pathological, as it has two modes. A more common occurrence is illustrated in Fig. 4.7, where there is one true mode (the "globally maximum" frequency) but also a secondary mode (a "locally maximum" frequency). A situation like this might be the result of sampling variability, in which case the secondary mode is just noise. In other cases, it might be the effect of a complex phenomenon and just "smoothing" the secondary mode is a mistake. We may list a few practical examples in which a secondary mode might result:

- The delivery lead time from a supplier, i.e., the time elapsing between issuing an order and receiving the shipment. Lead time may feature a little variability because of transportation times, but a rather long lead time may occur when the supplier runs out of stock. Ignoring this additional uncertainty may result in poor customer service.

- Consider the repair time of a manufacturing equipment. We may typically observe ordinary faults that take only a little time to be repaired, but occasionally we may have a major fault that takes much more time to be fixed.

- Quite often, in order to compare student grades across universities in different countries, histograms are prepared for each university and they are somehow matched in order to define fair conversion rules. Usually, this is done by implicitly assuming that there is a "standard" grade, to which some variability is superimposed. Truth is that the student population is far from uniform; we may have a secondary mode for the subset more skilled students, which actually constitute a different population than ordinary students.[12]

□

4.3.2 Dispersion measures

Location measures do not tell us anything about dispersion of data. We may have two distributions sharing the same mean, median, and mode, yet they are quite different. Figure 4.8, repeated from Chapter 1 (Fig. 1.6), illustrates the importance of dispersion in discerning the difference between distributions sharing location measures. One possible way to characterize dispersion is by measuring the *range* $X_{(n)} - X_{(1)}$, i.e., the difference between the largest and the smallest observations. However, the range has a couple of related shortcomings:

1. It uses only two observations of a possibly large dataset, with a corresponding potential loss of valuable information.

2. It is rather sensitive to extreme observations.

An alternative and arguably better idea is based on measuring deviations from the mean. We could consider the average deviation from the mean, i.e., something like

$$\frac{1}{n} \sum_{i=1}^{n} \left(X_i - \overline{X} \right)$$

However, it is easy to see that the above definition is useless, as the average deviation is identically zero by its very definition:

$$\frac{1}{n} \sum_{i=1}^{n} \left(X_i - \overline{X} \right) = \frac{1}{n} \sum_{i=1}^{n} X_i - \frac{1}{n} \sum_{i=1}^{n} \overline{X} = \overline{X} - \frac{n}{n} \overline{X} = 0 \qquad (4.3)$$

---

[12]Indeed, one of the most abject uses of statistics is using it to enforce some pattern on students' grades. As someone put it a while ago, there are lies, damned lies, and statistics.

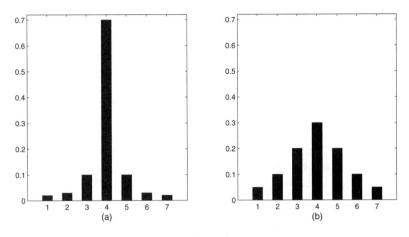

Fig. 4.8 Bar charts illustrating the role of dispersion: mean, median, and mode are the same, but the two distributions are quite different.

The problem is that we have positive and negative deviations canceling each other. To get rid of the sign of deviations, we might consider taking absolute values, which yields the *mean absolute deviation* (MAD):

$$\frac{1}{n} \sum_{i=1}^{n} |X_i - \overline{X}| \tag{4.4}$$

As an alternative, we may average the squared deviations, which leads to the most common measure of dispersion.

DEFINITION 4.6 (Variance) *In the case of a* population *of size n, variance is defined as*

$$\sigma^2 = \frac{1}{n} \sum_{i=1}^{n} (x_i - \mu)^2$$

In the case of a sample *of size n, variance is defined as*

$$S^2 = \frac{1}{n-1} \sum_{i=1}^{n} (X_i - \overline{X})^2$$

These definitions mirror the definition of mean for populations and samples. However, a rather puzzling feature of the definition of sample variance S^2 is the division by $n - 1$, instead of n. A convincing justification will be given in Section 9.1.2 within the framework of inferential statistics. For now, let us observe that the n deviations $(X_i - \overline{X})$ are not independent, since identity of Eq. (4.3) shows that when we know the sample mean and the first $n - 1$

deviations, we can easily figure out the last deviation.[13] In fact, there are only $n - 1$ independent pieces of information or, in other words, $n - 1$ *degrees of freedom*. Another informal argument is that since we do not know the true population mean μ, we have to settle for its estimate $\overline{X}$, and in estimating one parameter we lose one degree of freedom (1 df). This is actually useful as a mnemonic to help us deal with more complicated cases, where estimating multiple parameters results in the loss of more degrees of freedom.

Variance is more commonly used than MAD. With respect to MAD, variance enhances large deviations, since these are squared. Another reason, that will become apparent in the following, is that variance involves squaring deviations, and the function $g(z) = z^2$ is a nice differentiable one. MAD involves an absolute value $h(z) = |z|$, which is not that nice. However, taking a square does have a drawback: It changes the unit of measurement. For instance, variance of weekly demand should be measured in squares of items, and it is difficult to assign a meaning to that. This is why a strictly related measure of dispersion has been introduced.

DEFINITION 4.7 (Standard deviation) *Standard deviation is defined as the square root of variance. The usual notation, mirroring Definition 4.6, is σ for a population and S for a sample.*

The calculation of variance and standard deviation is simplified by the following shortcuts:

$$\sigma^2 = \frac{\sum_{i=1}^{n} x_i^2 - \frac{\left(\sum_{i=1}^{n} x_i\right)^2}{n}}{n} = \frac{1}{n}\left(\sum_{i=1}^{n} x_i^2 - n\mu^2\right) \tag{4.5}$$

$$S^2 = \frac{\sum_{i=1}^{n} X_i^2 - \frac{\left(\sum_{i=1}^{n} X_i\right)^2}{n}}{n-1} = \frac{1}{n-1}\left(\sum_{i=1}^{n} X_i^2 - n\overline{X}^2\right) \tag{4.6}$$

Example 4.12 Consider the sample:

$$62,\ 93,\ 126,\ 75,\ 34$$

We have

$$\sum_{i=1}^{5} X_i = 390, \qquad \sum_{i=1}^{5} X_i^2 = 35{,}150$$

---

[13]The careful reader will find this line of reasoning somewhat unconvincing, as the same observation could be applied to population variance. The true reason is that sample variance is used as an estimator of the true unknown variance, and the estimator is biased, i.e., is subject to a systematic error, if we divide by n rather than $n - 1$, as we prove in Section 9.1.2.

Table 4.7 Computing variance with raw and centered data.

X_i	$X_i - \overline{X}$
10,000,000,005	-22.5
10,000,000,010	-17.5
10,000,000,015	-12.5
10,000,000,020	-7.5
10,000,000,025	-2.5
10,000,000,030	2.5
10,000,000,035	7.5
10,000,000,040	12.5
10,000,000,045	17.5
10,000,000,050	22.5

Hence, sample variance is

$$S^2 = \frac{35{,}150 - \frac{390}{5}}{5 - 1} = 1182.50$$

and sample standard deviation is

$$S = \sqrt{1182.50} = 34.387$$

□

It is quite instructive to prove the above formulas. We consider here shortcut of Eq. (4.5), leaving the second one as an exercise:

$$
\begin{aligned}
\sigma^2 &\equiv \frac{1}{n} \sum_{i=1}^{n} (x_i - \mu)^2 = \frac{1}{n} \sum_{i=1}^{n} \left(x_i^2 - 2x_i\mu + \mu^2 \right) \\
&= \frac{1}{n} \left(\sum_{i=1}^{n} x_i^2 - 2\mu \sum_{i=1}^{n} x_i + n\mu^2 \right) = \frac{1}{n} \left(\sum_{i=1}^{n} x_i^2 - 2n\mu^2 + n\mu^2 \right) \\
&= \frac{1}{n} \left(\sum_{i=1}^{n} x_i^2 - n\mu^2 \right)
\end{aligned}
$$

These rearrangements do streamline calculations by hand or by a pocket calculator, but they can be computationally unfortunate when dealing with somewhat pathological cases.

Example 4.13 Consider the dataset in Table 4.7. The first column shows the raw data; the second column shows the corresponding *centered* data, which are obtained by subtracting the mean from the raw data. Of course, variance is the same in both cases, as shifting data by any amount does not affect dispersion. If we use the definition of variance, we get the correct result in both cases, $S^2 = 229.17$. However, if we apply the streamlined formula of Eq. (4.6), on a finite precision computer we get 0 for the raw data. This is a consequence of numerical errors, and we may see it clearly by considering just two observations with the same structure as the data in Table 4.7:

$$X_1 = \alpha + \epsilon_1, \qquad X_2 = \alpha + \epsilon_2$$

where ϵ_1 and ϵ_2 are much smaller than α. For instance, in the table we have $\alpha = 10{,}000{,}000{,}000$, $\epsilon_1 = 5$, and $\epsilon_2 = 10$. Then

$$\sum_{i=1}^{2} X_i^2 = X_1^2 + X_2^2 = (\alpha + \epsilon_1)^2 + (\alpha + \epsilon_2)^2$$

$$= 2\alpha^2 + 2\alpha(\epsilon_1 + \epsilon_2) + \left[\epsilon_1^2 + \epsilon_2^2\right]$$

and

$$2\overline{X}^2 = 2\left(\frac{X_1 + X_2}{2}\right)^2 = \frac{1}{2}(2\alpha + \epsilon_1 + \epsilon_2)^2$$

$$= 2\alpha^2 + 2\alpha(\epsilon_1 + \epsilon_2) + \left[\frac{\epsilon_1^2}{2} + \frac{\epsilon_2^2}{2} + \epsilon_1\epsilon_2\right]$$

We see that the two expressions are different, but since ϵ_1 and ϵ_2 are relatively small and get squared, the terms in the brackets are much smaller than the other ones. With a finite-precision computer arithmetic, they will be canceled in the calculations, so that the difference between the two expressions turns out to be zero. This is a *numerical error* due to *truncation*. In general, when taking the difference of similar quantities, a loss of precision may result. If we subtract the mean $\alpha + (\epsilon_1 + \epsilon_2)/2$, we compute variance with centered data:

$$X_1^c = \frac{\epsilon_1 - \epsilon_2}{2}, \quad X_2^c = \frac{\epsilon_2 - \epsilon_1}{2}$$

which yields the correct result

$$\frac{\epsilon_1^2}{2} + \frac{\epsilon_2^2}{2} - \epsilon_1\epsilon_2$$

with no risk of numerical cancelation. Although the effects need not be this striking in real-life datasets, it is generally advisable to work on centered data.

□

We close this section by pointing out a fundamental property of variance and standard deviation, due to the fact that they involve the sum of squares.

PROPERTY 4.8 *Variance and standard deviation can* never *be negative; they are zero in the "degenerate" case when there is no variability at all in the data.*

4.4   CUMULATIVE FREQUENCIES AND PERCENTILES

The median m is a value such that 50% of the observed values are smaller than or equal to it. In this section we generalize the idea to an arbitrary percentage. We could ask which value is such that 80% of the observations are smaller than or equal to it. Or, seeing things the other way around, we could ask what is the relative standing of an observed value. In Section 1.2.1 we anticipated quite practical motivations for asking such questions, which are of interest in measuring the service level in a supply chain or the financial risk of a portfolio of assets. The key concept is the set of cumulative (relative) frequencies.

DEFINITION 4.9 (Cumulative relative frequencies) *Consider a sample of n observations X_i, $i = 1, \ldots, n$, and group them in m classes, corresponding to each distinct observed value y_k, $k = 1, \ldots, m$. Classes are sorted in increasing order with respect to values: $y_k < y_{k+1}$. If f_k is the frequency of class k, the cumulative frequency of the corresponding value is the sum of all the frequencies up to and including that value:*

$$F_k = \sum_{j=1}^{k} f_j$$

By the same token, given relative frequencies $p_k = f_k/n$, we define cumulative relative frequencies:

$$P_k = \sum_{j=1}^{k} p_j$$

For the sake of simplicity, when no ambiguity arises, we often speak of cumulative frequencies, even though we refer to the relative ones.

Example 4.14 Consider the data in Table 4.8, which displays frequencies, relative frequencies, and cumulative frequencies for a dataset of 47 observations taking values in the set $\{1, 2, 3, 4, 5\}$. If we cumulate frequencies, we obtain cumulative frequencies:

$$F_1 = f_1 = 11$$
$$F_2 = f_1 + f_2 = 26$$
$$F_3 = f_1 + f_2 + f_3 = 36$$
$$F_4 = f_1 + f_2 + f_3 + f_4 = 43$$
$$F_5 = f_1 + f_2 + f_3 + f_4 + f_5 = 47$$

Table 4.8   Illustrating cumulative frequencies.

Value Y_k	Frequency f_k	Relative frequency p_k (%)	Cumulative frequency P_k (%)
1	11	23.40	23.40
2	15	31.91	55.32
3	10	21.28	76.60
4	7	14.89	91.49
5	4	8.51	100.00

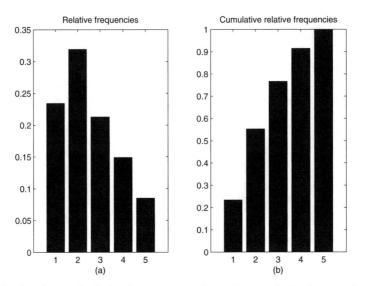

Fig. 4.9   Bar charts of relative frequencies and cumulative relative frequencies for the data in Table 4.8.

Cumulative relative frequencies are computed by adding relative frequencies:

$$P_1 = p_1 = 23.40\%$$
$$P_2 = p_1 + p_2 = 55.32\%$$
$$P_3 = p_1 + p_2 + p_3 = 76.60\%$$
$$P_4 = p_1 + p_2 + p_3 + p_4 = 91.49\%$$
$$P_5 = p_1 + p_2 + p_3 + p_4 + p_5 = 100.00\%$$

Since relative frequencies add up to 1, the last cumulative frequency must be 1 (or 100%). Since relative frequencies cannot be negative, cumulative frequencies form an increasing sequence; this is also illustrated in Fig. 4.9. Incidentally, the percentages in the first two rows of Table 4.8 may look wrong.

If we add up the first two relative frequencies, 23.40% and 31.91%, we obtain 55.31%, whereas the second cumulative relative frequency in the last column of the table is 55.32%. This is just the effect of rounding; indeed, such apparent inconsistencies are common when displaying cumulative frequencies.   ☐

Cumulative frequencies are related with a measure of relative standing of a *value* y_k, the *percentile rank*. A little example illustrates why one might be interested in percentile ranks.

Example 4.15 In French universities, grades are assigned on a numerical scale whose upper bound is 20 and the minimum for sufficiency is 10. A student has just passed a tough exam and is preparing to ask her parents for a well-deserved bonus, like a brand-new motorcycle, or a more powerful flamethrower, or maybe a trip to visit museums abroad. Unfortunately, her parents do not share her enthusiasm, as her grade is just 16, whereas the maximum is 20. Since this is just a bit above the midpoint of the range of sufficient grades (15), they argue that this is just a bit above average. She should do much better to earn a bonus! How can she defend her position?

As the saying goes, everything is relative. If she got the highest grade in the class, her claim is reasonable. Or maybe only 2 colleagues out of 70 earned a larger grade. What she needs to show is that she is near the top of the distribution of grades, and that a large percentage of students earned a worse grade.   ☐

The percentile rank of an *observation* X_i could be defined as the fraction of observations which are less than or equal to X_i:

$$\frac{\text{Number of observations less than or equal to } X_i}{\text{Total number of observations}} = \frac{b + e}{n}$$

where b is the number of observations *below* and e the number of observations *equal to* X_i, respectively. This definition is just based on the cumulative relative frequency corresponding to value X_i. However, there might be a little ambiguity. Imagine that all of the students in the example above have received the same grade, 16 out of 20. Using this definition, the percentile rank would be 100%. This is the same rank that our friend would get if she were the only student with a 16 out of 20, with everyone lagging far behind. A definition which does not discriminate between these two quite different cases is debatable indeed. We could argue that if everyone has received 16 out of 20, then the percentile rank for everyone should be 50%. Hence, we could consider the alternative definition of the percentile rank of X_i as

$$\frac{b + 0.5e}{n}$$

which accounts for observations equal to X_i in a slightly different way. Some well-known spreadsheets use still another definition, which eliminates the number of observations equal to X_i:

$$\frac{b}{b+a}$$

where b is the number of observations strictly below X_i, as before, and a is the number of observations strictly above X_i. We see that, sometimes, descriptive statistics is based on concepts that are a bit shaky; however, for a large dataset, the above ambiguity is often irrelevant in practice.

Now let us go the other way around. Given a value, we may be interested in its percentile rank, which is related to a cumulative frequency. Given a relative frequency, which is a percentage, we may ask what is the corresponding value. Essentially, we are inverting the mapping between values and cumulative frequencies. Values corresponding to a percentage are called *percentiles*. For instance, the median is just the 50% percentile, and we want to generalize the concept. Unfortunately, there is no standard definition of a percentile and software packages might yield slightly different values, especially for small datasets. In the following, we illustrate three possible approaches. None is definitely better than the other ones, and the choice may depend on the application.

Approach 1. Let us start with an intuitive definition. Say that we want to find the kth percentile. What we should do, in principle, is sort the n observations to get the order statistics $X_{(j)}$, $j = 1, \ldots, n$. Then, we should find the value corresponding to position $kn/100$. Since this ratio is not an integer in general, we might round it to the nearest integer. To illustrate, consider again the dataset (4.2), which we repeat here for convenience:

$$74.1, 76.4, 79.4, 79.9, 80.2, 82.1, 86.8, 89.3, 98, 103.5, 109.7, 121.2$$

What is the 42nd percentile? Since we have only 12 observations, one possible approach relies on the following calculation:

$$\frac{42 \times 12}{100} = 5.04 \approx 5$$

so that we should take the 5th element (80.2). Sometimes, it is suggested to add $\frac{1}{2}$ to the ratio above before rounding. Doing so, we are sure that at least 42% of the data are less than or equal to the corresponding percentile. Note that in the sample above we have distinct observed values. The example below illustrates the case of repeated values.

Example 4.16 Let us consider again the inventory management problem we considered in Section 1.2.1. For convenience, let us repeat here the cumulative frequencies of each value of observed demand:

Value	1	2	3	4	5
Cumulative frequency	0.15	0.55	0.80	0.90	1.00

Imagine that we want to order a number of items so that we satisfy the whole demand in at least 85% of the cases. If we trust the observed data, we should find a 85% percentile. Since we have 20 observations, we could look at the value $X_{(j)}$, where $j = 85 \times 20/100 = 17$. Looking at the disaggregated data, we see that $X_{(17)} = 4$, and there is no need for rounding. However, with large datasets it might be easier to work with cumulative frequencies. However, there is no value corresponding to a 85% cumulative frequency; what we may do, however, is take a value such that its cumulative frequency is *at least* 85%, which leads us to order four items. This example has two features:

- We want to be "on the safe side." We have a minimal service level, the probability of satisfying all customers, that we want to ensure. Hence, it makes sense to round up values. If the minimal service level were 79%, then $j = 79 \times 20/100 = 15.8$; looking at the order statistics, we see that $X_{(15)} = 3$ and $X_{(16)} = 4$, but we should order four items to be on the safe side.

- Percentiles in many cases, including this one, should correspond to *decisions*; since we can only order an integer number of items, we need a percentile that is an integer number. □

Approach 2. Rounding positions may be a sensible procedure, but it is not consistent with the definition of median that we have considered before. For an even number of observations, we defined the median as the average of two consecutive values. If we want to be consistent with this approach, we may define the kth percentile as a value such that:

1. At least $kn/100$ observations are less than or equal to it.

2. At least $(100 - k)n/100$ observations are greater than or equal to it.

For instance, if $n = 22$ and we are looking for the 80% percentile, we want a value such that at least $80 \times 22/100 = 17.6$ observations are less than or equal to it, which means that we should take $X_{(18)}$; on the other hand, at least $(100 - 80) \times 22/100 = 4.4$ values should be larger than or equal to it. Also this requirement leads us to consider the 18th observation, in ascending order. Hence, we see that in this case we just compute a position and then we round up. However, when $kn/100$ is an integer, two observations satisfy the above requirement, Indeed, this happens if we look for the 75% percentile and $n = 32$. Both $X_{(24)}$ and $X_{(25)}$ meet the two requirements stated above. So, we may take their average, which is exactly what happens when calculating the median of an even number of observations.

Approach 3. Considering the two methods above, which is the better one? Actually, it depends on our aims. Approach 2 does not make sense if the percentile we are looking must be a decision restricted to an integer value, as in Example 4.16. Furthermore, with approach 1 we are sure that the percentile will be an observed value, whereas with approach 2 we get a value that has not been observed. This is critical with integer variables, but if we are dealing with a continuous variable, it makes perfect sense. Indeed, there is still a third approach that can be used with continuous variables and is based on interpolating values, rather than rounding positions. The idea can be summarized as follows:

1. The sorted data values are taken to be the $100(0.5/n)$, $100(1.5/n)$, ..., $100([n - 0.5]/n)$ percentiles.

2. *Linear interpolation* is used to compute percentiles for percent values between $100(0.5/n)$ and $100([n - 0.5]/n)$.

3. The minimum or maximum values in the dataset are assigned to percentiles for percent values outside that range.

Let us illustrate linear interpolation with a toy example.

Example 4.17 We are given the dataset

$$3.4, \quad 7.2, \quad 8.3, \quad 9.6, \quad 12.5$$

According to the procedure above, 3.4 is taken as the 10% percentile, 7.2 is taken as the 30% percentile, and so on until 12.5, which is taken as the 90% percentile. If we ask for the 5% percentile, the procedure yields 3.4, the smallest observation. If we ask for the 95% percentile, the procedure yields 12.5, the largest observation.

Things are more interesting for the 42% percentile. This should be somewhere between 7.2, which corresponds to 30%, and 8.3, which corresponds to 50%. To see exactly where, in Fig. 4.10 we plot the cumulative frequency as a function of observed values, and we draw a line joining the two observations. How much should we move along the line from value 7.2 toward value 8.3? The length of the line segment is

$$8.3 - 7.2 = 1.1$$

and we should move by a fraction of that interval, given by

$$\frac{42 - 30}{50 - 30} = 0.6$$

Hence, the percentile we are looking for is

$$7.2 + 0.6 \times 1.1 = 7.86$$

☐

The above choice of low and high percentiles might look debatable, but it reflects the lack of knowledge about what may happen below $X_{(1)}$ and above

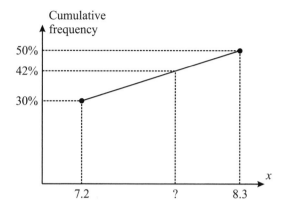

Fig. 4.10 Finding percentiles by interpolation.

$X_{(n)}$. Note that if we have a large dataset, $X_{(1)}$ will be taken as a lower percentile than in the example, reflecting the fact that with more observations we are more confident about the lowest value that observations may take; however, we seldom can claim that a value below $X_{(1)}$ cannot be observed. Similar considerations apply to $X_{(n)}$. In fact, there is little we can do about extreme values unless we superimpose a theoretical structure based on probability theory.

In practice, whatever approach we use, provided that it makes sense for the type of variable we are dealing with and the purpose of the analysis, it will not influence significantly the result for a large dataset. In the following chapters, when dealing with probability theory and random variables, we will introduce a strictly related concept, the quantile, which does have a standard definition.

4.4.1 Quartiles and boxplots

Among the many percentiles, a particular role is played by the *quartiles*, denoted by Q_1, Q_2, and Q_3, corresponding to 25%, 50%, and 75%, respectively. Clearly, Q_2 is simply the median. A look at these values and the mean tells a lot about the underlying distribution. Indeed, the *interquartile range*

$$\text{IQR} = Q_3 - Q_1$$

has been proposed as a measure of dispersion, and an alternative measure of skewness, called *Bowley skewness*, is

$$\frac{(Q_3 - Q_2) - (Q_2 - Q_1)}{(Q_3 - Q_1)} = \frac{(Q_3 - 2Q_2 + Q_1)}{(Q_3 - Q_1)}$$

The three quartiles are the basis of a common graphical representation of data, the *boxplot* (also known as a "whisker diagram"). A boxplot is shown in

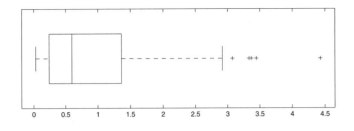

Fig. 4.11   A boxplot with outliers.

Fig. 4.11. In the picture you may notice a box. The line in the middle of the box corresponds to the median, whereas the two edges of the box correspond to the lower and upper quartiles. Dashed lines are drawn connecting the box to two *fences*. The two fences should be two bounds on the "normal" values of the observed variable. Any point beyond those fences is a potential outlier.

You will not be surprised to learn that there are alternative definitions of fences, and several variations on boxplots. One possible choice is to tentatively place the lower and upper fences at points

$$Q_1 - 1.5 \times \text{IQR}, \qquad Q_3 + 1.5 \times \text{IQR}$$

respectively. Points beyond such fences are regarded as outliers and are represented by a cross. If there is no outlier above the upper fence, this is placed corresponding to the largest observation; the lower fence is dealt with similarly. In Fig. 4.11 a dataset consisting of positive values is represented; since no observation is flagged as an outlier on the left part of the plot, the lower fence corresponds to the smallest observation, which is close to zero.

4.5   MULTIDIMENSIONAL DATA

So far, we have considered the organization and representation of data in one dimension, but in applications we often observe multidimensional data. Of course, we may list summary measures for each single variable, but this would miss an important point: the relationship between different variables.

We will devote all of Chapter 8 to issues concerning independence, correlation, etc. Here we want to get acquainted with those concepts in the simplest way. To begin with, let us consider bidimensional categorical data. To represent data of this kind, we may use a *contingency table*.

Example 4.18 Consider a sample of 500 married couples, where both husband and wife are employed. We collect information about yearly salary. For each person in the sample, we collect categorical information about the gender. The quantitative information about salary is transformed into categorical information by asking: Is salary less or more than \$30,000? We could express

Table 4.9   Contingency table for qualitative data

	Husband	
Wife	≤ \$30,000	> \$30,000
≤ \$30,000	212	198
> \$30,000	36	54

this as "high" and "low" income. We should not just disaggregate couples into two separate samples of 500 males and 500 females, as we could miss some information about the interactions between the two categorical variables. The contingency table in Table 4.9 is able to capture information about interactions. Armed with the contingency table, we may ask a few questions:

- What is the probability that a randomly selected female is high-income? We have 500 wives in the sample, and $36 + 54 = 90$ are high-income. Then, the desired probability[14] is

$$\frac{90}{500} = 18\%$$

- What is the probability that a randomly selected person, of whatever gender, is low-income? Note that we have 500+500 persons, since there are 500 pairs in the sample. We have 212 pairs in which both members are low-income, and $198 + 36$ pairs in which one of them is low-income. Hence, we should take the following ratio:

$$\frac{212 \times 2 + 198 + 36}{1000} = 65.8\%$$

- If we pick a couple at random, what is the probability that the wife is low-income, assuming that the husband is low-income? Apparently, this is a tough question, but we may find the answer using a little intuition. There are $212 + 36 = 248$ pairs in which the husband is low-income. We should restrict the sample to this subset and take the ratio

$$\frac{212}{248} = 85.48\%$$

since the wife is low-income in 212 out of these 248 pairs.

---

[14]Formally, we have not introduced probabilistic concepts yet, but we already know that relative frequencies may be interpreted as intuitive probabilities.

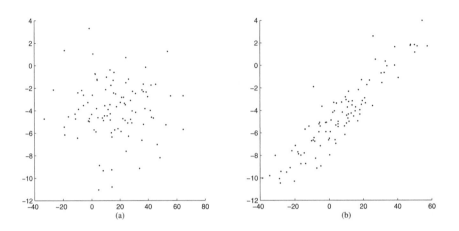

Fig. 4.12   Two scatterplots illustrating data dependence.

- If we pick a couple at random, what is the probability that the wife is low-income, assuming that the husband is high-income? Using the same idea as the previous question, we get

$$\frac{198}{198 + 54} = 78.57\%$$

In Section 5.3 we will see how the last two questions relate to the fundamental concept of conditional probability.   ☐

If we need to represent pairs of quantitative variables, we might aggregate them in classes and prepare corresponding contingency tables. A possibly more useful representation is the *scatterplot*, which is better suited to investigate the relationships between pairs of variables. In a bidimensional scatterplot, points are drawn corresponding to observations, which are pairs of values; coordinates are given by the values taken by the two variables in each observation. In Fig. 4.12 two radically different cases are illustrated. In scatterplot (a), we can hardly claim that the two variables have a definite relationship, since no pattern is evident; points look completely random. Scatterplot (b) is quite another matter, as it seems that there is indeed some association between the two variables; we could even imagine drawing a line passing through the data. This is what we do in Chapter 10, where we take advantage of this kind of association by building linear regression models.

Contingency tables and scatterplots work well in two dimensions, but what if we have 10 or even more dimensions? How can we visualize data in order to discern potentially interesting associations and patterns? These are challenging issues dealt with within multivariate statistics. One possibility is trying to generalize the analysis for two dimensions. For instance, we may arrange several scatterplots according to a matrix, one plot for each possible pair of variables. As you may imagine, these graphical approaches can be useful in low-dimensional cases but they are not fully satisfactory. More refined alternatives are based, e.g., on the following approaches:

- We can reduce data dimensionality by considering combination of variables or underlying factors.

- We can try to spot and classify patterns by cluster analysis.

Data reduction methods, including principal component analysis, factor analysis, and cluster analysis, are dealt with in Chapter 17.

Problems

4.1   You are carrying out a research about how many pizzas are consumed by teenagers, in the age range from 13 to 17. A sample of 20 boys/girls in that age range is taken, and the number of pizzas eaten per month is given in the following table:

4	12	7	11	9	7	8	13	16	11
4	7	5	7	11	7	7	41	9	14

- Compute mean, median, and standard deviation.

- Is there any odd observation in the dataset? If so, get rid of it and repeat the calculation of mean and median. Which one is more affected by an extreme value?

4.2   The following table shows a set of observed values and their frequencies:

Value	1	2	3	4	5	6	7	8
Frequency	5	4	7	10	13	8	3	1

- Compute mean, variance, and standard deviation.

- Find the cumulated relative frequencies.

4.3   You observe the following data, reporting the number of daily emergency calls received by a firm providing immediate repair services for critical equipment:

Day	1	2	3	4	5	6	7	8
N calls	5	4	6	2	3	8	2	4
Day	9	10	11	12	13	14	15	16
N calls	4	2	3	5	6	4	4	6

- Compute mean, mode, and quartiles.

- Find the cumulated relative frequencies.

4.4   Management wants to investigate the time it takes to complete a manual assembly task. A sample of 12 workers is timed, yielding the following data (in seconds):

> 21.3 15.2 13.6 16.1 15.0 19.2 21.0 14.3 15.6 20.1 21.1 22.2

- Find mean and median; do you think that the data are skewed?

- Find the standard deviation.

- What is the percentile rank of the person who took 20.1 seconds to complete the task?

- Find the quartiles, using the second approach that we have described for the calculation of percentiles.

- Suppose that management want to define an acceptable threshold, based on the 90% percentile; all workers taking more than this time are invited to a training session to improve their performance. Find this percentile using the interpolation method.

4.5   Professors at a rather unknown but large college have developed a habit of heavy drinking to forget about their students. The following data show the number of hangovers since the beginning of semester, disaggregated for male and female professors:

	N hangovers		
Gender	0	1	≥ 2
Male	61	23	40
Female	66	25	36

- Given that the professor is a female, what is the probability that she had a hangover twice or more during the semester?

- What is the probability that a professor is a male, given that he had a hangover once or less during the semester?

For further reading

- All introductory books on statistics offer a treatment of the essentials of descriptive statistics; a couple of examples are Refs. [3] and [4], which have also inspired some of the examples in this chapter.

- We did not cover at all the issues involved in designing and administering questionnaires for population surveys; they are dealt with at an introductory level in the text by Curwin and Slater [1].

- The graphical representation of multivariate data is dealt with, e.g., in the book by Härdle and Simar [2].

- Example 4.13 is based on H. Pottel, *Statistical flaws in Excel*. This unpublished paper may be downloaded from a few Webpages, including `http://www.mis.coventry.ac.uk/~nhunt/pottel.pdf`.

REFERENCES

1. J. Curwin and R. Slater, *Quantitative Methods for Business Decisions*, 6th ed., Cengage Learning EMEA, London, 2008.

2. W. Härdle and L. Simar, *Applied Multivariate Statistical Analysis*, 2nd ed., Springer, Berlin, 2007.

3. M.K. Pelosi and T.M. Sandifer, *Elementary Statistics: From Discovery to Decision*, Wiley, New York, 2003.

4. S.M. Ross, *Introduction to Probability and Statistics for Engineers and Scientists*, 4th ed., Elsevier Academic Press, Burlington, MA, 2009.

5

Probability Theories

Any book on quantitative methods includes a chapter on probability theory, and this one is no exception. However, the careful reader should wonder why this chapter's title mentions probability *theories*. In Section 5.1 we show that probability, like uncertainty, is a rather elusive concept. Descriptive statistics suggests the concept of probabilities as relative frequencies, but we may also interpret probability as plausibility related to a state of belief. The origin of the mathematical approach to probability can be traced back to Jacob Bernoulli, Thomas Bayes, and Pierre-Simon Laplace. Bernoulli's *Ars Conjectandi* (*The Art of Conjecture*) was published 8 years after his death in 1713, and Laplace published his *Théorie analytique des probabilités* in 1812. More recently, the axiomatic approach due to Andrei Nikolaevich Kolmogorov (1933) was proposed and has become a sort of standard approach to probability. We will follow the last approach in this and subsequent chapters, because it suits our purpose very well, but it is always healthy to keep in mind that "standard" does *not* mean "always the best." We come back to such issues in Chapter 14, while in this chapter we first introduce the axiomatic approach to probability theory in Section 5.2, laying down the fundamental concepts of events and probability measures, along with a set of basic rules of the game in order to work with probabilities in a sensible and consistent manner. In Section 5.3 we introduce conditional probabilities; we do so in a mathematically unsophisticated way, but we insist that conditioning is a powerful and essential concept that is used to model information availability to decision makers. Conditional probabilities also lead to a powerful result called *Bayes' theorem*, which we will come to appreciate in Section 5.4.

In this chapter we keep mathematical sophistication to a minimum, since our purpose is just to introduce the essential concepts that are used later to study random variables and inferential statistics. Some advanced topics in probability do require a more in-depth treatment based on a more sophisticated mathematical machinery. We provide references for the interested reader, and we will just give a little flavor of this later, in Section 7.10, in the context of random variables.

5.1 DIFFERENT CONCEPTS OF PROBABILITY

In Chapter 4 we have met relative frequencies, a fundamental concept in descriptive statistics. Intuitively, relative frequencies can be interpreted as "probabilities" in some sense, as they should tell us something about the likelihood of events. While this is legitimate and quite sensible in many settings, we should wonder whether this *frequentist* interpretation is the only meaning that we may possibly attach to the more or less intuitive notion of probability. In fact, when doing so we implicitly take for granted that

1. We have a suitable number of observations to estimate relative frequencies in a reliable way.

2. Past outcomes help us in making decisions for the future.

As you may imagine, none of the above should be taken for granted. There might be very rare, yet potentially relevant events whose likelihood is hard to evaluate precisely for the very reason that they are indeed rare. How many times did we observe a financial crisis due to subprime mortgages? Furthermore, market conditions do change in time, and past knowledge need not be 100% helpful in predicting the future. Whenever people are involved, rather than mechanical devices, repeatability of an experiment is not ensured.

Indeed, sometimes probability is more akin to the idea of "belief"; asking what is the probability that a war will erupt in some place under certain sociopolitical conditions is very different from asking what is the probability of some outcome in a game of chance based on dice throwing. Certainly, we should not like the idea of running many experiments to identify relative frequencies in the first case. Hence, we should pause a little and wonder whether there are different concepts of probabilities.

Consider a prototypical random experiment, dice throwing, and the following questions:

Q1. If we throw a die, what is the probability that the outcome is 5?

Q2. If we throw a die, what is the probability that the outcome is 5 or 2?

Q3. If we throw a die, what is the probability that the outcome is an even number?

Q4. If we throw two dice, what is the probability that the sum of the outcomes is 7?

The answers are rather easy to find, but we should reflect on the underlying principles that are used to come up with each answer.

A1. Assuming that the die is fair and no one is cheating, most of us would say that the answer is $\frac{1}{6}$, i.e., 1 in 6. Are we using relative frequencies in finding this answer? Not really, unless we want to throw that die a huge number of times to check the result empirically. The relative frequency that we would obtain will likely get close to $\frac{1}{6}$, but not exactly. Since the number of possible outcomes is 6 (ruling out the remote possibility that the die lands on an edge or a vertex), the intuitive justification for our answer is symmetry. We do not see a strong rationale for saying that the likelihoods of the possible results are different.[1] This symmetry is the foundation of the *classical* concept of probability. Actually, the die is not perfectly symmetric, as someone punched little holes on its faces, which are not perfectly equal. However, we do not know how to measure the impact of this lack of symmetry, if any. Of course, we could throw the die a huge number of times to see if there is a bias in favor of some outcome, but then the same procedure should be repeated for any kind of die, as it could depend on size, weight, and material. This does not sound too practical, but, since we are interested in management and decision making, there is a more important point. Say that there is indeed a small experimental discrepancy in the relative frequency of each outcome. Should we rely on that information in order to make a decision? Would it really make a difference?

A2. Since the two outcomes have the same probability, and they cannot occur at the same time, the intuitive answer is

$$1/6 + 1/6 = 2/6 = 1/3$$

Hence, we are just summing probabilities of elementary outcomes, which seems rather plausible in this case. Maybe, in more involved experiments, where we have to deal with complex events, we cannot just add probabilities like that. Nevertheless, the idea of adding probabilities looks sensible when outcomes are mutually exclusive, and it can be considered a basic rule of the game.

A3. Using more or less the same reasoning as before, the answer should be $\frac{1}{2}$. We may think of that as the sum of the probabilities of getting either 2, or 4, or 6. Alternatively, we may consider two mutually exclusive events, "even" and "odd," with the same probability. Whatever the choice, we

---

[1] Bernoulli referred to this concept as the *principle of insufficient reason.*

see that events need not be restricted to elementary outcomes. We might deal with events consisting of several elementary outcomes for at least a couple of reasons. First, maybe all we can observe is just "even" or "odd," because we are not able to see the exact result. In many practical problems, we are not allowed to observe everything and we must settle for some partial information. Moreover, we might be interested only in those two outcomes, because we are betting on them, and more detailed information is irrelevant for our purposes. Whatever the reason, we realize that events may consist of multiple outcomes, and we must find a sensible and consistent way to work with them.

A4. To begin with, it is reasonable to assume that the two dice do not influence each other. Then, there are $6 \times 6 = 36$ possible outcomes of the form (D_1, D_2), where both D_1 and D_2 can take any integer value between 1 and 6. Hence, we should just count the number of outcomes are such that $D_1 + D_2 = 7$. There are six such outcomes:

$$(1, 6) \quad (2, 5) \quad (3, 4) \quad (4, 3) \quad (5, 2) \quad (6, 1)$$

out of the 36 possible cases. Hence, the required probability is $\frac{6}{36} = 6$. What we see in action here is a classical approach with historical roots in gambling; we have many equally outcomes, and we just take the ratio of the number of "favorable" ones over their total number. In arriving at the total number of outcomes, and in assessing their likelihood, we assumed that the two dice are independent. As dice have no memory, we could even throw the same die twice. We have seen something similar in coin flipping (see Section 1.2.3); there, the probability of getting a particular result, say "head–head," when flipping a coin twice, is just the product of elementary probabilities:

$$\frac{1}{2} \times \frac{1}{2} = \frac{1}{4}$$

It seems that when considering independent events this is a plausible rule of the game. Yet, that discussion pointed out that sometimes events are not independent at all, and we may take advantage of this. So, we must make this concept a bit more precise.

The discussion of these four questions points out a few basic requirements on how we should work with probabilities. Moreover, we see that there are at least two different ways to regard probabilities: Descriptive statistics suggests the idea of probability as relative frequencies, whereas the classical approach relies on symmetry and counting arguments. Yet, these two concepts do not cover all of the possibilities. To see why, consider the following:

- Elementary counting does not work when there are an infinite number of outcomes, as in the case of real numbers on an interval.

- We can work with relative frequencies if past data are available and relevant, but this is not the case when forecasting sales for a brand new product, possibly representing a real technological breakthrough. If past data are not helpful at all, we might be forced to work with probabilities as beliefs, i.e., subjective assessments of likelihood.

Subjective probability does not sound like a rigorous and scientific concept. Yet, this is what we have to work with in many situations, and *subjective* does not imply *free from any rule*, as shown by the following experiment (described, e.g., in the text by Kahneman et al. [5]).

Example 5.1 Consider the following description of a person:

> *Linda is 31 years old, single, outspoken, and very bright. She majored in philosophy. As a student, she was deeply concerned with issues of discrimination and social justice, and she participated in antinuclear demonstrations.*

On the basis the information above, rank the following statements in decreasing order of likelihood, i.e., from the most probable to the least probable.

(a) Linda is a teacher in an elementary school.

(b) Linda works in a bookstore and takes yoga classes.

(c) Linda is active in a feminist movement.

(d) Linda is a psychiatric social worker.

(e) Linda is a member of the League of Women Voters.

(f) Linda is a bank teller.

(g) Linda is an insurance salesperson.

(h) Linda is a bank teller who is active in a feminist movement.

Please, do rank the statements before reading further!

Given the limited evidence we have, there are some statements that may be ranked in any order, depending on your subjective opinion. For instance, it is difficult to say if (f) is more likely than (g) or vice versa. Many people, when handed this question, rank (c) higher than (f) and (g) because the description suggests a rather precise kind of person. Again, this is consistent with the concept of subjective assessment of probability.

Yet, this does not mean that *any* ordering makes sense. A rather surprising experimental fact, reported in Ref. [5], is that many respondents consider (h) more likely than (f). It is easy to see that this makes no sense. If we consider the issue in terms of relative frequencies, the set of persons meeting the condition in (h) is clearly a subset of the set of persons meeting the condition in (f). Or, if we take a logical viewpoint, (h) implies (f), but not vice versa. Indeed, is no way (f) can be less plausible than (h).

Typically, those who rank (f) and (h) in the wrong way do not make the same mistake with (c) and (h). Arguably, the psychological trap is that Linda does not seem like the prototypical bank teller, but adding the second feature (she is an active feminist) to statement (h) tricks many into believing that this is more plausible than (f).   ☐

The example above shows that even if we deal with subjective probability, there must be some logical and consistent structure in the way we think. Indeed, in Section 1.2.2 we have seen that intuition may lead us to wrong conclusions. Let us consider a similar example here.

Example 5.2 A quick and easy test is able to predict the gender of a baby very early during childbearing. Unfortunately, the test is not 100% reliable:

- If the unborn child is a male, the result of the test is "male" with a probability of 90%.

- If the unborn child is a female, the result of the test is "female" with a probability of 70%.

Frances tries the test, and the result is "female." Mary tries the test, and the result is "male." Between Frances and Mary, which one should be more confident (or less uncertain) about the gender of her child?   ☐

In this case, too, many are tricked by wrong intuition and believe that the correct answer is Mary. If you see some similarity with the example of Section 1.2.2, please try the same line of reasoning to prove that the correct answer is Frances. (*Hint*: Say that we consider 200 unborn babies, and that exactly half of them are males; how many tests will predict "male"?)

In Section 5.4 we illustrate a systematic way to solve such puzzles. For now, we might have more than enough evidence that some discipline should be involved in dealing with probabilities. To this aim, we will consider the so-called *axiomatic* approach to the theory of probabilities. Even if it is not free from some criticism, this is the most common approach and is a starting point to educate our way of reasoning with probabilities.

5.2   THE AXIOMATIC APPROACH

The axiomatic approach aims at building a consistent theory of probability and is based on the following logical steps:

1. Defining the object of investigation, i.e., events

2. Defining an algebra of events, i.e., ways to combine events to describe nontrivial occurrences that we might be interested in

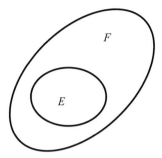

Fig. 5.1 E is a subset of F.

3. Defining the rules of the game that we need in order to assign a probability measure to each event in a coherent way

5.2.1 Sample space and events

To get going, we should first formalize a few concepts about running a random experiment and observing outcomes. The set of possible outcomes is called the *sample space*, denoted by Ω. For instance, in dice throwing the set of possible outcomes is

$$\Omega = \{1, 2, 3, 4, 5, 6\}$$

This is a finite set, but we might consider alternative random experiments where the set of possible outcome is an infinite set, such as the whole set of integer numbers. Combining random experiments, e.g., by throwing more dice, or the same one repeatedly, we may define rather complex sample spaces. An important feature of outcomes is that they are mutually exclusive, i.e., they cannot occur together.

As we pointed out before, we need not be only interested in simple events consisting of singletons, i.e., elementary outcomes. We have already met compound events such as

$$\mathsf{EVEN} = \{2, 4, 6\}, \qquad \mathsf{ODD} = \{1, 3, 5\}$$

Typically events correspond to statements such as "the outcome is larger than two" or "the outcome is between 3 and 5." Whatever the case, we see that the elements of these sets are also elements of the sample space Ω. We know from set theory that a set E is a subset of set F, denoted by $E \subseteq F$ when all of the elements of E belong to F (see Fig. 5.1).

DEFINITION 5.1 *An* event E *is a* subset *of the sample space, i.e.,* $E \subseteq \Omega$.

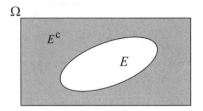

Fig. 5.2   An event and its complement.

5.2.2   The algebra of events

Given the definition of events, let us consider how we may build possibly complex events that have a practical relevance. Indeed, we often deal with the following concepts:

- The probability that an event does *not* occur

- The probability that at least one of two events occurs

- The probability that two events occur jointly

Since events are sets, it is natural to translate the concepts above in terms of set theory, relying on the usual difference, union, and intersection of sets.

The *difference* between sets A and B, denoted by $A\backslash B$, is a set consisting of the elements of A that do not belong to B. Given an event $E \subseteq \Omega$, its *complement* $E^c \equiv \Omega\backslash E$ occurs if and only if E does not. For instance, in dice throwing we have $\mathsf{EVEN}^c = \mathsf{ODD}$. Graphically, the complement of an event can be depicted as in Fig. 5.2. The probability of the complement E^c is just the probability that event E does not occur, but these two probabilities should be related in a plausible way.

The *union* of two sets A and B, denoted by $A \cup B$, is a set, consisting of the elements that belong to at least one of them (either A, or B, or both). Set union is depicted in Fig. 5.3. We immediately see that the probability of the union of two events is the probability that at least one of them (possibly both) occurs. Please note that we are *not* requiring that exactly one of them occurs. That would be an *exclusive* OR operation, which is perfectly legitimate per se, but set union is based on an *inclusive* OR operation.

Finally, the *intersection* of two sets A and B, denoted by $A \cap B$, is a set, consisting of the elements that belong to both A and B, as illustrated in Fig. 5.4. We immediately see that the probability of the intersection of two events is the probability that both of them occur jointly.[2]

---

[2]It may be worth clarifying what "jointly" really means. It does *not* necessarily mean "at the same time," since one can conceive a random experiment requiring multiple steps over time, and the two events could refer to things that happen in sequence. Nevertheless, they would be part of the same outcome of the overall experiment.

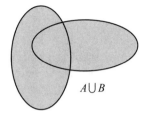

Fig. 5.3   The union of two sets.

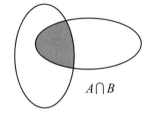

Fig. 5.4   The intersection of two sets.

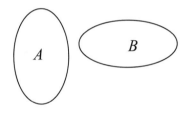

Fig. 5.5   Two disjoint sets.

The *empty set*, denoted by $\emptyset$, is a set with no element. Two sets are called *disjoint* if their intersection is the empty set, i.e., $A \cap B = \emptyset$ (see Fig. 5.5).

Given a sample space Ω, by the repeated application of these elementary set operations, we can build a huge collection of subsets of Ω. Let $\mathcal{F}$ the family of all sets we can build this way, working on events within a given sample space. We would like to assign probabilities to events, using sensible rules of the game, in such a way that the probabilities of complicated events are consistently related to the probabilities of the events that we used to build them. This is where the axioms of probability theory come into play. They are described in the next section in a simple and intuitive manner. We should mention that this intuition is all we need for the rest of the book, but a proper construction of probability theory is not that trivial when we deal with infinite sample spaces and possibly infinite collections of events. Generally speaking, going to infinite is always a tricky endeavor in the realm of mathematics. Still, the intuition we build, based on finite sample spaces, is perfectly adequate to our purposes.

5.2.3 Probability measures

The final step is associating each event $E \in \mathcal{F}$ with a probability measure $P(E)$, in some sensible way. As a starting point, it stands to reason that, for an event $E \subseteq \Omega$, its probability measure should be a number satisfying the following condition:

$$0 \leq P(E) \leq 1$$

This is certainly true if we think of probabilities in terms of relative frequencies, but it also applies to whatever likelihood concept we wish to consider, including subjective belief.[3] Furthermore, since the sample space Ω is, in a sense, the largest event including all of the other ones, we should also assume

$$P\{\Omega\} = 1$$

Loosely speaking, this condition says that something has to happen. Finally, let us consider the union of disjoint events. Again, intuition suggests that, in this case

$$P\left(E_1 \cup E_2\right) = P(E_1) + P(E_2)$$

Hence probabilities are *additive* for disjoint events. A simple example is question Q2 above, where the probability that a die yields 2 or 5 is just the sum of the two respective probabilities. The idea can be generalized to an arbitrary number of disjoint events but, when events have an intersection, additivity need not hold.

Example 5.3 Consider a deck of 52 poker cards. If we draw a card at random, the probability that it is a king is $\frac{4}{52} = \frac{1}{13}$. Similarly, there is a probability $\frac{1}{4}$ that it is spades. But what is the probability that it is the king or spades? If we just add probabilities, we get

$$\tfrac{4}{52} + \tfrac{13}{52} = \tfrac{17}{52}$$

but there is something wrong: We are counting twice the king of spades, which is the intersection of the set of kings and the set of spades. We get the correct result, $\frac{16}{52}$, if we subtract the intersection, so that common elements are correctly counted once. ▯

The example suggests that sensible rules should apply to manage complicated events, possibly amounting to a huge list. Actually, it turns out that the rules we need can be obtained as a consequence of the first three rules, which we take as the following *axioms* to define a probability measure.

---

[3]In this case, we could assign the value 1 to a statement that we are absolutely sure about, and the value 0 to something we do not believe at all, with every shade of gray in between.

DEFINITION 5.2 *A probability measure* $P(\cdot)$ *is a mapping from events E within a sample space Ω to real numbers such that*[4]

1. $0 \leq P(E) \leq 1$, *for all $E \in \mathcal{F}$*

2. $P(\Omega) = 1$

3. *For each sequence $E_1, E_2, E_3, \ldots$, of mutually exclusive (disjoint) events, i.e., such that $E_i \cap E_j = \emptyset$ for $i \neq j$, we have*

$$P\left(\bigcup_{i=1}^{\infty} E_i\right) = \sum_{i=1}^{\infty} P(E_i)$$

The last axiom may look a bit awkward, but it is just the generalization of additivity for probabilities of disjoint events to a possibly infinite (countable) number of events. From these axioms about events and probabilities, we can derive some properties that are intuitive, as well as some that are not.

Example 5.4 Given the probability $P(E)$ of event E, what is the probability of its complement E^c? Since these two events are obviously disjoint, and $E \cup E^c = \Omega$, using the axioms, we obtain

$$P(E) + P(E^c) = P(\Omega) = 1$$

Hence, the probability that an event does *not* happen $P(E^c) = 1 - P(E)$. Using this theorem, we may also see that $P(\emptyset) = 0$. □

Example 5.5 Example 5.3 above suggests that, if two events E_1 and E_2 are not disjoint, the following should hold:

$$P(E_1 \cup E_2) = P(E_1) + P(E_2) - P(E_1 \cap E_2)$$

To prove this, we may note that the union of sets E_1 and E_2 can be expressed in terms of disjoint sets:

$$E_1 \cup E_2 = E_1 \cup (E_2 \backslash E_1)$$

In plain English, this amounts to saying that the union of E_1 and E_2 can be rewritten as the union of two sets: E_1 and the part of E_2 that is disjoint from E_1 (it may help to check this by a simple drawing). Hence, we may use the third axiom:

$$P(E_1 \cup E_2) = P(E_1) + P(E_2 \backslash E_1) \tag{5.1}$$

Furthermore, we may express E_2, too, as the union of disjoint sets:

$$E_2 = (E_2 \backslash E_1) \cup (E_1 \cap E_2)$$

---

[4]Here we are a bit sloppy about the definition of the family $\mathcal{F}$ of events; more on this in Section 7.10.

In plain English, this amounts to saying that E_2 consists of the union of two subsets: the part of E_2 that is disjoint from E_1 and the intersection of E_1 and E_2. Then

$$P(E_2 \backslash E_1) = P(E_2) - P(E_1 \cap E_2)$$

which can be plugged into Eq. (5.1) to obtain the result immediately.   ◻

5.3 CONDITIONAL PROBABILITY AND INDEPENDENCE

Consider throwing a die twice. If we know that the result of the first draw is 4, does this change our probability assessment for the second draw? If the die is fair, and there is no cheating on the part of the person throwing it, the answer should be no. The two rolls are independent. In other cases, however, knowing that an event has occurred does tell us something about another event. We have seen such a case when dealing with the growth option example in Section 1.2.3. To formalize this, we should draw the line between two concepts:

1. The *a priori* or *unconditional* probability of an event, which should apply when we do not have any information about occurred events

2. The *conditional* probability, which results from a reassessment after collecting some (partial) knowledge represented by the occurrence of related events

DEFINITION 5.3 (Conditional probability) *The probability of event E, conditional on G, is denoted by $P(E \mid G)$ and is defined as*[5]

$$P(E \mid G) = \frac{P(E \cap G)}{P(G)}.$$

Example 5.6 In dice throwing we know that, a priori, $P(\{1\}) = P(\{2\}) = \frac{1}{6}$. But if we know that the event **EVEN** took place, we should update the *unconditional* probabilities, getting *conditional* probabilities. For instance, $P(\{1\} \mid \textbf{EVEN}) = 0$, as 1 is an odd number and is ruled out if we know that the event **EVEN** happened. We may also see intuitively that $P(\{2\} \mid \textbf{EVEN})$ should be $\frac{1}{3}$, as 2 is just one possible outcome out of three, if we know that the event **EVEN** occurred.

---

[5] *A word of caution:* The careful reader will immediately see that this definition is in trouble when $P(G) = 0$. It is tempting to say that this is not really an issue since, if G cannot occur, there is no point in conditioning with respect to that event. Unfortunately, we will see in Chapter 7, when dealing with continuous random variables, that events with zero probability can and do happen. Indeed, conditional probabilities are a more challenging subject than one would imagine at first sight, and this is why rigorous probability theory requires a nontrivial mathematical machinery.

We can obtain these results in a more systematic manner using the definition. For instance

$$P(\{2\} \,|\, \mathsf{EVEN}) = \frac{P(\{2\} \cap \mathsf{EVEN})}{P(\mathsf{EVEN})} = \frac{P(\{2\})}{P(\mathsf{EVEN})} = \frac{\frac{1}{6}}{\frac{1}{2}} = \frac{1}{3}$$

$\square$

The definition of conditional probability may look a bit weird at first, but it can be justified on the ground of the following intuition.

1. A priori, whatever happens must lie in the sample space Ω, and $P(\Omega) = 1$. If we know that G occurred, this is the new sample space, whose probability a priori was $P(G)$. In the example above, $G = \{2, 4, 6\}$ is the new sample space, if we know that event EVEN occurred. Dividing by the probability of G, which is typically less than 1, amounts to increasing all of the probabilities by a sensible renormalization factor. Indeed, such a renormalization implies that $P(G\,|\,G) = 1$, as it should be the case. This explains the term $P(G)$ at the denominator.

2. If G is the new sample space, event E may occur only if the intersection $E \cap G$ occurs. This explains the term $P(E \cap G)$ at the numerator.

There are cases in which the unconditional and the conditional probabilities are quite different. In other cases, information on an event G tells us nothing about another event E. In other words, the two events are independent.

DEFINITION 5.4 (Independence of two events) *Two events E and G are said to be independent if*

$$P(E \cap G) = P(E) \cdot P(G)$$

In other words, for independent events the joint probability can be expressed as the product of the individual probabilities.

This definition might seem a bit unrelated with conditional probability. However, it is easy to see that if E and G are independent events, then

$$P(E\,|\,G) \equiv \frac{P(E \cap G)}{P(G)} = \frac{P(E) \cdot P(G)}{P(G)} = P(E)$$

and, by the very same token, $P(G\,|\,E) = P(G)$. Hence, for independent events, unconditional and conditional probabilities are exactly the same and the occurrence of one event does not provide any useful information about the other one. Now it is a good idea is to check your understanding of independence with a couple of questions:

Q1. Are two disjoint events independent?

Q2. If $G \subset F$, are the two events independent?

Please: Answer before reading further!

When I ask students the first question, I typically emphasize the fact that the two events are disjoint and that they "have nothing to do with each other." This is usually enough to trick them into answering "Yes, disjoint events must be independent!" A bit of reflection should tell you that this is plain wrong. If E and G are disjoint and we know that G occurred, then we may rule out E. Hence, they *cannot* be independent. More formally (assuming that the two events have strictly positive probabilities):

$$E \cap G = \emptyset \quad \Rightarrow \quad P(E \mid G) = \frac{P(E \cap G)}{P(G)} = 0 \neq P(E).$$

The second question is a bit easier. If G is included in F, then the occurrence of G implies the occurrence of F. Formally (ruling out events with zero probability again), we can state

$$G \subseteq F \quad \Rightarrow \quad P(F \mid G) = \frac{P(F \cap G)}{P(G)} = \frac{P(G)}{P(G)} = 1 \neq P(F)$$

The moral of the story is that, even though the definition of independence concerns the possibility of factoring a joint probability into the product of independent probabilities, you have to think in terms of *information* to fully appreciate the issues involved. There is a good reason, though, to phrase the definition of independent events in terms of a product: It generalizes immediately to more than two events.

DEFINITION 5.5 (Independence of multiple events) *Consider a family of events* $\{E_1, E_2, \ldots, E_n\}$. *The events* $E_1, E_2, \ldots, E_n$ *are said to be independent if, given any arbitrary subset* $E_{j_1}, E_{j_2}, \ldots, E_{j_m}$ *of the family, with* $m \leq n$, *we have*

$$P(E_{j_1} \cap E_{j_2} \cap \cdots \cap E_{j_m}) = P(E_{j_1}) \cdot P(E_{j_2}) \cdots P(E_{j_m}) \qquad (5.2)$$

This definition might look overly involved. Is it not enough to require that the factorization condition applies to the whole family of n events and so be it? What the definition aims at capturing is that to have n independent events, knowledge about *any* subset of events should not tell us anything about the remaining ones. However, the following conjectures do seem plausible:

- If all of the events in a family are pairwise independent, can we say that they are independent in the sense of Definition 5.5?

- If Eq. (5.2) holds for the whole family, can we say that this implies a similar condition for the subsets of the family?

Intuition may be misleading, at times, and in fact the answer is no in both cases, as the following counterexamples show.

Example 5.7 Consider a random experiment consisting of the draw of an integer number between 1 and 4, where the four outcomes are equally likely. We see that the events $A \equiv \{1,2\}$, $B \equiv \{1,3\}$, $C \equiv \{1,4\}$ have the same probability, $\frac{1}{2}$. It is also easy to see that these events are pairwise independent:

$$P(A \cap B) = \tfrac{1}{4} = P(A) \cdot P(B)$$
$$P(A \cap C) = \tfrac{1}{4} = P(A) \cdot P(C)$$
$$P(B \cap C) = \tfrac{1}{4} = P(B) \cdot P(C)$$

However,

$$P(A \cap B \cap C) = P(\{1\}) = \tfrac{1}{4} \neq P(A) \cdot P(B) \cdot P(C)$$

To really get the point, it is useful to reason in terms of information and conditional probabilities. For instance, $P(A \mid B) = P(A) = \frac{1}{2}$, because knowing that B occurred does not provide us with any additional information about occurrence of event A. However, $P(A \mid B \cap C) = 1 \neq P(A) \cdot P(B \cap C)$, because if we know that event $B \cap C$ occurred, then necessarily the number 1 has been drawn, so A occurred for sure.   ◻

The example above shows that if we have three events, and all pairs are independent, the three of them are not necessarily independent. The next example goes the other way around, showing that even if the joint probability of three events factors into the product of their individual probabilities, they are not necessarily pairwise independent, so they cannot be considered independent.[6]

Example 5.8 Consider a three-dimensional space and events corresponding to a ball being placed at a point characterized by three coordinates (X, Y, Z). The possible points are $(1,0,0)$, $(0,1,0)$, $(0,0,1)$, $(1,1,0)$, $(1,1,1)$, with probabilities $\frac{1}{8}$, $\frac{1}{8}$, $\frac{3}{8}$, $\frac{2}{8}$, and $\frac{1}{8}$, respectively. We see immediately that:

$$P(\{X = 1\}) = P(\{Y = 1\}) = P(\{Z = 1\}) = \tfrac{1}{2}$$

and

$$P(\{X = 1\} \cap \{Y = 1\} \cap \{Z = 1\}) = \tfrac{1}{8} = P(\{X = 1\}) \cdot P(\{Y = 1\}) \cdot P(\{Z = 1\})$$

So, the probability of the joint event does factor into the product of individual probabilities. However, the events are not pairwise independent:

$$P(\{X = 1\} \cap \{Y = 1\}) = \tfrac{3}{8} \neq P(\{X = 1\}) \cdot P(\{Y = 1\})$$
$$P(\{X = 1\} \cap \{Z = 1\}) = \tfrac{1}{8} \neq P(\{X = 1\}) \cdot P(\{Z = 1\})$$
$$P(\{Y = 1\} \cap \{Z = 1\}) = \tfrac{1}{8} \neq P(\{Y = 1\}) \cdot P(\{Z = 1\})$$

◻

---

[6]Counterexamples of the first kind are rather common in textbooks, whereas counterexamples of the second kind are not. Example 5.8 is due to Crow [2].

A note on notation. In the last example we stuck to a notation inspired by set theory. As you can see, this can become a bit of a burden when events linked to numerical values of variables are involved. Since these are the kinds of events that we are mostly interested in, we can use a streamlined notation like $P(X = 1, Y = 1, Z = 1)$. In other words, joint events can also be denoted by getting rid of the intersection operator:

$$P(E_1 \cap E_2 \cap \cdots \cap E_n) \equiv P(E_1, E_2, \ldots, E_n)$$

5.4   TOTAL PROBABILITY AND BAYES' THEOREMS

Conditional probabilities are a very important and powerful concept. In this section we see how we may tackle problems like the one in Example 5.2, which we use as a guideline. To frame the problem clearly, let us define the following events:

- isM: the child is a male.

- isF: the child is a female.

- TM: the test predicts a male.

- TF: the test predicts a female.

Now the first question is: What do we know and what would we like to know? The problem statement provides us with the following conditional probabilities:

$$P(\text{TM} \,|\, \text{isM}) = 0.9, \qquad P(\text{TF} \,|\, \text{isF}) = 0.7$$

from which we may also infer

$$P(\text{TF} \,|\, \text{isM}) = 1 - 0.9 = 0.1, \qquad P(\text{TM} \,|\, \text{isF}) = 1 - 0.7 = 0.3$$

Using conditional probabilities, we also see that what we need is, in a sense, *inverting* the conditional information in the probabilities above, as we need to compare the two conditional probabilities:

$$P(\text{isM} \,|\, \text{TM}) \quad \text{and} \quad P(\text{isF} \,|\, \text{TF})$$

To see how we can accomplish our task, let us abstract a little and consider two events E and F. Intersection is a commutative operation:

$$(E \cap F) = (F \cap E) \quad \Rightarrow \quad P(E \cap F) = P(F \cap E)$$

Using the definition of conditional probability, we may write

$$P(E \cap F) = P(E \,|\, F) P(F)$$
$$P(F \cap E) = P(F \,|\, E) P(E)$$

but since the two left-hand sides are the same, we may also conclude that

$$P(E \mid F)P(F) = P(F \mid E)P(E)$$

We have proved the following theorem.

THEOREM 5.6 (Bayes' theorem) *Given two events E and F, we have*

$$P(F \mid E) = \frac{P(E \mid F)P(F)}{P(E)}$$

provided that $P(E) \neq 0$.

Immediate application to our problem yields

$$P(\mathsf{isM} \mid \mathsf{TM}) = \frac{P(\mathsf{TM} \mid \mathsf{isM})P(\mathsf{isM})}{P(\mathsf{TM})}, \qquad P(\mathsf{isF} \mid \mathsf{TF}) = \frac{P(\mathsf{TF} \mid \mathsf{isF})P(\mathsf{isF})}{P(\mathsf{TF})}$$

Let us assume for simplicity that $P(\mathsf{isM}) = P(\mathsf{isF}) = 0.5$. In the relationship above, we need the probabilities of events **TM** and **TF**. Let us focus on the first one: In how many ways can the result of the test predict a male? Well, there are two cases:

1. The child is indeed a male, and the test predicts the correct result; this happens with probability $P(\mathsf{TM} \mid \mathsf{isM})P(\mathsf{isM})$.

2. The child is in fact a female, but the test is wrong; this happens with probability $P(\mathsf{TM} \mid \mathsf{isF})P(\mathsf{isF})$.

Since the two events are mutually exclusive, we may add the two probabilities to obtain

$$P(\mathsf{TM}) = P(\mathsf{TM} \mid \mathsf{isM})P(\mathsf{isM}) + P(\mathsf{TM} \mid \mathsf{isF})P(\mathsf{isF})$$

A similar result holds for event **TF**. Let us pause a moment and generalize the result.

THEOREM 5.7 (Total probability theorem) *Given a sample space Ω and a family of mutually exclusive and collectively exhaustive events $H_1, H_2, \ldots, H_n$, the probability of event E can be expressed as*

$$P(E) = \sum_{i=1}^{n} P(E \cap H_i) = \sum_{i=1}^{n} P(E \mid H_i)P(H_i)$$

The events $H_1, H_2, \ldots, H_n$ form a *partition* of the sample space, as illustrated in Fig. 5.6. *Mutually exclusive* means that all of them are disjoint:

$$H_i \cap H_j = \emptyset, \qquad \text{for } i \neq j$$

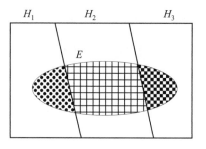

Fig. 5.6   An illustration of the total probability theorem.

Collectively exhaustive means that their union yields the whole sample space:

$$\bigcup_{i=1}^{n} H_i = \Omega$$

Given such a partition, we see that we may cut the event E into a collection of n mutually exclusive "slices" that, when patched together, yield back event E. The total probability theorem is a very convenient way to decompose the calculation of probabilities when we may slice the relevant event into disjoint pieces, as suggested in Fig. 5.6, and conditional probabilities are easy to compute. This is a very useful theorem in computing probabilities.

If we put Bayes' and total probability theorems together, we see that if $H_1, H_2, H_3, \ldots, H_n$ is a partition of the sample space, then for an event E we have the following equation:

$$P(H_i \,|\, E) = \frac{P(E \,|\, H_i)P(H_i)}{\sum_{j=1}^{n} P(E \,|\, H_j)P(H_j)}$$

Let us apply what we came up with to the gender prediction problem. The probability that Mary's child is indeed a male is

$$
\begin{aligned}
P(\text{isM} \,|\, \text{TM}) &= \frac{P(\text{TM} \,|\, \text{isM})P(\text{isM})}{P(\text{TM} \,|\, \text{isM})P(\text{isM}) + P(\text{TM} \,|\, \text{isF})P(\text{isF})} \\
&= \frac{0.9 \times 0.5}{0.9 \times 0.5 + 0.3 \times 0.5} = 0.75
\end{aligned}
$$

By the same token, the probability that Frances' child is indeed a female is

$$
\begin{aligned}
P(\text{isF} \,|\, \text{TF}) &= \frac{P(\text{TF} \,|\, \text{isF})P(\text{isF})}{P(\text{TF} \,|\, \text{isF})P(\text{isF}) + P(\text{TF} \,|\, \text{isM})P(\text{isM})} \\
&= \frac{0.7 \times 0.5}{0.7 \times 0.5 + 0.1 \times 0.5} = 0.875
\end{aligned}
$$

So, we see that Frances is the one who should be more confident about the gender of her child. We urge the reader to apply Bayes' theorem to the illness problem of Section 1.2.2 and find the result that we obtained there by an informal reasoning.

Bayes' theorem is fundamental in working with information and it is the starting point of a whole branch of statistics,[7] which we touch on in Chapter 14. To conclude the section, we consider a rather well-known puzzle.

Example 5.9 Consider a dumb but quite popular TV program, in which the participant sits in front of three boxes A, B, and C. One of the boxes contains a prize and the guy, who has no clue where the prize is, has to choose one. Say that he chooses A. The presenter *knows* where the prize is; he opens box C, showing that it does not contain the prize; then, he offers the participant the possibility of giving up the previous choice and switching to box B. Should the participant accept the offer?

When handed this question, the class typically divides into two camps:

1. One school of thought maintains that there is no point in switching from box A to box B. A priori, the probability of finding the prize was $\frac{1}{3}$; now, with two box remaining, the two probabilities are just $\frac{1}{2}$. Others go as far as to suggest that the presenter is cheating and trying to lure the participant into switching, in order to save the prize.

2. Another school of thought maintains that indeed the probabilities were symmetric a priori, but now the probability "mass" associated with box C should shift to box B; then, the probability that the prize is in box B is now $\frac{2}{3}$ and the participant would double the odds of winning by accepting the offer.[8]

Students hinting at the possibility that the presenter is cheating do have a point. We must state clearly the assumptions behind his behavior. In real games like this one, there are in fact many boxes with different prizes, and one would think that there is an incentive to try stealing the big one from the lucky participant. However, perhaps, a bigger incentive is to create suspense to keep the audience and make the game take more time, so that they can slip a few more juicy spots into the program. Therefore, let us assume that the presenter has no malicious intent and that his aim is just to stretch the game a little bit. Of course, whatever we conclude is as valid as this assumption, but this is a good feature of a formal analysis: Any assumption is stated clearly and we may assess its impact on our conclusions.

---

[7]Adherents to Bayesian statistics would object that this is the *only* branch of statistics, bridging the gap that the orthodox view leaves between probability theory and inferential statistics.

[8]The clash between the two camps was always huge fun for me. This has been recently spoiled by a move, "21," in which my "colleague" Kevin Spacey asks the question to a smart student who argues in a rather obscure way that the participant should indeed switch. So more and more students give the correct answer, but they typically do not know why!

The first step in tackling the problem is finding a sensible formalization. We are dealing with the following events:

- A, the prize is in box A.

- B, the prize is in box B.

- C, the prize is in box C.

- opC, the presenter opens box C after participant's choice.

What we need to do is evaluating the conditional probability $P(A \mid opC)$; note that

$$P(B \mid opC) = 1 - P(A \mid opC)$$

so calculating one of the two probabilities is quite enough.

The next step is to clearly state what we know, or we assume to know:

- A priori, the participant has no reason to believe that one box is more likely to contain the prize than the other ones:

$$P(A) = P(B) = P(C) = \tfrac{1}{3}$$

- The presenter is not cheating and knows where the prize is. Then, we can evaluate the following conditional probabilities:

  - $P(opC \mid A) = \tfrac{1}{2}$, because in such a case he could either open box B or C and nothing would change. So, let us assume that he chooses one of the two possibilities purely at random.

  - $P(opC \mid B) = 1$, because this is the only available option to him. He cannot open box A, because it is the selected one; he cannot open box B, because it would spoil the game.

  - $P(opC \mid C) = 0$, because he would necessarily open box B in this case, to avoid spoiling the game.

Now we are ready to apply Bayes' theorem:

$$P(A \mid opC) = \frac{P(opC \mid A) \cdot P(A)}{P(opC)}$$

What we miss in this expression is just $P(opC)$, which can be found by the total probability theorem:

$$P(opC) = P(opC \mid A) \cdot P(A) + P(opC \mid B) \cdot P(B) + P(opC \mid C) \cdot P(C)$$

If we put everything together, we obtain

$$P(A \mid opC) = \frac{P(opC \mid A) \cdot P(A)}{P(opC \mid A) \cdot P(A) + P(opC \mid B) \cdot P(B) + P(opC \mid C) \cdot P(C)}$$

$$= \frac{\tfrac{1}{2} \times \tfrac{1}{3}}{\tfrac{1}{2} \times \tfrac{1}{3} + 1 \times \tfrac{1}{3} + 0 \times \tfrac{1}{3}} = \frac{\tfrac{1}{2}}{\tfrac{1}{2} + 1} = \frac{1}{3}$$

Hence, the participant should switch to box B, since the odds of winning the prize would be $\frac{2}{3}$, rather than just $\frac{1}{3}$.   ☐

We should note that the conclusion of the example depends on all of the assumptions we made. This is a strength of a formal analysis, not a limitation: by stating a problem clearly, we point out which assumptions are critical as well as if and how our conclusion depends on them. If we are uncertain about the assumptions, it is no good reason not to consider their role explicitly.

Problems

5.1   Consider two events E and G, such that $E \subseteq G$. Then prove that $P(E) \leq P(G)$.

5.2   Assume that $P(A) = P(B)$ for two events A and B. Then prove that, given another event E

$$\frac{P(A\,|\,E)}{P(B\,|\,E)} = \frac{P(E\,|\,A)}{P(E\,|\,B)}$$

Find an interpretation of the result as a probability inversion formula.

5.3   In Example 5.9 we assumed that the presenter opens box C knowing where the prize is. Now, let us assume that he has no information on where the prize is. Does this change our conclusions?

For further reading

- Readable treatments on probability theory are offered by Ross [7, 8]; the second text [8]covers both probability and statistics and is easier going, whereas the first one is more heavily geared toward probability, even though it does not use too much of a sophisticated mathematical machinery.

- If you want to see a rigorous approach to probability theory, based on measure theory, one possible reference is the text by Resnick [6].

- In Ref. [1] a measure theoretic approach is followed as well, but at a less challenging level and with an unusual pedagogical twist; concepts are just reviewed and outlined, and then illustrated by plenty of solved problems that can really sharpen your understanding.

- Another relatively advanced text is Ref. [3], which does not emphasize formal approaches to probability theory, yet it is geared to advanced statistics based on the Bayesian paradigm, even though it relies on standard terminology and approach.

- A radical alternative is suggested in Ref. [4], which is a challenging reading based on an uncompromising Bayesian view of probability as plausibility.

REFERENCES

1. M. Capiński and T. Zastawniak, *Probability through Problems*, Springer-Verlag, Berlin, 2000.

2. E.L. Crow, A counterexample on independent events, *American Mathematical Monthly*, **74**:716–717, 1967.

3. M.H. DeGroot and M.J. Schervish, *Probability and Statistics,* 3rd ed., Addison Wesley, Boston, 2002.

4. E.T. Jaynes, *Probability Theory: The Logic of Science*, Cambridge University Press, Cambridge, UK, 2003.

5. D. Kahneman, P. Slovic, and A. Tversky, *Judgment under Uncertainty: Heuristics and Biases*, Cambridge University Press, Cambridge, UK, 1982.

6. S.I. Resnick, *A Probability Path*, Birkhäuser, Boston, 1999.

7. S.M. Ross, *Introduction to Probability Models,* 8th ed., Academic Press, San Diego, 2002.

8. S.M. Ross, *Introduction to Probability and Statistics for Engineers and Scientists,* 4th ed., Elsevier Academic Press, Burlington, MA, 2009.

6

Discrete Random Variables

In this chapter we start our investigation of random variables. Descriptive statistics deals with variables that can take values within a discrete or a continuous set. Correspondingly, we cover discrete random variables in this chapter, leaving continuous ones to Chapter 7. As we shall see, the mathematics involved in the study of continuous random variables requires concepts from calculus and is a bit more challenging than what is needed to cope with discrete ones. Hence, we prefer to proceed gradually, introducing intuitive concepts in the simpler case and emphasizing an intuitive link with descriptive statistics.

In Section 6.1 we introduce random variables formally, as associations of random events with numerical values. Then, in Section 6.2 we show how the distribution of a discrete random variable can be characterized by a probability mass function or a cumulative distribution function, which are related to concepts from descriptive statistics, i.e., histograms of relative frequencies and cumulative relative frequencies. Sections 6.3 and 6.4 proceed along the same conceptual path, introducing expected values of discrete random variables first, and then variance and standard deviations. Finally, in Section 6.5 we describe the main discrete probability distributions that are common in applications, along with some motivating examples relevant to business management.

As we pointed out, this is just a first step providing the reader with the basic knowledge about probability distributions, which is needed to tackle continuous random variables in the next chapter, where we also cover other concepts such as quantiles, skewness, and kurtosis; these apply to both discrete and continuous random variables, but we prefer treating them once within a more complete setting, after building some intuition based on the simpler, discrete case. Last but not least, in this chapter and the next one we just consider the univariate case; in Chapter 8 we introduce concepts about independence and correlation among multiple random variables, thus stepping into the domain of multivariate distributions.

6.1 RANDOM VARIABLES

In probability theory we work with events. The questions we may ask about events are quite limited, as they can either occur or not, and we may just investigate the probability of an event. In business management, more often than not we are interested in questions with a more quantitative twist, since events are linked to numerical values.

Example 6.1 Consider an airline adopting an overbooking strategy. The rationale behind overbooking is that aircraft capacity cannot be stored: If an aircraft with 50 seats flies with 10 empty ones, this does not mean that its capacity will be 60 on the next flight. Hence, many airlines accept bookings in excess of the actual capacity, in order to compensate for no-shows, i.e., passengers with a reservation, who do not show up at check-in. In real life, different tariffs are offered, defining the possibility of canceling the reservation and moving freely to another flight. In a simplified setting, the overbooking problem consists of determining the total number of bookings that we should accept. This decision does have an impact on the probability of an overbooking event, which occurs if any passenger cannot be accommodated at check-in. However, we are likely to be interested also in the *cost* of an overbooking strategy, as an overbooked passenger must be protected in some way, either by rerouting her to another flight or by offering overnight accommodation. Hence, we should associate numerical values with events.   ▯

Formally, a random variable is a mapping from events to numerical values.

DEFINITION 6.1 (Random variable) *A random variable is a function mapping outcomes within a sample space Ω to real numbers. This is sometimes denoted as follows:*

$$X : \Omega \to \mathbb{R}$$

We can also use the notation $X(\omega)$, with $\omega \in \Omega$, to emphasize the dependence on random outcomes.

In this definition, we should note that random variables are typically denoted by uppercase letters like X, whereas lowercase letters such as x are reserved to denote *realizations* of random variables, i.e., numerical values $x = X(\omega)$ corresponding to a specific event. Sometimes, an alternative notation is used, where $\tilde{\epsilon}$ refers to a random variable and ϵ refers to a realizations; this is handy with Greek letters, as you may imagine, and it is common in economics. The definition above is actually a bit informal and definitely incomplete. We should keep in mind that probability measures are associated with *events*, i.e., sets of elementary outcomes. Hence, the definition of a random variable depends on the family $\mathcal{F}$ of events $E \subseteq \Omega$. Hence, we should define a random variable referring to a probability space, consisting of the sample space Ω, the family of event $\mathcal{F}$, and the probability measure $\mathrm{P}(E)$, for events $E \in \mathcal{F}$. This may look a bit too abstract, but it is what we need to express information properly. For all of the practical purposes of this book, we need not be bothered about such technicalities, which are just outlined in Section 7.10 for the interested reader. All we need to know is that random experiments may result in numerical values that are practically relevant to us.

Example 6.2 Consider a lottery based on coin flipping, in which you win an amount €20 if the coin lands head, and you lose €10 otherwise. Then, we have a random variable X, such that

$$
X = \begin{cases} 20 & \text{if the outcome is head,} \\ -10 & \text{if the outcome is tail,} \end{cases}
$$

which readily translates to the probability distribution:

$$
X = \begin{cases} 20 & \text{with probability } 0.5, \\ -10 & \text{with probability } 0.5. \end{cases}
$$

Given our little lottery above, typical questions we may ask are

- If we keep repeating the lottery, what is the average win on the long term?

- What about its variability?

- More generally, how can we measure the risk of a lottery?

Considering the first two questions, we are naturally reminded of concepts like mean and variance, that we encountered in descriptive statistics. Indeed, we show later how these concepts can be adapted to random variables. However, there is a fundamental twofold gap when we move from descriptive statistics to probability theory and random variables.

1. Descriptive statistics is basically backward-looking: We calculate summary measures based on past observations, which could be referred to a whole population or to a possibly small sample. On the contrary, probability theory is concerned with what may happen *in the future*. Hence, probability theory is intrinsically forward-looking.

2. Furthermore, we implicitly assume that we know everything we need about a random variable, i.e., we assume that we are endowed with complete knowledge about the uncertainty of the phenomenon that we are representing.[1] In a sense, probability theory assumes knowledge about whole populations, not samples.

Clearly, there is a link, as one could use a sample to learn something about probabilities of future events, but we have already seen in Section 5.1 that this frequentist view is only one possibility. This is quite relevant when we consider the third question in the bullet list above. When dealing with risk management, one should ask if the past history contains all of the possible events. To get the point, imagine collecting statistics about credit crunches due to mortgage-backed securities before 2007.

In this chapter we assume that we know the whole set of values that a random variable may take, as well as their probabilities. This is the support of a discrete probability distribution.

DEFINITION 6.2 (Support of a discrete distribution) *The support of a discrete probability distribution is the set of values $x_i \in \mathbb{R}$, $i = 1, 2, 3, \ldots$ that the random variable X may take. We always assume that values in the support have been sorted, i.e., $x_i \leq x_{i+1}$.*

As we see from this definition, the values x_i may be real numbers in general; quite often, however, they are restricted to nonnegative integer numbers, i.e., $x_i \in \mathbb{Z}_+$. Yet, it is important to understand that this is *not* the feature that makes this distribution discrete. In the lottery above, in principle, the payoffs could be real numbers like 3π and -2π. The point is that in a discrete distribution the support is either a finite or a countably infinite set of values.[2] In the former case, the support is a set of numbers $x_i \in \mathbb{R}$, $i = 1, 2, \ldots, n$. Having said that, in the following the support of discrete distributions will mostly consist of integer numbers, which may include zero or not, depending on the application, but it is useful to keep the general framework in mind.

6.2 CHARACTERIZING DISCRETE DISTRIBUTIONS

When we deal with sampled data, it is customary to plot a histogram of relative frequencies in order to figure out how the data are distributed. When we consider a discrete random variable, we may use more or less the same concepts in order to provide a full characterization of uncertainty. For instance,

---

[1] In later chapters, we point out that we should speak of risk when the outcome of a random experiments is uncertain but the rules of the game are perfectly known, reserving the word *uncertainty* for the case where we do not really have all of the required information.

[2] The reader is referred back to Section 2.2 for an introduction to these concepts.

if we consider dice throwing, a full description of the related random variable is given by

$$P(X = i) = \tfrac{1}{6}, \qquad i = 1, 2, 3, \ldots, 6$$

This is an example of a probability mass function, which we formally define below and fully characterizes the probability distribution of a discrete random variable. However strange it may be, it turns out that a more flexible characterization is obtained by referring to cumulative relative frequencies, which leads to the definition of a cumulative distribution function. The essential advantage is that the latter function is more general and translates directly to the case of continuous random variables. In the next two sections we explore these concepts in detail.

6.2.1 Cumulative distribution function

The basic stuff of probability theory consists of events and their probability measures. Given a random variable X, consider the event $\{X \leq x\}$; incidentally, note how we use x to denote a *number*. The probability of this event is a function of x.

DEFINITION 6.3 (Cumulative distribution function) *Let X be a random variable. The function*

$$F_X(x) \equiv P(X \leq x)$$

for $x \in \mathbb{R}$, is called cumulative distribution function (CDF).

The notation $F_X(x)$ clarifies that the CDF is a function of x, associated with the random variable X, as suggested by the attached subscript. In passing, we may also note that the definition above is quite general, and it does not require that X is a discrete random variable; the CDF can be defined like this for continuous random variables as well. To build intuition, let us see how the CDF compares to cumulative relative frequencies in descriptive statistics.

Example 6.3 Let us consider a discrete random variable naturally related to dice throwing and build its CDF. To begin with, it is not possible to get a number strictly smaller than one. Hence

$$P(X \leq x) = 0, \qquad \text{for } x < 1.$$

Then, there is a jump in the function when we consider $x = 1$:

$$P(X \leq 1) = \tfrac{1}{6}$$

If we increase x, the CDF will stay there as long as $1 \leq x < 2$, since the discrete random variable cannot take any value in the open interval $(1, 2)$. Then, there is another jump for $x = 2$:

$$P(X \leq 2) = \tfrac{2}{6}$$

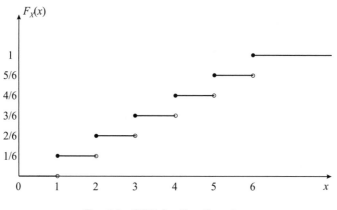

Fig. 6.1 CDF for dice throwing.

since the event $\{X \le 2\}$ includes the events $\{X = 1\}$ and $\{X = 2\}$. Similar jumps occur for $x = 3,4,5$. Finally, we have

$$P(X \le 6) = 1$$

For $x > 6$, the function will just stay there, since the support is bounded by $x = 6$. The resulting CDF is depicted in Fig. 6.1. Note that the figure is drawn according to the standard convention of discontinuous functions.[3] If we are a bit picky, we should say that the function is continuous from the right, but discontinuous from the left. In other words, $P(X \le 1) = \frac{1}{6}$, but if we approach $x = 1$ from the left, i.e, the value of the function we see is 0:

$$\lim_{x \uparrow 1} F_X(x) = 0, \qquad \lim_{x \downarrow 1} F_X(x) = \tfrac{1}{6}$$

☐

The plot in Fig. 6.1 suggests a few properties of the CDF for a discrete random variable:

1. The CDF is a nondecreasing function:

$$x_1 \le x_2 \qquad \Rightarrow \qquad F_X(x_1) \le F_X(x_2)$$

   To see this, observe that event $E_1 = \{X \le x_1\}$ is a subset of event $E_2 = \{X \le x_2\}$, which implies $P(E_1) \le P(E_2)$.

2. If we consider the support x_i, $i = 1, 2, 3, \ldots$, where $x_i < x_{i+1}$, we have

$$F_X(x) = 0 \qquad \text{if } x < x_1$$
$$\lim_{i \to +\infty} F_X(x_i) = 1$$

[3]See Section 2.4.

Table 6.1 A discrete probability distribution.

x	0	1	2	3
$p_X(x)$	0.15	0.20	0.35	0.30

Indeed, the value of the CDF is a probability, so it must stay within the interval $[0, 1]$.

3. The CDF for a discrete random variable is a piecewise constant function, with jumps corresponding to values included in the support.

6.2.2 Probability mass function

The CDF looks like a somewhat weird way of describing the distribution of a random variable. A more natural idea is just assigning a probability to each possible outcome in the support. Unfortunately, in the next chapter we will see that this idea cannot be applied to a continuous random variable. Nevertheless, in the case of a discrete random variable we may indeed associate a probability with the event $\{X = x_i\}$.

DEFINITION 6.4 (Probability Mass Function) *The* probability mass function *(PMF) of a discrete random variable is defined as the function*

$$p_X(x) = P(X = x)$$

The function is zero for values not included in the support. For values x_i in the support, the shorthand notation $p_i \equiv P(X = x_i)$ is often used.

Example 6.4 Consider the PMF in Table 6.1, and note that probabilities add up to 1. The PMF $p_X(x)$ is graphically illustrated in Fig. 6.2(a). The upward arrows are a common way to depict probability *masses* concentrated at discrete points within the distribution support. The height of each arrow corresponds to the probability of that value and provides us with a visual representation of likelihood. On the contrary, with continuous random variables the mass is distributed on continuous intervals. Figure 6.2(b) shows how we may get the CDF by summing up probabilities from the PMF:

$$F_X(x) = \begin{cases} 0 & \text{for } x < 0 \\ p_0 = 0.15 & \text{for } 0 \le x < 1 \\ p_0 + p_1 = 0.35 & \text{for } 1 \le x < 2 \\ p_0 + p_1 + p_2 = 0.70 & \text{for } 2 \le x < 3 \\ p_0 + p_1 + p_2 + p_3 = 1 & \text{for } 3 \le x \end{cases}$$

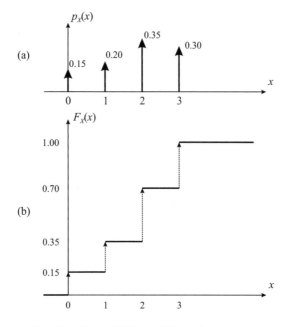

Fig. 6.2 From PMF to CDF and vice versa.

Each probability we add over points in the support contributes a jump to the CDF. We can also go the other way around, i.e., we may obtain probabilities by taking differences in the CDF:

$$p_0 = F_X(0)$$
$$p_1 = F_X(2) - F_X(1)$$
$$p_2 = F_X(3) - F_X(2)$$
$$p_3 = F_X(4) - F_X(3)$$

□

The example suggests general rules for moving from PMF and CDF and vice versa. We find the PMF by taking differences of consecutive values of the CDF over points x_i in the support:

$$p_X(x_i) \equiv \mathrm{P}(X = x_i) = \mathrm{P}(X \le x_i) - \mathrm{P}(X < x_i) = F_X(x_i) - F_X(x_{i-1}) \quad (6.1)$$

It is important to notice that if X is a discrete random variable and x_i is in its support, then

$$\mathrm{P}(X \le x_i) \ne \mathrm{P}(X < x_i)$$

Given the PMF, we just add up its values to find the CDF:

$$F_X(a) \equiv \mathrm{P}(X \le a) = \sum_{x_i \le a} p_X(x_i) \quad (6.2)$$

Indeed, the CDF and PMF provide us with the same information, which fully characterizes a discrete random variable.

6.3 EXPECTED VALUE

Both PMF and CDF provide us with all of the relevant information about a discrete random variable, maybe too much. In descriptive statistics, we use summary measures, such as mean, median, mode, variance, and standard deviation, to get a feeling for some essential features of a distribution, like its location and dispersion. In probability theory, there are corresponding concepts that we start exploring in this and the next section, where we consider concepts corresponding to mean and variance, leaving further developments to Sections 7.4 and 7.5. The single most relevant feature of a distribution is related to its "mean," a natural measure of location.

DEFINITION 6.5 (Expected value) *The expected value of a discrete random variable with PMF $p_X(x)$ is given by*

$$\mathrm{E}[X] \equiv \sum_{i=1}^{\infty} x_i\, p_X(x_i) = \sum_{i=1}^{\infty} x_i\, p_i \tag{6.3}$$

where we have used the shorthand notation $p_i = p_X(x_i)$.

This definition allows for an infinite support; in the case of a finite support, the sum will just run up to $i = N$. Quite often, the notation μ or μ_X is used to refer to the expected value of a random variable.

Example 6.5 Consider again the random variable whose PMF is shown in Table 6.1. Its expected value is

$$\mathrm{E}[X] = \sum_i x_i p_i = 0 \times 0.15 + 1 \times 0.20 + 2 \times 0.35 + 3 \times 0.30 = 1.8$$

Note that the expected value of a random variable taking integer values is not necessarily an integer number. □

6.3.1 Expected value vs. mean

Looking at Definition 6.3, the similarity with how the sample mean is calculated in descriptive statistics, based on relative frequencies, is obvious. However, there are a few differences that we must always keep in mind. This is why it is definitely advisable to avoid the term "mean" altogether, when referring to random variables. Using the term "expected value" may be tiresome at first, but it will enhance clarity of thinking, which will pay off later, when dealing with inferential statistics. Hence, it is useful to gain a thorough understanding of the differences between expected value in probability theory and mean in descriptive statistics.

- The first striking difference is that the expected value involves an infinite series, when the probability distribution has infinite support. This cannot happen in descriptive statistics, since what we observe is a bounded range of data. If the sum converges, it must be the case that very large values have a small probability. Allowing for the occurrence of an unlikely, but quite significant event is important in risk management.

- A related difference is that the probabilities need not come from empirical data. We will see in Section 6.5 that distributions may be obtained by conceptual random experiments driven by some underlying mechanism, which allows for an infinite set of outcome. We refer to such distributions as "theoretical" to set them apart from empirical distributions based on sampled data. A statistical mean can only be empirical.

- The expected value is a *number*. Given a probability distribution, the expected value is what it is. In descriptive statistics, the mean need not be a number. It will be a number only if the mean pertains to a population. If the mean comes from a random sample, then it will be *random variable*. Any time we sample, we get a different value.

Despite these remarkable differences, there is indeed a link between expected value and mean, which will become quite clear when dealing with inferential statistics; we may use a sample mean to *estimate* an expected value. We see that, in a sense, probability theory assumes complete knowledge about the underlying uncertainty, which can be equivalently encoded in the form of a PMF or CDF. Thus, in a sense, in probability theory we always work with the whole population.

In the following, we will learn to interpret the expected value in different ways, depending on our purpose:

- The expected value is a basic feature of a probability distribution, i.e., a location measure.

- The expected value can be regarded as a long-run average. This second interpretation is less obvious because, among other things, it assumes that the characteristics of a random process will be constant in the future. To really clarify what we mean, we should formally state the law of large numbers, specifying under which conditions this interpretation is sensible. Since the involved issues are not trivial, this is left to the advanced Section 9.8.

- The expected value can be interpreted as a forecast. This interpretation stresses the forward-looking nature of an expectation, as compared with the backward-looking nature of descriptive statistics. In Chapter 11, we clarify what makes a "good" forecast and what we need in order to transform forecasts into management decisions. From a practical perspective, we should always take into account the danger of building a forecast based on past history.[4]

6.3.2 Properties of expectation

We may think of the expected value as an operator mapping a random variable X into its expected value $\mu = \mathrm{E}[X]$. The expectation operator enjoys two very useful properties.

PROPERTY 6.6 (Linearity of expectation 1) *Given a random variable X with expected value $\mathrm{E}[X]$, we have*

$$\mathrm{E}[\alpha X + \beta] = \alpha\,\mathrm{E}[X] + \beta$$

for any numbers α and β.

This property is fairly easy to prove:

$$\mathrm{E}[\alpha X + \beta] = \sum_{i=1}^{\infty} p_i(\alpha x_i + \beta) = \alpha \sum_{i=1}^{\infty} p_i x_i + \beta \sum_{i=1}^{\infty} p_i = \alpha\,\mathrm{E}[X] + \beta$$

Informally, the property provides us with a quick rule for manipulating expectation as an operator, stating that numbers can be "taken outside" the expectation. The next property is a bit less trivial, as it involves the sum of multiple random variables.

PROPERTY 6.7 (Linearity of expectation 2) *Given m random variables X_i, $i = 1, \ldots, m$, we have*

$$\mathrm{E}\left[\sum_{i=1}^{m} X_i\right] = \sum_{i=1}^{m} \mathrm{E}[X_i]$$

A proof of this property is a bit involved, as it requires some tedious algebra, and it is omitted. What makes this property *conceptually* not trivial is that when dealing with multiple random variables, some care might be needed as they may have different distributions, and their mutual relationships may be quite complicated. Rather surprisingly, Property 6.7 states that the expected value of a sum of random variables is *always* the sum of their expected values.[5]

Taken together, the two properties state that expectation is a linear operator, in the sense that the expected value of a linear combination of random

---

[4]If you not get this point too clearly, imagine driving home just looking at the rearview mirror. A disclaimer: I wrote *imagine*, not *try!*

[5]We are assuming that pathological situations do not arise and that all of the involved expectations exist. This may not be the case if the series (6.3) diverges by going to infinity.

variables, is just the linear combination of their expected values:[6]

$$\mathrm{E}\left[\sum_{i=1}^{m} \alpha_i X_i\right] = \sum_{i=1}^{m} \alpha_i \mathrm{E}[X_i] \tag{6.4}$$

for arbitrary coefficients α_i, $i = 1, \ldots, m$.

6.3.3   Expected value of a function of a random variable

Typically, a random variable is just a risk factor that will affect some managerially more relevant outcome linked to cost or profit. This link may be represented by a function; hence, we are interested in functions of random variables. Given a random variable X and a function, like $g(x) = x^2$, or $g(x) = \max\{x, 0\}$, we define a new random variable $g(X)$. Not surprisingly, the expected value of a function of a random variable is

$$\mathrm{E}[g(X)] = \sum_{i=1}^{\infty} p_i g(x_i) \tag{6.5}$$

It is fundamental to really understand the expression in (6.5). We should consider each value x_i in the support, compute the corresponding value $g(x_i)$ of the function, multiply the result by probability p_i, and add everything up. This is the expected value of the function, which is *not* the function of the expected value. We should *not* calculate $\mathrm{E}[X]$ and then evaluate function $g(x)$ for $x = \mathrm{E}[X]$ since, in general

$$\mathrm{E}[g(X)] \neq g\left(\mathrm{E}[X]\right)$$

In other words, we *cannot* commute the two operators, i.e., the expectation $\mathrm{E}[\cdot]$ and the function $g(\cdot)$, as the following counterexample shows.

Example 6.6 Let X be a discrete random variable with support $\{-1, +1\}$. Both values have probability 0.5, so

$$\mathrm{E}[X] = [0.5 \times (-1) + 0.5 \times (+1)] = 0$$

Now consider function $g(x) = x^2$. The expected value of $g(X)$ is

$$\mathrm{E}[g(X)] = \mathrm{E}[X^2] = 0.5 \times (-1)^2 + 0.5 \times (+1)^2 = 0.5 + 0.5 = 1$$

This is definitely not the same as the function of the expected value:

$$g\left(\mathrm{E}[X]\right) = 0^2 = 0$$

$\square$

---

[6]See Section 3.3.3 for a definition of linear combination.

There is a case in which we may commute expectation and function, as suggested by Property 6.6: If we have linear affine function $h(x) = \alpha x + \beta$, then it is true that

$$\mathrm{E}[h(X)] = \mathrm{E}[\alpha X + \beta] = \alpha \mathrm{E}[X] + \beta = h(\mathrm{E}[X])$$

but this is a very peculiar situation. The following example illustrates the point again in a more practically relevant setting.

Example 6.7 As we have seen in Section 3.1, a European-style *call option* is a financial contract giving you the right, but *not* the obligation, to purchase a given asset (e.g., a stock share) at a fixed price (called the *strike price*), at a given date (the maturity of the option). Note that the investor holding the option is free to choose if she wants to exercise the option at maturity or not.[7] The other party of the contract, the option writer, is forced to sell the underlying asset if the option holder exercises the option. In contrast, the forward contracts we dealt with in Section 1.3.1 are more symmetric, since both sides of the contract are forced to carry out their obligations, i.e., to respectively buy and sell the underlying asset at the agreed forward price.

Say that we hold an option with strike price €40, written on a stock share whose price now is $S_0 = $ €35, maturing in 5 months. The stock price in five months, S_5, is a random variable. If it turns out that $S_5 > $ €40, then the holder can exercise the option and buy at 40€ the asset, which can then be sold for S_5, with a payoff $S_5 - $ €40. More generally, if S_T is the underlying asset price at maturity and K is the strike price, the payoff for the option holder is

$$g(S_T) = \max\{S_T - K, 0\}$$

Note that the option payoff cannot be negative, since the holder will not exercise the option if the asset price is below the strike price.[8]

Real-life probability distributions of stock share prices are rather complicated, but let us assume that the distribution of S_5 is fairly well approximated by a set of eight equally likely scenarios:

$$20, \ 25, \ 30, \ 35, \ 40, \ 45, \ 50, \ 55$$

What is the expected value of the option payoff at maturity? Since each value in this discrete support has probability $\frac{1}{8}$, symmetry suggests that $\mathrm{E}[S_5] = 37.5$. We see that $g(37.5) = \max\{37.5 - 40, 0\} = 0$, but this is not what we want. We should first calculate the option payoff in each scenario:

---

[7]In American-style options, the holder may exercise her right at any time before the expiration of the contract.

[8]This is not to say that the profit cannot be negative, as the holder must pay a price to get her hands on the option. Furthermore, the actual workings of financial markets are a bit more complicated, since prices move up and down in real time and there are transaction costs associated with buying and selling an asset. Hence, this expression of the option payoff just refers to the "value" of the contract for its holder.

$$0, 0, 0, 0, 0, 5, 10, 15$$

Then, the expected value of the option payoff is

$$\mathrm{E}[g(S_5)] = \frac{5 + 10 + 15}{8} = 3.75$$

                                                         □

For an arbitrary function $g(\cdot)$, we cannot say anything about the relationship between $\mathrm{E}[g(X)]$ and $g(\mathrm{E}[X])$. A notable exception is a convex function.[9]

THEOREM 6.8 (Jensen's inequality) *If $g(x)$ is a convex function, then*

$$g(\mathrm{E}[X]) \leq \mathrm{E}[g(X)].$$

It is quite instructive to see a proof for the simple case of a support consisting of two values, x_1 and x_2, with probabilities p and $1 - p$, respectively, since this sheds some light on the connection between discrete expectations and convex combinations. The expected value of X is

$$\mathrm{E}[X] = px_1 + (1 - p)x_2$$

which is a linear combination of values x_1 and x_2, with nonnegative weights that add up to 1. Hence, we are just taking a convex combination, and this also applies to a support consisting of more than two points. Using convexity of the function, we see that

$$\mathrm{E}[g(X)] = pg(x_1) + (1 - p)g(x_2) \leq g\left[px_1 + (1 - p)x_2\right] = g(\mathrm{E}[X])$$

6.4 VARIANCE AND STANDARD DEVIATION

The expected value of a random variable tells us something about the location of its distribution, but we need a characterization of dispersion and risk as well. In descriptive statistics, we consider squared deviations with respect to the mean. Here we do basically the same thing, with respect to the expected value.

DEFINITION 6.9 (Variance and standard deviation) *The variance of a random variable X is defined as*

$$\mathrm{Var}(X) \equiv \mathrm{E}[(X - \mathrm{E}[X])^2] = \mathrm{E}[(X - \mu_X)^2]$$

where we have used the shorthand notation $\mu_X = \mathrm{E}[X]$. Variance is often denoted by σ^2. Standard deviation σ is the square root of variance.

---

[9]Convexity was introduced in Section 2.11.

Once again, we note that variance and standard deviation for a random sample are random variables, whereas they are well-defined *numbers* for a random variable. This is much clearer if we write variance explicitly for a discrete random variable:

$$\sigma^2 = \sum_{i=1}^{\infty} p_i (x_i - \mu_X)^2$$

and compare it against variance for a sample consisting of n observations:

$$\frac{1}{n-1} \sum_{i=1}^{n} \left(X_i - \overline{X} \right)^2$$

The remarks we made in Section 6.3.1 regarding the expected value apply here as well: The support can be infinite, whereas a dataset cannot, and it takes a lot of knowledge, the whole PMF, to calculate variance. Furthermore, we should raise a couple of issues about variance, which can be better appreciated by thinking about the following question:

> We are interested in random fluctuations of weekly demand for the spare parts of a certain item. Variance of this demand is 10,000 items per week, squared. Is this a large or small variance?

To begin with, variance is an average squared deviation with respect to expected value, but it is hard to think of "items squared." This is why standard deviation is useful, as it is expressed with the same measurement units as the expected value. In the case above, standard deviation is 100 items per week. However, it is hard to tell if it is actually large. If expected value is 10,000 items per week, uncertainty is almost negligible. If expected value is 250 items per week, that standard deviation is fairly large. This is why sometimes we use the *coefficient of variation*, defined as

$$C_X \equiv \frac{\sqrt{\text{Var}(X)}}{|\text{E}[X]|} = \frac{\sigma}{|\mu|} \tag{6.6}$$

We need to take the absolute value of the expected value, as this could be negative. As a reference point, a variable with a fair amount of variability may have a coefficient of variation of about 1.

6.4.1 Properties of variance

The first thing we should observe is that variance cannot be negative, as it is the expected value of a squared deviation. It is zero for a random variable that is not random at all, i.e., a constant. In doing calculations, the following identity is quite useful:

$$\text{Var}(X) = \text{E}[X^2] - \text{E}^2[X] \tag{6.7}$$

This is the analog of Eqs. (4.5) and (4.6) in descriptive statistics, and proving it is a good exercise:

$$
\begin{aligned}
\mathrm{Var}(X) &\equiv \mathrm{E}[(X - \mu_X)^2] \\
&= \mathrm{E}[X^2 - 2X\mu_X + \mu_X^2] \\
&= \mathrm{E}[X^2] - 2\mu_X \mathrm{E}[X] + \mu_X^2 \\
&= \mathrm{E}[X^2] - 2\mu_X\mu_X + \mu_X^2 \\
&= \mathrm{E}[X^2] - \mathrm{E}^2[X]
\end{aligned}
$$

When dealing with expected value, we considered Properties 6.6 and 6.7, that essentially state that the expected value of a sum of random variables is the sum of their expected values. Does this carry over to variance? The answer is "not really." The first surprise is that, given numbers α and β, we have

$$
\mathrm{Var}(\alpha X + \beta) = \alpha^2 \, \mathrm{Var}(X) \tag{6.8}
$$

We see that a number α multiplying a random variable can be "taken outside" variance, but it gets squared. This is not surprising at all, since variance is a squared deviation. We also see that β does not play any role at all. Again, this makes sense, since shifting a probability distribution does change its expected value, a measure of location, but not its dispersion. It is easy to prove this property formally using the definition of variance and the properties of expectation:

$$
\begin{aligned}
\mathrm{Var}(\alpha X + \beta) &\equiv \mathrm{E}\left[(\alpha X + \beta - \mathrm{E}[\alpha X + \beta])^2\right] = \mathrm{E}\left[(\alpha X - \alpha\mu_X)^2\right] \\
&= \alpha^2 \mathrm{E}\left[(X - \mu_X)^2\right] = \alpha^2 \, \mathrm{Var}(X)
\end{aligned}
$$

Equation (6.8) suggests that there is something intrinsically "nonlinear" in variance, since the scale factor α gets squared. By the same token, we could wonder whether the variance of a sum of random variables is just the sum of variances:

$$
\mathrm{Var}\left[\sum_{i=1}^{m} X_i\right] = \sum_{i=1}^{m} \mathrm{Var}[X_i] \tag{6.9}
$$

Actually, this property *does not* hold in general.

Example 6.8 To see why Eq. (6.9) *cannot* hold in general, an intuitive example is helpful. Consider a simple financial portfolio allocation problem. We should allocate \$100,000 either to IBM or Microsoft stock shares. Say that 60% of our wealth is invested in IBM, and the rest in Microsoft:

$$
W_{\mathrm{IBM}} = \$60{,}000, \qquad W_{\mathrm{MS}} = \$40{,}000
$$

The return on our investment depends on the returns of the two stock shares, which are two random variables. Denoting the two random variables by R_{IBM} and R_{MS}, respectively, profit/loss is a random variable:

$$
\mathrm{P/L} = W_{\mathrm{IBM}} R_{\mathrm{IBM}} + W_{\mathrm{MS}} R_{\mathrm{MS}}
$$

with expected value

$$\mu_{\mathrm{P/L}} = W_{\mathrm{IBM}}\mu_{\mathrm{IBM}} + W_{\mathrm{MS}}\mu_{\mathrm{MS}}$$

Imagine, for the sake of the argument, that $\mu_{\mathrm{IBM}} = 5\%$ and $\mu_{\mathrm{MS}} = 10\%$. Since Microsoft has a higher expected return than IBM, why don't we invest all of our wealth in Microsoft shares? The easy answer is that, arguably, greater expected return comes with greater risk. How can we measure risk? The answer is definitely tricky, but a first attempt could be considering standard deviation of return, which is linked to variance. Now what is the variance of P/L? We will find the complete answer later, in Section 8.3.2, but we may immediately understand that we cannot assess risk of our portfolio without considering the relationships between the two returns.

- If IBM price increases when Microsoft price falls, and vice versa, there is an offsetting mechanism that should reduce risk, and this should be reflected somehow in variance of profit/loss.

- If the two prices have nothing to do with each other, i.e., if R_{IBM} and R_{MS} are independent random variables, whatever this means exactly, we will not see a big reduction in risk, but maybe some diversification benefit will result.

- On the contrary, if the two returns go hand in hand, then we do not have a well-diversified portfolio.[10] This should be reflected by a relative increase in overall variance with respect to the two previous cases.

Indeed, an old piece of advice suggests that we should not place all of our eggs into one basket.   ▯

The example shows that we are missing something, i.e., the mutual dependence of random variables, which we deal with in Chapter 8. Still, there is something that we may state here, without any proof. Equation (6.9) holds if all of the involved variables are independent.

PROPERTY 6.10 (Variance of a sum of independent variables) *Let X_i, $i = 1, \ldots, m$, be independent random variables with variance σ_i^2, and let α_i, $i = 0, \ldots, m$ be arbitrary real numbers. Let us also define random variables:*

$$Y = \sum_{i=1}^{m} X_i, \qquad Z = \alpha_0 + \sum_{i=1}^{m} \alpha_i X_i$$

---

[10]The Enron case is a good example of bad diversification from the perspective of their employees. On one hand, when disaster struck, they lost their jobs. As if this were not bad enough, the private pension fund of Enron was heavily invested in Enron stock shares. Hence, employees held a portfolio consisting of two assets, the job and the wealth invested in the pension fund, which can both be considered as claims to a stream of future cash flows. Unfortunately, the two assets were risky and the two risks were tightly related.

with variance σ_Y^2 and σ_Z^2, respectively. Then

$$\sigma_Y^2 = \sum_{i=1}^{m} \sigma_i^2 \tag{6.10}$$

$$\sigma_Y = \sqrt{\sum_{i=1}^{m} \sigma_i^2} \tag{6.11}$$

$$\sigma_Z^2 = \sum_{i=1}^{m} \alpha_i^2 \sigma_i^2 \tag{6.12}$$

Be sure to notice that we may add up variances, but not standard deviations.

6.5 A FEW USEFUL DISCRETE DISTRIBUTIONS

There is a wide family of discrete probability distributions, which we cannot cover exhaustively. Nevertheless, we may get acquainted with the essential ones, which will be illustrated by a few examples. First, we should draw the line between empirical and theoretical distributions. Since these terms may be a tad misleading, it is important to clarify their meaning.

- An *empirical distribution*, in the discrete case, consists of a set of values along with their probabilities. Hence, to describe it, we need the full PMF. This is why the term *nonparametric* distribution is also used.

- A *theoretical* or *parametric distribution* stems from a specific random experiment that may be conceptual, rather than empirical. This will be clearer in a moment, but the important point is that such a distribution can be fully characterized by a limited set of parameters, typically one or two. The shape of the distribution depends on the specific mechanism of the underlying random phenomenon.

The terminology for theoretical distributions is as such, because these distribution often allow for an infinite support, something that cannot stem from empirical experiments; an empirical distribution has a finite support by its very nature. Still, the parameters of a theoretical distribution can be fit against empirical data. Hence, we see that the boundary between the two classes is not that sharp, and is more related to their parametric vs. nonparametric nature. In describing the distributions below we will also use a few realistic examples to hone our skills.

6.5.1 Empirical distributions

Empirical distributions feature the closest link with descriptive statistics, since their PMF is typically estimated by collecting empirical relative frequencies.

For instance, if we consider a sample of 10 observations of a random variable X, and $X = 1$ occurs in three cases, $X = 2$ in five cases, and $X = 3$ occurs twice, we may estimate

$$p_1 = 0.3, \qquad p_2 = 0.5, \qquad p_3 = 0.2$$

Empirical distributions feature the largest degree of flexibility and may also be used with qualitative data. Furthermore, they may reflect quite complicated random phenomena, leading to possibly multimodal distributions, whereas we shall see that theoretical distributions typically have a single mode. Yet, this flexibility may backfire when it results in model overfitting, i.e., when the probability model reflects peculiarities in the sampled data that do not carry over to the overall population.

Another point that cannot be overemphasized is that empirical distributions have a finite support by definition, as they rule out values below the smallest observation and above the largest one. This may result in a wrong perception of risk. Sometimes, an empirical distribution is adjusted by adding "tails" derived from a theoretical model in order to avoid this problem, but this patch requires some ad hoc reasoning.

6.5.2   Discrete uniform distribution

The uniform distribution is arguably the simplest model of uncertainty, as it assigns the same probability to each outcome:

$$p(x_i) = p$$

This makes sense only if there is a finite number n of possible values that the random variable can assume. If they are consecutive integer numbers, we have an integer uniform distribution, which is characterized by the lower bound a and the upper bound b of its support. The condition:

$$\sum_{i=1}^{n} p_i = 1$$

immediately implies $p = 1/n$. We may think of a uniform distribution as a die with possibly many faces. A uniform distribution characterizes a case where we do not have any reason to believe that any outcome is more likely than the other ones; hence, we might associate a uniform distribution to a situation where we have very little information. There is little to say about this distribution, so it is time to tackle a nontrivial example.

Example 6.9 (Newsvendor problem) In Section 1.2.1 we briefly discussed the problem of purchasing items with a limited time window for sales under demand uncertainty. Now we are able to better understand it and even try a little numerical example. This kind of problem is typically labeled as newsvendor problem, as it resembles the challenge faced by late-nineteenth-century newsboys, who had to purchase newspapers before knowing demand, with considerable danger of scrapping a lot of sold newspaper.[11] More generally, the newsvendor problem is a prototype for all decisions involving fashion items, and its relevance is more and more significant as the rate at which products become obsolete is increasing.

To summarize the newsvendor problem:

- We purchase an item with unit cost c; this can also be the unit production cost in the "make" case. The item can be ordered only once, before the beginning of the time window.

- The product is sold at sales price s. The profit margin is $s - c$; if our order turns out to be small, i.e., if part of demand is not met, we lose profit opportunities.

- Unsold items are marked down and sold for $s_u < c$. If $s_u = 0$, then unsold items are just scrapped.

- Demand D is a random variable, whose probability distribution is known.

- The decision to be made is: How many items should we buy (or make)?

To gain a better feeling for the problem, let us consider the following numerical data:

- Unit cost is $c = 20$.

- Sales price is $s = 25$.

- Whatever is left, must just be scrapped; hence $s_u = 0$.

- Demand is uniformly distributed between 5 and 15; since the support consists of 11 values, we immediately see that the probability of each value is $p = \frac{1}{11}$; given the symmetry of the distribution, it is also easy to see that $E[D] = 10$.

Let us denote our decision variable, the order size, by q. At first sight, whenever we face a problem like this we should just come up with a good forecast of demand and buy that amount, $q = E[D] = 10$. Actually, this need not be the best strategy. Intuitively, such a naive solution disregards a lot of information, such as demand uncertainty and the involved economics, i.e., cost and price data. Before dwelling on mathematics, let us ask the following questions:

- If we go to a baker's shop to buy some bread at 6 p.m., are we likely to find what we want, i.e., real bread and not plastic? Experience suggests

---

[11] After the 1899 Newsboys strike, publishers agreed to buy back unsold copies.

that it is difficult to find real bread in the evening. Since this is a regular pattern, should we conclude that bakers are dumb demand forecasters?

- On the other hand, if we want to buy winter clothing, we always have the option to wait for markdown sales, since prices are marked down each and every winter, and the same applies to summer seasons. Should we conclude that owners of clothing shops are as dumb as bakers, but of the opposite breed?

- Or maybe none of them are dumb, and there is something we are missing?

It seems that we need more careful analysis, and the first question is: What makes a good solution to our problem? The easy answer is that we would like to maximize profit, but a moment of reflection shows that this is not a good answer and actually makes no sense at all. In fact, profit is random; more precisely it is a function of a random variable. It is our privilege to decide the order quantity q, but profit is a function of the random variable D as well. Formally, profit can be expressed as

$$\pi(q, D) = \begin{cases} (s - c)q & \text{if } q \leq D \\ sD - cq & \text{if } q > D \end{cases}$$

In fact, if $q \leq D$, we sell everything and the profit is the ordered quantity times profit margin; otherwise, profit is the difference between revenue from selling D items, minus the cost of purchasing q items (this applies to our case, since there is no salvage value associated with unsold items). We observe that our choice of q does not determine profit; rather, by choosing q we choose a *probability distribution* for profit. But how can we rank probability distributions? To do so, distributions should be associated with a single number. We fully address this issue in Chapter 13, where we deal with decision making under uncertainty, but the most natural choice is ranking based on expected value of profit. So, our problem can be formalized as

$$\max_{q \geq 0} \Pi(q) \equiv \mathrm{E}_D[\pi(q, D)]$$

Note that expected profit $\Pi(q)$ is a function of the decision variable q only, since random demand D is eliminated by expectation; the notation above clarifies that expectation is taken with respect to demand D. In our numerical example, expected profit is

$$\Pi(q) = \frac{1}{11}\left[\sum_{d=5}^{q}(sd - cq) + \sum_{d=q+1}^{15}(s - c)q\right]$$

In the sum, we use d to mark the difference between the random variable D and its realization d (a number). The first sum corresponds to demand

Table 6.2   Expected profit for a newsvendor problem.

q	5	6	7	8	9	10
$\Pi(q)$	25.00	27.73	28.18	26.36	22.27	15.91

q	11	12	13	14	15	—
$\Pi(q)$	7.27	−3.64	−16.82	−32.27	−50.00	—

scenarios where $D \leq q$, and the second one corresponds to demand scenarios where $D > q$; the case in which $D = q$ can be attributed to either sum, without changing the result. Each of the eleven possible outcomes is divided by the uniform probability $\frac{1}{11}$. Now we may calculate $\Pi(q)$ for each possible value of q.

- There is no point in buying fewer than five items since, according to our model of uncertainty, we are going to sell at least five.

- If $q = 5$, there is no risk at all, since we will sell everything, anyway, with a profit $\Pi(5) = 5 \times 5 = 25$.

- If $q = 6$, things are a little more involved. If it turns out that $D = 5$, we have a reduction in profit with respect to the previous case, since we lose 20 on one unsold item and profit is just 5. In any other scenario, we earn a profit of 30. To compare this profit distribution with the riskless profit of 25 when choosing $q = 5$, we compute the expected value:

$$\Pi(6) = \frac{5 + 10 \times 30}{11} \approx 27.73$$

showing that $q = 6$ is a better decision than $q = 5$, assuming that we have picked the correct criterion.

Proceeding this way, we get the results given in Table 6.2. We see immediately that the optimal solution is not $q = 10$, but $q = 7$. Is this really surprising? Looking at the data, we see that profit margin is $25 - 20 = 5$, which is the cost of a stockout occurring if we buy less than necessary. However, the cost of an unsold item is 20. Hence, it is not surprising that we came up with a rather conservative solution. With a large profit margin and/or a higher markdown sales price, the optimal order quantity would increase. Incidentally, this is why bakers are so conservative and fashion shops are so optimistic. Actually, rather than solving a problem by brute force, it would be nice to gain some insight, possibly by an analytical solution; since this is easily done assuming a continuous probability distribution, we defer this to Section 7.4.4.

This example shows a fundamental principle:

A point forecast, i.e., a single number, is typically not enough to come up with a good decision under uncertainty.

Before parting with this little example, it is very (and I stress *very*) important to understand a fundamental limitation of our modeling framework. We are assuming that our decisions do not influence uncertainty. Granted, there are cases in which this is a sensible assumption, or maybe a necessary one in order to make the problem tractable. Yet, a few histories of financial disasters suggest the opportunity of reflecting a bit on the appropriateness of such an assumption. We deal a bit with this thorny issue in Chapter 14; for now, let us observe the following:

- We are assuming that our choice of q does not influence demand. However, if we are too conservative and the problem is repeated over time, a lot of stockouts may erode our customer base.

- If the game is not repeated, then one might question the sensibility of using the expected value, which can be interpreted as a long-run average. Maybe, we should consider a risk measure as well, like standard deviation of profit. Still, if the approach is applied to a large number of items, we might argue that maximizing expected profit should lead to a good solution on average.[12]

- Marketing studies show that the number of items on the shelves may influence demand. Would you buy the last, lonely box of a perishable food item on the shelves?

- We are also assuming that sales price is given, but we might use it as a tool to influence demand and maximize profit.[13]

- Last but not least, we are also assuming that we can sell all of the leftover items at the markdown price s_u. This should not be taken for granted, since the markdown price could depend on how many items we are left with.

This might suggest opportunities for more complicated modeling, and indeed there are models for dynamic markdown pricing. However, when pulling such mathematical stunts, we should always keep in mind that a more complicated model requires more input data. If these data are not reliable, it is usually wiser to settle for a simpler and more robust model inspired by a parsimony principle. □

---

[12] As we shall see, this is essentially an application of the law of large numbers; yet, this is valid under a few assumptions, including independence, which in our setting means independence of demands for different items.
[13] The reader might refer back to Section 2.11.5.

6.5.3 Bernoulli distribution

The Bernoulli distribution is based on the idea of carrying out a random experiment, which may result in a success or a failure. Let p be the probability of success; then, $1 - p$ is the probability of failure. If we assign the value 1 to variable X in case of success, and 0 otherwise, we get the following PMF:

$$p_X(0) \equiv \mathrm{P}(X = 0) = 1 - p$$
$$p_X(1) \equiv \mathrm{P}(X = 1) = p$$

It is easy to calculate the expected value:

$$\mathrm{E}[X] = 1 \cdot p + 0 \cdot (1 - p) = p$$

and variance:

$$\mathrm{Var}(X) = \mathrm{E}[X^2] - \mathrm{E}^2[X] = \left[1^2 \cdot p + 0^2 \cdot (1 - p)\right] - p^2 = p(1 - p)$$

It is always useful to reflect a bit and get an intuitive feeling for the parameters of a probability distribution. If the support of the distribution is the set $\{0, 1\}$, the expected value is the same as the parameter p. Hence, the expected value of X is large when the probability of success is large, and this makes obvious sense. The same applies if you generalize the support to any other value. As a practical example, consider the launch of a new product and the corresponding profit if it is a success or a failure. Having only two values would arguably make a poor model of uncertainty, but it may be a simple starting point.[14]

Checking variance is a bit more interesting and leads us to a few useful observations:

- Variance cannot be negative, since $p \in [0, 1]$. A formula like $p(p - 1)$ would make no sense, since it would allow for a negative variance. Indeed, if we plot the formula for variance as a function of p, we get a concave parabola such that variance is positive for p in the allowable range, $0 \leq p \leq 1$, and negative outside.

- Variance is zero, the minimum possible value, when $p = 0$ and $p = 1$. This makes sense again; if success or failure are guaranteed, there is no uncertainty.

- The maximum variance is obtained when $p = 0.5$, which corresponds to the maximum level of uncertainty about the outcomes.

Now let us use this intuition in a toy example that is related to the Bernoulli random variable.

---

[14]You may wish to refer back to the examples in Section 1.2.3.

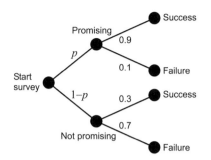

Fig. 6.3 Partial decision tree in Example 6.10.

Example 6.10 A firm has developed a new product and it must decide whether starting full scale production is worthwhile.[15] If the product is successful, profit will be €120,000; otherwise, it will just be €20,000. Probability of success is 0.6, or 60%, so there is quite some uncertainty. As an alternative, the firm could just sell the production license for €60,000 to another firm. However, there is still a third possibility; the firm could try to get some additional information by carrying out a market research survey, at a cost of €4,000. If the result of the survey is promising, the firm believes that the probability of success will be 0.9; otherwise, it will just be 0.3. Note that these are *conditional* probabilities, given the outcome of the survey.

- At what level of probability would the survey suggest a promising future for the product?

- What is the best course of action for the firm?

- What is the maximum cost of the survey that the firm should be willing to pay?

The first question seems a bit tricky. The point is that by carrying out a small-scale survey, interviewing potential customers to see if they like the new product, we will *not* increase the unconditional probability of success. The firm is not improving the product; it is only gathering information. The situation can be visualized by the tree in Fig. 6.3. The events we consider are P_{OK}, the product is successful; S_{good}, the survey result is promising; S_{bad}, the survey result is disappointing. If we do not carry out the survey, unconditional probability of success is 60%. If we carry out the survey, the unconditional probability of success, as seen from the root of the tree, is

$$
\begin{aligned}
P(P_{OK}) &= P(P_{OK}\,|\,S_{good}) \times P(S_{good}) + P(P_{OK}\,|\,S_{bad}) \times P(S_{bad}) \\
&= 0.9 \times p + 0.3 \times (1-p)
\end{aligned}
$$

[15]This example is adapted from the text by Curwin and Slater [4].

Setting this equal to 0.6 (60%) and solving for the unknown probability yields $p = 0.5$. This is the only probability ensuring consistency in our representation of uncertainty. Now, if we compare the expected profit from immediate product launch, we obtain

$$E[\pi] = 0.6 \times 120{,}000 + 0.4 \times 20{,}000 = 80{,}000 \qquad (6.13)$$

against the profit from selling a production license, we see that we should give the product a try.[16] If we carry out the survey, and the result is positive, expected profit is

$$E\left[\pi \mid S_{\text{good}}\right] = 0.9 \times 120{,}000 + 0.1 \times 20{,}000 = 110{,}000$$

from which the cost of the survey should be deducted. If the survey is not promising, then

$$E\left[\pi \mid S_{\text{bad}}\right] = 0.3 \times 120{,}000 + 0.7 \times 20{,}000 = 50{,}000$$

and we are better off selling the license. Hence, if we carry out the survey and make optimal use of the information that it provides, we obtain

$$E[\pi] = 0.5 \times 110{,}000 + 0.5 \times 60{,}000 - 4000 = 81{,}000 \qquad (6.14)$$

We see that the best course of action is:

- Carrying out the survey.

- If the result is promising, start production.

- If the result is not promising, sell the license.

Of course this is just a toy example, but it is important to really understand what does the trick. The survey does not improve our chances of success. Rather, it allows us to defer the decision while collecting additional information. We may see this by recalling what we have just observed about the Bernoulli random variable: Variance is decreased when the probability of success is driven away from 0.5, one way or the other. In this example, the unconditional standard deviation of profit if we do not carry out the survey will be

$$\sigma(\pi) = \sqrt{0.6 \times (120{,}000)^2 + 0.4 \times (20{,}000)^2 - (80{,}000)^2} = €48{,}989.79.$$

The conditional standard deviations depending on the survey outcome are

$$\sigma(\pi \mid S_{\text{good}}) = \sqrt{0.9 \times (120{,}000)^2 + 0.1 \times (20{,}000)^2 - (110{,}000)^2} = €30{,}000$$
$$\sigma(\pi \mid S_{\text{bad}}) = \sqrt{0.3 \times (120{,}000)^2 + 0.6 \times (20{,}000)^2 - (50{,}000)^2} = €45{,}825.76$$

---

[16]This is true if we care only about expected values. We will see in Chapter 13 that this applies to risk-neutral decision makers; risk-averse decision makers should also worry about other features.

We see that the effect of the survey is to reduce uncertainty, and the firm should be willing to pay at most €5000 for this reduction. To see this, observe that the expected profit in Eq. (6.14), without deducting the cost of the survey, is €85,000, which should be compared against €80,000, the expected profit of immediate product launch from Eq. (6.13).   ⬜

6.5.4 Geometric distribution

The *geometric distribution* is a generalization of the Bernoulli random variable. The underlying conceptual mechanism is the same, but the idea now is repeating identical and independent Bernoulli trials until we get the first success. The number of experiments needed to stop the sequence is a random variable X, with unbounded support $1, 2, 3, \ldots$. Finding its PMF is easy and is best illustrated by an example.

Example 6.11 The star of a horror movie is being chased by a killer monster, but she manages somehow to get home. The only problem is that she has n keys, and it goes without saying that she does not remember which one will open the door.[17] She starts picking one key at random and trying it, until she finds the right one. What is the probability that the door will open after k trials?

The answer depends on how cool our hero is after the long chase and whether she is lucid enough to set the nonworking keys apart. If she is not that lucid, then we are within the framework of the geometric distribution. The probability of success is $p = 1/n$ and, since wrong keys are put back in the same bunch with the other ones, all of the trials are identical and independent. If the door opens after k trials, this means that she failed $k - 1$ times. Since events are independent, we just take the product of individual probabilities. Denoting the number of trials by X, we obtain

$$
\begin{aligned}
\mathrm{P}(X = k) &= \underbrace{(1 - p) \times (1 - p) \times \cdots \times (1 - p)}_{k - 1 \text{ times}} \times p \\
&= (1 - p)^{k-1} p = \left(\frac{n - 1}{n} \right)^{k-1} \frac{1}{n}
\end{aligned}
$$

Note that, in principle, there is no upper bound on the number of trials.

If our hero keeps cool and discards the wrong keys, we should carry out a more careful analysis since now there is some memory in the process. The first failure has probability

$$
\frac{n - 1}{n}
$$

---

[17] You may have noticed that, in horror movies, keys *never* work; doors will not open and car engines will not start.

The second failure has probability

$$\frac{n-2}{n-1}$$

since now there is one key less. The pattern is the same for the next trials, and the last failure at the $(k-1)$th trial has probability

$$\frac{n-(k-1)}{n-(k-1)+1} = \frac{n-k+1}{n-k+2}$$

Finally, the probability that trial k is a success is

$$1 - \frac{n-k}{n-k+1} = \frac{1}{n-k+1}$$

Putting everything together, the desired probability is

$$\frac{n-1}{n} \times \frac{n-2}{n-1} \times \frac{n-3}{n-2} \times \cdots \times \frac{n-k+1)}{n-k+2} \times \frac{1}{n-k+1} = \frac{1}{n}$$

In this case, the distribution of the number of trials boils down to a uniform distribution, which may look a bit disappointing after all of this work. In fact, a much smarter idea is just realizing that if wrong keys are not reinserted in the bunch, the underlying random mechanism is equivalent to just throwing the n keys into n bins at random. The probability that the right key is in any of those bins is just $1/n$. Note that, without reinserting the keys, the support of the distribution is bounded by a worst case outcome of n trials.   ⬚

The example shows that the PMF of a geometric variable with parameter p is[18]

$$p_i \equiv \mathrm{P}(X = i) = (1-p)^{i-1}p \qquad i = 1, 2, 3, \dots$$

Very large values are quite unlikely, but not impossible. Figure 6.4 shows the PMF of a geometric variable for $p = 0.2$. We see that large values are associated with a very small probability. Empirically, you will only observe finite realizations but allowing for extreme, however unlikely, values is important for risk management. By the way, the careful reader should wonder whether it is true that probabilities add up to one for the geometric distribution. Indeed, recalling properties of the geometric series,[19] we see that

$$\sum_{i=1}^{\infty}(1-p)^{i-1}p = p\sum_{i=0}^{\infty}(1-p)^{i} = \frac{p}{1-(1-p)} = 1$$

---

[18]Sometimes, the geometric distribution is defined by counting only failures. If so, 0 would be included in the support and the PMF would be slightly changed accordingly.
[19]See Section 2.12.

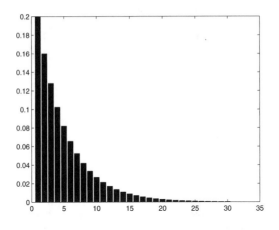

Fig. 6.4   PMF of a geometric distribution with parameter $p = 0.2$.

We could tackle another series to calculate the expected value of a geometric random variable. Straightforward application of the definition requires calculating the following:

$$\mathrm{E}[X] = \sum_{i=1}^{+\infty} i(1-p)^{i-1}p \tag{6.15}$$

This is a somewhat tedious calculation and is left as an exercise. In Chapter 8 we will discover a clever trick based on conditional expectation, allowing us to find the result quite easily. Using the same trick, it will prove very easy to show that the variance for a geometric random variable is

$$\mathrm{Var}(X) = \frac{1-p}{p^2}$$

6.5.5   Binomial distribution

The binomial distribution arises as yet another variation on Bernoulli trials. We run n independent and identical experiments and let X be a random variable counting the number of successes. The support of the resulting random variable is $\{1, 2, \ldots, n\}$, and its probability distribution depends on *two* parameters: the probability of success p and the number of experiments n. Since events are independent, it should be an easy affair to multiply probabilities of k successes and $n - k$ failures, to get the PMF. However, there is an additional twist which is best illustrated by a simple example.

Example 6.12 A random experiment consists of three Bernoulli trials with success probability p. What is the probability of getting exactly one success?

Since experiments are independent, the probability of a pattern in which we have one success and two failures is $(1-p)^2 p$, but this is *not* the answer to the question. In the geometric distribution we know that the success must occur in the last trial, but here any one of the three can be the success. Indeed, there are three outcomes for which $X = 1$:

$$(S, F, F), \ (F, S, F), \ (F, F, S)$$

where F and S denote failure and success, respectively. Hence, we see that $(1-p)^2 p$ is just the probability of *one* pattern in which there is one success, but since there are three, the correct probability is $P(X = 1) = 3(1-p)^2 p$.

$\square$

In the example above, there is an easy solution, since there are three sequences of three experiments such that there is exactly one success. But how many sequences of, say, 50 experiments may result in 18 successes? The answer is provided by binomial coefficients, which were introduced back in Section 2.2.4, when dealing with combinatorial analysis and permutations. Given n trials, the number of sequences containing k successes is given by the following binomial coefficient:

$$\binom{n}{k} \equiv \frac{n!}{(n-k)!k!}$$

In the example above

$$\binom{3}{1} \equiv \frac{3!}{(3-1)!1!} = \frac{3 \times 2 \times 1}{(2 \times 1) \times 1} = 3$$

Then, the PMF of a binomial random variable with parameters p and n is

$$P(X = k) = \binom{n}{k} p^k (1-p)^{n-k}, \qquad k = 1, \ldots, n \qquad (6.16)$$

Using properties of the binomial coefficients, it is easy to see that these probabilities add up to 1. An interesting feature of the binomial distribution is that it depends on two parameters. Figure 6.5 shows two PMFs for $n = 30$; plot (a) refers to the case $p = 0.2$ and plot (b) to the case $p = 0.4$. The support is the same for both distributions, but we see how a change in p shifts the PMF. The binomial distribution can be used as a model of uncertainty even when there is no underlying experiment based on Bernoulli trials. Indeed, the two parameters can be fine-tuned to fit empirical data. One such example is modeling demand for items that are sold in small amounts; when sales volume is high, a continuous random variable may be a simpler model.

A binomial random variable X can be regarded as the sum of n independent and identically distributed Bernoulli variables Y_i with parameter p. This

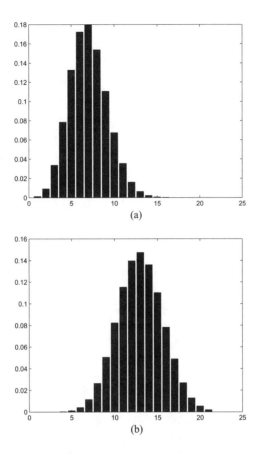

Fig. 6.5 The PMF of a two binomial distributions with parameter $n = 30$ and (a) $p = 0.2$, (b) $p = 0.4$.

is most useful to find expected value and variance by direct application of Property 6.7 and Eq. (6.9):

$$\mathrm{E}[X] \;=\; \mathrm{E}\left[\sum_{i=1}^{n} Y_i\right] \;=\; \sum_{i=1}^{n} \mathrm{E}[Y_i] \;=\; np$$

$$\mathrm{Var}(X) \;=\; \mathrm{Var}\left[\sum_{i=1}^{n} Y_i\right] \;=\; \sum_{i=1}^{n} \mathrm{Var}[Y_i] \;=\; np(1-p)$$

The following example illustrates the role and the limitations of binomial random variables for an interesting practical application, namely, overbooking strategies for airlines.[20]

---

[20]See also Example 6.1.

Example 6.13 An airline observes that 5% of booked passengers do not show up at check-in. Hence, they adopt an overbooking strategy, accepting a number of reservations that exceeds the number of available seats. When some passenger with a reservation cannot be accommodated, there is a cost, since the overbooked passenger must be rerouted, and maybe offered overnight accommodation.

- If an aircraft has 50 seats, and the airline accepts 52 reservations, what is the probability that all of the passengers checking in can be accommodated on the aircraft?

- If each overbooked passenger costs $300, what is the expected cost of the policy?

As a first check, we estimate the expected number of passengers checking in if there are 52 reservations:

$$52 \times (1 - 0.05) = 49.4 < 50$$

This means that the policy is sensible since the average number of actual passengers is less than the aircraft capacity. However, this does *not* mean that there will never be trouble. By the same token, it would be a gross mistake to say that the expected overbooking cost is zero. Remember that the expected value of a function is *not* the function of the expected value.

To analyze the problem we must model the underlying uncertainty. We may associate each booked passenger with a Bernoulli trial: She will show up (a success) with probability $p = 0.95$, and she will not show up with probability $1 - p = 0.05$. Here, we are using the information we have about a population of passengers as the probability that each *single* passenger does not check in. In doing so, the implicit assumption is that passengers cancel independently of one another; we immediately see that this is a simplification, since we are not taking behavior of families or groups into account. Still, doing so allows us to apply the binomial distribution to get a first feeling for the involved numbers.

Let us denote the number of passengers checking in by X, a binomial random variable with parameters $p = 0.95$ and $n = 52$. There will be no overbooking problem if the number of passengers checking in does not exceed aircraft capacity, and this happens with probability

$$P(\mathsf{OK}) = P(X = 0) + P(X = 1) + P(X = 2) + \cdots + P(X = 50)$$

Calculating the desired probability like this requires plenty of calculations. Since probabilities add up to one, it is definitely better to compute the probability above as

$$P(\mathsf{OK}) = 1 - P(X = 51) - P(X = 52)$$

Using the PMF of a binomial random variable, we find

$$P(X = 52) = 0.95^{52} = 0.0694$$

$$P(X = 51) = \binom{52}{1} \times 0.95^{51} \times 0.05 = 0.1901$$

$$P(\mathsf{OK}) = 1 - 0.0694 - 0.1901 = 74.05\%$$

By the same token, the expected cost of the overbooking strategy is

$$\$600 \times 0.0694 + \$300 \times 0.1901 = \$98.67$$

While the cost may seem relatively small, the probability of an overbooking is fairly large: In one case out of four the airline has to manage a situation, and this may be detrimental for its image. Actually, this really depends on how busy that flight is. Indeed, to be precise, we *did not* compute the unconditional probability $P(\mathsf{OK})$, but the conditional probability $P(\mathsf{OK} \mid \mathsf{res} = 52)$, i.e., the probability that there is no overbooking situation if there are 52 reservations.

A more sensible analysis should consider the number of reservations we receive, as well as the structure of reservation cancellations, which may involve pairs or larger groups of passengers, and not only single ones. In real life, airlines also price different fares with different rights to change a reservation. Proper capacity management in airlines is one of the most fruitful domains for quantitative analysis.[21]   ☐

6.5.6   Poisson distribution

The Poisson random variable arises naturally when we have to count the number of events occurring over a specific time interval. In later chapters, we see that this kind of distribution is intimately related to exponential random variables, which are dealt with in Section 7.6.3, and with the Poisson stochastic process, introduced in Section 7.9. For now, the best way to understand the Poisson distribution is by thinking about the random process of customer arrivals to a service facility. The main feature of such a process is the average arrival rate, i.e., the expected number of customers requesting service per unit time. Let us denote this rate by λ. If the arrival rate does not change in time and t is the length of a time interval, the expected number of customers arriving during that interval is λt.

Let X be the number of customers arriving during an interval of unit length. Its support is $\{0, 1, 2, 3, \ldots\}$, and its expected value is the arrival rate λ. The exact distribution of X depends on many things, including the random time elapsing between two consecutive arrivals. If interarrival times satisfy certain sensible properties, then the number of customers arriving during any

---

[21]See, e.g., Chapter 4 of Ref. [11] for an introduction to more complex models.

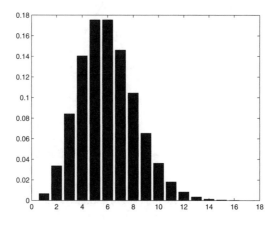

Fig. 6.6 The PMF of a Poisson random variable with parameter $\lambda = 5$.

time interval of unit length has Poisson distribution. The PMF of a Poisson random variable with parameter λ is

$$p_i \equiv P(X = i) = e^{-\lambda}\frac{\lambda^i}{i!}, \qquad i = 0, 1, 2, \ldots$$

Figure 6.6 shows the PMF of a Poisson random variable with parameter $\lambda = 5$. Recalling Taylor's expansion of the exponential function[22], it is easy to see that Poisson probabilities add up to 1:

$$\sum_{i=0}^{\infty} p_i = e^{-\lambda}\sum_{i=0}^{\infty}\frac{\lambda^i}{i!} = e^{-\lambda}e^{\lambda} = 1$$

Calculating the expected value is also fairly straightforward:

$$\mathrm{E}[X] \;=\; \sum_{i=0}^{\infty} i e^{-\lambda}\frac{\lambda^i}{i!} = \lambda e^{-\lambda}\sum_{i=1}^{\infty}\frac{\lambda^{i-1}}{(i-1)!} = \lambda e^{-\lambda}\sum_{k=0}^{\infty}\frac{\lambda^k}{k!} = \lambda$$

It can also be shown that $\mathrm{Var}(X) = \lambda$.

Example 6.14 A customer service center receives, on average, $\lambda = 3$ calls per hour. Using the PMF of a Poisson distribution, we may calculate the probability of receiving an arbitrary number of calls within one hour. Since

---

[22]See Example 2.31.

$e^{-3} = 0.0498$, we have:

$$p_0 = 0.0498 \times \frac{3^0}{0!} = 0.0498$$

$$p_1 = 0.0498 \times \frac{3^1}{1!} = 0.1494$$

$$p_2 = 0.0498 \times \frac{3^2}{2!} = 0.2240$$

$$p_3 = 0.0498 \times \frac{3^3}{3!} = 0.2240$$

$$p_4 = 0.0498 \times \frac{3^4}{4!} = 0.1680$$

$$\ldots$$

The probability of receiving two or more calls is

$$\sum_{k=2}^{\infty} P(X = k) = 1 - P(X = 0) - P(X = 1) = 1 - (0.0498 + 0.1494) = 0.8008$$

$\square$

At present, we cannot fully illustrate the conditions under which the number of events occurring with given rate can be modeled by a Poisson variable. They are linked to the memoryless property of the exponential random variable, which we defer to Section 8.5.2. Nevertheless, at an intuitive level, we may say that a Poisson distribution arises when

- The arrival process is stationary, i.e., the arrival rate does not change over time

- The arrival process is purely random, i.e., there is no regular pattern like the one arising from deterministic and regular arrivals

- The arrival numbers in two disjoint time intervals are two independent random variables

There is another way to shed some light on the Poisson distribution, which emphasizes its link with the binomial random variable. Consider a binomial random variable with parameters $p \to 0$ and $n \to +\infty$. In other words, we have a huge number of Bernoulli trials, but the probability of success for each one is tiny. Then, it can be shown that the distribution of such a Bernoulli random variable tends to the distribution of a Poisson random variable with parameter $\lambda = pn$.

Problems

6.1   Consider a generalization of the Bernoulli random variable, i.e., a variable taking values x_1 with probability p and x_2 with probability $1 - p$. Which values of p maximize and minimize variance?

6.2   Using Eq. (6.15), prove that the expected value of a geometric random variable X with parameter p is $E[X] = 1/p$. (*Hint:* Use the result in Example 2.40.)

6.3   Using the binomial expansion formula (2.3), prove that the PMF of the binomial distribution [see Eq. (6.16)] adds up to one.

6.4   You are about to launch a new product on the market. If it is a success you will make $16 million; otherwise, you lose $5 million. The probability of success is 65%. You could increase chances of success by delaying product launch in order to improve product design; this would take 6 months and would cost an additional $1 million. In order to account for the delay and the time value of money, we should discount cash flows at a rate of 3% (the rate refers to the 6-month period and is applied to profit/loss). What is the minimal improvement in success probability that makes the delay worthwhile?

6.5   According to an accurate survey, 40% of people checked at the exit of a well-known pub have made excessive use of alcoholic drinks. If we take a random sample of 25 persons, what is the probability that at least 4 of them are flagged?

6.6   Batteries produced by a company are known to be defective with a probability of 0.02. The company sells batteries in packages of eight and offers a money-back guarantee that at most one of them is defective. What is the probability that a package is returned? If a customer buys three packages, what is the probability that exactly one of them will be returned?

For further reading

- An elementary and quite readable introduction to random variables is offered in Ross [9]; at a similar level, Ross' earlier text [8] is a more detailed reference, rich in examples.

- The reader interested in a thorough treatment, while keeping mathematical complexity to a reasonable level, might consult Ref. [10]; see also the twin books by Grimmett and Stirzaker [5, 6].

- We did not bother too much to clarify which conditions should be required for a mapping from sample space Ω to numerical values to be a proper random variable. This is an issue of mostly theoretical interest, but in Ref. [3] you may find a readable treatment based on examples and problems. The more mathematically inclined readers will find Resnick's text [7] a challenging reading.

- On the application side, we have mentioned overbooking strategies and the newsvendor problem. An extensive reference on revenue management, including overbooking strategies, is Ref. [11]. The basic newsvendor problem and a few generalizations are covered, e.g., in Ref. [2].

- Other application-oriented sources are Refs. [1] and [4].

REFERENCES

1. D. Bertsimas and R.M. Freund, *Data, Models and Decisions: The Fundamentals of Management Science*, Dynamic Ideas, Belmont, MA, 2004.

2. P. Brandimarte and G. Zotteri, *Introduction to Distribution Logistics*, Wiley, New York, 2007.

3. M. Capiński and T. Zastawniak, *Probability through Problems*, Springer-Verlag, Berlin, 2000.

4. J. Curwin and R. Slater, *Quantitative Methods for Business Decisions*, 6th ed., Cengage Learning EMEA, London, 2008.

5. G. Grimmett and D. Stirzaker, *One Thousand Exercises in Probability*, Oxford University Press, Oxford, 2003.

6. G. Grimmett and D. Stirzaker, *Probability and Random Processes*, 3rd ed., Oxford University Press, Oxford, 2003.

7. S.I. Resnick, *A Probability Path*, Birkhäuser, Boston, 1999.

8. S.M. Ross, *Introduction to Probability Models,* 8th ed., Academic Press, San Diego, 2002.

9. S.M. Ross, *Introduction to Probability and Statistics for Engineers and Scientists,* 4th ed., Elsevier Academic Press, Burlington, MA, 2009.

10. D. Stirzaker, *Elementary Probability,* 2nd ed., Cambridge University Press, Cambridge, UK, 2003.

11. K.T. Talluri and G.J. Van Ryzin, *The Theory and Practice of Revenue Management*, Springer, New York, 2005.

7

Continuous Random Variables

In the previous chapter we have gained the essential intuition about random variables in the discrete setting. There, we introduced ways to characterize the distribution of a random variable by its PMF and CDF, as well as its expected value and variance. Now we move on to the more challenging case of a continuous random variable. There are several reasons for doing so:

- Some random variables are inherently continuous in nature. Consider the time elapsing between two successive occurrences of an event, like the request for service or a customer arrival to a facility. Time is a continuous quantity and, since this timespan cannot be negative, the support of a random variable modeling this kind of uncertainty is $[0, +\infty)$.

- Sometimes, continuous variables are used to model variables that are actually integers. As a practical example, consider demand for an item; a low-volume demand can be naturally modeled by a discrete random variable. However, when volumes are very high, it might be convenient to approximate demand by a continuous variable. To see the point, imagine a demand value like $d = 2.7$; in discrete manufacturing, you cannot sell 2.7 items, and rounding this value up and down makes a big difference; but what about $d = 10{,}002.7$? Quite often this turns out to be quite a convenient simplification, in both statistical modeling and in decision making.[1]

---

[1]We will appreciate this point when dealing with linear programming models in Chapter 12.

- Last but not least, in the next chapters on statistical applications, the most common probability distribution is, by far, the normal (or Gaussian) distribution, which is a continuous distribution whose support is the whole real line $\mathbb{R} = (-\infty, +\infty)$. As we will see, there are several reasons why the normal distribution plays a pivotal role in statistics.

This chapter extends the concepts that we have introduced for the discrete case, and it also presents a few new ones that are better appreciated in this broader context. The mathematical machinery for a full appreciation of continuous random variables is definitely more challenging than that required in the discrete case. However, an intuitive approach is adequate to pursue applications to business management. Cutting a few corners, the essential difficulty in dealing with continuous random variables is that we cannot work with the probability $P(X = x)$, as this is always zero for a continuous random variable. Unlike the discrete case, the probability mass is not concentrated at a discrete set of points, but it is distributed over a continuous set, which contains an infinite number of points even if the distribution support is a bounded interval like $[a, b]$. The role of the PMF is played here by a probability *density* function (PDF for short). Furthermore, the sums we have seen in the discrete context should be replaced by integrals.

Integrals were introduced in Section 2.13; we do not really need any in-depth knowledge, as integrals can be just interpreted as areas. In Section 7.1 we pursue an intuitive approach to see the link between such areas and probabilities. Then, we introduce density functions in Section 7.2, where we also see that the concept of cumulative distribution function (CDF) needs no adjustment when moving from discrete to continuous random variables. We see how expected values and variances are applied in this context in Section 7.3. Then, we expand our knowledge about the distribution of random variables by considering their mode, median, and quantiles in Section 7.4, and higher-order moments, skewness, and kurtosis in Section 7.5. All of these concepts apply to discrete random variables as well, but we have preferred to treat them once in the most general setting. As you can imagine, there is a remarkable variety of continuous distributions that can be applied in practice; we may use theoretical distributions whose parameters may be fit against empirical data, or we may just come up with an empirical distribution reflecting the data. In Section 7.6 we outline the main theoretical distributions – uniform, beta, triangular, exponential, and normal distributions – and we hint at how empirical distributions can be expressed. In Section 7.7 we take a first step toward statistical inference by considering sums of independent random variables; this will also lead us to the cornerstone central limit theorem, as well as a few more distributions that can be obtained from the normal and also play a pivotal role in inferential statistics; we will also have a preview of the often misunderstood law of large numbers. We illustrate a few applications in Section 7.8, with emphasis on quantiles of the normal distribution; what is remarkable, is that the very same concepts can be put to good use in diverse fields such as supply chain management and financial risk management. Finally, we consider sequences of random variables in time, i.e., stochastic processes, in Section 7.9. Section 7.10 can be skipped by most readers, as it has a more theoretical nature: Its aim is to clarify a point that we did not really investigate in the previous chapter, when we defined a random variable, i.e., the relationship between the random variable and the event structure of the underlying probability space.

7.1   BUILDING INTUITION: FROM DISCRETE TO CONTINUOUS RANDOM VARIABLES

The most natural way to characterize a discrete distribution is by its PMF, which can be depicted as a set of bars whose height is the probability of each value. What happens when we consider a random variable that may take any real value on an interval? A starting point to build intuition is getting back to descriptive statistics and relative frequency histograms. Imagine taking a sample of values, which are naturally continuous, and plotting the corresponding histogram of relative frequencies. The appearance of the histogram depends on how large the bins are. If they are rather coarse intervals, the histogram will look extremely jagged, like the one in Fig. 7.1(a). If we shrink the bins, we get thinner bars, looking like the histogram (b) in the figure. Please note that all of the histograms in Fig. 7.1 refer to the *same* set of data. You may also notice that the relative frequencies in the second case are lower than in the first one; this happens because there are more bins, and we assign fewer outcomes to each one. In the limit, if the bins get smaller and smaller and the sample size is large, we will get something like histogram (c) in the figure. This looks much like a continuous function, describing where sampled values are more likely to fall, whereas the PMF is just a set of values on a discrete set of points.

To make this idea a bit more precise, let us consider a continuous uniform distribution, which is arguably the simplest distribution we can think of. We got acquainted with the discrete uniform distribution in Section 6.5.2. If we consider a continuous uniform variable on the interval $[a, b]$, we should have a "uniform probability" over that interval, as depicted in Fig. 7.2. Given a point x in the interval, what is the probability that the random variable X takes that value, i.e., $P(X = x)$? Whatever this value is, it must be the same for all of the points in the interval. We know from Section 6.5.2 that in the discrete case $p = 1/n$, where n is the number of values in the support, but here we have an infinite number of values within the bounded interval $[a, b]$. Intuitively, if $n \to +\infty$, then $p = 1/n \to 0$. Moreover, if we assign any strictly positive value to p, the sum of probabilities will go to infinity, but we know that probabilities should add up to 1. It is tempting to think that the root of the trouble is that we are dealing with a support consisting of infinitely

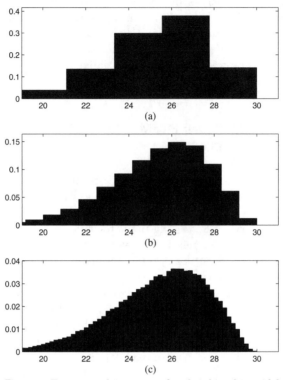

Fig. 7.1   Frequency histograms for shrinking bin widths.

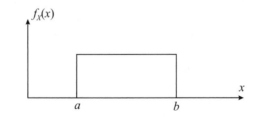

Fig. 7.2   A uniform distribution on the interval $[a, b]$.

many possible values. However, this is not really the case. In Section 6.5.6, we considered the Poisson distribution, which does have an infinite support. However, since probabilities vanish for large values of the random variable, their sum does converge to 1. This is not possible here, as we are considering a uniform variable. It seems that there is no way to assign a meaningful probability value to a single outcome in the continuous case.

However, there is a way out of the dilemma. We can assign sensible probabilities to *intervals*, rather than single values. To be concrete, consider a uni-

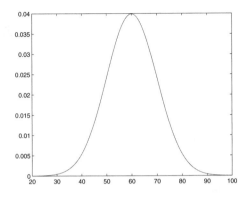

Fig. 7.3 A bell-shaped, nonuniform distribution.

form random variable on the interval $[0, 10]$, which is denoted by $X \sim U(0, 10)$. Common sense suggests the following results:

$$P(X \leq 3) = 0.3, \qquad P(5 \leq X \leq 8) = 0.3,$$

as in both cases we are considering an interval whose length is 3, i.e., 30% of the whole support. Notice that this probability depends on the width of the interval we consider, not on its location; indeed, this is what makes a distribution uniform, after all. More generally, it seems that if we consider an interval of width w included in $[0, 10]$, the probability that X falls there should be the ratio between w and the width of the whole support: $w/10$. By the way, we recall from elementary geometry that a point has "length" zero; hence, we begin to feel that in fact $P(X = x) = 0$ for any value x.

So far, so good, but what about a nonuniform distribution, like that in Fig. 7.3? If we consider several intervals of the same width, we cannot associate the same probability with them. Probability should be related to the height of the distribution, which is not constant. Hence, the length of subintervals will not do the trick. Nevertheless, to keep the shape of the distribution duly into account, we may associate probability of an interval with the *area* under the distribution, over that interval. The concept is illustrated in Fig. 7.4. If we shift the interval $[a, b]$ in the uniform case, we always get the same area, provided that the interval is a subset of the whole support; if we do the same in the bell-shaped case, we get quite different results. We start seeing that probabilities in the continuous case

1. Are distributed, whereas they are concentrated at a discrete set of points in the discrete case; we cannot work with a probability mass function associating single values x with the probability $P(X = x) = 0$.

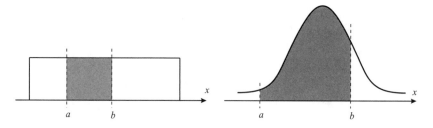

Fig. 7.4 Linking probability to areas.

2. May be associated with areas below a function that replaces the PMF, but plays a similar role; this function is the *probability density function* (PDF). We will denote the PDF of random variable X as $f_X(x)$.

To wrap up our intuitive reasoning, we should state one fundamental property of the PDF. When dealing with discrete variables, we know that probabilities add up to 1:

$$\sum_i p_X(x_i) = 1 \tag{7.1}$$

For continuous variables, it must be the case that

$$P(-\infty \leq X \leq +\infty) = 1$$

implying that the overall area below the PDF must be 1. But we also recall from Section 2.13 that this area can be expressed by an integral. Therefore, condition (7.1) should be replaced by

$$\int_{-\infty}^{+\infty} f_X(x)\,dx = 1 \tag{7.2}$$

If we are dealing with a uniform variable with support $[a, b]$, then

$$P(a \leq X \leq b) = 1$$

This is just a condition on the area of a rectangle with one edge of length $(b - a)$, and the other one corresponding to the value of the PDF. Therefore:

$$f_X(x) = \begin{cases} \dfrac{1}{b-a} & \text{if } x \in [a, b] \\ 0 & \text{otherwise} \end{cases}$$

If the support is the interval $[0, 10]$, then

$$\begin{aligned} P(X \leq 3) &= (3 - 0) \times \tfrac{1}{10} = 0.3 \\ P(5 \leq X \leq 8) &= (8 - 5) \times \tfrac{1}{10} = 0.3 \end{aligned}$$

as expected. With more general distributions, we have some more technical difficulties in calculating areas, but there are plenty of statistical tables and software packages taking care of this task for us.

7.2 CUMULATIVE DISTRIBUTION AND PROBABILITY DENSITY FUNCTIONS

A full characterization of discrete random variables can be given in terms of PMF or CDF. They are related, as the CDF can be obtained from the PMF by summing, and we can go the other way around by taking differences. For the reasons we have mentioned, in the continuous case the role of the PMF is assumed by a probability density function, whereas the CDF is defined in exactly the same way:[2]

$$F_X(x) \equiv P(X \leq x)$$

As we have pointed out, the PDF is a nonnegative function $f_X(x)$ that does *not* give the probability of a single value, but can be used to evaluate probabilities of intervals:

$$P(a \leq X \leq b) = \int_a^b f_X(x)\, dx$$

Then, it is easy to see the link between PDF and CDF:

$$F_X(x) = P(-\infty \leq X \leq x) = F_X(x) = \int_{-\infty}^x f_X(w)\, dw \qquad (7.3)$$

Since the PDF is nonnegative, the CDF is a nondecreasing function:

$$y > x \quad \Rightarrow \quad F_X(y) = \int_{-\infty}^y f_X(w)\, dw = \int_{-\infty}^x f_X(w)\, dw + \int_x^y f_X(w)\, dw$$

$$\geq \int_{-\infty}^x f_X(w)\, dw = F_X(x)$$

Furthermore

$$\lim_{x \to -\infty} F_X(x) = 0, \qquad \lim_{x \to +\infty} F_X(x) = 1$$

The last property is just another way to express the normalization condition on the PDF:

$$\int_{-\infty}^{+\infty} f_X(x)\, dx = 1$$

We may observe a few similarities between CDFs of discrete and continuous random variables:

- The CDF is a nondecreasing function ranging from 0 to 1.

- We obtain CDF by summing probabilities expressed by the PMF of discrete random variables, and by integrating the PDF of continuous

---

[2]Introductory treatments of random variables start from mass and density functions, as they are more intuitive and related to frequency histograms. However, a sound treatment should start from the CDF, which is a more general concept.

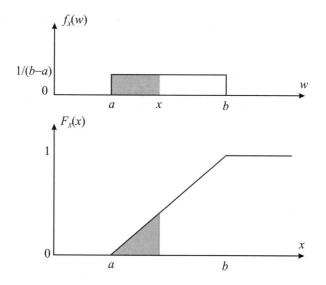

Fig. 7.5   From PDF to CDF in the case of a uniform distribution.

random variables; since the integral is, in a sense, a sort of continuous sum, this is not surprising.

The main difference is that, unlike the discrete case, the CDF of a continuous random variable is a continuous function, as illustrated in the following example.

Example 7.1 To illustrate the link between CDF and PDF, consider the PDF of a uniform distribution:

$$f_X(x) = \begin{cases} \dfrac{1}{b-a} & \text{if } x \in [a,b] \\ 0 & \text{otherwise} \end{cases}$$

Clearly, the event $\{X \leq a\}$ has zero probability; hence, we should expect that $F_X(x) = 0$ $x \leq a$. Indeed

$$\int_{-\infty}^{x} 0 \, dw = 0$$

For $a \leq x \leq b$, we have

$$F_x = \int_{a}^{x} \frac{1}{b-a} dw = \frac{w}{b-a} \Big|_{a}^{x} = \frac{x-a}{b-a} \tag{7.4}$$

Finally, $F_X(x) = 1$ for $x \geq b$. Both PDF and CDF of a uniform distribution are illustrated in Fig. 7.5.   □

From the example, we see that since the CDF is the integral of the PDF, it is a continuous function, even if the PDF has some kinky point. On the contrary, the CDF in the discrete case jumps because the probability mass is concentrated at a discrete set of points, rather than being distributed on a continuous support.

We may also better understand why the probability of any specific value is zero for continuous random variables:

$$P(X = x) = \int_x^x f_X(w)\, dw = 0$$

As a consequence, in the continuous case, we obtain

$$P(X \leq x) = P(X < x)$$

Another point worth mentioning is that

$$P(X > x) = P(X \geq x) = \int_x^{+\infty} f_X(w)\, dw = 1 - F_X(x)$$

As a final remark, we recall that in the discrete case we may go from CDF to PMF by taking differences. With continuous random variables, the equivalent operation is taking the derivative of the CDF. Indeed, we may also write $P(a \leq X \leq b)$ as

$$P(a \leq X \leq b) = P(X \leq b) - P(X \leq a) = F_X(b) - F_X(a)$$

which implies

$$\int_a^b f_x(x)\, dx = F_X(b) - F_X(a)$$

Recalling the fundamental theorem of calculus,[3] we conclude that

$$f_X(x) = \frac{dF_X(x)}{dx} \tag{7.5}$$

7.3   EXPECTED VALUE AND VARIANCE

Given a continuous random variable X and its PDF $f_X(x)$, its expected value is defined as follows:

$$E[X] \equiv \int_{-\infty}^{+\infty} x f_X(x)\, dx \tag{7.6}$$

Quite often, we use the short-hand notation $\mu_X = E[X]$. Again, this is a straightforward extension of the discrete case, where $E[X] \equiv \sum_i x_i p_X(x_i)$.

---

[3]See Theorem 2.22.

Example 7.2 As an illustration, let us consider the expected value of a uniform random variable on $[a, b]$. Symmetry suggests that the expected value should be the midpoint of the support. Indeed

$$
\begin{aligned}
\mathrm{E}[X] &= \int_{-\infty}^{+\infty} x f_X(x)\, dx = \int_a^b x\, \frac{1}{b-a}\, dx = \left.\frac{x^2}{2(b-a)}\right|_a^b \\
&= \frac{b^2 - a^2}{2(b-a)} = \frac{(b+a)(b-a)}{2(b-a)} = \frac{a+b}{2}
\end{aligned}
$$

$\Box$

By the same token, we define variance of a continuous random variable as:

$$
\mathrm{Var}(X) \equiv \mathrm{E}\left[(X - \mu_X)^2\right] = \int_{-\infty}^{+\infty} (x - \mu_X)^2 f_X(x)\, dx.
$$

Common shorthand notations for variance are σ^2 and σ_X^2; its square root σ_X is standard deviation. More generally, we define the expected value of a function $g(X)$ of a random variable as

$$
\mathrm{E}\left[g(X)\right] = \int_{-\infty}^{+\infty} g(x) f_X(x)\, dx
$$

The considerations we made about expected values of discrete random variables apply here as well. Since integration is a linear operator,[4] just like the sum, expectation is linear in the continuous case, too. All of the properties of expectation and variance, that we have introduced for the discrete case, carry over to the continuous case. In particular, we recall the following very useful properties:

$$
\begin{aligned}
\mathrm{Var}(X) &= \mathrm{E}\left[X^2\right] - (\mathrm{E}[X])^2 \\
\mathrm{E}(\alpha X + \beta) &= \alpha \mathrm{E}[X] + \beta \\
\mathrm{Var}(\alpha X + \beta) &= \alpha^2 \mathrm{Var}(X)
\end{aligned}
$$

where α and β are arbitrary real numbers.

7.4 MODE, MEDIAN, AND QUANTILES

In the chapter on descriptive statistics, we have introduced concepts like mode, median, and percentiles. We have also remarked that some concepts, in particular the percentiles, are somewhat shaky in the sense that there are slightly different definitions and ways of calculating them using observed data. In this section we examine probabilistic counterparts of these concepts, and how they are related to discrete and continuous random variables.

---

[4]See Property 2.23.

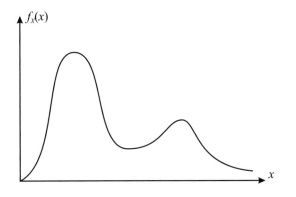

Fig. 7.6   A bimodal PDF.

7.4.1   Mode

The mode of a probability distribution is a point at which the PMF or the PDF is maximized. The concept is easy to grasp, but we should point out that the mode for a continuous random variable is *not* the value with maximum probability, since probabilities of all possible values are just zero. On the contrary, the interpretation for discrete variables is closer to the intuitive concept that we introduced along with descriptive statistics.

The mode need not be unique in principle, as multiple maxima are possible. Most of the theoretical distributions we examine later in this chapter have a single mode, but, in practice, we may find multimodal distributions in the sense illustrated in Fig. 7.6. We have a well-defined mode, but there is a secondary local maximum. If the distribution is built by fitting against empirical data, it may be the case that the secondary mode is just sampling noise. However, we should never discard the possibility that we really need a sort of mixed distribution to model different dynamics of a phenomenon.

7.4.2   Median and quantiles for continuous random variables

Roughly speaking, the median is a value splitting a dataset into two equal parts. When dealing with continuous random variables, we find that the median is a value m_X such that

$$F_X(m_X) = 0.5$$

Geometrically, the median splits the PDF in two parts with an area equal to 0.5. In descriptive statistics, the median can be regarded as a specific case of percentile that corresponds to a 50% probability. In probability theory, the term is usually replaced by quantiles.

DEFINITION 7.1 (Quantiles of continuous random variable) *Given the CDF $F_X(x)$ of a continuous random variable and a probability level $\alpha \in$*

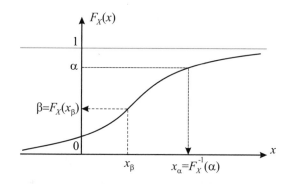

Fig. 7.7 Probability and quantiles for a continuous random variable.

$[0, 1]$, *we define the* **quantile** x_α *of the distribution as the number satisfying the equation*

$$F_X(x_\alpha) = \alpha \tag{7.7}$$

Geometrically, the quantile x_α is a number leaving an area α to its left, under the PDF. Conceptually, computing a quantile requires inversion of the CDF, as illustrated in Fig. 7.7. Be sure to understand this figure, as quantiles play a prominent role in many applications to follow:

1. Given a value x_β, we may find the corresponding probability $\beta = P\{X \le x_\beta\}$ by evaluating the CDF $F_X(x_\beta)$.

2. Given a probability α, we may find the corresponding quantile $\alpha = F_X^{-1}(\alpha)$, which is a *value*, by inverting the CDF.

A natural question is if the CDF is in fact an invertible function. In most cases, when dealing with continuous random variables, the CDF is a strictly increasing and continuous function; hence, inverting the function poses no difficulty. When support is infinite, we cannot really find quantiles corresponding to probabilities 0 and 1, and we should set $x_0 = -\infty$ and $x_1 = +\infty$. There is no guarantee of finding a unique quantile, as the CDF may be a nondecreasing function that is constant on certain intervals, rather than a strictly increasing function. This may happen if the support of the distribution consists of disjoint intervals.

Example 7.3 Consider values $x_a < x_b < x_c < x_d$ and a continuous random variable X whose support consists of the disjoint intervals $[x_a, x_b]$ and $[x_c, x_d]$. Since X cannot assume values between x_b and x_c, the CDF is constant on the interval $[x_b, x_c]$, and $F_X(x_b) = F_X(x_c) = \alpha$, for some probability value α. Clearly, the quantile x_α seems undefined, since the function is noninvertible on that interval. □

The example may look somewhat pathological, but in fact this is what happens with discrete random variables. This is why quantiles need to be defined in a more general way.

Table 7.1 PMF and CDF for the discrete probability distribution of Example 7.4.

x	1	2	3	4	5
$p_X(x)$	0.25	0.30	0.25	0.10	0.10
$F_X(x)$	0.25	0.55	0.80	0.90	1.00

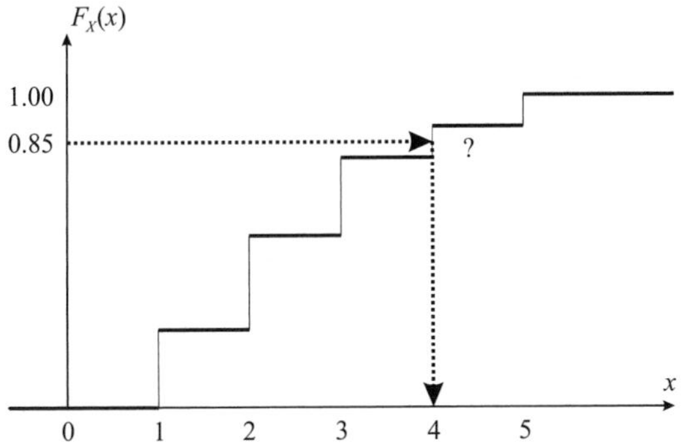

Fig. 7.8 The CDF for a discrete random variable is not invertible.

7.4.3 Quantiles for discrete random variables

Computing quantiles for a discrete random variable by applying Definition 7.1 would require inverting the CDF. However, this is a piecewise constant function, featuring jumps at each value of the distribution support, which makes its inversion impossible in general.

Example 7.4 Consider random demand for a spare part, sold in low volumes, over the next time period. There is no inventory at present, and we must determine the purchased quantity, in such a way that the probability of satisfying the whole demand is above a minimal threshold. Assume that randomness in demand can be modeled by the PMF of Table 7.1, and that we would like to meet demand with a probability of 0.85. In the parlance of supply chain management, we should say that our service level is 85%.[5] The purchased amount should correspond to the quantile with 0.85 probability level. A look at Table 7.1 shows that there is no value of x such that $F_X(x) = 0.85$. Indeed, the function is not invertible, as illustrated in Fig. 7.8. What would one do in practice? The sensible solution, since $x = 3$ gives only a 80% service level, is to select $x = 4$ to meet the required constraint. □

---

[5]This is just one possible definition of service level, related to the probability of not having a stockout; one can also refer to the size of the stockout; see, e.g., Chapter 5 of Ref. [4].

What we have done in the example above makes sense: The quantile is related to a decision, and we make it in such a way to stay on the safe side. In fact, Definition 7.1 can be generalized as follows.

DEFINITION 7.2 (Generalized definition of quantiles) *Let $F_X(x)$ be the CDF of random variable X. Given a probability level $\alpha \in [0,1]$, we define the* **quantile** x_α *of the distribution as the smallest number x_α such that $F_X(x_\alpha) \geq \alpha$. Formally*

$$x_\alpha = \min x, \quad \text{s.t.} \ \ F_X(x) \geq \alpha \tag{7.8}$$

We immediately see that if the CDF is invertible, this definition boils down to the previous one. Indeed, Eq. (7.8) corresponds to the so-called *generalized inverse function*. The reader is urged to check that applying this definition, we do find the decision we chose in Example 7.4. Unlike percentiles, quantiles of probability distributions have a precise definition that makes perfect sense, as we will see in the applications described later in Section 7.8. Before proceeding with theoretical concepts, it is worth pausing a little and check a remarkable example.

7.4.4 An application: the newsvendor problem again

In Example 6.9 we have considered and solved numerically a hypothetical instance of the newsvendor problem. The procedure was based on brute force and did not provide us with any valuable insight into the structure of the problem itself. Furthermore, if we approximate the distribution of demand by a continuous distribution, which makes sense for high sale volumes, we cannot try any possible value. An analytical solution would definitely be more elegant and useful. An easy way to find it is based on marginal analysis.[6] Say that we have purchased $q - 1$ items. Should we buy one more?

We recall that profit margin of one sold unit is given by $m = s - c$, i.e., selling price minus purchasing cost. If a unit remains unsold at the end of the sale time window, we incur a cost $c_u = c - s_u$, i.e., purchasing cost minus markdown price, the cost of unsold items. If we buy unit number q, we might incur a cost c_u; if we do not buy it, we might miss the profit margin m. We should figure out if we *expect* that this marginal unit is profitable or not. To attach probabilities to events, observe that the marginal unit will contribute m if demand D is at least q; otherwise, it will reduce profit by c_u. Hence, the *expected marginal profit* is

$$m\mathrm{P}(D \geq q) - c_u\mathrm{P}(D < q)$$

Note that if demand is a discrete random variable, we should be careful with the inequalities, as in that case the probability $\mathrm{P}(D \geq q)$ is different from

---

[6]See Chapter 5 of Ref. [4] for a more rigorous analysis and for some generalizations of the basic newsvendor model.

$P(D > q)$; if demand has continuous distribution, we may be sloppy with the inequalities. If the expected marginal profit is positive, we should buy one more item; otherwise, we should not. Hence, we must find an optimal quantity q^* such that

$$mP(D \geq q^*) - c_u P(D < q^*) > 0, \quad mP(D \geq q^* + 1) - c_u P(D < q^* + 1) < 0$$

The task is considerably simplified by assuming that demand is a continuous random variable. Then, expected marginal profit can be regarded as the first-order derivative of expected profit, and the optimal solution can be found by enforcing the first-order necessary condition for optimality:

$$
\begin{aligned}
0 &= m(1 - P\{D \leq q^*\}) - c_u P\{D \leq q^*\} \\
&= m(1 - F_D(q^*)) - c_u F_D(q^*)
\end{aligned}
\tag{7.9}
$$

where $F_D(\cdot)$ is the CDF of demand. This in turn implies that

$$F_D(q^*) = P(D \leq q^*) = \frac{m}{m + c_u} \tag{7.10}$$

To be precise, we should also check the second-order derivative, since stationarity alone cannot tell a minimum from a maximum. We may recall that the derivative of the CDF is just the PDF. Hence, if we differentiate Eq. (7.9), we obtain

$$-m f_D(q^*) - c_u f_D(q^*)$$

which is certainly negative, since the PDF is positive.[7]

By using Eq. (7.10), it is not only easy to find the optimal ordering quantity; we also gain a fundamental insight into the problem structure:

1. First we compute the ratio $m/(m+c_u)$, which can be interpreted as the probability of satisfying the whole demand, i.e., the service level.

2. Then we find the quantile corresponding to that service level.

We should note that, since m and c_u are positive, the ratio is bounded by the interval $[0, 1]$. Since the CDF is monotonically increasing, we order more (and raise the service level) when m is large with respect to c_u. We order less when profit margin is small, or when the cost of unsold items is too large. This is not too surprising and explains the different patterns that we observe at different retail shops, depending on what they sell. An obvious question is: When is it optimal to order the expected value of demand? If the distribution

---

[7]Clearly, we are ruling out the pathological case in which the PDF is zero. Negative profit margin or cost of unsold items would make no sense. By writing expected profit explicitly, we could check that it is a concave function; hence, stationarity is a necessary and sufficient condition for optimality.

is symmetric, the expected value is just the median. We will order the median if the optimal service level is 50%, which happens when $m = c_u$.

Example 7.5 Let us solve Example 6.9 by assuming, this time, a continuous uniform distribution on the interval $[5, 15]$. With the data of the problem, we have $m = 25 - 20 = 5$ and $c_u = 20 - 0 = 20$. Hence, the optimal service level is $5/(5 + 20) = 20\%$. Using Eq. (7.4) for the CDF of the continuous uniform distribution, we find

$$\frac{q^* - 5}{15 - 5} = 0.2 \quad \rightarrow \quad q^* = 5 + 0.2 \times 10 = 7$$

By sheer luck, we get the same integer solution we found by brute force in the discrete case. In general, the value q^* is not integer, and we have to check whether it is optimal to round it up or down.   ▯

   The classical newsvendor problem has an easy and elegant solution, but we should not forget the limitations of the underlying model:

- We are assuming that any leftover item can be sold at the markdown price s_u, independently of the amount of unsold items; in practice, markdown management procedures may be necessary.[8]

- We are assuming that the expected value of profit is a sensible objective. While this might make sense in the long run, sometimes risk aversion should be taken into account (see Section 13.2).

Despite all of these limitations, the basic newsvendor model is am extremely useful model for building intuition and can be generalized in many ways. A rather surprising fact is that the reasoning above, based on marginal analysis, can be applied in completely different settings.

Example 7.6 We consider here the simplest model for revenue management in the airline industry (we considered overbooking strategies in Example 6.13). This model is known as *Littlewood's two-class, single-capacity control model*.[9] Assume that we may sell tickets at two different prices, p_1 and p_2, with $p_1 > p_2$. The two prices correspond to two different classes of customers: Class 2 consists of passengers reserving their flight well in advance, maybe because they are planning their holidays. Class 1 consists of business travelers, flying because of business engagements that might pop up at the last minute. Business travelers are typically willing to pay much more for a flight, and this is why they are charged a higher price. Clearly, we are uncertain about the two demands for seats, but we imagine that a probability distribution for the two demands is known (we also assume that demands of the two classes are

---

[8]See, e.g, Chapter 10 of Ref. [10].
[9]See Chapter 2 of Ref. [17] for more information and extensions.

independent random variables). According to our assumptions, the demand D_2 from passengers of class 2 is realized before demand D_1 from class 1. If we satisfy all of the requests of class 2 (up to capacity C of the aircraft), we may regret our decision if it turns out that

$$D_1 > C - D_2$$

In such a case, had we reserved more seats to class 1 by rejecting demand from class 2 beyond some threshold level, we would have made more money by selling some seats at a price $p_1 > p_2$. Hence, we can define a "protection level," which is the capacity reserved for passengers of class 1. Of course, if the protection level is set too high, we could regret our decision as well, if we turn down too many requests from class 2, only to find later that demand D_1 is low and seats are left empty.

Rather surprisingly, the problem can be tackled much along the lines of the newsvendor problem by marginal analysis. Let us denote the reservation level by y_1, i.e., the number of seats reserved to class 1. Let us also denote by $F_1(\cdot)$ the CDF for demand from class 1. Say that a customer of class 2 requests a seat when the remaining capacity is y_1. Should we accept or reject the offer? To answer the question, we should trade off a sure revenue p_2 against an uncertain revenue p_1. This revenue is uncertain, because we will make p_1 only if demand D_1 is at least y_1. We should accept the request if

$$p_2 > p_1 \mathrm{P}(D_1 \geq y_1)$$

and we should reject it otherwise. If we assume a continuous distribution for D_1, the optimal protection level y_1^* is exactly where the two amounts above are the same:

$$p_2 = p_1 \mathrm{P}(D_1 \geq y_1) = p_1(1 - F_1(Y_1^*)) \quad \Rightarrow \quad y_1^* = F_1^{-1}\left(1 - \frac{p_2}{p_1}\right)$$

To check the solution, observe that y_1^* is large when p_2 is small with respect to p_1; in the limit, if p_2/p_1 goes to zero, the protection level is the aircraft capacity and all of the requests from class 2 will be rejected.   □

We should stress that both basic newsvendor and Littlewood's models are rather crude simplifications of reality. Nevertheless, they do provide us with essential intuition and can be extended to cope with more realistic models of actual problems.

7.5 HIGHER-ORDER MOMENTS, SKEWNESS, AND KURTOSIS

Expected value and variance do not tell us the whole story about a random variable. To begin with, they do not say anything about the possible lack of symmetry. From descriptive statistics, we know that to characterize symmetry of a distribution, or lack thereof, we need a coefficient measuring its skewness. Furthermore, we may have distributions according to which extreme events, like huge losses due to a stock market crash, are pretty rare, and distributions in which they are not that unlikely. The probability of extreme realizations of random variables depends on the probability mass associated with the tails of the distribution. Needless to say, when dealing with risk management we do need measures taking these features into account. They all rely on the definition of a more general concept, moments of a random variable.

DEFINITION 7.3 (Moments of a random variable) *The moment of order k of random variable X is defined as* $\mathrm{E}[X^k]$. *The central moment of order k is defined as* $\mathrm{E}[(X - \mu_X)^k]$.

We immediately see that expected value is just the first-order moment, whereas variance is the second-order *central* moment. To characterize deviations with respect to the expected value, we need an even-order moment, to avoid cancelation between positive and negative deviations. But in order to capture lack of symmetry, we need just that, which is captured by a central moment of odd order. Furthermore, in order to capture fat tails, we need higher-order moments than just the second one. This motivates the following definitions.

DEFINITION 7.4 (Skewness and kurtosis) *Skewness is defined as*

$$\gamma \equiv \frac{\mathrm{E}\left[(X - \mu_X)^3\right]}{\sigma^3} \tag{7.11}$$

Kurtosis is defined as

$$\kappa = \frac{\mathrm{E}\left[(X - \mu_X)^4\right]}{\sigma^4} \tag{7.12}$$

Looking at these definition, one could wonder why we should divide the moment of order $k = 3, 4$ by a corresponding power of the standard deviation σ. The point is that the above properties should not depend on change of scale or shifts in the underlying distribution: If we define a random variable $Y = \alpha + \beta X$, its skewness and kurtosis should be the same as X. It is easy to see that skewness is not changed by this linear affine transformation:

$$\mathrm{E}\left[\left(\frac{Y - \mathrm{E}[Y]}{\sqrt{\mathrm{Var}(Y)}}\right)^3\right] = \mathrm{E}\left[\left(\frac{\alpha + \beta X - \alpha - \beta\mu_X}{\sqrt{\mathrm{Var}(\alpha + \beta X)}}\right)^3\right]$$

$$= \mathrm{E}\left[\left(\frac{\beta(X - \mu_X)}{\sqrt{\beta^2 \mathrm{Var}(X)}}\right)^3\right] = \mathrm{E}\left[\frac{(X - \mu)^3}{\sigma^3}\right]$$

The same applies to kurtosis. In Fig. 7.9 we illustrate two asymmetric distributions. The PDF on the left is skewed to the right and has positive skew; in such a case, the median is smaller than the expected value. The PDF

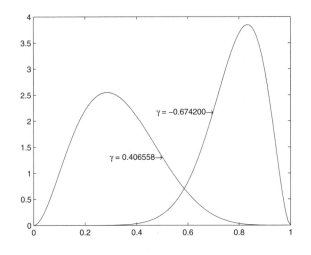

Fig. 7.9 Schematic representation of positive and negative skewness.

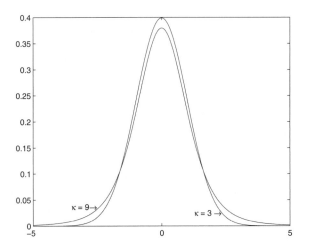

Fig. 7.10 Schematic representation of kurtosis.

on the right is skewed to the left and has negative skew. Figure 7.10 shows two distributions with different tail behavior. The distribution with kurtosis $\kappa = 9$ has fatter tails and a corresponding lower mode. This makes sense, as the overall area below any PDF must always be 1; if tails are fatter, some probability mass is removed from the central portion of the distribution.

7.6   A FEW USEFUL CONTINUOUS PROBABILITY DISTRIBUTIONS

In the following sections we describe some continuous probability distributions. The main criterion of classification is theoretical vs. empirical distributions. The former class consists of distributions that are characterized by a very few parameters; indeed, they can also be labeled as parametric distributions. Theoretical distributions will never fit empirical data exactly, but they provide us with very useful tools, as they can be justified by some assumption about the underlying randomness. Furthermore, they have PDFs in analytical form, which may help us in finding analytical solutions to a wide set of problems. On the contrary an empirical distribution will, of course, fit observed data very well, but there is a hidden danger in doing so: We might overfit the distribution, obtaining a PDF or a CDF that does fit the peculiarities of the observed sample, but does not describe the properties of the population very well.

Empirical distribution will be the last example we cover here. First we consider the few main theoretical distributions, to provide the reader with the essential feeling for them. We start from the simplest case, the uniform distribution; then we consider the triangular and the beta distributions, which may be used as rough-cut models when little information is available on the underlying uncertainty. Then we describe the exponential and the normal distributions. They play a central role in probability theory because of their properties and because they can be used as building blocks to obtain other distributions. We defer the treatment of a few distributions obtained from the normal to Section 7.7.2, as they require some background on sums of random variables.

7.6.1   Uniform distribution

We have already met the uniform distribution in Section 7.1, where we specified its PDF and CDF. To say that a random variable X is uniformly distributed on the interval $[a, b]$, the notation $X \sim U(a, b)$ is used. We have already shown that the expected value is the midpoint on the support:

$$E[X] = \frac{a + b}{2}$$

Since the uniform distribution is symmetric, the median and the expected value are the same, and skewness is zero. It can be shown that variance is

$$\mathrm{Var}(X) = \frac{(b - a)^2}{12}$$

A peculiarity of the uniform distribution is that it has no well defined mode, since the PDF is constant. All of the remaining theoretical distributions have a single mode.

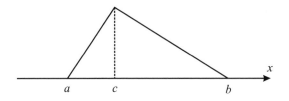

Fig. 7.11 PDF of a triangular distribution.

It is reasonable to say that the uniform distribution is a very dry model of uncertainty, as it just provides us with bounds on the possible realizations of X. It is often stated that the uniform distribution should be used whenever we have no idea about the underlying uncertainty. Actually this is a bit debatable, and the following argument has been proposed to counter this view. Suppose that the only thing we know about variable X is that it can take values between 0 and 1. Apparently, a uniform distribution $U(0, 1)$ is an obvious choice. But now consider the variable $Y = X^\alpha$, for some value $\alpha > 0$. We cannot say anything about Y, either, and the variable is bounded between 0 and 1. However, we cannot say that both X and Y are uniformly distributed. Indeed, representing almost complete ignorance is not as easy as it may seem. Nevertheless, a uniform distribution is often used in Bayesian statistics as a noninformative prior.[10] Another quite relevant application of a $U(0, 1)$ distribution is random-number generation for Monte Carlo simulation.[11] When we have to simulate randomness by a computer program, we first generate a $U(0, 1)$ variable, which is then transformed to whatever we need to model uncertainty.

7.6.2 Triangular and beta distributions

Triangular distribution is a possible model of uncertainty when limited knowledge is available. Three parameters characterize it: the extreme points of the support $[a, b]$ and that the mode c, where $a \leq c \leq b$. The PDF for a triangular random variable is depicted in Fig. 7.11. The expected value and variance for a triangular distribution are

$$E[X] = \frac{a + b + c}{3}, \qquad \text{Var}(X) = \frac{a^2 + b^2 + c^2 - ab - ac - bc}{18}$$

respectively.

Imagine a project planning problem, which involves tasks of quite uncertain duration. If we are able to assign the support, i.e., lower and upper bound on the time to complete a task, and a mode, we might consider using a triangular

---

[10]See Section 14.7.
[11]See Section 9.7.

distribution. A distribution that is widely used in such applications, but featuring a better academic pedigree, is the beta distribution. This distribution has support on the interval $[0, 1]$ and depends on two parameters, α_1 and α_2. Its PDF is

$$f_X(x) = \frac{x^{\alpha_1 - 1}(1 - x)^{\alpha_2 - 1}}{\mathrm{B}(\alpha_1, \alpha_2)}, \qquad x \in [0, 1]$$

To be precise, when $\alpha_1, \alpha_2 < 1$, the support is the open interval $(0, 1)$, as PDF goes to infinity at its extreme points. In the following, we will just consider the case $\alpha_1, \alpha_2 > 1$. The definition of the PDF involves a normalization factor $\mathrm{B}(\alpha_1, \alpha_2)$, the beta function, defined as

$$\mathrm{B}(\alpha_1, \alpha_2) = \int_0^1 x^{\alpha_1 - 1}(1 - x)^{\alpha_2 - 1} \, dx$$

The beta distribution can be adapted to many practical cases by shifting and scaling, which results in an arbitrary support $[a, b]$. The expected value and variance are

$$\mathrm{E}[X] = \frac{\alpha_1}{\alpha_1 + \alpha_2}, \qquad \mathrm{Var}(X) = \frac{\alpha_1 \alpha_2}{(\alpha_1 + \alpha_2)^2 (\alpha_1 + \alpha_2 + 1)}$$

respectively. The mode, for $\alpha_1, \alpha_2 > 1$, is

$$m = \frac{\alpha_1 - 1}{\alpha_1 + \alpha_2 - 2}$$

Figure 7.12 shows three examples of beta distributions, for different settings of its parameters. Looking at the PDF, we see that the distribution is symmetric when $\alpha_1 = \alpha_2$. Indeed, skewness is expressed as follows:

$$\gamma = \frac{2(\alpha_2 - \alpha_1)}{\alpha_1 + \alpha_2 + 2} \sqrt{\frac{\alpha_1 + \alpha_2 + 1}{\alpha_1 \alpha_2}}$$

7.6.3 Exponential distribution

The exponential distribution is one the main tools used to model uncertainty, and it is related to other distributions, as well as to an important family of stochastic processes that we will investigate later. An exponential random variable can only take nonnegative values, i.e., its support is $[0, +\infty)$, and it owes its name to the functional form of its density:

$$f_X(x) = \begin{cases} \lambda e^{-\lambda x} & \text{if } x \geq 0 \\ 0 & \text{if } x < 0 \end{cases}$$

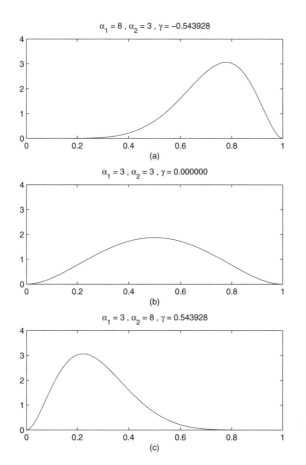

Fig. 7.12   PDF of symmetric and skewed beta distributions.

Here $\lambda > 0$ is a given parameter, and the notation $X \sim \exp(\lambda)$ is often used.[12] Straightforward integration[13] yields the CDF

$$F_X(x) = \int_0^x \lambda e^{-\lambda t}\, dt = 1 - e^{-\lambda x} \tag{7.13}$$

and the expected value is

$$E[X] = \int_0^\infty x \lambda e^{-\lambda x}\, dx = \frac{1}{\lambda}$$

---

[12]Note that this notation refers to the parameter characterizing the distribution, rather than to the expected value.

[13]See Example 2.43; the example also shows that the area below the PDF of the exponential distribution is actually 1.

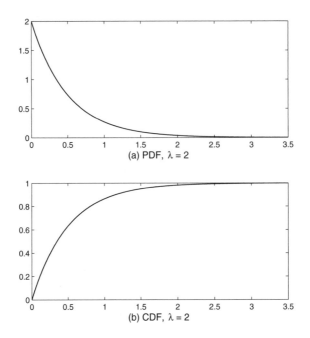

Fig. 7.13 PDF and CDF of an exponential distribution with $\lambda = 2$.

It is worth noting that the expected value is quite different from the mode, which is zero. It can be shown that variance for the exponential distribution is $1/\lambda^2$, implying that the coefficient of variation is $c_X = 1$. Figure 7.13 shows the PDF and the CDF for an exponential distribution with parameter $\lambda = 2$.

Unlike the uniform distribution, there are typically good physical reasons for adopting this distribution to model a random quantity. A common use of exponential distribution is to model time elapsing between two random events, e.g., the interarrival time between two consecutive service requests. Note that λ is, within this interpretation, a rate at which events occur, e.g., average number of service requests per unit time; the mean interarrival time is $1/\lambda$. In fact, we often speak of exponential random variables with *rate* λ. There are a few important points worth mentioning:

- The exponential distribution is linked to the Poisson distribution, which we covered in Section 6.5.6. Imagine that the successive interarrival times of service requests are independent[14] and exponentially distributed

---

[14]We did not cover independence between random variables yet; however, recalling the concept of independent events, the concept should be rather clear.

with rate λ, and count the number of such requests arriving during a time interval of length t. Then, the number of requests we count is a discrete random variable following a Poisson distribution with parameter λt. In Section 7.9 we will see that this phenomenon corresponds to a common stochastic process, which is unsurprisingly known as the *Poisson process*.

- If we sum n independent exponential variables with rate λ, we obtain a new probability distribution that is called *Erlang*. This distribution is also widely used in applications to model time between events.

- Probably the most important feature of an exponential random variable is its "lack of memory." We will consider this property in more detail in Section 8.5.2, but we can realize its intuitive meaning and its practical relevance by considering the waiting time for the arrival of a bus at a bus stop. If we know that the time between two consecutive arrivals is uniformly distributed between, say, 2 and 10 minutes, and we have been waiting for 9 minutes, we may have a pretty clear idea about the time we still have to wait. The more we have waited in so far, the less we are supposed to wait in the future. On the contrary, if this time is exponentially distributed, the fact that we waited for a long time does not change the distribution; the distribution when we get to the bus stop and the distribution after waiting 20 minutes are the same. A full understanding of this requires concepts about independence and conditional distributions, which are provided in Chapter 8, but it is important to see the practical implication of this property. Imagine that we use the exponential distribution to model time between failures of an equipment. Lack of memory implies that even if the machine has been in use for a long time, this does not mean that it is more likely to have a failure in the near future. Note again the big difference with a uniform distribution. If we know that time between failures is uniformly distributed between, say, 50 and 70 hours, and we also know that 69 hours have elapsed since the last failure, we must expect the next failure within one hour. If the time between failures is exponentially distributed and 69 hours have elapsed, we cannot conclude anything, since from a probabilistic point of view the machine is brand new. If we think of purely random failures, due to bad luck, the exponential distribution may be a plausible model, but definitely it is not if wear is the main driving factor of failures.

7.6.4 Normal distribution and its quantiles

The normal distribution is by far the most common, and misused, distribution in the theory of probability. It is also known as Gaussian distribution, but the term "normal" illustrates its central role quite aptly. Its PDF has a seemingly

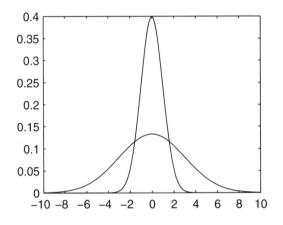

Fig. 7.14   PDF of two normal distributions.

awkward form

$$f_X(x) = \frac{1}{\sqrt{2\pi}\,\sigma} \exp\left\{ -\frac{1}{2}\left(\frac{x-\mu}{\sigma}\right)^2 \right\}, \qquad -\infty < x < +\infty \qquad (7.14)$$

depending on two parameters, μ and σ^2. Actually, we met such a function a while ago,[15] and we noted its peculiar bell shape. Figure 7.14 shows two PDFs for $\mu = 0$, and $\sigma = 1$, $\sigma = 3$. Actually, it is quite easy to interpret the PDF (7.14):

- The initial factor $1/\sqrt{2\pi}\sigma$ is just a normalization factor, and its role is only to ensure that the area below the PDF is 1.

- The expected value is just the parameter μ; indeed, this parameter has the effect of shifting the PDF left and right.

- The variance is just the parameter σ^2; indeed, this parameter has the effect of changing the scale, i.e., spreading or concentrating the bell, as we can see in Fig. 7.14.

We often use the notation $X \sim \mathcal{N}(\mu, \sigma^2)$ to indicate that X has normal distribution; note that the second parameter corresponds to variance, rather than standard deviation. It is very easy to see that for the normal distribution expected value (mean), mode, and median are just the same. The PDF is clearly symmetric with respect to the expected value, so skewness is zero. On the contrary, a somewhat surprising fact is that kurtosis for a normal variable is $\kappa = 3$, and it does not depend on the specific value of the parameters.

---

[15]See Example 2.10.

Indeed, in some books the definition of kurtosis, which we gave in Definition 7.4, is replaced by

$$\kappa_e = E\left[\frac{(X - \mu_X)^4}{\sigma^4}\right] - 3$$

This is a surprising definition for the uninitiated, and we prefer the alternative one. The point is that the tail behavior of the normal distribution is a sort of benchmark, and it may be useful to express kurtosis of other distribution with reference to this base case. The appropriate name for κ_e is *excess kurtosis*.

The last point shows that all of the possible normal distributions are essentially the same in terms of tail behavior. In fact, there is something more to notice. We can transform any normal random variable into any other normal variable, with different parameters, just by a linear affine transformation. Consider a generic normal $X \sim \mathcal{N}(\mu, \sigma^2)$, and consider the variable

$$Z = \frac{X - \mu}{\sigma} \qquad (7.15)$$

In terms of PDF, we are just shifting the graph and changing its scale, without changing its basic form. Using the familiar rules concerning expected values and variance, we observe the following:

$$\begin{aligned} E[Z] &= E\left[\frac{X - \mu}{\sigma}\right] = \frac{E[X] - \mu}{\sigma} = 0 \\ \text{Var}(Z) &= \text{Var}\left(\frac{X - \mu}{\sigma}\right) = \frac{\text{Var}(X)}{\sigma^2} = 1 \end{aligned}$$

A normal variable $Z \sim \mathcal{N}(0, 1)$, with zero expected value and unit variance is called *standard normal*. The transformation (7.15) is called *standardization*. Actually, it applies to any distribution, as it yields a variable with zero expected value and unit variance, but it plays an important role for the normal distribution. We may also go the other way around: Given a standard normal Z, we may invert (7.15) to get an arbitrary normal by *destandardization*:

$$X = \mu + \sigma Z \qquad (7.16)$$

The normal distribution has many nice properties, which we will discover in the following text and justify its popularity. One unpleasing feature, though, is that its CDF cannot be calculated analytically. As we know from chapter 2, integrating the density (7.14) requires finding its antiderivative. As it turns out, this is impossible and we must resort to numerical methods to evaluate the integral and, therefore, the CDF. This poses no practical difficulty as plenty of software is available to carry out this task efficiently and with more than adequate precision. We should mention that, traditionally, any text involving probability and statistics provides the reader with tables to carry out calculations by hand.[16]   The trouble is that we cannot have a set of tables for any possible normal distribution. However, we can easily carry out the job once, for the standard normal, and then apply standardization and destandardization to work with an arbitrary normal.[17]   Tables for the standard normal provide us with values of the following CDF:

$$\Phi(z) = \mathrm{P}(Z \leq z) = \frac{1}{\sqrt{2\pi}} \int_{-\infty}^{z} e^{-x^2/2} \, dx$$

Sometimes, only the right area is tabulated:

$$\mathrm{P}(0 \leq Z \leq z) = \frac{1}{\sqrt{2\pi}} \int_{0}^{z} e^{-x^2/2} \, dx$$

Of course, this does not change anything because of the symmetry of the PDF. Given a way to compute $\Phi(z)$, we can deal with probabilities for an arbitrary normal variable $X \sim \mathcal{N}(\mu, \sigma^2)$. To find the probability $\mathrm{P}(X \leq \beta)$, we should just apply standardization:

$$\mathrm{P}(X \leq \beta) = \mathrm{P}\left(\frac{X - \mu}{\sigma} \leq \frac{\beta - \mu}{\sigma}\right) = \mathrm{P}\left(Z \leq \frac{\beta - \mu}{\sigma}\right) = \Phi\left(\frac{\beta - \mu}{\sigma}\right)$$

Example 7.7 Consider $X \sim \mathcal{N}(3, 16)$, i.e., a normal variable with expected value 3 and standard deviation 3. Let us compute $\mathrm{P}(2 < X < 7)$:

$$
\begin{aligned}
\mathrm{P}(2 < X < 7) &= \mathrm{P}\left(\frac{2-3}{4} \leq \frac{X-3}{4} \leq \frac{7-3}{4}\right) \\
&= \mathrm{P}\left(Z \leq \frac{7-3}{4}\right) - \mathrm{P}\left(Z \leq \frac{2-3}{4}\right) \\
&= \Phi(1) - \Phi\left(-\frac{1}{4}\right) \\
&= 0.8413 - 0.4013 \\
&= 0.4400
\end{aligned}
$$

When using statistical tables, we cannot carry out the above calculation directly, as typically we are provided with values $\Phi(z)$ only for $z \geq 0$. However, we may easily to take advantage of symmetry to compute $\Phi(-\frac{1}{4})$:

1. We need the area of the PDF to the left of $z = -\frac{1}{4}$.

---

[16]This book is an exception, since tables are a heritage of the past, and software tools are easier to use in practice, not to mention much more precise. Some tools, like R, are freely available on the Web (see http://www.r-project.org). Nevertheless, statistical tables are offered on the book Webpage for the readers' convenience.

[17]The idea is also helpful when designing numerical algorithms to compute or invert the CDF of the normal; we just need one procedure, for the standard normal, and a simple transformation of its output provides us with what we need.

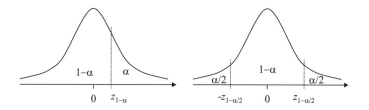

Fig. 7.15 Using quantiles of the standard normal distribution.

2. Because of symmetry with respect to the expected value $E[Z] = 0$, this is just the area to the right of $z = \frac{1}{4}$.

3. But this is just the probability:

$$P\left(Z \geq \frac{1}{4}\right) = 1 - P\left(Z \leq \frac{1}{4}\right) = 1 - \Phi\left(\frac{1}{4}\right)$$

4. Hence:

$$\phi\left(-\frac{1}{4}\right) = 1 - \Phi\left(\frac{1}{4}\right) = 1 - 0.5987 = 0.4013$$

$\Box$

The kind of gimmicks of the example above are not required anymore, if you have a decent piece of software, but they are still worth learning to really know the ropes of working with normal variables. This is also important because one of the most common tasks in statistics is the use of quantiles of normal distributions. Numerical inversion of the CDF for the standard normal, or reading statistical tables the other way around, yields the quantiles:

$$P(Z \leq z_q) = q$$

for a probability level $q \in (0,1)$. Actually, the usual notation in statistical applications is $z_{1-\alpha}$, where α is a rather small number, like 0.1 or 0.05; geometrically, the quantile $z_{1-\alpha}$ leaves an area $1 - \alpha$ of PDF to its left, and α is the area of the right tail. This is illustrated in Fig. 7.15. From the figure, we also see that if we want to leave two symmetric tails on the left and on the right, such that their total area is α, we should consider quantile $z_{1-\alpha/2}$ and observe that

$$P(-z_{1-\alpha/2} \leq Z \leq z_{1-\alpha/2}) = 1 - \alpha$$

Now, we know that there is a way to find quantiles z_q for the standard normal, but how can we find a quantile x_q for a generic normal variable? The quick-and-dirty recipe mirrors destandardization:

$$x_q = \mu + \sigma z_q$$

To see why this works, observe the following:

$$P\left(Z \leq z_q\right) = q \iff P\left(\mu + \sigma Z \leq \mu + \sigma z_q\right) = q \iff P\left(X \leq \mu + \sigma z_q\right) = q$$

Example 7.8 Consider a normal variable X with expected value $\mu = 100$ and standard deviation $\sigma = 20$. What is its 95% quantile? We are looking for a number $x_{0.95}$ such that

$$P(X \leq x_{0.95}) = 0.95$$

Statistical software provides us with the corresponding quantile for the standard normal distribution: $z_{0.95} = 1.6449$. Hence

$$x_{0.95} = \mu + \sigma z_{0.95} = 100 + 20 \times 1.6449 = 132.8971$$

▯

Example 7.9 (A well-known rule for the normal distribution) Given a normal variable $X \sim \mathcal{N}(\mu, \sigma^2)$, we might wonder how many realizations are expected to fall in an interval of the form $\mu \pm k\sigma$. We find

1. $P(\mu - \sigma \leq X \leq \mu + \sigma) \approx 68.26\%$

2. $P(\mu - 2\sigma \leq X \leq \mu + 2\sigma) \approx 95.44\%$

3. $P(\mu - 3\sigma \leq X \leq \mu + 3\sigma) \approx 99.74\%$

We see that almost all of the realizations are expected to fall "within three standard deviations of the mean." In other words, the width of the interval including almost all of them is six standard deviations; indeed, a managerial philosophy has been called *six sigma* because of this.

This also shows that the normal distribution has rather thin tails, and this is why it serves as a benchmark in terms of kurtosis. If we observe events that go much beyond the three-sigma wall we should question the applicability of a model based on the normal distribution. A well-known example is the stock market crash of October 19, 1987. This date did deserve the name of "Black Monday," as the Dow Jones Industrial Average index dropped from 2246 to 1738, a decline of almost 25% in one day. Fitting a normal distribution against index returns shows that this event was about 20 standard deviations below average. In fact, it is rather common to observe such extreme events on financial markets. On the one hand, alternative distributions have been proposed, with fatter tails, to better account for such phenomena. On the other hand, more radical approaches have been proposed, modeling the dynamic behavior of stock market participants, which are not completely rational decisionmakers. The very applicability of probability modeling to this kind of system have been questioned.[18]   ▯

---

[18]We will outline a few related issues in Chapter 14. There are many interesting accounts of what happened on the Black Monday and the hard lessons that were learned on that infamous day; see, among others, Refs. [2] and [3].

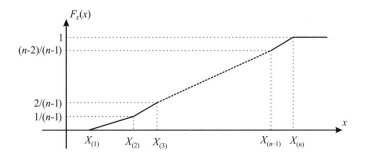

Fig. 7.16 The CDF for an empirical distribution.

7.6.5 Empirical distributions

Sometimes, no theoretical distribution seems to fit available data, and we resort to an empirical distribution. A standard way to build an empirical distribution is based on order statistics, i.e., sorted values from a sample. Assume that we have a sample of n values and order statistics $X_{(i)}$, $i = 1, \ldots, n$, where $X_{(i)} \leq X_{(i+1)}$. The value $X_{(1)}$ is the smallest observation and $X_{(n)}$ is the largest one.

Assume that we want to rule out values above and below the observed range. Then, we can write

$$\mathrm{P}(X < X_{(1)}) = F_X(X_{(1)}) = 0, \qquad \mathrm{P}(X > X_{(n)}) = 1 - F_X(X_{(n)}) = 0$$

The last condition implies $F_X(X_{(n)}) = 1$. Note that this is rather arbitrary; sometimes, small tails are appended to the extreme points of the observed range in order to avoid overfitting. This is, however, an arbitrary and ad hoc procedure. To assign values of the CDF for intermediate order statistics, we may simply divide the range from 0 to 1 in equal intervals, which results in the following rule:

$$F_X(X_{(k)}) = \frac{k-1}{n-1}$$

For values falling between order statistics, linear interpolation is the simplest choice and results in a CDF like the one illustrated in Fig. 7.16. Given the CDF, the PDF is obtained. Clearly, choosing linear interpolation results in a kinky CDF with nondifferentiable points. In order to get a smoother curve, we could also interpolate with higher-order polynomials.[19]

Once again, we should remark that fitting an empirical distribution has some hidden traps. It is easy to trust available data too much and to obtain a distribution that reflects peculiarities in the sample, which do not necessarily

---

[19] A common choice is based on *cubic splines*, which are third-order polynomials associated with each interval, selected in such a way that the function is continuous, as well as its first- and second-order derivative.

carry over to the whole population. Furthermore, it is sometimes possible to come up with mixtures of theoretical distributions that do fit the data and eliminate the need for arbitrary choices, e.g., as far as the support and the tail behavior are concerned.

Example 7.10 One standard reason for rejecting simple theoretical distributions is that empirical frequencies may display multiple modes. Then, it is tempting to fit whatever we have observed, but this may result in a poor understanding of the underlying phenomena. Consider, for instance, the time needed to complete surgical operations. If we take statistics about these times, quite likely we will observe multiple modes. But this might be linked to the different kinds of operations being executed, which may range from quite simple to very complex ones. It is much better to fit discrete probabilities against the different classes of operations, and then to model the variability of the time within each class around the expected value for each one. ▯

7.7   SUMS OF INDEPENDENT RANDOM VARIABLES

A recurring task in applications is summing random variables. If we have n random variables X_i, $i = 1, \dots, n$, we may build another random variable

$$Y = \sum_{i=1}^{n} X_i$$

What can we say about the distribution of Y? The answer depends on two important features of the terms in the sum:

- Is the distribution of all of the X_i the same?

- Are the involved variable independent?

We will clarify what we mean by "independent random variables" formally in Chapter 8, but what we know about independent events and conditional probabilities is enough to get the overall idea: Two variables are independent if knowing the realization of one of them does not help us in predicting the realization of the other one.

DEFINITION 7.5 (i.i.d. variables) *We say that the variables X_i, $i = 1, \dots, n$, are **i.i.d.** if they are independent and identically distributed.*

Arguably, the case of i.i.d. variables is the easiest we may think of. Unfortunately, even in this case, characterizing the distribution of the sum on random variables is no trivial task. It might be tempting to think that the distribution of Y should, at least qualitatively, similar to the distribution of the X_i, but a simple counterexample shows that this is not the case.

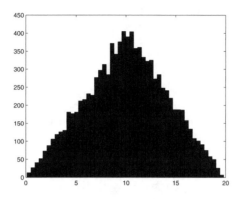

Fig. 7.17   Sampling the sum of two i.i.d. uniform variables.

Example 7.11 Consider two independent random variables, uniformly distributed between 0 and 10: $U_1, U_2 \sim U(0, 10)$. The support of their sum, $Y = U_1 + U_2$, is clearly the interval $[0, 20]$, but what about the distribution? The analytical answer would require a particular form of integral, but we may guess the answer by sampling this distribution with the help of statistical software. Figure 7.17 shows the histogram obtained by sampling 10,000 observations of the sum. A look at the plot suggests a triangular distribution. In fact, it can be shown that the distribution of Y is triangular, with support on interval $[0, 20]$ and mode $m = 10$.

By the same token, if we sum two i.i.d. exponential variables, we do not get an exponential. There is an important case in which distribution is preserved by summing.

PROPERTY 7.6 *The sum of jointly normal random variables is a normal random variable.*

It is important to note that this property does not assume independence: It applies to normal variables that are not independent and have different parameters. The term "jointly" may be puzzling, however. The point is that characterizing the joint distribution of random variables is not as simple as it may seem. It is not enough to specify the distribution of each single variable, as this provides us with no clue about their joint behavior. The term above essentially says that we are dealing with a multivariate normal distribution, which we define later, in Section 8.4.

So, the results concerning the general distribution of the sum of random variables are somewhat discouraging, but we recall that something more can be said if we settle for the basic features of a random variable, i.e., expected value and variance. We stated a couple of properties when dealing with discrete random variables, that carry over to the continuous case.

PROPERTY 7.7 (Expected value of a sum of random variables) *The expected value of the sum of random variables is the sum of their expected values, assuming that they exist:*

$$\mathrm{E}\left[\sum_{i=1}^{n} X_i\right] = \sum_{i=1}^{n} \mathrm{E}[X_i]$$

PROPERTY 7.8 (Variance of a sum of independent random variables) *The variance of the sum of independent random variables is the sum of their variances, assuming that they exist:*

$$\mathrm{Var}\left(\sum_{i=1}^{n} X_i\right) = \sum_{i=1}^{n} \mathrm{Var}\left(X_i\right)$$

It is important to notice that the two properties above do not require variables to be identically distributed. The property about variance does require independence, however.

Example 7.12 Consider two independent normal variables, $X_1 \sim \mathcal{N}(10, 25)$ and $X_2 \sim \mathcal{N}(-8, 16)$. Then, the sum $Y = X_1 + X_2$ is a normal random variable, with expected value

$$\mathrm{E}[Y] = \mathrm{E}[X_1] + \mathrm{E}[X_2] = 10 - 8 = 2$$

and standard deviation

$$\sigma_Y = \sqrt{\mathrm{Var}(X_1) + \mathrm{Var}(X_2)} = \sqrt{25 + 16} = 6.4031$$

Note that we *cannot* add standard deviations; doing so would lead to a wrong result $(5 + 4 = 9)$. ⬚

The example illustrates the fact that, when variables are independent, we may sum variances, but *not* standard deviations:

$$\sigma_{X_1 + X_2} = \sqrt{\sigma_{X_1}^2 + \sigma_{X_2}^2} \neq \sigma_{X_1} + \sigma_{X_2}$$

Another important remark concerns the sum and the difference between independent random variables. The property implies $\mathrm{Var}\left(X + Y\right) = \mathrm{Var}\left(X\right) + \mathrm{Var}\left(Y\right)$, but what about their difference? We must apply the property carefully:

$$\begin{aligned}\mathrm{Var}\left(X - Y\right) &= \mathrm{Var}\left(X + (-Y)\right) = \mathrm{Var}\left(X\right) + \mathrm{Var}\left(-Y\right)\\ &= \mathrm{Var}\left(X\right) + (-1)^2 \mathrm{Var}\left(Y\right) = \mathrm{Var}\left(X\right) + \mathrm{Var}\left(Y\right)\end{aligned}$$

We see that the variance of a difference is *not* the difference of the variance; it is also worth noting that this would easily lead to nonsense, as by taking differences of variances we could find a negative variance.

7.7.1 The square-root rule

Consider a sequence of i.i.d. random variables observed over time, X_t, $t = 1, \ldots, T$. Let μ and σ be the expected value and standard deviation of each X_t, respectively. Then, if we consider the sum over the T periods, $Y = \sum_{t=1}^{T} X_t$, we have

$$\mathrm{E}[Y] \;=\; \sum_{t=1}^{T} \mathrm{E}[X_t] \;=\; \mu T \tag{7.17}$$

$$\sigma_Y \;=\; \sqrt{\sum_{t=1}^{T} \mathrm{Var}[X_t]} \;=\; \sigma\sqrt{T} \tag{7.18}$$

We see that the expected value scales linearly with time, whereas the standard deviation scales with the *square root* of time. Sometimes students and practitioners are confused by the result concerning standard deviations. It is important to draw the line between the sum of T random variables and the product of T and one random variable.

Example 7.13 Consider demand for an item, over a time interval consisting of T time buckets, say, weeks. The time interval could be delivery lead time, i.e., the time elapsing between the instant at which we issue a replenishment order and the time instant at which we receive the corresponding shipment from the supplier. In practice, demand during lead time is a relevant variable for inventory management decisions. Say that μ and σ are expected value and standard deviation of weekly demand, respectively, and assume that demands in different weeks are independent.

Then, the expected value of demand during lead time is $T\mu$, but its standard deviation is $\sqrt{T}\sigma$ and *not* $T\sigma$. The typical way to get the wrong result is by considering demand during lead time as a random variable

$$Y = TX$$

where X is a random variable corresponding to demand during one week. It is true that $\mathrm{Var}(Y) = T^2\mathrm{Var}(X)$, but this is the wrong reasoning; by doing so, we assume that demand during a week is realized, and then it is replicated for T weeks. But this does not correspond to the real phenomenon. ⬛

The square-root rule shows that, if T is very small, then the volatility term $\sqrt{T}\sigma$ dominates the expected value term, as the square root of T goes to zero more slowly than T itself does, when the latter goes to zero. This has some implications for measuring financial risk, as we shall see later, but there is another hidden trap here. It is tempting to apply the rule by considering fractional values of T, but this may lead to nonsense. An example will illustrate the point.

Example 7.14 Let us assume that the yearly demand for an item is normally distributed with expected value 1000 and standard deviation 250. If the lead time is 2 months, what is the distribution of lead time demand D_{LT}? If we assume that the year consists of 12 identical months of 30 days, and we assume that demands in different months are independent, we could consider the application of the above rules with $T = \frac{2}{12} = \frac{1}{6}$. In terms of expected value and standard deviation, this would imply

$$\text{E}\left[D_{\text{LT}}\right] = \frac{1000}{6} \approx 166.67; \qquad \sigma_{D_{\text{LT}}} = \frac{250}{\sqrt{6}} \approx 102.06$$

This might make some sense, but can we say that lead time distribution is normally distributed? If we recall the three-sigma rule, we note that

$$\text{E}\left[D_{\text{LT}}\right] - 3\sigma_{D_{\text{LT}}} = 166.67 - 3 \times 102.06 = -139.51 < 0$$

In fact, if we assume normality, the probability of negative demand is far from negligible. This example shows that if we assume normality of monthly demand, we may deduce normality of demand during a year, but we cannot go the other way around. □

7.7.2   Distributions obtained from the normal

As we pointed out, if we sum i.i.d. random variables, we may end up with a completely different distributions, with the normal as a notable exception. However, there are ways to combine independent normal random variables that lead to new distributions that have remarkable applications, among other things, in inferential statistics. In fact, statistical tables are available for the random variables we describe below, providing us with quantiles we need to carry out statistical tests, as we will see in later chapters.

The chi-square distribution   Consider a set of independent standard normal variables Z_i, $i = 1, \ldots, n$. Consider random variable X defined as

$$X = Z_1^2 + Z_2^2 + \cdots + Z_n^2$$

Obviously, X cannot have normal distribution, as it cannot take negative values. This distribution is called *chi-square*, with n degrees of freedom. This is often denoted as $X \sim \chi_n^2$. The following results can be proved:

$$\text{E}[X] = n, \qquad \text{Var}(X) = 2n$$

Figure 7.18 shows the PDF for chi-quare variables with 4 and 8 degrees of freedom. The second one corresponds to the PDF with the lower mode, and the higher expected value and variance.

Student's t distribution   Consider a standard normal variable Z and a chi-square variable χ_n^2 with n degrees of freedom. Also assume that they are

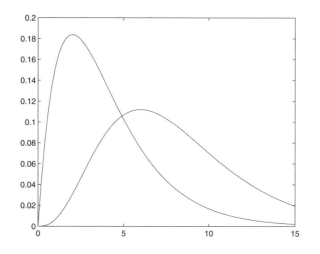

Fig. 7.18   PDF of two chi-square variables, χ_4^2 and χ_8^2.

independent. Then, the random variable

$$T_n = \frac{Z}{\sqrt{\chi_n^2/n}}$$

has Student's t distribution with n degrees of freedom.[20] We can show that

$$\mathrm{E}[T_n] = 0, \qquad \mathrm{Var}[T_n] = \frac{n}{n-2}$$

incidentally, we see that variance need not be always defined, as it may go to infinity. Figure 7.19 shows the PDFs of T_1 and T_5 variables, along with the PDF of standard normals. The PDF with the highest mode, drawn with a continuous line, corresponds to the standard normal; T_1, represented with a dash–dotted line, features the lowest mode and the fattest tails. Indeed, the t distribution does look much like a standard normal, but it has fatter tails. When the number of degrees of freedom increases, the t distribution gets closer and closer to the standard normal. To see this quantitatively, we observe that kurtosis for the t distribution is

$$\kappa = \frac{3n - 6}{n - 4}$$

---

[20] Just in case you are wondering about a professor called Student, this is actually a pseudonym used in 1908 by William Sealy Gosset to publish his results about this distribution. Gosset's employer, the Guinness brewery, forbade members of its staff to publish scientific papers, for fear of trade secret leaks.

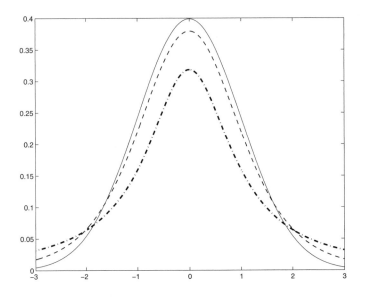

Fig. 7.19 Comparing the PDFs of two t distributions, T_1 and T_5 against the standard normal.

when n goes to infinity, kurtosis tends to 3, which is the kurtosis for a normal variable. Indeed, traditional statistical tables display quantiles for t variables up to $n = 30$, suggesting the use of quantiles for the standard normal for larger values of n. This approximation is not needed anymore, but it is useful to keep in mind that t distribution does tend to the standard normal for large values of n.

*The **F** distribution* Consider two independent random variables with chi-square distributions $\chi^2_{n_1}$ and $\chi^2_{n_2}$, respectively. The random variable

$$Y = \frac{\chi^2_{n_1}/n_1}{\chi^2_{n_2}/n_2}$$

is said to have F distribution with n_1 and n_2 degrees of freedom, which is denoted by $Y \sim F(n_1, n_2)$. Note that the degrees of freedom cannot be interchanged, as the former refers to the denominator of the ratio, the latter to its denominator.

There is a relationship between F and t distributions, which can be grasped when considering a $F(1, n)$ variable. This involves a χ^2_1 variable, with 1 degree of freedom, which is just a standard normal squared. Hence, what we have is

$$F_{1,n} = \frac{Z^2}{\chi^2_n} = (t_n)^2$$

i.e., the square of a t variable with n degrees of freedom. Furthermore, when n_2 is large enough, we get a $F_{n,\infty}$ variable. By the law of large numbers,

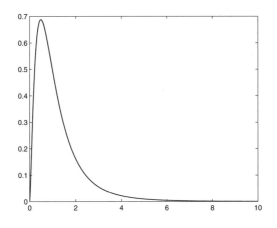

Fig. 7.20 PDF of a $F(5, 10)$ random variable.

that we will state precisely later, when n_2 goes to infinity, the ratio $\chi^2_{n_2}/n_2$ variable converges to the numerical value 1. Hence a $F_{n,\infty}$ variable is just a χ^2_n variable divided by n:

$$F_{n,\infty} = \frac{\chi^2_n}{n}$$

Figure 7.20 shows the PDF of a F random variable with 5 and 10 degrees of freedom. After this list of weird distributions obtained from the standard normal, the reader might well wonder why one should bother. The answer will be given when dealing with inferential statistics and linear regression, but we can offer at least some intuition:

- We know from descriptive statistics that the sample variance involves squaring observations; when the population is normally distributed, this entails essentially squaring normal random variables, and the distribution of sample variance is linked to chi-square variables.

- Furthermore, when we standardize a normal random variable, we take a normal variable (minus the expected value), and we divide it by its standard deviation; this leads to the t distribution.

- Finally, a common task in inferential statistics is comparison of two variances. A typical way to do this is to take their ratio and check whether it is small or large. When variances come from sampling normal distributions, we are led to consider the ratio of two chi-square variables.

Comparing the PDFs of the three distributions, we see that the t distribution is symmetric, whereas the other two have nonnegative support. This should be kept in mind when working with quantiles from these distributions.

The lognormal distribution Unlike the previous distributions, the lognormal does not stem from statistical needs, but it is worth mentioning anyway because of its role in financial applications, among others. A random variable Y is lognormally distributed if $\log Y$ is normally distributed; put another way, if Y is normal, then e^Y is lognormal. Since the exponential is a nonnegative function, a lognormal random variable cannot take negative values. In fact, it has often been used (and misused) as a model of random stock prices since, unlike the normal, it cannot yield negative prices.[21] Another noteworthy feature of lognormal random variables is that a product of lognormals is a lognormal variable; this is a consequence of the similar property of sums of normal variables and the properties of logarithms.

The following formulas illustrate the relationships between the parameters of a normal and a lognormal distribution. If $X \sim \mathcal{N}(\mu, \sigma^2)$ and $Y = e^X$, then

$$E[Y] = e^{\mu + \sigma^2/2}$$
$$\text{Var}(Y) = e^{2\mu + \sigma}(e^{\sigma^2} - 1)$$

In particular, we see that

$$E\left[e^X\right] = e^{\mu + \sigma^2/2} \geq e^\mu = e^{E[X]}$$

Since the exponential is a convex function, this is a consequence of Jensen's inequality.[22] Figure 7.21 shows the PDF of a lognormal variable with parameters $\mu = 0$ and $\sigma = 1$.

7.7.3   Central limit theorem

As we noted, it is difficult to tell which distribution we obtain when summing a few i.i.d. variables. Surprisingly, we can tell something pretty general when we sum a *large* number of such variables. We can get a clue by looking at Fig. 7.22. We see the histogram obtained by sampling the sum of independent exponential random variables with rate $\lambda = 0.5$ or, in other words, expected value 2; the sample size is 10,000. In plot (a) we see the histogram for just one exponential variable; we observe the exponential shape that we expect. Plot (b) shows the histogram when $n = 10$ independent exponentials are summed; finally, plot (c) shows what happens for $n = 100$. The last histogram looks suspiciously like a normal density. Indeed, the celebrated central limit theorem confirms the intuition.

THEOREM 7.9 (Central limit theorem) *Let $X_1, X_2, \ldots, X_n$, be a sequence of i.i.d. random variables with expected μ and standard deviation σ.*

---

[21] We recall that stock shares are limited liability assets; hence, they cannot take negative values. In other words, the worst-case return is -100%.
[22] See Theorem 6.8.

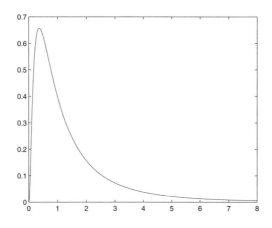

Fig. 7.21   PDF of a lognormal variable with parameters $\mu = 0$ and $\sigma = 1$.

Then, for n large, the following holds:

$$\mathrm{P}\left(\frac{X_1 + X_2 + \cdots X_n - n\mu}{\sqrt{n}\sigma} \leq x\right) \approx \mathrm{P}(Z \leq x)$$

where Z is standard normal.

This theorem essentially states that the sum of n i.i.d. variables tends to a normal distribution with expected value $n\mu$ and standard deviation $\sqrt{n}\sigma$; by standardization we get Z.[23] The central limit theorem contributes to explain why the normal distribution plays a pivotal role: When we sum many random contributions, we tend to end up with a normal distribution. For instance, demand for items sold in high volumes can often be modeled by a normal distribution, resulting from the sum of many individual demands, whereas this model is inappropriate for low-volume items.

7.7.4   The law of large numbers: a preview

The sample mean plays a key role in descriptive statistics and, as we shall see, in inferential statistics as well. In this section we take a first step to characterize its properties and, in so doing, we begin to appreciate an often cited principle: the law of large numbers.

---

[23]From a theoretical perspective, we should say that we have convergence *in distribution*. We outline the fundamental concepts of stochastic convergence in Section 9.8; many readers, however, may ignore the related subtleties.

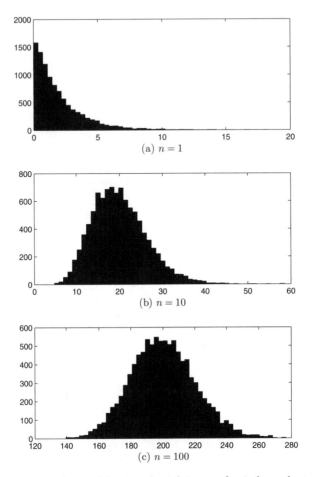

Fig. 7.22 Histograms obtained by sampling the sum of n independent exponentials with rate $\lambda = 0.5$, for $n = 1, 10, 100$.

Consider a sample consisting of i.i.d. variables X_i, $i = 1, \ldots, n$, with expected value μ and variance σ^2. The sample mean

$$\overline{X} = \frac{1}{n} \sum_{i=1}^{n} X_i$$

is a random variable and it is natural to wonder what is its distribution. From what we have seen, we know that a general answer does not exist. However, if the sample comes from a normal population, the sample mean is normally distributed, because the sum of normals is normal as well. Furthermore, we can rely on the central limit theorem to conclude that sample mean will tend to be normal when the sample is large enough.

Another intuitive property of the sample mean is that it should get closer and closer to the true expected value μ, when n progressively increases. Indeed, on the basis of the properties of sums of random variables, we obtain

$$\mathrm{E}[\overline{X}] = \mathrm{E}\left[\frac{1}{n}\sum_{i=1}^{n} X_i\right] = \frac{1}{n}\sum_{i=1}^{n} \mathrm{E}\left[X_i\right] = \frac{1}{n}\sum_{i=1}^{n} \mu = \mu$$

Furthermore, relying on the independence assumption, we also see that

$$\mathrm{Var}(\overline{X}) = \mathrm{Var}\left(\frac{1}{n}\sum_{i=1}^{n} X_i\right) = \frac{1}{n^2}\sum_{i=1}^{n} \mathrm{Var}\left(X_i\right) = \frac{1}{n^2}\sum_{i=1}^{n} \sigma^2 = \frac{\sigma^2}{n}$$

This is a remarkable result: The larger the sample size, the lower the variance of the sample mean. In the limit, this variance goes to zero; but a random variable with zero variance is just a number. Then, we may suspect that we should write something like this:

$$\lim_{n\to+\infty} \frac{1}{n}\sum_{i=1}^{n} X_i = \mu \tag{7.19}$$

This gets close to a precise statement of the law of large numbers. Actually, stating this law precisely requires to specify all of the hidden assumptions as well. Furthermore, the limit above has no clear meaning: What is the limit of a sequence of *random variables*? How can a random variable *tend to a number*? A sound statement of the law of large numbers requires some concepts of stochastic convergence. We will outline these concepts in the advanced Section 9.8; however, most readers may skip the involved technicalities.

7.8 MISCELLANEOUS APPLICATIONS

In this section we outline a few applications from logistics and finance. The three examples will definitely look repetitive, and possibly boring, but this is exactly the point: Quantitative concepts may be applied to quite different situations, and this is why they are so valuable. In particular, we explore here three cases in which quantiles from the normal distributions are applied.

7.8.1 The newsvendor problem with normal demand

We know from Section 7.4.4 that the optimal solution of a newsvendor problem with continuous demand is the solution of the equation

$$F_D(q^*) = \frac{m}{m + c_u}$$

i.e., the quantile of demand distribution, corresponding to probability $m/(m+c_u)$. If we assume normal demand, with expected value μ and standard deviation σ, then the optimal order quantity (assuming that we want to maximize expected profit) is

$$q^* = \mu + z_{m/(m+c_u)}\sigma$$

Assume that items are purchased from a supplier for \$10 per item and then are sold at \$15, and that the salvage value of unsold items is \$3. The expected value of demand over the sales window is 10,000 items, and its standard deviation is 2000 items. Then we find

$$\frac{m}{m + c_u} = \frac{15 - 10}{(15 - 10) + (10 - 3)} = 0.4167$$

Note that service level is lower than 50%, so the corresponding quantile from the standard normal distribution is negative, and we should buy less than expected demand. Indeed, statistical software yields

$$z_{0.4167} = -0.2104 \qquad \Rightarrow \qquad q^* = 10{,}000 - 0.2104 \times 2000 \approx 9579$$

Note that, since the profit margin is low with respect to the cost of unsold items, we should be conservative; the larger the risk, measured by standard deviation, the less we buy.

7.8.2   Setting the reorder point in inventory control

Say that we are in charge of managing the inventory of a component, whose supply lead time is 2 weeks. Weekly demand is modeled by a normal random variable with expected value 100 and standard deviation 20 (let us pretend that this makes sense). If we apply a reorder point policy based on the EOQ model, we should order a fixed quantity whenever the inventory level falls below a reorder point R.[24] How can we set R in order to achieve a 95% service level?

The service level in this case is the probability of *not* having a stockout during the delivery lead time. Note that we may run out of stock during the time window between the instant at which we issue the order to our supplier and the time instant at which items are received and inventory is replenished. Hence, we should consider the probability that demand during lead time does not exceed the reorder point R, which should be set in such a way that

$$P(D_{LT} \le R) = 0.95$$

If we assume that weekly demand is normal, then we should just compute a quantile from the normal distribution again. If demands in two consecutive

---

[24]This is a rather imprecise statement, as ordering decisions should consider backorders and on-order inventory See, e.g., Ref. [4] for a complete treatment.

weeks are independent, then the distribution of the demand during lead time is normal with parameters

$$\mu_{LT} = 100 + 100 = 200$$
$$\sigma_{LT} = \sqrt{20^2 + 20^2} = \sqrt{2} \times 20 = 28.2843$$

Since $z_{0.95} = 1.6449$, we should set

$$R = \mu_{LT} + z_{0.95} \cdot \sigma_{LT} = 200 + 1.6449 \times 28.2843 \approx 247$$

Note that, if there were no risk, we would just set $R = 200$. The additional 47 items we keep on stock are a *safety stock*. To reduce safety stock, and save money related to holding inventory, we should reduce demand uncertainty and/or lead time. This is precisely one of the cornerstones of the so-called Toyota approach, which was originally applied within the automotive industry to car manufacturing; its extension to other industries resulted in the well-known *just-in-time* philosophy.

7.8.3   An application to finance: value at risk (VaR)

Most financial investments entail some degree of risk. Imagine a bank holding a portfolio of assets; the bank should set aside enough capital to make up for possible losses on the portfolio. To determine how much capital the bank should hold, precise guidelines have been proposed, e.g., by the Basel committee. Risk measures play a central role in such regulations, and a commonly proposed risk measure is *value at risk* [VaR; please note the capitalization of letters to avoid ambiguity with *variance* (Var)]. It must be mentioned that bank regulation has been the subject of quite some controversy, in the wake of financial disasters following the subprime mortgage crisis of 2007–2008. In particular, VaR has been criticized as an inadequate risk measure, offering a false sense of security. It has even been suggested that VaR should not be taught at all in business schools.[25] However, like it or not, VaR is used; hence, students and practitioners should be fully aware of what it is and what it is not. We will investigate VaR and its limitations further in Section 13.2.3. The first step, however, is understanding VaR in a simple setting.

Informally, VaR allows to say something like

> *We are X percent sure that we will not lose*
> *more than V dollars over the next N days.*

We immediately see that VaR is actually a quantile of the distribution of losses. To clarify the idea, consider a portfolio consisting of a single stock: We own $10 million in Microsoft shares, and we want to estimate one-day

---

[25]See N.N. Taleb and P. Triana, Bystanders to this financial crime were many, *Financial Times* Dec. 8, 2008.

VaR, with 99% confidence level. The simplest textbook calculation goes like this: Assume that the daily return of the stock is normally distributed. We know from the square-root rule of Section 7.7.1 that, in such a short timespan, volatility (standard deviation) dominates drift (expected return). Then, assume that daily return is a normal variable with expected value 0% and standard deviation $\sigma_d = 2\%$. Daily profits and losses can be expressed as the daily variation δW in our wealth:

$$\delta W = N \cdot S \cdot \delta R$$

where N is the number of stocks, S their initial price, and δR the random return. We have a loss when $\delta R < 0$. If we plot the PDF of profit, losses correspond to the left tail; if we plot the PDF for loss, they are on the right tail; in the case of the normal distribution, given its symmetry, this makes no real difference. To find VaR, we need to solve the equation

$$P\left(L \leq \text{VaR}_{0.99}\right) = 0.99$$

where $L = -\delta W$ is loss. Given what we have seen repeatedly about quantiles of the normal distribution, we see that we should compute

$$\text{Var}_{0.99} = N \cdot S \cdot \sigma_d \cdot z_{0.99} = \$10{,}000{,}000 \times 0.02 \times 2.33 = \$466{,}000$$

Now what about the 10-day VaR? The following reasoning is often proposed: Since volatility scales with the square root of time, it follows that

$$\text{VaR}_{99\%}(10 \text{ days}) = \sqrt{10} \times \text{VaR}_{99\%}(1 \text{ day}) = \$1{,}473{,}621$$

Obviously, this reasoning assumes independence in daily returns, which should certainly not be taken for granted. Furthermore, when we consider time horizons, the drift should not be neglected. More generally, we know that the normal distribution has thin tails and should not be considered a safe model of uncertainty in finance. So, the above calculation should be regarded just as a starting point. Nevertheless, it proves our point: Quantitative concepts can be used (and misused) in a variety of unrelated settings.

7.9 STOCHASTIC PROCESSES

So far, we have considered a single random variable. However, more often than not, we have to deal with multiple random variables. There are two cases in which we have to do so:

- We might observe different random variables, say, X_i, $i = 1, \ldots, n$, at the same time. In such a case, we speak of *cross-sectional data*. As practical examples, think of the return of several financial assets over an investing horizon; alternatively, consider the demand for several items, which could be complementary or substitute goods.

- We might observe a single random variable, over multiple time periods, say, X_t, $t = 1, \ldots, T$. In such a case, we speak of *longitudinal data*. For instance, we may observe the weekly return of a financial asset over a timespan of a few months, or daily demand for an item.

In practice, we may also have the two views in combination, i.e., multiple variables observed over a timespan of several periods. In such a case, we speak of *panel data*. Given the scope of this book, we will not consider panel data. If we observe cross-sectional data, and the corresponding random variables are independent, then we may just study one variable at a time, and that's it. However, this is just a very special and simple case. We need more sophisticated tools to deal with dependence, which is the topic of the next chapter. In this section, we want to consider random variables over time.

DEFINITION 7.10 (Stochastic process) *A time-indexed collection of random variables is called a* **stochastic process**. *If time is discretized, we have a* **discrete-time** *process:*

$$X_t, \qquad t = 0, 1, 2, 3, 4, \ldots \qquad (7.20)$$

If time is continuous, we have a **continuous-time** *(also known as* **continuous-parameter***) process:*

$$X(t), \qquad t \in [0, +\infty) \qquad (7.21)$$

In general, one could consider a collection of random variables depending on space, rather than time. Then, we could speak of discrete- or continuous-parameter processes. In the applications we consider in the book, the parameter will always be time. In some loose sense, the stochastic process is a generalization of deterministic functions of time, in that for any value of t it yields a random variable (which is a function itself) rather than a number. If we observe a sequential realization of the random variables over time, we get a *sample path* of the process. In this introductory book, we will essentially deal with discrete-time processes, but it is a good idea to consider at least a simple example of a relevant continuous-time process.

Example 7.15 (Poisson process) The Poisson process is an example of a counting process, i.e., a stochastic process $N(t)$ counting the number of events that occurred in the time interval $[0, t]$. Such a process starts from zero and has unit increments over time. We may use such a process to model order or customer arrivals. The Poisson process is obtained when we make specific assumptions about the interarrival times of customers. Let X_k, $k = 1, 2, 3, 4, \ldots$, be the interarrival time between customer $k - 1$ and customer k; by convention, X_1 is the arrival time of the first customer after the start time $t = 0$. We obtain a Poisson process if we assume that variables X_k are mutually independent and all exponentially distributed with parameter λ, which is in this case the arrival rate, i.e., the average number of customers

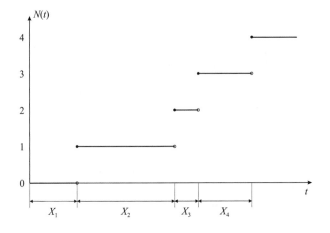

Fig. 7.23 Sample path of the Poisson process.

arriving per unit time. A sample path is illustrated in Fig. 7.23; we see that the process "jumps" whenever a customer arrives, so that sample paths are piecewise constant.

We have already mentioned the link between Poisson and exponential distributions and the Poisson process. If we consider a time interval $[t_1, t_2]$, with $t_1 < t_2$, then the number of customers who arrived in this interval, i.e., $N(t_2) - N(t_1)$, has Poisson distribution with parameter $\lambda(t_2 - t_1)$. Furthermore, if we consider another time interval $[t_3, t_4]$, where $t_3 < t_4$, which is disjoint from the previous one, i.e., $(t_2 < t_3)$, then the random variables $N(t_2) - N(t_1)$ and $N(t_4) - N(t_3)$ are independent. We say that the Poisson process has *stationary* and *independent* increments.

The Poisson process is a useful model for representing the random arrival of customers who have no mutual relationships at all. This is a consequence of the lack of memory of the exponential distribution.

The model can be generalized to better fit reality. For instance, if we observe the arrival process of customers at a big retail store, we easily observe variations in the arrival rate. If we introduce a time-varying rate $\lambda(t)$, we get the so-called *inhomogeneous Poisson process*. Furthermore, if we consider not only customer (or order) arrivals, but the demanded quantities as well, we see the opportunity of associating another random variable, the quantity per order, with each customer. The cumulative quantity demanded $D(t)$ in the time interval $[0, t]$ is another stochastic process, which is known as a *compound Poisson process*. The sample paths of this process would be qualitatively similar to those in Fig. 7.23, but the size of the jumps would be random. This is a possible model for demand, when sale volumes are not large enough to warrant use of a normal distribution.   ⬜

Naive thinking would draw us to the conclusion that, in order to characterize a stochastic process, we should give the distribution of X_t for all the

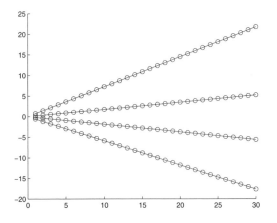

Fig. 7.24   Sample paths of the stochastic process $X_t = t \cdot \tilde{\epsilon}$, $\tilde{\epsilon} \sim N(0,1)$.

relevant time instants t. This is what we call the *marginal distribution*. The following example shows that marginal distributions do not tell the whole story.

Example 7.16 (A Gaussian process) A common class of stochastic processes consists of sequences of random variables whose marginal distribution is normal, which is why they are termed *Gaussian processes*. To be precise, we should say that a Gaussian process requires that the random variables $X_{t_1}, X_{t_2}, \ldots, X_{t_m}$ have a *jointly* normal distribution for any possible choice of time instants $t_1, t_2, \ldots, t_m$, but for the sake of simplicity we will put in the same bag any process for which the marginal distribution of X_t is normal. However, it is important to realize that in doing so we are considering processes that may be very different in nature. Consider the stochastic process

$$X_t = t \cdot \tilde{\epsilon}, \qquad t = 0, 1, 2, 3, \ldots$$

where $\tilde{\epsilon}$ is standard normal variable. In our loose sense, we may say that this is a Gaussian process, since X_t is normal with expected value 0 and variance t^2. However, it is a somewhat degenerate process, since uncertainty is linked to the realization of a *single* random variable. If we know the value of X_t for a single time instant, then we can figure out the whole sample path. Figure 7.24 illustrates this point by showing a few sample paths of this process. A quite different process is obtained if all variables X_t are normal with parameters μ and σ^2 and mutually *independent*. Figure 7.25 shows a sample path of the process $X_t = t \cdot \tilde{\epsilon}_t$, where $\tilde{\epsilon}_t \sim N(0,1)$. However, the marginal distributions of the individual random variables X_t are exactly the same for both processes.

□

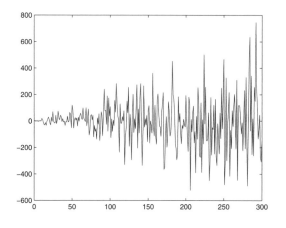

Fig. 7.25   Sample path of process $X_t = t \cdot \tilde{\epsilon}_t$, where $\tilde{\epsilon}_t \sim N(0,1)$.

The example shows that we really need some way to characterize the interdependence of jointly distributed random variables, which is the topic of next chapter. Discrete-time stochastic processes will be discussed further in Chapter 11 on time series models.

7.10 PROBABILITY SPACES, MEASURABILITY, AND INFORMATION

Successful investing in stock shares is typically deemed a risky and complex endeavor. However, the following piece of advice seems to offer a viable solution:[26]

>   *Buy a stock. If its price goes up, sell it. If it goes down, don't buy it.*

In this section we dig a little deeper into concepts related to measurability of random variables and their relationship to the flow of information and its impact on decisions. Despite their more theoretical character,[27] the concepts we consider here are often met when reading books on quantitative finance, where it is common to read about filtrations and adapted processes. We will not try a full and rigorous treatment, which would require a quite sophisticated

---

[26] In his history of the Great Crash of 1929, John K. Galbraith attributes this fundamental advice to an American comedian; see J.K. Galbraith, *The Great Crash 1929*, originally published in 1954, reprinted by Mariner Books, 2009.

[27] This supplement may be safely skipped. The only section in which we use related concepts is Section 13.4, where we illustrate multistage stochastic linear programming models.

machinery; still, we will be able to understand what is wrong with the above suggestion from a probabilistic perspective, and the concepts that we illustrate should look less intimidating after getting an intuitive feel for them.

We pointed out that a random variable is actually a mapping

$$X : \Omega \to \mathbb{R}$$

from a set Ω of outcomes of a random experiment to the set of real numbers $\mathbb{R}$. Indeed, random variables are often denoted by $X(\omega)$ to emphasize this point. However, not all conceivable mappings are legitimate random variables. To see this, we need to clarify the concept of probability space.

DEFINITION 7.11 (Probability space) *A probability space is a triple, usually denoted by* $(\Omega, \mathcal{F}, P)$*, where* Ω *is the sample space, consisting of outcomes of a random experiment;* $\mathcal{F}$ *is a family of subsets of* Ω*, the events, with suitable closure properties; and* P *is a probability measure mapping events into the interval* $[0, 1]$*.*

To fully get the message behind this definition, we should observe that for a given sample space we may define different probability measures, which is not surprising, but we may also define different families of events. If we roll a die, the obvious sample space is $\Omega = \{1, 2, 3, 4, 5, 6\}$, and we may consider a family of events obtained by arbitrary combinations of set operations like union, complement, and intersection. This would make a rather large family of all subsets of size 1, 2, etc., also including Ω itself and its complement, the empty set $\emptyset$:

$$\mathcal{F}_1 = \{\Omega, \emptyset, \{1\}, \{2\}, \ldots, \{6\}, \{1, 2\}, \{1, 3\}, \ldots \{5, 6\}, \{1, 2, 3\}, \ldots\} \quad (7.22)$$

However, we may constrain events a bit in order to reflect information or lack thereof. For instance, we might consider the following family of events:

$$\mathcal{F}_2 = \{\Omega, \emptyset, \{1, 3, 5\}, \{2, 4, 6\}\} \quad (7.23)$$

It is easy to check that if we try taking complements and unions of elements in $\mathcal{F}_2$, we still get an element of $\mathcal{F}_2$. Since intersection is just a combination of these two operations, we see that $\mathcal{F}_2$ is closed under elementary set operations. This family of events, with respect to $\mathcal{F}_1$, is definitely less rich, and this reflects lack of information. It is the set of events we would deal with if the only information available about the roll of the die were "even" or "odd."

When assigning a probability measure to subsets of Ω, we need to make sure that we are able to do the same for any event that we may obtain by elementary set operations. In other words, we should not get a subset that is not an event. This requirement may be expressed by requiring that $\mathcal{F}$ be a field.

DEFINITION 7.12 (Field) *A family* $\mathcal{F}$ *of subsets of* Ω *is called a* **field** *if the following conditions hold:*

1. $\Omega \in \mathcal{F}$

2. $E \in \mathcal{F} \Rightarrow (\Omega \setminus E) \in \mathcal{F}$

3. $E, G \in \mathcal{F} \Rightarrow (E \cup G) \in \mathcal{F}$

A field is also called an *algebra of sets*. The conditions in the definition state that $\mathcal{F}$ is closed under elementary set operations. Note that the first and second conditions imply that the empty set $\emptyset$ belongs to the field $\mathcal{F}$.

Example 7.17 Given $\Omega = \{1, 2, 3, 4\}$, consider the family of subsets

$$\mathcal{G} = \{\Omega, \emptyset, \{1\}, \{2, 3, 4\}, \{1, 2\}, \{3, 4\}, \}$$

This is not a field since, for instance

$$\{1\} \cup \{3, 4\}, \notin \mathcal{G}$$

Actually, to cope with continuous random variables and more generally with probability distributions with infinite support, a stronger concept is needed: a σ-*field*, also called a σ-*algebra*. To define this stronger concept, the third condition is extended to a countable union of events:

$$E_1, E_2, E_3, \ldots \in \mathcal{F} \Rightarrow \bigcup_{i=1}^{\infty} E_i \in \mathcal{F}$$

Indeed, whenever the concept infinity comes into play, pathological cases can occur. The very concept of σ-field is necessary to avoid weird cases in which it is impossible to assign a probability measure to an event. We will not be concerned with these anomalies, since we limit our treatment to a finite sample space.

Even describing a finite field by enumerating all of its subsets may be a daunting task. However, we may describe it implicitly by considering a finite *partition* $\mathcal{P}$ of Ω, i.e., a finite family of subsets E_i, $i = 1, \ldots, n$, such that

$$E_i \cap E_j = \emptyset, \text{ for } i \neq j, \qquad \bigcup_{i=1}^{n} E_i = \Omega$$

Given a partition $\mathcal{P}$, we may consider the σ-field $\sigma(\mathcal{P})$ generated by combining subsets in the partition in any possible way. In the case of (7.22), the partition consists of all singleton sets, whereas in the case of (7.23) we have the two subsets of even and odd outcomes.

Let us now turn to random variables. Given a probability space, we may define random variables as mappings of outcomes into real numbers, but we should clarify how we associate a measure probability with a random variable. Actually, we associate a probability measure with underlying events. Indeed, the probability that we assign to random variables should be associated with the underlying *events* in the field $\mathcal{F}$. Consider a discrete random variable X and numeric value a. How can we define the probability $P(X = a)$? We should consider the subset of outcomes $\omega \in \Omega$ such that $X(\omega) = a$, which essentially amounts to inverting the function $X(\omega)$:[28]

$$X^{-1}(a) \equiv \{\omega \in \Omega : X(\omega) = a\}$$

Then, the probability we seek is just the probability measure of the subset $X^{-1}(a)$. However, this is only possible if any such subset is an event in the σ-field $\mathcal{F}$.

Example 7.18 Consider the sample space $\Omega = \{1, 2, 3, 4\}$ and a partition

$$\mathcal{P} = \{\{1\}, \{2, 3, 4\}\}$$

Let $\mathcal{F} = \sigma(\mathcal{P})$ be the field generated by this partition and define the mapping $X(\omega)$:

$$X(\omega) = 1 + \omega$$

This mapping is *not* a random variable with respect to the field $\mathcal{F}$. In fact, we cannot assign the probability $P(X = 3)$, since

$$\{\omega \in \Omega : X(\omega) = 3\} = \{2\} \notin \mathcal{F}$$

To be a random variable, a mapping $Y(\omega)$ should be *constant* for the three outcomes in $\{2, 3, 4\}$. For instance

$$Y(\{1\}) = -1, \ Y(\{2\}) = 1, \ Y(\{3\}) = 1, \ Y(\{4\}) = 1$$

is a legitimate random variable. ▯

What is wrong with Example 7.18 is not the mapping $X(\omega)$ per se; it is its association with the field $\mathcal{F}$. If we had a richer field, generated by the partition of Ω into its singletons, there would be no issue. Technically speaking, we say that X is not $\mathcal{F}$-measurable.

DEFINITION 7.13 (Measurable random variable) *We say that a random variable $X(\omega)$ is $\mathcal{F}$-measurable if*

$$\{\omega \in \Omega : X(\omega) = x\} \in \mathcal{F}$$

for all values of x.

---

[28] The careful reader will immediately guess that what we are saying works only for a discrete random variable, since for continuous random variables the probability of observing a specific value is zero. Indeed, in a rigorous treatment we should consider events $\{\omega \in \Omega : X(\omega) \leq a\}$, but to keep things as intuitive as possible, we will refrain from doing so.

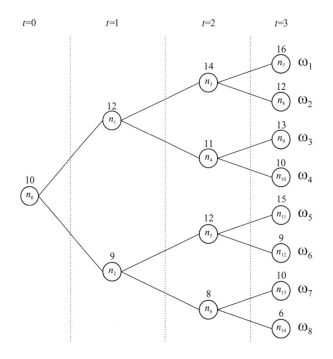

$t=0$ $t=1$ $t=2$ $t=3$

Fig. 7.26   An event tree.

In other words, the inverse function for any value x must be an event in the field $\mathcal{F}$, so that we may associate a probability measure with it. This can be done or not depending on the random variable and the richness of the field of events $\mathcal{F}$. If we go back to dice throwing, it is clear that if our field is given by (7.23), we can assign only one value to all even outcomes, and another value to all odd outcomes. The field is, in a sense, smaller than $\mathcal{F}_1$, since all the events in $\mathcal{F}_2$ are events in $\mathcal{F}_1$, but the converse is not true; this represents a limitation in the available information.

The link between event fields, measurability, and information can be further clarified if we consider a dynamic problem. Let us consider a stochastic process in the form of the event tree depicted in Fig. 7.26. To be concrete, let us interpret this as a stochastic process describing the price of a stock share. At time $t = 0$, the stock price is $X_0 = 10$. Then, the price may go up or down, resulting in a stochastic process X_t, $t = 0,1,2,3$. In this case the sample space consists of outcomes ω_i, $i = 1, 2, \ldots, 8$, and each outcome corresponds to a scenario, i.e., a possible path of stock prices. For instance, outcome ω_3 is associated with scenario $(10, 12, 11, 13)$. If we are at any terminal node in the scenario tree, we know which scenario has occurred, since we can observe the whole history of stock prices. However, if we are, e.g., on node n_4, we do not know whether we are observing scenario ω_3 or ω_4, since they cannot be distinguished. Nevertheless, we do have some information, since by observing the past history of stock prices we can rule out any other scenario. In the root node n_0 we have the least information, since any scenario is possible.

All of this is reflected in the event fields with which random variables X_t, $t = 0,1,2,3$, are associated. We can capture information by suitable partitions of the sample space

$$\Omega = \{\omega_1, \omega_2, \omega_3, \omega_4, \omega_5, \omega_6, \omega_7, \omega_8\}$$

At time $t = 0$ we cannot say anything, and our field of events is

$$\mathcal{F}_0 = \{\emptyset, \Omega\} = \{\emptyset, \{1, 2, 3, 4, 5, 6, 7, 8\}\}$$

At time $t = 1$ we can at least rule out half of the scenarios. This is reflected by the more refined partition

$$\mathcal{P}_1 = \{\{1, 2, 3, 4\}, \{5, 6, 7, 8\}\}$$

which generates the field

$$\mathcal{F}_1 = \{\emptyset, \{1, 2, 3, 4\}, \{5, 6, 7, 8\}, \Omega\}$$

At time $t = 2$, there is a further branching, refining the partition

$$\mathcal{P}_2 = \{\{1, 2\}, \{3, 4\}, \{5, 6\}, \{7, 8\}\}$$

which generates an even richer field:

$$\mathcal{F}_2 \;=\; \{\, \emptyset, \{1,2\}, \{3,4\}, \{5,6\}, \{7,8\}, \{1,2,3,4\}, \{1,2,5,6\}, \ldots, \{5,6,7,8\}$$
$$\{1,2,3,4,5,6\}, \ldots, \{3,4,5,6,7,8\}, \Omega \,\}$$

Finally, at time $t = 3$, we have the finest partition, consisting of singletons

$$\mathcal{P}_3 = \{\, \{1\}, \{2\}, \{3\}, \{4\}, \{5\}, \{6\}, \{7\}, \{8\} \,\}$$

which generates the richest field $\mathcal{F}_3$, consisting of all possible subsets Ω. We note that, as time goes by, we get larger and larger fields.

DEFINITION 7.14 (Filtration) *An increasing sequence of σ-fields*

$$\mathcal{F}_0 \subset \mathcal{F}_1 \subset \mathcal{F}_2 \subset \cdots$$

defined on a common sample space Ω is called a **filtration**.

A filtration defines precisely how information is collected by observing a stochastic process. This concept can be defined for continuous-time processes with a continuous state space, and it requires a sophisticated mathematical machinery. However, the essential message is quite simple:

> *The sequence of decisions we make by observing the stochastic process at times $t = 0, 1, 2, \ldots$ must reflect the available information and cannot be anticipative.*

The piece of advice with which we have opened this section is clearly not implementable: It would require knowledge of the future. From a technical perspective, consider decision variables Z_t^b and Z_t^s representing the number of stock shares that we buy and sell, respectively, at time t. At time $t = 0$ we have a unique decision, since we can just buy or sell here and now. Seen from time $t = 0$, the variables for the next time instants are random variables, as they depend on the decision that we will make on the basis of the observed path of the stochastic process and our expectations for the future, which are represented by the scenario tree. The random variables at time t must be $\mathcal{F}_t$-measurable. If we are at node n_4 in the event tree of Fig. 7.26, we cannot say "buy if scenario is ω_3" and "do not buy if scenario is ω_4." The decision, whatever it is, must be the same for the two scenarios. Otherwise, the random variable corresponding to the decision at that node would not be constant on the event $\{\omega_3, \omega_4\}$, and it would not be measurable; we would be in trouble just as in Example 7.18. Technically speaking, we say that decisions must be *adapted* to the filtration $\mathcal{F}_t$. The piece of financial advice we have considered, unfortunately, is not adapted to the filtration and would require clairvoyance to be practically implementable.

Problems

7.1   A random variable X has normal distribution with $\mu = 250$ and $\sigma = 40$. Find the probability that X is larger than 200.

7.2   Consider a normal variable with $\mu = 250$ and $\sigma = 20$, and find the probability that X falls in the interval between 230 and 260.

7.3   We should set the reorder point R for an item, whose demand during lead time is uncertain. We have a very rough model of uncertainty – the lead time demand is uniformly distributed between 5000 and 20000 pieces. Set the reorder point in such a way that the service level is 95%.

7.4   You are working in your office, and you would like to take a very short nap, say, 10 minutes. However, every now and then, your colleagues come to your office to ask you for some information; the interarrival time of your colleagues is exponentially distributed with expected value 15 minutes. What is the probability that you will not be caught asleep and reported to you boss?

7.5   A friend of yours is an analyst and is considering a probability model to capture uncertainty in monthly demand of an item featuring high-volume sales. He argues that the central limit applies and, after a thorough check of data, proposes a normal distribution with expected value 12,000 and standard deviation 7000 items. Is this a reasonable model?

7.6   Let X be a normal random variable with expected value $\mu = 6$ and standard deviation $\sigma = 1$. Consider random variable $W = 3X^2$. Find the expected value $E[W]$ and the probability $P(W > 120)$.

7.7   You have just issued a replenishment order to your supplier, which is not quite reliable. You have ordered 400 items, but what you will receive is a normal random variable with that expected value, and standard deviation 40 (let us assume that using a continuous random variable is a sensible approximation of the discrete random variable modeling the integer number of received items). After receiving the shipment, you will have to serve a number of customer requests. The amount that your customers ask for is a random variable with expected value 3 and standard deviation 0.3. How many customer requests should you receive in order to have a probability of stockout larger than 10%?

7.8   Let $X \sim \chi_n^2$ be a chi-square variable with n degrees of freedom. Prove that $E[X] = n$.

7.9   You work for a manufacturing firm producing items with a limited time window for sale. Items are sold by a distributor facing uncertain demand over the time window, which we model by a normal distribution with expected value 10,000 and standard deviation 2500. The distributor decides how many items to order using a newsvendor model. From the distributors perspective, each item costs $10 and is sold at the recommended price of $14. Unsold items are bought back by the manufacturer. Assume that the manufacturer would like to see at least 15,000 items on the shelves (in order to promote her brand name); at what price should she be willing to buy unsold items back?

7.10   You are in charge of deciding the purchased amount of an item with limited time window for sales and uncertain demand. The unit purchase cost is $10 per item, the selling price is $16, and unsold items at the end of the sales window have a salvage value of $7 per item. Demand is also influenced by the level of competition. If there is none, demand is uniformly distributed between 1200 and 2200 items. However, if a strong competitor enters the game, demand is uniformly distributed between 100 and 1100 items.

- If the probability that the competitor enters the market is assumed to be 50%, how many items should you order to maximize expected profit? (Let us assume that selling prices are the same in both scenarios.)

- What if this probability is 20%? Does purchased quantity increase or decrease?

7.11   In some applications we are interested in the distribution of the maximum among a set of realization of random variables. Let us consider a set of n i.i.d. variables U_i, $i = 1, \ldots, n$, with uniform distribution on the unit interval $[0, 1]$. Let X be their maximum: $X = \max\{U_1, U_2, \ldots, U_n\}$. Prove that the CDF of X is $F_X(x) = x^n$.

For further reading

Most useful references for this chapter are essentially the same as in the previous chapter, some of which are repeated here just for readers' convenience.

- For an elementary introduction to continuous random variables, see Ref. [14] or [13]. The first reference also illustrates the link between probability theory and statistics, whereas the second one is more concerned with probability theory, including stochastic processes.

- At a more advanced level is Ref. [16]. Stochastic processes are dealt with at a reasonably sophisticated level in Ref. [7], which is accompanied by Ref. [6].

- Issues in measurability of random variables are dealt with in Refs. [5], [12], and [15], which definitely make more challenging reading.

- On the application side, the link between probability theory and mathematical finance is well illustrated in Ref. [11]. A useful reading on value at risk is provided in Ref. [9].

- The role of stochastic models in manufacturing is well documented in Ref. [8]; applications to supply chain management are also described at an elementary level in Ref. [4]; see Ref. [1] or [18] for a more advanced treatment.

REFERENCES

1. S. Axsäter, *Inventory Control,* 2nd ed., Springer, New York, 2006.

2. J.C. Bogle, Black Monday and black swans, *Financial Analysts Journal,* **64**(2):30–40, 2008.

3. R. Bookstaber, *A Demon of Our Own Design: Markets, Hedge Funds, and the Perils of Financial Innovation,* Wiley, New York, 2008.

4. P. Brandimarte and G. Zotteri, *Introduction to Distribution Logistics,* Wiley, New York, 2007.

5. M. Capiński and T. Zastawniak, *Probability through Problems,* Springer-Verlag, Berlin, 2000.

6. G. Grimmett and D. Stirzaker, *One Thousand Exercises in Probability,* Oxford University Press, Oxford, 2003.

7. G. Grimmett and D. Stirzaker, *Probability and Random Processes,* 3rd ed., Oxford University Press, Oxford, 2003.

8. W. Hopp and M. Spearman, *Factory Physics,* 3rd ed., McGraw-Hill, New York, 2008.

9. P. Jorion, *Value at Risk: The New Benchmark for Controlling Derivatives Risk,* McGraw-Hill, New York, 1997.

10. R.L. Phillips, *Pricing and Revenue Optimization,* Stanford University Press, Stanford, CA, 2005.

11. S.R. Pliska, *Introduction to Mathematical Finance: Discrete Time Models,* Blackwell Publishers, Malden, MA, 1997.

12. S.I. Resnick, *A Probability Path,* Birkhäuser, Boston, 1999.

13. S.M. Ross, *Introduction to Probability Models,* 8th ed., Academic Press, San Diego, 2002.

14. S.M. Ross, *Introduction to Probability and Statistics for Engineers and Scientists,* 4th ed., Elsevier Academic Press, Burlington, MA, 2009.

15. S.M. Ross and E.A. Peköz, *A Second Course in Probability,* Probability Bookstore, Boston, 2007.

16. D. Stirzaker, *Elementary Probability,* 2nd ed., Cambridge University Press, Cambridge, UK, 2003.

17. K.T. Talluri and G.J. Van Ryzin, *The Theory and Practice of Revenue Management,* Springer, New York, 2005.

18. P.H. Zipkin, *Foundations of Inventory Management,* McGraw-Hill, New York, 2000.

8

Dependence, Correlation, and Conditional Expectation

So far, when dealing with a sequence of random variables, we always assumed that they were independent. In this chapter, at last, we investigate the issue of dependence. To get the basic intuition, consider the hypothetical demand data in Table 8.1.

Table 8.1 Demand data for two items: Are they independent?

D_1	110	120	90	80	100	80	105	95	120	100
D_2	54	53	49	44	51	40	53	47	60	49

Can we say that the two random variables D_1 and D_2 are independent? Looking at the raw data may be a bit confusing, but things get definitely clearer if we consider the sample means, $\overline{D}_1 = 100$ and $\overline{D}_2 = 50$. We observe that whenever D_1 is above average, D_2 tends to be, too; vice versa, whenever D_1 is below average, D_2 tends to be, too. Hence, we have a sort of concordance between the two random variables, which should not be there if the two variables were independent. Capturing this concordance leads us to the definition covariance and correlation, which is the main purpose of this chapter.

Before doing so, we introduce the formal concepts of joint and marginal distributions in Section 8.1. The mathematics here is a bit more complicated than elsewhere, and the section can be skipped by those who just wish an intuitive understanding. In Section 8.2, we summarize properties of independent random variables. Then, in Section 8.3, we characterize the interdependence between two random variables in terms of covariance and correlation. We illustrate the role of these concepts in risk management, and we also point out their limitations. In the book, we will not cover multivariate distributions too much, as this is a topic that goes beyond an introductory text, but in Section 8.4 we treat one fundamental case: the multivariate normal distribution. Finally, in Section 8.5, we take advantage of the possible links between random variables to compute expectations by conditioning. Indeed, conditional expectation is an essential concept in many applications.

8.1 JOINT AND MARGINAL DISTRIBUTIONS

In order to fully appreciate the issues involved in characterizing the dependence of random variables, as well as to appreciate the role of independence, we should have some understanding of how to characterize the joint distribution of random variables.[1] For the sake of simplicity, we will deal only with the case of two random variables with a joint distribution, leaving the general case as a relatively straightforward extension. The pathway to define all the relevant concepts for two jointly distributed random variables X and Y is similar to the case of a random single variable. The unifying concept is the cumulative distribution function (CDF), as it applies to both discrete and continuous variables. If we have two random variables, we consider joint events such as $\{X \leq x\} \cap \{Y \leq y\}$, and associate a probability measure with them.

DEFINITION 8.1 (Joint cumulative distribution function) *The joint CDF for random variables X and Y is defined as*

$$F_{X,Y}(x,y) \equiv \mathrm{P}\left(\{X \leq x\} \cap \{Y \leq y\}\right)$$

In the following, we will often use the streamlined notation $\mathrm{P}(X \leq x, Y \leq y)$ *to denote a joint event.*

The joint CDF is a function of two variables, x and y, and it fully characterizes the joint distribution of the random variables X and Y.

We may also define the joint probability mass function (PMF) for discrete variables, and the joint probability density function (PDF) for continuous variables. For instance, let us refer to a pair of discrete variables, where X may take values x_i, $i = 1,2,3,\ldots$, and Y may take values y_j, $j = 1,2,3,\ldots$. The joint PMF is

$$p_{X,Y}(x_i, y_j) = \mathrm{P}(X = x_i, Y = y_j)$$

---

[1]This section can be skipped by less mathematically inclined readers.

We see immediately that, given the PMF, we may recover the CDF

$$F_{X,Y}(x_i, y_j) = \sum_{l=1}^{i} \sum_{k=1}^{j} p_{X,Y}(x_l, y_k)$$

Going the other way around, we may find the PMF given the CDF

$$\begin{aligned} p_{X,Y}(x_i, y_j) &= F_{X,Y}(x_i, y_j) - F_{X,Y}(x_{i-1}, y_j) \\ &\quad -F_{X,Y}(x_i, y_{j-1}) + F_{X,Y}(x_{i-1}, y_{j-1}) \end{aligned} \tag{8.1}$$

We also recall that a PMF makes sense in the discrete case as these probabilities are well defined, whereas for the continuous case they are zero. When dealing with continuous random variables, we have introduced a PDF, which allows us to express probabilities associated with sets, rather than single values; to do so, we need to integrate the PDF over the set of interest. The idea can be generalized to jointly distributed random variables, but in this case we integrate over a two-dimensional domain.[2] Hence, we have a joint PDF $f_{X,Y}(x, y)$ such that

$$P((X, Y) \in C) = \iint_{(x,y) \in C} f_{X,Y}(x, y) dx\, dy \tag{8.2}$$

In general, we cannot be sure that such a function exists. To be precise, we should say that the two random variables are jointly continuous if the joint PDF exists. Given the joint PDF, we may find the joint CDF by integration:

$$F_{X,Y}(a, b) = P\left(X \in (-\infty, a], Y \in (-\infty, b]\right) = \int_{-\infty}^{a} \int_{-\infty}^{b} f_{X,Y}(x, y) dy\, dx \tag{8.3}$$

Example 8.1 It is instructive to see the connection between Eqs. (8.1) and (8.2), in the context of jointly distributed, continuous random variables. Consider the rectangular area

$$C = [x_{i-1}, x_i] \times [y_{j-1}, y_j]$$

which is depicted as the darkest rectangle in Fig. 8.1. The probability

$$P((X, Y) \in C)$$

is the area below the joint PDF, over the rectangle. How can we find this area in terms of the joint CDF? Looking at Eq. (8.3), we see that the CDF gives the areas below the PDF, over infinite regions to the southwest with respect to each point. These are displayed as quadrants in the figure. It is

---

[2]Integrals in two dimensions are outlined in Section 3.9.3.

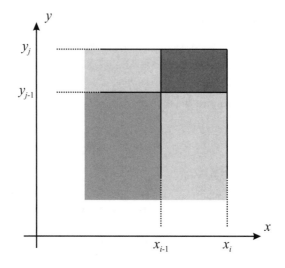

Fig. 8.1 Interpreting Eq. (8.1).

easy to see that we can find the probability of C by taking the right sums and differences of the area over the quadrants. Let us denote the quadrant to the southwest with respect to point (x, y) as $Q(x, y)$. Then, in terms of set operations (difference, union, and intersection), we may write

$$C = \{Q(x_i, y_j) \setminus [Q(x_{i-1}, y_j) \cup Q(x_i, y_{j-1})]\} \cup [Q(x_{i-1}, y_j) \cap Q(x_i, y_{j-1})]$$

where, of course, $Q(x_{i-1}, y_j) \cap Q(x_i, y_{j-1}) = Q(x_{i-1}, y_{j-1})$, which is depicted as the quadrant with the intermediate shading in the figure. Translating set operations to sums and differences, we get (8.1).   ⬜

In the example we see how to obtain a probability in terms of the CDF, for jointly continuous random variables. To find the PDF in terms of the CDF, we should consider the limit case of a rectangle with edges going to zero. Doing so yields

$$f_{X,Y}(a, b) = \frac{\partial^2 F}{\partial x\, \partial y}(a, b)$$

In the single-variable case, we can find the PDF by taking the derivative of the CDF; since double integrals are involved in the bidimensional case, it should not come as a surprise that a second-order, mixed derivative is involved here.

Having defined joint characterizations of random variables, a first question is: Can we relate the joint CDF, PMF, and PDF to analogous functions describing the single variables? In general, whatever refers to a single variable, within the context of a multivariate distribution, is called *marginal*. So, to be more specific, given the joint CDF $F_{X,Y}(x, y)$, how can we find marginal CDFs $F_X(x)$ and $F_Y(y)$ pertaining to each individual variable? In principle, the task for the CDF is fairly easy. We obtain the marginal CDFs for the two random variables as follows:

$$F_X(x) = P(X \le x) = P(X \le x, Y \le +\infty) = F_{X,Y}(x, +\infty)$$

By the same token, $F_Y(y) = F_{X,Y}(+\infty, y)$.

In the discrete case, to obtain the marginal PMFs from the joint PMF, we just use the total probability theorem[3] and the fact that events $\{Y = y_j\}$ are disjoint:

$$
\begin{aligned}
p_X(x_i) &= P(X = x_i) = P\left(\bigcup_j \{X = x_i, Y = y_j\}\right) \\
&= \sum_j P(X = x_i, Y = y_j) = \sum_j p_{X,Y}(x_i, y_j)
\end{aligned}
$$

If we want to obtain marginal PDFs from the joint PDF, we may work in much the same way as in the discrete case:

$$P(X \in A) = P(X \in A, Y \in (-\infty, +\infty)) = \int_A \int_{-\infty}^{+\infty} f_{X,Y}(x, y) dy\, dx$$

If we introduce the marginal density:

$$f_X(x) = \int_{-\infty}^{+\infty} f_{X,Y}(x, y) dy$$

we have

$$P(X \in A) = \int_A f_X(x) dx$$

The marginal PDF for Y is obtained in the same way, integrating with respect to x.

We see that, given a joint distribution, we may find the marginals. It is tempting to think that, given the two marginals, we may recover the joint distribution. This is *not* true, as the marginal distributions do not say anything about the link among random variables. In Example 7.16 we have seen that, in the context of a discrete-time stochastic process X_t, quite different processes may share the same marginal distribution for all time periods t. The following example, taking advantage of the concepts we have just learned, shows that quite different joint distributions may yield the same pair of marginals.

Example 8.2 Consider the following PDFs (in a moment we will check that they are legitimate densities):

$$
\begin{aligned}
f_{X,Y}(x, y) &= 1, & 0 \le x, y \le 1 \\
g_{X,Y}(x, y) &= 1 + (2x - 1)(2y - 1), & 0 \le x, y \le 1
\end{aligned}
$$

---

[3]See Theorem 5.7.

They look quite different, but it is not too difficult to see that they yield the same marginals. The first case is easy:

$$f_X(x) = \int_0^1 f_{X,Y}(x,y)\, dy = \int_0^1 1\, dy = 1$$

By symmetry, we immediately see that $f_Y(y) = 1$ as well. Hence, the two marginals are two uniform distributions on the unit interval $[0,1]$. Now let us tackle the second case. As we learned in Section 3.9.3, when integrating with respect to y, we should just treat x as a constant:

$$
\begin{aligned}
g_X(x) &= \int_0^1 [1 + (2x-1)(2y-1)]dy \\
&= \int_0^1 1\, dy \;+\; (2x-1)\int_0^1 (2y-1)\, dy \\
&= 1 + (2x-1)[y^2 - y]\Big|_0^1 = 1
\end{aligned}
$$

As before, it is easy to see by symmetry that $f_Y(y) = 1$ as well. Again, the two marginals are two uniform distributions, but the link between the two random variables is quite different.

Before closing the example, it is easy to see that $f_{X,Y}(x,y)$ is a legitimate density as it is never negative and

$$\int_0^1 \int_0^1 1\, dx\, dy = 1$$

Actually, this is just the area of a unit square. Checking the legitimacy of $g_{X,Y}(x,y)$ is a bit more difficult. The easy part is checking that the integral over the unit square is 1. Given the marginals above, we may write

$$\int_0^1 \int_0^1 g_{X,Y}(x,y)dy\, dx = \int_0^1 g_X(x)dx = \int_0^1 1\, dx = 1$$

But we should also check that the function is never negative. One way of doing so would be to find its minimum over the unit square. The optimization of a function of multiple variables is fully addressed in Chapter 12, but what we know from Section 3.9.1 suggests that a starting point is the stationarity conditions

$$\frac{\partial g}{\partial x} = 2(2y-1) = 0, \qquad \frac{\partial g}{\partial y} = 2(2x-1) = 0$$

These conditions imply $x = y = 0.5$ but, unfortunately, this is neither a minimum nor a maximum. We invite the reader to check that the Hessian has two eigenvalues of opposite sign. However, with a little intuition, we may see that this density involves the product of two linear terms, $2x - 1$ and $2y - 1$, and these range on the interval $[-1, 1]$, given the bounds on x and y.

Hence, their product will never be smaller than -1, and the overall PDF will never be negative (its minimum is 0, for $x = 1, y = -1$ and $x = -1, y = 1$).

□

The example points out that there is a gap between two marginals and a joint distribution. The missing link is exactly what characterizes the dependence between the two random variables. This missing link is the subject of a whole branch of probability theory, which is called *copula theory*; a copula is a function capturing the essential nature of the dependence between random variables, separating it from the marginal distributions. Copula theory is beyond the scope of this book; in this chapter and in the following, we will just rely on a simple and partial characterization of dependence between random variables, based on covariance and correlation.

8.2 INDEPENDENT RANDOM VARIABLES

In the previous section we formally introduced the concept of the joint cumulative distribution function (CDF). In the case of two random variables, X and Y, this is a function $F_{X,Y}(x, y)$ of two arguments, giving the probability of the joint event $\{X \leq x, Y \leq y\}$:

$$F_{X,Y}(x, y) = P(X \leq x, Y \leq y)$$

The joint CDF tells the whole story about how the two random variables are linked. Then, on the basis of the joint CDF, we may also define a joint probability mass function (PMF) $p_{X,Y}(x, y)$ for discrete random variables, and a joint probability density function (PDF) $f_{X,Y}(x, y)$ for continuous random variables. These concepts translate directly to the more general case of n jointly distributed random variables.

If we have the joint PMF or PDF, we may compute the usual things, such as expected values, variances, and expected values of function of the random variables. The most general statement, when two random variables X and Y are involved, concerns the expected value of a function $g(X, Y)$ of the two random variables. It can be computed as

$$E[g(X,Y)] = \begin{cases} \sum_i \sum_j g(x_i, y_j) p_{X,Y}(x_i, y_j), & \text{in the discrete case} \\[2ex] \int_{-\infty}^{+\infty} \int_{-\infty}^{+\infty} g(x, y) f_{X,Y}(x, y)\, dy\, dx & \text{in the continuous case} \end{cases}$$

To find the expected value $E[X]$, all we have to do is plugging $g(x, y) = x$ in the formula above; if we are interested in variance $\text{Var}(Y)$, then we plug $g(x, y) = (y - \mu_Y)^2$, where $\mu_Y = E[Y]$. We will not really have to compute such things in the remainder of the book, but the reader may appreciate the potential difficulty of computing multiple sums or integrals.

Apart from this computational difficulty, the very task of characterizing the full joint distribution of random variables may be difficult. It is even more difficult to infer this information based on empirical data. This is why one often settles for a limited characterization of dependence in terms of correlation. Before doing so, it is quite useful to understand how independence may simplify the analysis significantly. We will state, without any proof, a few fundamental properties that hold under the assumption of independence. All of these results are a consequence of a well-known fact: If events are independent, the probability of a joint event is just the product of the probabilities of all of the individual events. In the case of the joint CDF, this implies:

$$F_{X,Y}(x, y) = P(X \leq x) \cdot P(Y \leq y) = F_X(x) \cdot F_Y(y)$$

In other words, the joint CDF can be factored into the product of the two marginal CDFs, $F_X(x)$ and $F_Y(y)$. The same factorization applies to the joint PMFs and PDFs:

$$p_{X,Y}(x, y) = p_X(x)p_Y(y), \qquad f_{X,Y}(x, y) = f_X(x)f_Y(y)$$

It is by using this factorization that we can prove, e.g., the properties concerning the variance of a linear combination of independent random variables, that we have already used a few times and is recalled here for readers' convenience:

$$\text{Var}\left(\sum_{i=1}^{n} \lambda_i X_i\right) = \sum_{i=1}^{n} \lambda_i^2 \text{Var}(X_i)$$

for random variables X_i and coefficients λ_i, $i = 1, \ldots, n$. Later, in Section 8.3.2, we discuss the general case where independence is not assumed. In passing, we may also state the following theorem.

THEOREM 8.2 *Consider a function of random variables X and Y and assume that it can be factorized as the product of two terms $g(X)h(Y)$. If the two random variables are independent, then $E[g(X)h(Y)] = E[g(X)] \cdot E[h(Y)]$.*

In particular, it is important to notice that the expected value of a sum is always the sum of the expected value, but this commutation cannot be applied to products in general:

$$E[XY] \neq E[X]E[Y] \tag{8.4}$$

However, equality holds if X and Y are independent.

8.3 COVARIANCE AND CORRELATION

If two random variables are not independent, it is natural to investigate their degree of dependence, which means finding a way to measure it and to take advantage of it. The second task leads to statistical modeling, which we will investigate later in the simplest case of linear regression. The first task is not as easy as it may seem; the joint density would tell us the whole story, but it is difficult to manage. More importantly, it is difficult to estimate from empirical data. We would like to come up with a limited set of summary measures that are easy to estimate, but fully capturing dependence with a single number is a tricky issue, as we will learn shortly.

A clue on what we could do to measure dependence can be obtained by checking Table 8.1 again. We observed that in most joint realizations of the two random variables, both values tend to be either larger or smaller than the respective averages. This leads to the definition of covariance.

DEFINITION 8.3 (Covariance) *The covariance between random variables X and Y is defined as*

$$\text{Cov}(X, Y) \equiv \text{E}[(X - \text{E}[X])(Y - \text{E}[Y])]$$

Quite often, the covariance between two random variables is denoted by σ_{XY}.

Covariance is the expected value of the product of two deviations from the mean, and its sign depends on the signs of the two factors. We have positive covariance when the events $\{X > \text{E}[X]\}$ and $\{Y > \text{E}[Y]\}$ tend to occur together, as well as the events $\{X < \text{E}[X]\}$ and $\{Y < \text{E}[Y]\}$, because the signs of the two factors in the product tend to be the same. If the signs tend to be different, we have a negative covariance.

Example 8.3 If two products are complements, it is natural to expect positive covariance between their demands; negative covariance can be expected if they are substitutes. Similarly, if we observe over time the demand for an item whose long- or midterm consumption is steady, a day of high demand should typically be followed by a day with low demand. As a concrete example, consider the weekly demand for diapers after a week of intense promotional sales. □

From a computational perspective, it is very handy to express covariance as follows:

$$
\begin{aligned}
\text{Cov}(X, Y) &\equiv \text{E}\Big[(X - \mu_X) \cdot (Y - \mu_Y)\Big] \\
&= \text{E}\Big[XY - \mu_X Y - X\mu_Y + \mu_X \mu_Y\Big] \\
&= \text{E}[XY] - \mu_X \mu_Y \tag{8.5}
\end{aligned}
$$

We easily see that if two variables are independent, then their covariance is zero, since independence implies $\text{E}[XY] = \text{E}[X] \cdot \text{E}[Y]$, courtesy of Theorem 8.2. However, the converse is *not* true in general, as we may see from the following counterexample.

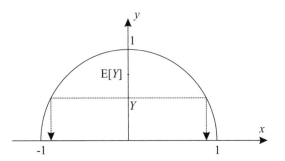

Fig. 8.2   A counterexample about covariance.

Example 8.4 (Two dependent random variables may have zero co-variance) Let us consider a uniform random variable on the interval $[-1, 1]$; its expected value is zero and on its support the density function is constant and given by $f_X(x) = \frac{1}{2}$. Now, define random variable Y as

$$Y = \sqrt{1 - X^2}.$$

Clearly, there is a very strong interdependence between X and Y because, given the realization of X, Y is perfectly predictable. However, their covariance is zero! We have seen that

$$\text{Cov}(X, Y) = \text{E}[XY] - \text{E}[X]\text{E}[Y]$$

but $\text{E}[X] = 0$ and

$$\text{E}[XY] = \int_{-1}^{1} x\sqrt{1 - x^2} \cdot \tfrac{1}{2} \, dx = 0$$

because of the symmetry of the integrand function, which is an odd function, in the sense that $f(-x) = -f(x)$. One intuitive way to explain the weird finding of this example is the following. First note that points with coordinates (X, Y) lie on the upper half of the unit circumference $X^2 + Y^2 = 1$. But if $Y < \text{E}[Y]$, we may have either $X > \text{E}[X]$ or $X < \text{E}[X]$. This is illustrated in Fig. 8.2. A similar consideration applies when $Y > \text{E}[Y]$.   ▯

The example shows that covariance is not really a perfect measure of dependence, as it may be zero in cases in which there is a very strong dependence. In fact, covariance is rather a measure of *concordance* between a pair of random variables. In other words, covariance measures a *linear* association between random variables. A strong nonlinear link, as the one in Fig. 8.2 may not be detected at all, or only partially. This point will be much clearer when we deal with simple linear regression in Chapter 10.

8.3.1   A few properties of covariance

The essential properties of covariance are the following:

Property 1. $\text{Cov}(X, X) = \text{Var}(X)$. This property shows that covariance is a generalization of variance and explains its name. Using shorthand notation, $\sigma_{XX} = \sigma_X^2$.

Property 2. $\text{Cov}(X, Y) = \text{Cov}(Y, X)$. This property points out an important issue: Covariance is a measure of association, but it has nothing to do with cause–effect relationships. This is an important point to keep in mind when building statistical models based on empirical data. Causality is not necessarily proved by statistical techniques which just exploit associations.

Property 3. $\text{Cov}(\alpha X, Y) = \alpha \, \text{Cov}(X, Y)$, where α is any number. This property states that numbers can be "taken outside" covariance. It is instructive to note that, applying this property, we obtain

$$
\begin{aligned}
\text{Var}(\alpha X) &= \text{Cov}(\alpha X, \alpha X) = \alpha \, \text{Cov}(X, \alpha X) \\
&= \alpha^2 \text{Cov}(X, X) = \alpha^2 \text{Var}(X)
\end{aligned}
$$

as expected.

Property 4. $\text{Cov}(X, Y + Z) = \text{Cov}(X, Y) + \text{Cov}(X, Z)$. This is a sort of "distributive" property that comes handy when dealing with sums of random variables. While the first three properties are trivial to prove, it is instructive to prove this last one using (8.5) and linearity of expectation:

$$
\begin{aligned}
\text{Cov}(X, Y + Z) &= \text{E}[X(Y + Z)] - \text{E}[X]\text{E}[Y + Z] \\
&= \text{E}[XY] + \text{E}[XZ] - \text{E}[X]\text{E}[Y] - \text{E}[X]\text{E}[Z] \\
&= \text{Cov}(X, Y) + \text{Cov}(X, Z)
\end{aligned}
$$

8.3.2   Sums of random variables

In Section 7.7 we dealt with sums of random variables, under the restrictive assumption of independence. Finally, armed with covariance, we may tackle the general case.

THEOREM 8.4 (Variance of the sum/difference of two random variables) *Given two random variables X and Y, the variance of their sum and difference is*

$$
\begin{aligned}
\text{Var}(X + Y) &= \text{Var}(X) + \text{Var}(Y) + 2\,\text{Cov}(X, Y) \\
\text{Var}(X - Y) &= \text{Var}(X) + \text{Var}(Y) - 2\,\text{Cov}(X, Y)
\end{aligned}
$$

respectively.

This theorem is somewhat reassuring; if we take the difference of two random variables, rather than the sum, the minus does count after all, but it plays a role depending on covariance. It is instructive to consider a pair of alternative proofs of the first relationship (the second one is easily obtained from the first by using property 3 of covariance). One possibility is using the definition of variance directly:

$$\begin{aligned}
\mathrm{Var}(X+Y) &\equiv \mathrm{E}\left[(X+Y-\mu_Y-\mu_Y)^2\right] \\
&= \mathrm{E}\left[(X-\mu_X)^2+(Y-\mu_Y)^2+2(X-\mu_X)(Y-\mu_Y)\right] \\
&= \mathrm{Var}(X)+\mathrm{Var}(Y)+2\,\mathrm{Cov}(X,Y)
\end{aligned}$$

Another possibility is to take advantage of property 4 of covariance:

$$\begin{aligned}
\mathrm{Var}(X+Y) &= \mathrm{Cov}(X+Y,X+Y) \\
&= \mathrm{Cov}(X,X)+\mathrm{Cov}(X,Y)+\mathrm{Cov}(Y,X)+\mathrm{Cov}(Y,Y) \\
&= \mathrm{Var}(X)+\mathrm{Var}(Y)+2\,\mathrm{Cov}(X,Y)
\end{aligned}$$

Using the second proof technique, it is fairly easy to come up with the following statement.

THEOREM 8.5 (Variance of the sum of random variables) *Given a collection of random variables X_i, $i = 1, \ldots, n$, we obtain*

$$\mathrm{Var}\left(\sum_{i=1}^{n} X_i\right) = \sum_{i=1}^{n} \mathrm{Var}(X_i) + 2\sum_{i=1}^{n}\sum_{j<i}\mathrm{Cov}(X_i, X_j)$$

We see that, in the case of mutually independent variables, covariances will be zero, and the variance of the sum does boil down to the sum of variances. A very interesting application of the theorem arises in the area of financial portfolio management.

Example 8.5 (A classical model of risk in financial portfolio management) Consider the task of an investor who must allocate her wealth to two risky financial assets. The returns of the two assets are modeled by random variables R_1 and R_2, respectively. Assume that the investor knows the two expected returns, μ_1 and μ_2; the two variances of the return, σ_1^2 and σ_2^2; and the covariance, σ_{12}. What the investor has to decide is the fraction of wealth that she should allocate to each asset; let us denote the weights of the two assets in the portfolio by w_1 and w_2, respectively. The two weights must add up to 1

$$w_1 + w_2 = 1$$

and they cannot be negative if short-selling is ruled out (see Example 1.3). Hence, the return of the portfolio is a weighted sum of random variables:

$$R_p = w_1 R_1 + w_2 R_2$$

There are numerous issues involved in portfolio decisions, but a preliminary requirement is to figure out the expected return of the portfolio and some measure of risk, in order to trade them off. It is easy to see that the expected return is

$$\mu_p = \mathrm{E}\left[w_1 R_1 + w_2 R_2\right] = w_1 \mu_1 + w_2 \mu_2$$

One possible measure of risk is standard deviation of portfolio return. We will see a few shortcomings of this risk measure later, but it is certainly a useful input for the decision of our investor. To find standard deviation, one has to find variance first. This is a bit trickier than the expected value:

$$\sigma_p^2 = \mathrm{Var}\left(w_1 R_1 + w_2 R_2\right) = w_1^2 \sigma_1^2 + 2 w_1 w_2 \sigma_{12} + w_2^2 \sigma_2^2$$

Taking a careful look at variance, as a function of portfolio weights w_1 and w_2, should ring a bell: This is a quadratic form, a concept that we introduced in Section 3.8. We also know that quadratic forms can be expressed in a very compact way by using vectors and matrices. So, let us group portfolio weights into the following vector:

$$\mathbf{w} = \left[\begin{array}{c} w_1 \\ w_2 \end{array}\right]$$

Let us also group variances and covariances into the following matrix:

$$\boldsymbol{\Sigma} = \left[\begin{array}{cc} \sigma_{11} & \sigma_{12} \\ \sigma_{21} & \sigma_{22} \end{array}\right]$$

where $\sigma_{ii} = \sigma_i^2$, $i = 1, 2$. Matrix $\boldsymbol{\Sigma}$ is called the *covariance matrix*. Given property 2 of covariance, $\sigma_{ij} = \sigma_{ji}$; so $\boldsymbol{\Sigma}$ is a symmetric matrix. Now it is easy to check that variance of portfolio return can be expressed as

$$\sigma_p^2 = \mathbf{w}^T \boldsymbol{\Sigma} \mathbf{w}$$

$\square$

The finding of the example above can be easily generalized to a collection of n random variables grouped into the following vector:

$$\mathbf{X} = \left[\begin{array}{c} X_1 \\ X_2 \\ \vdots \\ X_n \end{array}\right]$$

Let us also introduce the vector of expected values and the covariance matrix

$$\boldsymbol{\mu} = \left[\begin{array}{c} \mu_1 \\ \mu_2 \\ \vdots \\ \mu_n \end{array}\right], \quad \boldsymbol{\Sigma} = \left[\begin{array}{ccccc} \sigma_{11} & \sigma_{12} & \sigma_{13} & \cdots & \sigma_{1n} \\ \sigma_{21} & \sigma_{22} & \sigma_{23} & \cdots & \sigma_{2n} \\ \sigma_{31} & \sigma_{32} & \sigma_{33} & \cdots & \sigma_{3n} \\ \vdots & \vdots & \vdots & \ddots & \vdots \\ \sigma_{n1} & \sigma_{n2} & \sigma_{n3} & \cdots & \sigma_{nn} \end{array}\right]$$

where $\sigma_{ii} = \sigma_i^2$, $i = 1, \ldots, n$. If we form a linear combination

$$Z = \sum_{i=1}^{n} \alpha_i X_i = \boldsymbol{\alpha}^T \mathbf{X}$$

then we have

$$\mathrm{E}[Z] \; = \; \sum_{i=1}^{n} \alpha_i \mu_i \; = \; \boldsymbol{\alpha}^T \boldsymbol{\mu} \tag{8.6}$$

$$\mathrm{Var}(Z) \; = \; \sum_{i,j=1}^{n} \alpha_i \alpha_j \sigma_{ij} \; = \; \boldsymbol{\alpha}^T \boldsymbol{\Sigma} \boldsymbol{\alpha} \tag{8.7}$$

Since we know that variance is nonnegative, we may immediately deduce the following property of the covariance matrix.

PROPERTY 8.6 *The covariance matrix $\boldsymbol{\Sigma}$ is a symmetric, positive semidefinite matrix.*

This property has several implications. We may be interested in minimizing risk as measured by variance, with respect to a vector of decision variables $\mathbf{x}$, subject to constraints. The objective function is the quadratic form $\mathbf{x}^T \boldsymbol{\Sigma} \mathbf{x}$, which, by the property above, is a convex function.[4] Furthermore, we know that a symmetric matrix has orthogonal eigenvectors.[5] This can be exploited in data reduction methods such as principal component analysis, which is the subject of Chapter 17.

8.3.3   The correlation coefficient

The covariance is a generalization of variance. Hence, it is not surprising that it shares a relevant shortcoming: Its value depends on the unit of measurement of the underlying quantities. We recall that it is impossible to say whether a variance of 10,000 is large or small; a similar consideration applies to standard deviation, which at least is measured in the same units as expected values, so that we may consider the coefficient of variation $C_X = \sigma_X / \mu_X$. We may try a normalization as well, in order to define an adimensional version of covariance.

DEFINITION 8.7 (Correlation coefficient) *The correlation coefficient between random variables X and Y is defined as*

$$\rho_{XY} \equiv \frac{\mathrm{Cov}(X,Y)}{\sqrt{\mathrm{Var}(X)}\sqrt{\mathrm{Var}(Y)}} = \frac{\sigma_{XY}}{\sigma_X \sigma_Y}$$

[4]See Section 3.8.
[5]See Theorem 3.10.

The coefficient of correlation is adimensional, and it can be easily interpreted on the basis of the following theorem.

THEOREM 8.8 *The correlation coefficient ρ_{XY} takes values in the range $[-1, 1]$. If $\rho_{XY} = \pm 1$, then X and Y are related by $Y = a + bX$, where the sign of b is the sign of the correlation coefficient.*

PROOF Consider the following linear combination of X and Y:

$$Z = \frac{X}{\sigma_X} + \frac{Y}{\sigma_Y}$$

We know that variance cannot be negative; hence

$$\text{Var}\left(\frac{X}{\sigma_X} + \frac{Y}{\sigma_Y}\right) = \text{Var}\left(\frac{X}{\sigma_X}\right) + \text{Var}\left(\frac{Y}{\sigma_Y}\right) + 2\text{Cov}\left(\frac{X}{\sigma_X}, \frac{Y}{\sigma_Y}\right)$$
$$= 1 + 1 + 2\rho_{X,Y} \geq 0$$

where σ_X and σ_Y are the standard deviations of X and Y, respectively. This inequality immediately yields $\rho_{X,Y} \geq -1$. By the same token, consider a slightly different linear combination:

$$\text{Var}\left(\frac{X}{\sigma_X} - \frac{Y}{\sigma_Y}\right) = 1 + 1 - 2\rho_{X,Y} \geq 0 \quad \Rightarrow \quad \rho_{X,Y} \leq 1$$

We also know that if $\text{Var}(Z) = 0$, then Z must be a constant. In the first case, variance will be zero if $\rho = -1$. Then, we may write

$$Z = \frac{X}{\sigma_X} + \frac{Y}{\sigma_Y} = \alpha$$

for some constant α. Rearranging the equality, we have

$$Y = -X\frac{\sigma_Y}{\sigma_X} + \alpha\sigma_Y$$

This can be rewritten as $Y = a + bX$, and since standard deviations are non negative, we see that the slope is negative. Considering the second linear combination yields a similar relationship for the case $\rho = 1$, in which the slope b is positive. ∎

Given the theorem, it is fairly easy to interpret a specific value of correlation:

- A value close to 1 shows a strong degree of positive correlation.

- A value close to -1 shows a strong degree of negative correlation.

- If correlation is zero, we speak of *uncorrelated variables*.

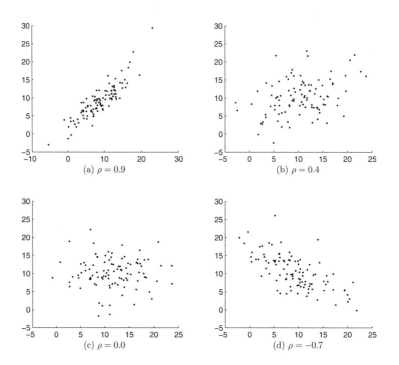

Fig. 8.3 Samples of jointly normal variables for different values of the correlation coefficient ρ.

A visual illustration of correlation is given in Fig. 8.3. Each scatterplot shows a sample of 100 joint observations from a joint normal with $\mu_1 = \mu_2 = 10$, $\sigma_1 = \sigma_2 = 5$, and different values of correlation. The effect of correlation is quite evident if we think to "draw a line" going through each cloud; the slope of the line corresponds to the sign of the correlation. In the limit case of $\rho = \pm 1$, the observations would exactly lie on a line. We stress again that uncorrelated variables need not be independent. A notable case, in which lack of correlation implies independence, is the multivariable normal distribution, which we cover next.

8.4 JOINTLY NORMAL VARIABLES

A detailed coverage of multivariate distributions is beyond the scope of the book, but we should at least consider a generalization of normal distribution. A univariate normal distribution is characterized by its expected value μ and by its variance σ^2. In the multivariate case, we have a vector of expected values $\boldsymbol{\mu}$ and a covariance matrix $\boldsymbol{\Sigma}$. We consider a random vector taking values in $\mathbb{R}^n$:

$$\mathbf{X} = \begin{bmatrix} X_1 \\ X_2 \\ \vdots \\ X_n \end{bmatrix}$$

We say that $\mathbf{X}$ has *jointly normal* or *multivariate normal* distribution if its joint density is given by

$$f_{\mathbf{X}}(\mathbf{x}) = \frac{1}{\left(\sqrt{2\pi}\right)^n |\boldsymbol{\Sigma}|^{1/2}} \exp\left\{-\frac{1}{2}(\mathbf{X}-\boldsymbol{\mu})^T \boldsymbol{\Sigma}^{-1}(\mathbf{X}-\boldsymbol{\mu})\right\} \tag{8.8}$$

where $\exp(\cdot)$ is the exponential function and $|\boldsymbol{\Sigma}|$ is the determinant of the covariance matrix. This expression may look a bit intimidating, but it is easy to see that, for $n = 1$, it boils down to the familiar density of a univariate normal. The notation $\mathbf{X} \sim \mathcal{N}(\boldsymbol{\mu}, \boldsymbol{\Sigma})$ is used to refer to a multivariate normal variable.

To get a better feeling for the multivariate normal density, it may be instructive to write it down more explicitly for a bivariate case. We have

$$\boldsymbol{\mu} = \begin{bmatrix} \mu_1 \\ \mu_2 \end{bmatrix}, \qquad \boldsymbol{\Sigma} = \begin{bmatrix} \sigma_2^1 & \rho\sigma_1\sigma_2 \\ \rho\sigma_1\sigma_2 & \sigma_2^2 \end{bmatrix}$$

Let us write the determinant explicitly:

$$|\boldsymbol{\Sigma}| = \begin{vmatrix} \sigma_1^2 & \rho\sigma_1\sigma_2 \\ \rho\sigma_1\sigma_2 & \sigma_2^2 \end{vmatrix} = \sigma_1^2\sigma_2^2 - \rho^2\sigma_1^2\sigma_2^2 = \sigma_1^2\sigma_2^2(1-\rho^2)$$

The inverse of the covariance matrix is[6]

$$\boldsymbol{\Sigma}^{-1} = \frac{1}{\sigma_1^2\sigma_2^2(1-\rho^2)} \begin{bmatrix} \sigma_2^2 & -\rho\sigma_1\sigma_2 \\ -\rho\sigma_1\sigma_2 & \sigma_1^2 \end{bmatrix}$$

and a few calculations yield

$$f_{\mathbf{X}}(x_1, x_2) = \frac{1}{2\pi\sigma_1\sigma_2\sqrt{1-\rho^2}} \exp\left\{-\frac{z_1^2 - 2\rho z_1 z_2 + z_2^2}{2(1-\rho^2)}\right\}$$

where

$$z_1 = \frac{x_1 - \mu_1}{\sigma_1}, \qquad z_2 = \frac{x_2 - \mu_2}{\sigma_2}$$

Clearly, z_1 and z_2 represent standardizations of variables x_1 and x_2.

What is the shape of this density function? If the two variables were uncorrelated, i.e., if $\rho = 0$, the level curves of the density function would

---

[6]See the rule described in Example 3.12.

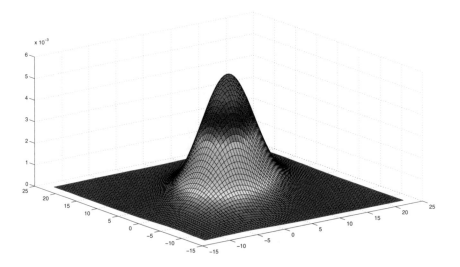

Fig. 8.4 Surface plot of a joint normal PDF.

just be concentric circles in terms of the standardized variables. In terms of the original variables, we would have a set of concentric ellipses, with a horizontal axis and a vertical axis. The effect of correlation is to rotate the ellipses. Figure 8.4 shows a surface plot of the density function of a bivariate normal with $\mu_1 = \mu_2 = 5$, $\sigma_1 = \sigma_2 = 6$, and $\rho = 0.6$. We see the familiar bell shape, but it is the contour plot of Fig. 8.5 that illustrates the effect of positive correlation.

Let us check what happens in the density above if we set $\rho = 0$, destandardizing z_i to get a clearer picture:

$$
\begin{aligned}
f_{\mathbf{X}}(x_1, x_2) &= \frac{1}{2\pi\sigma_1\sigma_2} \exp\left\{-\frac{z_1^2 + z_2^2}{2}\right\} \\
&= \frac{1}{\sqrt{2\pi}\sigma_1} \exp\left\{-\frac{(x_1 - \mu_1)^2}{2\sigma_1^2}\right\} \cdot \frac{1}{\sqrt{2\pi}\sigma_2} \exp\left\{-\frac{(x_2 - \mu_2)^2}{2\sigma_2^2}\right\} \\
&= f_{X_1}(x_1) f_{X_2}(x_2)
\end{aligned}
$$

We see that the joint density can be factored into the product of two marginal densities, which are themselves normal. Indeed, the following theorem holds in general.

THEOREM 8.9 *If* $\mathbf{X}$ *is a vector of jointly normal random variables and if they are pairwise uncorrelated, then they are independent.*

The theorem states that for jointly normal variables, lack of correlation implies independence. We noted that independence implies lack of correlation, but

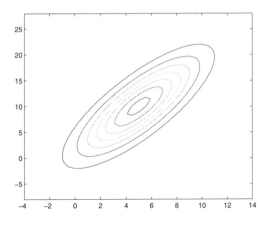

Fig. 8.5   Contour plot of a joint normal PDF.

the converse is not true in general (see Example 8.4). The multivariate normal is a significant exception.

8.5   CONDITIONAL EXPECTATION

We are already familiar with the concept of conditional probability when events are involved. When dealing with random variables X and Y, we might wonder whether knowing something about Y, possibly even its realized value, can help us in predicting the value of X. To introduce the concepts in the simplest way, it is a good idea to work with a pair of discrete random variables with discrete support. So, let us consider a variable X that can take values x_i, $i = 1, \ldots, k$, and a variable Y that can take values y_j, $j = 1, \ldots, l$. Given the joint PMF, we know all of the relevant probabilities

$$p_{XY}(x, y) = P(X = x_i, Y = y_j)$$

and we may also consider conditional probabilities, such as

$$P(X = x_i \mid Y = y_j) = \frac{P(X = x_i, Y = y_j)}{P(Y = y_j)}$$

assuming, of course, that $P(Y = y_j) \neq 0$. Generalizing a bit, let us define the conditional PMF:

$$p_{X|Y}(x, y) \equiv P(X = x \mid Y = y) = \frac{p_{X,Y}(x, y)}{p_Y(y)}$$

It is essential to note that, if the two random variables are independent, then

$$p_{X|Y}(x,y) = \frac{p_{X,Y}(x,y)}{p_Y(y)} = \frac{p_X(x)p_Y(y)}{p_Y(y)} = p_X(x)$$

i.e., knowledge of Y is no use in predicting X. If the two variables are not independent, one natural question concerns the expected value of X if we know that $Y = y_j$. Such a conditional expectation is obtained as follows:

$$E[X \mid Y = y_j] = \sum_{i=1}^{k} x_i P(X = x_i \mid Y = y_j) = \sum_{i=1}^{k} x_i p_{X|Y}(x_i|y_j)$$

Example 8.6 Let X and Y be two binary random variables whose joint distribution is characterized by the PMF:

$$p_{X,Y}(0,0) = 0.1, \quad p_{X,Y}(0,1) = 0.3, \quad p_{X,Y}(1,0) = 0.4, \quad p_{X,Y}(1,1) = 0.2$$

Let us find the distribution of X conditional on $Y = 0$ or $Y = 1$. The first step is computing the marginal PMF of Y:

$$p_Y(0) = p_{X,Y}(0,0) + p_{X,Y}(1,0) = 0.1 + 0.4 = 0.5$$
$$p_Y(1) = p_{X,Y}(0,1) + p_{X,Y}(1,1) = 0.3 + 0.2 = 0.5$$

Then we find $p_{X|Y}(x,0)$ first:

$$P(X = 0 \mid Y = 0) = \frac{p_{X,Y}(0,0)}{p_Y(0)} = \frac{1}{5}$$

$$P(X = 1 \mid Y = 0) = \frac{p_{X,Y}(1,0)}{p_Y(0)} = \frac{4}{5}$$

By the same token:

$$P(X = 0 \mid Y = 1) = \frac{p_{X,Y}(0,1)}{p_Y(1)} = \frac{3}{5}$$

$$P(X = 1 \mid Y = 1) = \frac{p_{X,Y}(1,1)}{p_Y(1)} = \frac{2}{5}$$

Now we may compute the conditional expected values:

$$
\begin{aligned}
E[X \mid Y = 0] &= 0 \times P(X = 0 \mid Y = 0) + 1 \times P(X = 1 \mid Y = 0) \\
&= 0 \times \tfrac{1}{5} + 1 \times \tfrac{4}{5} = 0.8 \\
E[X \mid Y = 1] &= 0 \times P(X = 0 \mid Y = 1) + 1 \times P(X = 1 \mid Y = 1) \\
&= 0 \times \tfrac{3}{5} + 1 \times \tfrac{2}{5} = 0.4
\end{aligned}
$$

Incidentally, the *unconditional* expected value of X is

$$E[X] = 0 \times 0.1 + 0 \times 0.3 + 1 \times 0.4 + 1 \times 0.2 = 0.6$$

We see that knowledge of Y does change our expectation about X. The two random variables are not independent.   □

The case of two jointly continuous random variables is conceptually similar, and it goes through the definition of the following PDF:

$$f_{X|Y}(x, y) = \frac{f_{X,Y}(x, y)}{f_Y(y)}$$

for y such that $f_Y(y) \neq 0$. It is no surprise that we cannot divide by a probability $P(Y = y)$, as this is identically zero, but the concept is quite similar to the discrete case.

Conditioning is a useful concept that can be exploited, among other things:

1. To simplify the calculations of expectations

2. To characterize properties of some probability distributions

3. To characterize properties of certain stochastic processes

We illustrate these points in the following sections.

8.5.1   Computing expectations by conditioning

In this section we take advantage of a fundamental theorem concerning *iterated expectation*. Before formalizing the idea, let us illustrate it by a simple example.

Example 8.7 You are lost in an underground mine and stand in front of two tunnels. One of the two tunnels will lead you to the surface after a 5-hour walk; the other one will just lead back where you are now, a 2-hour walk. What is the expected value of the time it will take to resurface?

This depends on how smart you are. You do not know which is the right tunnel, so you have a 50% percent chance of selecting the wrong one; yet, the least you can do is to not repeat the mistake, by marking the tunnel you select first. In this case, the calculation is fairly simple:

$$E[X] = 5 \times P(\mathsf{OK}) + (2 + 5) \times P(\mathsf{NOK}) = 6$$

Here X is the random time to get to the surface, OK denotes the event that you take the right tunnel, and NOK corresponds to the wrong one.

A more interesting case occurs if you are memoryless, i.e., you do not remember the choice you made if you get back to square 1, Apparently, this is a very complicated case, as there is the possibility of infinite cycling. Using conditioning, the calculation is quite easy. The key point is that if you take the right tunnel, after 5 hours you are done. If you take the wrong one, after 2 hours you are get back to the starting point, and the time to surface is random; however, its expectation is exactly the same as in the first trial, due to lack of memory. Formally

$$
\begin{aligned}
\mathrm{E}[X] &= \mathrm{E}[X \mid \mathsf{OK}]\mathrm{P}(\mathsf{OK}) + \mathrm{E}[X \mid \mathsf{NOK}]\mathrm{P}(\mathsf{NOK}) \\
&= 5 \times 0.5 + (2 + \mathrm{E}[X]) \times 0.5
\end{aligned}
$$

Solving for $\mathrm{E}[X]$, we find

$$
\mathrm{E}[X] = \frac{5 \times 0.5 + 2 \times 0.5}{0.5} = 7
$$

Not surprisingly, it should take more time to resurface if you are memoryless.

$\Box$

The example suggests that, when computing expectations, we may use an analog to the theorem of total probabilities.

THEOREM 8.10 (Law of iterated expectations) *The expected value* $\mathrm{E}[X]$ *can be expressed in terms of the conditional expectation* $\mathrm{E}[X \mid Y]$ *as*

$$
\mathrm{E}[X] = \mathrm{E}\left[\mathrm{E}[X \mid Y]\right]
$$

We should note that the outermost expectation is with respect to Y, *as* $\mathrm{E}[X \mid Y]$ *is a function of the random variable* Y.

PROOF   We prove the result only in the case of two discrete random variables X and Y with finite support; X may take values x_i, $i = 1, \ldots, k$, and Y may take values y_j, $j = 1, \ldots, l$. Using the total probability theorem and the fact that the events $\{Y = y_j\}$ are disjoint, we may write

$$
\mathrm{P}(X = x_i) = \sum_{j=1}^{l} \mathrm{P}(X = x_i, Y = y_j) = \sum_{j=1}^{l} \mathrm{P}(X = x_i \mid Y = y_j)\mathrm{P}(Y = y_j)
$$

Then we may rewrite $\mathrm{E}[X]$ as follows:

$$
\begin{aligned}
\mathrm{E}[X] &\equiv \sum_{i=1}^{k} x_i \mathrm{P}(X = x_i) = \sum_{i=1}^{k} x_i \left\{ \sum_{j=1}^{l} \mathrm{P}(X = x_i \mid Y = y_j)\mathrm{P}(Y = y_j) \right\} \\
&= \sum_{i=1}^{k} \sum_{j=1}^{l} x_i \mathrm{P}(X = x_i \mid Y = y_j)\mathrm{P}(Y = y_j) \\
&= \sum_{j=1}^{l} \sum_{i=1}^{k} x_i \mathrm{P}(X = x_i \mid Y = y_j)\mathrm{P}(Y = y_j) \\
&= \sum_{j=1}^{l} \mathrm{P}(Y = y_j) \sum_{i=k}^{l} x_i \mathrm{P}(X = x_i \mid Y = y_j) \\
&= \sum_{j=1}^{l} \mathrm{P}(Y = y_j)\mathrm{E}[X \mid Y = y_j] \\
&= \mathrm{E}[\mathrm{E}[X \mid Y]]
\end{aligned}
$$

In the proof, we have used the possibility of swapping sums, which is an obvious consequence of commutative property of addition for finite sums; with infinite sums, some more caution should be used. ▌

Example 8.8 In Section 6.5.4 we considered the geometric distribution with parameter p and we have claimed that its expected value is

$$E[X] = \frac{1}{p}$$

A very straightforward way to obtain this result exploits conditioning on the outcome of the first trial (the reader should recall the physical motivation of this distribution). If the first trial is a success, which occurs with probability p, we have $X = 1$ because we have just attained our success and we stop the sequence of trials immediately. Otherwise, we have already failed once, and we must try again. However, since experiments are independent, we are just back to square 1, and the expected number of trials to go is the same as before. Formally

$$E[X] = E[X \mid \mathsf{OK}] \cdot P(\mathsf{OK}) + E[X \mid \mathsf{NOK}] \cdot P(\mathsf{NOK}) = 1 \cdot p + (1 + E[X])(1 - p)$$

from which we immediately get $E[X] = 1/p$, which confirms our previous result. The real bonus, though, comes when computing variance. As a preliminary step, we have

$$
\begin{aligned}
E[X^2] &= E[X^2 \mid \mathsf{OK}] \cdot P(\mathsf{OK}) + E[X^2 \mid \mathsf{NOK}] \cdot P(\mathsf{NOK}) \\
&= 1^2 \cdot p + E[(1 + X)^2](1 - p) \\
&= p + (1 + 2E[X] + E[X^2])(1 - p) \\
&= p + \left(1 + \frac{2}{p}\right)(1 - p) + E[X^2](1 - p)
\end{aligned}
$$

which yields

$$E[X^2] = \frac{2 - p}{p^2}$$

Then we immediately obtain

$$Var(X) = E\left[X^2\right] - E^2[X] = \frac{2 - p}{p^2} - \frac{1}{p^2} = \frac{1 - p}{p^2}$$

▯

8.5.2 The memoryless property of the exponential distribution

We have introduced the exponential distribution in Section 7.6.3, where we also pointed out its link with the Poisson distribution and the Poisson process. The standard use of exponential variables to model random time between events relies on its memoryless property, which we are now able to appreciate.

Consider an exponential random variable X with parameter λ, and say that X models the random life of some equipment (or a lightbulb) whose average life is $1/\lambda$. What is the probability that the random life of this device will exceed a threshold t? Given the CDF of the exponential distribution,[7] we see that

$$P(X > t) = e^{-\lambda t}$$

This makes sense, as this probability goes to zero when t increases, with a speed that is high when expected life is short. Now suppose that, after lighting the bulb, we notice that it is alive and kicking at time t; we could wonder what its expected residual life is, given this information. In general, after a long timespan of work, the death of a piece of equipment gets closer and closer.[8] To formalize the problem, we should consider the conditional probability that the overall life of the lightbulb is larger than $t + s$:

$$
\begin{aligned}
P(X > t + s \mid X > t) &= \frac{P\left(\{X > t+s\} \cap \{X > t\}\right)}{P(X > t)} \\
&= \frac{P(X > t + s)}{P(X > t)} \\
&= \frac{e^{-\lambda(t+s)}}{e^{-\lambda t}} \\
&= e^{-\lambda s} \\
&= P(X > s)
\end{aligned}
$$

We see a rather surprising result: The elapsed time t does not influence the residual life s and, after a timespan of length t, the lightbulb is statistically identical to a brand-new one; hence, it is "memoryless."

The memoryless character of the exponential distribution can be a good reason to use or not to use it to model random phenomena. For instance, it is suitable for modeling certain "purely random" phenomena, but not situations such as failures due to wear. When its use is warranted, simple and manageable results are often obtained. One well-known property is the PASTA property, which stands for *Poisson arrivals see time averages*. In plain words, and cutting a few corners, if observers join a system according to a Poisson process, what they observe is the system as if it were in steady state. A full appreciation of this important property requires technical machinery that is beyond the scope of the book, but a simple example will shed some light on its relevance.

Example 8.9 (The confectioner's shop puzzle) A mature lady loves confectionery. Every now and then, she feels the urgent need for a few pastries.

---

[7]See Eq. (7.13).
[8]Even more so in case of early burnout, which is typical of former engineering students like the author.

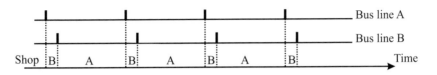

Fig. 8.6 Bus arrival schedule for Example 8.9.

Lucky her, just in front of her home there is a stop, where bus lines A and B stop. Both will get her to a confectioner's shop, which will call shop A and shop B. The lady is quite unpredictable: Her pastry frenzy can burst out at any time in the day. So, let us say that the time at which she arrives at the bus stop is uniformly distributed over the working hours of a day. What is the probability that she will satisfy her need at shop A, rather than shop B? The easy answer is that it depends on the arrival frequencies of the two bus lines. If one frequency is much higher than the other one, this should have an effect. Let us assume that the arrivals of the two buses are quite regular, and that the frequencies are the same. To fix ideas, let us say that the time between bus arrivals is exactly 10 minutes for each bus line.

Please: Think a bit before reading further and give your answer!

Since the two frequencies are the same, it is very tempting to say that the distribution should be 50–50, i.e., she could end up to either shop with the same probability. Now, I say that the actual probabilities are quite different. She reaches shop A with probability 90%, and shop B with probability 10%. How could you explain the mystery?

Please: Think a bit before reading further and give your answer!

While it is very tempting to assume a 50–50 distribution, we are neglecting the very peculiar nature of deterministic arrivals. On each line, a bus arrives exactly every 10 minutes, but we should consider how the arrivals are phased in time. Consider the time schedule illustrated in Fig. 8.6. The shift between the arrival times of the two bus lines is such that it is much more likely to get to shop A. To see this, imagine that the lady arrives just after the first bus of line A stopped; them she will catch the line B bus and reach shop B. If she arrives just after the first bus of line B stopped, she will reach shop A. The picture suggests that this second event is much more likely due to the odd shift between bus arrivals. However, if the bus arrival processes were Poisson processes with same rate, i.e., if the time between arrivals were two independent exponential variables with the same expected value, the probabilities would be 50–50. In fact, when the lady arrives, the time elapsed from the last arrival on each line would provide her with no information about which bus will arrive first. ⬚

The example is a bit pathological but quite instructive: The interaction between random and deterministic phenomena may be less trivial than one might expect.

8.5.3   Markov processes

In Section 7.9 we introduced stochastic processes as sequences of random variables; assuming a discrete-time stochastic process, we have a sequence of the form X_t, $t = 1, 2, 3, \ldots$. We have also pointed out that we cannot characterize a stochastic process in terms of the marginal distributions of each variable X_t. In principle, we should assign the joint distribution of all the involved variables: a daunting task, indeed. For this reason, whenever it is practically acceptable, we work with processes in which mutual dependence among random variables is at least limited to a simple structure, if not absent at all. The Poisson process, thanks to the memoryless property of the exponential distribution, is a simple case. Another class of relatively simple, yet practically relevant processes features a "limited" amount of memory, which is reflected by a manageable dependence structure.

DEFINITION 8.11 (Discrete-time Markov processes) *Consider a discrete-time stochastic process X_t, $t = 1, 2, 3, \ldots$. If the condition*

$$\mathrm{E}[X_{t+1} \mid X_t, \ X_{t-1}, \ X_{t-2}, \ X_{t-3}, \ldots] = \mathrm{E}[X_{t+1} \mid X_t]$$

holds for any value of the time index t, the process is called a **Markov process**.

We see that a Markov process features a limited amount of memory, as the only relevant past observation in the conditional expectation above is the last observation. A simple example of Markov process occurs when the set of possible values of X_t is a finite set. We speak in such a case of a *discrete-time Markov chain.*

Example 8.10 Financial markets are characterized by *volatility*, which is linked to the standard deviation of returns. One interesting feature of volatility is that we observe periods of relative calm, in which volatility is reasonable, followed by periods of nervousness, where volatility is quite large. Imagine that we want to build a model in which markets can be in one of two states, low and high; the time bucket we consider is a single trading day. High volatility tends to persist; hence, we cannot just assign a probability that, on one day, markets will be in one of the two states. We should build a regime-switching model, accounting for the fact that each state tend to persist: After a day of high volatility, we are more likely to observe another day of high volatility; the same holds for a day with low volatility. A naive regime-switching model is illustrated in Fig. 8.7. The idea is that if the last day was in the low state, the next day will feature the same level of volatility with probability 0.8. However,

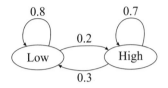

Fig. 8.7 A simple regime switching model.

there is a probability 0.2 that we will observe a day with high volatility. If we get to the high state one day, the next day will feature high volatility again with probability 0.7, whereas we have a 0.3 probability of moving back to the low volatility state. Formally, we have the following conditional probabilities that represent *transition* probabilities:

$$P\,(\text{low} \mid \text{low}) = 0.8, \quad P\,(\text{high} \mid \text{low}) = 0.2$$
$$P\,(\text{high} \mid \text{high}) = 0.7, \quad P\,(\text{low} \mid \text{high}) = 0.3$$

Since we deal with discrete states with a qualitative nature, we deal with conditional probabilities rather than conditional expectations of numerical variables, but this is a two-state, discrete-time Markov chain.

Now, a good question is: If we get to the high state, what is the expected number of days that we will spend in that state? The answer can be found by referring back to the geometric distribution. Doing so, it is easy to see that the expected "sojourn" time in the high volatility state is

$$E[T_h] = \frac{1}{0.7} = 1.43$$

Arguably, the values above are not quite realistic. Furthermore, according to this model, the number of days we spent in each state has no influence in the future evolution. The only relevant piece of information is the last visited state. In fact, the geometric random variable plays, in the context of discrete distributions, the same role that the exponential plays in the context of the continuous ones; both are memoryless distributions. Indeed, in continuous-time Markov chains, the sojourn time in each state is exponentially distributed. The applicability of a memoryless distribution to model a real-life case must be carefully and critically evaluated. Nevertheless, they are so easy to deal with that it is often better to build a complex model with multiple states, approximating a more realistic distribution, than coming up with a more realistic, but intractable model.   ▯

Problems

8.1   You have to decide how much ice cream to buy in order to meet demand at two retail stores. Demand is modeled as follows:

$$D_1 = 100X + \epsilon_1, \qquad D_2 = 120X + \epsilon_2$$

where X, ϵ_1, and ϵ_2 are independent normal variables with expected value and variance given by (28,16), (200,100), (300,150), respectively. Random variable X can be regarded as a common risk factor linked to temperature (the two retail stores are close enough that their sales levels are influenced by the same temperature value), whereas the other variables are specific factors, possibly related to competition in the zone of each store, as well as to pure random variability. Ice cream is stored at a central warehouse, which is close to the two retail stores, so that whatever is needed can be immediately transported by a van.

- Find how much ice cream you should order, in such a way that service level at the warehouse level will be 95%.

- Would this quantity increase or decrease in case of positive correlation between ϵ_1 and ϵ_2?

8.2 You are in charge of component inventory control. Your firm produces end items P_1 and P_2, which share a common component C. You need two components C for each piece of type P_1 and three components for each piece of type P_2. Over the next time period, demand is uncertain and modeled by a normal distribution. Demand for item P_1 has expected value 1000 and standard deviation 250; the corresponding values for P_2 are 700 and 180. Assuming that the two demands are independent, determine the desired inventory level for component C in such a way that its service level is 92%.

8.3 You have invested \$10,000 in IFM stock shares and \$20,000 in Peculiar Motors stock shares. Compute the one-day value at risk, at 95% level, assuming normally distributed daily returns. Daily volatility is 2% for IFM and 4% for Peculiar Motors, and their correlation is 0.68.

8.4 Consider two random variables X and Y, not necessarily independent. Prove that $\text{Cov}(X - Y, X + Y) = 0$.

For further reading

- Modeling multivariate distributions requires a relatively sophisticated mathematical machinery. For introductory treatments, you may refer to Ref. [4], [6], or [10].

- Conditional expectation is also a concept that may be treated at an intuitive level, or using measure-theoretic probability; we have just outlined probability spaces in Section 7.10. For a full account of this approach to conditional expectation, see Ref. [2] or, at an advanced level, Ref. [9].

- We have just hinted at portfolio management as a domain in which the concept of this chapter may be applied. For a readable treatment, see Ref. [1] or [5].

- Covariance and correlation, as we pointed out, are not able to fully account for dependence. A more general approach is copula theory. For a general introduction, see Ref. [8]; financial applications are described in Ref. [3]; in Chapter 2 of Ref. [7] you may also find a quite readable and nice introduction to the issues of modeling dependence.

- Markov processes and Markov chains are a widely used tool in stochastic modeling; for an illustration through a diversified set of examples, see the text by Tijms [11].

REFERENCES

1. Z. Bodie, A. Kane, and A. Marcus, *Investments,* 8th ed., McGraw-Hill, New York, 2008.

2. M. Capiński and T. Zastawniak, *Probability through Problems*, Springer-Verlag, Berlin, 2000.

3. U. Cherubini, E. Luciano, and W. Vecchiato, *Copula Methods in Finance*, Wiley, New York, 2004.

4. M.H. DeGroot and M.J. Schervish, *Probability and Statistics,* 3rd ed., Addison Wesley, Boston, 2002.

5. E.J. Elton and M.J. Gruber, *Modern Portfolio Theory and Investment Analysis,* 5th ed., Wiley, New York, 1995.

6. G. Grimmett and D. Stirzaker, *Probability and Random Processes,* 3rd ed., Oxford University Press, Oxford, 2003.

7. A. Meucci, *Risk and Asset Allocation*, Springer, New York, 2005.

8. R.B. Nelsen, *An Introduction to Copulas,* 2nd ed., Springer, New York, 2006.

9. S.I. Resnick, *A Probability Path*, Birkhäuser, Boston, 1999.

10. S.M. Ross, *Introduction to Probability Models,* 8th ed., Academic Press, San Diego, 2002.

11. H.C. Tijms, *A First Course in Stochastic Models*, Wiley, Chichester, West Sussex, UK, 2003.

9

Inferential Statistics

In the last few chapters we have modeled uncertainty using the tools of probability theory. Problems in probability theory may require a fair level of mathematical sophistication, and often students are led to believe that the involved calculations are the real difficulty. However, this is not the correct view; the real issue is that whatever we do in probability theory assumes a lot of knowledge. When dealing with a random variable, we need a function describing its whole probability distribution, like CDF, PMF, or PDF; in the multivariate case, the full joint distribution might be required, which can be a tricky object to specify. More often than not, this knowledge is not available and it must be somehow inferred from available data, if we are lucky enough to have them.

It is quite instructive to compare the definition of expected value in probability theory, assuming a continuous distribution, and sample mean in descriptive statistics:

$$\mathrm{E}[X] \equiv \int_{-\infty}^{+\infty} x f_X(x)dx, \qquad \overline{X} \equiv \frac{1}{n}\sum_{i=1}^{n} X_i$$

The first expression looks definitely more intimidating than the second one, which is an innocent-looking average. Yet, the real trouble is getting to know the PDF $f_X(x)$; calculating the integral is just a technicality. In this chapter we relate these two concepts, as the sample mean can be used to estimate the expected value, when we do not know the underlying probability density. Indeed, the sample mean does not look troublesome, yet it is: If we select the sample randomly, the sample mean is a random variable, possibly with a large variance. We need to clarify the difference between a *parameter*, like the expected value $\mu = E[X]$, which is a number that we want to estimate, an *estimator*, like the sample variance $\overline{X}$, which is a random variable, and an *estimate*, which is a specific realization of that random variable for a sample. This raises many issues:

- How reliable is the estimate we get?

- How can we state something about estimation errors?

- How can we determine a suitable sample size?

- How can we check the truth of a hypothesis about the unknown expected value?

To find an answer to these and many other questions, we need the tools of inferential statistics.

This chapter has been structured in two parts, corresponding to different kinds of reader:

- The first few sections provide the reader with the essentials of point and interval estimation of a parameter and hypothesis testing; these procedures have been automated in many software packages, but any business student and practitioner should have a reasonable background in order to apply these techniques with a minimum of critical sense.

- Then, we dig a bit deeper into issues such as stochastic convergence, the law of large numbers, and parameter estimation; readers who are just interested in the essentials of statistical inference can safely skip these sections, which are a bit more challenging and aimed at bridging the gap with advanced books on inferential statistics.

It is also worth pointing out that, given the aims of this chapter, we have taken an *orthodox* approach to statistics, as this is definitely simpler and corresponds to what is generally taught in business classes around the world. However, in Chapter 14 we stress that this is just one possible approach, and we provide readers with a glimpse of Bayesian statistics.

In Section 9.1 we clarify the meaning of terms like "random sample" and "statistic," laying down the foundations for the remainder of the chapter. Then, in Sections 9.2 and 9.3, we cover two classical topics, confidence intervals and hypothesis testing, within the framework of the basic problem of inferential statistics: estimating the expected value of a probability distribution. These three sections provide readers with the essential knowledge that anyone involved in business management should have, if anything, to understand essential issues and difficulties in analyzing data.

Then, we broaden our perspective a bit, while keeping the required mathematics to a rather basic level. In Section 9.4 we consider estimating other parameters of interest, like variance, probabilities, correlation, skewness, and kurtosis; we also consider the comparison of two populations in terms of their means. In all of these techniques, we assume that the probability distribution of the underlying population is known qualitatively, and we want to estimate some quantitative parameters. We will mostly refer to the simple case of a normal population. However, assuming that we are dealing with the parameters of a given distribution is itself a hypothesis that should be tested. Can we be sure that the underlying distribution is normal? Nonparametric statistics can be used, among other things, to check the fit of an assumed probability distribution against empirical data. We outline some basic tools, like the chi-square test, in Section 9.5. We conclude the first part of this chapter with the basics of analysis of variance (ANOVA), in Section 9.6, and Monte Carlo simulation, in Section 9.7. It should be mentioned that both of these topics would require a whole book for a thorough coverage, but we can provide readers with an essential understanding of why they are useful and what kind of knowledge is required for their sensible use.

The last three sections, as we mentioned, are more challenging and can be skipped at first reading. Their aim is to bridge the gap between the elementary "cookbook" treatment that any introduction to business statistics offers, and the more advanced and mathematically demanding references. In Section 9.8 we outline the essential concepts of stochastic convergence; they are needed for an understanding of the law of the large numbers and also provide us with a justification of many estimation concepts that find wide application in statistical modeling and econometrics. We consider a more general framework for parameter estimation in Section 9.9, where we discuss desirable properties of estimators, as well as general strategies to obtain them, like the method of moments and maximum likelihood. By a similar token, we outline a more general approach to hypothesis testing in Section 9.10, dealing with a few issues that are skipped in the very elementary treatment of Section 9.3.

9.1 RANDOM SAMPLES AND SAMPLE STATISTICS

Inferential statistics relies on random samples. There are many ways to take a sample:

- Given a large population, we may administer a questionnaire to a relatively small sample of randomly selected individuals; alternatively, we may structure the sample in such a way that it is representative of the overall population.

- Given a stream of manufactured items, we may take a random sample of them to check for defects; alternatively, we may carefully plan experiments in order to assess the impact of factors related to product design or manufacturing technology.

- Sometimes, rather than working with a real system, we have to rely on a computer-based model reflecting randomness in the real world, and carry out Monte Carlo simulation experiments to assess system performance.

- Finally, sometimes we have a set of data and we must do the best with it, as it would be too costly or impossible to collect others. Destructive tests, like experiments with car crashes, are rather costly, but at least they can be carefully planned in order to squeeze out as much information as possible. When dealing with a time series of stock prices, we cannot set experimental conditions at will to check the impact of controlled factors. We have to rely on the observed history, pretending that it came from random sampling, and that's it.

The last point should be stressed: Quite often we *assume* that observed data have been generated by a process that we represent as a statistical model. If we want to use tools from probability and statistics, this is a necessary step. However, we should always keep in mind that whatever conclusion we come up with, it is only as good as the underlying model, which can be a rather drastic simplification of reality. Hence, it is important to formalize what we mean by *random sample*, in order to have a clear picture of the assumptions that statistical tools rely on; the validity of these assumptions should be carefully checked on a casewise basis.

DEFINITION 9.1 (Random sample) *A random sample is a sequence* $X_1, X_2, \ldots, X_n$ *of independent and identically distributed (**i.i.d.**) random variables. Each element X_i in the sample is referred to as an* **observation**, *and n is the* **sample size**.

It is very important to stress the role of *independence* in this definition. All of the concepts introduced in the chapter depend critically on this assumption. It may well be the case that there is correlation within a real-life sample, and a blindfolded application of naive statistical procedures may lead to erroneous conclusions and a possible business disaster. Furthermore, we also assume that the data are somewhat homogeneous, since they are identically distributed. Clearly, if the data have been observed under completely different settings, the conclusions we draw from their analysis may be severely flawed.

Example 9.1 Consider performing a quality check on manufactured items. This typically consists of the measurement of a few quantities for each item, that should conform to some specifications. There could be a small variability in such measures that results from pure randomness and does not significantly affect quality perceived by customers. Let us denote by X_k the measured value for item k. If we consider this as the realization of a random variable, can we say that these variables are independent? Well, it may be the case that, due to tool wear, the machine starts producing a sequence of items that do not meet the specifications; then, the values we observe are not really independent.

The conditional probability that item $k + 1$ is defective, given that item k is defective, can be larger than the unconditional probability of producing a defective item. □

Example 9.2 Sometimes, the assumptions that variables are identically distributed may be wrong. Imagine taking measures of whatever you like within a population consisting of men and women. If gender has a significant impact on the measured variable, it may be a gross mistake to attribute variability to randomness. □

Example 9.3 Consider daily demand d_t, $t = 1, \ldots, T$, for an item. Considering this as a sequence of i.i.d. variables may be a gross mistake as well. It may be the case that demands on two consecutive days are negatively correlated; if customers buy many items on day t, they might not buy any more until their inventory has depleted. Furthermore, at retail stores it may well be the case that sales on Saturdays are much larger than sales on Tuesdays. Such seasonal patterns are commonly observed, in which case data should be deseasonalized before they are analyzing.[1] □

If we take for granted that we are dealing with i.i.d. variables, we may under- or overestimate true variability. Assuming that we have a legitimate random sample, we use available data to estimate some quantity of interest, possibly a summary measure. By far, the most common such measure is the sample mean, but there are other measures that we may be interested in, like variance and correlation. Formally, given a random sample, we compute one or more *sample statistics* (not to be confused with Statistics itself).

DEFINITION 9.2 (Statistic) *A* **statistic** *is a random variable whose value is determined by a random sample.*

In other words, a statistic is a function of a random sample and, as such, it is a *random variable*. This is quite important to realize; we use the most familiar statistic, sample mean, to illustrate basic concepts and possible pitfalls.

9.1.1 Sample mean

The sample mean is a well-known concept from descriptive statistics:

$$\overline{X} = \frac{1}{n} \sum_{i=1}^{n} X_i \tag{9.1}$$

If data come from a legitimate random sample, sample mean is a statistic. A natural use of sample mean is to *estimate* the expected value of the underlying

---

[1]See Chapter 11 on time series.

Table 9.1 Sample means from normal distributions.[a]

Sample number	$\mu = 10, \sigma = 20$ $n = 10$	$\mu = 10, \sigma = 20$ $n = 1{,}000{,}000$	$\mu = 10, \sigma = 0.2$ $n = 10$
1	14.7812	10.0208	10.1338
2	5.2892	10.0004	10.1093
3	14.6516	9.9837	10.0633
4	13.4969	10.0188	9.9147
5	10.9173	10.0149	10.0321
6	15.3640	9.9980	9.9755
7	14.6960	10.0019	10.0624
8	-0.6176	9.9670	10.0643
9	17.7920	10.0194	9.9893
10	25.5372	10.0092	9.9911

[a] Column headers report expected value μ and standard deviation σ for the underlying distribution, as well as the sample size n; 10 random samples are drawn for each setting of these parameters.

random variable, that is unknown in practice. It is important to understand what we are doing: We are using a random variable as an *estimator* of some unknown parameter, which is a number. The realization of that random variable, the *estimate*, is, of course, a number, but we will typically get different numbers whenever we repeat the sampling procedure. They may even be quite different numbers.

Example 9.4 In Table 9.1 we show the realized value of the sample mean in ten samples from a normal distribution.[2] The first column shows a lot of variability in the sample mean. This is hardly a surprise, considering that the standard deviation, $\sigma = 20$, is twice as much as the expected value, $\mu = 10$, and the sample size is rather small, $n = 10$. If sample size is increased considerably ($n = 1{,}000{,}000$), we get the second column, which is affected by much less variability. This also happens if we take a small sample from a distribution with small standard deviation, $\sigma = 0.2$, as shown in the third column.   □

The example shows that if we use a random variable to estimate a parameter, variability of the estimator is an obvious issue. We consider the desirable properties of an estimator in greater depth in Section 9.9, but two obvious features are as follows:

---

[2] The samples have been generated using a pseudorandom number generator on a computer. Such generators are the basis of Monte Carlo simulation, which is outlined in Section 9.7.

1. The estimator should be *unbiased*, i.e., its expected value should be the value of the unknown parameter that we wish to estimate.

2. The variability of the estimator should be as small as possible, for a given sample size.

Clearly, we should be interested in the distribution of any sample statistic we use. In more detail, given the probability distribution of the i.i.d. random variables X_i, $i = 1, \ldots, n$, in the sample, the least we can do is to determine the expected value and variance of sample statistics which are relevant for our analysis; if possible, we should also find their exact distribution.

Example 9.5 (Expected value and variance of the sample mean)
Consider a sample consisting of i.i.d. variables X_i with expected value μ and variance σ^2. In Section 7.7 we used properties of sums of independent random variables to prove that

$$E[\overline{X}] = E\left[\frac{1}{n}\sum_{i=1}^{n}X_i\right] = \frac{1}{n}\sum_{i=1}^{n}E[X_i] = \frac{1}{n}\sum_{i=1}^{n}\mu = \mu \qquad (9.2)$$

$$\text{Var}(\overline{X}) = \text{Var}\left(\frac{1}{n}\sum_{i=1}^{n}X_i\right) = \frac{1}{n^2}\sum_{i=1}^{n}\text{Var}(X_i)$$

$$= \frac{1}{n^2}\sum_{i=1}^{n}\sigma^2 = \frac{\sigma^2}{n} \qquad (9.3)$$

We repeat the derivation of these formulas to emphasize the role of the i.i.d. assumption. Independence plays no role in deriving Eq. (9.2), but we do assume that variables are identically distributed. This equation shows that sample mean is indeed an unbiased estimator of the unknown parameter μ. Equation (9.3) shows that the larger the sample size, the lower the variance of the sample mean, as both Table 9.1 and intuition suggest. However, intuition is not enough because it does not clarify the essential role of independence in this. In the limit, if all of the X_i were perfectly correlated, they would be the same number, and increasing the sample size would be of little use.

This example is also useful for understanding the basic framework of *orthodox statistics*. The expected value μ is an *unknown number*, and we will use random samples to draw inferences about it. On the contrary, in the *Bayesian* framework,[3] probability distributions are associated with the parameters; these distributions may be used to model a priori knowledge or subjective opinions that we might have. Subjective opinions play no role in the orthodox framework. ∎

The example shows that it is fairly easy to come up with essential properties of the sample mean, but what about its distribution? Unfortunately, there is

---

[3]See Chapter 14.

little we can say in general, as we know that summing identically distributed variables does not yield an easy distribution in general. From Section 7.7 we know that if we sum uniform or exponential variables, we do not get uniform or exponential variables, respectively, even if we assume independence. However, we recall two essential results:

- If we sample from a normal distribution, i.e., if variables X_i are normal, then the sample mean will be exactly normal, since the sum of jointly normal variables is itself normal.

- If we sample from a generic distribution and the sample size is large enough, the central limit theorem[4] tells us that the sample mean will tend to a normal random variable. In particular, the distribution of the standardized statistic

$$Z = \frac{\overline{X} - \mu}{\sigma/\sqrt{n}}$$

tends to a standard normal.

The central limit theorem is fairly general and plays a key role in inferential statistics, but in some cases it is better to exploit the structure of the problem at hand.

Example 9.6 (Sampling from a Bernoulli population) Consider a qualitative variable referring to a property that may or may not hold for an individual of a population. If we sample from that population, we are typically interested in the fraction p of the population, for which the property holds. Let X_i be a random variable set to 1 if the observed individual i enjoys the property, 0 otherwise. We immediately see that we are sampling a Bernoulli population; i.e., the random variables X_i are Bernoulli distributed, with a parameter $p \equiv P(X = 1)$ that we wish to estimate.

Since we are summing n independent Bernoulli variables X_i with expected value p, what we get is related to a binomial random variable Y. From Section 6.5.5, we know that

$$E[Y] = E\left[\sum_{i=1}^{n} X_i\right] = np \tag{9.4}$$

$$\text{Var}[Y] = \text{Var}\left[\sum_{i=1}^{n} X_i\right] = np(1-p) \tag{9.5}$$

Equation (9.4) tells us that we may estimate the fraction p by just counting the fraction of the sample that enjoys the following property:

$$\hat{p} = \frac{1}{n}\sum_{i=1}^{n} X_i$$

---

[4]See Section 7.7.3.

Equation (9.5) tells us how the estimate of the expected value also yields an estimate of variance, $\hat{p}(1 - \hat{p})$ ▯

The case of a Bernoulli population is somewhat peculiar, since there is only one parameter that yields both expected value and variance. In general, if variance is a distinct parameter σ^2, we are in trouble when trying to exploit Eq. (9.3), since it relies on another unknown parameter σ that we have to estimate.

9.1.2 Sample variance

The typical estimator of variance is *sample variance*:

$$S^2 = \frac{1}{n-1} \sum_{i=1}^{n} (X_i - \overline{X})^2 \tag{9.6}$$

This formula can be understood as a sample counterpart of the definition of variance: It is basically an average squared deviation with respect to sample mean. When doing calculations by hand, the following rearrangement can be useful:

$$
\begin{aligned}
S^2 &= \frac{1}{n-1} \sum_{i=1}^{n} (X_i - \overline{X})^2 \\
&= \frac{1}{n-1} \left(\sum_{i=1}^{n} X_i^2 - 2\overline{X} \sum_{i=1}^{n} X_i + n\overline{X}^2 \right) \\
&= \frac{1}{n-1} \left(\sum_{i=1}^{n} X_i^2 - n\overline{X}^2 \right) \tag{9.7}
\end{aligned}
$$

The *sample standard deviation* is just S, the square root of sample variance. We recall from Example 4.13 that rewriting variance like this may result in bad numerical behavior with some peculiar datasets; nevertheless, it is quite convenient in proving useful relationships.

An obviously weird feature of sample variance is that we take an average of n terms, yet we divide by $n-1$. On the basis of the limited tools of descriptive statistics there is no convincing way of understanding why, when dealing with a population, we are told to divide by n, whereas we should divide by $n - 1$ when dealing with a sample. The right way to appreciate the need for this correction is to check for unbiasedness of sample variance.

THEOREM 9.3 *Sample variance is an unbiased estimator of true variance, i.e.,* $\mathrm{E}[S^2] = \sigma^2$.

PROOF   In the proof, we exploit rewriting (9.7) of sample variance:

$$E\left[S^2\right] = \frac{1}{n-1}\left(\sum_{i=1}^{n} E\left[X_i^2\right] - nE\left[\overline{X}^2\right]\right)$$

$$= \frac{n}{n-1}\left\{\sigma^2 + \mu^2 - \left(\frac{\sigma^2}{n} + \mu^2\right)\right\}$$

$$= \sigma^2$$

In the second line above, we applied the usual property, $\mathrm{Var}(Y) = E\left[Y^2\right] - E^2[Y]$, to each X_i in the sum and to the sample mean $\overline{X}$.   ∎

To fully understand the result, observe that

$$E\left[\sum_{i=1}^{n} X_i^2 - n\mu^2\right] = n\left(\sigma^2 + \mu^2\right) - n\mu^2 = n\sigma^2$$

but

$$E\left[\sum_{i=1}^{n} X_i^2 - n\overline{X}^2\right] = n\left(\sigma^2 + \mu^2\right) - n\left(\frac{\sigma^2}{n} + \mu^2\right) = n\sigma^2 - \sigma^2$$

In the second case, we should not divide by n when measuring deviations against $\overline{X}$ rather than μ, and this results in a bias that is corrected by the $n-1$ factor. From an intuitive perspective, we could say that the need to estimate the unknown expected value implies that we "lose one degree of freedom" in the n available data in the sample.

Finding the expected value of sample variance is fairly easy, but characterizing its full distribution is not. One simple case is when the sample is normal. Intuitively, we see from Eq. (9.7) that sample variance involves squares of normal variables. Given what we know about the chi-square and Student's t distribution,[5] the following theorem, which summarizes basic results on the distribution of the estimators that we have considered, should not come as a surprise.[6]

THEOREM 9.4 (Distributional properties of sample statistics) *Let* $X_1, \ldots, X_n$ *be a random sample from a normal distribution with expected value* μ *and variance* σ^2. *Then*

1. *The sample mean* $\overline{X}$ *has normal distribution with expected value* μ *and variance* σ^2/n.

---

[5]This chapter relies heavily on some probability distributions that are derived from the normal, like the chi-square, t, and F distributions. They were described in Section 7.7.2, to which the reader may refer before proceeding further.
[6]See, e.g., Ref. [15] for a proof.

2. *The random variable* $(n-1)S^2/\sigma^2$ *has chi-square distribution with* $n-1$ *degrees of freedom.*

3. *Sample mean and sample variance are independent random variables.*

4. *The random variable*

$$T = \frac{\overline{X} - \mu}{S/\sqrt{n}}$$

has t distribution with $n-1$ *degrees of freedom.*

Statement 1 is quite natural, since we know that the sum of jointly normal variables is itself a normal variable. Statement 2 can be understood by noting that in the sample mean we square and sum independent normal variables and recalling that by squaring and summing independent standard normals we get a chi-square; the distribution has $n-1$ degrees of freedom, which is also reasonable, given what we observed about sample variance. Statement 3, on the contrary, is somewhat surprising, since sample mean and sample variance are statistics depending on the same random variables, but it is essential in establishing the last distributional result, which will play a fundamental role in the following. We should note that if the true variance were known, we could work with the statistic

$$Z = \frac{\overline{X} - \mu}{\sigma/\sqrt{n}}$$

which is a standard normal. If the random sample is not normal, the results above do not hold. Indeed, many results that are routinely used in inferential statistics are valid only for normal samples. Luckily, variations on the theme of the central limit theorem provide us with asymptotic results, i.e., properties that apply to *large* samples. These results justify the application of statistical procedures, which are obtained for normal samples, to large nonnormal samples. In what follows, we will rely on these procedures,[7] but it is important to keep in mind that they just yield approximated results, and that due care must be exercised when dealing with small samples.

9.2 CONFIDENCE INTERVALS

The sample mean is a *point estimator* for the expected value, in the sense that it results in an estimate that is a single number. Since this estimator it is subject to some variance, it would be nice to have some measure of how

---

[7]A thorough investigation of these issues is beyond the scope of an introductory textbook; see the references at the end of the chapter. The important message is that you should stray from the cookbook recipe approach that is widely taken when dealing with inferential statistics.

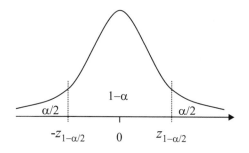

Fig. 9.1 Illustrating the link between quantiles of the standard normal and confidence intervals.

much we can trust that single number. In other words, we would like to get an emf interval estimate, which typically comes in the form of a *confidence interval*. Roughly speaking, a confidence interval is a range in which the true, unknown parameter should lie with some probability. As we shall see, this statement must be taken with due care if we wish a rigorous interpretation, but it is a good declaration of intent. This probability is known as the *confidence level*, and it should be relatively large, say, 95% or 99%. In this section, we derive a confidence interval for the expected value of a normal distribution or, in more colloquial terms, for the mean of a normal population. Later, we apply the idea to the parameter of a Bernoulli population and to the variance of a normal population.

From Section 9.1.1 we know that the statistic

$$Z = \frac{\overline{X} - \mu}{\sigma/\sqrt{n}}$$

has standard normal distribution, if the sample comes from a normal population with parameters μ and σ^2. Note that, if we were so lucky as to get a "perfect" sample, we would have $\overline{X} = \mu$ and $Z = 0$. In real life, the sample mean $\overline{X}$ will be smaller or larger than the true expected value μ, and Z will be negative or positive, accordingly. Figure 9.1 illustrates the question, in terms of the PDF of a standard normal and its quantiles. There, $z_{1-\alpha/2}$ is the quantile with probability level $1 - \alpha/2$, i.e., a number such that $P(Z \le z_{1-\alpha/2}) = 1 - \alpha/2$. Correspondingly, $\alpha/2$ is the area of the right tail, to the right of quantile $z_{1-\alpha/2}$. Given the symmetry of the normal distribution, we observe that $P(Z \le -z_{1-\alpha/2}) = \alpha/2$; i.e., we also have a left tail with probability (area) $\alpha/2$. Then, by construction, the probability that the Z statistic falls between those two quantiles is as follows:

$$P\left(-z_{1-\alpha/2} \le \frac{\overline{X} - \mu}{\sigma/\sqrt{n}} \le z_{1-\alpha/2}\right) = 1 - \alpha$$

This is just the area under the PDF, between the quantiles. If you prefer, it is the total area, which amounts to 1, minus the areas associated with the left and right tails, each one amounting to $\alpha/2$. Now, let us rearrange the first inequality:

$$-z_{1-\alpha/2} \leq \frac{\overline{X} - \mu}{\sigma/\sqrt{n}} \quad \Rightarrow \quad \mu - z_{1-\alpha/2}\frac{\sigma}{\sqrt{n}} \leq \overline{X}$$

Doing the same with the second inequality and putting both of them together, we conclude that

$$P\left(\overline{X} - z_{1-\alpha/2}\frac{\sigma}{\sqrt{n}} \leq \mu \leq \overline{X} + z_{1-\alpha/2}\frac{\sigma}{\sqrt{n}}\right) = 1 - \alpha \qquad (9.8)$$

We should note carefully that μ is an unknown number that is bracketed by two random variables with probability $1 - \alpha$. Then, we conclude that the interval

$$\left(\overline{X} - z_{1-\alpha/2}\frac{\sigma}{\sqrt{n}}, \quad \overline{X} + z_{1-\alpha/2}\frac{\sigma}{\sqrt{n}}\right) \qquad (9.9)$$

is a confidence interval for the expected value μ of a normal distribution, with confidence level $1 - \alpha$. Formula (9.9) is very easy to apply, but there is a little fly in the ointment: It assumes that the standard deviation σ is known. Arguably, if the expected value is unknown, it seems quite unreasonable that standard deviation is known. In fact, however weird it may sound, there are situations in which this happens.

Example 9.7 Imagine that we are measuring a physical quantity with an instrument affected by some measurement uncertainty. Then, what we read from the instrument is a random variable X; if there is no measurement bias, $E[X] = \mu$, where μ is the true value of the quantity we are measuring. We may regard the measurement as $X = \mu + \epsilon$, where ϵ is some noise corrupting what we read on the instrument. Unbiasedness amounts to stating $E[\epsilon] = 0$. If the instrument has been well calibrated using appropriate procedures, we may have a pretty good idea of the standard deviation of each observation, which is actually the standard deviation σ_ϵ of noise. Hence, we (almost) know the standard deviation of random variable X, but not its expected value.   ▯

Strictly speaking, even in the example above we do not really the know standard deviation, but we may have a reliable estimate. Even more important: The knowledge about the standard deviation has been obtained preliminarily, by a procedure that is independent from the random sample we use to estimate the expected value. Nevertheless, in business settings, we typically do not know the standard deviation at all. Hence, we have to settle for sample standard deviation S. Note that, unlike the case in the example above, $\overline{X}$ and S come from the *same* sample, and we cannot just plug S and replace σ into (9.9). Luckily, we already know that the statistic

$$T = \frac{\overline{X} - \mu}{S/\sqrt{n}}$$

has t distribution with $n - 1$ degrees of freedom. Hence, if we denote by $t_{1-\alpha/2,n-1}$ the $(1 - \alpha/2)$-quantile of the t distribution, using the same procedure described above yields the confidence interval

$$\left(\overline{X} - t_{1-\alpha/2,n-1}\frac{S}{\sqrt{n}}, \quad \overline{X} + t_{1-\alpha/2,n-1}\frac{S}{\sqrt{n}} \right) \qquad (9.10)$$

From a qualitative point of view, this confidence interval has the same form as (9.9). The only difference is the use of quantiles $t_{1-\alpha/2,n-1}$ from the t distribution instead of $z_{1-\alpha/2}$. From Fig. 7.19 we recall that a t distribution features fatter tails than the standard normal; the fewer the degrees of freedom, the fatter the tails. Hence,

$$t_{1-\alpha/2,n-1} > z_{1-\alpha/2}$$

and the bottom line result is a wider confidence interval. This makes sense, since when estimating variance σ^2 we add some uncertainty to the overall process; more so, when the sample size n is small.

Example 9.8 Let us consider the random sample:

$$\{43, 79, 26, 137, 45, 55, 93, 52, 46, 17\}$$

under the assumption that it comes from a normal distribution, and let us compute a 95% confidence interval for the expected value. We first compute the statistics

$$n = 10, \qquad \overline{X} = 59.3, \qquad S \approx 35.2422$$

From statistical tables, or a suitable piece of software, we obtain

$$t_{1-\alpha/2,n-1} = t_{0.975,9} \approx 2.2622$$

By straightforward application of Eq. (9.10), we obtain the confidence interval (34.0893, 84.5107). ▯

Formula (9.10) is so easy to apply, and it has been implemented in so many software packages, that it is tempting to apply it without much thought. It is also natural to wonder why one should bother to understand where it comes from. The next sections address these issues.

9.2.1 A counterexample

Consider the queueing system in Fig. 9.2. Customers arrive according to a random process. If there is a free server, service is started immediately; otherwise, the customer joins the queue and waits. Service time is random as well, and whenever a server completes a service, the first customer in the queue (if any) starts her service. To keep it simple, let us assume that the system is open 24/7, so that there is no issue with closing times. Now, what

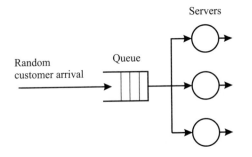

Fig. 9.2 A queueing system with multiple servers.

is the first and foremost feature to measure quality of service experienced by customers? If you have some experience in long queues, I guess that waiting time is what comes to your mind. In fact, a common management problem is determination of the number of servers in such a way to strike a compromise between cost and service quality. Hence, we would like to assess the average waiting time, given the number of servers. Of course, an average does not fully capture our problem, as we would also like to make sure that a prescribed maximum waiting time is rarely, if ever, exceeded, but let us stick to the simplest performance measure we may think of.

A first step in the analysis is modeling uncertainty. A common assumption is that customers arrive according to a Poisson process, i.e., that time between two consecutive arrivals is exponentially distributed with a rate λ. If service times were exponentially distributed, it would be good news, as there are easy formulas giving the average waiting time for a queueing system involving exponential distributions only. Unfortunately, we know that the exponential distribution is memoryless,[8] and this makes it a poor candidate for modeling service times. Furthermore, customer arrivals may have a time-varying rate; indeed, at retail stores we observe hours at which there is a burst of arrivals, whereas other hours are pretty quiet. Luckily, we may rely on computer simulation programs to analyze complex queueing systems, which can make a faithful model of reality. We will get a clue about how such models work in Section 9.7. For now, let us just say that we have a way to collect a sequence of observed waiting times. Let W_k be the waiting time of the kth customer, and say that we simulate the system long enough, collecting waiting times for customers $k = 1, \ldots, n$. To make "long enough" more precise, we could apply formula (9.10) to calculate a confidence interval. If the interval is too wide, we can simulate more customers to get a satisfactory estimate of average waiting time.

---

[8] See Section 8.5.2.

Now I have two questions for you. The first one is pretty easy; the second one is a bit more challenging. Question 1 is: Can we apply the procedure above to come up with the confidence interval we need?

Please! Sit down, think a while, and give your answer before reading further. This is the easy question, after all.....

When I ask this question in class, almost all students agree. Then the tougher question 2 comes: How *many* mistakes have you made, if your answer was yes?

1. To begin with, if we start our simulation with an empty system, what is the waiting time of the first customer? Of course, it is zero; the first few customers will find an empty system, and their waiting times do not tell us much. In fact, this transient phase may affect the statistics we collect. We should discard the first observations to warm the system up and avoid this issue. We should start collecting waiting times only when the system is in steady state. From a more general perspective, confidence intervals assume a sample of identically distributed random variables, but the initial waiting times have a different probability distribution.

2. A more general issue is that the waiting times are unlikely to be normally distributed, and the resulting confidence interval will only be an approximation; as we have said, however, this is a fairly good one for a large sample.

3. Actually, the really serious mistake is that waiting times of successive customers are *not* independent. If customer k undergoes a long waiting time, this means that this unlucky customer arrived when the system is congested and there is a long queue. Hence, we might expect the waiting time for customer $k + 1$ to be large as well. Formally, waiting times of successive customers are positively correlated.

The important message is that *you should never take independence for granted.* In this specific case, it can be shown by a deeper analysis that the blindfold application of standard statistical procedures results in an underestimation of the variability of waiting times. Hence, the width of the confidence interval is underestimated as well, and the net result is that we are overconfident in our conclusions.

In practice, the way out requires *batching* observations. If the system is stable and the queue does not explode to infinity, we should expect that if there is a moment of congestion, it will be resolved after a while and the queue length will revert back to a normal level. Hence, we should also expect that the waiting times of two faraway customers, say, W_k and W_{k+1000}, are practically independent. In other words, intuition suggests that the random variables W_k and W_{k+s} should have some positive correlation, but this tends to fade out for increasing values of s. Then, we group observations into m

batches, each one consisting of n customers, amounting to a total sample of nm customers. Now consider the m batch means

$$\overline{W}^j = \sum_{k=(j-1)n+1}^{jn} W_k, \qquad j = 1, \ldots, m$$

where batch $j = 1$ consists of customers $1, 2, \ldots, n$, batch $j = 2$ consists of customers $n+1, n+2, \ldots, 2n$, etc. Each sample mean $\overline{W}^j$ is, at least approximately, independent from the other ones, and we may apply the standard procedure on them. The further good news is that, courtesy of the central limit theorem, they should be more or less normal,[9] providing further justification for the approach.

9.2.2   An important remark about confidence levels

A further point concerns the correct interpretation of the confidence level. Consider the 95% confidence interval we calculated in Example 9.8. We *cannot* say that the confidence interval (34.0893, 84.5107) contains the unknown expected value with probability 95%. What we *can* say is that if we repeat the sampling procedure many times, and we compute a confidence interval for each sample, about 95% of the confidence intervals will contain the true unknown expected value. But there is nothing we can say about a specific confidence interval. To clarify the point, let us consider the simpler case of a normal random variable X with expected value 0. We can say that $P(X \leq 0) = 0.5$. But if we observe a *realization* $x = -0.55$, we certainly cannot say that $P(-0.55 \leq 0) = 0.5$. More generally, if μ is the expected value of a random variable X and x is a realization of that variable, the expression $P(x \leq \mu)$ is meaningless, since it involves two numbers. The condition $x \leq \mu$ is either true or false, and since we do not know μ, there is nothing we can say about it. By the same token, the statement about the confidence level $1 - \alpha$ of a confidence interval applies a priori, i.e., to a pair of random variables that provide us with the lower and upper bounds of the interval. But it would be wrong to claim that a confidence interval provides us with some probabilistic information about the expected value. Actually, this applies within the framework of orthodox statistics, whereby the expected value μ is a number. Within a Bayesian framework, we do associate a probability distribution with an unknown parameter; this distribution can be the result of merging a priori beliefs with empirical evidence from sampled observations.

   To reinforce a clear view of what a confidence interval is, we may consider the general definition of interval estimators and estimates.[10]

---

[9]To be precise, this statement should be taken with care, as the central limit theorem requires independence as well. Nevertheless, if there are enough servers to ensure stability of queues and batches are fairly large, what we observe in practice is not far from normality.
[10]See the text by Casella and Berger [3].

DEFINITION 9.5 (Interval estimate and estimator) *An* **interval estimate** *of a real valued parameter* θ *is a pair of functions* $L(\mathbf{x})$ *and* $U(\mathbf{x})$, *where* $\mathbf{x} \in \mathbb{R}^n$ *and* n *is the size of the sample, such that* $L(\mathbf{x}) \leq U(\mathbf{x})$, *for any* $\mathbf{x}$ *in the range of interest. If the sample* $\mathbf{X} = \mathbf{x}$ *is observed, the inference* $L(\mathbf{x}) \leq \theta \leq U(\mathbf{x})$ *is made. The* random *interval* $[L(\mathbf{X}), U(\mathbf{X})]$ *is called an* **interval estimator**.

This definition includes standard confidence intervals as a specific case, and it clearly points out that the interval *estimator* is a pair of random variables. The interval *estimate* consists of the realization of the two random variables. Any probabilistic statement must refer to estimators, and not to estimates.

9.2.3 Setting the sample size

From a qualitative perspective, the form of the confidence interval (9.10) suggests the following observations:

- When the sample is very large, we may use the quantiles $z_{1-\alpha/2}$ from the standard normal distribution, since the t distribution tends to a standard normal, when the degrees of freedom go to infinity.

- The larger the confidence level $1 - \alpha$, the larger the confidence interval; in other words, a wider interval is required to be "almost sure" that it includes the true value (in the sense that we have just clarified!).

- The confidence interval is large when the underlying variability σ of the observations is large.

- The confidence interval shrinks when we increase the size of the sample. Actually, it might be the case that the interval gets larger by adding a few observations, if these additional data result in a larger sample standard deviation S, but this is a pathological behavior that we may observe if we add a few observations to a small sample.

The last statement is quite relevant, and is related to an important issue. So far, we have considered a given sample and we have built a confidence interval. However, sometimes we have to go the other way around: Given a required precision, how large a sample should we take? One way of formalizing the issue is the following. Say that we require the following condition on the maximum absolute error:

$$|\overline{X} - \mu| \leq \epsilon \tag{9.11}$$

Clearly, we can obtain only a probabilistic guarantee: Condition (9.11) should hold with confidence level $1 - \alpha$. This requirement can be rewritten as a pair of inequalities, since $\overline{X} - \mu$ could be positive or negative:

$$\overline{X} - \mu \leq \epsilon$$
$$\overline{X} - \mu \geq -\epsilon$$

The two inequalities may be rearranged as

$$\overline{X} - \epsilon \leq \mu \leq \overline{X} + \epsilon$$

from which we see that ϵ is just the half-length of the confidence interval. Hence, we should find a sample size n as follows:

$$t_{1-\alpha/2,n-1}\frac{S}{\sqrt{n}} = \epsilon \quad \Rightarrow \quad n \approx \left(\frac{t_{1-\alpha/2,n-1}S}{\epsilon}\right)^2 \qquad (9.12)$$

The careful reader will certainly be dissatisfied with the last equation: How can we find the sample size n *before* sampling, if this requires knowledge of sample standard deviation S, which is obviously known only *after* sampling? Furthermore, n depends on n itself, since (9.12) defines n as a function the quantile $t_{1-\alpha/2,n-1}$ that we should use. The standard way out of this dilemma is as follows:

- Assume that n should be large enough to warrant use of quantiles $z_{1-\alpha/2}$ of the standard normal, which suppresses the circular dependence on n.

- Take an exploratory sample to find a rough estimate S, which can be used to find the total sample size n. Clearly, there is no guarantee that, after completion of the overall sampling, the initial estimate S will turn out to be close to the new estimate based on the whole sample. So, we should take the exploratory sample, add the tentative number of observations, and then check again, possibly repeating the procedure.

Example 9.9 A sample of size $n = 25$, taken from a normal population, yields

$$\overline{X} = 121.8290, \qquad S = 48.4709$$

The resulting 95% confidence interval is $(101.8212, 141.8368)$ which is too large for your purposes, and you would like to take a sample large enough to guarantee, with confidence level 95%, that the absolute error is smaller than 1. The required sample size would be

$$n = \left(\frac{z_{0.975}S}{\epsilon}\right)^2 = \left(\frac{1.96 \times 48.4709}{1}\right)^2 \approx 9026$$

□

The example illustrates the potentially high price of precision when variability is large. To get a better feeling for this issue, let us say that when sample size is n, the half-width of the confidence interval is H; how large should a sample size n' be, in order to ensure $H' = H/10$, while keeping the same confidence level? Note that this requirement corresponds to an improvement of one order of magnitude in precision. Using the formula for the half-width we see that

$$H' = \frac{H}{10} = z_{1-\alpha/2}\frac{S}{\sqrt{n}} \times \frac{1}{10} = z_{1-\alpha/2}\frac{S}{\sqrt{100n}}$$

which implies $n' = 100n$. Hence, to improve precision by one order of magnitude, the sample size must increase by two orders of magnitude. The reason is that the effect of increasing sample size is "killed" by the square-root function $\sqrt{n}$, which is concave.

We emphasize again that in this section we have just derived a confidence interval for the expected value of a normal distribution. Asymptotic results and the central limit theorem allow us to apply Eq. (9.10) as an approximation for large samples, but it is sometimes necessary or advisable to take a different route when estimating other parameters, such as variance, or when dealing with different distributions. In Section 9.4 we illustrate some generalizations, while keeping the treatment at an elementary level. In Section 9.9 we consider parameter estimation within a more systematic and general framework.

9.3 HYPOTHESIS TESTING

The need for testing a hypothesis about an unknown parameter arises from many problems related to inferential statistics. There are general and powerful ways to build appropriate procedures for testing hypotheses, which we outline in Section 9.10. Since they do require some level of mathematical sophistication, we offer here an elementary treatment that is strongly linked to how we have built confidence intervals for the expected value of a normal distribution. To get a grasp of the underlying issues, we begin with hypotheses about the expected value of a normal distribution. We will generalize a bit in Section 9.4, where we consider testing the difference in the means of two populations, variance, proportions, and correlation. In Section 9.6 we show how Analysis of Variance allows to deal with more than two populations. A solid understanding of these topics is more than adequate for most business practitioners. In order to get a feeling for the involved issues, let us introduce a little numerical example.

Example 9.10 Consider the data listed in Table 9.2 and the following claims:

1. The sample comes from a normal distribution, i.e., from a normal population.

2. The expected value of this distribution, or the population mean, if you prefer, is $\mu_0 = 5$.

How can we check the truth of these claims? At present, we have absolutely no tool with which to verify that a sample comes from a normal distribution. We will consider the issue later, in Section 9.5, but we should note that this does not involve just a few parameters, but the whole distribution. Since a nonparametric test looks a bit hard, let us take normality for granted, at least for now. Since we already know a bit about estimating an expected value, we can try checking the second claim. Probably, the first step should be the

Table 9.2 Sample data for hypothesis testing.

5.00	1.82	15.95	−13.74	9.28	13.96	12.31	10.78	5.40	11.77

calculation of the standard sample statistics:

$$n = 10, \qquad \overline{X} = 7.253, \qquad S = 8.5757$$

We see that the sample mean $\overline{X}$ is quite different from $\mu_0 = 5$. This could lead us to reject the claim. However, we also see that variability is quite large and that the sample size is rather small. The discrepancy between $\overline{X}$ and μ_0 could be explained in either of two ways:

1. The claim is false, i.e., $\mu_0 \neq 5$.

2. The claim is true, but the sample mean differs from the expected value because of randomness in sampling.

In this case we are somewhat embarrassed. If the sample statistics were, say, $\overline{X} = 20$ and $S = 0.5$, we would feel rather confident that the claim is false. If, on the contrary, we had $\overline{X} = 5.1$, we would probably say that the small discrepancy is not statistically significant. But what does *statistically significant* exactly mean, anyway?

Such questions like this one are quite relevant in practice, and to check similar claims properly, we need to rely on sensible statistical procedures. In hypothesis testing we state some hypothesis concerning an unknown parameter, in this case the expected value. This hypothesis should be checked against an alternative. For instance, in our first example, we should test $H_0: \mu = 5$, against the alternative $H_a: \mu \neq 5$. More generally, we may test the hypothesis

$$H_0: \mu = \mu_0,$$

for a given μ_0, against the *alternative hypothesis*

$$H_a: \mu \neq \mu_0.$$

The first hypothesis H_0 is called the *null hypothesis*. The somewhat odd name is justified by the fact that in many practical settings, if the null hypothesis is true, then we should just sit and do nothing. It is when the hypothesis is rejected that we should act before is too late, or we have discovered an interesting fact or some unexpected pattern in data. Clearly, this is just an interpretation that need not always be true. Note that the null and the alternative hypotheses are complementary and jointly cover all of the possible values of μ_0. In the following, we will also consider other kinds of hypotheses, such as

$$H_0 : \mu \geq \mu_0 \qquad \text{against} \qquad H_a : \mu < \mu_0$$

or

$$H_0 : \mu \leq \mu_0 \qquad \text{against} \qquad H_a : \mu > \mu_0$$

Note again that the two hypotheses are mutually exclusive and cover all of the possible values of μ. The following example illustrates why the above forms may arise in practical contexts.

Example 9.11 Consider the manufacturing process for producing shafts that must fit into a hole. Ideally, the diameter of all of the shafts should exactly be as specified by product design; in practice, a slight discrepancy is unavoidable. However, we do not want a diameter larger than specified, as the shaft will not fit the hole; on the other hand, we do not want a diameter smaller than specified, either, as this may result in dangerous vibrations. Taking a random sample of items, we should check that the average diameter is sufficiently close to the nominal value. If the sample mean is too large or too small, then we should probably act quickly in order to bring back the manufacturing process under control. ⬚

Example 9.12 Unlike the previous example, there are cases in which the situation is not symmetric. Consider, for instance, the concentration of a water pollutant; if the measured concentration is below the maximum level allowed, no one will complain. We should act only if we find that the concentration exceeds a certain danger threshold; if there is less pollutant than acceptable, no one would add any just to stick to that threshold (hopefully). By the same token, there are cases in which we should act when the average is below some level. No consumer association would complain, if they find more candies than expected in a box. (The viewpoint of the producer is different!) ⬚

Whatever hypothesis we test, it might be true or false. If we accept a true hypothesis or we reject a false one, then we did well. Unfortunately, there are two types of errors that we could make:

- We commit a *type I* error if we reject a true hypothesis.

- We commit a *type II* error if we accept a false hypothesis.

The probability of an error is something that we have to accept, as in real life we will (almost) *never* know the truth; the best we can do is keeping the probability of errors as small as possible. Which kind of error is the most relevant one depends on the practical situation at hand, and the cost of making one of the two mistakes. Indeed, in practice, this also influences the way in which the null hypothesis is stated. Such considerations notwithstanding, we will see shortly that it is definitely easier to control the probability of a type

I error. Keeping this probability low means that we want some quite strong evidence against the null hypothesis to reject it. An unfortunate consequence is that if we reject the null hypothesis only when we are pretty sure, the probability of a type II error tends to increase.[11] The best way to get the point is keeping in mind the following sound principle:[12]

You are innocent until proven guilty.

Clearly, if we take such an attitude, we realize that sometimes we might lack enough evidence to reject the null hypothesis, even if it is very suspicious. This means that the probability of a type II error will increase. Indeed, it is sometimes better to say that we "fail to reject" the null hypothesis, as we are not really accepting it.[13] Incidentally, this raises an immediate issue: How can we *prove* something using hypothesis testing? Of course, we cannot really prove anything using sampled data, but we can collect strong statistical evidence supporting a claim. However, this claim *should not* be taken as the null hypothesis. We should take the negation of our claim as the null hypothesis H_0; then, if we are able to reject H_0, this will provide good support for our idea.[14]

To make all of the above ideas operational, we have to be more specific. As we said, we consider first a test concerning the expected value of a normal distribution, where we want to keep the probability of a type I error reasonably low. Such an error occurs if the null hypothesis is true and we reject it by mistake. Let us start with the null hypothesis $H_0: \mu = \mu_0$, tested against the alternative $H_a: \mu \neq \mu_0$. If the null hypothesis is true, and the population is normal, then the test statistic

$$\text{TS} \equiv \frac{\overline{X} - \mu_0}{S/\sqrt{n}}$$

has t distribution with $n - 1$ degrees of freedom. We already used this result when deriving a confidence interval, but please note a fundamental difference:

---

[11] The probability of *not* making a type II error is called the *power* of the test; in other words, the power of the test is the probability that it will reject a false null hypothesis. A full treatment of this topic is beyond the scope of the book, and we will just state a few considerations in Section 9.3. However, it stands to reason that an increase in the power of the test implies an increase on the probability of a type I error as well.

[12] This principle should not be taken for granted and is a relatively recent conquest. In dark ages and dark places, *you* have the burden of proving that you are not guilty.

[13] This is a rather controversial issue. In fact, what is usually taught in statistics textbooks is a sort of hybrid between what have been conflicting views in the development of inferential statistics. See, e.g., Ref. [9] for an illustration.

[14] This is somewhat similar in spirit to mathematical proofs by *reductio ad absurdum*: We negate what we want to prove, and then we derive a contradiction, which is a proof of our thesis. In our case, the contradiction is with respect to empirical data, in a probabilistic sense.

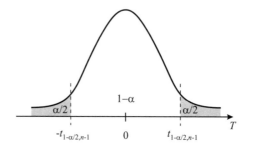

Fig. 9.3 A rejection region for hypothesis testing.

In this case we *know* the value μ_0, which is provided by the null hypothesis. Now let us follow an intuitive route:

- Reasonably, we should reject H_0 if the sample mean $\overline{X}$ is "far" from μ_0 (both larger or smaller), which implies that the statistic TS will be "far" from 0, i.e., too large (positive) or too small (negative).

- Then we could find a positive critical value t^* such that we reject the null hypothesis if TS $< -t^*$ or TS $> t^*$, and we fail to reject H_0 if $|\text{TS}| \leq t^*$. Given the symmetry of the t distribution, there is no point in defining two different thresholds for negative or positive values.

- The critical value t^* defines two regions:

  - A *rejection region* $\mathcal{C}$, such that if the statistic is in that region, TS $\in \mathcal{C}$, then we reject the null hypothesis.

  - An *"acceptance"* region, such that we fail to reject the null hypothesis if TS $\notin \mathcal{C}$.

  In our case, the rejection region is the union of intervals $(-\infty, -t^*)$ and $(t^*, +\infty)$, i.e., a subset of the real line. It is useful to notice that the rejection region, for the TS statistic above, consists of two symmetric tails of a probability distribution. More generally, the rejection region can be defined in terms of a subset of $\mathbb{R}^n$, where n is the sample size; then, we reject H_0 if $(X_1, X_2, \ldots, X_n) \in \mathbf{C}$.

- Actually, because of sampling errors, it might well be the case that the test statistic falls in the rejection region by pure chance; this, however, should be relatively unlikely. We want a suitably small probability of type I error α, say, $\alpha = 0.05$. This value is often called the *significance level*. Then it is easy to see that by setting $t^* = t_{1-\alpha/2, n-1}$, we obtain a rejection region associated with a probability of type I error given by α. The idea is illustrated in Fig. 9.3. If the null hypothesis is true, the test statistic will fall in the acceptance region with probability $1 - \alpha$:

$$P_{\mu_0}\left(-t_{1-\alpha/2,n-1} \leq \frac{\sqrt{n}(\overline{X} - \mu_0)}{S} \leq t_{1-\alpha/2,n-1}\right) = 1 - \alpha$$

where we use the notation P_{μ_0} to emphasize that we compute this probability under the probability measure assumed in H_0. The rejection region consists of two tails. Each tail is associated with a probability $\alpha/2$; even if the null hypothesis is true, because of sampling variability we have a probability $\alpha/2$ that the standardized sample mean falls on the right tail ($\overline{X}$ is much larger than μ_0), and a probability $\alpha/2$ that it falls on the left tail ($\overline{X}$ is much smaller than μ_0). If, in such a circumstance, we reject H_0, the probability of a type I error is α, putting the two tails together.

Wrapping everything up, the procedure prescribes the following conditions, for a given significance level α:

$$\text{"Accept" } H_0 \text{ if } \left|\frac{\sqrt{n}(\overline{X} - \mu_0)}{S}\right| \leq t_{1-\alpha/2,n-1}$$

$$\text{Reject } H_0 \text{ if } \left|\frac{\sqrt{n}(\overline{X} - \mu_0)}{S}\right| > t_{1-\alpha/2,n-1}$$

A test like this is called a *two-sided* or *two-tail* test, as the rejection region consists of two tails. Let us illustrate with an example.

Example 9.13 Consider again the data in Table 9.2 and the null hypothesis $H_0\colon \mu = 5$, against the alternative $H_a\colon \mu \neq 5$. As we observed, the statistics

$$n = 10, \qquad \overline{X} = 7.253, \qquad S = 8.5757$$

seem to contradict the claim, but we should test this carefully. Assume that we choose a significance level $\alpha = 0.1$. The test statistic is

$$\text{TS} = \frac{\sqrt{10} \times (7.253 - 5)}{8.5757} = 0.8308$$

Indeed, TS $\neq 0$, but we do not know yet if this is really significant. To find the critical value drawing the line between acceptance and rejection, we may consult statistical tables to find $t_{1-\alpha/2,n-1} = t_{0.95,9} = 1.8331$. Since the sample size is $n = 10$, we have 9 degrees of freedom. Once again, note that if the rejection region consists of two tails, we should split its total area α on the two tails; this is quite similar to what we do when calculating confidence intervals. Since $|$ TS $| = 0.8308 < 1.8331$, we *cannot* reject the hypothesis with that significance level; see Fig. 9.4. What would happen with a smaller probability of type I error? Well, we would fail to reject the null hypothesis again, since decreasing the significance level means that we are even more

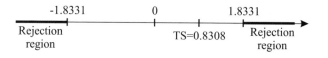

Fig. 9.4 Checking rejection, given a test statistic.

conservative. For instance, if we set $\alpha = 0.05$, we should compare the test statistic against the quantile $t_{0.975,9} = 2.2622$. Then, the rejection region consists of two smaller tails, and we would fail to reject again, since TS is still in the acceptance region. In order to reject the null hypothesis, we should increase the probability of a type I error. If we set $\alpha = 0.5$, the quantile marking the rejection region is $t_{0.75,9} = 0.7027$. In this case, $|\text{TS}| = 0.8308 > 0.7027$, i.e., the test statistic is in the rejection region. However, we have a very large probability of type I error; if $\alpha = 50\%$, we are basically flipping a coin.

One could wonder which value of α draws the line between acceptance and rejection. Using suitable software for evaluating the CDF of t distribution, we find that

$$P(\text{TS} \leq 0.8308) = 0.7862$$

Hence

$$P(\text{TS} > 0.8308) = 1 - 0.7862 = 0.2138$$

If we choose $\alpha > 2 \times 0.2138 = 42.76\%$, then we reject the null hypothesis. Note that we should double the probability above because this is a two-tail test. This leads to the p-value concept, which is discussed later.

To summarize, we failed to reject H_0. Can we say that we *accept* it? Well, this is a bit of a philosophical question, but probably one would not say that a sample mean $\overline{X} = 7.253$ really supports $H_0 : \mu = 5$. All we can say is that the evidence is not hard enough to reject it. By the way, I can tell you that in this hypothetical case the null hypothesis was indeed true; the sample was obtained by running a generator of pseudorandom variates, sampling a normal distribution with expected value 5 and standard deviation 10; in real life, you will never know.   ▯

The careful reader will probably find some close connection between what we have learned about confidence intervals and hypothesis testing. Indeed, it turns out that the testing procedure above could also be carried out in terms of confidence intervals. What we could do is:

- Compute a confidence interval with confidence level $1 - \alpha$

- Reject the null hypothesis if μ_0 does not fall inside the confidence interval

We prefer to avoid this way of reasoning in order to enforce the basic concepts about hypothesis testing. Furthermore, as we show in the next section, this way of thinking is not that helpful if the form of the null hypothesis is different, leading to a one-tail test.

9.3.1   One-tail tests

When the null hypothesis is of the form $H_0 : \mu = \mu_0$, we consider a two-tail rejection region. In many problems, the null hypotheses has the form $H_0 : \mu \geq \mu_0$ or $H_0 : \mu \leq \mu_0$ are more appropriate. As one could expect, this leads to a rejection region consisting of one tail. Before illustrating the technicalities involved, it is useful to consider a practical example.

Example 9.14 A firm manufactures a product whose average useful life is 1250 hours. This average comes from an extended experience in the past, and it can be considered as a very reliable estimate of the expected value of life. The firm is currently engaged in a product improvement program, based on a different manufacturing process. A sample of 30 items produced by this new process is tested, resulting in a sample mean of 1315 hours and a sample standard deviation of 70 hours. Can we say that the new process is really better than the old one?

This is a typical case lending itself to hypothesis testing, but here we are not interested in a two-tail test. What we would like to check is whether the difference between 1315 and 1250 hours is statistically significant and cannot be attributed to a lucky sample. In other words, we would like to reject the hypothesis that the expected life is still 1250, but we certainly would not be happy if the sample mean were less than that, as this would not support the claim that the firm did a good job. The correct way to state the problem, from the point of view of the firm itself, requires the null hypothesis

$$H_0 : \mu \leq 1250 \tag{9.13}$$

against the alternative:

$$H_a : \mu > 1250 \tag{9.14}$$

The firm would like to prove that the product has really been improved, but we should *not* state a null hypothesis like $H_0 : \mu \geq 1250$. This is quite tempting, and in fact many students fall into this trap and make different sorts of mistakes:

- Some go as far as to state hypotheses involving the sample mean 1315, which makes no sense since this is a sample statistic, not an expected value.

- Moreover, we should not run a two-tail test, as we would also reject the null hypothesis when the sample mean is 1000 hours, which does not really suggest that the product has been improved.

- Finally, a null hypothesis like $H_0 : \mu \geq 1250$ is true regardless of whether they improved the product (average life is larger than 1250) or not (average life is still 1250); hence, a test of such a hypothesis does not discriminate between the alternatives that we are interested in comparing.

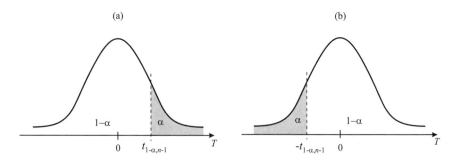

Fig. 9.5   Rejection regions for one-tail hypothesis testing.

We should keep in mind that, with standard test procedures, we can either *reject* or *fail to reject*. We do not really prove that a claim is true; we may just argue that there is strong evidence against a claim. So, if we want to support a statement, we should try to reject its negation. This is why, in this problem, the claim that the firm would like to prove plays the role of the alternative hypothesis (9.14).   ⏹

What kind of rejection region should be associated with the test in the example? The reasoning is almost the same as in two-tail testing, but it is easy to see that we should reject the null hypothesis if the test statistic TS is large, as this supports the one-sided alternative. If the test statistic is large, this means that the average life we observe is much larger than μ_0, and it is hard to explain this discrepancy by sampling variability alone. Indeed, the rejection region is the right tail, as shown in Fig. 9.5(a). Formally, the test should be run as follows:

$$\text{``Accept'' } H_0 \text{ if } \quad \frac{\sqrt{n}(\overline{X} - \mu_0)}{S} \leq t_{1-\alpha,n-1}$$

$$\text{Reject } H_0 \text{ if } \quad \frac{\sqrt{n}(\overline{X} - \mu_0)}{S} > t_{1-\alpha,n-1}$$

Note that here we should use a quantile with probability level $(1 - \alpha)$ instead of $(1 - \alpha/2)$. In this case, the rejection region consists of one tail, and the probability of a type I error is not split in two. In Example 9.14, the test statistic is

$$\text{TS} = \frac{\sqrt{30} \times (1315 - 1250)}{70} = 5.086 \tag{9.15}$$

The sample size is 30, so we are not really allowed to use quantiles from the standard normal distribution, but we are close; by remembering the $\pm 3\sigma$ rule, we see that the null hypothesis will be rejected for any sensible significance level when the test statistic is larger than 3, when using quantiles of the standard normal. If we insist in using the t distribution, say, with $\alpha = 0.5\%$, which is pretty small, the correct quantile is $t_{0.995,29} = 2.756 < \text{TS}$. So we see that we reject even if we require a very small probability of type I error.

To complete the picture, let us consider the case of

$$H_0 : \mu \geq \mu_0$$

against the alternative

$$H_a : \mu < \mu_0$$

Here, the rejection region is the left tail [see Fig. 9.5.(b)]:

$$\text{``accept'' } H_0 \text{ if } \quad \frac{\sqrt{n}(\overline{X} - \mu_0)}{S} \geq t_{\alpha, n-1}$$

$$\text{reject } H_0 \text{ if } \quad \frac{\sqrt{n}(\overline{X} - \mu_0)}{S} < t_{\alpha, n-1}$$

The quantile $t_{\alpha, n-1}$, with a sensibly small significance level ($\alpha < 50\%$), is actually negative. Given the symmetry of the t distribution, $t_{\alpha, n-1} = -t_{1-\alpha, n-1}$, and the rejection region is also characterized by the condition

$$\frac{\sqrt{n}(\overline{X} - \mu_0)}{S} < -t_{1-\alpha, n-1}$$

Before we proceed further, we should insist again that, strictly speaking, what we have said so far applies only to the expected value of a normal distribution. If the sample is normal, then we can say that the TS statistic has t distribution. In other cases, we have to carefully examine the distribution of the test statistic that we are considering, in order to properly set up the test; yet, the kind of reasoning is the same. It is also important to see the role of the two hypotheses in setting up the test:

- The alternative hypothesis determines the rejection region, which in the case we consider here might be the right tail, the left tail, or both. If the test statistic falls in one of these "extreme" regions, there is strong evidence against the null hypothesis.

- The null hypothesis specifies the probability distribution of the test statistic. In this simple setting, the null hypothesis just specifies the expected value μ_0. This is not so obvious for one-tail tests: If the null hypothesis is $H_0 : \mu \leq \mu_0$, why should we just consider μ_0 and not any smaller value, that would be compatible with the null hypothesis anyway? To see a heuristic justification, we should keep in mind the conservative nature of hypothesis testing. If TS falls in the rejection region assuming that $\mu = \mu_0$, i.e., if

$$\text{TS} = \frac{\sqrt{n}(\overline{X} - \mu_0)}{S} > t_{1-\alpha, n}$$

then it will also fall in the rejection region for any assumed value less than μ_0. By assuming $\mu = \mu_0$, we take the worst case from the point of view of the alternative hypothesis; this approach is consistent with the idea of keeping the probability of a type I error small. In Section 9.10 we offer a more rigorous look at hypothesis testing.

As a final observation, we have illustrated testing procedures for the mean of a normal population, when standard deviation σ is not known and must be estimated by its sample counterpart S. This is the standard case in business applications, but if σ were known, it would be easy to adapt the approach. We should just consider a test statistic $Z = \sqrt{n}(\overline{X} - \mu_0)/\sigma$, which is standard normal, and apply the procedures described above using quantiles of the standard normal distribution.

9.3.2   Testing with p-values

In the manufacturing example of the previous section we found such a large value for the test statistic that we are quite confident that the null hypothesis should be rejected, whatever significance level we choose. In other cases, finding a suitable value of α can be tricky. Recall that the larger the value of α, the easier it is to reject the null hypothesis. This happens because the rejection region, whether one- or two-tail, increases with α. We could find a case in which we "accept" the null hypothesis if $\alpha = 0.05$, but we reject it if $\alpha = 0.06$. This is clearly a critical situation, because the right confidence level is nowhere engraved on a rock. A useful concept from this perspective is the p-value.

It is easier to understand the concept referring to a one-tail test with H_0: $\mu \leq \mu_0$, vs. H_a: $\mu > \mu_0$. The rejection region, as shown in Fig. 9.5(b), is the right tail. If the value of the test statistic is TS $= t$, the p-value is defined as

$$p \equiv P(T_{n-1} > t), \tag{9.16}$$

where T_{n-1} is a t variable with $n - 1$ degrees of freedom. It is important to realize that we compute the p-value *after* having observed the sample: The random variable TS has been realized, and its numeric value is t. It is easy to see that we would reject the null hypothesis for any significance level $\alpha > p$, and we would fail to reject it for any value $\alpha < p$. Hence, calculating a p-value is a way to draw the line between rejection and failure to reject.

Example 9.15 Let us consider again the manufacturing case above, where H_0: $\mu \leq 1250$, H_a: $\mu > 1250$, $S = 70$, and $n = 30$. But now let us assume that the sample mean turns out to be $\overline{X} = 1260$. In this case, the value of the test statistic TS is

$$t = \frac{\sqrt{30} \times (1260 - 1250)}{70} = 0.7825$$

Table 9.3   Hypothesis testing about the mean of a normal population, when variance is unknown (TS = test statistic; α = significance level).

H_0	H_a	TS	Test with level α	p-value if TS = t
$\mu = \mu_0$	$\mu \neq \mu_0$	$\dfrac{\sqrt{n}(\overline{X} - \mu_0)}{S}$	Reject if $|\,\mathrm{TS}\,| > t_{1-\alpha/2, n-1}$	$2\mathrm{P}(T_{n-1} \geq |t|)$
$\mu \leq \mu_0$	$\mu > \mu_0$	$\dfrac{\sqrt{n}(\overline{X} - \mu_0)}{S}$	Reject if $\mathrm{TS} > t_{1-\alpha, n-1}$	$\mathrm{P}(T_{n-1} \geq t)$
$\mu \geq \mu_0$	$\mu < \mu_0$	$\dfrac{\sqrt{n}(\overline{X} - \mu_0)}{S}$	Reject if $\mathrm{TS} < -t_{1-\alpha, n-1}$	$\mathrm{P}(T_{n-1} \leq t)$

By using software or statistical tables of the CDF for a t distribution with 29 degrees of freedom, we obtain $\mathrm{P}(T_{n-1} \leq 0.7825) = 0.7799$. Hence

$$p = \mathrm{P}(T_{n-1} > 0.7825) = 1 - 0.7799 = 0.2201$$

This means that, in order to reject, we should admit a large probability, 22.01%, of committing a type I error. This seems a bit too large and, indeed, the difference between 1260 and 1250 does not seem that significant, for this sample size and this variability.

If the sample mean were $\overline{X} = 1290$, repeating the above calculation would yield $p = 0.002$. Hence, we would reject the null hypothesis for any significance level $\alpha > 0.2\%$. This is strong evidence that the difference between 1290 and 1250 is statistically significant and cannot be attributed to sampling variability alone. By the way, we should sit down and reflect about the difference between what is *statistically* significant and what is significant from the *business* perspective. If the values we are comparing refer to the useful life of some product, we should not take it for granted that the standard customer will notice the difference. The decision to switch to the new manufacturing process or not depends on the cost of the improved process with respect to the old one, and the awareness level of customers.   ▯

To complete the overall picture, with two-sided tests we reject in two cases: when $\mathrm{TS} < -t_{1-\alpha/2, n-1}$ and when $\mathrm{TS} > t_{1-\alpha/2, n-1}$. Hence, the p-value is the probability that the absolute value of a random variable T_{n-1} is larger than the absolute value of the test statistic. Exploiting symmetry of t distribution, we can write this probability as

$$p = 2\mathrm{P}(T_{n-1} \geq |t|) \tag{9.17}$$

when TS = t. The remaining one-tail test is easy to figure out, and Table 9.3 summarizes what we have found so far about hypothesis testing.

A remark on p-values   In closing this section, it is important to point out a common misunderstanding related to p-values, which is due to an incorrect way of reading (9.16) and (9.17). Since p-values are evaluated using probabilities, it is tempting to consider them as probabilities. However, this is wrong: p-values are random variables, not probabilities. True, we do calculate p-values using probabilities, but they depend on the numerical realization t of the test statistic TS, which is a random variable. If we take two random samples, we will find different p-values. So, we cannot consider them as probabilities of type I errors, which are given by the significance level α, which is specified *before* taking the sample. What p-values provide us with is a feeling for statistical significance. When a p-value is very small, this suggests that there is really strong evidence against the null hypothesis. In fact, many statistical software tools print something like `P > |t| = 0.000` or `Corresponding p-value < 0.005`. These are an indication of strong reasons for rejecting the null hypothesis, as the test statistic falls on the fair tails of the distribution. A large p-value suggests that we should take a large value of α to reject the null hypothesis, implying a large probability of committing a type I error. Since this is not safe, in such a case it is wise to admit that we cannot reject H_0.

9.4   BEYOND THE MEAN OF ONE POPULATION

In this section we generalize what we have seen so far concerning the expected value of a normal population. We consider the following problems:

- Estimating and testing the difference in the mean of two populations, which is used to assess if there are significant differences between them

- Hypothesis testing for variances of normal populations

- Testing Bernoulli populations, i.e., estimating and testing proportions

- Estimating covariances and related issues

- Estimating and testing a coefficient of correlation

- Estimating skewness and kurtosis

As you can see, this is a rather long list of topics. Because of space limitations, we will take a somewhat "cookbook" approach, cutting some corners to keep the treatment to a reasonable size. Nevertheless, we think that the previous discussions have provided the reader with a sensitivity for the pitfalls of inferential statistics techniques. Our aim is to illustrate the tools provided by statistical software packages and their rationale. We refer readers to the references for a more thorough and solid treatment.

9.4.1 Testing hypotheses about the difference in the mean of two populations

Sometimes, we have to run a test concerning two (or more) populations. For instance, we could wonder if two markets for a given product are really different in terms of expected demand. Alternatively, after the re-engineering of a business processes, we could wonder whether the new performance measures are significantly different from the old ones. In both cases, the rationalization of the problem calls for assessing the difference between two expected values, $\mu_1 - \mu_2$, where μ_1 and μ_2 are the expected values of two random variables. As we have seen, finding a confidence interval and hypothesis testing are both related to the distributional properties of a relevant statistic. Given our aim, it is quite natural to take samples from the two populations, with respective sizes n_1 and n_2, and exploit the statistic

$$\overline{X}_1 - \overline{X}_2 \tag{9.18}$$

i.e., the difference between the two sample means. Exactly what we should do depends on a number of questions:

- Is the number of observations, from both populations, large or small?

- Are the two variances known? If they are not, can we assume that they are equal?

- Are the samples from the two populations independent?

- Are the two populations normal?

In the following, we consider a subset of the possible cases, assuming normal populations, in order to illustrate the issues involved.

The case of large and independent samples   If the two samples are both large and mutually independent, the statistic (9.18) is, at least approximately, normally distributed. If the populations are normal, it is in fact normally distributed. Furthermore, independence allows to estimate the standard deviation of the difference by

$$S_{\overline{X}_1 - \overline{X}_2} = \sqrt{\frac{S_1^2}{n_1} + \frac{S_2^2}{n_2}} \tag{9.19}$$

where S_1^2 and S_2^2 are the two sample variances. The standardized statistic

$$Z = \frac{(\overline{X}_1 - \overline{X}_2) - (\mu_1 - \mu2)}{S_{\overline{X}_1 - \overline{X}_2}}$$

is (approximately) standard normal. Applying the same reasoning that we have used with a single normal population, the following confidence interval can be built:

$$(\overline{X}_1 - \overline{X}_2) \pm z_{1-\alpha/2} S_{\overline{X}_1 - \overline{X}_2}$$

On the basis of these estimates, it is also easy to test if the two populations are significantly different; in this case, the test boils down to checking whether the origin lies within the confidence interval.

Example 9.16 We want to compare the average yearly wage for two groups of professionals. Two independent samples are taken, with size $n_1 = 55$ and $n_2 = 65$, respectively; the observed sample means are $\overline{X}_1 = €71,000$ and $\overline{X}_2 = €67,000$, and sample standard deviations are $S_1 = €3,000$ and $S_2 = €5,000$. Can we say that the observed difference is statistically significant?

The first step is calculating the sample standard deviation according to Eq. (9.19):

$$S_{\overline{X}_1 - \overline{X}_2} = \sqrt{\frac{3000^2}{55} + \frac{5000^2}{65}} = 740.44$$

Then, we specify the null hypothesis,

$$H_0 : \mu_1 - \mu_2 = 0$$

and the alternative

$$H_a : \mu_1 - \mu_2 \neq 0$$

Accordingly, the test statistic is

$$Z = \frac{4000}{740.44} = 5.4022$$

Since $Z > 3$, the null hypothesis is rejected with any sensible value of the significance level. For instance, if we select $\alpha = 5\%$, the relevant quantile is $z_{0.975} = 1.96$, which is smaller than the test statistic. If we want a 95% confidence interval for the difference in the two population means, we obtain

$$4000 \pm 1.96 \times 740.44 \quad \Rightarrow \quad (€2548.76, €5451.24)$$

We see that 0 is not included in the confidence interval, suggesting again that the difference is statistically significant. ▯

The case of small and independent samples   With small samples (say, $n_1, n_2 < 30$), the procedure is not so simple, unless we know the two population variances σ_1^2 and σ_2^2. Since this is hardly the case, we cannot rely on the normality of the test statistic. A relatively easy case occurs when we may assume that the variances of the two populations are the same, since this allows pooling observations to estimate the common standard deviation as

$$S_p = \sqrt{\frac{(n_1 - 1)S_1^2 + (n_2 - 1)S_2^2}{n_1 + n_2 - 2}}$$

Note that the pooled estimator is based on a weighted combination of the two sample variances, where weights are related to the respective degrees of freedom $n_1 - 1$ and $n_2 - 1$. Then, we use the standard deviation of the statistic $\overline{X}_1 - \overline{X}_2$, i.e.

$$S_{\overline{X}_1 - \overline{X}_2} = S_p \sqrt{\frac{1}{n_1} + \frac{1}{n_2}}$$

to build the confidence interval

$$(\overline{X}_1 - \overline{X}_2) \pm t_{n_1+n_2-2,1-\alpha/2} \cdot S_{\overline{X}_1 - \overline{X}_2}$$

We may test hypotheses by a similar token. Here, we rely on a t distribution, which requires that the two populations be normal, and the total degrees of freedom are $n_1 + n_2 - 2$, since we estimate two means.

If the two variances are different, we could try again to resort to the t distribution, at least as a reasonable approximation, but it is not clear how many degrees of freedom we should use. A (nontrivial) distributional result justifies the following estimate:

$$\hat{f} = \frac{\left(\dfrac{S_1^2}{n_1} + \dfrac{S_2^2}{n_2} \right)^2}{\dfrac{1}{n_1 - 1} \left(\dfrac{S_1^2}{n_1} \right)^2 + \dfrac{1}{n_2 - 1} \left(\dfrac{S_2^2}{n_2} \right)^2}$$

Since in general $\hat{f}$ is not an integer, we may round it down (which makes sense, because with fewer degrees of freedom a confidence interval is larger and more conservative) and build the confidence interval

$$(\overline{X}_1 - \overline{X}_2) \pm t_{\hat{f},1-\alpha/2} \sqrt{\frac{S_1^2}{n_1} + \frac{S_2^2}{n_2}}$$

The case of paired observations: paired t testing   All of the procedures described above rely on the independence between the two samples. Now assume, on the contrary, that the samples are strictly related. Such a case occurs when the observations are actually paired. For instance, assume that we sample random financial scenarios, indexed by k, and we evaluate the performance of two portfolio management policies on each scenario, resulting in observations $X_k^{(1)}$ and $X_k^{(2)}$. In this case, we cannot say that the two observations are independent; arguably, both policies could result in a bad performance when applied in a recessive scenario. However, if we are just interested in checking if one of the two policies has a significant advantage over the other one, we can work directly with the observed differences

$$D_k \equiv X_k^{(1)} - X_k^{(2)}$$

Table 9.4 Testing the effectiveness of a preventive maintenance policy.

Plant	Before	After	D_k
1	18.5	21	-2.5
2	24.5	22	2.5
3	30.5	23	7.5
4	16	14.5	1.5
5	23.5	25.5	-2.0
6	25	21.5	3.5
7	18	23.5	-5.5
8	20	17.5	2.5
9	15	15.5	-0.5
10	32	28.5	3.5

and the statistics

$$\overline{D} = \frac{1}{n}\sum_{k=1}^{n} D_k$$

$$S_D = \sqrt{\frac{1}{n-1}\left(\sum_{k=1}^{n} D_k^2 - n\overline{D}^2\right)}$$

We see that, by pairing observations, we are back to the case of single population, and a confidence interval for the difference is

$$\overline{D} \pm t_{n-1,1-\alpha/2}\frac{S_D}{\sqrt{n}}$$

We may also test the difference, running a test which is aptly called the *paired t test*.

Example 9.17 A large corporation runs 10 production plants around the world, which suffer from excessive downtimes, i.e., wasted time because of machine breakdowns. A new preventive maintenance policy is applied, and the corporation would like to check whether it has been effective. To this aim, the data illustrated in Table 9.4 are collected. For each plant, we know the monthly number of hours actually lost, before and after the introduction of the new preventive maintenance policy. The last column shows the reduction of lost hours, where a negative sign implies that actually more production capacity was lost after the new policy was implemented. Of course, this may occasionally happen and does not imply that the policy is ineffective. If we denote by μ_1 the expected lost hours before and by μ_2 the expected lost hours after changing the maintenance process, we should test the null hypothesis

$$H_0: \mu_1 \leq \mu_2$$

versus the alternative

$$H_a: \mu_1 > \mu_2$$

The test statistics are

$$\overline{D} = 1.05, \quad S_D = 3.7301, \quad TS = \frac{1.05}{3.7301} = 0.2815$$

It is easy to see that we cannot reject the null hypothesis for any safe significance level. For instance, if we choose $\alpha = 0.1$, the relevant quantile is $t_{0.9,9} = 1.3830$. The rejection region is the right tail, but TS < 1.3830. The p-value is

$$P(T_9 > 0.2815) = 0.3923$$

which is definitely too large to conclude that the new maintenance procedure has been effective. □

9.4.2   Estimating and testing variance

It is easy to prove that sample variance S^2 is an unbiased estimator of variance σ^2, but if we want a confidence interval for variance, we need distributional results on S^2, which depend on the underlying population. For a normal population we may take advantage of Theorem 9.4. In particular, we recall that the sample variance is related to the chi-square distribution as follows:

$$\frac{(n-1)S^2}{\sigma^2} \sim \chi^2_{n-1}$$

where χ^2_{n-1} denotes a chi-square distribution with $n-1$ degrees of freedom. The conceptual path we should follow is the same that have seen for the expected value, with a slight difference: Unlike the t distribution, the chi-square distribution is not symmetric and has only positive support (see Fig. 7.18). To build a confidence interval with confidence level $(1 - \alpha)$, we need the two quantiles, $\chi^2_{\alpha/2,n-1}$ and $\chi^2_{1-\alpha/2,n-1}$, defined by the conditions[15]

$$P\left(K_2 \leq \chi^2_{\alpha/2,n-1}\right) = \frac{\alpha}{2}, \qquad P\left(K_2 \leq \chi^2_{1-\alpha/2,n-1}\right) = 1 - \frac{\alpha}{2}$$

where $K_2 \sim \chi^2_{n-1}$. Both quantiles are positive, and $\chi^2_{\alpha/2,n-1} \leq \chi^2_{1-\alpha/2,n-1}$. Then we have

$$P\left(\chi^2_{\alpha/2,n-1} \leq \frac{(n-1)S^2}{\sigma^2} \leq \chi^2_{1-\alpha/2,n-1}\right) = 1 - \alpha$$

---

[15] We always use a notation whereby quantiles are associated with the probability that they leave on the left. There are books where $\chi^2_{\alpha/2,n-1}$ is the larger quantile, since the probability $\alpha/2$ is associated with the area right tail, but we prefer a more uniform notation.

which after rearranging leads to the confidence interval

$$\left(\frac{(n-1)S^2}{\chi^2_{1-\alpha/2,n-1}}, \frac{(n-1)S^2}{\chi^2_{\alpha/2,n-1}} \right) \tag{9.20}$$

We may also test a hypothesis about variance, just as we did for the expected value:

$$H_0: \sigma^2 = \sigma_0^2, \qquad \text{vs.} \qquad H_a: \sigma^2 \neq \sigma_0^2$$

The distributional result above implies that, under the null hypothesis, we have

$$\frac{(n-1)S^2}{\sigma_0^2} \sim \chi^2_{n-1} \tag{9.21}$$

From Section 7.7.2 we also recall that the expected value of a χ^2_{n-1} variable is $n-1$. Then, if the test statistic has a value that is sensibly smaller or sensibly larger than $n-1$, we should question the null hypothesis. The test procedure is the following:

$$\text{"accept" } H_0 \text{ if } \quad \chi^2_{\alpha/2,n-1} \leq \frac{(n-1)S^2}{\sigma_0^2} \leq \chi^2_{1-\alpha/2,n-1}$$

$$\text{reject } H_0 \quad \text{otherwise}$$

Example 9.18 The following sample:

20.7533	46.6777	-35.1769	27.2435	16.3753
-16.1538	1.3282	16.8525	81.5679	65.3887

was generated by a pseudorandom variate generator; the underlying distribution was normal, with $\mu = 10$ and $\sigma = 20$. Now let us forget what we know and build a confidence interval for standard deviation, with confidence level 95%. The sample standard deviation is

$$S = 35.3977$$

To apply Eq. (9.20) we need the following quantiles from the chi-square distribution with 9 degrees of freedom:

$$\chi^2_{\alpha/2,n-1} = 2.7004, \quad \chi^2_{1-\alpha/2,n-1} = 19.0228$$

These quantiles, unlike those from the normal distribution, are not symmetric and yield the confidence interval

$$\left(\frac{9 \times 35.3977^2}{19.0228}, \frac{9 \times 35.3977^2}{2.7004} \right) = (592.81, \ 4176.05)$$

Taking square roots, we notice that the confidence interval for the standard deviation

$$(24.35, \ 64.62)$$

does not include the true value $\sigma = 20$, which may happen with probability 5%. If we want to test the null hypothesis

$$H_0 : \sigma = 20$$

against the alternative hypothesis $H_a : \sigma \neq 20$, with significance level $\alpha = 0.05$, we calculate the test statistic according to Eq. (9.21):

$$\frac{9 \times 35.3977^2}{20^2} = 28.19$$

This value looks pretty large. In fact, $28.19 > \chi^2_{1-\alpha/2,n-1} = 19.0228$, and the null hypothesis is (incorrectly) rejected. Since this was a two-tail test, we could have equivalently observed that $\sigma_0 = 20$ was not included in the confidence interval above. Again, we must be aware that type I errors are a real possibility. □

If we have to compare the variances of two populations, we should run a test such as

$$H_0 : \sigma_1^2 = \sigma_2^2 \qquad \text{vs.} \qquad H_a : \sigma_1^2 \neq \sigma_2^2$$

If both populations are normal and we take two *independent* samples of size n_1 and n_2, respectively, from Theorem 9.4 we know that

$$\frac{(1-n_1)S_1^2}{\sigma_1^2} \sim \chi^2_{n_1-1}, \qquad \frac{(1-n_2)S_2^2}{\sigma_2^2} \sim \chi^2_{n_2-1}$$

where S_1^2 and S_2^2 are the sample variances for the two samples. Hence, we have two independent chi-square variables. In Section 7.7.2 we have seen that the ratio of two independent chi-square variables is related to the F distribution; more precisely, $(S_1^2/\sigma_1^2)/(S_2^2/\sigma_2^2)$ has F distribution with $n_1 - 1$ and $n_2 - 1$ degrees of freedom. Then, under the null hypothesis, we have

$$\frac{S_1^2}{S_2^2} \sim F(n_1 - 1, n_2 - 1)$$

Using the familiar logic, we should run the following test:

$$\text{``accept'' } H_0 \text{ if } \quad F_{\alpha/2,n_1-1,n_2-1} \leq \frac{S_1^2}{S_2^2} \leq F_{1-\alpha/2,n_1-1,n_2-1}$$

$$\text{reject } H_0 \quad \text{otherwise}$$

Example 9.19 A reliable production process should yield items with low variability of key measures related to their quality. Imagine that we compare two technologies to check if they are significantly different in terms of such variability. A sample of $n_1 = 10$ items obtained by process 1 yields $S_1^2 = 0.15$, whereas process 2 yields $S_2^2 = 0.29$, for sample size $n_2 = 12$. Can we say that there is a significant difference in variance?

Let us choose $\alpha = 10\%$. The test statistic is

$$F = \frac{0.15}{0.29} = 0.5172$$

By using suitable statistical software, we find the following quantiles of the F distribution with 9 and 11 degrees of freedom:

$$F_{0.05,9,11} = 0.3223, \qquad F_{0.95,9,11} = 2.8962$$

Since $F = 0.5172 \in [0.3223, 2.8962]$, we cannot reject the null hypothesis. Superficially, it seems that there is quite a difference in the two variances, as one is almost twice as much as the other one, but the sample sizes are too small to draw a reliable conclusion. If we had $n_1 = 100$ and $n_2 = 120$, then we would use

$$F_{0.05,99,119} = 0.7258$$

and we could reject the null hypothesis. $\square$

Once again, we see that if we have distributional results on relevant statistics, we can follow the usual drill to come up with confidence intervals, hypothesis tests, p-values, etc.

9.4.3   Estimating and testing proportions

So far, we have mostly relied on normality of observations or, at least approximately, on normality of the sample mean for large samples. However, there are cases in which we should be a bit more specific and devise approaches which are in tune with the kind of observations we are taking. Such a case occurs when dealing with a property that may or may not hold for each observed element of a population. In Example 9.6 we considered sampling a population and estimating the fraction of its members that enjoy a generic property. Conceptually, we are sampling a Bernoulli population with unknown parameter p. Letting $X_i = 1$ if observation i is "yes," 0 otherwise, a natural estimator of p is

$$\hat{p} = \frac{1}{n} \sum_{i=1}^{n} X_i$$

Since $n\hat{p}$ has binomial distribution, we should use quantiles from the binomial distribution to build confidence intervals and to test hypotheses about p. This is not difficult, as these quantiles have been tabulated. However, if the sample is large enough, we may rely on the central limit theorem to conclude that

$$\frac{\sum_{i=1}^{n} X_i - np}{\sqrt{np(1-p)}} = \frac{\hat{p} - p}{\sqrt{p(1-p)/n}} \sim \mathcal{N}(0, 1)$$

at least approximately. An alternative view is that, essentially, we are approximating a binomial distribution by a normal, but we are relating both its expected value and variance to parameters p and n; since n is known, we are relating two features to one unknown parameter, without losing sight of the structure of the binomial distribution. Then, for a large sample, we have

$$P\left(-z_{1-\alpha/2} \leq \frac{\hat{p} - p}{\sqrt{\dfrac{p(1-p)}{n}}} \leq z_{1-\alpha/2}\right) \approx 1 - \alpha$$

where $z_{1-\alpha/2}$ is the usual quantile of the standard normal. Note that the familiar drill for the normal distribution does not work in this case. The problem is that the unknown parameter p occurs in a complicated way, since it also gives variance. To find a confidence interval in the usual form, we should substitute $\hat{p}$ for p in the denominator of the ratio above. This yields the *approximate* confidence interval

$$\left(\hat{p} - z_{1-\alpha/2}\sqrt{\frac{\hat{p}(1-\hat{p})}{n}}, \ \hat{p} + z_{1-\alpha/2}\sqrt{\frac{\hat{p}(1-\hat{p})}{n}}\right)$$

This confidence interval looks much like the confidence interval for the mean of a normal population, with sample variance S^2 substituted by $\hat{p}(1 - \hat{p})$, which is an estimate of the variance of a Bernoulli random variable. This is so natural that one tends to forget that there are two approximations involved here. The first one has distributional nature and is justified by the central limit theorem; the second one relies on the estimate of variance of a Bernoulli random variable.

Using the same machinery, we may run hypothesis tests. A natural hypothesis that we may wish to test is

$$H_0 : p \leq p_0 \qquad \text{vs.} \qquad H_a : p > p_0$$

Under the null hypothesis, we may argue that the test statistic

$$\frac{\sum_{i=1}^{n} X_i - np_0}{\sqrt{np_0(1-p_0)}} \tag{9.22}$$

has approximately standard normal distribution. We have to rely on the central limit theorem here, too; however, since we are plugging the number p_0 from the null hypothesis, there is no other trouble. Clearly, we are inclined to reject H_0 if the count of "yes" answers in the sample is too large, i.e., if

$$\frac{\sum_{i=1}^{n} X_i - np_0}{\sqrt{np_0(1-p_0)}} \geq z_{1-\alpha} \tag{9.23}$$

for a significance level α.

Example 9.20 According to process specifications, a certain machine should produce no more than 5% defective parts. Then, if we take a sample of 300 parts, the fraction of defective items should be something like $300 \times 0.05 = 15$. Now assume that, as a matter of fact, we observe 19 defective items. Is this finding compatible with the above percentage? We should test the null hypothesis

$$H_0 : p \leq 0.05$$

against the alternative hypothesis $H_a : p > 0.05$. Using the normal approximation, the test statistic (9.22) is

$$\frac{19 - 15}{\sqrt{300 \times 0.05 \times 0.95}} = 1.0596$$

Comparing this value against the quantile $z_{0.95} = 1.6449$, we see that we cannot reject the null hypothesis at 5% significance level. If we use suitable software, we find that the quantile at 95% for the binomial distribution with parameters $n = 300$ and $p = 0.05$ is $b = 21$. So, we should observe at least 22 defective items to reject the null hypothesis. We may check the quality of the normal approximation by finding this threshold number with normal quantiles. Using (9.23) we find

$$\sum_{i=1} n X_i \geq n p_0 + z_{1-\alpha} \sqrt{n p_0 (1 - p_0)}$$

$$= 300 \times 0.05 + 1.6449 \times \sqrt{300 \times 0.05 \times 0.95} = 21.2092$$

which is compatible with the exact quantile of the binomial distribution.   ⬜

The example suggests that the normal approximations works fairly well, but care must be exercised when dealing with small sample sizes and probabilities. We should stress that common wisdom suggests that a sample size should be at least 30 to use normal approximations, but this rule of thumb does not apply here as it disregards the impact of p. It is often suggested that the product np should be at least 20 to rely on the normal approximation.

9.4.4   Estimating covariance and related issues

Just as we have defined sample variance, we may define *sample covariance* S_{XY} between random variables X and Y:

$$S_{XY} = \frac{1}{n-1} \sum_{i=1}^{n} (X_i - \overline{X})(Y_i - \overline{Y}) \tag{9.24}$$

where n is the size of the sample, i.e., the number of observed *pairs* (X_i, Y_i). Sample covariance can also be rewritten as follows:

$$S_{XY} = \frac{1}{n-1} \left(\sum_{i=1}^{n} X_i Y_i - n \overline{X}\,\overline{Y} \right)$$

To see this, we note the following:

$$\sum_{i=1}^{n}(X_i - \overline{X})(Y_i - \overline{Y}) = \sum_{i=1}^{n}X_iY_i - \sum_{i=1}^{n}X_i\overline{Y} - \sum_{i=1}^{n}\overline{X}Y_i + \sum_{i=1}^{n}\overline{X}\,\overline{Y}$$

$$= \sum_{i=1}^{n}X_iY_i - n\overline{X}\,\overline{Y} - n\overline{X}\,\overline{Y} + n\overline{X}\,\overline{Y}$$

$$= \sum_{i=1}^{n}X_iY_i - n\overline{X}\,\overline{Y} \qquad (9.25)$$

This rewriting mirrors the relationship $\sigma_{XY} = E[XY] - \mu_X\mu_Y$ from probability theory. It is important to realize that our sample must consist of *joint* realizations of variables X and Y. If we want to investigate the impact of temperature on ice cream demand, we must have pairs of observations taken in the same place at the same time; clearly, mixing observations is no use.

This definition is consistent with sample variance, since $S_X^2 = S_{XX}$. Yet, one thing that may not look that obvious is why we should divide by $n - 1$. When dealing with variance, a common justification is that we lose one degree of freedom because we estimate one parameter, the unknown expected value. This simple rule does not sound convincing here, as we use two sample means $\overline{X}$ and $\overline{Y}$ as estimators of the respective expected values. Indeed, the best idea is to really prove that the estimator above is unbiased.

THEOREM 9.6 *The sample covariance (9.24) is an unbiased estimator of covariance, i.e.,* $E[S_{XY}] = \sigma_{XY}$.

PROOF   Using (9.25) and the fact that the pairs (X_i, Y_i) are identically distributed, we obtain

$$E\left[\sum_{i=1}^{n}X_iY_i - n\overline{X}\,\overline{Y}\right] = \sum_{i=1}^{n}E[X_iY_i] - nE[\overline{X}\,\overline{Y}]$$

$$= nE[XY] - \frac{1}{n}E\left[\sum_{i=1}^{n}X_i\sum_{j=1}^{n}Y_j\right]$$

$$= nE[XY] - \frac{1}{n}\sum_{i=1}^{n}\sum_{j=1}^{n}E[X_iY_j] \qquad (9.26)$$

Since the pairs are also independent, for $i \neq j$ we have

$$E[X_iY_j] = E[X_i] \cdot E[Y_j] = \mu_X\mu_Y$$

There are $n^2 - n = n(n - 1)$ such terms in the last double sum in Eq. (9.26), hence

$$\sum_{i=1}^{n}\sum_{j=1}^{n}E[X_iY_j] = nE[XY] - n(n - 1)\mu_X\mu_Y$$

So we see that

$$
\mathrm{E}\left[\sum_{i=1}^{n} X_i Y_i - n\overline{X}\,\overline{Y}\right] = n\mathrm{E}[XY] - \frac{1}{n}\left(n\mathrm{E}[XY] - n(n-1)\mu_X\mu_Y\right)
$$

$$
= (n-1)[\mathrm{E}[XY] - \mu_X\mu_Y] = (n-1)\sigma_{XY}
$$

which proves the result.   ∎

Now let us consider a quite practical question. If we have a large set of jointly distributed random variables, what is the required effort if we want to estimate their covariance structure? Equivalently, how many correlations we need? The covariance matrix is symmetric; hence, if we have n random variables, we have to estimate n variances and $n(n-1)/2$ covariances. This amounts to

$$
n + \frac{n(n-1)}{2} = \frac{n(n+1)}{2}
$$

entries. Hence, if $n = 1,000$, we should estimate 500,500 entries in the covariance matrix. A daunting task, indeed! If you think that such a case will never occur in practice, please consider Example 8.5, on portfolio management. You might well consider 1,000 assets for inclusion in the portfolio. In such a case, can we estimate the covariance matrix? What you know about statistical inference tells that you might need a lot of data to come up with a reliable estimate of a parameter. If you have to estimate a huge number of parameters, you need a huge collection of historical data. Unfortunately, many of them would actually tell us nothing useful: Would you use data from the 1940s to characterize the distribution of returns for IBM stock shares now? We need a completely different approach to reduce our estimation requirements.

Example 9.21 (Single-factor models for portfolio management) The returns of a stock share are influenced by many factors. Some are general economic factors, such as inflation and economic growth. Others are peculiar factors of a single firm, depending on its management strategy, product portfolio, etc. In between, we may have some factors that are quite relevant for a group of firms within a specific industrial sectors, much less for others; think of the impact of oil prices on energy or telecommunications.

Rather than modeling uncertain returns individually, we might try to take advantage of this structure of general and specific factors. Let us take the idea to an extreme and build a simple model whereby there is one factor common to all of stock shares, and a specific factor for each single firm. Formally, we represent the random return for the stock share of firm i as

$$
R_i = \alpha_i + \beta_i R_m + \epsilon_i
$$

where

- α_i and β_i are parameters to be estimated.

- R_m is a random variable representing the common risk factor; the subscript m stands for *market*; indeed, financial theory suggests that a suitable common factor could be the return of a market portfolio consisting of all stock shares, with a proportion depending on their relative capitalization with respect to the whole market.

- ϵ_i is a random variable representing individual risk, which in financial parlance is referred to as **idiosyncratic** risk; a natural assumption about these variables is that $\mathrm{E}[\epsilon_i] = 0$ (otherwise, we would include the expected value into α_i). Another requirement is that the common factor really captures whatever the stock returns have in common, and that the specific factors are independent. Typically, we do not require independence, but only lack of correlation. These requirements can be formalized as:

$$\mathrm{Cov}(\epsilon_i, R_m) = 0 \qquad\qquad (9.27)$$
$$\mathrm{Cov}(\epsilon_i, \epsilon_j) = 0, \qquad i \neq j \qquad (9.28)$$

In the next chapter, where we deal with linear regression, we will see that condition (9.27) is actually ensured by model estimation procedures based on least squares. On the contrary, condition (9.28) is just an assumption, resulting in a so-called *diagonal model*, since the covariance matrix of specific factors is diagonal.

The model above is called single-factor model for obvious reasons, but what are its advantages from a statistical perspective? Let us check how many unknown parameters we should estimate in order to evaluate expected return and variance of return for an arbitrary portfolio. To begin with, observe that for a portfolio with weights w_i, we have

$$R_p = \sum_{i=1}^{n} w_i(\alpha_i + \beta_i R_m + \epsilon_i) = \sum_{i=1}^{n} w_i\alpha_i + R_m \sum_{i=1}^{n} w_i\beta_i + \sum_{i=1}^{n} w_i\epsilon_i$$

Then,

$$\mathrm{E}[R_p] = \sum_{i=1}^{n} w_i\alpha_i + \mathrm{E}[R_m]\sum_{i=1}^{n} w_i\beta_i + \sum_{i=1}^{n} w_i\mathrm{E}[\epsilon_i] = \sum_{i=1}^{n} w_i\alpha_i + \mu_m \sum_{i=1}^{n} w_i\beta_i$$

where μ_m is expected return of the market portfolio (more generally, the expected value of whatever common risk factor we choose). From this, we see that we need to estimate:

- n parameters α_i, $i = 1, \ldots, n$

- n parameters β_i, $i = 1, \ldots, n$

- The expected value μ_m

These add up to $2n + 1$ parameters. Variance is a bit trickier, but we may use the diagonality condition (9.28) to eliminate covariances and obtain

$$\text{Var}(R_p) = \text{Var}(R_m) \left(\sum_{i=1}^{n} w_i \beta_i \right)^2 + \sum_{i=1}^{n} w_i^2 \text{Var}(\epsilon_i) + 2 \sum_{i \neq j} w_i w_j \text{Cov}(\epsilon_i, \epsilon_j)$$

$$= \sigma_m^2 \left(\sum_{i=1}^{n} w_i \beta_i \right)^2 + \sum_{i=1}^{n} w_i^2 \sigma_i^2$$

where σ_m^2 is the variance of the common risk factor and σ_i^2 is the variance of each idiosyncratic risk factor, $i = 1, \ldots, n$. They amount to $n + 1$ additional parameters that we should estimate, bringing the total to $3n + 2$ parameters. In the case of $n = 1,000$ assets, the we have a grand total of $3,002$ parameters; this is a large number, anyway, but pales when compared with the $500,500$ entries of the full covariance matrix. $\square$

This example is quite instructive. Please keep in mind that we are not estimating parameters for fun. This is a book about management, and estimates are supposed to be used as an input to decision-making procedures; as you may expect, wrong estimates may lead to poor decisions. As it often turns out in practice, the more sophisticated the decision process, the larger the impact of wrong estimates.

The single-factor model we just outlined has a pervasive impact in both portfolio management and corporate finance. In fact, it is the starting point of a well-known equilibrium model, the *capital asset pricing model* (CAPM). To put it simply, CAPM is based on rewriting the single-factor model in terms of excess returns with respect to the risk-free return r_f:

$$R_i - r_f = \alpha_i + \beta_i (R_m - r_f) + \epsilon_i$$

By taking expected values, we obtain

$$\mu_i - r_f = \alpha_i + \beta_i(\mu_m - r_f) \tag{9.29}$$

It is useful to interpret the expected excess return $\mu_i - r_f$ as a *risk premium*, since it is the return above the risk-free rate that an investor expects if she holds the risky asset i; by the same token, $\mu_m - r_f$ is the risk premium from holding the market portfolio, i.e., a broadly diversified portfolio reflecting the relative market capitalization of firms. Then, according to CAPM, the following conditions hold at equilibrium in Eq. (9.29):

$$\alpha_i = 0 \tag{9.30}$$

$$\beta_i = \frac{\text{Cov}(R_i, R_m)}{\text{Var}(R_m)} \tag{9.31}$$

These conditions state that there is *no specific* risk premium for asset i; the risk premium is only due to the correlation of its return with the general market return. This is a controversial result, relying on many debatable assumptions, and we cannot really discuss it in any detail. However, there are two possibilities about CAPM: Either it is valid model, or it is not.

1. If it is a valid model, the practical implication is that you should not pay any financial analyst, since the best that you can do is to invest in a market portfolio, surrogated by a wide market index. There is no hidden alpha to take advantage of by active stock picking, and only efficiency in passive portfolio management matters. In fact, this is why exchange traded funds (ETFs) are so popular; they are passive funds tracking broad market indexes at low cost. Of course, if one believes CAPM, statistics should be used to support the thesis empirically.

2. If it is not a valid model, then you should try to manage a portfolio actively. This means that you should try to use statistics to estimate parameters in order to gain differential knowledge that can be used to make money. Furthermore, from an organizational perspective this task may be decomposed in two parts, one pertaining to general market conditions (involving macroeconomic factors), and one pertaining to specific information about a firm.

This brief discussion should convince you about the fundamental role of quantitative methods in practice, and their impact on organization of a portfolio management firm.

More generally, this motivates the use of statistical models, which we begin discussing in Chapter 10 on simple linear regression. As you may imagine, it is hard to believe that one single factor can really capture everything stock shares may have in common. As a result, we do not have a diagonal model, since there is some commonality left unexplained by a single factor. This leads us to consider multiple factor models, which need multivariate statistical methods that are outlined in Part IV.

9.4.5 Correlation analysis: testing significance and potential dangers

To estimate the correlation coefficient ρ_{XY} between X and Y, we may just plug sample covariance S_{XY} and sample standard deviations S_X, S_Y into its definition, resulting in the *sample coefficient of correlation*, or sample correlation for short:

$$R_{XY} = \frac{S_{XY}}{S_X S_Y} = \frac{\sum_{i=1}^{n}(X_i - \overline{X})(Y_i - \overline{Y})}{\sqrt{\sum_{i=1}^{n}\left(X_i - \overline{X}\right)^2} \cdot \sqrt{\sum_{i=1}^{n}\left(Y_i - \overline{Y}\right)^2}} \qquad (9.32)$$

The factors $n-1$ in S_{XY}, S_X, and S_Y cancel each other, and it can be proved that $-1 \leq r_{XY} \leq +1$, just like its probabilistic counterpart ρ_{XY}.

Once again, we stress that the estimator that we have just defined is a *random variable* depending on the random sample we take. Whenever we estimate a nonzero correlation coefficient, we should wonder whether it is statistically significant. Even if the "true" correlation is $\rho_{XY} = 0$, because of sampling variability it is quite unlikely that we get a sample correlation $R_{XY} = 0$. How can we decide if a nonzero sample correlation is really meaningful? A simple strategy is to test the null hypothesis

$$H_0: \ \rho_{XY} = 0$$

against the alternative hypothesis

$$H_a: \ \rho_{XY} \neq 0$$

However, we need a statistic whose distribution under the null hypothesis is fairly manageable. One useful result is that, if the sample is normal, the statistic

$$T = R_{XY}\sqrt{\frac{n-2}{1-R_{XY}^2}}$$

is approximately distributed as a t variable with $n-2$ degrees of freedom, for a suitably large sample. In the following example we show how to take advantage of this distributional result.

Example 9.22 Say that you are convinced that there is a positive correlation between normal random variables X and Y. You take a sample of 12 joint observations. Which is the minimum value of correlation that you would find statistically significant? In this case, we should consider a one-sided test, with null hypothesis

$$H_0: \ \rho_{XY} \leq 0$$

tested against the alternative one

$$H_a: \ \rho_{XY} > 0$$

Assume that we test with the standard significance level $\alpha = 5\%$. The degrees of freedom we should consider are $n - 2 = 10$, so the rejection region is $T > t_{0.05,10} = 1.812$. Now we have to transform this rejection region for the T statistic into a rejection region for the sample correlation. Setting

$$R_{XY}\sqrt{\frac{10}{1-R_{XY}^2}} = 1.812$$

we solve for R_{XY}:

$$\sqrt{1 - R_{XY}^2} = R_{XY} \cdot \frac{\sqrt{10}}{1.812}$$

$$\Rightarrow 1 - R_{XY}^2 = R_{XY}^2 \cdot \frac{10}{1.812^2}$$

$$\Rightarrow R_{XY} = \frac{1}{\sqrt{1 + \frac{10}{1.812^2}}} = 0.4972$$

where, of course, we should take the positive root. ☐

Correlation analysis is very useful, but as with any other tool, we must be well aware of its pitfalls and limitations to use it properly. They are a direct consequence of a few observations we made in Chapter 8, where we analyzed the link between independence and lack of correlation. We repeat them here for convenience, along with some additional warning.

Correlation measures association, not causation: Lurking variables   A common misunderstanding is the confusion of correlation with *causation*. When X and Y are correlated, it is tempting to conclude that X "causes" Y. This may be true, but it is knowledge of the phenomenon that allows us to draw such a conclusion. Correlation per se does not measure anything except a *symmetric* association. In fact, the definitions of covariance and correlation are symmetric, and this also applies to their sample counterparts: $R_{XY} = R_{XY}$. Therefore, it is not at all clear which variable is the cause and which one is the effect.

Sometimes, we might detect a spurious correlation between X and Y, which is actually the effect of another variable Z. To see how this may happen, imagine that there is indeed a causal relationship between Z and Y, where Z is the cause and Y the effect, which is reflected by a positive correlation between Z and Y. If Z is positively correlated with X, because of a pure noncausal association, we will detect a positive correlation between X and Y as well; however, this positive correlation does not reflect any cause-effect relationship. In such a case, we call Z a *lurking variable*.

Example 9.23 Consider the relationship between how much we spend in advertisements, measured by X, and demand, measured by Y. If we detect a positive correlation between X and Y, we might be satisfied by our last marketing campaign. But now suppose that Z measures the discount we offer to support promotional sales. Quite often, a marketing campaign involves increased advertising and a reduction in price to expand the customer base. So, it *might* be the case that the real cause of the increase in demand is the reduction in price. ☐

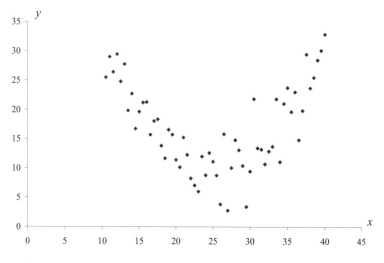

Fig. 9.6   Scatterplot of a nonlinear relationship between X and Y.

Lurking variables are a quite common issue and we may be easily lead to wrong conclusions. We will see other examples in Section 16.2.1, when dealing with multiple regression models.

Correlation measures only linear associations   We have already pointed out that, in general, lack of correlation does not imply independence. When the relationship between X and Y is nonlinear, the coefficient of correlation could not reflect this link at all. An example is shown in Fig. 9.6, where there is indeed a link between the two variables, but sample correlation is practically zero. This happens because when Y is larger than its mean, X can be larger or smaller than its mean (see also Example 8.4). To overcome this difficulty, there are a few tricks that can be used:

- One possible strategy is a nonlinear transformation of variables. Sometimes, rather than considering X, we may take $\sqrt{X}$ or $\log X$. These nonlinear transformations are commonly used to develop statistical models.

- Another possibility is to account for nonlinearity explicitly, relating X and Y by a nonlinear model; nonlinear regression is outlined in Section 16.4.

The King-Kong effect   Another danger of correlation analysis is the impact of a single "odd" observation. The issue is illustrated in Fig. 9.7. In Fig. 9.7(a) we see a random sample from a bivariate normal distribution with zero correlation ($\rho_{XY} = 0$). The sample correlation is very small, $R_{XY} = 0.0275$, and this is not significant and clearly compatible with sampling variability. Now assume that we add a single observation, $(X_k, Y_k) = (40, 40)$, which is quite far apart, as shown in Fig. 9.7(b). Now sample correlation jumps to

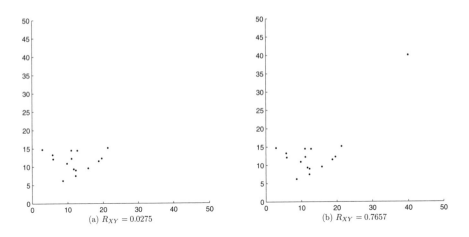

Fig. 9.7   An illustration of the King Kong effect.

$R_{XY} = 0.7657$, because of the impact of the two deviations $(X_k - \overline{X})$ and $(Y_k - \overline{Y})$ on sample correlation. This is often called the "King Kong" or "Big Apple" effect.[16] In practice, the King-Kong effect might be the effect of an *outlier*, i.e., a measurement error or, possibly, an observation that actually belongs to a different population.

9.4.6   Estimating skewness and kurtosis

We have defined skewness and kurtosis as:[17]

$$\gamma \equiv E\left[\frac{(X - \mu_X)^3}{\sigma^3}\right], \quad \kappa = E\left[\frac{(X - \mu_X)^4}{\sigma^4}\right]$$

These definitions are related to higher-order moments of random variables. Just like expected value and variance, these are probabilistic definitions, and we should wonder *if* and *how* these measures should be estimated on the basis of sampled data. The "if" should not be a surprise. If we know that the sampled population is normal, there is no point in estimating skewness and kurtosis, since we know that $\gamma = 0$ and $\kappa = 3$ for a normal distribution. By the same token, for other distributions, skewness and kurtosis are related to parameters of the distribution. For instance, a beta distribution is defined in terms of two parameters, α_1 and α_2; any other feature depends on these parameters. Hence, we should really estimate these parameters, rather than

---

[16] Statistics on towns in the United States may be affected by the inclusion of New York, which has peculiar characteristics.
[17] See Section 7.4.

the expected value, variance, skewness, or kurtosis. Indeed, we will take this more general view in Section 9.9.

On the contrary, if we do not take a specific distribution for granted, we might be interested in an estimate of skewness and kurtosis. One possible way to estimate these characteristics relies on the definition of higher-order moments.[18] If we denote the central moment of order k by

$$m_k = \mathrm{E}\left[(X - \mu_X)^k\right]$$

we have

$$\gamma = \frac{m_3}{(m_2)^{3/2}}, \quad \kappa = \frac{m_4}{(m_2)^2}$$

Then, we could consider *sample moments* of order k

$$M_k = \frac{1}{n} \sum_{i=1}^{n} (X_i - \overline{X})^k$$

and define sample skewness as follows:

$$G = \frac{M_3}{(M_2)^{3/2}} = \frac{\dfrac{1}{n} \sum_{i=1}^{n} (X_i - \overline{X})^3}{\left[\dfrac{1}{n} \sum_{i=1}^{n} (X_i - \overline{X})^2\right]^{3/2}}$$

By the same token, we would consider the following definition of sample kurtosis:

$$K = \frac{M_4}{(M_2)^2} = \frac{\dfrac{1}{n} \sum_{i=1}^{n} (X_i - \overline{X})^4}{\left[\dfrac{1}{n} \sum_{i=1}^{n} (X_i - \overline{X})^2\right]^2}$$

Unfortunately, these are not unbiased estimators of γ and κ. The following corrections are typically used:[19]

$$G_1 = \frac{\sqrt{n(n-1)}}{n-2} G, \quad K_1 = \frac{n-1}{(n-2)(n-3)} \left[(n+1)K - 3(n-1)\right] + 3$$

Neither of these adjustments eliminates bias, however; some software packages offer the choice between these two forms of estimator. We stress again that if we assume a given distribution, estimating skewness and kurtosis is a false

---

[18] See Definition 7.3.
[19] See, e.g., Ref. [10] for a comparison of measures of sample skewness and sample kurtosis.

problem. If we do not take a specific distribution for granted, we might use either form of sample skewness and sample kurtosis to get a feeling for the sensibility of our hypothesis. For instance, if we are checking normality, we should find a sample skewness close to 0 and a sample kurtosis close to 3. If the estimates are quite far from what we expect, we should be inclined to rule out normality; however, even if we find $G \approx= 0$ and $K \approx 3$, this would not be enough to conclude normality. In fact, there are nonparametric tests that can be used to check the plausibility of an assumption about the whole distribution, without referring to specific parameters.

9.5 CHECKING THE FIT OF HYPOTHETICAL DISTRIBUTIONS: THE CHI-SQUARE TEST

So far, we have been concerned with parameters of probability distributions. We never questioned the fit of the distribution itself against empirical data. For instance, we might assume that a population is normally distributed, and we may estimate and test its expected value and variance. However, normality should not be taken for granted, just like any other claim about the underlying distribution. Sometimes, specific knowledge suggests strong reasons that justify the assumption; otherwise, this should be tested in some way. When we test whether experimental data fit a given probability distribution, we are not really testing a hypothesis about a parameter or two; in fact, we are running a nonparametric test. The chi-square test is one example of such a test.

The idea is fairly intuitive and basically relies on the idea of a relative frequency histogram, although the technicalities do require some care. The first step is to divide the range of possible observed values in J disjoint intervals, corresponding to bins of a frequency histogram. Given a probability distribution, we can compute the probability p_j, $j = 1, \ldots, J$, that a random variable distributed according to that distribution falls in each bin. If we have n observations, the number of observations that should fall in interval j, if the assumed distribution is indeed the true one, should be $E_j = np_j$. This number should be compared against the number O_j of observations that actually fall in interval j; a large discrepancy would suggest that the hypothesis about the underlying distribution should be rejected. As does any statistical test, the chi-square test relies on a distributional property of a statistic. It can be shown that for a large number of samples, the statistic

$$\chi^2 = \sum_{j=1}^{J} \frac{(O_j - E_j)^2}{E_j}$$

has (approximately) a chi-square distribution. We should reject the hypothesis if χ^2 is too large, i.e., if $\chi^2 > \chi^2_{1-\alpha,m}$, where

- $\chi^2_{1-\alpha,m}$ is a quantile of the chi-square distribution

- α is the significance level of the test

- m is the number of degrees of freedom

What we are missing here is m, which depends on the number of parameters of the distribution that we have estimated using the data. If no parameter has been estimated, i.e., if we assumed a specific parameterized distribution prior to observing data, the degrees of freedom are $J - 1$; if we estimated p parameters, we should use $J - p - 1$.

The idea of the test, as we stressed, is pretty intuitive. However, it relies on approximated distributional results that may be critical. Another tricky point is that the result of the test may depend on the number and placement of bins. Rules of thumb have been proposed and are typically embedded in statistical software. Nevertheless, we should mention that there are other general strategies to test goodness of fit, like the Kolmogorov–Smirnov test, as well as *ad hoc* testing procedures for specific distributions.

9.6   ANALYSIS OF VARIANCE

Analysis of variance (ANOVA) is the collective name of an array of methods that find wide applications in inferential statistics. In essence, we compare groups of observations in order to check if there are significant differences between them, which may be attributed to the impact of underlying factors. One such case occurs when we compare sample means taken from m populations, in order to test the hypothesis that the respective expected values are all the same. Note that, so far, we only considered two populations; with ANOVA we may check an arbitrary number of populations. The ability to analyze the impact of factors is also useful to assess the significance of statistical models, as well as to design statistical experiments. The approach relies on the comparison of different estimates of variance, which should not be significantly different, if factors are not relevant; if we find a statistically significant difference in estimates, then we may reject the hypothesis that factors have no impact.

In this section, we take a somewhat limited view, which is nevertheless able to convey the essentials of ANOVA. We consider two simple and specific cases:

1. One-way ANOVA, whereby we assume that there is one factor at work.

2. Two-way ANOVA, whereby we assume that there are two factors at work.

We will take another view at ANOVA in the context of linear regression in Section 10.3.4.

9.6.1   One-way ANOVA

In Section 9.4.1 we considered a test concerning the hypothesis that the means of two (normal) populations are the same:

$$H_0: \mu_1 = \mu_2 \quad \text{vs.} \quad H_a: \mu_1 \neq \mu_2$$

It is easy to imagine situations in which we want to check a similar claim for more than two populations. To set the stage for the following treatment, let us assume that we have m normal populations, $i = 1, \ldots, m$, and that we take a sample of n elements from each population. If the number of observations for each population is the same, we have a *balanced* design; otherwise, we have an *unbalanced* design. Formally, we are considering the following random variables:

$$X_{ij} \sim \mathcal{N}(\mu_i, \sigma^2), \quad i = 1, \ldots, m; \ j = 1, \ldots, n$$

where the subscripts i and j refer to populations and observations, respectively. As usual, all observations are assumed independent. We denote by μ_i the unknown expected value of population i. Formally, the null hypothesis we want to test is

$$H_0: \mu_1 = \mu_2 = \cdots = \mu_m$$

against the alternative H_a that not all expected values are the same. Another key assumption concerns population variances. They are unknown, but it is assumed that all of them have the same value σ. This might seem a bold assumption, but keep in mind that we want to check the equality of the expected values or, more informally, if there is any significant difference among the populations; hence, in terms of null hypothesis, it is natural to assume the same variance.

Since we have m samples of size n, we have a grand total of nm independent, normally distributed observations. If we standardize, square, and add all of them, we obtain a chi-square random variable with nm degrees of freedom

$$\sum_{i=1}^{m} \sum_{j=1}^{n} \frac{(X_{ij} - \mu_i)^2}{\sigma^2} \sim \chi_{nm}^2 \tag{9.33}$$

Since expected values are unknown, we should replace them with sample means for each population

$$\overline{X}_{i\cdot} = \frac{1}{n} \sum_{j=1}^{n} X_{ij}, \quad i = 1, 2, \ldots, m \tag{9.34}$$

where the notation $\overline{X}_{i\cdot}$ points out that this is a sample mean obtained by summing over the second subscript j. If we plug these sample means into Eq. (9.33), we get the random variable

$$\frac{\text{SS}_w}{\sigma^2} \sim \chi_{nm-m}^2$$

where we define

$$SS_w \equiv \sum_{i=1}^{m} \sum_{j=1}^{n} \left(X_{ij} - \overline{X}_{i\cdot} \right)^2$$

as the *sum of squares within samples*, since deviations are taken with respect to each expected value within each population. This is again a chi-square variable, but with $nm - m$ degrees of freedom, since we have estimated m expected values. Given the properties of chi-square variables, we obtain

$$\frac{E\left[SS_w\right]}{nm - m} = \sigma^2$$

which means that $SS_w/(nm - m)$ is an unbiased estimator of σ^2, regardless of whether the null hypothesis H_0 is true.

Now we build another estimator of σ^2, which is unbiased *only* if H_0 holds, i.e., if the expected values are the same: $\mu_i = \mu$, for $i = 1, \ldots, m$. In such a case, we could estimate μ by taking the overall sample mean

$$\overline{X}_{\cdot\cdot} = \frac{1}{nm} \sum_{i=1}^{m} \sum_{j=1}^{n} X_{ij} \tag{9.35}$$

Then, to estimate σ^2, we could take a different route. Let us define the *sum of squares between samples*

$$SS_b \equiv n \sum_{i=1}^{m} \left(\overline{X}_{i\cdot} - \overline{X}_{\cdot\cdot} \right)^2$$

To see the rationale behind the definition, let us observe that, under the null hypothesis, the variables

$$\frac{\overline{X}_{i\cdot} - \mu}{\sqrt{\sigma^2/n}}, \qquad i = 1, \ldots, m$$

are standard normal. If we square and sum these variables, we get the following chi-square variable:

$$n \sum_{i=1}^{m} \frac{\left(\overline{X}_{i\cdot} - \mu \right)^2}{\sigma^2} \sim \chi_m^2$$

If we plug Eq. (9.35) into the sum above to replace the unknown expected value μ, under the null hypothesis, we obtain

$$n \sum_{i=1}^{m} \frac{\left(\overline{X}_{i\cdot} - \overline{X}_{\cdot\cdot} \right)^2}{\sigma^2} \sim \chi_{m-1}^2$$

This implies that, under H_0

Table 9.5 Sample data for one-way ANOVA.

i	X_{ij}						$\sum_j X_{ij}$
1	116	155	32	126	110	61	600
2	150	150	126	66	128	157	777
3	115	136	124	83	107	64	629

- $E\left[SS_b\right]/\sigma^2 = m - 1$
- $SS_b/(m-1)$ is an unbiased estimator of σ^2

To summarize, we have two estimators for the unknown variance: $SS_w/(nm - m)$ is always unbiased; $SS_b/(m-1)$ is unbiased only if the means are all the same. Then, under the null hypothesis, the ratio of the two estimators should be close to 1. Moreover, it can be shown that $SS_b/(m-1)$ tends to overestimate σ^2 if H_0 is not true. Then, we consider the following test statistic:

$$\text{TS} = \frac{SS_b/(m-1)}{SS_w/(nm-m)}$$

which under H_0 is a F variable with $m-1$ and $nm - m$ degrees of freedom. We reject the hypothesis when the test statistic is too large. More precisely, if $F_{1-\alpha,m-1,nm-m}$ is the $(1-\alpha)$-quantile of the F distribution, we obtain a test with significance level α if we reject when $\text{TS} > F_{1-\alpha,m-1,nm-m}$.

Example 9.24 Let us apply one-way ANOVA to the data listed in Table 9.5, where we have three samples of size $n = 6$, taken from $m = 3$ populations. The first step is to calculate the sums

$$\bar{X}_{1\cdot} = 600/6 = 100.00, \quad \bar{X}_{2\cdot} = 777/6 = 129.50, \quad \bar{X}_{3\cdot} = 629/6 = 104.83$$
$$\bar{X}_{\cdot\cdot} = (100 + 129.5 + 104.83)/3 = 111.44$$

We observe that the three sample means do look rather different. Now we should test the null hypothesis

$$H_0: \mu_1 = \mu_2 = \mu_3$$

We proceed calculating the following sums of squares:[20]

$$SS_b = 6 \times \left[(100 - 111.44)^2 + (129.5 - 111.44)^2 \right.$$
$$\left. + (104.83 - 111.44)^2\right] = 3004.11$$
$$SS_w = 19436.33$$

---

[20] Calculating SS_w is definitely tedious by hand; this task can be streamlined by taking advantage of the identity $SS_w = \sum_i \sum_j X_{ij}^2 - nm\bar{X}_{\cdot\cdot}^2 - SS_b$.

Thus, we find the following alternative estimates of the unknown variance σ^2:

$$\text{SS}_b/2 = 1502.06, \qquad \text{SS}_w/(18-3) = 1295.76$$

which do look different, at first sight. The test statistic is

$$\text{TS} = \frac{1502.06}{1295.76} = 1.1592$$

and, assuming a significance level $\alpha = 5\%$, it should be compared with the following quantile of the F distribution with 2 and 15 degrees of freedom:

$$F_{0.95,2,15} = 3.6823$$

We see that the test statistic does not fall into the rejection region and, therefore, the apparent difference in sample means is not statistically significant. Actually, using the CDF of the F distribution, we obtain the p-value

$$p = \text{P}(F_{2,15} > \text{TS}) = 0.3403$$

which is pretty large. In order to reject the null hypothesis, we should accept a very large probability of a type I error. $\square$

This procedure can be easily adapted to the unbalanced case, where the m samples have not the same size. It is often argued, however, that a balanced design is preferable for nonnormal populations, as the resulting test is a bit more robust to lack of normality.

9.6.2 Two-way ANOVA

In one-way ANOVA we are testing if observations from different populations have a different mean, which can be considered as the one factor affecting such observations. In two-way ANOVA we consider the possibility that two factors affect observations. As a first step, it is useful to reconsider one-way ANOVA in a slightly different light. What we are implicitly assuming is that each random variable can be expressed as the sum of an unknown value plus a random disturbance

$$X_{ij} = \mu_i + \epsilon_{ij}$$

where $\text{E}[\epsilon_{ij}] = 0$. Then, $\text{E}[X_{ij}] = \mu_i$, which is the only factor affecting the expected value of the observations. If we denote the average expected value by μ, where

$$\mu = \frac{1}{m} \sum_{i=1}^{m} \mu_i$$

we may write

$$\text{E}[X_{ij}] = \mu + \alpha_i$$

where $\alpha_i = \mu_i - \mu$ and $\sum_{i=1}^{m} \alpha_i = 0$. Hence, the average value of α_i is zero, but the null hypothesis of one-way ANOVA is much stronger, since it amounts to saying that there is no effect due to α_i, and this is true if $\alpha_i = 0$ for all i.

We may generalize the idea and consider two factors

$$E[X_{ij}] = \mu + \alpha_i + \beta_j$$

where

$$\sum_{i=1}^{m} \alpha_i = \sum_{j=1}^{n} \beta_j = 0$$

In this case, we are taking into consideration the presence of two factors, which are *not* interacting. If we want to account for interaction, we should extend the model to

$$E[X_{ij}] = \mu + \alpha_i + \beta_j + \gamma_{ij}$$

If we organize observations in rows indexed by i and columns indexed j, we may test the following hypotheses:

1. There is no row effect, i.e., $\alpha_i = 0$, for all i.

2. There is no column effect, i.e., $\beta_j = 0$, for all j.

3. There is no effect due to interaction, i.e., $\gamma_{ij} = 0$, for all i and j.

Let us consider the first case, assuming that there is no interaction and that variance is σ^2, for all i and j:

$$H_0 : \text{all } \alpha_i = 0 \quad \text{vs.} \quad H_a : \text{not all } \alpha_i \text{ are equal to } 0$$

As in one-way ANOVA, we build different estimators of σ^2, one of which is unbiased only if the null hypothesis is true. To obtain an estimator that is always valid, let us consider

$$\sum_{i=1}^{m} \sum_{j=1}^{n} \frac{(X_{ij} - E[X_{ij}])^2}{\sigma^2} = \sum_{i=1}^{m} \sum_{j=1}^{n} \frac{(X_{ij} - \mu - \alpha_i - \beta_j)^2}{\sigma^2} \tag{9.36}$$

This is a chi-square variable with nm degrees of freedom, if observations are normal and independent. To estimate the unknown parameters, we consider the appropriate sample means

$$\hat{\mu} = \overline{X}_{..} ; \quad \hat{\alpha}_i = \overline{X}_{i.} - \overline{X}_{..} ; \quad \hat{\beta}_j = \overline{X}_{.j} - \overline{X}_{..}$$

We should recall that, since the sum of the parameters α_i is zero, we need to estimate only $m - 1$ of them; by the same token, we need to estimate only $n - 1$ parameters β_j. So, we need to estimate a grand total of

$$1 + (m - 1) + (n - 1) = m + n - 1$$

parameters. Then, if we plug the above estimators into Eq. (9.36), we find that

$$\sum_{i=1}^{m}\sum_{j=1}^{n}\frac{\left(X_{ij}-\hat{\mu}-\hat{\alpha}_i-\hat{\beta}_j\right)^2}{\sigma^2}=\sum_{i=1}^{m}\sum_{j=1}^{n}\frac{\left(X_{ij}-\overline{X}_{i\cdot}-\overline{X}_{\cdot j}+\overline{X}_{\cdot\cdot}\right)^2}{\sigma^2}$$

is chi-square with

$$nm-(m+n-1)=(m-1)(n-1)$$

degrees of freedom. Then, if we define the *sum of squared errors* as

$$SS_e\equiv\sum_{i=1}^{m}\sum_{j=1}^{n}\left(X_{ij}-\overline{X}_{i\cdot}-\overline{X}_{\cdot j}+\overline{X}_{\cdot\cdot}\right)^2$$

we have

$$E\left[\frac{SS_e}{(m-1)(n-1)}\right]=\sigma^2$$

Therefore, we have built an unbiased estimator of variance. Now, we build another estimator, which is unbiased only under the null hypothesis. In fact, under H_0, we have:

$$E\left[\overline{X}_{i\cdot}\right]=\mu+\alpha_i=\mu$$

Since $\text{Var}(\overline{X}_{i\cdot})=\sigma^2/n$, the sum of squared standardized variables

$$\sum_{i=1}^{m}\frac{\left(\overline{X}_{i\cdot}-\mu\right)^2}{\sigma^2/n}$$

is a chi-square variable with m degrees of freedom, if the null hypothesis is true. Replacing μ by its estimator $\overline{X}_{\cdot\cdot}$, we lose one degree of freedom. So, if we define the *row sum of squares*

$$SS_r\equiv n\sum_{i=1}^{m}\left(\overline{X}_{i\cdot}-\overline{X}_{\cdot\cdot}\right)^2$$

we have

$$E\left[SS_r/(m-1)\right]=\sigma^2$$

Therefore, we have another estimator of variance, but this one is unbiased only under the null hypothesis. When H_0 is not true, this estimator tends to overestimates σ^2. Then, we may run a test based on the test statistic

$$TS=\frac{SS_r/(m-1)}{SS_e/[(m-1)(n-1)]}$$

which, under H_0, has F distribution with $(m-1)$ and $(m-1)(n-1)$ degrees of freedom. Given a significance level α, we reject the null hypothesis that there is no row effect if

$$\text{TS} \geq F_{1-\alpha,m-1,(m-1)(n-1)}$$

Clearly, a similar route can be taken to check for column effects, where we define a *column sum of squares*

$$\text{SS}_c \equiv m \sum_{j=1}^{n} \left(\overline{X}_{\cdot j} - \overline{X}_{\cdot\cdot}\right)^2$$

which is related to a chi-square variable with $n-1$ degrees of freedom, and we reject the null hypothesis that there is no column effect if

$$\text{TS} = \frac{\text{SS}_c/(n-1)}{\text{SS}_e/[(m-1)(n-1)]} \geq F_{1-\alpha,n-1,(m-1)(n-1)}$$

The case with interactions is a bit trickier, but it follows the same conceptual path.

9.7 MONTE CARLO SIMULATION

Monte Carlo simulation is a widely used tool in countless branches of physics, engineering, economics, finance, and business in general. Roughly speaking, the aim is to simulate a system on a computer, in order to evaluate its performance under random scenarios. The name was actually invented by physicists and aptly reflects the role of randomness. Indeed, Monte Carlo simulation is one of the more successful, as well as potentially dangerous, application areas of probability and statistics. While "Monte Carlo" is typically associated with "simulation," a more in-depth view highlights two sides of the coin:

- Monte Carlo *sampling* refers to a mathematical method that may be used, e.g., to discretize a continuous representation of uncertainty in order to generate a set of scenarios. For instance, we may consider a normal distribution and generate a sample of observations that are supposed to approximate the original continuous distribution. Monte Carlo sampling is also a numerical method to evaluate challenging multidimensional integrals.

- Monte Carlo *simulation* relies on Monte Carlo sampling, and it is used to predict the performance of a system in which uncertainty and dynamic interactions of system components preclude the application of analytical modeling techniques. In other words, it is impossible to build a mathematical model giving us a clue of how well a system will work, and we are forced to rely on empirical experimentation with a simulated model of reality.

Whereas Monte Carlo sampling is more of a statistical tool, Monte Carlo simulation may also involve nontrivial modeling of complex systems by special purpose software environments and programming languages.

Example 9.25 (Monte Carlo sampling as a numerical integration tool) Imagine that we need the value of the following integral:

$$\int_0^1 g(x)dx$$

From Section 2.13 we know that we should find the antiderivative of function g, but this is sometimes impossible to do. A seemingly weird way of tackling the problem is obtained if we rewrite the integral in probabilistic terms:

$$\int_0^1 g(x)dx = \int_0^1 g(x) \cdot 1 \, dx = \int_0^1 g(x)f_U(x)dx = \mathrm{E}[g(U)]$$

What we have done is consider the harmless factor 1 as the PDF of a uniform random variable $U \sim \mathcal{U}(0,1)$. Then, the integral can be regarded as the expected value of a function of a random variable. Furthermore, if we were able to draw a sample of independent random observations $U_1, U_2, \ldots, U_n$ from the uniform distribution, we could apply the arsenal of statistical inference to the estimator

$$\frac{1}{n}\sum_{i=1}^n g(U_i)$$

If the sample is large enough, this sample mean should converge to the true value of the integral above, and we may also check whether the sample is indeed large enough by evaluating a confidence interval.

Actually, there are numerical integration methods that are faster and more accurate than Monte Carlo sampling, when we have to integrate a function of one variable. But consider the expected value of a function of a vector $\mathbf{X}$ of random variables taking values in $\mathbb{R}^m$, with a possibly complicated joint PDF as follows:

$$\mathrm{E}[g(\mathbf{X})] = \int_{\mathcal{D}} g(\mathbf{x})f_{\mathbf{X}}(\mathbf{x})d\mathbf{x}$$

This may be an integral in a very high-dimensional space, over a domain $\mathcal{D} \subset \mathbb{R}^m$ that is the support of the distribution. Standard numerical methods are not applicable when m is even moderately large, whereas Monte Carlo sampling requires only to generate random variables with a given probability distribution and calculate a sample mean.   ☐

The basic ingredient of Monte Carlo sampling is the generation of observations of random variables on a computer. Conceptually, this is an impossible task, since there is no randomness in a computer algorithm; however, using methods outlined in Section 9.7.2, we are able to generate *pseudo*random variables that are practically satisfactory. Monte Carlo sampling can be used for scenario generation[21] in many risk management applications. From a practical perspective, the required machinery of Monte Carlo sampling can be conceptually challenging, but it is easy to implement, quite often in spreadsheet form. On the contrary, Monte Carlo simulation of complex dynamic systems may require nontrivial programming skills.

Example 9.26 (Monte Carlo simulation of an inventory control policy) We must control the inventory of an item subject to random demand. Daily demand d_t, $t = 1,2,3,\ldots$, can be just a sequence of i.i.d. variables, or a very complicated process subject to intertemporal dependence and seasonal patterns.[22] We consider a simple control policy, which is widely known as (s, S) policy:

- Every T days we check the inventory level; if it is below a threshold level s, we order enough items to bring inventory back to a target level S, where $S \geq s$; more precisely, we order $S - I_t$ items, where I_t is the current inventory level when it is reviewed; items are received after a delivery lead time LT.

- If the inventory level is above the threshold s, we do nothing.

The role of the threshold level s is to avoid ordering too often, since we would incur large fixed ordering and transportation costs. The policy parameters s and S should be set in order to find a satisfactory balance between inventory holding costs, ordering costs, and penalties due to stockouts. The reader should compare this approach with the economic order model that we introduced in Section 2.1. Since we check inventory with a review period T, the (s, S) policy is called *periodic review*; on the contrary, the (Q, R) policy, based on a fixed order quantity Q and a reorder level R, is a *continuous review* policy, since we should check inventory level whenever we draw an item from inventory. A possible advantage of periodic review policies is that, by choosing a common ordering period, we may synchronize and aggregate orders for different items with a common supplier, resulting in a reduction in transportation costs.

In very simple cases, mathematical models can be devised to predict the average cost as a function of the control parameters s and S; then, we may tune them in the best way. Unfortunately, many complicating factors may preclude this approach:

---

[21] See also Section 13.4.3.
[22] The time series models that we discuss in Chapter 11 can be used to model and simulate virtually any demand pattern.

- The demand process can be difficult to deal with analytically, because of correlation over time and other nontrivial patterns.

- Items may be perishable and have limited shelf life.

- Delivery lead time LT may be uncertain; note that if LT is deterministic and smaller than the review period T, the issued order will be received before the next inventory review; however, in general, we should take the queue of incoming deliveries into account. Indeed, inventory control does not rely on physical, on-hand inventory; rather, we should consider the inventory position, which includes both on-hand and on-order inventory, as well as possible backlog associated with customers that we have not yet satisfied because of a stockout.

- The customer behavior when a stockout occurs may be nontrivial; some customers will wait for delivery, possibly reneging if the waiting time is too large; other customers may be lost; sometimes, they will switch to a substitute item.

- There could be an interaction between different items held in stock.

- There could be an interaction between our and suppliers' inventories.

In these cases, we should resort to performance evaluation by Monte Carlo simulation. ☐

The reader should appreciate the difference between the two examples. On a conceptual level, the simulation of a complex supply chain to estimate its average cost is just the estimation of an expected value. We are just integrating a complicated function that cannot be represented by a formula. However, from a practical perspective, the second example requires considerable modeling skills and cannot be tackled by plain spreadsheets. The dynamic interaction of components and actors calls for much more. We will not consider implementation issues at all, but we should mention that a massive array of software environments and languages is available to tackle huge simulation models. They are also equipped with graphical animation tools that are way beyond the scope of this book. What is relevant, from our point of view, is the fundamental role that probability and statistics play in a proper simulation study. Too often, users fascinated by flashy computer graphics forget this point.

9.7.1   Discrete-event vs. discrete-time simulation

The time we experience in everyday life is continuous. Engineers simulating, e.g., the flight behavior of an aircraft, have to build a continuous-time model accounting for quite complex dynamics. To make the model amenable to numerical simulation, suitable discretization schemes have to be devised;

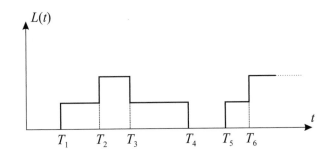

Fig. 9.8 Sample path of a discrete-event queueing system.

indeed, nothing is continuous in the digital world of computers. The way in which this discretization is done influences the computational burden and accuracy of predictions obtained by running the model. Over the years, we all have noticed the improvement in weather forecasts due to more sophisticated models and numerical schemes, as well as dramatically faster computer hardware, allowing for fine discretizations. The knowledge involved in suitable discretizations of continuous systems is really cutting-edge.

Luckily, time discretization is often much more natural in many business models. In the inventory control problem of Example 9.26, there is really no need to model the exact timing of demand, if we review inventory in the evening and we receive shipments in the morning. The system will change its state only at naturally discrete-time instants, resulting in very simple dynamics:

$$I_t = I_{t-1} + x_t - d_t$$

i.e., inventory at the end of day t is just the inventory at the end of the day before, plus what we receive from suppliers on day t, minus demand on day t. We see that simulating such a system involves state changes at regular time steps. In other cases, system dynamics is not so easy and regularly paced. Consider again the queueing system of Section 9.2.1. Assume that there is only one server in the system and that its service time for each customer is random. If we model the arrival process by a Poisson process, we have to simulate the system in continuous time; according to this model, the time elapsing between successive arrivals is exponentially distributed, and the exponential distribution is a continuous one. A possible sample path of such a system is illustrated in Fig. 9.8, which shows how the main state variable, the queue length $L(t)$, evolves in time. To be precise, $L(t)$ counts the number of customers in the system, including the one currently being served. From this sample path, it is easy to figure out the sequence of events taking place:

- At time T_1 the first customer arrives; the server is idle, so service starts immediately.

- At time T_2 the second customer arrives; the server is busy, so this customers is enqueued; now there are two customers in the system.

- At time T_3 the first customer completes her service, and the second one starts being served; the number of customers in the system is decreased.

- At time T_4 the second service is completed; there is no one in the queue, and server gets idle.

- At time T_5 the third customer arrives, and so on.

We notice that the relevant state variable follows a piecewise constant path, whereby abrupt changes are due to occurrence of events. Queueing systems are the standard example of *discrete-event* dynamics. To simulate a discrete-event system we need the following ingredients:

- A system clock, whose current value is denoted by t_{cl}, representing simulated time.

- An ordered list of events – the clock is advanced to the earliest scheduled event.

- Data structures to represent the system state.

- A set of procedures to manage each type of event; events change the system state and possibly schedule the occurrence of later events.

In a standard queueing system, there are two natural events: the arrival of a new customer and the completion of a service. The way in which these events should be handled can be outlined as follows.

Arrival of a customer. When a customer arrives, we check the server state:

- If it is idle, we start service immediately; we also schedule the next completion time, by drawing a random variable T_s according to the chosen probability distribution and adding it to the current clock time t_{cl}; we insert a completion event in the sorted list of events, with clock time $T_s + t_{cl}$.

- If the server is busy, we insert the customer in the queue.

In any case, we schedule the next arrival by drawing an interarrival time T_a, and inserting an arrival event in the sorted list of events, with clock time $T_a + t_{cl}$.

Completion of a service. When a service is completed, we should check the queue:

- If it is empty, the state of the server is changed to idle.

- Otherwise, we pick the first customer in the queue, possibly according to some priority scheme, draw a random variable T_s accounting for service time, and insert a completion event in the sorted list of events, with time $T_s + t_{cl}$.

After the current event is processed, we pick the next one in the list, advance the clock t_{cl} accordingly, and process the event as prescribed.

Conceptually, discrete-event simulations are not so difficult for very simple systems. However, when there are many entities and resources in the system, with complicated interactions, suitable simulation languages and environments are needed. In particular, the collection of relevant statistics about waiting times, queue lengths, and resource utilizations can be awkward without suitable tools. Clearly, it takes some skill to implement a simulation model, but this kind of technicality should not hide some more relevant difficulties:

- Which system features are really important, and which can be neglected? Any model is a simplification of reality; too many details will make model building more difficult; even worse, data requirements will become unmanageable. There is no point in building a complex model if the data we feed are not reliable. This type of decision is more of an art than a science, as you can imagine.

- The output of any Monte Carlo simulation should be regarded as a random variable. Which statistical techniques should we use to analyze the output? How should we plan the simulation experiments? A good knowledge of inferential statistics is an essential ingredient of a serious and successful simulation study.

9.7.2   Random-number generation

Any Monte Carlo approach relies on the ability of generating variables that look reasonably random. Clearly, no computer algorithm can be truly random, but all we need is a way of generating pseudorandom variables that would trick statistical tests into believing that they are truly random. The starting point of any such strategy is the generation of uniform variables on the unit interval [0,1]; then, there is a wide array of methods to transform uniform variables into whatever distribution we need.

In the past, the standard textbook method used to generate $\mathcal{U}(0, 1)$ variates was based on linear congruential generators (LCGs). A LCG generates a sequence of nonnegative integer numbers Z_i as follows; given an integer number Z_{i-1}, we generate the next number in the sequence by computing

$$Z_i = (aZ_{i-1} + c) \bmod m$$

where a (the multiplier), c (the shift), and m (the modulus) are properly chosen parameters and "mod" denotes the remainder of integer division (e.g., 15 mod 6 = 3). Then, to generate a uniform variate on the unit interval,

Table 9.6 Sample sequence generated by a linear congruential generator.

i	Z_i	U_i	i	Z_i	U_i
1	6	0.3750	11	0	0.0000
2	1	0.0625	12	3	0.1875
3	8	0.5000	13	2	0.1250
4	11	0.6875	14	13	0.8125
5	10	0.6250	15	4	0.2500
6	5	0.3125	16	7	0.4375
7	12	0.7500	17	6	0.3750
8	15	0.9375	18	1	0.0625
9	14	0.8750	19	8	0.5000
10	9	0.5625	20	11	0.6875

we return the number (Z_i/m). Clearly, we also need an initial number Z_0 to start the sequence; this number is the *seed* of the sequence.

Example 9.27 Let us take $a = 5$, $c = 3$, and $m = 16$. If we start from the seed $Z_0 = 7$, the first integer number we generate is

$$Z_1 = (5 \times 7 + 3) \bmod 16 = 38 \bmod 16 = 6$$

Hence

$$U_1 = \tfrac{6}{16} = 0.3750$$

Then we obtain

$$Z_2 = (5 \times 6 + 3) \bmod 16 = 33 \bmod 16 = 1, \quad U_2 = \tfrac{1}{16} = 0.0625$$

Proceeding this way, we get the sequence illustrated in Table 9.6.   ▯

It is clear that there is nothing random in the sequence generated by a LCG. Starting the sequence from the same seed will always yield the same sequence. Furthermore, we actually generate rational numbers, rather than real ones; this is not a serious problem, provided that m is large enough. But there is another reason to choose a large value for m: The generator is periodic! In fact, we may generate at most m distinct integer numbers Z_i, in the range from 0 to $m - 1$, and whenever we repeat a previously generated number, the sequence repeats itself (which is not very random at all). We see from the previous example that we return to the initial seed $Z_0 = 7$ after 16 steps. Since the maximum possible period is m, we should make it very large in order to have a large period. The proper choice of a and c ensures that the period is maximized and that the sequence looks random. Designing a good random-number generator is not easy; luckily, when we purchase good numerical software, someone has already solved the issue for us. Actually, recent developments have led to better generators, but all we need is an idea of what seeds are and how they influence random-number generation.

The second issue we have to tackle is how to generate a pseudorandom variable with an arbitrary distribution. One general approach is based on the *inverse transform* method. Suppose that we are given the CDF $F(x) = P\{X \leq x\}$, and that we want to generate random variables according to F. If we are able to invert $F(x)$ easily, we may apply the following approach:

1. Draw a random number $U \sim \mathcal{U}(0, 1)$.

2. Return $X = F^{-1}(U)$.

It is easy to see that the random variate X generated by this method is actually characterized by the distribution function F:

$$P\{X \leq x\} = P\{F^{-1}(U) \leq x\} = P\{U \leq F(x)\} = F(x)$$

where we have used the monotonicity of F and the fact that U is uniformly distributed.

Example 9.28 A typical distribution that can be simulated easily by the inverse transform method is the exponential distribution. If $X \sim \exp(\mu)$, where $1/\mu$ is the expected value of X, its distribution function is

$$F(x) = 1 - e^{-\mu x}$$

Direct application of the inverse transform yields

$$x = -\frac{1}{\mu} \ln(1 - U)$$

Since the distributions of U and $(1 - U)$ are actually the same, it is customary to generate exponential variates by drawing a random number U and returning $-\ln(U)/\mu$. ▯

The inverse transform method is only one of the general recipes that are used in practice. Sometimes, ad hoc methods are applied to specific distributions. As far we are concerned, we just need to know that statistical software is available to sample virtually any conceivable distribution.

9.7.3 Methodology of a simulation study

Building a simulation model requires much more than programming events and generating pseudorandom variables. A simulation study involves the following steps as well:

Input analysis. This phase requires devising probability distributions that match empirical data, if available. We should not just assume that, say, a random time has exponential distribution. We should check the fit of any distribution, e.g., by running nonparametric tests like the chi-square goodness-of-fit test.

Model verification. Model verification means checking that the program does what it is supposed to do, i.e., that there are no bugs. Note that this does not imply that the model is a good one; it means simply checking that the program behaves according to its specification.

Model validation. This is a check on the model itself. Even when the program works as specified, it is no guarantee of success, if the specifications themselves are flawed. We should check that the inputs, as well as the business rules encoded in the simulation program, are realistic. Model validation is fairly easy if we are simulating an existing system; we should just compare observed and predicted performance. In other cases, things are not that easy.

Output analysis. This is carried out along the lines of confidence intervals. The caveats of Section 9.2.1 apply.

Finally, we should mention the role of properly designed experiments to check the influence of design parameters. ANOVA techniques do play a major role here.

9.8 STOCHASTIC CONVERGENCE AND THE LAW OF LARGE NUMBERS

In this section we start considering in some more depth the issues involved in inferential statistics. The aim is to bridge the gap between the elementary treatment that is commonly found in business-oriented textbooks, and higher-level books geared toward mathematical statistics. As we said, most readers can safely skip these sections. Others can just have a glimpse of what is required to tackle a deeper study of statistical and econometric models. Furthermore, the usual cookbook treatment is geared towards normal populations, which may result in a distorted and biased perspective. A normal distribution is characterized by two parameters, μ and σ^2, which have an obvious interpretation as expected value and variance, respectively. As a result, we tend to identify the estimation of parameters with the estimation of expected values or variances. However, this is a limited view. For instance, the beta distribution[23] is characterized by two parameters, α_1 and α_2; if we know those parameters, we may compute whatever moment we want, expected value, variance, skewness, etc. But to estimate these parameters, we cannot

---

[23] See Section 7.6.2.

just rely on a sample mean. Thus, we need a more general framework to tackle parameter estimation. Indeed, there are a few alternative strategies to obtain estimators. We should understand what makes a good estimator, in order to assess the tradeoffs when alternatives are available. Finally, we have derived confidence intervals and hypothesis testing procedures in a somewhat informal and ad hoc manner. Actually, there are general strategies that help to assess desirable properties of them as well.

We deal with the above issues in the rest of the chapter, and a good starting point to motivate our work is a better investigation of the law of large numbers, which we previewed in Section 7.7.4. Intuitively, if we take a large random sample, the sample mean should get closer and closer to the true expected value. Therefore, we should expect a convergence result loosely stated as follows:

$$\lim_{n \to +\infty} \frac{1}{n} \sum_{i=1}^{n} X_i \overset{?}{\to} \mu$$

The question mark stresses the fact that this convergence is quite critical: In which sense can we say that a *random variable* converges to a *number*? More generally, we could consider convergence of a sequence of random variables X_n to a random variable X. The relevance of a "stochastic convergence" concept to analyze parameter estimation problems for large samples should be obvious. Furthermore, we have illustrated the central limit theorem in Section 7.7.3. There, we loosely stated that a certain random variable "converges" to a normal distribution; we should make this statement more precise. As it turns out, there is not a single and all-encompassing concept of stochastic convergence, which may be defined in a few different ways; they may be more or less useful, depending on the application; furthermore, some convergence concepts may be weaker, but easier to verify.

9.8.1 Convergence in probability

The first stochastic convergence concept that we illustrate is not the strongest one, but it can be easier to grasp.

DEFINITION 9.7 (Convergence in probability) *A sequence of random variables, $X_1, X_2, \ldots$, converges in probability to a random variable X if, for every $\epsilon > 0$*

$$\lim_{n \to +\infty} P\left(\left| X_n - X \right| \geq \epsilon\right) = 0$$

The definition may look intimidating, but it is actually intuitive: X_n tends to X if, for an arbitrarily small ϵ, the probability of the event $\{|X_n - X| \geq \epsilon\}$ goes to zero. In other words

$$\lim_{n \to +\infty} P\left(\left| X_n - X \right| < \epsilon\right) = 1$$

i.e., the difference between the two realizations cannot be larger than a small and arbitrary constant ϵ. Convergence in probability is denoted as follows:

$$X_n \xrightarrow{p} X$$

or

$$\plim_{n \to +\infty} X_n = X$$

We may also apply the definition to express convergence to a specific number a, rather than a random variable.

Example 9.29 Consider the following sequence of random variables:

$$X_n = \begin{cases} 0 & \text{with probability } 1 - \dfrac{1}{n} \\ n & \text{with probability } \dfrac{1}{n} \end{cases}$$

We see that, for n going to infinity, X_n may take a larger and larger value n, but its probability is vanishing. Indeed, the probability mass gets concentrated on the value 0 and

$$\plim_{n \to \infty} X_n = 0$$

Using this concept, we may clearly state (without proof) one form of the law of large numbers.

THEOREM 9.8 (Weak law of large numbers) *Let* $X_1, X_2, \ldots$ *be a sequence of i.i.d. random variables, with* $\mathrm{E}[X_i] = \mu$ *and finite variance* $\mathrm{Var}(X_i) = \sigma^2 < +\infty$. *Then*

$$\overline{X}_n = \frac{1}{n} \sum_{i=1}^{n} X_i \xrightarrow{p} \mu$$

As you may imagine, there must be a different, strong form of this law; as we shall see, it uses a stronger form of stochastic convergence. Despite its relative weakness, convergence in probability is relevant in the theory of estimation, in that it allows us to define an important property of an estimator. Consider a sequence of estimators Y_n of an unknown parameter θ; each Y_n is a statistic depending on a sample of size n. We say that this is a *consistent estimator* if

$$\plim_{n \to \infty} Y_n = \theta$$

The weak law of large numbers implies that the sample mean is a consistent estimator of expected value.

9.8.2 Convergence in quadratic mean

Consider the ordinary limits of expected value and variance of the sample mean:

$$\lim_{n\to\infty} \mathrm{E}\left[\overline{X}\right] = \mu, \quad \lim_{n\to\infty} \mathrm{Var}\left(\overline{X}\right) = \lim_{n\to\infty} \frac{\sigma^2}{n} = 0$$

When variance goes to zero like this, intuition suggests that the probability mass gets concentrated and some kind of convergence occurs.

DEFINITION 9.9 (Convergence in quadratic mean to a number) *If* $\mathrm{E}[X_n] = \mu_n$ *and* $\mathrm{Var}(X_n) = \sigma_n^2$, *and the ordinary limits of the sequence of expected values and variances are a real number* β *and* 0, *respectively, then the sequence* X_n *converges in quadratic mean to* β.

Convergence in quadratic mean is also referred to as *mean-square convergence*, denoted as $X_n \xrightarrow{m.s.} \beta$. Actually, the definition above is just a very specific case of the more general concept of rth mean convergence to a random variable.

DEFINITION 9.10 (Convergence in rth mean) *Given* $r > 1$, *if the condition* $\mathrm{E}\left[|X_n^r|\right] < \infty$ *holds for all* n, *we say that the sequence* X_n *converges in rth mean to* X *if*

$$\lim_{n\to\infty} \mathrm{E}\left[(X_n - X)^r\right] = 0$$

Not surprisingly, Definition 9.9 of mean-square convergence is equivalent to requiring

$$\lim_{n\to\infty} \mathrm{E}\left[(X_n - \beta)^2\right] = 0$$

Convergence in quadratic mean implies convergence in probability, but the converse is not true, as illustrated by the following counterexample.

Example 9.30 Let us consider again the sequence of random variables of Example 9.29

$$X_n = \begin{cases} 0 & \text{with probability } 1 - \dfrac{1}{n} \\ n & \text{with probability } \dfrac{1}{n} \end{cases}$$

Whatever n we take, we have

$$\mathrm{E}[X_n] = 0 \cdot \left(1 - \frac{1}{n}\right) + n \cdot \frac{1}{n} = 1$$

However, this contradicts the finding of Example 9.29, i.e., $X_n \xrightarrow{p} 0$. Actually, we solve the trouble by noting this sequence does *not* converge in quadratic mean, since variance is not finite for n going to infinity:

$$\begin{aligned} \lim_{n\to\infty} \mathrm{Var}(X_n) &= \lim_{n\to\infty} \left\{\mathrm{E}[X_n^2] - \mathrm{E}^2[X_n]\right\} \\ &= \lim_{n\to\infty} \left\{\left(1 - \frac{1}{n}\right) \cdot 0^2 + \frac{1}{n} \cdot n^2 - 1^2\right\} \\ &= +\infty \end{aligned}$$

□

9.8.3 Convergence in distribution

We have already met convergence in distribution when dealing with the central limit theorem.

DEFINITION 9.11 (Convergence in distribution) *The sequence of random variables* $X_1, X_2, \ldots$, *with CDFs* $F_n(x)$, *converges in distribution to random variable* X *with CDF* $F(x)$ *if*

$$\lim_{n \to \infty} |F_n(x) - F(x)| = 0$$

for all points at which $F(x)$ *is continuous.*

Convergence in distribution can be denoted as

$$X_n \overset{d}{\to} X$$

When convergence in distribution occurs, we may speak of a *limiting* distribution. The central limit theorem states that the limiting distribution of

$$Z_n = \frac{\sum_{i=1}^{n} X_n - \mu}{\sigma/\sqrt{n}}$$

is the standard normal distribution. It is important to notice that convergence in distribution does not at all imply convergence to a constant. Furthermore, it can be shown that convergence in distribution is weaker than convergence in probability to a random variable, in the sense that

$$(X_n \overset{p}{\to} X) \quad \Rightarrow \quad (X_n \overset{d}{\to} X)$$

but the converse is not true.

9.8.4 Almost-sure convergence

The last type of convergence that we consider is a strong one in the sense that it implies convergence in probability, which in turn implies convergence in distribution.

DEFINITION 9.12 (Almost-sure convergence) *The sequence of random variables* $X_1, X_2, \ldots$, *converges almost surely to random variable* X *if, for every* $\epsilon > 0$, *the following condition holds:*

$$P\left(\lim_{n \to \infty} |X_n - X| < \epsilon\right) = 1$$

Almost-sure convergence is denoted by $X_n \overset{a.s.}{\longrightarrow} X$. Sometimes, almost-sure convergence is referred to as convergence *with probability 1*; correspondingly, the notation "w.p.1" rather than "a.s." may be found .

Comparing definition 9.12 of almost-sure convergence with definition 9.7 of convergence in probability may be confusing, as they seem quite similar. The difference is a swap of the limit operator with the probability operator. This commutation is not innocuous at all, and it does make a difference. To understand the definition, it is useful to recall the definition of random variables as functions $X(\omega)$ from the sample space Ω to real numbers and to reflect on the concept of *pointwise convergence* of functions. For instance, consider a sequence of deterministic functions $f_n(x)$, where $x \in [0, 1]$. We say that this sequence of functions converges pointwise to function $f(x)$ if, for all $x \in [0, 1]$, we have

$$\lim_{n \to \infty} f_n(x) = f(x)$$

Definition 9.12 weakens this condition a bit, in the sense that we require that the condition

$$\lim_{n \to \infty} X_n(\omega) = X(\omega)$$

holds only for *almost any* $\omega \in \Omega$. The set of outcomes ω for which the sequence does not converge must be a set of null measure; equivalently, we have convergence for a set with probability measure 1. The following nice examples illustrate these concepts.[24]

Example 9.31 Let the sample space be $\Omega = [0, 1]$, where probability is uniform. Consider the sequence of random variables

$$X_n(\omega) = \omega + \omega^n$$

and the random variable $X(\omega) = \omega$. If $\omega \in [0, 1)$, we see that

$$\lim_{n \to \infty} X_n(\omega) = \omega = X(\omega)$$

This does not happen for $\omega = 1$, since $X_n(1) = 2$ for any n. Hence, we have convergence of $X_n(\omega)$ to $X(\omega)$ for all outcomes except a single one, which is a set of null measure. Hence

$$X_n(\omega) \xrightarrow{a.s.} X(\omega)$$

▯

Example 9.32 Let the sample space be $\Omega = [0, 1]$, again with uniform probability, as in the previous example. The indicator function associated with an interval is denoted by $I_{[a,b]}(x)$ and takes value 1 if $x \in [a, b]$, 0 otherwise.

---

[24]These two examples are taken from the text by Casella and Berger [3, p. 234].

Now define a sequence of random variables as follows:

$$X_1(\omega) = \omega + I_{[0,1]}(\omega)$$

$$X_2(\omega) = \omega + I_{[0,\frac{1}{2}]}(\omega)$$

$$X_3(\omega) = \omega + I_{[\frac{1}{2},1]}(\omega)$$

$$X_4(\omega) = \omega + I_{[0,\frac{1}{3}]}(\omega)$$

$$X_5(\omega) = \omega + I_{[\frac{1}{3},\frac{2}{3}]}(\omega)$$

$$X_6(\omega) = \omega + I_{[\frac{2}{3},1]}(\omega)$$

$$\vdots$$

To see the logic behind this sequence, notice that the interval $[0, 1]$ is sliced in two parts to define X_2 and X_3; then in three parts to define X_4, X_5, and X_6; etc. These slices, for increasing n, are smaller and smaller. Whatever ω we choose, $X_1(\omega) = s + 1$. The other variables in the sequence take either value s, or value $s + 1$, depending on whether ω is in the interval of the indicator function associated with each variable in the sequence.

Now let us consider the random variable $X(\omega) = \omega$. The sequence X_n converges in probability to X, i.e., $X_n \xrightarrow{p} X$. To see this, consider the probability

$$P(|X_n - X| \geq \epsilon). \tag{9.37}$$

The random variables $X_n(\omega)$ and $X(\omega)$ differ on a subinterval of Ω that is smaller and smaller for n increasing; in other words, the measure of this interval goes to zero. In fact, for a suitably small ϵ, the probability in Eq. (9.37) is just the probability that X_n falls in this subinterval, but since this probability goes to zero, X_n converges in probability to X. However, the sequence X_n does *not* converge almost surely to X. To see this, fix an arbitrary ω; for increasing n, the values in the sequence alternate between ω and $\omega + 1$, and there is no pointwise convergence. ☐

The counterexample shows that convergence in probability does not imply almost-sure convergence, whereas it can be shown that almost-sure convergence does imply convergence in probability. Not surprisingly, if we apply almost-sure convergence, we get a version of the law of large numbers that is stronger than the one in Theorem 9.8.

THEOREM 9.13 (Strong law of large numbers) *Let $X_1, X_2, \ldots$ be a sequence of i.i.d. random variables, with $E[X_i] = \mu$ and finite variance $\text{Var}(X_i) = \sigma^2 < +\infty$. Then, we have*

$$\overline{X}_n \equiv \frac{1}{n} \sum_{i=1}^{n} X_i \xrightarrow{a.s.} \mu$$

A comparison of this theorem with the similar Theorem 9.8 about the weak law of large numbers is puzzling, as they do look the same. Actually, finiteness of variance is a stronger condition than necessary, and alternative statements can be found in the literature; for instance, we could just require $E\left[\|X_n\|\right] < \infty$. The price we pay for relaxing assumptions is in terms of quite involved theorem proofs, which depend on the type of stochastic convergence involved. We leave such technicalities aside and just remark that the convergence concept in the strong law of large numbers is indeed stronger than the convergence concept used in the weak law.

9.8.5 Slutsky's theorems

A few useful theorems in calculus allow us to manipulate limits in an intuitive manner and justify the rules for calculating derivatives that we stated back in Section 2.8. For instance, the limit of the product of two sequences is the product of their limits, if they exist; furthermore, if a function $g(\cdot)$ is continuous, than g can be interchanged with the limit operator, i.e.

$$\lim_{x \to x_0} g(x) = g\left(\lim_{x \to x_0} x\right) = g(x_0)$$

These properties are generalized to stochastic limits by a group of results called Slutsky's theorems.

THEOREM 9.14 *For a continuous function $g(\cdot)$, we have*

$$\operatorname*{plim}_{n \to \infty} g(X_n) = g\left(\operatorname*{plim}_{n \to \infty} X_n\right)$$

The practical implication of this theorem is that if we are able to find a consistent estimator of a parameter, we also have a consistent estimator of a function of that parameter.

THEOREM 9.15 *If the sequence of random variables X_n converges in distribution to random variable X, i.e., $X_n \xrightarrow{d} X$, and the sequence Y_n converges in probability to a constant c, i.e., $\operatorname{plim}_{n \to \infty} Y_n = c$, then*

$$X_n + Y_n \xrightarrow{d} X + c$$
$$X_n Y_n \xrightarrow{d} cX$$
$$X_n/Y_n \xrightarrow{d} X/c, \qquad if\ c \neq 0$$

Example 9.33 Standard procedures for confidence intervals and hypothesis testing are derived assuming a normal sample, but they are often applied to nonnormal populations. We should make sure that this makes sense, at least for large samples. The central limit theorem implies that, if $\overline{X}$ is the sample mean of a sequence of i.i.d. variables, the statistic

$$Z = \frac{\overline{X} - \mu}{\sigma/\sqrt{n}}$$

has an approximately standard normal distribution. However, this result assumes knowledge of σ; if we replace σ by the sample standard deviation S, things are not obvious at all. We typically resort to the t distribution, but this again assumes a normal sample. This is where Slutsky's theorems come in handy. Let us consider the statistic

$$T = \frac{\overline{X} - \mu}{S/\sqrt{n}}$$

and rewrite it as follows:

$$T = \frac{\sqrt{n}(\overline{X} - \mu)/\sigma}{S/\sigma}$$

The numerator converges in distribution to a standard normal, courtesy of the central limit theorem:

$$\frac{\sqrt{n}\,(\overline{X} - \mu)}{\sigma} \xrightarrow{d} \mathcal{N}(0, 1)$$

As to the denominator, it can be shown that sample variance S^2 is a consistent estimator of variance σ^2, i.e., $S^2 \xrightarrow{p} \sigma^2$. Then, using Slutsky's theorems, we see that

$$\frac{S^2}{\sigma^2} \xrightarrow{p} 1$$

and therefore

$$T = \frac{\overline{X} - \mu}{S/\sqrt{n}} \xrightarrow{d} \mathcal{N}(0, 1)$$

$\square$

9.9 PARAMETER ESTIMATION

Introductory treatments of inferential statistics focus on normal populations. In that case, the two parameters characterizing the distribution, μ and σ^2, coincide with expected value, the first-order moment, and variance, the second-order central moment. Hence, students might believe that parameter estimation is just about calculating sample means and variances. It is easy to see that this is not the case.

Example 9.34 Consider a uniform random variable $X \sim \mathcal{U}[a, b]$. We know from section 7.6.1 that

$$E[X] = \frac{a+b}{2}, \qquad \text{Var}(X) = \frac{(b-a)^2}{12}$$

Clearly, sample mean $\overline{X}$ and sample variance S^2 do not provide us with direct estimates of parameters a and b. However, we might consider the following way of transforming the statistics to estimates of parameters. If we substitute μ and σ^2 with their estimates, we get

$$a + b = 2\overline{X}$$
$$-a + b = 2\sqrt{3}S$$

Note that in taking the square root of variance, we should only consider the positive root to get a positive standard deviation. Solving this system yields

$$\hat{a} = \overline{X} - \sqrt{3}S, \quad \hat{b} = \overline{X} + \sqrt{3}S$$

□

This example suggests a general strategy to estimate parameters:

- Estimate moments by a random sample.

- Set up a system of equations relating parameters and moments, and solve it.

Indeed, this is the starting point of a general parameter estimation approach called *method of moments*. However, a more careful look at the example should raise an issue. Consider the order statistics of a random sample of n observations of a uniform distribution:

$$U_{(1)} \leq U_{(2)} \leq U_{(3)} \leq \cdots \leq U_{(n)}$$

If we take the above approach, *all* of these observations play a role in estimating a and b. But this is a bit counterintuitive; to characterize a uniform distribution we need a lower and an upper bound on its realizations. So, it seems that only $U_{(1)}$ and $U_{(n)}$, i.e., the smallest and the largest observations should play a role. We might suspect that there are alternative strategies for finding *point estimators*. In this section we outline two approaches: the method of moments and the method of maximum likelihood. Since there are alternative ways to build estimators, it is just natural to wonder how we can compare them. Therefore, we should first list the desirable properties that make a good estimator.

9.9.1 Features of point estimators

We list here a few desirable properties of a point estimator $\hat{\theta}$ for a parameter θ. When comparing alternative estimators, we may have to trade off one property for another. We are already familiar with the concept of unbiased estimator. An estimator $\hat{\theta}$ is **unbiased** if

$$E[\hat{\theta}] = \theta$$

We have shown that sample mean is an unbiased estimator of expected value; in Chapter 10 we will show that ordinary least squares, under suitable hypotheses, yield unbiased estimators of the parameters of a linear statistical model. Biasedness is related to the expected value of an estimator, but what about its variance? Clearly, ceteris paribus, we would like to have an estimator with low variance.

DEFINITION 9.16 (Efficient unbiased estimator) *An unbiased estimator $\hat{\theta}_1$ is more efficient than another unbiased estimator $\hat{\theta}_2$ if*

$$\text{Var}(\hat{\theta}_1) < \text{Var}(\hat{\theta}_2)$$

Note that we must compare unbiased estimators in assessing efficiency. Otherwise, we could obtain a nice estimator with zero variance by just choosing an arbitrary constant. It can be shown that, under suitable hypotheses, the variance of certain estimators has a lower bound. This bound, known as the *Cramér–Rao bound*, is definitely beyond the scope of this book, but the message is clear: We cannot go below a minimal variance. If the variance of an estimator attains that lower bound, then it is efficient. Another property that we have already hinted at is consistency. An estimator is **consistent** if $\text{plim}\,\hat{\theta} = \theta$.

Example 9.35 The statistic

$$\frac{1}{n}\sum_{i=1}^{n}\left(X_i - \overline{X}\right)^2$$

is not an unbiased estimator of variance σ^2, as we know that we should divide by $n-1$ rather than n. However, it is a consistent estimator, since when $n \to \infty$, there is no difference between dividing by $n-1$ or n. Incidentally, by dividing by n we lose unbiasedness but we retain consistency while reducing variance. □

9.9.2   The method of moments

We have already sketched an application of this method in Example 9.34. To state the approach in more generality, let us introduce the sample moment of order k:

$$M_k \equiv \frac{1}{n}\sum_{i=1}^{n}X_i^k$$

The sample moment is the sample counterpart of moment $m_k = \mathrm{E}[X^k]$. Let us assume that we need an estimate of k parameters $\theta_1, \theta_2, \ldots, \theta_k$. The method of moments relies on the solution of the following system of equations:

$$\begin{cases} m_1 = g_1(\theta_1, \theta_2, \ldots, \theta_k) \\ m_2 = g_2(\theta_1, \theta_2, \ldots, \theta_k) \\ \quad\vdots \\ m_k = g_k(\theta_1, \theta_2, \ldots, \theta_k) \end{cases}$$

In general, this is a system of nonlinear equations that may be difficult to solve, and we cannot be sure that a unique solution, if any, exists. However, assuming that there is in fact a unique solution, we may just replace moments m_k by sample moments M_k, and solve the system to obtain estimators $\hat{\theta}_1, \hat{\theta}_2, \ldots, \hat{\theta}_k$.

Example 9.36 Let $X_1, X_2, \ldots X_n$ be a random sample from a normal population with parameters $(\theta_1, \theta_2) \equiv (\mu, \sigma)$. We know that

$$m_1 = \mu, \qquad m_2 = \sigma^2 + \mu^2$$

Plugging sample moments and solving the system yields

$$\hat{\mu} = M_1 = \overline{X}$$

$$\hat{\sigma} = \sqrt{M_2 - \overline{X}^2} = \sqrt{\frac{1}{n} \sum_{i=1}^{n} X_i^2 - \overline{X}^2} = \sqrt{\frac{1}{n} \sum_{i=1}^{n} (X_i - \overline{X})^2}$$

We note that we do not obtain an unbiased estimator of variance.

9.9.3 The method of maximum likelihood

The method of maximum likelihood is an alternative approach to find estimators in a systematic way. Imagine that a random variable X has a PDF characterized by a single parameter θ; we indicate this as $f_X(x; \theta)$. If we draw a sample of n i.i.d. variables from this distribution, the joint density is just the product of individual PDFs:

$$f_{X_1, \ldots, X_n}(x_1, \ldots, x_n; \theta) = f_X(x_1; \theta) \cdot f_X(x_2; \theta) \cdots f_X(x_n; \theta) = \prod_{i=1}^{n} f_X(x_i; \theta)$$

This is a function depending on the parameter θ. If we are interested in estimating θ, *given* a sample $X_i = x_i$, $i = 1, \ldots, n$, we may swap the role of variables and parameters and build the *likelihood function*

$$L(\theta) = L(\theta; x_1, \ldots, x_n) = f_{X_1, \ldots, X_n}(x_1, \ldots, x_n; \theta) \tag{9.38}$$

The shorthand notation $L(\theta)$ is used to emphasize that this is a function of the unknown parameter θ, for a given sample of observations. On the basis this framework, intuition suggests that we should select the parameter yielding the largest value of the likelihood function. For a discrete random variable, the interpretation is more natural: We select the parameter maximizing the probability of what we have indeed observed. For a continuous random variable, we cannot really speak of probabilities, but the rationale behind the method of maximum likelihood should be apparent: We try to find the best explanation of what we observe.

Example 9.37 Let us consider the PDF of an exponential random variable with parameter λ:

$$f_X(x; \lambda) = \lambda e^{-\lambda x}$$

Given observed values $x_1, \ldots, x_n$, the likelihood function is

$$L(\lambda) = \prod_{i=1}^{n} \lambda e^{-\lambda x} = \lambda^n \exp\left\{-\lambda \sum_{i=1}^{n} x_i\right\}$$

Quite often, rather than attempting direct maximization of the likelihood function, it is convenient to maximize its logarithm, i.e., the *loglikelihood function*

$$l(\theta) = \ln L(\theta) \tag{9.39}$$

The rationale behind this function is easy to grasp, since by taking the logarithm of the likelihood function, we transform a product of functions into a sum of functions

$$l(\lambda) = n \ln \lambda - \lambda \sum_{i=1}^{n} x_i$$

The first-order optimality condition yields

$$\frac{n}{\lambda} - \sum_{i=1}^{n} x_i = 0 \quad \Rightarrow \quad \hat{\lambda} = \frac{1}{\dfrac{1}{n} \displaystyle\sum_{i=1}^{n} X_i} = \frac{1}{\overline{X}}$$

The result is rather intuitive, since $E[X] = 1/\lambda$. In problem 9.17, the reader is invited to check that the method of moments yields the same estimator.

$\square$

If we must estimate multiple parameters, the idea does not change much. We just solve a maximization problem involving multiple variables.

Example 9.38 Let us consider maximum-likelihood estimation of the parameters $\theta_1 = \mu$ and $\theta_2 = \sigma^2$ of a normal distribution. Straightforward manipulation of the PDF yields the loglikelihood function

$$l(\mu, \sigma^2) = -\frac{n}{2} \ln(2\pi) - \frac{n}{2} \ln \sigma^2 - \frac{1}{2\sigma^2} \sum_{i=1}^{n} (x_i - \mu)^2$$

The first-order optimality condition with respect to μ is

$$\frac{\partial l}{\partial \mu} = \frac{1}{\sigma^2} \sum_{i=1}^{n} (x_i - \mu) = 0 \quad \Rightarrow \quad \hat{\mu} = \frac{1}{n} \sum_{i=1}^{n} x_i$$

Hence, the maximum-likelihood estimator of μ is just the sample mean. If we apply the first-order condition with respect to σ^2, plugging the estimator $\hat{\mu}$, we obtain

$$\frac{\partial l}{\partial \sigma^2} = -\frac{n}{2} \frac{1}{\sigma^2} + \frac{1}{2\sigma^4} \sum_{i=1}^{n} (x_i - \hat{\mu})^2 \quad \Rightarrow \quad \hat{\sigma}^2 = \frac{1}{n} \sum_{i=1}^{n} (x_i - \hat{\mu})^2$$

$\square$

In the two examples above, maximum likelihood yields the same estimator as the method of moments. Then, one could well wonder whether there is really any difference between the two approaches. The next example provides us with a partial answer.

Example 9.39 Let us consider a uniform distribution on interval $[0, \theta]$. On the basis of a random sample of n observations, we build the likelihood function

$$L(\theta) = \begin{cases} \dfrac{1}{\theta^n} & \text{for } 0 \leq x_i \leq \theta, \, i = 1, \ldots, n \\ 0 & \text{otherwise} \end{cases}$$

This may look a bit weird at first sight, but it makes perfect sense, as the PDF is zero for $x > \theta$. Indeed, θ cannot be smaller than the largest observation:

$$\theta \geq \max\{x_1, x_2, \ldots, x_n\}$$

In this case, since there is a constraint on θ, we cannot just take the derivative of $L(\theta)$ and set it to 0 (first-order optimality condition). Nevertheless, it is easy to see that likelihood is maximized by choosing the smallest θ, subject to constraint above. Hence, using the notation of order statistics, we find

$$\hat{\theta} = X_{(n)}$$

$\square$

While the application of maximum likelihood to exponential and normal variables looks a bit dull, in the last case we start seeing something worth noting. The estimator is quite different from what we have obtained by using the method of moments. The most striking feature is that this estimator does *not* use the whole sample, but just a single observation. We leave it as an exercise for the reader that, in the case of a uniform distribution on the interval $[a, b]$, maximum-likelihood estimation yields

$$\hat{a} = X_{(1)}, \quad \hat{b} = X_{(n)}$$

Indeed, the smallest and largest observations are *sufficient statistics* for the parameters a and b of a uniform distribution.[25] We refrain from stating a formal definition of sufficient statistics, but the idea is rather intuitive. Given a random sample $\mathbf{X}$, a sufficient statistic is a function $T(\mathbf{X})$ that captures all of the information we need from the sample in order to estimate a parameter. As a further example, sample mean is a sufficient statistic for the parameter μ of a normal distribution. This concept has far-reaching implications in the theory of inferential statistics.

A last point concerns unbiasedness. In the first two examples, we have obtained biased estimators for variance, even though they are consistent. It can be shown[26] that an unbiased estimator of θ for the uniform distribution on $[0, \theta]$ is

$$\frac{n+1}{n} X_{(n)} \qquad (9.40)$$

Again, we see that maximum likelihood yields a less than ideal estimator, even though for $n \to \infty$ there is no real issue. It turns out that maximum-likelihood estimators (MLEs) do have limitations, but a few significant advantages as well. Subject to some technical conditions, the following properties can be shown for MLEs:

- They are consistent.

- They are asymptotically normal.

- They are asymptotically efficient.

- They are invariant, in the sense that, given a function $g(\cdot)$, the MLE of $\gamma = g(\theta)$ is $g(\hat{\theta})$, where $\hat{\theta}$ is the MLE of θ.

As a general rule, finding MLEs requires the solution of an optimization problem by numerical methods, but there is an opportunity here. Whatever constraint we want to enforce on the parameters, depending on domain-specific knowledge, can be easily added. We obtain a constrained optimization problem that can be tackled using the theory that we illustrate in Chapter 12.

9.10   SOME MORE HYPOTHESIS TESTING THEORY

Arguably, one of the most important contributions of the theory of maximum-likelihood estimation is that it provides us with a systematic approach to find estimators. Similar considerations apply to interval estimation and hypothesis testing. We have very simple and intuitive ways of computing confidence intervals and testing hypotheses about the mean of a normal population, but

---

[25] See Ref. [5, p. 375] for a proof.
[26] See problem 9.19.

we might well be clueless when dealing with less trivial problems. Ad hoc methods may be difficult to find, and we need general strategies. Furthermore, a sound theory is useful to assess desirable properties of a test. For instance, we just considered the probability of a type I error, but we disregarded type II errors completely. Actually, we cannot just focus on type I errors. Indeed, it is quite easy to obtain a test with zero probability of a type I error; we have just to choose an empty rejection region, $\mathcal{C} = \emptyset$. Unfortunately, by doing so, we obtain a test where the probability of a type II error is 1. We should also keep an eye on type II errors, which leads us to considering both the size and the power of a test. In the next two subsections, we first introduce these two concepts, and then we get a glimpse of a more general strategy to obtain hypothesis testing procedures.

9.10.1 Size and power of a test

In the elementary theory of hypothesis testing we consider a null hypothesis such as

$$H_0 : \mu = \mu_0$$

against the alternative $H_a : \mu \neq \mu_0$. Given a sample $\mathbf{X} = (X_1, \ldots, X_n)$, we considered a rejection region $\mathcal{C}$ related to two tails of a standard normal or t distribution. In this case, it is quite easy to evaluate the probability of a type I error

$$\alpha = \mathrm{P}_{\mu_0}(\mathbf{X} \in \mathcal{C})$$

This is just the probability that, even though the null hypothesis is true, the test statistic falls in the rejection region. The notation P_{μ_0} refers to the fact that we associate a probability distribution with the unique value μ_0 considered in the null hypothesis. But which probability distribution should we consider when the null hypothesis is $H_0 : \mu \leq \mu_0$?

To generalize the analysis, let us consider a vector of parameters $\boldsymbol{\theta}$ and a null hypothesis of the form

$$H_0 : \boldsymbol{\theta} \in \Theta_0$$

versus the alternative

$$H_a : \boldsymbol{\theta} \in \Theta_0^c$$

Here Θ_0 is an arbitrary region, and Θ_0^c is its complement. The vector of parameters could be subject to additional restrictions; for instance, a subset of parameters could take only nonnegative values. Let us denote the set of feasible values of parameters by Θ; clearly, $\Theta = \Theta_0 \cup \Theta_0^c$. In the first case above the set Θ_0 is a singleton, $\Theta_0 = \{\mu_0\}$, and we speak of a *simple hypothesis*; otherwise, we have a *composite hypothesis*.

DEFINITION 9.17 (Size of a test) *We say that a test with rejection region $\mathcal{C}$ has **size** α if*

$$\alpha = \max_{\boldsymbol{\theta} \in \Theta_0} \mathrm{P}_{\boldsymbol{\theta}}(\mathbf{X} \in \mathcal{C})$$

Of course, the size of a test is essentially another name for the significance level. From its definition, we see that α is related to the worst-case distribution in terms of type I errors. Since we want to be conservative, it is natural to give priority to type I errors by keeping α reasonably small. However, given a test size, we should find a test minimizing the probability of a type II error. If $\boldsymbol{\theta} \in \Theta_0^c$, the probability of type II error is

$$P_{\boldsymbol{\theta}}(\mathbf{X} \notin \mathcal{C}) = 1 - P_{\boldsymbol{\theta}}(\mathbf{X} \in \mathcal{C})$$

Note that

$$P_{\theta}(\mathbf{X} \in \mathcal{C}) = \begin{cases} \text{probability of a type I error,} & \text{if } \boldsymbol{\theta} \in \Theta_0 \\ 1 - \text{probability of a type II error,} & \text{if } \boldsymbol{\theta} \in \Theta_0^c \end{cases}$$

DEFINITION 9.18 (Power of a test) *The* **power function** *for a test with rejection region $\mathcal{C}$ is a function of $\boldsymbol{\theta}$, defined as*

$$\beta(\boldsymbol{\theta}) = P_{\boldsymbol{\theta}}(\mathbf{X} \in \mathcal{C})$$

Note that the power of a test is a *function*, as it depends on the true value the unknown parameter vector; hence, in a practical setting, we are not able to find a single value giving the power of the test. Still, we may observe that the ideal power function is 0 for $\boldsymbol{\theta} \in \Theta_0$ and 1 for $\boldsymbol{\theta} \in \Theta_0^c$. This ideal power function cannot be obtained in practice, but we can look for tests that have maximal power when $\boldsymbol{\theta} \in \Theta_0^c$. The theory of optimal, or most powerful, tests is developed in the context of mathematical statistics and it relies on systematic ways to devise testing procedures. We outline one of them in the next section.

9.10.2   Likelihood ratio tests

We have introduced likelihood functions as a useful tool for parameter estimation. They also play a role in hypothesis testing and lead to the so-called *likelihood ratio test* (LRT). The test is based on the likelihood ratio statistic

$$\lambda(\mathbf{x}) = \frac{\sup_{\boldsymbol{\theta} \in \Theta_0} L(\boldsymbol{\theta}; \mathbf{x})}{\sup_{\boldsymbol{\theta} \in \Theta} L(\boldsymbol{\theta}; \mathbf{x})} \tag{9.41}$$

A LRT is a test whose rejection region is of the form

$$\mathcal{C} = \{\mathbf{x} \mid \lambda(\mathbf{x}) \leq c\}$$

for a suitable constant $c < 1$. The rationale of the approach is rather intuitive. If the statistic $\lambda(\mathbf{x})$ is small, the restriction of $\boldsymbol{\theta}$ to Θ_0 does not seem justified.

Example 9.40 Let us consider again the one-sided test on the expected value of a normal population

$$H_0 : \mu \leq \mu_0, \quad \text{vs.} \quad H_a : \mu > \mu_0$$

For simplicity, let us assume that the variance σ^2 is known. In the first part of the chapter, we have considered a one-tail test, with a somewhat heuristic justification. Let us see if we can find a more formal justification based on a LRT. The likelihood function, for a sample of size n, is

$$L(\mu; \sigma^2, \mathbf{x}) = \frac{1}{(2\pi\sigma)^{n/2}} \exp\left[-\frac{\sum_{i=1}^{n}(x_i - \mu)^2}{2\sigma^2}\right]$$

Note that this is a function of μ, for given $\mathbf{x}$ and σ^2. In the likelihood ratio, we have $\Theta = (-\infty, +\infty)$ and $\Theta_0 = (-\infty, \mu_0)$. We know from Example 9.38 that the denominator of the likelihood ratio in Eq. (9.41) is maximized for $\mu = \bar{x}$. As to the numerator, there are two cases:

1. $\bar{x} \le \mu_0$. In this case, the constraint defining Θ_0 is nonbinding[27]

   in the maximization of the likelihood function at the numerator, and $\lambda(\mathbf{x}) = 1$. This case is not interesting.

2. $\bar{x} > \mu_0$. In this case, the constraint defining Θ_0 is binding and in fact the statistic seems to contradict the null hypothesis; the likelihood function at the numerator is maximized for $\mu = \mu_0$.

Assuming that the interesting case $\bar{x} > \mu_0$ applies, the likelihood ratio may be rewritten as follows:

$$
\begin{aligned}
\lambda(\mathbf{x}) &= \frac{\dfrac{1}{(2\pi\sigma)^{n/2}} \exp\left[-\dfrac{\sum_{i=1}^{n}(x_i - \mu_0)^2}{2\sigma^2}\right]}{\dfrac{1}{(2\pi\sigma)^{n/2}} \exp\left[-\dfrac{\sum_{i=1}^{n}(x_i - \bar{x})^2}{2\sigma^2}\right]} \\[2em]
&= \frac{\exp\left[-\dfrac{\sum_{i=1}^{n}(x_i - \bar{x} + \bar{x} - \mu_0)^2}{2\sigma^2}\right]}{\exp\left[-\dfrac{\sum_{i=1}^{n}(x_i - \bar{x})^2}{2\sigma^2}\right]} \\[2em]
&= \frac{\exp\left[-\dfrac{\sum_{i=1}^{n}(x_i - \bar{x})^2 + \sum_{i=1}^{n}(\bar{x} - \mu_0)^2}{2\sigma^2}\right]}{\exp\left[-\dfrac{\sum_{i=1}^{n}(x_i - \bar{x})^2}{2\sigma^2}\right]} \\[2em]
&= \exp\left[-\dfrac{\sum_{i=1}^{n}(\bar{x} - \mu_0)^2}{2\sigma^2}\right] \\[1em]
&= \exp\left[-\dfrac{n(\bar{x} - \mu_0)^2}{2\sigma^2}\right].
\end{aligned}
$$

---

[27] We define binding and nonbinding constraints later, in Section 12.1. However, the idea is pretty simple. A constraint of the form $g(\mathbf{x}) \le 0$ is binding at the optimal solution $\mathbf{x}^*$ if $g(\mathbf{x}^*) = 0$; it is nonbinding if there is some slack, i.e., if $g(\mathbf{x}^*) < 0$.

Now, we should reject when $\lambda(\mathbf{x}) < c$, which implies rejection when the statistic

$$\frac{\sqrt{n}(\overline{X} - \mu_0)}{\sigma}$$

is large. This is exactly what we do in standard test procedures; the rejection region is the left tail of a standard normal distribution. Furthermore, the probability of error of type I is set to α by selecting the critical value $z_{1-\alpha}$. The case with σ^2 unknown is dealt with by a similar token, even though the calculations are a bit more involved. ∎

The LRT approach is not only a systematic way to devise testing procedures, but it also plays a key role in establishing theoretical results on most powerful tests; we refer the interested reader to the advanced references on mathematical statistics.

Problems

9.1   The director of a Masters' program wants to assess the average IQ of her students. A sample of 18 students yields the following results:

$$130, 122, 119, 142, 136, 127, 120, 152, 141$$
$$132, 127, 118, 150, 141, 133, 137, 129, 142$$

- Build a 95% confidence interval for the average IQ.

- Assuming that IQ is normally distributed, how would you estimate the probability that IQ is larger than 130? What if you do not want to assume normality? Compare the two approaches.

9.2   You have to compute a confidence interval for the expected value of a random variable. Using a standard procedure, you take a random sample of size $N = 20$, and the sample statistics are $\overline{X} = 13.38$ and $S = 4.58$.

- Compute a confidence interval at level 95%.

- If you take additional observations, raising sample size to $N = 50$, how large do you expect the new confidence interval be? What hypothesis are you making in your reasoning?

9.3   Find the 97% confidence interval, given a sample mean of 128.37, sample standard deviation of 37.3, and sample size of 50. What is the width of the confidence interval? Suppose that you want to cut the confidence interval by 50% (i.e., the new width should be half the previous one). How many additional observations would you use?

9.4   In standard confidence intervals, you use the sample mean as an estimator of expected value. Now suppose that a friend of yours suggests the following alternative estimator:

$$\tilde{X} = \frac{1}{N}\left(\frac{1}{2}X_1 + \frac{3}{2}X_2 + \frac{1}{2}X_3 + \frac{3}{2}X_4 + \cdots + \frac{1}{2}X_{N-1} + \frac{3}{2}X_N\right)$$

where we assume that N is an even number. Prove that

$$\mathrm{E}\left[\tilde{X}\right] = \mu, \quad \mathrm{Var}\left(\tilde{X}\right) = 1.25 \times \frac{\sigma^2}{N}$$

where $\mathrm{E}[X] = \mu$ and $\mathrm{Var}(X) = \sigma^2$. Would you use this estimator? Why or why not?

9.5   TakeItEasy produces special shoes for runners, whose average life is 1250 km. In order to improve the product, they experiment with a new design, and test prototypes with a sample of 30 runners. The sample mean of product life is 1315 km, with a standard deviation of 70 km. Can we say that TakeItEasy has actually improved their product?

9.6   Air quality is measured by the concentration of a dangerous pollutant. The mayor of a city has engaged in a program to improve traffic conditions in order to decrease the concentration of that pollutant. Of course, there is a lot of day-to-day variability in measures. In the past, the average concentration was 29 (measured in some units). In a sample of 20 days after completion of the program, the sample mean has been 26.9, with a standard deviation of 8. Can we say that the mayor's program has been effective?

9.7   You want to compare the reliability of two machines that insert chips onto electronic cards. The main problem is the occurrence of jams in the feeding mechanism, as this requires stopping production to fix the trouble. To this aim, you observe the number of jams when producing standard batches of electronic cards, resulting in the following table:

	Machine 1	Machine 2
Sample size	46	54
Sample mean	8.2	9.4
Sample standard deviation	2.1	2.9

Is there a significant difference between the two machines?

9.8   A study was done to measure the impact of fatigue on human performance when carrying out a certain task. The performance is measured by an appropriate index, the larger the better, which is measured at the beginning of the shift and after 3 hours of work. Ten workers are observed, resulting in the following table:

beginning	after 3 hours
68	57
64	53
69	72
88	63
72	73
80	50
85	53
116	80
77	65
78	49

Is there a significant effect due to weariness?

9.9   The following dataset is a random sample from a normal distribution:

$$103.23, \ 111.00, \ 86.45, \ 105.17, \ 101.91$$
$$92.15, \ 97.40, \ 102.06, \ 121.47, \ 116.62$$

Find a 95% confidence interval for variance.

9.10   In order to estimate the fraction of defective parts, you take a sample of size 1000 and find that 63 are not acceptable. Find a 99% confidence interval for the fraction of defective parts.

9.11   In one-way ANOVA we define the sum of squares SS_b and SS_w. Prove the identity

$$SS_w = \sum_i \sum_j X_{ij}^2 - nm\overline{X}_{..}^2 - SS_b$$

9.12   Apply one-way ANOVA to check equality of means for the following sample:

$i = 1$	$i = 2$	$i = 3$
82.31	240.80	181.55
160.98	228.27	188.83
230.84	278.73	334.07
522.06	278.16	326.81
449.25	172.16	327.55

9.13   A m-Erlang distribution with rate λ is obtained when summing m independent exponential random variables with rate λ. This distribution may be used to model more realistic random service times in queueing systems. Devise an efficient method to generate a sample of pseudorandom variables from the m-Erlang distribution.

9.14   Define an algorithm to generate pseudorandom variables characterized by the following density function:

$$f(x) = \begin{cases} x - 1, & \text{if } 1 \le x \le 2 \\ 3 - x, & \text{if } 2 \le x \le 3 \\ 0, & \text{otherwise} \end{cases}$$

9.15   Consider the simulation of a continuous review (Q, R) inventory control policy. Define the relevant events for the system, and outline a procedure for the management of each event. To deal with a specific case, assume that customers arrive according to a Poisson process and order a random amount of items drawn from a given probability distribution. Lead time is assumed to be deterministic. First assume that customers are impatient, i.e., if their order cannot be satisfied immediately from stock, they are lost. Then, adapt the event management procedures to the case of patient customers (allowing for backlogged demand).

9.16   Consider a sequence of random variables

$$X_n = \begin{cases} 0 & \text{with probability } 1 - \dfrac{1}{n^2} \\ n & \text{with probability } \dfrac{1}{n^2} \end{cases}$$

Does this sequence converge in probability to a number? What about convergence in quadratic mean?

9.17   Consider an exponential distribution with rate λ. On the basis of a random sample of size n, apply the method of moments to estimate λ.

9.18   Apply the method of maximum likelihood to estimate the parameters of a uniform distribution on the interval $[a, b]$.

9.19   Prove the result of Eq. (9.40). You should use the result established in Problem 7.11 to find the expected value of $X_{(n)}$, given a sample of n independent observations from the uniform distribution on the interval $[0, \theta]$.

For further reading

- A basic treatment of inferential statistics is offered by many books, including Ref. [13].

- An excellent reference at the intermediate level is Ref. [15], which has influenced the initial part of this chapter.

- Advanced-level treatments are offered, e.g., by Refs. [3], [5], and [8]. The first reference, in particular, has influenced the presentation in the advanced sections of this chapter.

- Indeed, inferential statistics can involve quite sophisticated mathematics, if brought to a higher level, but it also raises interesting "philosophical" controversies. If you want to dig deeper into hypothesis testing, Ref. [9] deals with the meaning of significance levels and p-values. See also Refs. [4] and [12], which gives a historical perspective as well.

- We have very briefly touched on the fundamental theme of simulation; an extensive treatment can be found in Ref. [11], whereas Ref. [14] offers a more succinct exposition. Monte Carlo simulation is also fundamental in finance; see Ref. [2], among many other sources.

- We have hinted at factor models for financial portfolio management; see Ref. [1] for a very readable introduction and motivation.

- A thorough analysis of stochastic convergence can be found in Ref. [7]. This is also a fundamental theme in more application-oriented books, like those on econometrics; see, e.g., Ref. [6] or [16] (which has also influenced the treatment in this chapter).

REFERENCES

1. Z. Bodie, A. Kane, and A. Marcus, *Investments*, 8th ed., McGraw-Hill, New York, 2008.

2. P. Brandimarte, *Numerical Methods in Finance and Economics: A Matlab-Based Introduction*, 2nd ed., Wiley, New York, 2006.

3. G. Casella and R.L. Berger, *Statistical Inference*, 2nd ed., Duxbury, Pacific Grove, CA, 2002.

4. R. Christensen, Testing Fisher, Neyman, Pearson, and Bayes, *The American Statistician*, **59**:121–126, 2005.

5. M.H. DeGroot and M.J. Schervish, *Probability and Statistics*, 3rd ed., Addison-Wesley, Boston, 2002.

6. W.H. Greene, *Econometric Analysis*, 6th ed., Prentice-Hall, Upper Saddle River, NJ, 2008.

7. G. Grimmett and D. Stirzaker, *Probability and Random Processes*, 3rd ed., Oxford University Press, Oxford, 2003.

8. R.V. Hogg, J.W. McKean, and A.T. Craig, *Introduction to Mathematical Statistics*, 6th ed., Prentice-Hall, Upper Sadle River, NJ, 2005.

9. E. Hubbard and M.J. Bayarri, Confusion over measures of evidence (p's) versus errors (α's) in classical statistical testing, *The American Statistician*, **57**:171–178, 2003.

10. D.N. Joanes and C.A. Gill, Comparing measures of sample skewness and kurtosis, *Journal of the Royal Statistical Society, Series D, The Statistician*, **47**:183–189, 1998.

11. A.M. Law and W.D. Kelton, *Simulation Modeling and Analysis*, 3rd ed., McGraw-Hill, New York, 1999.

12. D.J. Murdoch, Y.-L. Tsai, and J. Adcock, P-values are random variables, *The American Statistician*, **62**:242–245, 2008.

13. M.K. Pelosi and T.M. Sandifer, *Elementary Statistics: From Discovery to Decision*, Wiley, New York, 2003.

14. S.M. Ross, *Simulation*, 4th ed., Elsevier Academic Press, Burlington, MA, 2006.

15. S.M. Ross, *Introduction to Probability and Statistics for Engineers and Scientists*, 4th ed., Elsevier Academic Press, Burlington, MA, 2009.

16. J.H. Stock and M.W. Watson, *Introduction to Econometrics*, 2nd ed., Addison-Wesley, Boston, 2006.

10

Simple Linear Regression

In this chapter we take advantage of all the probabilistic and statistical knowledge we have built in the previous chapters to get into the realm of empirical model building. Models come in many forms, but what we want to do here is finding a relationship between two variables, say, x and y, based on a set of n joint observations (x_i, y_i), $i = 1, \ldots, n$. We got acquainted with correlation in Chapter 8, and if two variables are correlated we can try to put such knowledge to good use for decision making and forecasting. The first step in building a model is choosing a functional form representing the link between the variables of interest. The simplest relationship that comes to mind is linear:

$$y = a + bx$$

This is called a *simple linear regression model*. It is obviously *linear*, but one could and should wonder whether a more complicated, nonlinear functional form is better suited to our task. It is *simple* since there is only one variable x that we use to "explain" the variable y; multiple linear regression models rely on possibly several explanatory variables. We cover these more advanced models in Chapter 16, since they rely on a definitely more challenging technical machinery. Yet, even the innocent-looking simple linear regression model hides a lot of issues, which are best understood in a simple setting. A deep understanding of these issues is needed to tackle nonlinear and multiple regression models.

It is tempting to interpret x as an input or a cause, and y as an output or an effect. Granted, there are many practical cases in which this interpretation does make sense, but we should never forget that a regression model relies on association and not causation. The same caveats that we have pointed out when dealing with correlation apply to regression models. By a similar token, when referring to functions it is customary to call x the independent variable and y the dependent variable; obviously, these terms can be a bit misleading in a statistical framework. In the following we will refer to x by the terms *explanatory variable* or *regressor*; y will be called *response* or *regressed variable*.

To build and use a linear regression model, we must accomplish the following steps:

1. We must devise a suitable way to choose the coefficients a and b.

2. We should check if the model makes sense and is reliable enough.

3. We should use the model by

   - Building knowledge to understand a phenomenon

   - Generating forecasts and scenarios for decision making under uncertainty

We accomplish the first step in Section 10.1, where we lay down the foundations of the least-squares method. Section 10.2 deals with the second step, which requires building a statistical framework for linear regression. This is needed to state precise assumptions behind our modeling endeavor, which should be thoroughly checked before using the model; we also need to draw statistical inferences and the test hypotheses about the estimated coefficients in the model. We do so in Section 10.3, for the simpler case of a nonstochastic regressor, i.e., when the explanatory variable x is treated as a number rather than a random variable. Then, in Section 10.4, we tackle the third step. There are different uses of a linear regression model, and statistics in general. We might be interested in understanding a physical or social phenomenon; in such a case a model is used for knowledge discovery purposes and to ascertain the impact of explanatory variables. In a business management setting, we are more likely to be interested in using the model to generate forecasts and scenarios as an input to a decision-making procedure. However, we should not undervalue model building *per se*: Gathering data and identifying a model is often a good way to reach a common understanding of a multifaceted problem, possibly involving different members of an organization with a limited view of the overall process. Finally, in Section 10.5 we relax a few limiting assumptions made in Section 10.3, and outline extensions such as the weighted least-squares method, and in Section 10.6 we take a look at the links between linear algebra and linear regression. The last two sections can be safely omitted by readers who just need a basic understanding of linear regression.

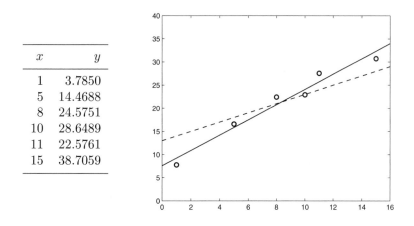

x	y
1	3.7850
5	14.4688
8	24.5751
10	28.6489
11	22.5761
15	38.7059

Fig. 10.1 Data for linear regression.

10.1 LEAST-SQUARES METHOD

Consider the data tabulated and depicted in Fig. 10.1. These *joint* observations are displayed as circles, and a look at the plot suggests the possibility of finding a linear relationship between x and y. A linear law relating the two variables, such as[1]

$$y = a + bx$$

can be exploited to understand what drives a social or physical phenomenon, and is one way of exploiting correlation between data. The sample correlation coefficient for these data is 0.9623, and even if the sample is a toy one, it is significant at 1% confidence level using the test of Section 9.4.5 (the p-value is 0.0021). Intuitively, the sign of the slope b should be the same as the sign of the correlation. If we relate expenditure in advertisement and revenue, we would expect a positive correlation. If we relate price and demand, we would expect negative correlation. Before we go on, it is important to insist on two points:

- A linear relationship need not be the best model. We know that correlation picks up only linear associations, but there could be a more complicated relationship. Nevertheless, linear models are the natural starting point in our investigation.

- It is tempting to interpret x as a cause and y as an effect; however, correlation only measures association. For instance, when relating price and

---

[1]Strictly speaking, we are using a linear *affine* function; the "linear" term should be reserved for the case when $a = 0$, but we will not bother too much about subtlety.

demand we cannot say in general which one is the independent variable, as this depends on the specific market and its level of competition.

Clearly, there is no way to find a "perfect fit" line passing through all of the observed points, unless a linear model is an exact representation of reality. This will never be the case in real life, because of uncertainty and/or unmodeled factors.

Now what makes a good model, and how can we measure its fit against empirical data? Let us compare the two alternative lines depicted in Fig. 10.1. It is a fairly safe bet most readers would choose the continuous line rather than the dashed one. Our eye is measuring the distance between the observed points (x_i, y_i) and the corresponding points $(x_i, \hat{y}_i)$ that the linear model would predict. To find a prediction $\hat{y}_i$, associated with a setting of the regressor x_i, simply requires plugging the latter into the regression model:

$$\hat{y}_i = a + bx_i$$

To formalize this distance for each individual observation, we can define a *residual*:[2]

$$e_i \equiv y - \hat{y}_i = y_i - (a + bx_i)$$

In order to evaluate the overall fit of a model we must aggregate the single residuals into one distance measure. It should be clear that the average residual is not quite useful, as positive and negative deviations cancel each other. Borrowing the variance idea from statistics, we may consider squared deviations to get rid of the sign and define the *sum of squared residuals* (SSR) as follows:

$$\text{SSR} \equiv \sum_{i=1}^{n} e_i^2$$

According to the *least-squares* approach, the best-fit line is the one that minimizes SSR with respect to the regression coefficients a and b. Of course, there is no difference between minimizing SSR or the average squared residual, as multiplying the objective function by $1/n$ does not change the optimal solution.

Finding the best coefficients a and b calls for the solution of a least-squares problem and, in the case of simple linear regression, is a straightforward exercise in calculus: We just need to enforce the first-order optimality conditions, one per regression coefficient.[3] A first condition is obtained by requiring sta-

---

[2]It is only natural to call e_i an *error*. However, this term is reserved for a related quantity that will be introduced later.

[3]They are sufficient conditions, as the objective function is convex with respect to the decision variables.

tionarity of the partial derivative of SSR with respect to the intercept a:

$$\frac{\partial \text{SSR}}{\partial a} = -\sum_{i=1}^{n} 2 \left(y_i - a - bx_i\right)$$

$$= -2 \left(\sum_{i=1}^{n} y_i - \sum_{i=1}^{n} a - \sum_{i=1}^{n} bx_i\right)$$

$$= -2 \left(\sum_{i=1}^{n} y_i - na - b \sum_{i=1}^{n} x_i\right) = 0$$

Rearranging the condition yields

$$a^* = \frac{1}{n} \sum_{i=1}^{n} y_i - \frac{b}{n} \sum_{i=1}^{n} x_i = \overline{y} - b\overline{x} \tag{10.1}$$

where $\overline{x}$ and $\overline{y}$ are the average values of x and y. It is useful to interpret this condition: It tells us that the *barycenter* $(\overline{x}, \overline{y})$ of the experimental data lies on the regression line, which does make good sense. This condition also implies that the average residual for the best-fit model is zero:

$$\overline{e} = \frac{1}{n} \sum_{i=1}^{n} [y_i - (a + bx_i)] = \overline{y} - a - b\overline{x} = \overline{y} - (\overline{y} - b\overline{x}) - b\overline{x} = 0 \tag{10.2}$$

The second optimality condition is obtained by enforcing stationarity of the partial derivative with respect to the slope b:

$$\frac{\partial \text{SSR}}{\partial b} = -2 \sum_{i=1}^{n} x_i \left(y_i - a - bx_i\right) = 0$$

We can get rid of the leading factor -2, which is irrelevant, and plug the optimal value a^* into this condition:

$$\sum_{i=1}^{n} x_i \left[y_i - \left(\frac{1}{n} \sum_{j=1}^{n} y_j - \frac{b}{n} \sum_{j=1}^{n} x_j\right) - bx_i \right]$$

$$= \frac{1}{n} \left(n \sum_{i=1}^{n} x_i y_i - \sum_{i=1}^{n} x_i \cdot \sum_{i=1}^{n} y_i + b \sum_{i=1}^{n} x_i \cdot \sum_{i=1}^{n} x_i - nb \sum_{i=1}^{n} x_i^2 \right) = 0$$

Solving for b yields

$$b^* = \frac{n \sum_{i=1}^{n} x_i y_i - \sum_{i=1}^{n} x_i \cdot \sum_{i=1}^{n} y_i}{n \sum_{i=1}^{n} x_i^2 - \left(\sum_{i=1}^{n} x_i\right)^2} \tag{10.3}$$

This way of expressing the optimal slope is arguably the most immediate for carrying out the calculations. However, we may also rewrite slope in a way that is easier to remember, dividing numerator and denominator by n and making averages of x and y explicit:

$$b^* = \frac{\displaystyle\sum_{i=1}^{n} x_i y_i - n\bar{x}\bar{y}}{\displaystyle\sum_{i=1}^{n} x_i^2 - n\bar{x}^2} \tag{10.4}$$

With least squares, we find explicit expressions for a^* and b^*. In the following we will see how these expressions may be interpreted intuitively to improve our understanding; however, it is better to illustrate the approach first by a small numerical example. To streamline notation, we will drop the asterisk ($*$) and denote the optimal value of coefficients by a and b.

Example 10.1 Let us apply the least-squares method to the data of Fig. 10.1. The necessary calculations are reported in Table 10.1. Applying (10.3), we have

$$b = \frac{6 \times 1388.1444 - 50 \times 132.7598}{6 \times 536 - 50^2} = 2.3616$$

Plugging this value into (10.1) yields

$$a = \frac{132.7598}{6} - 2.3616 \times \frac{50}{6} = 2.4470$$

So, the linear regression model is

$$y = 2.3616x + 2.4470 \tag{10.5}$$

and it corresponds to the continuous line in Fig. 10.1, which in fact is the best-fit line in the least-squares sense. □

10.1.1 Alternative approaches for model fitting

In the least-squares method, we square residuals and solve the corresponding optimization problem analytically. We should wonder what is so special with squared residuals. We might just as well take the absolute values of the residuals and solve

$$\min_{a,b} \sum_{i=1}^{n} |y_i - (a + bx_i)|$$

Another noteworthy point is that in so doing we are essentially considering average values of squared or absolute residuals; please note again that minimizing the sum or the average of squared residuals is the same thing, as

Table 10.1   Calculations for Example 10.1.

x_i	y_i	x_i^2	$x_i y_i$
1	3.7850	1	3.7850
5	14.4688	25	72.3440
8	24.5751	64	196.6008
10	28.6489	100	286.4890
11	22.5761	121	248.3371
15	38.7059	225	580.5885
Sums 50	132.7598	536	1388.1444

dividing the function by n does not change the optimal solution. When minimizing an average, there might be a good fit for most observations, but a rather large discrepancy for a very few ones. These could be outliers that can be omitted, but if this is not acceptable, an alternative fitting approach is to minimize the worst-case residual:

$$\min_{a,b} \left\{ \max_{i=1,\ldots,n} |y_i - (a + bx_i)| \right\}$$

This is a min–max problem, where the decision variables are a and b; for given values of slope and intercept, we get a different set of n residuals, and we pick the largest one in absolute value. The aim is minimizing the maximum residuals with respect to regression coefficients.[4]

In principle, there is nothing wrong with these alternative fitting models. In forecasting, we do consider mean absolute deviations, as we will discover in Chapter 11. In function approximation, there is a whole theory concerned with the minimization of the maximum approximation error. The real trouble stems from the fact that the two models described above do not lend themselves to a closed-formula treatment, since the absolute value is not a differentiable function. We will see in Section 12.2 how to transform them into linear programming problems, which can be readily solved by commercially available software packages. However, we just get a numerical solution precluding any further interpretation. On the contrary, a closed formula allows us to cast least-squares fitting within a proper statistical framework, which is essential, as pointed out in the next section.

---

[4]What we are doing is just applying different concepts of the vector norm, which were introduced in Section 3.3.2.

10.1.2   What is linear, exactly?

If we label a model like $y = a + bx$ as linear, no eyebrow should be raised. Now, consider a regression model involving a squared explanatory variable:

$$y = a + b_1 x + b_2 x^2 \qquad (10.6)$$

Is this linear? Actually it is, in terms of the factor that matters most: fitting model coefficients. True, the model is nonlinear in terms of the explanatory variable x, but the actual unknowns when we apply least squares are the coefficients a, b_1, and b_2. In this respect, the model is linear and we may easily apply the multivariate least-squares approach, which we will illustrate in Chapter 16. If we want to tell the difference between the two models above, we may say that the first one is linear and *first-order*, whereas the second one is linear and *second-order*. In a polynomial model like (10.6), the order of the model is the largest degree occurring in the polynomial.

10.2   THE NEED FOR A STATISTICAL FRAMEWORK

So far, we have regarded linear regression as a numerical problem, which is fairly easy to solve by least squares. However, this is a rather limited view. To begin with, it would be useful to gain a deeper insight into the meaning of the regression coefficients, in particular the slope b. The careful reader might have noticed that the numerator in formula (10.4) looks suspiciously like a covariance between variables x and y, whereas the denominator suggests the idea of variance of x. The interpretation must be handled with care since, so far, we considered variables x and y as numbers and not random variables. Still, let us pursue this line of intuition by manipulating the expression of slope a bit. In the following, we will use the rather obvious identities

$$\sum_{i=1}^{n}(x_i - \bar{x}) \equiv 0 \qquad \text{and} \qquad \sum_{i=1}^{n}(y_i - \bar{y}) \equiv 0$$

Equation (10.4) can be rewritten as

$$b = \frac{\displaystyle\sum_{i=1}^{n} x_i y_i - \sum_{i=1}^{n} x_i \bar{y}}{\displaystyle\sum_{i=1}^{n} x_i^{2} - \sum_{i=1}^{n} x_i \bar{x}} = \frac{\displaystyle\sum_{i=1}^{n} x_i \left(y_i - \bar{y}\right)}{\displaystyle\sum_{i=1}^{n} x_i \left(x_i - \bar{x}\right)}$$

In this ratio, using the identities above, we may subtract zero from both numerator and denominator and rearrange, which yields

$$
b \;=\; \frac{\displaystyle\sum_{i=1}^{n} x_i\,(y_i - \bar{y}) - \sum_{i=1}^{n} \bar{x}\,(y_i - \bar{y})}{\displaystyle\sum_{i=1}^{n} x_i\,(x_i - \bar{x}) - \sum_{i=1}^{n} \bar{x}\,(x_i - \bar{x})} \;=\; \frac{\displaystyle\sum_{i=1}^{n} (x_i - \bar{x})\,(y_i - \bar{y})}{\displaystyle\sum_{i=1}^{n} (x_i - \bar{x})^2} \tag{10.7}
$$

This is yet another way of writing the slope,[5] and by dividing numerator and denominator by $n - 1$, we get an expression in terms of sample statistics:

$$
b \;=\; \frac{\dfrac{1}{n-1}\displaystyle\sum_{i=1}^{n} (x_i - \bar{x})\,(y_i - \bar{y})}{\dfrac{1}{n-1}\displaystyle\sum_{i=1}^{n} (x_i - \bar{x})^2} \;=\; \frac{S_{xy}}{S_x^2} \;=\; \frac{r_{xy} S_x S_y}{S_x^2} \;=\; \frac{r_{xy} S_y}{S_x}
$$

Formally, the quantities S_x, S_y, S_{xy}, and r_{xy} are similar to sample standard deviations, covariances, and correlation coefficients; even if the interpretation is debatable, as we are dealing with numbers and not with random samples, the notation is quite handy and tells us a lot. The slope coefficient is closely related to the sample correlation between x and y. Indeed, we know that correlation picks up the linear association between random variables. Positive correlation results in an upward-sloping regression line, whereas negative correlation results in a downward-sloping line. The exact slope depends on the standard deviation of x and y, but this is somewhat arbitrary as it is influenced by how we measure x and y.

The interpretation of slope in terms of correlation should be properly cast within a statistical framework. Indeed, a regression model relies on observed data that we can associate with a sampling mechanism, provided we postulate a data-generating process, i.e., an underlying stochastic model that we observe through a sample of noisy observations from outside. Within this framework, regression coefficients are *estimates* of the unknown parameters of the data-generating process. If we rely on concepts from inferential statistics and analyze linear regression as an estimation problem, we are able to

- Quantify uncertainty about estimates, i.e., calculate confidence intervals for the underlying parameters

- Test hypotheses about the effect of explanatory variables

---

[5] Incidentally, this form is somewhat annoying to use when using a pocket calculator, since it involves centering all of the data. If we recall the roundoff issues of Example 4.13, it is no surprise that this form is numerically more precise when used with a computer; centering data is usually a good idea to improve numerical stability.

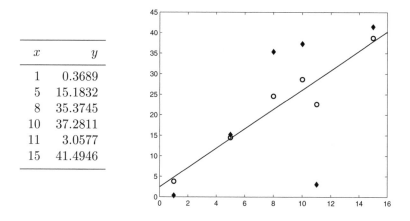

x	y
1	0.3689
5	15.1832
8	35.3745
10	37.2811
11	3.0577
15	41.4946

Fig. 10.2 A different dataset yielding the same regression line as in Fig. 10.1.

- Quantify uncertainty about predictions, i.e., calculate prediction intervals in order to come up with robust decisions

An example will illustrate why all of the above is so important.

Example 10.2 Consider the dataset tabulated in Fig. 10.2, and plotted there as filled diamonds; the figure also includes the dataset of Fig. 10.1, represented by the empty circles. The two datasets look quite different, yet we invite the reader to check that the two regression lines are the same (disregarding slight numerical discrepancies). Now imagine that we want to predict the response with the explanatory variable set to $x_{01} = 13$ or $x_{02} = 17$. The prediction, based on the two identical linear regression models, would just be the same. We have just to plug values of x into the regression line (10.5) to obtain

$$\hat{y}_{01} = 2.3616 \times 13 + 2.4470 = 33.1478$$
$$\hat{y}_{02} = 2.3616 \times 17 + 2.4470 = 42.5942$$

This is straightforward, but in which case should we trust the linear regression model more? Intuition suggests that the second dataset is affected by a larger level of uncertainty, and that the resulting predictions should be taken with due care. Indeed, for the first dataset the sum of squared residuals for the best-fit line is $SSR_1 = 53.1495$, whereas in the second case it is $SSR_2 = 1000$. Furthermore, the value $x = 13$ lies within the range of the observed sample on which the fitting is based, whereas $x = 17$ does not. Intuition again suggests that if we are *extrapolating* outside the range of observed values we should be more uncertain about our predictions, but the least-squares method per se does not provide us with tools to address these issues.   □

The first step in building a statistical framework is a clear statement of the assumptions about the underlying data-generating process, i.e., the process by which observations of the response variable are generated for given values of the explanatory variable. We have two possibilities:

1. If the explanatory variable is a *number* x, we speak of a *nonstochastic* regressor and may write a linear data-generating model as

$$Y_i = \alpha + \beta x_i + \epsilon_i, \qquad i = 1, \dots, n \qquad (10.8)$$

2. If the explanatory variable is a *random variable* X, we speak of a *stochastic* regressor and may write

$$Y_i = \alpha + \beta X_i + \epsilon_i, \qquad i = 1, \dots, n \qquad (10.9)$$

Even if the regressor is nonstochastic, the response Y_i is a random variable because of the terms ϵ_i, which are random variables and are called *errors*. Even though errors are modeled as random variables, they need not be random in principle. Errors may represent truly random factors, but they could also represent other variables that would explain the response, but cannot be observed directly. They can also represent modeling errors, i.e., deviations from a simple linear functional form. If so, they have more to do with our ignorance and modeling mistakes than with true randomness.[6] We will just put everything under one roof and random errors will play the role of a catchall. However, this does not mean that everything goes: We do model errors as random variables, but they are subject to very precise conditions for our modeling mechanism to work. It is essential to validate a regression model a posteriori by checking whether there is a blatant violation of these assumptions.

The difference between nonstochastic and stochastic regressors has implications on how the data-generating model should be characterized. Before dwelling on the required formal assumptions, we show with a few examples that the difference is substantial and not academic.

Example 10.3 Imagine that we are a monopolist producer of a good and we want to set its selling price. The price will arguably affect demand, and we could conceive a linear demand model such as

$$D = \alpha + \beta p + \epsilon$$

---

[6]This dichotomy of views is well known in physics and has been the subject of heated controversy between Albert Einstein and Niels Bohr. According to the first view, Nature is basically deterministic but it looks random because of our ignorance; according to the second view, randomness is an intrinsic feature of Nature. Einstein expressed the first view by stating "God does not play dice with universe," to which Bohr replied "Stop telling God what to do."

where D is demand and p is price. In this market setting, price is a managerial lever under our complete control and, of course, it can be perfectly observed, without any measurement uncertainty. Hence, price in this case is a nonstochastic regressor and there is a clear causal relationship between price and demand. We could try to estimate the demand function in order to find the optimal price to maximize profit.[7]   □

Example 10.4 Now consider a model relating the random return R_k, that we may earn from investing in stock share k, to the random return R_m from the market portfolio as a whole. The return from the market portfolio can be typically proxied by a broad market index such as S&P500. If we regress R_k on R_m, namely

$$R_k = \alpha + \beta R_m + \epsilon$$

we build a model with a stochastic regressor, as we cannot indisputably claim control over the market. Furthermore, in this case we also see an important point again: Linear regression models, like correlation analysis, are about *association*, not *causation*. Unlike the monopolistic producer setting a price, in this case it is hard to say which variable is the cause and which one is the effect. On one hand, we could say that the general market mood does affect return on asset k; on the other hand, asset k is itself part of the market portfolio and contributes to return R_m. A regression model like this has many uses[8] and can be exploited to assess the systematic risk of an asset, i.e., the risk component that cannot be diversified away by holding a widely diversified portfolio.   □

Example 10.5 Stochastic and nonstochastic regressors may coexist in the same model, when there are multiple explanatory variables. Consider a model predicting the rating or the share obtained by broadcasting a movie on a TV network. The rating depends on many factors, including characteristics of the movie itself, such as its genre and the number of star actors and actresses featured in it, as well as the broadcast slot, i.e., the time, day, and month of the broadcast. These are variables that a TV network may control, because it selects which movie to show and when. However, the rating also depends on what competitor networks show at the same time, i.e., their counterprogramming decisions. Clearly, the level of competition is not under our control,[9] and should be considered a stochastic regressor.   □

Whether regressors are stochastic or not, Y_i is a random variable since errors make our observations noisy, and α and β in the underlying data-generating process are *unknown parameters* that we try to estimate. What

[7]See Section 2.11.5.
[8]See Example 9.21.
[9]Well, this applies to healthy competitive markets. There are always exceptions to any rule, unfortunately.

we actually observe should be written, in the case of nonstochastic regressors, as

$$Y_i = a + bx_i + e_i \qquad (10.10)$$

where the regression coefficients a and b are found by least squares and the residuals e_i are evaluated by comparing predicted and observed values of the response variable. Note the difference between this expression and (10.8). We use Roman letters (a, b, e_i) when referring to coefficients and residuals, which are estimates and observed variables, respectively; we use Greek letters (α, β, ϵ_i) when referring to underlying things that we cannot observe directly. Even if the regressors are nonstochastic, the regression coefficients a and b are random variables,[10] since they depend on the random variables Y_i through Eqs. (10.1) and (10.3). They are estimators of the unknown parameters α and β. It is very useful to see how this conceptual framework is exactly the same we have used when estimating the expected value μ, an unknown number, by the sample mean $\overline{X}$, an observed random variable. In this case, we should also notice that the errors ϵ_i are *not* directly observable, since we do not know the true parameters. We must settle for a proxy of errors, the residuals e_i.

Least squares provide us with estimators of the parameters in the data-generating processes (10.8) or (10.9). Whenever we estimate parameters, we would like to have unbiased estimators with as little variance as possible. In order to investigate the properties of the least-squares estimators, we need a precise statement of conditions on regressor variables and errors. In the next sections we treat in fair detail the case of nonstochastic regressors, which is easier to deal with. We will outline the case of stochastic regressors later, in Section 10.5. It turns out that the fundamental results for the case of stochastic regressors are not that different, but the underlying assumptions must be expressed in a more complicated way.

The validity of the regression model also depends on our assumptions about the errors ϵ_i. Very roughly speaking, it is generally assumed that they are independent of each other and that their expected value is zero. Independence among errors simplifies results considerably and ensures that we are not really missing something systematic in our model; if their expected value were not zero, we could just include it in the intercept α. A fundamental issue is whether errors are somehow related to the value of x_i or X_i:

- We speak of a *homoskedastic* case if the variance of errors is the same for each observation, and in particular it does not depend on the value of the regressor variable: $\text{Var}(\epsilon_i) = \sigma_\epsilon^2$, for all $i = 1, \ldots, n$.

- Otherwise, we have a *heteroskedastic* case.

---

[10] Strictly speaking, we should use uppercase letters A and B, but this would make notation a bit unpleasant. An alternative would be to use $\hat{\alpha}$, $\hat{\beta}$, and $\hat{\epsilon}_i$ to emphasize their role as estimators and proxies of unobserved variables, respectively.

Strictly speaking, the estimation approach we are describing, where errors are i.i.d. variables, should be referred to as ordinary least squares (OLS); the slope and intercept given by (10.1) and (10.3) are OLS estimators. Variations on the theme are used to deal with violations of the assumptions above.

Finally, in order to derive more specific results on confidence intervals and in testing hypotheses about the underlying parameters, we might use further assumptions concerning the exact distribution of the errors. If we assume normality, rather simple results may be found; alternatively, the central limit theorem can be invoked to prove asymptotic normality of estimators, under technical conditions. All of these assumptions should not be taken for granted, and they should be carefully checked in order to assess the validity of the regression model. Whether least-squares estimators are biased or not, as well as their efficiency, depends on them.

10.3   THE CASE OF A NONSTOCHASTIC REGRESSOR

In this section we want to tackle a few statistical issues concerning the estimation of the unknown parameters of the data-generating model, featuring a nonstochastic regressor and homoskedastic errors:

$$Y_i = \alpha + \beta x_i + \epsilon_i, \qquad i = 1, \ldots, n$$

As we said, the values x_i are numbers and the errors ϵ_i are independent and identically distributed random variables satisfying the following assumptions:

- Their expected value is zero: $\mathrm{E}[\epsilon_i] = 0$.

- The variance for all errors is the same and does not depend on x_i, i.e., $\mathrm{Var}(\epsilon_i) = \sigma_\epsilon^2$.

- When needed, we will also assume that errors are normally distributed: $\epsilon_i \sim N(0, \sigma_\epsilon^2)$. In such a case, independence can be substituted by lack of correlation: $\mathrm{E}[\epsilon_i \epsilon_j] = 0$, for $i \neq j$.

Our task mirrors what we did when estimating expected value by sample mean:

1. We want to assess the properties of least-squares estimators a and b in terms of bias and estimation error.

2. We want to find confidence intervals for estimators.

3. We want to test hypotheses about the regression parameters.

4. We want to assess the performance of our model in terms of explanatory power and prediction capability.

Since the results that we obtain depend on the assumptions above, it is important to check them a posteriori by analyzing the residuals $e_i = Y_i - (a + bx_i)$, which are a proxy of the unobservable errors. Simple graphical checks are illustrated in Section 10.3.5.

10.3.1 Properties of estimators

The least-squares estimators a and b, as given by (10.1) and (10.3), are random variables, since they depend on the observed response variables Y_i, which in turn depend on the errors. Given the underlying assumptions about regressors and errors, we see that

$$\mathrm{E}[Y_i] = \alpha + \beta x_i, \qquad \mathrm{Var}(Y_i) = \sigma_\epsilon^2$$

In order to assess the viability of the estimators, we must assess some features of their probability distribution, which in turn requires rewriting them in terms of the underlying random variables ϵ_i. Let us start with the estimator of the slope parameter β. Using Eq. (10.7), where we plug random variables Y_i in place of numbers y_i, we get the (ordinary) least-squares estimator b:

$$b = \frac{\sum_{i=1}^{n} (x_i - \bar{x})\left(Y_i - \overline{Y}\right)}{\sum_{i=1}^{n} (x_i - \bar{x})^2} \tag{10.11}$$

Given the underlying model, we see that the sample mean of the response variables is given by

$$\overline{Y} = \alpha + \beta\bar{x} + \bar{\epsilon}$$

where $\bar{x}$ is the average of the x_i, and $\bar{\epsilon}$ is the sample mean of the errors, over the n observations. Then we rewrite b as

$$
\begin{aligned}
b &= \frac{\sum_{i=1}^{n} (x_i - \bar{x}) \cdot [\alpha + \beta x_i + \epsilon_i - (\alpha + \beta\bar{x} + \bar{\epsilon})]}{\sum_{i=1}^{n} (x_i - \bar{x})^2} \\[2mm]
&= \frac{\sum_{i=1}^{n} \left[\beta (x_i - \bar{x})^2 + (x_i - \bar{x})(\epsilon_i - \bar{\epsilon})\right]}{\sum_{i=1}^{n} (x_i - \bar{x})^2} \\[2mm]
&= \beta + \frac{\sum_{i=1}^{n} (x_i - \bar{x})(\epsilon_i - \bar{\epsilon})}{\sum_{i=1}^{n} (x_i - \bar{x})^2} \tag{10.12}
\end{aligned}
$$

We see that b is given by the sum of β and a random term depending on the errors ϵ_i. To prove unbiasedness, we need to show that the expected value of this random term is zero:

$$
\begin{aligned}
E[b] &= E\left[\beta + \frac{\sum_{i=1}^{n}(x_i - \bar{x})(\epsilon_i - \bar{\epsilon})}{\sum_{i=1}^{n}(x_i - \bar{x})^2}\right] = \beta + E\left[\frac{\sum_{i=1}^{n}(x_i - \bar{x})(\epsilon_i - \bar{\epsilon})}{\sum_{i=1}^{n}(x_i - \bar{x})^2}\right] \\
&= \beta + \frac{\sum_{i=1}^{n}(x_i - \bar{x}) \cdot E[\epsilon_i - \bar{\epsilon}]}{\sum_{i=1}^{n}(x_i - \bar{x})^2} = \beta + 0
\end{aligned}
$$

In the manipulations above, we have used the fact that β and x_i are numbers and can be taken outside the expectation; then we rely on the assumption that the expected value of the errors is zero, as well as the expected value of their sample mean.

The same line of reasoning can be adopted to prove unbiasedness of a. We rewrite (10.1) relying on the assumptions:

$$
\begin{aligned}
a &= \overline{Y} - b\bar{x} = \frac{1}{n}\sum_{i=1}^{n}Y_i - \frac{b}{n}\sum_{i=1}^{n}x_i \\
&= \frac{1}{n}\sum_{i=1}^{n}(\alpha + \beta x_i + \epsilon_i) - \frac{b}{n}\sum_{i=1}^{n}x_i \qquad (10.13) \\
&= \alpha + \frac{1}{n}\sum_{i=1}^{n}(\beta - b)x_i + \frac{1}{n}\sum_{i=1}^{n}\epsilon_i \qquad (10.14)
\end{aligned}
$$

We see that a can be broken down in the sum of three pieces; taking the expected value yields

$$
\begin{aligned}
E[a] &= E\left[\alpha + \frac{1}{n}\sum_{i=1}^{n}(\beta - b)x_i + \frac{1}{n}\sum_{i=1}^{n}\epsilon_i\right] \\
&= \alpha + \frac{1}{n}\sum_{i=1}^{n}E[\beta - b]x_i + \frac{1}{n}\sum_{i=1}^{n}E[\epsilon_i] = \alpha
\end{aligned}
$$

In these manipulations we have broken down the expected value of a sum into the sum of the expected values, as usual, and we have taken numbers outside the expectation. The second term is zero because b is an unbiased estimator of β and the last term is zero because of our assumptions about the errors. We should note that these results do not depend on the specific distribution of errors, which need not be normal.

Having an unbiased estimators is good news, but we also need to assess their variability. We are concerned about the estimation errors $(b - \beta)$ and $(a - \alpha)$. Generally speaking, variability of an estimator can be measured by the *standard error of estimate*, which we denote by SE($\cdot$):

$$\text{SE}(b) \equiv \sqrt{\text{E}\left[(b - \beta)^2\right]}, \qquad \text{SE}(a) \equiv \sqrt{\text{E}\left[(a - \alpha)^2\right]}$$

Since our estimators are unbiased, we can recast SE into a more familiar form; recalling that $\text{E}[Z^2] = \text{Var}(Z) + \text{E}^2[Z]$, for any random variable Z, we get

$$\text{SE}(b) = \sqrt{\text{Var}(b - \beta) + \text{E}^2[b - \beta]} = \sqrt{\text{Var}(b - \beta)} = \sqrt{\text{Var}(b)}$$

since β is a number. We see that, because of unbiasedness, the standard error of the estimate is just the standard deviation of the estimator. A similar relationship holds for SE(a), a, and α.

Let us evaluate SE of the least-squares estimators, starting with b. We may rewrite Eq. (10.12) slightly:

$$b = \beta + \frac{\sum_{i=1}^{n} (x_i - \bar{x})(\epsilon_i - \bar{\epsilon})}{\sum_{i=1}^{n} (x_i - \bar{x})^2} = \beta + \frac{\sum_{i=1}^{n} (x_i - \bar{x})\epsilon_i}{\sum_{i=1}^{n} (x_i - \bar{x})^2} \tag{10.15}$$

This holds because $\sum_i (x_i - \bar{x})\bar{\epsilon} = 0$. Then we may calculate the variance of b directly:

$$
\begin{aligned}
\text{Var}(b) &= \text{Var}\left[\beta + \frac{\sum_{i=1}^{n} (x_i - \bar{x})\epsilon_i}{\sum_{i=1}^{n} (x_i - \bar{x})^2}\right] = \frac{\sum_{i=1}^{n} \text{Var}\left[(x_i - \bar{x})\epsilon_i\right]}{\left(\sum_{i=1}^{n} (x_i - \bar{x})^2\right)^2} \\
&= \frac{\sum_{i=1}^{n} (x_i - \bar{x})^2 \sigma_\epsilon^2}{\left(\sum_{i=1}^{n} (x_i - \bar{x})^2\right)^2} = \frac{\sigma_\epsilon^2}{\sum_{i=1}^{n} (x_i - \bar{x})^2}
\end{aligned}
$$

In the manipulations above we have taken advantage of the nature of the x_i (numbers) and of the errors ϵ_i (mutually independent and with constant standard deviation σ_ϵ). So, the standard error of b is

$$\text{SE}(b) = \frac{\sigma_\epsilon}{\sqrt{\sum_{i=1}^{n} (x_i - \bar{x})^2}} \tag{10.16}$$

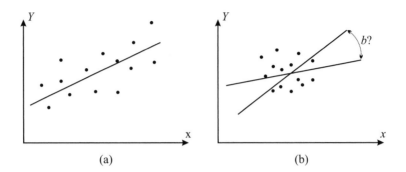

Fig. 10.3 It is difficult to tell the right slope when data are concentrated.

An apparent missing piece of information in this formula is σ_ϵ; however, we may estimate this on the basis of residuals, as shown later in Section 10.3.2. Now it is useful to interpret the result we have obtained.

- As expected, the reliability of our estimate of the slope depends on the intrinsic variability of the phenomenon we are modeling. If the noisy contribution from errors is low, then the n observations are very close to the line $Y = \alpha + \beta x$ and estimating the slope is a fairly easy task. Indeed, we see that $\text{SE}(b)$ is proportional to σ_ϵ.

- Another fairly intuitive observation is that the more observations we have, the better. This is pretty evident in the standard deviation of the sample mean, where a factor $\sqrt{n}$ pops up. We do not see here an explicit contribution of the number of observations, but nevertheless each additional observation adds a squared contribution to the denominator of the ratio, reducing $\text{SE}(b)$.

- A less obvious observation is that our ability to estimate the slope depends on where the observations are located. The denominator of the ratio includes a term looking like a variance, and it is in fact a measure of the (nonrandom) variability of the observations x_i. If the points x_i are close to each other, i.e., are close to their average $\bar{x}$, we have a small denominator. It is difficult to see the impact of small variations of x on Y, because this effect is "buried" in background noise. If the observed range of x is wide enough, assessing the impact of x on Y is easier. This is illustrated intuitively in Fig. 10.3.

Formula (10.16) suggests that, in order to get a good estimate of slope, we should have observations over a large range of the explanatory variable x. However, it is worth noting that a linear relationship might just be an acceptable approximation of a nonlinear phenomenon over a limited range of values. Hence, by taking a wide sample we might run into a different kind of trouble, namely a poor fit resulting from the nonlinearity of the observed phenomenon.

In order to assess the standard estimation error for a, we may follow the same route that we took for the slope. We use (10.14) to express variance of the estimator:

$$
\begin{aligned}
\mathrm{Var}(a) \ &= \ \mathrm{Var}\left[\alpha + (\beta - b)\,\bar{x} + \frac{1}{n}\sum_{i=1}^{n}\epsilon_i\right] \\
&= \ [\mathrm{SE}(b)\cdot\bar{x}]^2 + \frac{1}{n^2}\sum_{i=1}^{n}\mathrm{Var}(\epsilon_i) + 2\mathrm{Cov}\left((\beta - b)\bar{x}, \frac{1}{n}\sum_{i=1}^{n}\epsilon_i\right) \\
&= \ \mathrm{SE}^2(b)\cdot\bar{x}^2 + \frac{\sigma_\epsilon^2}{n} - \frac{2\bar{x}}{n}\mathrm{Cov}\left(b - \beta, \sum_{i=1}^{n}\epsilon_i\right)
\end{aligned}
\tag{10.17}
$$

This expression looks quite complex, but it is easy to read. The first term relates $\mathrm{SE}(a)$ to $\mathrm{SE}(b)$; the second term is related to the variance of the sample mean of errors $\bar{\epsilon}$; finally, we can show that the last term is zero. To see this, let us rewrite $b - \beta$ using (10.15) to make the contribution of errors explicit:

$$
\begin{aligned}
\mathrm{Cov}\left(b - \beta, \sum_{i=1}^{n}\epsilon_i\right) \ &= \ \mathrm{Cov}\left(\frac{\sum_{i=1}^{n}(x_i - \bar{x})\,\epsilon_i}{\sum_{i=1}^{n}(x_i - \bar{x})^2}, \sum_{j=1}^{n}\epsilon_j\right) \\
&= \ \frac{\sum_{i=1}^{n}\sum_{j=1}^{n}\mathrm{Cov}\left((x_i - \bar{x})\,\epsilon_i, \epsilon_j\right)}{\sum_{i=1}^{n}(x_i - \bar{x})^2} = \frac{\sum_{i=1}^{n}\mathrm{Cov}\left((x_i - \bar{x})\,\epsilon_i, \epsilon_i\right)}{\sum_{i=1}^{n}(x_i - \bar{x})^2} \\
&= \ \frac{\sum_{i=1}^{n}\sigma_\epsilon^2\,(x_i - \bar{x})}{\sum_{i=1}^{n}(x_i - \bar{x})^2} = 0
\end{aligned}
$$

where we have exploited the mutual independence between the errors ϵ_i and the fact that their variance does not depend on the observations. Now, plugging the expression of $\mathrm{SE}(b)$, we obtain

$$
\mathrm{SE}(a) = \sigma_\epsilon\sqrt{\frac{\bar{x}^2}{\sum_{i=1}^{n}(x_i - \bar{x})^2} + \frac{1}{n}} = \sqrt{[\mathrm{SE}(b)\bar{x}]^2 + \frac{\sigma_\epsilon^2}{n}}
\tag{10.18}
$$

This formula, too, lends itself to a useful interpretation, which is apparent when looking at the two nonzero terms in Eq. (10.17).

- The term $\mathrm{SE}(b)\bar{x}$ tells us that getting the wrong slope will influence the error that we make in estimating the intercept, but this depends on the

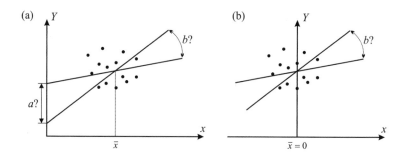

Fig. 10.4 Schematics of the impact of slope on estimating the intercept.

average value of the explanatory variable. This may be understood by looking at Fig. 10.4. If $\bar{x} = 0$, rotating the regression line around the barycenter of data has no impact on the intercept, whereas the impact is large when $\bar{x}$ is large.

- The second term is related to a shift in the intercept due to the variability of the sample error. If $\bar{\epsilon} = 0$, then there is no such shift; otherwise the contribution to the estimation error depends on the sign of this sample mean. The magnitude of this effect depends on the intrinsic variability, measured by σ_ϵ, and by the sample size n.

Now that we know the expected value and the variance of our least-squares estimators, a natural question is how they are distributed. The answer is fairly straightforward in our setting, nonstochastic regressors and homoskedastic errors, provided that we also assume that errors are normally distributed, i.e., $\epsilon_i \sim N(0, \sigma_\epsilon^2)$. A look at Eqs. (10.12) and (10.14) shows that, in our settings, the least-squares estimators are essentially linear combinations of errors, i.e., linear combinations of normal variables. Since a linear combination of normal variables is itself a normal variable, we immediately see that estimators are normally distributed, too. We have proved the following theorem.

THEOREM 10.1 *If regressor variables x_i are nonstochastic and errors are i.i.d. normal variables, then least-squares estimators are normal random variables: $b \sim N(\beta, \mathrm{SE}^2(b))$ and $a \sim N(\alpha, \mathrm{SE}^2(a))$, with standard errors given by (10.16) and (10.18).*

This result can actually be generalized. If we do not assume that errors are normally distributed, we can invoke the central limit theorem to show that least-squares estimators are asymptotically normal.[11] In Section 10.5 we also see that this holds for stochastic regressors and for heteroskedastic errors,

---

[11] From a technical perspective, we should add conditions concerning outliers, expressed in terms of higher-order moments.

but the involved formulas, as well as the underlying theory, are a bit more complicated.

10.3.2 The standard error of regression

Equations (10.16) and (10.18) help us in assessing the uncertainty about the estimate of unknown regression parameters. A missing piece in this puzzle, however, is the standard deviation σ_ϵ of the random errors, which are not directly observable. The only viable approach we have is to rely on the residuals $e_i = Y_i - \hat{Y}_i$ as a proxy for the errors ϵ_i. The intuition is that, if we trust the estimated model, $\hat{Y}_i$ is the expected value of Y_i and, in order to assess the variability of the errors, it is reasonable to consider the variability of the observations with respect to their expected value. Another way to get the picture is by noting that the assumptions behind the statistical model imply that $\sigma_\epsilon^2 = \text{Var}(Y_i)$; however, we cannot just take the usual sample variance of the observed Y_i, since the expected value of Y_i is not a constant, but it depends on the value of x_i. The following result, which holds under our assumptions, helps us in estimating the standard deviation of errors.

THEOREM 10.2 *If errors are i.i.d. and normally distributed with standard deviation σ_ϵ, then the ratio of SSR (sum of squared residuals) and σ_ϵ^2 is a chi-square random variable with $n - 2$ degrees of freedom:*

$$\frac{\text{SSR}}{\sigma_\epsilon^2} \sim \chi_{n-2}^2$$

Given what we know about the chi-square distribution,[12] we immediately conclude that an unbiased estimator of the standard deviation of errors is given by the *standard error of regression* (SER), defined as follows:

$$\text{SER} \equiv \hat{\sigma}_\epsilon = \sqrt{\frac{\sum_{i=1}^n \left(Y_i - \hat{Y}_i\right)^2}{n - 2}} \tag{10.19}$$

Example 10.6 Let us find the SER for the data in Fig. 10.1. We recall that, in that case, the regressed model was

$$\hat{Y} = 2.4470 + 2.3616x$$

Plugging values of the regressors and computing the residuals yields the results listed in Table 10.2. Then we obtain

$$\text{SER} = \sqrt{\frac{\text{SSR}}{6 - 2}} = 3.6452$$

---

[12] See Section 7.7.2.

Table 10.2 Residuals for the data of Fig. 10.1.

Y_i	$\hat{Y}_i$	e_i
3.7850	4.8085	-1.0235
14.4688	14.2548	0.2140
24.5751	21.3394	3.2357
28.6489	26.0626	2.5863
22.5761	28.4241	-5.8480
38.7059	37.8704	0.8355

We urge the reader to check that, for the data in Fig. 10.2, we get SER = 15.8114. This much larger SER shows that, even though the regressed models are the same, the underlying errors are quite different. ☐

The definition of SER in Eq. (10.19) is essentially a sample standard deviation, with a couple of twists:

1. Deviations are not taken against a constant value, but with respect to a sort of moving target depending on the regressors x_i.

2. Unlike the usual sample standard deviation, we divide by $n - 2$ rather than $n - 1$. The cookbook recipe way of remembering this is that we lose 2 degrees of freedom because deviations depend on two estimated parameters, rather than just the usual sample mean $\overline{X}$. Also note that if we had just $n = 2$ observations, we could say nothing about errors, because the observed points would exactly fit one line, and there would be no deviation between Y_i and $\hat{Y}_i$.

10.3.3 Statistical inferences about regression parameters

Now we are armed with the necessary knowledge to draw statistical inferences about the regression parameters. Mirroring what we did with the estimation of expected value, we should

- Calculate confidence intervals for slope and intercept

- Check the significance of regression coefficients by testing suitable hypotheses

The technicalities involved here are essentially the same as those involved in dealing with estimation of the expected value, and we avoid repeating the reasoning. In the case of the expected value, everything revolves around the statistic

$$T = \frac{\overline{X} - \mu}{S/\sqrt{n}}$$

where $S/\sqrt{n}$ is the standard deviation of the estimator $\overline{X}$ or, in other words, its standard error, since the estimator is unbiased. In the case of the slope in a regression model, the relevant statistic is

$$T = \frac{b - \beta}{\mathrm{SE}(b)}$$

Given our distributional results so far, it is no surprise that this statistic has t distribution with $n - 2$ degrees of freedom, assuming that errors are normally distributed. A similar result applies to the intercept. Hence, for a large sample, our estimators tend to be normally distributed. Courtesy the central limit theorem, it can be shown that least-squares estimators are approximately normal even if errors are nonnormal, provided that the other assumptions hold.

To compute confidence intervals, we fix a confidence level $1 - \alpha$, get the corresponding quantile $t_{1-\alpha/2,n-2}$ from t distribution, and compute

$$b \pm t_{1-\alpha/2,n-2} \, \mathrm{SE}(b)$$

for the slope, and

$$a \pm t_{1-\alpha/2,n-2} \, \mathrm{SE}(a)$$

for the intercept, relying on formulas (10.16), (10.18), and (10.19).

Example 10.7 Let us compute confidence intervals for the regression parameters for the data in Fig. 10.2. We already know that

$$a = 2.4470, \quad b = 2.3616, \quad \mathrm{SER} = 15.8114$$

furthermore $\bar{x} = 8.3333$. Using the formulas for standard errors, we find

$$\mathrm{SE}(b) = \frac{15.8114}{\sqrt{\displaystyle\sum_{i=1}^{6} (x_i - 8.3333)^2}} = 1.4474$$

and

$$\mathrm{SE}(a) = 15.8114 \times \sqrt{\frac{(8.3333)^2}{\displaystyle\sum_{i=1}^{6} (x_i - 8.3333)^2} + \frac{1}{6}} = 13.6803$$

If confidence level is 95%, the quantile we need is

$$t_{0.975,4} = 2.7764$$

The confidence interval for the slope is

$$2.3616 \pm 2.7764 \times 1.4474 = (-1.6571, \, 6.3802)$$

and the confidence interval for the intercept is

$$2.4470 \pm 2.7764 \times 13.6803 = (-35.5357, \ 40.4296)$$

We see that these are pretty large confidence intervals; even worse, they contain the origin, and we are not really even sure about the sign of the underlying parameters! This is no surprise, given the extremely scarce and noisy data, but it is what we can honestly say (always keeping the underlying assumptions in mind). We urge the reader to check that for the less noisy data of Fig. 10.1, we get confidence intervals

$$(1.4351, \ 3.2880), \qquad (-6.3096, \ 11.2036)$$

for slope and intercept, respectively. Also in this case, with very few data, we cannot trust the regression model too much, but at least we have a clear idea of the sign of the effect of the explanatory variable on the response variable.

$\square$

Hypothesis testing proceeds much along the usual lines. The most common test we carry out is a t test concerning the slope, i.e., the significance of the effect of an explanatory variable. Even if an explanatory variable has no effect on the response variable, i.e., its slope coefficient is $\beta = 0$, the estimated value will not be zero because of random errors. It is then natural to test the null hypothesis

$$H_0 : \beta = 0$$

against the alternative one

$$H_a : \beta \neq 0$$

It is customary to run a two-tail test, although we could use a one-tail test if we have a clear idea about the sign of the effect. Then the test statistic boils down to

$$T = \frac{b}{\text{SE}(b)}$$

which is calculated by any software package implementing linear regression. The p-value of the t test of the slope, given that the test statistic T assumes the value t, is

$$p = 2\text{P}\{|T_{n-2}| > t\}$$

where T_{n-2} is a t random variable with $n-2$ degrees of freedom. Note that we are assuming a two-tail test, and that the result is exact if errors are normal.

Example 10.8 Continuing Example 10.7, we see that the test statistic for the noisy dataset of Fig. 10.2 is

$$t = \frac{2.3616}{1.4474} = 1.6316$$

The quantile for a two-tail test with $\alpha = 5\%$ is the same quantile we used to calculate the confidence interval

$$t_{0.975,4} = 2.7764$$

from which we immediately see that the test statistic does *not* fall in the rejection region and we cannot reject the null hypothesis that the actual slope β is zero. The p-value of the test is

$$2P\{T_4 > t\} = 0.1781$$

where T_4 is a t variable with 4 degrees of freedom. We could reject the null hypothesis if we accepted a probability of type I error that is just a little less than 20%. Hence, we cannot really say that the effect of the explanatory variable is significant on the basis of the sparse and noisy data we have. For the less noisy first dataset, the test statistic is $t = 7.0772$ and the p-value is 0.0021, less than 1%.　　　　　　　　　　　　　　　　　　　　　　　　□

Testing the intercept requires the same conceptual framework, provided we use the standard error $SE(a)$. Usually, we are more interested in testing slopes, as they measure the impact of an explanatory variable on the response. However, there are situations in which testing the intercept is even more important.

Example 10.9 (Capital asset pricing model) In Example 9.21 we considered a factor model for returns of a financial asset. This is essentially a regression model that can be cast in the following form:

$$R_k - r_f = \alpha_k + \beta_k(R_m - r_f) + \epsilon_k$$

Here R_k is the (random) return from holding the risky asset k for some holding period, R_m is the (random) return from holding the market portfolio m for some holding period, and r_f is the risk-free return over the same period (a number). This model is expressed in terms of excess return, i.e., the difference between a random return and the risk-free return.

A fundamental piece of financial theory is the capital asset pricing model (CAPM). This is much more than a regression model, as it is an equilibrium model concerning expected returns. Essentially, the model states that, at equilibrium

$$E[R_k - r_f] = \beta_k E[R_m - r_f] \tag{10.20}$$

In other words, according to CAPM, $\alpha_k = 0$. The practical implication, if we believe the model, is that the risk premium from holding asset k, i.e., the expected excess return, depends only on the systematic risk, i.e., the risk from holding the market portfolio. The unsystematic risk, i.e., the specific risk related to firm k, is not rewarded, as it can be diversified away by holding a properly diversified portfolio. Then, the risk is just measured by the asset beta:

$$\beta_k = \frac{\text{Cov}(R_k, R_m)}{\text{Var}(R_m)}$$

We cannot discuss the exact conditions leading to CAPM, but it is just natural to consider an empirical test of the theory. In this case, we are interested in testing the null hypothesis

$$H_0 : \alpha_k = 0$$

against the alternative $H_a : \alpha_k \neq 0$, to see if empirical data reject the theory. Apparently, it would be easy to check this by running suitable regressions, observing returns over consecutive time periods. Unfortunately, this is not that easy, and in fact empirical testing of CAPM has raised a fair share of controversy. From a financial perspective, it is not so obvious what makes a "market" portfolio, even though one could try surrogating that by a broad market index like S&P500. From a statistical perspective, there is a quite critical point: Who says that errors in consecutive time periods are independent? □

The last remark of this example is a good reminder that assumptions should never be taken for granted. More sophisticated regression approaches have been devised to cope with correlated errors.

10.3.4   The R^2 coefficient and ANOVA

Testing the significance of a slope coefficient is a first diagnostic check of the suitability of a regression model. However, per se, this test does not tell us much about the predictive power of the model as a whole. The reader may better appreciate the point by thinking about a multiple regression model involving several explanatory variable: A t test checks the impact of each regressor individually, but it does not tell us anything about the overall model. Indeed, the standard error of regression is a measure of the uncertainty of predictions and can help in this respect, but it suffers from the usual problem with standard deviation: Its magnitude depends on how we measure things. The typical performance measure that is used to check the usefulness of a regression model is the R^2 coefficient, also known as *coefficient of determination*. The R^2 coefficient can be understood in two related ways:

- It can be seen as the fraction of total variability that the model is able to explain.

- It can be seen as the squared coefficient of correlation between the explanatory variable x and the response Y, or between the prediction $\hat{Y}$ and the observed response Y.

Whatever the interpretation, the values of the R^2 coefficient are bounded between 0 and 1; the closer to 1, the higher the explanatory power of the model.

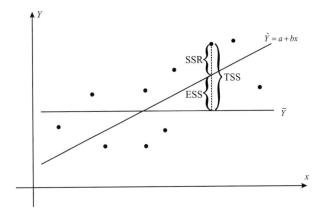

Fig. 10.5   Geometric interpretation of the R^2 coefficient.

Let us pursue the former interpretation first, with the graphical aid of Fig. 10.5. When we look at raw data, we see that the response variable Y is subject to variability with respect to its average $\overline{Y}$. This variability can be measured by the sum of squared deviations with respect to $\overline{Y}$, which is called the *total sum of squares* (TSS):

$$\text{TSS} \equiv \sum_{i=1}^{n}(Y_i - \overline{Y})^2$$

Not all of this variability is noise, since part of it can be attributed to variability in the explanatory variable x. This "predictable" variability is explained by the regression model in terms of the predicted response $\hat{Y} = a + bx$. So, we may measure the explained variability by the *explained sum of squares*:

$$\text{ESS} \equiv \sum_{i=1}^{n}(\hat{Y}_i - \overline{Y})^2$$

From Fig. 10.5, we see that there is some residual part of variability that cannot be accounted for by model predictions. This unexplained variability is related to residuals, more precisely to the sum of squared residuals, that we already defined as

$$\text{SSR} \equiv \sum_{i=1}^{n}(Y_i - \hat{Y}_i)^2$$

This view is reinforced by the following theorem.

THEOREM 10.3 *For a linear regression model, the total variability can be partitioned into the sum of explained and unexplained variability as follows:*

$$\text{TSS} = \text{ESS} + \text{SSR}$$

PROOF   The total variability can be rewritten by adding and subtracting $\hat{Y}_i$ within squared terms and by expanding them:

$$
\begin{aligned}
\text{TSS} &= \sum_{i=1}^{n}(Y_i - \overline{Y})^2 = \sum_{i=1}^{n}(Y_i - \hat{Y}_i + \hat{Y}_i - \overline{Y})^2 \\
&= \underbrace{\sum_{i=1}^{n}(Y_i - \hat{Y}_i)^2}_{\text{SSR}} + \underbrace{\sum_{i=1}^{n}(\hat{Y}_i - \overline{Y})^2}_{\text{ESS}} + \underbrace{2\sum_{i=1}^{n}(Y_i - \hat{Y}_i)(\hat{Y}_i - \overline{Y})}_{\nu}
\end{aligned}
$$

Now we prove that the last term $\nu = 0$. The first step is to rewrite model predictions in terms of deviations with respect to averages, using the least-squares estimator of intercept:

$$
\hat{Y}_i = a + bx_i = \overline{Y} - b\overline{x} + bx_i = \overline{Y} + b(x_i - \overline{x})
$$

Plugging this expression into the expression of ν, and using the least-squares estimator of slope, we get

$$
\begin{aligned}
\nu &= \sum_{i=1}^{n}[Y_i - \overline{Y} - b(x_i - \overline{x})]b(x_i - \overline{x}) \\
&= b\left[\sum_{i=1}^{n}(Y_i - \overline{Y})(x_i - \overline{x}) - b\sum_{i=1}^{n}(x_i - \overline{x})^2\right] \\
&= b\left[b\sum_{i=1}^{n}(x_i - \overline{x})^2 - b\sum_{i=1}^{n}(x_i - \overline{x})^2\right] = 0
\end{aligned}
$$

On the basis of this result, we may define the R^2 coefficient as the ratio of explained variability and total variability:

$$
R^2 \equiv \frac{\text{ESS}}{\text{TSS}} = 1 - \frac{\text{SSR}}{\text{TSS}} \tag{10.21}
$$

We see that R^2 is also 1 minus the fraction of unexplained variability. When the coefficient of determination is close to 1, little variability is left unexplained by the model.

Recalling the meaning of slope b, we would expect the regression model to have good explanatory power when there is a strong correlation between the explanatory and the output variables. Strictly speaking, since we are dealing with a nonstochastic regressor, we are a bit sloppy when talking about this correlation, and it would be better to consider the correlation between $\hat{Y}_i$ and

Y_i. Hence, let us consider the squared sample correlation:

$$r_{Y\hat{Y}}^2 = \frac{\left[\sum_{i=1}^{n}(Y_i - \overline{Y})\left(\hat{Y}_i - \overline{\hat{Y}}\right)\right]^2}{\sum_{i=1}^{n}(Y_i - \overline{Y})^2 \cdot \sum_{i=1}^{n}\left(\hat{Y}_i - \overline{\hat{Y}}\right)^2}$$

It is not difficult to prove that this expression boils down to R^2. To begin with, we see that the average prediction and the average observation of the response variable are the same:

$$\overline{\hat{Y}} = \frac{1}{n}\sum_{i=1}^{n}\hat{Y}_i = \frac{1}{n}\sum_{i=1}^{n}(a + bx_i) = a + b\bar{x} = \overline{Y}$$

Hence, we may rewrite the numerator of $r_{Y\hat{Y}}^2$ as

$$\sum_{i=1}^{n}(Y_i - \overline{Y})(\hat{Y}_i - \overline{Y}) = \sum_{i=1}^{n}(Y_i - \hat{Y}_i + \hat{Y}_i - \overline{Y})(\hat{Y}_i - \overline{Y})$$

$$= \underbrace{\sum_{i=1}^{n}(Y_i - \hat{Y}_i)(\hat{Y}_i - \overline{Y})}_{=\nu=0} + \sum_{i=1}^{n}(\hat{Y}_i - \overline{Y})^2$$

We see that we get the same null term ν that we encountered in proving Theorem 10.3. Hence, we can rewrite the squared correlation as

$$r_{Y\hat{Y}}^2 = \frac{\left[\sum_{i=1}^{n}(\hat{Y}_i - \overline{Y})^2\right]^2}{\sum_{i=1}^{n}(Y_i - \overline{Y})^2 \cdot \sum_{i=1}^{n}\left(\hat{Y}_i - \overline{Y}\right)^2} = \frac{\sum_{i=1}^{n}(\hat{Y}_i - \overline{Y})^2}{\sum_{i=1}^{n}(Y_i - \overline{Y})^2} = R^2$$

Thus, we have another aspect to consider in interpreting the coefficient of determination.

The R^2 coefficient gives us an evaluation of the overall fit of the model. One word of caution is in order, however, when we think of using multiple explanatory variables. It is easy to understand that the more variable we add, the better the fit: R^2 cannot decrease when we add more and more variables. However, there is a tradeoff, since adding degrees of freedom means that we are increasing the uncertainty of our estimates. Furthermore, the more variables, the more parameters we use to improve the fit within the sampled data, but this need not translate out of sample. We will return to these issues in Chapter 16.

The interpretation of R^2 in terms of explained variability suggests another way of checking the significance of the regression model as a whole, rather than in terms of a t test on the slope coefficient. As we mentioned, this is important when dealing with multiple regression, which involves multiple coefficients; these may be tested individually, but we would also like to have an evaluation of the overall model. This test applies concepts that we introduced when dealing with analysis of variance (ANOVA) in Section 9.6. Again, there are a couple of ways to interpret the idea.

1. If the model has explanatory power, then the explained variability ESS should be large with respect to the unexplained variability SSR. We could check their ratio, and if this is large enough, then the model is significant. In fact, there are many settings in which one has to compare variances. To carry out the test, we need to build a test statistic with a well-defined probability distribution, and this calls for a little adjustment.

2. Remembering what we did in Section 9.6, we may consider a joint test on the parameters of a multiple regression model:

$$Y_i = \alpha + \beta_1 x_i + \beta_2 x_2 + \cdots + \beta_m x_m$$

The model is significant if at least one coefficient is nonzero. Hence, we may run the following test:

$$H_0 \colon \beta_1 = \beta_2 = \cdots = \beta_m = 0$$
$$H_a \colon \text{at least one } \beta_i \neq 0$$

To devise a test, let us use some intuition first. If we assume that errors are normally distributed, then the sums of squares are related to the sum of squared normals, and we know that this leads to a chi-square distribution. If we want to check the ratio of sums of squares, we have to deal with a ratio of chi-square random variables, and we know that this involves an F distribution. The reasoning here is quite informal, as we should check independence of the two sums of squares, the F distribution involves two independent chi-square variables. Another issue is related to the degrees of freedom, which are typically associated with sums of squares and define which kind of chi-square variable is involved. To get a clue, let us consider the relationship between TSS, ESS, and SSR:

$$\sum_{i=1}^{n}(Y_i - \overline{Y})^2 = \sum_{i=1}^{n}(\hat{Y}_i - \overline{Y})^2 + \sum_{i=1}^{n}(Y_i - \hat{Y}_i)^2$$

We know that the left-hand term has $n - 1$ degrees of freedom, since only $n - 1$ terms can be assigned freely. We suggested that the last term, which is related to the standard error of regression, has $n - 2$ degrees of freedom. Then, the remaining term ESS must have just 1 degree of freedom. To further

Table 10.3 Typical ANOVA table summarizing source of variation, sum of squares, degrees of freedom, mean squares, F statistic, and the corresponding p-value.

Source	SS	df	MS	F	p
Regression	ESS	$\mathrm{df}_{\exp}$	$\mathrm{MS}_{\exp} = \dfrac{\mathrm{ESS}}{\mathrm{df}_{\exp}}$	$\dfrac{\mathrm{MS}_{\exp}}{\mathrm{MS}_{\mathrm{err}}}$	p-value
Error	SSR	$\mathrm{df}_{\mathrm{err}}$	$\mathrm{MS}_{\mathrm{err}} = \dfrac{\mathrm{SSR}}{\mathrm{df}_{\mathrm{err}}}$		
Total	TSS	$\mathrm{df}_{\exp} + \mathrm{df}_{\mathrm{err}}$			

justify this intuition, please note that we may rewrite all of the involved terms in the ESS as

$$\hat{Y}_i - \overline{Y} = b(x_i - \bar{x})$$

This shows that any term can be written as a *single* function of the Y_i, so this sum of squares must have 1 degree of freedom. Correcting the sums of squares for their degrees of freedom, we define the *explained mean square* and the mean square due to errors:

$$\mathrm{MS}_{\exp} = \frac{\mathrm{ESS}}{1}, \qquad \mathrm{MS}_{\mathrm{err}} = \frac{\mathrm{SSR}}{n-2}$$

and consider the following test statistic:

$$F = \frac{\mathrm{MS}_{\exp}}{\mathrm{MS}_{\mathrm{err}}}$$

If the true slope is zero, i.e., under the null hypothesis, this variable is a ratio of chi-square variables and has F distribution, more precisely an $F(1, n - 2)$ distribution. We are encouraged to reject the null hypothesis if the test statistic is large, as this suggests that the ESS is large with respect to SSR. Using quantiles from the F distribution, we may test the regression model. Again, we should emphasize that this depends on a normality assumption and that we have just provided an intuitive justification, not a serious proof. It is customary to express the ANOVA test and the related F test in the tabular form shown in Table 10.3.

Example 10.10 To illustrate the procedure, let us consider the ANOVA for the two datasets of Figs. 10.1 and 10.2. The two respective ANOVA tables are illustrated in Table 10.4. In both cases, we have 5 degrees of freedom in TSS, as there are $n = 6$ observations; 1 degree of freedom is associated with ESS, and $n - 2 = 4$ with SER. Apart from little numerical glitches, we see that the ESS is the same for both regression models, since the regression coefficients

Table 10.4 ANOVA tables for the datasets illustrated in Figs. 10.1 and 10.2.

(a)

Source	SS	df	MS	F	p
Regression	665.5175	1	665.5175	50.0864	0.0021
Error	53.1495	4	13.2874		
Total	718.6670	5			

(b)

Source	SS	df	MS	F	p
Regression	665.5153	1	665.5153	2.6621	0.1781
Error	1000.0012	4	250.0003		
Total	1665.5164	5			

and the value of the explanatory variables are the same. What makes the difference is the SER, i.e., the sum of squared residuals, which is much larger in the second case because of the much larger variability in the second dataset. This variability is left unexplained by the regression model. In this simple regression model, $MS_{exp} = ESS$, since 1 degree of freedom is associated with explained variability, whereas $MS_{err} = SSR/4$. The test statistic is 50.0864 in the first dataset. This is a rather large value, for a $F(1,4)$ distribution, and it is quite unlikely that it is just the result of sampling variability. Indeed, the p-value is 0.0021, which means that the model is significant at less than 1%. On the contrary, for the second dataset the test statistic is small, 2.6621, and the corresponding p-value is 17.81%. This means that, to reject the null hypothesis that the regression model is not significant, we must accept a very large probability of a type I error.   ▯

One should wonder whether this amounts to a test genuinely different from the t test on the slope. In the simple regression case, the two tests are actually equivalent. Indeed, in Section 7.7.2 we pointed out that an $F_{1,n}$ variable is just a t_n variable squared. So, the t test on the only slope in a simple regression model or the F test are equivalent. Indeed, the p-values above are exactly the same that we got in Example 10.8 when running the t test. Furthermore, for the first dataset, we have

$$F = 53.1495 = 7.0772^2 = t^2$$

This does not generalize to multiple regression models, where there are multiple coefficients and the degrees of freedom will play an important role.

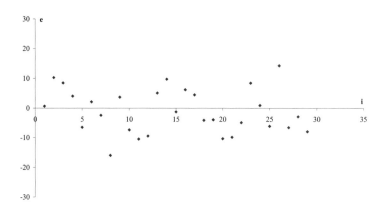

Fig. 10.6 Plot of residuals coherent with the regression model assumptions; e (ordinate) refers to residual and i (abscissa) is an observation index.

10.3.5   Analysis of residuals

All of the theory we have built so far relies on specific assumptions about the errors ϵ_i, which we recall here once again for convenience:

- They have expected value zero.

- They are mutually independent and identically distributed; in particular they have the same standard deviation, which does not depend on x_i.

Since errors are not observable directly, we must settle for a check based on residuals. The check can exploit sound statistical procedures, which are beyond the scope of this book; for our purposes, it is enough if we reinforce the important concepts by illustrating a few visual checks that we can carry out just by plotting residuals.

The assumption about the expected value is automatically met, since the way we build the least-squares estimators implies that the average residual is zero, as shown in Eq. (10.2). What we need to check is that there is no evident autocorrelation and lack of stationarity. The plot of the residuals e_i should look like Fig. 10.6, where we see that they do reasonably resemble pure noise. On the contrary, in Fig. 10.7 we see a pattern that is typically associated with positively correlated errors. If we observe a positive residual for observation number i, the next observation $i + 1$ is likely to display a positive residual as well. The same holds for negative errors, and we see "waves" of positive vs. negative residuals. Such a pattern may be observed for at least a couple of reasons. The first one is that there is indeed some correlation between consecutive observations in time. In such a case subscript i really refers to time and to the order in which observations were taken; the obvious case is when time is the explanatory variable. Another possible reason has really

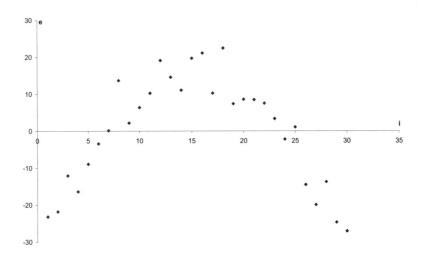

Fig. 10.7 Plot of residuals suggesting autocorrelation in the errors.

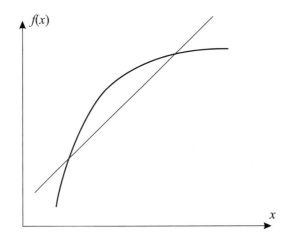

Fig. 10.8 Using linear regression with a nonlinear underlying function results in "autocorrelated" residuals.

little to do with statistics: We may also observe a pattern like this when there are nonlinearities in the observed phenomenon. Consider, for instance, the nonlinear function in Fig. 10.8, and imagine using linear regression to approximate it, using a few sample points (possibly affected by noise). The nonlinear curve is somehow cut by the regression line, and this results in a nonrandom pattern in the residuals: They have one sign in the middle range of the interval of x, and the opposite sign near the extreme points. Of course, this second case has more to do with the appropriateness of the

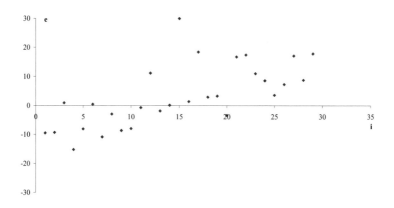

Fig. 10.9 Plot of residuals suggesting that the mean of the error process is not stationary.

selected functional form, and we are somewhat improperly using statistical concepts as a diagnostic tool. In passing, we should also clarify the meaning of the subscript i associated with an observation. If the explanatory variable is time, subscript i refers to the position in the chronological sequence of observations. But if we are regressing sales against price, we might wish to sort observations according to the value of the explanatory variable, so that subscript i does *not* refer to the order in which we took our samples. Doing so, we miss the possibility of seeing potential patterns due to the impact of time, however. Hence, it is advisable to plot residuals in any sensible order to ensure a multiple check.

Another check concerns the stationarity of the error process in terms of mean and standard deviation. By construction, the average residual is zero over the whole dataset, but a plot of residuals like the one in Fig. 10.9 suggests lack of stationarity. If i is actually related to the temporal sequence of observations, we could consider running a multiple regression in which time is an explanatory variable. Last but not least, we assumed homoskedasticity, but a plot of residuals such as Fig. 10.10 suggests that variance is not constant over our observations. In Section 10.5 we outline two possible ways to cope with heteroskedasticity. We close this section by just giving a hint of more quantitative tests.

- To check autocorrelation, we could measure the correlation between e_i and e_{i+1}. Again, we should be careful about the meaning of the subscript i: If it is time, checking correlation between consecutive samples makes sense; otherwise, we should sort observations according to values of the explanatory variable.

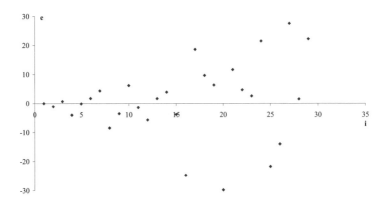

Fig. 10.10   Plot of residuals suggesting that the variance of the error process is not stationary.

- To check whether time should be included as an explanatory variable, we may measure the "correlation" between e_i and i, where i refers to the chronological order of observations.

- Finally, to check homoskedasticity, we may measure the "correlation" between ϵ_i^2 and i.

10.4   USING REGRESSION MODELS

Regression models can be used in a variety of ways, but the essential possibilities are

1. Using a regression model to assess the impact of explanatory variables

2. Using a regression model to forecast the value of the response variable for given values of the explanatory variables

In the first case, we are actually concerned with the estimate of slope; the idea is that understanding the phenomenon can lead to knowledge and to better policies. Apparently, there is little difference from the second case since, after all, a good estimate of parameters should lead to good forecasts. In fact, there are at least a couple of subtle differences between the two mindsets. If the only practical relevance of a regression model is forecasting, we may be quite happy with a model that does not yield any plausible explanation of a phenomenon, but it is empirically good at forecasting future values of the response variables. In the next chapter we will see that, in the limit, one can resort to time series models that do not try at all to explain anything; their only role is forecasting, without any ambition to find explanations. A completely different approach is taken, for instance, by sociologists, since in that case a true understanding of the driving forces behind an observed result is the true added value of a regression model. Furthermore, forecasting when regressors are themselves stochastic may seem to be a quite tricky business, as regressors themselves should be forecasted. For instance, in finance one can build a regression model to link the return of many stock shares to a few macroeconomic factors, which might be easier to forecast. In this section we just deal with forecasting when either regressors are nonstochastic and under our control, or when the explanatory variable is time.

Apparently, forecasting is quite easy. Given the regression model, just plug the value of x_0, and the predicted response is

$$\mathrm{E}\left[\hat{Y}_0\right] = a + bx_0$$

Note that this is just a point forecast, i.e., a single value. If we knew the underlying statistical model and its parameters, the response would be a random variable

$$Y_0 = \alpha + \beta x_0 + \epsilon_0$$

Assuming that errors are normally distributed, the response would be normal as well, with

$$\mathrm{E}[Y_0] = \alpha + \beta x_0, \qquad \mathrm{Var}(Y_0) = \sigma_\epsilon$$

This uncertainty is just linked to the variance of the future realization of the error ϵ_0. However, we should take the uncertainty of our estimators into account, too. We should consider $\hat{Y}_0$ as a random variable, since it is based on estimators a and b. Since we know how to characterize the uncertainty of our estimates, we might build a confidence interval for $\hat{Y}_0$. However, when we consider the realization of the error term in the future, we are dealing with a *prediction interval*, which means dealing with the distribution of a random variable, rather than the estimation of an unknown parameter, which is a number. Indeed, we may even go as far as to build a probability distribution for Y_0. Assuming that regressors are nonstochastic and that errors are normally distributed, this distribution will be t with $n-2$ degrees of freedom, or a normal distribution for a sufficiently large sample.

Since errors are independent, we may just add the variances of $\hat{Y}_0$ and ϵ_0. The former depends on the variances of a and b, but we should not forget their covariance. So, let us write everything explicitly:

$$\hat{Y}_0 = a + bx_0 = \overline{Y} - b(\bar{x} - x_0)$$

$$= \overline{Y} - (\bar{x} - x_0) \times \frac{\sum_{i=1}^{n}(x_i - \bar{x})Y_i}{\sum_{i=1}^{n}(x_i - \bar{x})^2}$$

To streamline notation, let

$$\gamma = \frac{1}{\sum_{i=1}^{n}(x_i - \bar{x})^2}$$

which is a number under our assumptions. Then

$$\hat{Y}_0 = \sum_{i=1}^{n} Y_i \left[\frac{1}{n} - \gamma(x_i - \bar{x})(\bar{x} - x_0) \right]$$

Taking its variance, we obtain

$$\text{Var}(\hat{Y}_0) = \sum_{i=1}^{n} \text{Var}(Y_i) \left[\frac{1}{n} - \gamma(x_i - \bar{x})(\bar{x} - x_0) \right]^2$$

$$= \sigma_\epsilon^2 \left[\sum_{i=1}^{n} \frac{1}{n^2} + \gamma^2(\bar{x} - x_0)^2 \sum_{i=1}^{n}(x_i - \bar{x})^2 - \frac{2\gamma}{n}(\bar{x} - x_0) \underbrace{\sum_{i=1}^{n}(x_i - \bar{x})}_{=0} \right]$$

$$= \sigma_\epsilon^2 \left[\frac{1}{n} + \frac{(x_0 - \bar{x})^2}{\sum_{i=1}^{n}(x_i - \bar{x})^2} \right] \tag{10.22}$$

Taking into account the variance of the error affecting Y_0, we conclude that

$$\text{SE}(Y_0) = \sqrt{\text{Var}(Y_0 - a - bx_0)} = \sigma_\epsilon \sqrt{1 + \frac{1}{n} + \frac{(x_0 - \bar{x})^2}{\sum_{i=1}^{n}(x_i - \bar{x})^2}} \tag{10.23}$$

Here we are using the standard error of Y_0, which we should interpret as a standard error of prediction, rather than a standard error of estimate. As usual, it is important to interpret the result we have obtained. We see that variance depends on three contributions:

1. The first term is just σ_ϵ^2 and is due to the realization of the future error term.

2. The second term, σ_ϵ^2/n, depends on the underlying uncertainty and is mitigated by increasing the sample.

3. The last term is a bit more complicated. The denominator of the ratio tells us that the more observation are spread out, the better; indeed we know that this affects our ability to estimate slope. The numerator depends on the distance between x_0, the value of the explanatory variable

for which we predict the response, and the average x in the available dataset. We see that the farther we go, the less certain we are. Indeed, we cannot extrapolate to much outside the available data. Actually, the situation can be even worse if the underlying phenomenon is nonlinear, and the linear regression model is just a suitable approximation for a limited range of the explanatory variable.

In practice, we must estimate σ_ϵ^2 by SSR, and when computing prediction intervals we should use quantiles from t_{n-2} distribution. With large samples, quantiles from the standard normals can be used.

Example 10.11 A firm observes the following data correlating sales price and sales volume (measured in thousands of items):

Price (€)	2.6	2.8	3.3	3.5	3.8
Sales	2.63	3.35	1.86	1.35	0.47

The firm wants to predict sales if the sales price is raised to €4. Since sales are not completely predictable, the firm would also like a prediction interval with a 95% confidence level to come up with a robust decision.

Least squares yield

$$a = 8.6096, \qquad b = -2.0867$$

Not surprisingly, slope is negative, since we expect sales to drop after a price increase. Note, however, that the dataset shows that sales were higher when price was €2.8 than when it was €2.6; other factors, or just pure randomness, may be at play. Then, a point forecast is easily obtained:

$$\hat{Y}_0 = 8.6096 - 4 \times 2.0867 = 0.2626$$

Thus, we expect to sell something like 263 items. To build a prediction interval, we need to estimate SER, based on the residuals $e_i = Y_i - \hat{Y}_i$:

$$e_1 = 2.6300 - 3.1840 = -0.5540$$
$$e_2 = 3.3500 - 2.7667 = 0.5833$$
$$e_3 = 1.8600 - 1.7233 = 0.1367$$
$$e_4 = 1.3500 - 1.3060 = 0.0440$$
$$e_5 = 0.4700 - 0.6800 = -0.2100$$

Then

$$\text{SER} = \hat{\sigma}_\epsilon = \sqrt{\frac{\sum_{i=1}^{5}(Y_i - \hat{Y}_i)^2}{3}} = 0.4871$$

We may also check that $R^2 = 0.857$, which looks satisfactory. Applying (10.23), where $\bar{x} = 3.2$, we get

$$\text{SE}(Y_0) = 0.4871 \times \sqrt{1 + \frac{1}{5} + \frac{(4 - 3.2)^2}{\sum_{i=1}^{5}(x_i - 3.2)^2}} = 0.6631$$

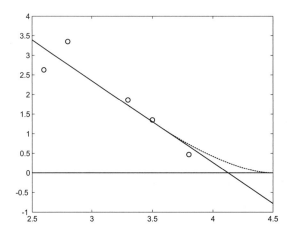

Fig. 10.11   Bad regression-based sales forecast in Example 10.11.

Since $t_{0.975,3} = 3.1824$, the 95% prediction interval is as follows, expressed in items:

$$262.6 \pm 3.1824 \times 663.1 \approx (-1848,\ 2373)$$

This is a pretty large amount of uncertainty, but there is something far worse: The prediction interval does not make any sense, as it includes negative values. On second thought, this is not quite surprising, since SER = 487.1 is large with respect to $\hat{Y}_0 = 262.6$. Note, however, that the large value of R^2 may induce a false sense of security. To visualize what is really going wrong, we may have a look at Fig. 10.11. We see that SER is not that large in absolute terms, but it is relatively large when the price goes up, because it affects much smaller sales volumes. Furthermore, since sales cannot be negative,[13] a negatively sloped regression line is a sensible model within a limited range of prices. When prices are very high, a nonlinear model like the one hinted at by the dashed line in the figure would make more sense.   ⬜

10.5   A GLIMPSE OF STOCHASTIC REGRESSORS AND HETEROSKEDASTIC ERRORS

In this section we outline what happens when we relax a bit our assumptions about the underlying statistical model:

$$Y_i = \alpha + \beta X_i + \epsilon_i \tag{10.24}$$

---

[13]Well... sales cannot be negative, unless you are very, very bad with marketing!

The first thing to note is that now the explanatory variable is random. This is certainly going to make things a tad more complicated, but we do not want to change our assumptions substantially, which is possible by taking advantage of the law of iterated expectations.[14] The trick is to restate assumptions *conditionally* on X_i, whenever necessary:

- (X_i, Y_i) are i.i.d. realizations of a joint probability distribution; this also implies that errors are independent (but it does not imply that they are identically distributed).

- $\mathrm{E}[\epsilon_i \mid X_i] = 0$, i.e., the conditional expectation of errors is zero; note that by the law of iterated expectations this implies that the unconditional expectation of the errors is zero:

$$\mathrm{E}[\epsilon_i] = \mathrm{E}[\mathrm{E}[\epsilon_i \mid X_i]] = 0$$

  but the converse is not true.

- A technical assumption, which we will disregard in the following but that is needed to derive some results concerning estimators, is that outliers are unlikely, in the sense that (X_i, Y_i) have finite fourth-order moments; we recall that fat tails are measured by kurtosis, which is related to fourth-order moments.

The assumption of homoskedasticity should be expressed in terms of conditional variance

$$\mathrm{Var}(\epsilon_i \mid X_i) = \sigma_\epsilon^2$$

but we do not want to take this for granted.

   This framework looks more complicated, but the essential results we have shown in the simpler setting hold with the assumptions mentioned above. To see an example of the technicalities involved, let us see how unbiasedness of the estimate of slope can be proved. Here we concentrate on the slope, since it is usually the parameter of interest, but the approach can be used to deal with the intercept as well. The starting point is again Eq. (10.12), which should be written as

$$b = \beta + \frac{\sum_{i=1}^{n} \left(X_i - \overline{X} \right) \epsilon_i}{\sum_{i=1}^{n} \left(X_i - \overline{X} \right)^2} = b + W \tag{10.25}$$

We would like to prove that $E(W) = 0$, where W is the ratio of the two sums in the equation above. If we take the expectations, in this case we cannot

---

[14]See Theorem 8.10.

just move explanatory variable outside, since they are not numbers anymore. However, we can use the assumptions and the law of iterated expectations, by conditioning on X_i, $i = 1, \ldots, n$:

$$E[W] = E[E[W \mid X_1, X_2, \ldots, X_n]]$$

The conditional expectation allows us to treat explanatory variables as numbers, performing the same tricks again:

$$E[W \mid X_1, X_2, \ldots, X_n] = \frac{\displaystyle\sum_{i=1}^{n} (X_i - \overline{X}) \, E[\epsilon_i \mid X_1, X_2, \ldots, X_n]}{\displaystyle\sum_{i=1}^{n} (X_i - \overline{X})^2} = 0$$

Note that we take X_i outside the conditional expectation, and then use the fact that

$$E[\epsilon_i \mid X_1, X_2, \ldots, X_n] = E[\epsilon_i \mid X_i] = 0$$

due to assumptions concerning the independence between observations and the conditional expectation of errors. An equivalent way of seeing this result is that the assumptions imply the conditional unbiasedness of the estimator

$$E[(b - \beta) \mid X_1, X_2, \ldots, X_n] = 0$$

which in turn implies unbiasedness by the application of iterated expectations.

Things are not that easy when we consider standard errors without assuming homoskedasticity. To see why, imagine that we consider variance of the numerator in W:

$$\text{Var}[(X_i - \overline{X})\epsilon_i].$$

Since regressors are stochastic, we cannot take them outside variance. We cannot factorize the product and simplify the expression, since ϵ_i need not be independent of X_i. Moreover, we cannot automatically assume that the variance of a ratio is the ratio of the variances. If we want to evaluate $\text{SE}(b)$, we must settle for more complicated formulas. An important asymptotic result is the normality of the estimator of slope.

THEOREM 10.4 *The asymptotic distribution of the estimator b, under the previously stated assumptions, is characterized by the following limit:*

$$\sqrt{n}(b - \beta) \xrightarrow{d} \mathcal{N}\left(0, \; \frac{\text{Var}(\nu_i)}{[\text{Var}(X_i)]^2}\right)$$

where $\nu_i = (X_i - \mu_X)\epsilon_i$ and $\xrightarrow{d}$ refers to convergence in distribution.

PROOF   A fully detailed and rigorous proof would be somewhat technical and tedious, but we may at least appreciate the role of stochastic convergence concepts, including Slutsky's theorem, which we illustrated in Section 9.8.5. Equation (10.25) implies

$$\sqrt{n}(b - \beta) = \sqrt{n} \times \frac{\frac{1}{n} \sum_{i=1}^{n} \left(X_i - \mu_X + \mu_X - \overline{X}\right) \epsilon_i}{\frac{1}{n} \sum_{i=1}^{n} \left(X_i - \overline{X}\right)^2}$$

$$= \frac{\frac{1}{\sqrt{n}} \sum_{i=1}^{n} \nu_i}{\frac{1}{n} \sum_{i=1}^{n} \left(X_i - \overline{X}\right)^2} - \frac{\left(\overline{X} - \mu_X\right) \frac{1}{\sqrt{n}} \sum_{i=1}^{n} \epsilon_i}{\frac{1}{n} \sum_{i=1}^{n} \left(X_i - \overline{X}\right)^2} \qquad (10.26)$$

where $\nu_i = (X_i - \mu_X)\epsilon_i$ and $\mathrm{E}[\nu_i] = \mathrm{E}[\mathrm{E}[(X_i - \mu_X)\epsilon_i \,|\, X_i]] = 0$.

To take advantage of Slutsky's theorem, we need to assess which terms above converge, in probability or in distribution, to some relevant quantity. What we know about sample mean and sample variance implies

$$\frac{1}{n} \sum_{i=1}^{n} \left(X_i - \overline{X}\right)^2 \xrightarrow{p} \sigma_X^2$$

$$\left(\overline{X} - \mu_X\right) \xrightarrow{p} 0$$

The central limit theorem tells us that

$$\frac{1}{\sqrt{n}} \sum_{i=1}^{n} \epsilon_i = \sqrt{n}\,\overline{\epsilon} \xrightarrow{d} \mathcal{N}(0, \sigma_\epsilon^2)$$

Then, applying Slutsky's theorem, we see that the second term in (10.26) converges in probability to zero. Applying the central limit theorem to the numerator of the first term yields

$$\frac{\frac{1}{\sqrt{n}} \sum_{i=1}^{n} \nu_i}{\sigma_\nu} \xrightarrow{d} \mathcal{N}(0, 1)$$

where σ_ν is the standard deviation of ν_i. Then, we also see that

$$\frac{\left(\frac{1}{\sqrt{n}} \sum_{i=1}^{n} \nu_i\right) \Big/ \sigma_\nu}{\left(\frac{1}{n} \sum_{i=1}^{n} \left(X_i - \overline{X}\right)^2\right) \Big/ \sigma_X^2} \xrightarrow{d} \mathcal{N}(0, 1)$$

from which the theorem follows immediately. We should note that this proof is not quite rigorous, as applying the central limit theorem requires finiteness of variance, which can be ensured by proper assumptions. ∎

The theorem implies that, for a large sample, the estimator is unbiased, asymptotically normal, and consistent. As usual, we do not really know the variance of b in the statement of the theorem, but we can estimate it by the following formula, based on observed residuals and the substitution of variances with their sample counterparts:

$$\hat{\sigma}_b^2 = \frac{1}{n} \times \frac{\dfrac{1}{n-2}\displaystyle\sum_{i=1}^{n}(X_i - \overline{X})^2 e_i^2}{\left[\dfrac{1}{n}\displaystyle\sum_{i=1}^{n}(X_i - \overline{X})^2\right]^2}$$

When drawing statistical inferences, we use the same procedure as in the homoskedastic case, but we use the standard error $\mathrm{SE}(b) = \hat{\sigma}_b$. Many software packages implement these formulas, which are *robust to heteroskedasticity* and require a minimal set of assumptions.

An alternative to these robust formulas is obtained if we assume some more specific structure on the nature of heteroskedasticity. For instance, let us assume that

$$\mathrm{Var}(\epsilon_i) = \sigma_i^2 = \frac{\sigma_\epsilon^2}{w_i}$$

i.e., variances of errors are known up to a proportionality constant σ_ϵ^2. Then we may rewrite (10.24) multiplying both sides of the equation by $\sqrt{w_i}$:

$$\sqrt{w_i}Y_i = \sqrt{w_i}\alpha + \sqrt{w_i}\beta X_i + \sqrt{w_i}\epsilon_i$$

Using this trick, we see that $\mathrm{Var}(\sqrt{w_i}\epsilon_i) = \sigma_\epsilon^2$, i.e., we are back to the homoskedastic case. Now the sum of squared residuals is

$$\sum_{i=1}^{n}\left(\sqrt{w_i}Y_i - \sqrt{w_i}a + \sqrt{w_i}bX_i\right)^2 = \sum_{i=1}^{n} w_i \left(Y_i - a + bX_i\right)^2$$

The resulting approach is called *weighted least squares* and should be contrasted to the *ordinary* least squares (OLS) that we introduced in this chapter. The result is, after all, rather intuitive: We should attribute more weight to observations with a large value of w_i, i.e., observations affected by a smaller noise.

What we have accomplished by introducing weighted least squares may sound purely academic, as it seems quite hard to have a detailed knowledge of variances σ_i^2 or weights w_i. However, there are more general approaches that can be used to approximate this knowledge. One such procedure is the estimation of an equation describing variance, i.e., a model relating σ_i^2 to one or more variables by a regression equation. The procedure may be sketched as follows:

1. Fit a regression model using ordinary least squares and evaluate residuals e_i.

2. Regress the squared residuals on the explanatory variable to obtain an equation predicting variance for each observation as a function of the explanatory variable, $\hat{\sigma}_i^2 = f(x_i)$; in the case of multiple regression, analysis of residuals may suggest the most appropriate variables to use to estimate such a variance function.

3. Given the estimated variance function, obtain estimates of weights $w_i = 1/\hat{\sigma}_i^2$.

4. Apply weighted least squares.

10.6   A VECTOR SPACE LOOK AT LINEAR REGRESSION

The aim of this section is to broaden our view about linear regression by analyzing it in the light of some concepts from linear algebra. In fact, linear regression can be regarded as a sort of orthogonal projection within suitably chosen vector spaces. To see this, let us group observations and residuals into vectors as follows:

$$\mathbf{Y} = \begin{bmatrix} Y_1 \\ Y_2 \\ \vdots \\ Y_n \end{bmatrix}, \quad \mathbf{x} = \begin{bmatrix} x_1 \\ x_2 \\ \vdots \\ x_n \end{bmatrix}, \quad \mathbf{e} = \begin{bmatrix} e_1 \\ e_2 \\ \vdots \\ e_n \end{bmatrix}$$

Let us also denote by $\mathbf{u} = [1, 1, \ldots, 1]^T$, i.e., a column vector whose entries all set to 1. Ideally, we would like to find coefficients a and b such that vector $\mathbf{Y}$ can be expressed as a linear combination of $\mathbf{x}$ and $\mathbf{u}$:

$$\mathbf{Y} = a\mathbf{u} + b\mathbf{x}$$

However, there is little chance to express a vector in $\mathbb{R}^n$ with a basis consisting of two vectors only, and we settle for a vector that is as close as possible, by minimizing the norm of the vector of residuals:

$$\min_{a,b} \|\mathbf{e}\|^2, \quad \mathbf{e} = \mathbf{Y} - (a\mathbf{u} + b\mathbf{x})$$

We are looking for a vector in the linear subspace spanned by $\mathbf{x}$ and $\mathbf{u}$, which is as close as possible to $\mathbf{Y}$. In other words, we are *projecting* $\mathbf{Y}$ on this subspace. From linear algebra, we know that this projection has some orthogonality properties, in the sense that the difference between $\mathbf{Y}$ and its projection and the projection itself should be orthogonal vectors. Intuitively, if we consider a plane, and a point outside the plane, the path of minimal length between the point and the plane lies along a line that is orthogonal to the plane itself.

Therefore, we should expect that[15]

$$\mathbf{e} \cdot (a\mathbf{u} + b\mathbf{x}) = 0$$

where the product dot · denotes the usual inner product between vectors in $\mathbb{R}^n$. Since inner product is a linear operator, we should just check orthogonality between $\mathbf{e}$ and the basis vectors $\mathbf{u}$ and $\mathbf{x}$. However, from least-squares theory, we know that the average of residuals is indeed zero, so

$$\mathbf{e} \cdot \mathbf{u} = \sum_{i=1}^{n} e_i = 0$$

Checking orthogonality between $\mathbf{e}$ and $\mathbf{x}$ proceeds along familiar lines, based on the form of least-squares estimators:

$$
\begin{aligned}
\mathbf{e} \cdot \mathbf{x} &= \sum_{i=1}^{n} e_i x_i = \sum_{i=1}^{n} [Y_i - (a + bx_i)] x_i \\
&= \sum_{i=1}^{n} [Y_i - \overline{Y} - b(x_i - \bar{x})] x_i = \sum_{i=1}^{n} [(Y_i - \overline{Y})x_i - b(x_i^2 - x_i\bar{x})] \\
&= \sum_{i=1}^{n} (Y_i - \overline{Y})(x_i - \bar{x}) - b \sum_{i=1}^{n} (x_i^2 - x_i\bar{x}) \\
&= \sum_{i=1}^{n} (Y_i - \overline{Y})(x_i - \bar{x}) - b \left(\sum_{i=1}^{n} x_i^2 - n\bar{x}^2 \right) = 0
\end{aligned}
$$

Hence, we see that ordinary least squares may be interpreted in terms of orthogonal projection on linear subspaces. This view of linear regression in linear algebraic terms will prove most useful in understanding multiple linear regression.

So far, in this section, we did not refer to any probabilistic or statistical concept. Thus, it may be useful to interpret linear regression in probabilistic terms. Note that, in practice, regression is carried out on *sampled* data; however, we may also consider best approximation problems between random variables. Let us consider a random variable Y, which we want to approximate by a linear affine transformation of another variable X, i.e.

$$Y \approx a + bX$$

for coefficients a and b that we must determine in a suitable way. What are the desirable properties of such an approximation? To begin with, the two expected values should be the same:

$$E[Y] = a + bE[X] \tag{10.27}$$

---

[15] See Section 3.3.2 for an introduction to inner products and norms.

which is just the probabilistic counterpart of (10.1). To find another condition, we should require that the approximation is good in a probabilistic sense. If we introduce the error[16] $\epsilon = Y - (a + bX)$, we may require that its variance is small; therefore, we solve the problem

$$\min_{a,b} \text{Var}(\epsilon)$$

subject to condition (10.27). Actually, from this requirement we find a condition on b only. To see this, let us express variance of error:

$$\text{Var}(\epsilon) = \text{Var}(Y - a - bX) = \text{Var}(Y) + b^2 \text{Var}(X) - 2b \text{Cov}(Y, X)$$

Minimization with respect to b yields

$$b = \frac{\text{Cov}(Y, X)}{\text{Var}(X)}$$

This is just the probabilistic counterpart of (10.7).

Since we find a best approximation of Y in terms of X by linear regression, the error should not carry any information related to X. In other words ϵ and X should be uncorrelated. Indeed

$$\begin{aligned} \text{Cov}(\epsilon, X) &= \text{Cov}(Y - a - bX, X) = \text{Cov}(Y, X) - b \text{Cov}(X, X) \\ &= \text{Cov}(Y, X) - b \text{Var}(X) = 0 \end{aligned}$$

This also implies that ϵ is uncorrelated with the overall approximation $a + bX$. Now, can we bridge the gap between this probabilistic view of regression, and the view above, based on inner products and orthogonal projection? To this aim, we should define a suitable inner product between random variables, as well as an orthogonality concept. A comparison between the two views suggests that we may link orthogonality and lack of correlation. In other words, random variables X_1 and X_2 are "orthogonal" if $\text{Cov}(X_1, X_2) = 0$. This in turn suggests the definition of an inner product between random variables:

$$\langle X_1, X_2 \rangle = \text{Cov}(X_1, X_2)$$

Let us see if this makes sense. From linear algebra, we recall the properties that any legitimate inner product should enjoy:

1. $\langle X_1, X_2 \rangle = \langle X_2, X_1 \rangle$

2. $\langle X_1 + X_2, Z \rangle = \langle X_1, Z \rangle + \langle X_2, Z \rangle$

3. $\langle aX_1, X_2 \rangle = a\langle X_1, X_2 \rangle$, for any scalar a

---

[16] Since we are not dealing with a sample-oriented, statistical model with unobservables, we use the error, rather than the residual, which is its observable surrogate.

4. $\langle X, X \rangle \geq 0$, and $\langle X, X \rangle = 0$ only if $X = 0$

The first three properties correspond to properties of covariance that we listed in Section 8.3.1. The last property says first that variance cannot be negative, which is fine. The problem comes from the second point: Variance is zero for any constant random variable, not only for the constant zero. This suggests that the might be some more work to do, which is beyond the scope of an introductory book, but it turns out that there is way to solve this issue and properly define an inner product of random variables.[17]

Leaving the last technicality aside, it does seem that we can define a linear space of random variables, on which we may take linear combinations, and the inner product is just covariance. Within this framework, the two views described above are indeed strictly related. They both amount to orthogonal projection within linear spaces with a properly defined inner product. The inner product also defines a norm

$$\|x\|^2 = \langle x, x \rangle$$

where x may be a vector or a random variable. By least squares, we do minimize the squared norm of an error/residual, which is orthogonal to the projected element, either vector or random variable. We close this section by a short example reinforcing this general framework.

Example 10.12 (Pythagorean theorem for random variables) We know that if vectors $\mathbf{x}$ and $\mathbf{y}$ are orthogonal, then

$$\|\mathbf{x} + \mathbf{y}\|^2 = \|\mathbf{x}\|^2 + \|\mathbf{y}\|^2$$

If we apply this to the legs and the hypothenuse of a right triangle, we get the familiar form of Pythagorean theorem. If we apply the idea to a linear space of random variables, whereby the squared norm is variance, we get something quite familiar. If two random variables are uncorrelated, we obtain

$$\|X + Y\|^2 = \text{Var}(X + Y) = \text{Var}(X) + \text{Var}(Y) = \|X\|^2 + \|Y\|^2$$

$\square$

Problems

10.1   Consider the following sales data:

Week	1	2	3	4	5	6	7	8	9	10
Sales	30	20	45	35	30	60	40	50	45	65

Build a linear regression model to predict sales and calculate R^2.

---

[17]See, e.g., the text by Luenberger [3].

10.2   Given the observed data

x	45	50	55	60	65	70	75
y	24.2	25.0	23.3	22.0	21.5	20.6	19.8

build a 95% confidence interval for the slope.

10.3   A firm sells a perishable product, with a time window for sales limited to 1 month. The product is ordered once per month, and the delivery lead time is very small, so that the useful shelf life is really 1 month. Each piece is bought at €10 and its sold for €14; if the product expires, it can be scrapped for €2 per unit. Over the last 4 months a positive trend in sales has been observed:

Month	Jan	Feb	Mar	Apr
Sales	102	109	123	135

Hence, the firm resorts to linear regression to forecast sales over the next period. How many items should the firm buy, in order to maximize expected profit in May?

For further reading

- An introductory treatment of linear regression is given, e.g., in Ref. [6].

- The treatment in Ref. [7] is slightly higher-level, but quite readable and careful about distributional results for estimators.

- Robust formulas to heteroskedasticity are discussed in Ref. [8], which also illustrates many of applications to economics.

- The use of linear regression for forecasting purposes can be appreciated by reading Ref. [4], which also deals with time series models, which we cover in the next chapter.

- Other extensions of basic linear regression modeling are dealt with, e.g., in Ref. [5], including generalized and weighted least squares.

- For more in-depth treatment, the reader may have a look at Ref. [1] or [2], among others.

- Readers interested in the vector space view of regression may find the reading of Luenberger's text [3] enlightening.

REFERENCES

1. N.R. Draper and H. Smith, *Applied Regression Analysis,* 3rd ed., Wiley, New York, 1998.

2. R.R. Hocking, *Methods and Applications of Linear Models: Regression and the Analysis of Variance,* 2nd ed., Wiley, Hoboken, NJ, 2003.

3. D.G. Luenberger, *Optimization by Vector Space Methods,* Wiley, New York, 1969.

4. S. Makridakis, S.C. Wheelwright, and R.J. Hyndman, *Forecasting: Methods and Applications,* 3rd ed., Wiley, New York, 1998.

5. D.C. Montgomery, C.L. Jennings, and M. Kulahci, *Introduction to Time Series Analysis and Forecasting,* Wiley, New York, 2008.

6. M.K. Pelosi and T.M. Sandifer, *Elementary Statistics: From Discovery to Decision,* Wiley, New York, 2003.

7. S.M. Ross, *Introduction to Probability and Statistics for Engineers and Scientists,* 4th ed., Elsevier Academic Press, Burlington, MA, 2009.

8. J.H. Stock and M.W. Watson, *Introduction to Econometrics,* 2nd ed., Addison-Wesley, Boston, 2006.

11

Time Series Models

Forecasting is a common task in business management. In Chapter 10, on simple linear regression models, we have met a kind of statistical model that can be used as a forecasting tool, provided that

- We are able to find potential explanatory variables

- We have enough data on all the relevant variables, in order to obtain reliable estimates of model parameters

Even though, strictly speaking, linear regression captures association and not causation, the idea behind such a model is that knowledge about explanatory variables is useful to predict the value of the explained variable. Unfortunately, there are many cases in which we are not able to find a convincing set of explanatory variables, or we lack data about them, possibly because they are too costly to collect. In some extreme cases, not only do we lack enough information about the explanatory variables, but we even lack information about the predicted variable. One common case is forecasting sales for a brand-new kind of product, with no past sales history. Then, we might have to settle for a *qualitative*, rather than *quantitative* forecasting approach. Qualitative forecasting may take advantage of qualified expert opinion; various experts may be pooled in order to obtain both a forecast and a measure of its uncertainty.[1] Actually, these two families of methods can be and, in fact, are often integrated. Even when plenty of data are available, expert

---

[1]Some qualitative forecasting methods, like the Delphi method, have been developed to manage the judgmental forecasts of a pool of experts. If poorly applied, these methods

opinions are a valuable commodity, since statistical models are intrinsically *backward*-looking, whereas we should look forward in forecasting.

In this book, we stick to quantitative approaches, such as linear regression, leaving qualitative forecasting to the specialized literature. Within the class of quantitative forecasting methods, an alternative to regression models is the family of *time series* models. The distinguishing feature of time series models is that they aim at forecasting a variable of interest, based only on observations of the variable itself; no explanatory variable is considered.

Example 11.1 Stock trading on financial markets is one of those human endeavors in which good forecasts would have immense value. One possible approach is based on *fundamental analysis*. Given a firm, its financial and industrial performance is evaluated in order to assess the prospect for the price of its shares. In this kind of analysis, it is assumed that there is some rationality in financial markets, and we try to explain stock prices, at least partially, by a set underlying factors, which may also include macroeconomic factors such as inflation or oil price. On the contrary, *technical analysis* is based only on patterns and trends in the stock price itself. No explanatory variable is sought. The idea is that financial markets are mostly irrational, and that psychology should be used to explain observed behavior. In the first case, some statistical model, possibly a complicated linear regression model, could be arranged. In the second case, time series approaches are used.[2]   ⬚

(*Note:* Time series models can be built to forecast a wide variety of variables, such as interest rates, electric power consumption, inflation, unemployment, etc. To be concrete, in most, if not all of our examples, we will deal with demand forecasting. Yet, the approaches we outline are much more general than it might seem.)

To illustrate the nature of time series models formally, let us introduce the fundamental notation, which is based on a discrete time representation in *time buckets* or periods (e.g., weeks) denoted by t:

- Y_t is the realization of the variable of interest at time bucket t. If we are observing weekly demand for an item at a retail store, Y_t is the demand observed at that store *during* time bucket t.

- $F_{t,h}$ is the forecast generated *at the end* of time bucket t with horizon of h time buckets; hence, $F_{t,h}$ is a prediction of demand at time $t + h$,

---

may lead to a consensus forecast destroying valuable information; in fact, the disagreement between experts may be leveraged to find a measure of inherent uncertainty.

[2]For the sake of simplicity, we have stated the two approaches in a somewhat extreme and overly simplistic manner. In practice, we may well build a statistical model in which behavioral factors are accounted for by suitable explanatory variables; in fact, quantitative models of this kind have been proposed for active portfolio management.

where $h = 1, 2, 3, \ldots$. It is important to clarify the roles of subscripts t and h in our definition of forecast:

- t indicates *when* the forecast is made, and it defines the information set on the basis of which the forecast is built; for instance, at time t we have information about demand during all the time buckets up to and including t.

- h defines how many steps ahead in the future we want to forecast; the simplest case is $h = 1$, which implies that after observing demand in time bucket t, we are forecasting future demand in time bucket $t + 1$.

We should distinguish quite clearly *when* and *for when* the forecast is calculated. One might wonder why we should forecast with a horizon $h > 1$. The answer is that sometimes we have to plan actions in considerable advance. Consider, for instance, ordering an item from a supplier whose lead time is three time buckets; clearly, we cannot base our decision on a demand forecast with $h = 1$, since what we order now will not be delivered during the next time bucket.

In time series models, the information set consists only of observations of Y_t; no explanatory variable is considered. Trivial examples of forecasting formulas could be

$$F_{t,1} = Y_t \quad \text{or} \quad F_{t,1} = \frac{1}{t} \sum_{k=1}^{t} Y_k$$

In the first case, we just use the last observation as a forecast for the next time bucket. In the second case, we take the average of all of the past t observations, from time bucket 1 up to time bucket t. In a sense, these are two extremes, because we either use a very tiny information set, or a set consisting of the whole past history. Maybe one observation is too prone to spikes and random shocks, which would probably add undesirable noise, rather than useful information, to our decision process; on the other hand, the choice of using all of them does not consider the fact that some observations far in the past could be hardly relevant. Furthermore, we are not considering the possibility of systematic variations due to trends or seasonality. In the following, we describe both heuristic and more formal approaches to forecasting.

There is an enormous variety of forecasting methods, and plenty of software packages implementing them. What is really important is to understand a few recurring and fundamental concepts, in order to properly evaluate competing approaches. However, there is an initial step that is even more important: *framing* the forecasting process within the overall *business process*. We insist on this in Section 11.1. The most sophisticated forecasting algorithm is utterly useless, if it is not in tune with the surrounding process. Whatever approach we take, it is imperative to monitor forecast errors; in Section 11.2 we define several error measures that can be used to choose among alternative models, to fine-tune coefficients governing their functioning, and to check and improve performance. Section 11.3 illustrates the fundamental ideas of time series decomposition, which highlights the possible presence of factors like trend and seasonality. Section 11.4 deals with a very simple approach, moving average, which has a limited domain of applicability but is quite useful in pointing out basic tradeoffs that we have to make when fine-tuning a forecasting algorithm. Section 11.5 is actually the core of the chapter, where we describe the widely used family of exponential smoothing methods; they are easy to use and quite flexible, as they can account for trend and seasonality in a straightforward and intuitive way. Finally, Section 11.6 takes a more formal route and deals with autoregressive and moving-average models within a proper statistical framework. This last section is aimed at more advanced readers and may be safely skipped.

11.1   BEFORE WE START: FRAMING THE FORECASTING PROCESS

When learning about forecasting algorithms, it is easy to get lost in technicalities and forget a few preliminary points.

Forecasting is not about a single number. We are already familiar with inferential statistics and confidence intervals. Hence, we should keep in mind that a single number, i.e., a point forecast, may be of quite little use without some measure of uncertainty. As far as possible, forecasting should be about building a whole probability distribution, not just a number to bet on.

Choose the right time bucket. Imagine that you order raw materials at the end of every week; should you bother about daily forecasting? Doing so, you add unnecessary complexity to your task, as weekly forecasts are what you really need. Indeed, it is tempting to use large time buckets in order to aggregate demand with respect to time and reduce forecast errors; this is not advisable if your business process requires forecasts with small time buckets.

What should we forecast? If this sounds like a dumb question, think again. Imagine that you are a producer of T-shirts, available in customary sizes and a wide array of colors. Forecasting sales of each single item may be a daunting task. As a general rule, forecasts are more reliable if we can aggregate items. Rather than forecasting demand for each combination of size and color, we could aggregate sizes and consider only forecasting for a set of colors. In fact, demand for a specific combination of color and size may be rather volatile, but the fraction of population corresponding to each size is much more stable. We may forecast aggregate sales for each T-shirt model, and then use common factors across models to disaggregate and obtain forecasts for each single combination. Of course, in the end, we want to forecast sales for each individual item, but in the *process* of building a forecast we use suitable aggregation and disaggregation strategies. This approach is also quite powerful in pooling demand data across items, thus improving the quality of estimates of common factors, but it must be applied to compatible product families. We cannot apply size factors that are standard for adults to T-shirts depicting cartoon characters (maybe).

Forecasts are a necessary evil. Common sense says that forecasts are always wrong, but we cannot do without them. This is pretty true, but sometimes we can do something to make our life easier and/or mitigate the effect of forecast errors. If our suppliers have a long delivery lead time, we must forecast material requirements well in advance. Since the larger the forecast horizon, the larger the uncertainty in forecasting, it is wise to try whatever we can to shorten lead time (possibly by choosing geographically closer suppliers). The same applies to manufacturing lead times. Sometimes, seemingly absurd approaches may improve forecasting. Consider the production process for sweaters. Just like with T-shirts, forecasting a specific combination of color, model, or size is quite hard. Two key steps in producing a sweater are dying the fabric with whatever color we like, and knitting. The commonsense approach is to dye first and knit later. However, the time needed to knit is much longer than the time needed to dye. This means that one has to forecast sales of a specific combination much in advance of sales, with a corresponding criticality. By swapping the two steps, knitting first, one may postpone the final decisions, and rely on more accurate forecasts. This is a typical postponement decision,[3] whereby the impact on cost and quality must be traded off against the payoff from better matching of supply and demand.

The general point is that the forecasting process should support the surrounding business process and should help in making decisions. Forecast errors may be numerically large or small, but what really matters is their economic consequence.

As a last observation, let us consider another seemingly odd question: Can we observe what we are forecasting? To be concrete, let us consider demand again. Can we observe demand at a retail store? In an age of bar codes and point-of-sale data acquisition, the answer could be "yes." Now imagine that you wish a pot of your favorite cherry yoghurt, but you find the shelf empty; what are you going to do? Depending on how picky you are, you

---

[3]The case we are outlining is well known: Benetton reversed the sequence of manufacturing steps in order to ease severe difficulties with forecasting. Hewlett-Packard is another company that has successfully adopted product design for postponement, in order to improve supply chain performance.

might settle for a different packaging, a different flavor, or a different brand, or you could just go home quite angry. But unless you are *very* angry and start yelling at any clerk around, no one will know that potential demand has been lost. What we may easily measure are *sales*, not demand. Hence, we use sales as a proxy of demand, but this may result in underforecasting. If this looks like a peculiar case, consider a business-to-business setting. A potential customer phones and needs 10 electric motors now; your inventory level is down, but you will produce a new batch in 5 days. Unfortunately, the customer really needs those motors right now; so, he hangs up and phones a competitor. Chances are, this potential demand is never recorded anywhere in the information system. Again, we risk underestimating demand. Statistical techniques have been devised to correct for these effects, but sometimes it is more a matter of organization than sophisticated math.

11.2   MEASURING FORECAST ERRORS

Before we delve into forecasting algorithms, it is fundamental to understand how we may evaluate their performance. This issue is sometimes overlooked in practice: Once a forecast is calculated and used to make a decision, it is often thrown away for good. This is a mistake, as the performance of the forecasting process should be carefully monitored.

In the following, we will measure the quality of a forecast by a *forecasting error*, defined as

$$e_t = Y_t - F_t' \tag{11.1}$$

This is the forecast error for the variable at time t. The notation F_t' refers to the forecast for time bucket t. It is not important *when* the forecast was made, but *for when*. If the horizon is $h = 1$, then $F_t' = F_{t-1,1}$, whereas $F_t' = F_{t-2,2}$ if $h = 2$. Also, notice that by this definition the error is positive when demand is larger than the forecast (i.e., we underforecasted), whereas the error is negative when demand is smaller than the forecast (i.e., we over-forecasted). It is important to draw the line between two settings in which we may evaluate performance:

1. In *historical simulation*, we apply an algorithm on past data and check what the performance would have been, had we adopted that method. Of course, to make sense, historical simulation must be non-anticipative, and the information set that is used to compute $F_{t,h}$ must not include any data after time bucket t. This may sound like a trivial remark, but, as we will see later, there might be an indirect way to violate this nonanticipativity principle.

2. In *online monitoring*, we gather errors at the end of each time bucket to possibly adjust some coefficients and fine-tune the forecasting algorithm. In fact, many forecasting methods depend on coefficients that can be set once and for all or dynamically adapted.

Whatever the case, performance is evaluated over a *sample* consisting of T time buckets, and of course we must aggregate errors by taking some form of average. The following list includes some of the most commonly used indicators:

- Mean error:

$$\text{ME} = \frac{1}{T} \sum_{t=1}^{T} e_t \qquad (11.2)$$

- Mean absolute deviation:

$$\text{MAD} = \frac{1}{T} \sum_{t=1}^{T} |e_t| \qquad (11.3)$$

- Root-mean-square error:

$$\text{RMSE} = \sqrt{\frac{1}{T} \sum_{t=1}^{T} e_t^2} \qquad (11.4)$$

- Mean percentage error:

$$\text{MPE} = \frac{1}{T} \sum_{t=1}^{T} \frac{e_t}{Y_t} \qquad (11.5)$$

- Mean absolute percentage error:

$$\text{MAPE} = \frac{1}{T} \sum_{t=1}^{T} \frac{|e_t|}{Y_t} \qquad (11.6)$$

As clearly suggested by Eq. (11.2), ME is just the plain average of errors over the sample, so that positive and negative errors cancel each other. As a consequence, ME does not discriminate between quite different error patterns like those illustrated in Table 11.1; the average error is zero in both case (a) and (b), but we cannot say that the accuracy is the same. Indeed, ME measures *systematic deviation* or *bias*, but not the accuracy of forecasts. A significantly positive ME tells us that we are systematically underforecasting demand; a significantly negative ME means that we are systematically overstating demand forecasts. This is an important piece of information because it shows that we should definitely revise the forecasting process, whereas a very small ME does not necessarily imply that we are doing a good job. To measure accuracy, we should get rid of the error sign. This may be accomplished by taking the absolute value of errors or by squaring them. The first idea leads to MAD; the second one leads to RMSE, where the square root is taken in

Table 11.1   Mean error measures bias, not accuracy.

Case	e_1	e_2	e_3	e_4	e_5	e_6	ME
(a)	0	0	0	0	0	0	0
(b)	+5	-5	-10	+10	-5	+5	0

Table 11.2   RMSE penalizes large errors more than MAD does.

Case	e_1	e_2	e_3	e_4	e_5	e_6	ME	MAD	RMSE
(a)	+1	+1	+1	+1	+1	+1	1	1	1
(b)	0	-3	0	0	+3	0	1	1	1.73

order to express error and demand in the same units. In case (a) of Table 11.1, MAD is 0; on the contrary, the reader is invited to verify that MAD is $\frac{40}{6} = 6.67$ in case (b), showing that MAD can tell the difference between the two sequences of errors. RMSE, with respect to MAD, tends to penalize large errors, as we may appreciate from Table 11.2. Cases (a) and (b) share the same ME and MAD, but RMSE is larger in case (b), which indeed features larger errors.

The reader will certainly notice some similarity between RMSE and standard deviation. Comparing the definition in Eq. (11.4) and the familiar definition of sample variance and sample standard deviation in inferential statistics, one could even be tempted to divide by $T - 1$, rather than by T. This temptation should be resisted, however, since it misses a couple of important points.

- In inferential statistics we have a random sample consisting of independent observations of identically distributed variables. In forecasting there is no reason to think that the expected value of the variable we are observing is stationary. We might be chasing a moving target, and this has a profound impact on forecasting algorithms.

- Furthermore, in inferential statistics we use the *whole* sample to calculate the sample mean $\overline{X}$, which is then used to evaluate the sample standard deviation; a bias issue results, which must corrected. Here we use past demand information to generate the forecast F_t'; then we evaluate the forecasting error with respect to a *new* demand observation Y_t, which was *not* used to generate F_t'.

By the same token, MAD as defined in Eq. (11.3) should not be confused with MAD as defined in Eq. (4.4), where we consider deviations with respect to a sample mean. In fact, in forecasting literature the name *mean absolute error* (MAE) is sometimes used, rather than MAD, in order to avoid this ambiguity.

A common feature of ME, MAD, and RMSE is that they measure the forecast error using the same units of measurement as demand. Now imagine that MAD is 10; is that good or bad? Well, very hard to say: If MAD is 10 when demand per time bucket is something like 1000, this is pretty good. If MAD is 10 and demand is something like 10, this is pretty awful. This is why MPE and MAPE have been proposed. By dividing forecast errors by demand,[4] we consider a *relative* error. Apparently, MPE and MAPE are quite sensible measures. In fact, they do have some weak spots:

- They cannot be adopted when demand during a time bucket can be zero. This may well be the case when the time bucket is small and/or demand is sporadic. In modern retail chains, replenishment occurs so often, and assortment variety is so high that this situation is far from being a rare occurrence.

- Even when demand is nonzero , these indices can yield weird results if demand shows wide variations. A somewhat pathological example is shown in Table 11.3. The issue here is that forecast 1 is almost always perfect, and in one unlucky case it makes a big mistake, right when demand is low; the error is 9 when demand is 1, so percentage error is an astonishing 900%, which yields MAPE=90% when averaged over the 10 time buckets. Forecast 2, on the contrary, is systematically wrong, but it is just right at the right moment; the second MAPE is only 18%, suggesting that the second forecaster is better than the first one, which is highly debatable.

A possible remedy to the two difficulties above is to consider a ratio of averages, rather than an average of ratios. We may introduce the following performance measures:

$$\text{ME\%} = \frac{\text{ME}}{\overline{\overline{Y}}}$$

$$\text{MAD\%} = \frac{\text{MAD}}{\overline{Y}}$$

$$\text{RMSE\%} = \frac{\text{RMSE}}{\overline{Y}}$$

The idea is to take performance measures, expressed in absolute terms, and make them relative by dividing by the average demand

$$\overline{Y} = \frac{1}{T} \sum_{t=1}^{T} Y_t$$

---

[4]It is important to keep in mind that errors should be divided by demand Y_t, and not by forecast F_t'. This second choice may lead to forecast manipulations when a forecaster's wages are tied to forecast errors.

Table 11.3 Illustrating potential dangers in using MPE and MAPE.

(a)

Period	1	2	3	4	5	6	7	8	9	10
Demand	10	10	10	10	1	10	10	10	10	10
Forecast 1	10	10	10	10	10	10	10	10	10	10
Error 1	0	0	0	0	-9	0	0	0	0	0
Forecast 2	12	12	12	12	1	12	12	12	12	12
Error 2	-2	-2	-2	-2	0	-2	-2	-2	-2	-2

(b)

	ME	MAD	MPE (%)	MAPE (%)
Forecast 1	-0.9	0.9	-90	90
Forecast 2	-1.8	1.8	-18	18

Source: From Ref. [2, page 110].

In these measures, it is very unlikely that average demand over the T time buckets is zero; if this occurs, there is a big problem, but not with forecasting.

As we see, there is a certain variety of measures, and we must mention that others have been proposed. As common in management, we should keep several indicators under control, in order to get a complete picture. We close the section with a few additional remarks:

- Checking forecast accuracy is a way to gauge the inherent uncertainty in demand; this is helpful when we want to come up with something more than just a point forecast.

- We should never forget that a forecast is an input to a decision process. Hence, alternative measures might consider the effect of a wrong forecast, most notably from the economic perspective. However, how this is accomplished depends on the strategies used to hedge against forecast errors.

11.2.1 In- and out-of-sample checks

As will be clear from the following, when we want to apply certain forecasting algorithms, we might need to fit one or more parameters used to calculate a forecast. This is typically done by a proper initialization. When the algorithm depends on estimates of parameters, if we start from scratch, initial performance will be poor because the algorithm has to learn about the demand pattern first. However, we would like to ascertain how the algorithm performs in steady state, not in this initial transient phase.

If we are carrying out a historical simulation, based on a sample of past data, we may use a portion of available information to fit parameters. This portion of data that we use for fitting and initial learning is the *fit sample*. Arguably, the larger the fit sample, the better the initialization. However, this does not leave us with any data to test performance. In fact, it would be not quite correct to use knowledge of data to fit parameters, and then predict the very same data that we have used to initialize the algorithm. Performance evaluation should be carried out *out-of-sample*, i.e., predicting data that have not been used in any way for initialization purposes.[5]

This will be much clearer in the following, but it is important to state this principle right from the beginning. The available sample of data should be split into

1. A *fit sample* used for initialization

2. A *test sample* to evaluate performance in a realistic and sensible way

This approach is also known as *data splitting*, and it involves an obvious tradeoff, since a short fit sample would leave much data available for testing, but initial performance could be poor; on the other hand, a short test sample makes performance evaluation rather unreliable.

11.3 TIME SERIES DECOMPOSITION

The general idea behind time series models is that the data-generating process consists of two components:

1. A *pattern*, which is the "regular" component and may be quite variable over time, but in a fairly predictable way

2. An *error*, which is the "irregular" and unpredictable component[6]

Some smoothing mechanism should be designed in order to filter errors and expose the underlying pattern. The simplest decomposition scheme we may adopt is

$$Y_t = \mu + \epsilon_t \tag{11.7}$$

where ϵ_t is a random variable with expected value 0. Additional assumptions, for the sake of statistical tractability, may concern independence and normality of these random shocks. Faced with such a simple decomposition, we could just take an average of past observations in order to come up with

---

[5]We should mention that out-of-sample testing should also be carried out with a regression model.

[6]The term "unpredictable" should be better qualified. This component is unpredictable *conditional on our knowledge*. Actually, some components of the observed process could be predictable, if we had better information. As a concrete example, consider demand spikes due to trade promotions, as observed by a manufacturer who is not informed in advance by retailers.

an estimate of μ, which is the expected value of Y_t. In real life, we do not have constant expected values, as market conditions may change over time. Hence, we could postulate a model like

$$Y_t = \mu_t + \epsilon_t \tag{11.8}$$

Here μ_t could be a "slowly" varying function of time, associated with market cycles, on which fast swings due to noise are superimposed. Alternatively, we could think of μ_t as a stochastic process whose sample paths are piecewise constant. In other words, every now and then a shock arrives, possibly due to the introduction or withdrawal of similar products by competitors (or ourselves), and μ_t jumps to a new value. Hence, we should not only filter random errors but also track the variation of μ_t, adapting our estimate. Equation (11.8) may look like a sort of catchall, which could fit any process. However, it is typically much better to try and see more structure in the demand process, discerning predictable and unpredictable variability. The elements of predictable variability that we will be concerned with in this chapter are[7]

- Trend, denoted by Θ_t

- Seasonality, denoted by S_t

Trend is quite familiar from linear regression, and, strictly speaking, it is related to the slope of a line. Such a line represent a tendency of demand, on which random shocks and possibly seasonality are superimposed. In the following text, the intercept and slope of this line will be denoted as B (the *level* or *baseline demand*) and T (the trend in the strict sense), respectively; hence, the equation of the line is

$$\Theta_t = B + tT$$

Seasonality is a form of predictable variability with precise regularity in timing. For instance, we know that in Italy more ice cream is consumed in summer than in winter, and that demand at a large retail store is larger on Saturdays than on Tuesdays. It is important to separate seasonality from the trend component of a time series; high sales of ice cream in July–August should not be mistaken for an increasing trend. An example of time series clearly featuring trend and seasonality is displayed in Fig. 11.1. There is an evident increasing trend, to which a periodic oscillatory pattern is superimposed. There is a little noise component, but we notice that demand peaks occur regularly in time, which makes this variability at least partially predictable.

Generally speaking, time series decomposition requires the specification of a functional form

$$Y_t = f(\Theta_t, S_t, \epsilon_t)$$

---

[7]It is also standard procedure to include a *cycle* component, which we will formally disregard, as it can be subsumed by other components. The name "cycle" stems from its link with economic cycles.

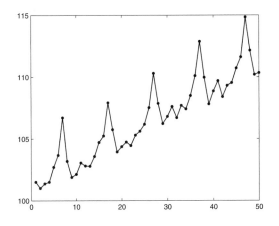

Fig. 11.1   Time series featuring trend and seasonality.

depending on each component: trend, seasonality, and noise. In principle, we may come up with weird functional forms, but the two most common patterns are as follows:

- The *additive* decomposition

$$Y_t = \Theta_t + S_t + \epsilon_t \tag{11.9}$$

- The *multiplicative* decomposition

$$Y_t = \Theta_t \cdot S_t \cdot \epsilon_t \tag{11.10}$$

Within an additive decomposition, a possible assumption is that the noise term ϵ_t is normally distributed, with zero mean. This is not a harmless assumption for a multiplicative scheme, as it results in negative values that make no sense when modeling demand. A possible alternative is assuming that ϵ_t is lognormally distributed, with expected value 1. We recall from section 7.7.2 that a lognormal distribution results from taking the exponential of a normal variable, and it has positive support; hence, if we transform data by taking logarithms in Eq. (11.10), we obtain an additive decomposition with normal noise terms. It is also important to realize that the seasonal component is periodic

$$S_t = S_{t+s}$$

for some seasonality cycle s. If time buckets correspond to months, yearly seasonality corresponds to $s = 12$; if we partition a year in four quarters, then $s = 4$. The seasonal factor S_t in a multiplicative scheme tells us the extent to which the demand in time bucket t exceeds its long-run average, corrected by trend. For instance, if there is no trend, a multiplicative factor $S_t = 1.2$ tells that demand in time bucket t is 20% larger than the average. It is reasonable to normalize multiplicative seasonality factors in such a way that their average value is 1, so that they may be easily interpreted.

Example 11.2 Consider a year divided in four 3-month time buckets. We could associate S_1 with winter, S_2 with spring, S_3 with summer, and S_4 with autumn. Then, if things were completely static, we would have

$$S_1 = S_5 = S_9 = \ldots; \ S_2 = S_6 = S_{10} = \ldots; \ \ldots$$

Now assume that, in a multiplicative model, we have

$$S_1 = 1.2, \quad S_2 = 1.3, \quad S_3 = 0.8$$

What should the value of S_4 be? It is easy to see that we should have $S_4 = 0.7$.
$\square$

Example 11.3 (Additive vs. multiplicative seasonal factors) The kind of seasonality we have observed in Fig. 11.1 is additive. In fact, the width of the oscillatory pattern does not change over time. We are just adding and subtracting periodic factors. In Fig. 11.2 we can still see a trend with the superimposition of a seasonal pattern. However, we notice that the width of the oscillations is increasing. In fact, the two following conditions have a remarkably different effect:

- Demand in August is 20 items above normal (Fig. 11.1)

- Demand in August is 20% above normal (Fig. 11.2)

In the case of an additive seasonality term, the average of the seasonal factors in a whole cycle should be 0. With a multiplicative seasonality and an increasing trend, the absolute increase in August demand, measured by items, is itself increasing, leading to wider and wider swings.
$\square$

The impact of noise differs as well, if we consider additive rather than multiplicative shocks. The most sensible model can often be inferred by visual inspection of data. To fix ideas and keep the treatment to a reasonable size, we will always refer to the following hybrid scheme, because of its simplicity:

$$Y_t = (B + tT)S_t + \epsilon_t \tag{11.11}$$

If there were no seasonality, the pattern would just be a line with intercept B and slope T. Since we assume $\mathrm{E}[\epsilon_t] = 0$, on the basis of *estimates* of parameters $\hat{B}$, $\hat{T}$, and $\hat{S}_t$, our demand forecast is

$$\hat{Y}_t = (\hat{B} + t\hat{T})\hat{S}_t \tag{11.12}$$

Here we use the notation $\hat{Y}_t$, which is closer to the one used in linear regression, rather than $F_{t,h}$. We do so because, so far, it is not clear when and on

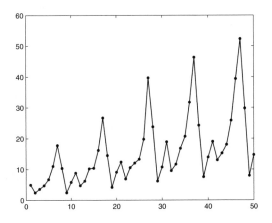

Fig. 11.2   Time series featuring trend and multiplicative rather than additive seasonality.

the basis of which information the forecast is generated. Given a sample of demand observations, there are several ways to fit the parameters in this decomposition.

Example 11.4 Consider a sample of two seasonal cycles, consisting of four time buckets each; the seasonal cycle is $s = 4$, and we have demand observations Y_t, $t = 1, 2, \ldots, 8$. To decompose the time series, we could solve the following optimization model:

$$\min \quad \sum_{t=1}^{8} \left(\hat{Y}_t - Y_t \right)^2$$

$$\begin{aligned}
\text{s.t.} \quad & \hat{Y}_1 = (B + 1 \times T)S_1 \\
& \hat{Y}_2 = (B + 2 \times T)S_2 \\
& \hat{Y}_3 = (B + 3 \times T)S_3 \\
& \hat{Y}_4 = (B + 4 \times T)S_4 \\
& \hat{Y}_5 = (B + 5 \times T)S_1 \\
& \hat{Y}_6 = (B + 6 \times T)S_2 \\
& \hat{Y}_7 = (B + 7 \times T)S_3 \\
& \hat{Y}_8 = (B + 8 \times T)S_4 \\
& S_1 + S_2 + S_3 + S_4 = 4 \\
& S_t \geq 0
\end{aligned}$$

The optimization is carried out with respect to level B, trend T, and the four seasonal factors that repeat over time. Note that seasonal factors are normalized, so that their average value is 1; furthermore, either level or trend could be negative (but not both), whereas seasonal factors are restricted to nonnegative values, otherwise negative demand could be predicted.   □

The optimization problem in the example looks much like a least-squares problem, but, given the nonlinearity in the constraints, it is not as easy to solve as the ordinary least-squares problems we encounter in linear regression. Alternative procedures, much simpler and heuristic in nature, have been proposed to decompose a time series. We will not pursue these approaches in detail, as they are based on the unrealistic idea that the parameters in the demand model are constant over time.

In fact, market conditions do change, and we must track variations in the underlying unknown parameters, updating their estimates when new information is received. Hence, we should modify the static decomposition scheme of Eq. (11.12) as follows:

$$\hat{Y}_{t+h} = F_{t,h} = (\hat{B}_t + h\hat{T}_t)\hat{S}_{t+h-s}, \qquad h = 1, 2, 3, \ldots \qquad (11.13)$$

Since we update estimates dynamically, now we make the information set, i.e., demand observations up to and including time bucket t, and the forecast horizon h explicit. In this scheme we have three time-varying estimates:

- The estimate of the *level* component $\hat{B}_t$ at the end of time bucket t; if there were no trend or seasonality, the "true" parameter B_t would be an average demand, possibly subject to slow variations over time.

- The estimate of the *trend* component $\hat{T}_t$ at the end of time bucket t, which is linked to the slope of a line. The slope can also change over time, as inversions in trend are a natural occurrence. Note that when forecasting demand Y_{t+h} at the end of time bucket t, the estimate of trend at time t should be multiplied by the forecast horizon h.

- The estimate of the multiplicative *seasonality factor* $\hat{S}_{t+h-s}$, linked to percentage up- and downswings. This is the estimate at the end of time bucket t of the seasonal factor that applies to all time buckets similar to $t + h$, i.e., to a specific season within the seasonal cycle. We clarify in Section 11.5.4 that such an estimate is obtained at time bucket $t + h - s$, where s is the length of the seasonal cycle, but we may immediately notice that at the end time bucket t we cannot using anything like $\hat{S}_{t+h}$, since this is a *future* estimate.

This dynamic decomposition leads to the heuristic time series approaches that are described in the following two sections.

11.4   MOVING AVERAGE

Moving average is a very simple algorithm, which serves well to illustrate some tradeoffs that we will face later. As a forecasting tool, it can be used when we assume that the underlying data generating process is simply

$$Y_t = B_t + \epsilon_t \qquad (11.14)$$

This is the model we obtain from (11.13) if we do not consider trend and seasonality.[8] In plain words, the idea is that demand is *stationary*, with average B_t. In principle, the average should be constant over time. If so, we should just forecast demand by a plain average of all available observations. The average has the effect of filtering noise out and revealing the underlying "signal." In practice, there are slow variations in the level B_t. Therefore, if we take the sample mean of all available data

$$\overline{Y} = \frac{1}{T} \sum_{t=1}^{T} Y_t$$

we may suffer from two drawbacks:

1. We might be considering data that do not carry any useful information, as they pertain to market conditions that no longer apply.

2. We assign the same weight $1/T$ to all demand observations, whereas more recent data should have larger weights; note that, in any case, weights must add up to 1.

A moving average includes only the most recent k observations:

$$\hat{B}_t = \frac{1}{k} \sum_{i=t-k+1}^{t} Y_i \qquad (11.15)$$

The coefficient k is a *time window* and characterizes the moving average. To get a grip of the sum, in particular of the +1 term in the lower limit, imagine that $k = 2$; then, at time t, after observing Y_t, we would take the average

$$\hat{B}_t = \frac{Y_{t-1} + Y_t}{2}$$

We see that the sum should start with time bucket $t - 1$, not $t - 2$. In a moving average with time window k, each observation within the last k ones has weight $1/k$ in the average. This is illustrated in Fig. 11.3. The estimate of the level is used to build a forecast. Since demand is assumed stationary, the horizon h plays no role at all, and we set

$$F_{t,h} = \hat{B}_t, \qquad h = 1, 2, 3, \ldots$$

---

[8]We should mention that moving averages are also used in some static decomposition algorithms, as a preliminary step to smooth data and expose trend and seasonality components.

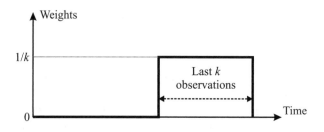

Fig. 11.3   Time window in a moving-average scheme.

Example 11.5 Let us apply a moving average with time window $k = 3$ for the dataset

$$\mathbf{Y} = [12, 20, 14, 15, 13, 18, 22, 10]$$

and compute MAD, assuming a forecast horizon of $h = 1$. We can make a first forecast only at the end of time bucket $t = 3$, after observing $Y_3 = 14$:

$$F_{3,1} = \hat{B}_3 = \frac{Y_1 + Y_2 + Y_3}{3} = \frac{12 + 20 + 14}{3} = 15.33$$

Here, $\hat{B}_3$ is the estimate of the level parameter B_t at the end of time bucket $t = 3$. Then, stepping forward, we drop $Y_1 = 12$ from the information set and include $Y_4 = 15$. Proceeding this way, we obtain the following sequence of estimates and forecasts:

$$F_{4,1} = \hat{B}_4 = \frac{Y_2 + Y_3 + Y_4}{3} = \frac{20 + 14 + 15}{3} = 16.33$$

$$F_{5,1} = \hat{B}_5 = \frac{Y_3 + Y_4 + Y_5}{3} = \frac{14 + 15 + 13}{3} = 14.00$$

$$F_{6,1} = 15.33, \quad F_{7,1} = 17.67, \quad F_{8,1} = 16.67$$

As we noticed, forecasts do not depend on the horizon; since demand is stationary, any forecast $F_{t,h}$ based on the information set up to and including time bucket t will be the same for $h = 1, 2, \ldots$. For instance, say that at the end of $t = 5$ we want to forecast demand during time bucket $t = 10$; the forecast would be simply

$$F_{5,5} = F_{5,1} = \hat{B}_5 = 14.00$$

To compute MAD, we must match forecasts and observations properly. The first forecast error that we may compute is

$$e_4 = Y_4 - F_4' = Y_4 - F_{3,1} = 15 - 15.33 = -0.33$$

By averaging absolute errors over the sample, we obtain the following MAD:

$$\mathrm{MAD} = \frac{|15 - 15.33| + |13 - 16.33| + \cdots + |10 - 17.67|}{5} = 4.4$$

Note that we have a history of 8 time buckets, but errors should be averaged only over the 5 periods on which we may calculate an error. The last forecast $F_{8,1}$ is not used to evaluate MAD, as the observation Y_9 is not available.   ▯

A standard question asked by students after seeing an example like this is:

Should we round forecasts to integer values?

Since we are observing a demand process taking integer values, it is tempting to say that indeed we should round demand forecasts. Actually, there would be two mistakes in doing so:

- The point forecast is an estimate of the *expected value* of demand. The expected value of a discrete random variable may well be noninteger.

- We are confusing forecasts and decisions. True, we cannot purchase 17.33 items to meet demand; it must be either 17 or 18. However, what if items are purchased in boxes containing 5 items? What about making a robust decision hedging against demand uncertainty? What about existing inventory on hand? The final decision will depend on a lot of factors, and the forecast is just one of the many inputs needed.

11.4.1   Choice of time window

In choosing the time window k, we have to consider a tradeoff between

- The ability to smooth (filter) noise associated with occasionally large or small demand values

- The ability to adapt to changes in market conditions that shift average demand

If k is large, the method has a lot of inertia and is not significantly influenced by occasional variability; however, it will be slow to adapt to systematic changes. On the contrary, if k is low, the algorithm will be very prompt, but also very nervous and prone to chase false signals. The difference in the behavior of a moving average as a function of time window length is illustrated in Fig. 11.4. The demand history is depicted by empty circles, whereas the squares show the corresponding forecasts (calculated one time bucket before; we assume $h = 1$). The demand history shows an abrupt jump, possibly due to opening another distribution channel. Two simulations are illustrated, with time windows $k = 2$ and $k = 6$, respectively. Note that, in plot (a), we start forecasting at the end of time bucket $t = 2$ for $t = 3$, since we assume $h = 1$, whereas in plot (b) we must wait for $t = 6$. We may also notice that, when $k = 2$, each forecast is just the vertical midpoint between the two last observations. We see that, with a shorter time window, the forecast tends to chase occasional swings; on the other hand, with a longer time window, the adaptation to the new regime is slower.

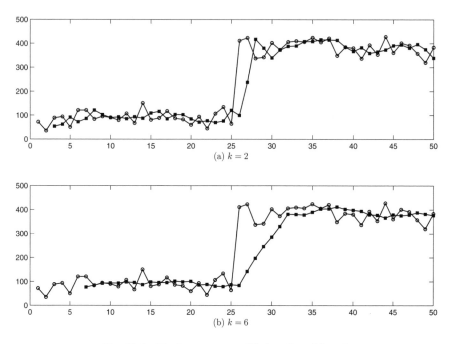

Fig. 11.4   Moving average with $k = 2$ and $k = 6$.

Moving average is a very simple approach, with plenty of applications beyond demand forecasting.[9] However, this kind of average can be criticized because of its "all or nothing" character. As we see in Fig. 11.3, the most recent observations have the same weight, $1/k$, which suddenly drops to 0. We could try a more gradual scheme in which

- Different weights are attributed to more recent observations

- A small but non-zero weight is associated with older observations

In other words, weights should decrease gradually to zero for older observations. This is exactly what the family of exponential smoothing algorithms accomplishes.

11.5   HEURISTIC EXPONENTIAL SMOOTHING

Exponential smoothing algorithms are a class of widely used forecasting methods that were born on the basis of heuristic insight. Originally, they lacked

[9]Moving averages of stock prices are used in many trading strategies based on technical analysis, whether you believe in them or not.

a proper statistical background, unlike the more sophisticated time series models that we outline in Section 11.6. More recently, attempts to justify exponential smoothing have been put forward, but the bottom line is that they proved their value over time. Indeed, there is no consistent evidence that more sophisticated methods have the upper hand, when it comes to real-life applications, at least in demand forecasting. However, heuristic approaches are less well suited to deal with other domains, such as financial markets, which indeed call for more sophistication. Apart from their practical relevance, exponential smoothing algorithms have great pedagogical value in learning the ropes of forecasting. Methods in this class have a nice intuitive appeal and, unlike moving averages, are readily adapted to situations involving trend and seasonality. One weak point that they suffer from is the need for ad hoc approaches to quantify uncertainty in forecasts; always keep in mind that a point forecast has very limited value in robust decision making, and we need to work with prediction intervals or, if possible, an estimate of a full-fledged probability distribution. In the next two sections, we illustrate the basic idea of exponential smoothing in the case of stationary demand. We also point out a fundamental issue with exponential smoothing: initialization. Then, we extend the idea to the cases of trend, multiplicative seasonality, and trend plus seasonality.

11.5.1 Stationary demand: three views of a smoother

In this section, we deal with the case of stationary demand, as represented by Eq. (11.14). In *simple exponential smoothing* we estimate the level parameter B_t by a mix of new and old information:

$$\hat{B}_t = \alpha Y_t + (1-\alpha)\hat{B}_{t-1} \qquad (11.16)$$

where α is a coefficient in the interval $[0,1]$. In (11.16), the new information consists of the last observation of demand Y_t, and the old information consists of the estimate $\hat{B}_{t-1}$, which was computed at the end of time bucket $t-1$, after observing Y_{t-1}. The updated estimate $\hat{B}_t$ is a weighted average of new and old information, depending on the smoothing coefficient α. To understand its effect, it is useful to examine the two extreme cases:

- If we set $\alpha = 1$, we forget all of the past immediately, as the new estimate is just the last observation.[10] If α is large, we are very fast to catch new trends, but we are also quite sensitive to noise and spikes in demand.

- If we set $\alpha = 0$, new information is disregarded altogether; if α is very small, there is a lot of inertia and the learning speed is quite low.

We see that α plays a role similar to the time window k in a moving average. By increasing α we make the forecaster more responsive, but also more nervous

---

[10]In fact, exponential smoothing with $\alpha = 1$ is equivalent to moving average with $k = 1$.

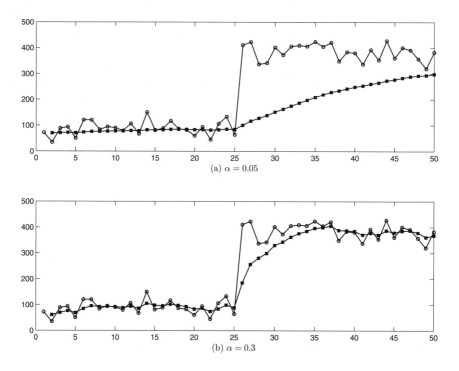

Fig. 11.5 Effect of the smoothing coefficient α, when demand jumps to a new level.

and sensitive to noise; by decreasing α, we increase inertia, and noise is better smoothed. The tradeoff is illustrated in Fig. 11.5; a relatively small smoothing coefficient ($\alpha = 0.05$) makes the adaptation to new market conditions very slow in case (a); by increasing α, the method is more responsive, as shown in case (b). In Fig. 11.6 we see the case of a sudden spike in demand. When α is small, on one hand the spike has a smaller effect; on the other hand the effect is more persistent, making the forecast a bit biased for a longer time period, as shown in plot (a). When α is large, there is an immediate effect on the forecast, which has a larger error and a larger bias, but this fades away quickly as the spike is rapidly forgotten, as shown in plot (b). In fact, the smoothing coefficient is also known as the *forgetting factor*.

Since we assume a stationary demand, the horizon h does not play any role and, just like with moving average, we have

$$F_{t,h} = \hat{B}_t, \qquad h = 1, 2, 3 \ldots$$

Equation (11.16) illustrates the way simple exponential smoothing should be implemented on a computer, but it does not shed much light on why this method earned such a name. Let us rewrite exponential smoothing in terms

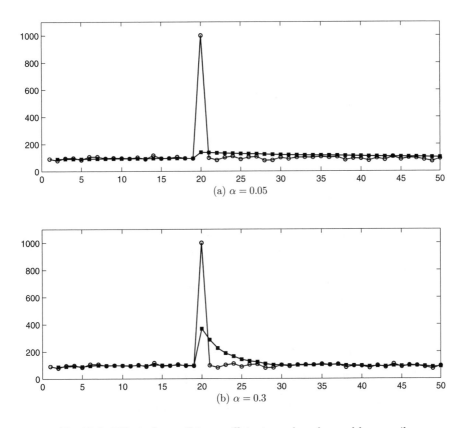

Fig. 11.6 Effect of smoothing coefficient α, when demand has a spike.

of forecasts *for* a time bucket, assuming that $h = 1$, so that

$$\hat{B}_{t-1} = F_{t-1,1} = F'_t$$

If we collect terms involving α, we get

$$F_{t,1} = \hat{B}_t = \hat{B}_{t-1} + \alpha(Y_t - \hat{B}_{t-1})$$

But $Y_t - \hat{B}_{t-1} = Y_t - F'_t = e_t$, and we get the second view of exponential smoothing

$$F_{t,1} = F_{t-1,1} + \alpha e_t \qquad (11.17)$$

This shows that the new forecast is the old one, corrected by the last forecast error, which is smoothed by the coefficient $\alpha \le 1$. The larger is α, the stronger is the correction. This shows why this algorithm is a smoother: It dampens the correction induced by the error, which could be just the result of a transient

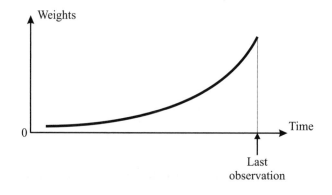

Fig. 11.7 Exponentially decaying weights in exponential smoothing.

spike. Note that the forecast is increased when the error is positive, i.e., we underforecasted demand, and decreased otherwise.

To understand where the term "exponential" comes from, we still need a third view, which is obtained by unfolding Eq. (11.16) recursively. Applying the equation at time $t-1$, we obtain

$$\hat{B}_{t-1} = \alpha Y_{t-1} + (1-\alpha)\hat{B}_{t-2}$$

Plugging this equation into (11.16) yields

$$\hat{B}_t = \alpha Y_t + \alpha(1-\alpha)Y_{t-1} + (1-\alpha)^2\hat{B}_{t-2}$$

If we apply the same reasoning to $\hat{B}_{t-2}$, $\hat{B}_{t-3}$, etc., we find

$$\hat{B}_t = \alpha Y_t + \alpha(1-\alpha)Y_{t-1} + \alpha(1-\alpha)^2 Y_{t-2} + (1-\alpha)^3\hat{B}_{t-3}$$
$$= \alpha Y_t + \alpha(1-\alpha)Y_{t-1} + \alpha(1-\alpha)^2 Y_{t-2} + \alpha(1-\alpha)^3 Y_{t-3} + (1-\alpha)^4\hat{B}_{t-4}$$
$$= \sum_{k=0}^{\infty} \alpha(1-\alpha)^k\, Y_{t-k} \tag{11.18}$$

This third view clearly shows that exponential smoothing is just another average. We leave it as an exercise for the reader to prove that weights add up to one, but they are clearly an exponential function of k, with base $(1-\alpha) < 1$. The older the observation, the lower its weight. Figure 11.7 is a qualitative display of exponentially decaying weights, and it should be compared with the time window in Fig. 11.3. The exponential decay is faster when α is increased, as the base $(1-\alpha)$ of the exponential function in Eq. (11.18) is smaller.

11.5.2 Stationary demand: initialization and choice of α

One obviously weird feature of Eq. (11.18) is that it involves an infinite sequence of observations. However, in real life we do not have an infinite number of observations; the sum must be truncated somewhere in the past, right before we started collecting information. The oldest term in the average, in practice, corresponds to the *initialization* of the algorithm. To see this, let us assume that we have observations $Y_1, \ldots, Y_T$. If we apply Eq. (11.16) at time bucket $t = 1$, we have

$$\hat{B}_1 = \alpha Y_1 + (1 - \alpha)\hat{B}_0$$

The term $\hat{B}_0$ is an initial estimate, or we should better say a *guess*, since it should be evaluated before the very first observation is collected. By applying the same idea that lead us to Eq. (11.18), we find

$$\hat{B}_T = \alpha Y_T + \alpha(1 - \alpha)Y_{T-1} + \cdots + (1 - \alpha)^T \hat{B}_0$$

Hence, after collecting T observations, the weight of the initial estimate $\hat{B}_0$ in the average is $(1 - \alpha)^T$. If α is small, this initial term has a slowly decaying impact, for increasing T, and it may play a critical role, as a bad initialization will be forgotten too slowly.

To deal with initialization, we should clearly frame the problem:

1. If we are applying exponential smoothing online, to a brand-new product with no past sales history, we must make some educated guess, possibly depending on past sales of a similar product. Arguably, α should be larger at the beginning, since we must learn rapidly, and then it could be reduced when more and more information is collected.

2. If we are working online, but we are forecasting demand for an item with a significant past sales history, we may use old observations to initialize the smoother, but we should do it properly.

3. By the same token, initialization must be carried out properly when working offline and evaluating performance by *historical simulation*.

To see what "properly" means in case 2 above, imagine that we are at the end of time bucket $t = T$, and we have an information set consisting of T past observations:

$$Y_1, Y_2, \ldots, Y_{T-2}, Y_{T-1}, Y_T$$

We are working online and want to forecast demand for $t = T + 1$, and one possible way to initialize the smoother is to compute the sample mean of these observations, setting

$$\hat{B}_T = \overline{Y} = \frac{1}{T}\sum_{k=1}^{T} Y_T$$

Then, our forecast for Y_{T+1} would be $F_{T,1} = \hat{B}_T$. However, in doing so, we are forgetting the very nature of exponential smoothing, since the sample mean assigns the same weight to all of the observations. We should try to set up a situation as similar as possible to that of a forecasting algorithm that has been running forever. The way to do so is to apply exponential smoothing on the past data, from Y_1 to Y_T, using the sample mean $\overline{Y}$ as an initialization of $\hat{B}_0$. Apparently, we are playing dirty here, since we use past data to come up with an initialization that is shifted back to the end of time bucket $t = 0$. However, there is nothing wrong with this, as long as we do *not* compute forecast errors to evaluate performance.

The last remark is also essential when carrying out historical simulation to evaluate performance. Assume again that we have a sample of T observations Y_t, $t = 1, \ldots, T$, and that the size of the fit sample is $\tau < T$. We partition the sample as follows:

$$\underbrace{Y_1, Y_2, \ldots, Y_\tau,}_{\text{fit sample}} \underbrace{Y_{\tau+1}, Y_{\tau+2}, \ldots, Y_T}_{\text{test sample}}$$

Then, the correct procedure is:

1. Use observations in the fit sample to initialize $\hat{B}_0$

2. Apply exponential smoothing on the fit sample, updating parameter estimates as prescribed, without calculating errors

3. Proceed on the test sample, collecting errors

4. Evaluate the selected forecast error measures

The reader could wonder why we should carry out step 2; a tempting shortcut is to proceed directly with the application of the smoother on the test sample. A good way to get the important message is by referring back to Monte Carlo simulation of a queueing system (Section 9.2.1). In that case, we should discard the initial part of the simulation, which is just a transient phase, if the queueing system starts empty, and gather statistics only in steady state. The same consideration applies here. By running the smoother on the fit sample, we warm the system up and forget the initialization. We illustrate the idea in the example below.

Example 11.6 Consider a product whose unit purchase cost is $10, is sold for $13, and, if unsold within a shelf life of one week, is scrapped, with a salvage value of $5. We want to decide how many items to buy, based on the demand history reported in Table 11.4. If we were convinced that demand is stationary, i.e., its expected value and standard deviation do not change, we could just fit a probability distribution like the normal, based on sample mean and sample standard deviation. Then, the newsvendor model would provide us with the answer we need. However, if expected demand might shift in time, exponential smoothing can be applied. To see this, imagine taking the standard sample statistics when the demand shows an abrupt jump, like the demand history depicted in Fig. 11.5. We recall that the underlying demand model is

$$Y_t = B_t + \epsilon_t$$

Table 11.4 Application of exponential smoothing to a newsvendor problem.

t	Y_t	$\hat{B}_t$	e_t	e_t^2
0	–	118.4000	–	–
1	99	116.4600	–	–
2	94	114.2140	–	–
3	122	114.9926	–	–
4	138	117.2933	–	–
5	139	119.4640	–	–
6	70	114.5176	−49.4640	2446.6879
7	103	113.3658	−11.5176	132.6552
8	58	107.8293	−55.3658	3065.3768
9	61	103.1463	−46.8293	2192.9796
10	100	102.8317	−3.1463	9.8994
11	154	107.9485	51.1683	2618.1948
12	73	104.4537	−34.9485	1221.3998
13	113	105.3083	8.5463	73.0396
14	92	103.9775	−13.3083	177.1111
15	139	107.4797	35.0225	1226.5770

Imagine that, at some time bucket $t = t^*$, the level jumps from value B' to a larger value B''. If we knew the "true" value of the level B_t at each time bucket, demand uncertainty would be related only to the standard deviation σ_ϵ of the unpredictable component ϵ_t. If we calculate sample mean and standard deviation mixing the two components, the jump in B_t would result in an estimate of σ_ϵ that is much larger than true value; the sample mean would be halfway between two values B' and B''. In practice, we need to update dynamically the estimates of both expected value and standard deviation; hence, let us apply exponential smoothing, with $\alpha = 0.1$ and a fit sample of 5 time buckets.

The calculations are displayed in Table 11.4. On the basis of the fit sample, the initial estimate of level $\hat{B}_0$ is

$$\hat{B}_0 = \frac{99 + 94 + 122 + 138 + 139}{5} = 118.4$$

The first demand observation is used to update the estimate of level:

$$\hat{B}_1 = 0.1 \times 99 + 0.9 \times 118.4 = 116.46$$

Note that we should *not* compute an error for the first time bucket, comparing $Y_1 = 99$ against $F_{0,1} = \hat{B}_0 = 118.4$, since the demand observation Y_1 itself has been used to initialize the estimate; incidentally, using a fit sample of size 1, the first error would be zero. It would also be tempting to run the smoother starting from time bucket $t = 6$, the first time bucket in the test sample. However, it is better to "forget" the initialization and the consequent transient phase, by running the smoother through the full fit sample, without collecting errors. We proceed using the same mechanism and, after observing $Y_5 = 139$, we update the estimate again:

$$\hat{B}_5 = \alpha Y_5 + (1 - \alpha)\hat{B}_4 = 0.1 \times 139 + 0.9 \times 117.2933 = 119.4640$$

Now we may start forecasting and calculating errors:

$$e_6 = 70 - 119.4640 = -49.4640$$

Please note that we should compare $Y_6 = 70$ with the forecast in the line above in the table, $F_{5,1} = 119.4640$. Of course, you could arrange the table in such a way that each demand observation and the related forecast are on the same line. At the end of the test sample, the last forecast is

$$F_{15,1} = 107.4797$$

but this is not used in assessing performance, since observation Y_{16} is not available; however, we use it as an estimate $\hat{\mu}$ of the expected value of demand, valid for time bucket $t = 16$. In so doing, we implicitly assume that our estimates are unbiased.

Now we also need an estimate $\hat{\sigma}$ of the standard deviation. Since we cannot use sample standard deviation, we resort to using RMSE as an estimator. In order to evaluate RMSE, we square errors, as shown in the last column of Table 11.4. We obtain

$$\text{RMSE} = \sqrt{\frac{(-49.4640)^2 + (-11.5176)^2 + \cdots + (35.0225)^2}{10}} = 36.2821$$

Hence, we assume that demand in time bucket $t = 16$ is normally distributed with the following expected value and standard deviation, respectively:

$$\hat{\mu} = 107.4797, \qquad \hat{\sigma} = 36.2821$$

Now we apply the standard newsvendor model. The service level is calculated from the economics of the problem:[11]

$$\frac{m}{m + c} = \frac{3}{3 + 5} = 0.3750$$

corresponding to a quantile $z_{0.375} = -0.3186$. The order quantity should be

$$\hat{\mu} + z_{0.375}\hat{\sigma} = 107.4797 - 0.3186 \times 36.2821 = 95.9202 \approx 96$$

[11] See Section 7.4.4.

After collecting more observations, the above estimates would be revised; hence, the order quantity is not constant over time.   ▯

Apart from initialization, another issue in exponential smoothing is the choice of the smoothing coefficient α. The example suggests one possible approach, based on historical simulation: Choose the coefficient by minimizing a selected error measure. However, a possibly better strategy is dynamic adaptation. When dealing with a new product, we said that we could start with a larger α in order to forget a possibly bad initial guess rapidly; then, we could reduce α in order to shift emphasis toward noise filtering. Another element that we should take into consideration is bias. In Fig. 11.5 we have seen that the smoother can be slow to adapt to new market conditions, when an abrupt change occurs. The effect is that forecasts become systematically biased, which is detected by a mean error which is significantly different from zero. A strategy that has been proposed in the literature is to increase α when mean error is significantly different from zero. However, bias can also be the effect of a wrong demand model. Simple exponential smoothing assumes stationary demand, but a systematic error will result if, for instance, a strong upward or downward trend is present.

11.5.3   Smoothing with trend

Demand may exhibit additive trend components that, in a static case, could be represented by the following demand model

$$D_t = B + tT + \epsilon_t$$

where B is the level and T is the trend. Looking at the demand model, linear regression seems a natural candidate to estimate these two parameters. However, level and trend might change over time, suggesting the opportunity of a dynamic demand model and an adaptation of exponential smoothing. Given estimates $\hat{B}_t$ and $\hat{T}_t$ of level and trend, respectively, at the end of time bucket t, after observing Y_t, the demand forecast with horizon h is

$$F_{t,h} = \hat{B}_t + h\hat{T}_t \tag{11.19}$$

Note that, unlike simple moving average and exponential smoothing, the forecast horizon does play a role here, as it multiplies trend. The following adaptation of exponential smoothing is known as *Holt's linear method*:

$$\hat{B}_t = \alpha Y_t + (1 - \alpha)(\hat{B}_{t-1} + \hat{T}_{t-1}) \tag{11.20}$$

$$\hat{T}_t = \beta(\hat{B}_t - \hat{B}_{t-1}) + (1 - \beta)(\hat{T}_{t-1}) \tag{11.21}$$

Here α, β are two smoothing coefficients in the range $[0, 1]$. Comparing Eq. (11.20) against simple exponential smoothing, we note that we cannot update $\hat{B}_t$ solely on the basis of $\hat{B}_{t-1}$ as the two values are not directly comparable when trend is involved. In fact, a forecast for Y_t, based on the information set up to $t - 1$, would be

$$F_{t-1,1} = \hat{B}_{t-1} + \hat{T}_{t-1}$$

Then, by applying the same error correction logic of Eq. (11.17), we should adapt the estimate of level as follows:

$$\hat{B}_t = (\hat{B}_{t-1} + \hat{T}_{t-1}) + \alpha[Y_t - (\hat{B}_{t-1} + \hat{T}_{t-1})]$$

which leads to Eq. (11.20). Equation (11.21) uses a smoothing coefficient β to update the estimate of trend. In this case, the new information consists of the difference between the two last estimates of level, $\hat{B}_t - \hat{B}_{t-1}$. We might wonder if we should not use the growth in demand, $Y_t - Y_{t-1}$, in updating trend estimates. Indeed, this is a possible alternative; however, the difference in demand might be oversensitive to noise, whereas the difference in estimated levels is more stable. This might be a disadvantage, however, when trend changes, since adaptation could be too slow. This can be countered by using a larger coefficient β, which is sensible since the level estimates are by themselves noise smoothers.

We may appreciate the flexibility of the exponential smoothing framework, which can be easily and intuitively adapted to different demand models. However, a careful look at Eq. (11.19) suggests a potential danger when h is large: We should not extrapolate a trend too much in the future. In particular, with a negative trend, this could even result in a negative demand forecast. The above formulas apply to an *additive* trend, but when demand is low and trend is negative, a model with *multiplicative* trend should be adopted. In such a model, the demand forecast is

$$F_{t,h} = \hat{B}_t \cdot (\hat{T}_t)^h$$

and a decreasing trend is represented by $\hat{T}_t < 1$; this corresponds to a percentage decrease in demand, and it prevents negative forecasts. We will not give update formulas for multiplicative trend, but they can be found in the specialized literature.

Exponential smoothing with trend presents the same initialization issues that we have encountered with stationary demand. Given a fit sample, there are alternative initialization methods:

- Linear regression on the fit sample

- Heuristic approaches based on time series decomposition

Whatever approach we take, the minimal fit sample consists of two demand observations. From a mathematical perspective, it is impossible to estimate two parameters with just one observation; a simpler view is that with one observation there is no way to estimate a trend. Also note that, if we select this minimal fit sample, we will just fit a line passing through two points. On this fit sample, errors will be zero, because with two parameters and two observations we are able to find an exact fit. This reinforces the view that errors should not be computed on the fit sample.

11.5.4 Smoothing with multiplicative seasonality

In this section we consider the case of pure seasonality. Forecasts are based on the demand model of Eq. (11.13), in which the trend parameter is set to $\hat{T}_t = 0$:

$$F_{t,h} = \hat{B}_t \cdot \hat{S}_{t+h-s} \tag{11.22}$$

where s is the length of the seasonal cycle, i.e., a whole cycle consists of s time buckets.[12] To get a grip of this model, imagine a yearly cycle consisting of 12 monthly time buckets, and say that we are at the end of December of year X. Then, $t = 0$ is December of year X and time bucket $t = 1$ corresponds to January of year $X + 1$. On the basis of the estimate $\hat{B}_0$, how can we forecast demand in January? Of course, we cannot take $\hat{B}_0 \cdot \hat{S}_0$, since this involves the seasonal factor of December. We should multiply $\hat{B}_0$ by the seasonal factor of January, but this is *not* $\hat{S}_1$, because $\hat{S}_1$ is the estimate of this seasonal factor after observing Y_1. Since $s = 12$ and $h = 1$, the correct answer is

$$F_{0,1} = \hat{B}_0 \cdot \hat{S}_{0+1-12} = \hat{B}_0 \cdot \hat{S}_{-11}$$

where $\hat{S}_{-11}$ is the estimate of the seasonal factor of January at the end of January in year X. We should use seasonal factors estimated one year ago! After observing Y_1, i.e., January demand in year $X + 1$, we can update $\hat{S}_1$, which will be used to forecast demand for January in year $X + 2$. It is easy to devise exponential smoothing formulas to update estimates of both level and seasonality factors:

$$\hat{B}_t = \alpha \frac{Y_t}{\hat{S}_{t-s}} + (1 - \alpha)\hat{B}_{t-1} \tag{11.23}$$

$$\hat{S}_t = \gamma \frac{Y_t}{\hat{B}_t} + (1 - \gamma)\hat{S}_{t-s} \tag{11.24}$$

Here α and γ are familiar smoothing coefficients in the range $[0, 1]$. Equation (11.23) is an adaptation of simple exponential smoothing, in which demand Y_t is *deseasonalized* by dividing it by the current estimate of the correct seasonal factor, $\hat{S}_{t-s}$. To see why this is needed, think of ice cream demand in summer; we should not increase the estimate of level after observing high sales, as this

---

[12]To be precise, Eq. (11.22) applies only if the forecast horizon does not exceed the cycle length, i.e., if $h \leq s$; a more general formulation is $F_{t,h} = \hat{B}_t \cdot \hat{S}_{t+h-\lfloor\{(h-1)/s\}+1\rfloor \cdot s}$, where $\lfloor x \rfloor$ is the "floor" operator, rounding down x to the nearest integer.

is just a seasonal effect. The need for deseasonalization is quite common in many application domains. Equation (11.24) takes care of updating the seasonal factor of time bucket t, based on the previous estimate $\hat{S}_{t-s}$, and the new information on the seasonal factor, which is obtained by dividing the last observation by the revised level estimate.

Initialization requires fitting the parameters $\hat{B}_0$ and $\hat{S}_{s-j}$, for $j = 1, 2, \ldots, s$. We need $s+1$ parameters, but actually the minimal fit sample consists of just s observations, i.e., a whole cycle. In fact, we lose one degree of freedom because the average value of multiplicative seasonality factors must be 1. Of course, the more cycles we use, the better the fit will be. Assuming that the fit sample consists of k full cycles, we have $l = k \cdot s$ observations. A reasonable way to fit parameters in this case is the following:

1. The initial estimate of level is set to average demand over the fit sample:

$$\hat{B}_0 = \frac{1}{l} \sum_{j=1}^{l} Y_j \qquad (11.25)$$

Note that we cannot take this plain average if the fit sample does not consist of full cycles; when doing so in such a case, we would overweight some seasons within the cycle.

2. Seasonal factors are estimated by dividing average demand of time buckets corresponding to each season within the fit sample, divided by the level estimate:

$$\hat{S}_{j-s} = \frac{1}{k\hat{B}_0} \sum_{\tau=0}^{k-1} Y_{j+\tau s} \qquad \text{for} \quad j = 1, \ldots, s \qquad (11.26)$$

where $k = l/s$ is the number of full cycles in the fit sample. To understand the sum above, say that we have $k = 3$ full cycles, each one consisting of $s = 4$ time buckets. Then, the seasonal factor for the first season within the cycle is estimated as

$$\hat{S}_{-3} = \frac{Y_1 + Y_5 + Y_9}{3\hat{B}_0}$$

Example 11.7 Table 11.5 shows demand data for 6 time buckets. We assume $s = 3$, so what we see is a history consisting of two whole cycles. We want to initialize the smoother with a fit sample consisting of one cycle, and evaluate MAPE on a test sample consisting of the second cycle, applying exponential smoothing with coefficients $\alpha = 0.1$ and $\gamma = 0.2$. Initialization

Table 11.5 Applying exponential smoothing with multiplicative seasonality.

t	Y_t	$\hat{B}_t$	$\hat{S}_t$	F'_t	e_t	$|e_t|/Y_t$
-2	–	–	1.4126	–	–	–
-1	–	–	0.9821	–	–	–
0	–	74.3333	0.6054	–	–	–
1	105	74.3333	1.4126	105.0000	–	–
2	73	74.3333	0.9821	73.0000	–	–
3	45	74.3333	0.6054	45.0000	–	–
4	117	75.1829	1.4413	105.0000	12.0000	10.26%
5	81	75.9125	0.9991	73.8343	7.1657	8.85%
6	39	74.7635	0.5886	45.9560	−6.9560	17.84%

yields the following parameters:

$$\hat{B}_0 = \frac{105 + 73 + 45}{3} = 74.3333$$

$$\hat{S}_{-2} = \frac{105}{74.3333} = 1.4126$$

$$\hat{S}_{-1} = \frac{73}{74.3333} = 0.9821$$

$$\hat{S}_0 = \frac{45}{74.3333} = 0.6054$$

A useful check is

$$1.4126 + 0.9821 + 0.6054 = 3$$

If we apply Eqs. (11.23) and (11.24) after the first observation, we obtain

$$\hat{B}_1 = 0.1 \times \frac{105}{1.4126} + 0.9 \times 74.3333 = 74.3333$$

$$\hat{S}_1 = 0.2 \times \frac{105}{74.3333} + 0.8 \times 1.4126 = 1.4126$$

We note that estimates do not change! A closer look reveals that the first forecast would have been

$$F_{0,1} = \hat{B}_0 \cdot \hat{S}_{-2} = 74.3333 \times 1.4126 = 105$$

We do not update estimates, because the forecast was perfect. By a similar token, errors are zero and parameters are not updated for time buckets $t = 2$ and $t = 3$. On second thought, this is no surprise: We have used a model with four parameters, and actually 3 degrees of freedom, to match three observations. Of course, we obtain a perfect match, and errors are zero throughout the fit sample. All the good reasons for not calculating errors there! Things get interesting after observing Y_4, and the calculations in Table 11.5 yield

$$\text{MAPE} = 12.31\%$$

As a further exercise, let us compute forecasts $F_{6,2}$ and $F_{6,3}$:

$$F_{6,2} = \hat{B}_6 \cdot \hat{S}_{6+2-3} = \hat{B}_6 \cdot \hat{S}_5 = 74.7635 \times 0.9991 = 74.6927$$
$$F_{6,3} = \hat{B}_6 \cdot \hat{S}_{6+3-3} = \hat{B}_6 \cdot \hat{S}_6 = 74.7635 \times 0.5886 = 44.0083$$

⬜

Once again, the example shows the danger of calculating errors within the fit sample; performance evaluation must always be carried out out-of-sample to be fair. Of course, we will not incur such a blatant mistake with a larger fit sample. In the literature, a *backward initialization* is often suggested. The idea is to run the smoother backward in time, starting from the last time bucket, in order to obtain initialized parameters. Doing so with a large data set will probably avoid gross errors, but a clean separation between fit and test samples is arguably the wisest idea.

11.5.5 Smoothing with trend and multiplicative seasonality

The last exponential smoothing approach we consider puts everything together and copes with additive trend and multiplicative seasonality. The *Holt–Winter method* is based on Eq. (11.13), which we repeat for convenience:

$$F_{t+h} = (\hat{B}_t + h\hat{T}_t) \cdot \hat{S}_{t+h-s} \tag{11.27}$$

The overall scheme uses three smoothing coefficients and it proceeds as follows

$$
\begin{aligned}
\hat{B}_t &= \alpha \frac{Y_t}{\hat{S}_{t-s}} + (1-\alpha)(\hat{B}_{t-1} + \hat{T}_{t-1}) \\
\hat{T}_t &= \beta(\hat{B}_t - \hat{B}_{t-1}) + (1-\beta)(\hat{T}_{t-1}) \\
\hat{S}_t &= \gamma \frac{Y_t}{\hat{B}_t} + (1-\gamma)\hat{S}_{t-s}
\end{aligned}
$$

All of the remarks we have made about simpler versions of exponential smoothing apply here as well, including initialization. To initialize the method, we need $s + 2$ parameters, which must be estimated on the basis of at least $s + 1$ demand observations, since seasonal factors are not independent. Heuristic initialization approaches, based on time series decomposition, are described in the references at the end of the chapter. Alternatively, we could use a formal approach based on least-squares-like optimization models, as illustrated in Example 11.4.

11.6   A GLANCE AT ADVANCED TIME SERIES MODELING

The class of exponential smoothing methods was born out of heuristic intuition, even though methodological frameworks were later developed to provide them with a somewhat more solid justification. Despite these efforts, exponential smoothing methods do suffer from at least a couple of drawbacks:

- They are not able to cope with correlations between observations over time; for instance, we might have a demand process with positive or negative autocorrelation, formally defined below, and this must be considered by forecasting procedures.

- They do not offer a clear way to quantify uncertainty; using RMSE as an estimate of standard deviation, as we did in Example 11.6, may be a sensible heuristic, but again it may be inadequate when autocorrelation is an issue.

It is also worth noting that simple linear regression models share some of the limitations above, as standard OLS estimates do assume uncorrelated errors. If necessary, we may resort to an alternative body of statistical theory that deals with formal time series models. In this section we give a flavor of the theory, by outlining the two basic classes of such models: *moving-average* and *autoregressive processes*. They can be integrated in the more general class of ARIMA (autoregressive integrated moving average) processes, also known as Box–Jenkins models. As we shall see, time series modeling offers many degrees of freedom, maybe too many; when observing a time series, it may be difficult to figure out which type of model is best-suited to capture the essence of the underlying process. Applying time series requires the following steps:

1. *Model identification.* We should first select a model structure, i.e., its type and its order.

2. *Parameter estimation.* Given the qualitative model structure, we must fit numerical values for its parameters. We will not delve into the technicalities of parameter estimation for time series models, but this step relies on the statistical tools that we have developed in Section 9.9.

3. *Forecasting and decision making.* In the last step, we must make good use of the model to come up with a forecast and a quantification of its uncertainty, in order to find a suitably robust decision.

All of the above complexity must be justified by the application at hand. Time series models are definitely needed in quantitative finance, maybe not in demand forecasting. We should also mention that, quite often, a quantitative forecast must be integrated with qualitative insights and pieces of information; if so, a simpler and more intuitive model might be easier to twist as needed.

In its basic form, time series theory deals with *weakly stationary processes*, i.e., time series Y_t with the following properties, related to first- and second-order moments:

1. The expected value of Y_t does not change in time: $E[Y_t] = \mu$.

2. The covariance between Y_t and Y_{t+k} depends only on time lag k.

The second condition deserves some elaboration.

DEFINITION 11.1 (Autocovariance and autocorrelation)
Given a weakly stationary stochastic process Y_t, the function

$$\gamma_Y(k) = \mathrm{Cov}(Y_t, Y_{t+k})$$

is called **autocovariance of the process with time lag** k. *The function*

$$\rho_Y(k) = \frac{\gamma_Y(k)}{\sigma^2}$$

is called the **autocorrelation function** *(ACF).*

The definition of autocorrelation relies on the fact that variance is constant as well:

$$\rho_Y(k) = \rho(Y_t, Y_{t+k}) = \frac{\mathrm{Cov}(Y_t, Y_{t+k})}{\sqrt{\mathrm{Var}(Y_t)}\sqrt{\mathrm{Var}(Y_{t+k})}} = \frac{\gamma_Y(k)}{\sigma^2}$$

In practice, autocorrelation may be estimated by the sample autocorrelation function (SACF), given a sample path Y_t, $t = 1, \ldots, T$:

$$R_k = \frac{\displaystyle\sum_{t=k+1}^{T} \left(Y_t - \overline{Y}\right)\left(Y_{t-k} - \overline{Y}\right)}{\displaystyle\sum_{t=1}^{T}\left(Y_t - \overline{Y}\right)^2} \tag{11.28}$$

where $\overline{Y}$ is the sample mean of Y_t. The expression in Eq. (11.28) may not look quite convincing, since the numerator and the denominator are sums involving a different number of terms. In particular, the number of terms in the numerator is decreasing in the time lag k. Thus, the estimator looks biased and, for a large value of k, R_k will vanish. However, this is what one expects in real life. Furthermore, although we could account for the true number of terms involved in the numerator, for large k the sum involves very few terms and is not reliable. Indeed, the form of sample autocorrelation in Eq. (11.28) is what is commonly used in statistical software packages, even

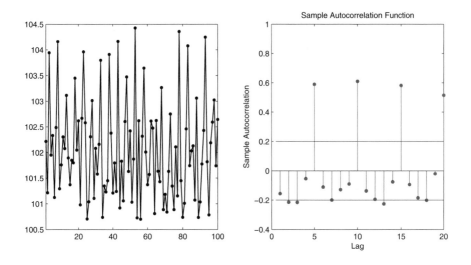

Fig. 11.8 A seasonal time series (left) and its autocorrelogram (right).

though alternatives have been proposed.[13] If T is large enough, under the null hypothesis that the true autocorrelations ρ_k are zero, for $k \geq 1$, the statistic $\sqrt{T} R_k$ is approximately normal standard. Since $z_{0.99} = 1.96 \approx 2$, a commonly used approximate rule states that if

$$| R_k | > \frac{2}{\sqrt{T}} \qquad (11.29)$$

the sample autocorrelation at lag k is statistically significant. For instance, if $T = 100$, autocorrelations outside the interval $[-0.2, 0.2]$ are significant. We should keep in mind that this is an approximate result, holding for a large number T of observations. We may plot the sample autocorrelation function at different lags, obtaining an *autocorrelogram* that can be most useful in pointing out hidden patterns in data.

Example 11.8 (Detecting seasonality with autocorrelograms) Consider the time series displayed in figure 11.8. A cursory look at the plot may not suggest much structure in the data, but the autocorrelogram does. The autocorrelogram displays two horizontal lines defining a band, outside which autocorrelation is statistically significant; notice that $T = 100$ and the two horizontal lines are set at -0.2 and 0.2, respectively; this is consistent with

[13]The denominator in Eq. (11.28) is related to the estimate of autocovariance. In the literature, sample autocovariance is typically obtained by dividing the sum by T, even though this results in a biased estimator. If we divide by $T - k$, we might obtain an autocovariance matrix which is not positive semidefinite; see pp. 220-221 of Brockwell and Davis [3].

Eq. (11.29). We notice that sample autocorrelation is stronger at time lags 5, 10, 15, and 20. This suggests that there is some pattern in the data. In fact, the time series was generated by sampling the following process:

$$Y_t = B \cdot S_{\mod(t,5)+1} \cdot e^{0.02\epsilon_t}, \qquad t = 1, \dots, 100$$

where:

- ϵ_t is a sequence of independent, standard normal variables; taking the exponential of ϵ_t makes sure that we have positive values.

- $\mod(t, 5)$ denotes the remainder of integer division of t by 5; actually, this is just a process with a seasonal cycle of length 5, with parameters corresponding to the following level and seasonal factors:

$$B = 100,$$
$$S_1 = 0.6410, S_2 = 0.3205, S_3 = 0.9615, S_4 = 2.4359, S_5 = 0.6410$$

We see that an autocorrelogram is a useful tool to spot seasonality and other hidden patterns in data. ▯

Another important building block in time series modeling is *white noise*, denoted by ϵ_t in the following. This is just a sequence of i.i.d. random variables. If they are normal, we have a Gaussian white noise. As we have mentioned, the first step in modeling a time series is the identification of the model structure. In the following we outline a few basic ideas that are used to this aim, pointing out the role of autocorrelation in the analysis.

11.6.1 Moving-average processes

A *finite-order moving-average process of order q*, denoted by MA(q), can be expressed as

$$Y_t = \mu + \epsilon_t - \theta_1 \epsilon_{t-1} - \theta_2 \epsilon_{t-2} - \dots - \theta_q \epsilon_{t-q}$$

where random variables ϵ_t are white noise, with $\mathrm{E}[\epsilon_t] = 0$ and $\mathrm{Var}(\epsilon_t) = \sigma^2$. These variables play the role of *random shocks* and drive the process. It is fairly easy to see that the process is weakly stationary. A first observation is that expected value and variance are constant:

$$\mathrm{E}[Y_t] = \mu + \mathrm{E}[\epsilon_t] - \theta_1 \mathrm{E}[\epsilon_{t-1}] - \dots - \theta_q \mathrm{E}[\epsilon_{t-q}] = \mu$$
$$\mathrm{Var}(Y_t) = \mathrm{Var}(\epsilon_t) + \theta_1^2 \mathrm{Var}(\epsilon_{t-1}) + \dots + \theta_q^2 \mathrm{Var}(\epsilon_{t-q}) = \sigma^2 \left(1 + \theta_1^2 + \dots + \theta_q^2\right)$$

The calculation of autocovariance is a bit more involved, but we may take advantage of the uncorrelation of white noise:

$$
\begin{aligned}
\gamma_Y(k) &= \mathrm{Cov}(Y_t, Y_{t+k}) \\
&= \mathrm{E}\left[(\epsilon_t - \theta_1 \epsilon_{t-1} - \dots - \theta_q \epsilon_{t-q})(\epsilon_{t+k} - \theta_1 \epsilon_{t+k-1} - \dots - \theta_q \epsilon_{t+k-q})\right] \\
&= \begin{cases} \sigma^2 \left(-\theta_k + \theta_1 \theta_{k+1} + \dots + \theta_{q-k} \theta_q\right) & k = 1, 2, \dots, q \\ 0 & k > q \end{cases}
\end{aligned}
$$

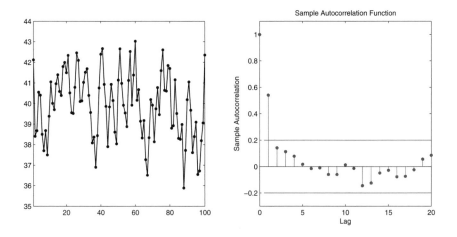

Fig. 11.9 Sample path and corresponding SACF for the moving-average process $Y_t = 40 + \epsilon_t + 0.8\epsilon_{t-1}$.

As a consequence, the autocorrelation function is

$$\rho_Y(k) = \frac{\gamma_Y(k)}{\gamma_Y(0)} = \begin{cases} \dfrac{-\theta_k + \theta_1\theta_{k+1} + \cdots + \theta_{q-k}\theta_q}{1 + \theta_1^2 + \cdots + \theta_q^2}, & k = 1, 2, \ldots, q \\ 0, & k > q \end{cases} \quad (11.30)$$

Thus, the autocorrelation function depends only on the lag k. We also notice that the autocorrelation function cuts off for lags larger than the order of the process. This makes sense, since the process Y_t is a moving average of the driving process ϵ_t. Hence, by checking whether the sample autocorrelation function cuts off after a time lag, we may figure out whether a time series can be modeled as a moving average, as well as its order q. Of course, the sample autocorrelation will not be exactly zero for $k > q$; nevertheless, by using the autocorrelogram and its significance bands, we may get some clue.

Example 11.9 Let us consider a simple MA(1) process

$$Y_t = 40 + \epsilon_t + 0.8\epsilon_{t-1}$$

where ϵ_t is a sequence of uncorrelated standard normal variables (Gaussian white noise). In Fig. 11.9 we show a sample path obtained by Monte Carlo simulation, and the corresponding sample autocorrelogram. The sample autocorrelation looks significant at time lag 1, which is expected, given the nature of the process. Note that, by applying Eq. (11.30), we find that the autocorrelation function, for a MA(1) process $Y_t = \mu + \epsilon_t - \theta_1\epsilon_{t-1}$, is

$$\rho_Y(1) = \frac{-\theta_1}{1 + \theta_1^2} \quad (11.31)$$

$$\rho_Y(k) = 0, \qquad k > 1 \quad (11.32)$$

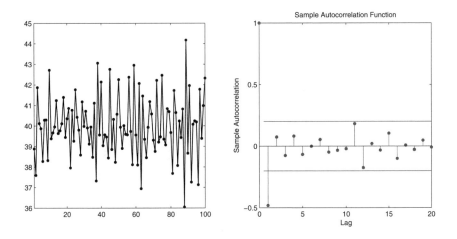

Fig. 11.10 Sample path and corresponding SACF for the moving-average process $Y_t = 40 + \epsilon_t - 0.8\epsilon_{t-1}$.

Figure 11.10 shows the sample path and autocorrelogram of a slightly different MA(1) process:

$$Y_t = 40 + \epsilon_t - 0.8\epsilon_{t-1}$$

The change in sign in θ_1 has an effect on the sample path; an upswing tends to be followed by a downswing, and vice versa. The autocorrelogram shows a cutoff after time lag 1, and a negative autocorrelation.

If we increase the order of the process, we should expect more significant autocorrelations. In Fig. 11.11, we repeat the exercise for the MA(2) process

$$Y_t = 40 + \epsilon_t + 0.9\epsilon_{t-1} + 0.5\epsilon_{t-2}$$

We notice that, in this case, the autocorrelation function cuts off after time lag $k = 2$. 　　　　　　　　　　　　　　　　　　　　　　　□

We should mention that sample autocorrelograms are a statistical tool. It may well be the case that, for the moving-average processes in the example, we get a different picture. This is a useful experiment to carry out with the help of statistical software.

11.6.2   Autoregressive processes

In finite-order moving-average processes, only a finite number of past realizations of white noise influence the value of Y_t. This may be a limitation for those processes in which all of the previous realizations have an effect, even though this possibly fades in time. This consideration led us from forecasting using simple moving averages to exponential smoothing. In principle,

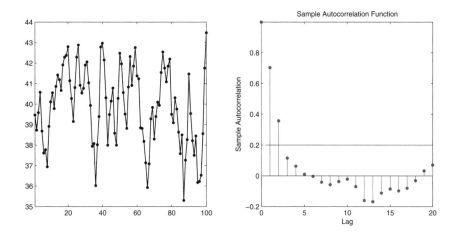

Fig. 11.11 Sample path and corresponding SACF for the moving-average process $Y_t = 40 + \epsilon_t + 0.9\epsilon_{t-1} + 0.5\epsilon_{t-2}$.

we could consider an infinite-order moving-average process, but having to do with an infinite sequence of θ_{t-k} coefficients does not sound quite practical. Luckily, under some technical conditions, such a process may be rewritten in a compact form involving time-lagged realizations of the Y_t itself. This leads us to the definition of an *autoregressive process* of a given order. The simplest such process is the autoregressive process of order 1, AR(1):

$$Y_t = \delta + \phi Y_{t-1} + \epsilon_t \qquad (11.33)$$

One could wonder under which conditions this process is stationary, since we cannot use the same arguments as in the moving-average case. A *heuristic* argument to find the expected value $\mu = \mathrm{E}[Y_t]$ is based on taking expectations and dropping the time subscript in Eq. (11.33):

$$\mu = \delta + \phi\mu \quad \Rightarrow \quad \mu = \frac{\delta}{1 - \phi}$$

The argument is not quite correct, as it leads to a sensible result if the process is indeed stationary, which is the case if $|\phi| < 1$. Otherwise, intuition suggests that the process will grow without bounds. The reasoning can be made precise by using the infinite-term representation of Y_t, which is beyond the scope of this book. Using the correct line of reasoning, we may also prove that

$$\gamma_Y(k) = \frac{\sigma^2 \phi^k}{1 - \phi^2}, \qquad k = 0, 1, 2, \ldots$$

In particular

$$\mathrm{Var}(Y_t) = \gamma_Y(0) = \frac{\sigma^2}{1 - \phi^2}$$

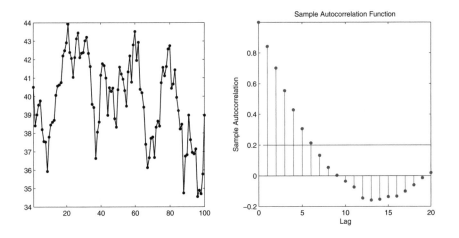

Fig. 11.12 Sample path and corresponding SACF for the autoregressive process $Y_t = 8 + 0.8Y_{t-1} + \epsilon_t$.

and we may also observe that, for a stationary AR(1) process,

$$\rho_Y(k) = \frac{\gamma_Y(k)}{\gamma_Y(0)} = \phi^k, \qquad k = 0, 1, 2, \dots \tag{11.34}$$

We notice that autocorrelation is decreasing, but it fades away with no sharp cutoff.

Example 11.10 In Figs. 11.12 and 11.13, we show a sample path and the corresponding sample autocorrelogram for the two AR(1) processes

$$Y_t = 8 + 0.8Y_{t-1} + \epsilon_t \quad \text{and} \quad Y_t = 8 - 0.8Y_{t-1} + \epsilon_t$$

respectively. Notice that the change in sign in the ϕ coefficient has a significant effect on the sample path, as well as on autocorrelations. In the first case, autocorrelation goes to zero along a relatively smooth path.[14] The sample path of the second process features evident up- and downswings; we also notice an oscillatory pattern in the autocorrelation. □

The autocorrelation behavior of AR processes does not present the cutoff properties that help us determine the order of a MA process. The tool that has been developed for AR process identification is the *partial autocorrelation function* (PACF). The rationale behind PACF is to measure the degree of association between Y_t and Y_{t-k}, removing the effects of intermediate lags,

---

[14]This need not be the case, as we are working with *sample* autocorrelations. Nevertheless, at least for significant values, we observe a monotonic behavior consistent with Eq. (11.34).

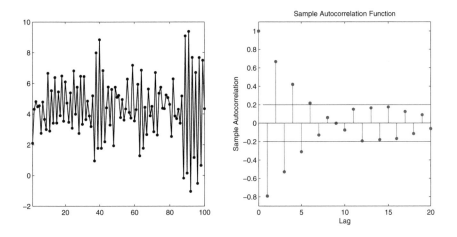

Fig. 11.13 Sample path and corresponding SACF for the autoregressive process $Y_t = 8 - 0.8Y_{t-1} + \epsilon_t$.

i.e., $Y_{t-1}, \ldots, Y_{t-k+1}$. We cannot dwell too much on PACF, but we may at least get a better intuitive feeling as follows.

Example 11.11 (Partial correlation) Consider three random variables X, Y, and Z, and imagine regressing X and Y on Z:

$$\hat{X} = a_1 + b_1 Z$$
$$\hat{Y} = a_2 + b_2 Z$$

Note that we are considering a probabilistic regression, not a sample-based regression. From Section 10.6, we know that

$$b_1 = \frac{\text{Cov}(X, Z)}{\text{Var}(Z)}, \qquad b_2 = \frac{\text{Cov}(Y, Z)}{\text{Var}(Z)}$$

Furthermore, we have regression errors

$$X^* = X - \hat{X} = X - (a_1 + b_1 Z)$$
$$Y^* = Y - \hat{Y} = Y - (a_2 + b_2 Z)$$

which may be regarded as the random variables X and Y, after the effect of Z is removed. The correlation $\rho(X, Y)$ may be large because of the common factor Z (the "lurking" variable). If we want to get rid of it, we may consider the partial correlation $\rho(X^*, Y^*)$. □

Following the intuition provided by the example, we might consider estimating the partial autocorrelation between Y_t and Y_{t-k} by the following linear regression:

$$Y_t = b_0 + b_1 Y_{t-1} + b_2 Y_{t-2} + \cdots + b_{k-1} Y_{t-k+1} + b_k Y_{t-k}$$

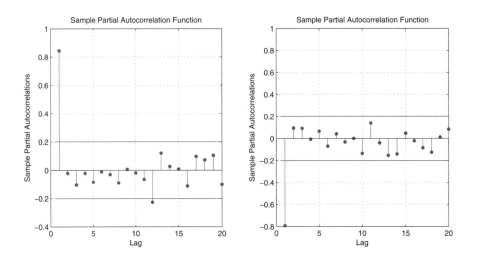

Fig. 11.14 Sample partial autocorrelation function for the autoregressive processes of Example 11.10.

By including intermediate lagged variables $Y_{t-1}, \ldots, Y_{t-k+1}$, we capture their effect by the regression coefficients $b_1, \ldots, b_{k-1}$. Then, we could use b_k as an estimate of partial autocorrelation. Actually, this need not be the sounder approach, but software packages provide us with ready-to-use functions to estimate the PACF by its sample counterpart (SPACF). In Fig. 11.14 we show the SPACF for the two AR(1) processes of Example 11.10. We see that the SPACF cuts off after lag 1, even though statistical sampling errors suggest that there is a significant value at larger lags in the first case. SPACF can be used to assess the order of an AR model.

11.6.3 ARMA and ARIMA processes

Autoregressive and moving-average processes may be merged into ARMA (autoregressive moving-average) processes like:

$$Y_t = \delta + \phi_1 Y_{t-1} + \cdots + \phi_p Y_{t-p} + \epsilon_t - \theta_1 \epsilon_{t-1} + \cdots + \theta_q \epsilon_{t-q} \qquad (11.35)$$

The model above is referred to as ARMA(p, q) process, for self-explanatory reasons. Conditions ensuring stationarity have been developed for ARMA processes, as well as identification and estimation procedures. Clearly, the ARMA modeling framework affords us plenty of opportunities to fit historical data. However, it applies only to stationary data. It is not too difficult to find real-life examples of data processes that are nonstationary. Just think of stock market indices; most investors really wish that the process is not stationary.

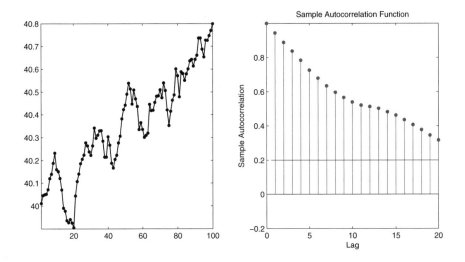

Fig. 11.15 A sample path and the corresponding SACF for the random walk $Y_t = Y_{t-1} + 0.05\eta_t$.

Example 11.12 (A nonstationary random walk) A quite common building block in many financial models is the *random walk*. An example of random walk is

$$Y_t = Y_{t-1} + 0.05\eta_t \tag{11.36}$$

where η_t is a sequence of independent and standard normal random variables. This is actually an AR process, but from Section 11.6.2 we know that it is nonstationary, as $\phi = 1$. A sample path of the process is shown in Fig. 11.15. In this figure, the nonstationarity of the process is pretty evident, but this need not always be the case. In figure 11.16 we show another sample path for the same process. A subjective comparison of the two sample paths would not suggest that they are just two realizations of the same stochastic process. However, the two autocorrelograms show a common pattern: Autocorrelation fades out slowly. Indeed, this is a common feature of nonstationary processes. Figure 11.17 shows the SPACF for the second sample path. We see a very strong partial autocorrelation at lag 1, which cuts off immediately. Again, this is a pattern corresponding to the process described by Eq. 11.36.   □

Since the theory of stationary MA and AR processes is well developed, it would be nice to find a way to apply it to nonstationary processes as well. A commonly used trick to remove nonstationarity in a time series is *differencing*, by which we consider the time series

$$Y'_t = Y_t - Y_{t-1} \tag{11.37}$$

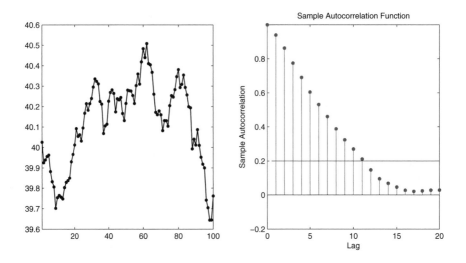

Fig. 11.16 Another sample path and the corresponding SACF for the random walk $Y_t = Y_{t-1} + 0.05\eta_t$.

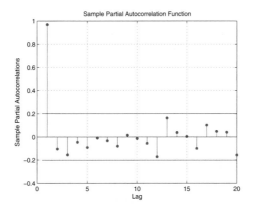

Fig. 11.17 Sample partial autocorrelation function for the random walk $Y_t = Y_{t-1} + 0.05\eta_t$.

Applying differencing to the sample path of Fig. 11.15 results in the sample path and SACF illustrated in figure 11.18. The shape of the SACF is not surprising, since the differenced process is just white noise.

Example 11.13 (What is nonstationarity, anyway?)   A time series with trend

$$Y_t = \alpha + \beta t + \epsilon_t \tag{11.38}$$

where ϵ_t is white noise, is clearly nonstationary and features a deterministic trend. A little digression is in order to clarify the nature of nonstationarity

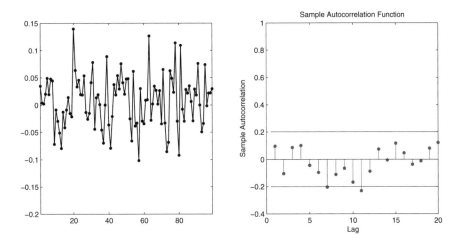

Fig. 11.18 The effect of differencing on the sample random walk of Fig. 11.15.

in a random walk

$$Y_t = Y_{t-1} + \epsilon_t \tag{11.39}$$

The sample paths in Example 11.12 show that in the random walk does not feature a deterministic trend. Recursive unfolding of Eq. (11.39) results in

$$Y_t = Y_0 + \sum_{k=1}^{t} \epsilon_k$$

Therefore

$$E[Y_t \mid Y_0] = Y_0$$

Hence, we must have a different kind of nonstationarity in the random walk of Eq. (11.39) than in the process described by Eq. (11.38). To investigate the matter, let us consider the expected value of the increment $Y_t' = Y_t - Y_{t-1}$, conditional on Y_{t-1}:

$$E[Y_t' \mid Y_{t-1}] = E[\epsilon_t] = 0$$

Therefore, given the last observation Y_{t-1}, we cannot predict whether the time series will move up or down. Now, let us consider a stationary AR(1) process

$$Y_t = \phi Y_{t-1} + \epsilon_t$$

where $\phi \in (-1, 1)$. The increment in this case is

$$Y_t' = (\phi - 1)Y_{t-1} + \epsilon_t$$

Since $(\phi - 1) < 0$, we have

$$\begin{cases} E[Y_t' \mid Y_{t-1}] < 0 & \text{if } Y_{t-1} > 0 \\ E[Y_t' \mid Y_{t-1}] > 0 & \text{if } Y_{t-1} < 0 \end{cases}$$

This suggests that a stationary AR(1) process is *mean reverting*, in the sense that the process tends to return to its expected value; the nonstationary random walk does not enjoy this property. ⧠

If we introduce the *backshift operator B*, defined by

$$BY_t = Y_{t-1}$$

we may express the first difference in Eq. (11.37) as

$$Y_t' = Y_t - BY_t = (1 - B)Y_t$$

Sometimes, differencing must be repeated in order to obtain a stationary time series. We obtain *second-order differencing* by repeated application of (first-order) differencing:

$$
\begin{aligned}
Y_t'' &= Y_t' - Y_{t-1}' \\
&= (Y_t - Y_{t-1}) - (Y_{t-1} - Y_{t-2}) \\
&= Y_t - 2Y_{t-1} + Y_{t-2} \\
&= (1 - 2B + B^2)Y_t \\
&= (1 - B)^2 Y_t
\end{aligned}
$$

This suggests that we may formally apply the algebra of polynomials to the backshift operator, in order to find differences of arbitrary order. By introducing polynomials

$$\Phi(B) = 1 - \phi_1 B - \cdots - \phi_p B^p$$
$$\Theta(B) = 1 - \theta_1 B - \cdots - \theta_q B^q$$

we may rewrite the ARMA model of Eq. (11.35) in the compact form

$$\Phi(B)Y_t = \delta + \Theta(B)\epsilon_t \tag{11.40}$$

We may extend the class of stationary ARMA models, in order to allow for nonstationarity. We find the more general class of ARIMA (autoregressive integrated moving average) processes, also known as *Box–Jenkins models*. An ARIMA(p, d, q) process can be represented as follows:[15]

$$\Phi(B)(1 - B)^d Y_t = \delta + \Theta(B)\epsilon_t \tag{11.41}$$

where $\Phi(B)$ and $\Theta(B)$ are polynomials of order p and q, respectively, and d is a differencing order such that the process Y_t is stationary, whereas, if we take differences of order $(d - 1)$, the process is still nonstationary. The name

---

[15]To be precise, this representation requires that the two polynomials $\Phi(B)$ and $\Theta(B)$ satisfy some additional technical conditions that are beyond the scope of this book.

"integrated" stems from the fact that we obtain the nonstationary process by integrating a stationary one, i.e., by undoing differencing. In most business applications the order d is 0 or 1. A full account of this class of models is beyond the scope of this book; we refer the reader to the bibliography provided at the end of this chapter, where it is also shown how Box–Jenkins models can be extended to cope with seasonality.

11.6.4 Using time series models for forecasting

Time series models may be used for forecasting purposes. As usual, we should find not only a point forecast, but also a prediction interval. Given an information set consisting of observations up to Y_t, we wish to find a forecast $F_{t,h} = \hat{Y}_{t+h}(t)$, at time t, with horizon $h \geq 1$, that is "best" in some well specified sense. A reasonable criterion is the minimization of the mean squared error

$$\mathrm{E}\left[\left(Y_{t+h} - \hat{Y}_{t+h}(t)\right)^2\right]$$

It can be shown that this is obtained by the conditional expectation

$$\mathrm{E}[Y_{t+h} \,|\, Y_t, Y_{t-1}, Y_{t-2}, \ldots]$$

To be concrete, assume that we have estimated the parameters of a model like

$$Y_t = \delta + \phi_1 Y_{t-1} + \cdots \phi_p Y_{t-p} + \epsilon_t - \theta_1 \epsilon_{t-1} - \cdots - \theta_q \epsilon_{t-q}$$

If we want to forecast with horizon $h = 1$, we should step forward to

$$Y_{t+1} = \delta + \phi_1 Y_t + \cdots \phi_p Y_{t-p+1} + \epsilon_{t+1} - \theta_1 \epsilon_t - \cdots - \theta_q \epsilon_{t-q+1} \qquad (11.42)$$

To build a point forecast, we note the following:

- If all of the observations $Y_{t-p+1}, \ldots, Y_t$ are available,[16] we should just plug their values into Eq. (11.42).

- Of course we do not know the future random shock ϵ_{t+1}; however, if random shocks are uncorrelated, we may just plug its expected value $\mathrm{E}[\epsilon_{t+1}] = 0$ into the equation.

- Unfortunately, we cannot directly observe the past shocks $\epsilon_{t-q+1}, \ldots, \epsilon_t$. In chapter 10 on simple linear regression, we faced a similar issue related to the difference between unobservable errors and observable residuals. We have to estimate $\hat{\epsilon}_{t-q+1}, \ldots, \hat{\epsilon}_t$ on the basis of prediction errors.

---

[16] In principle, with a high-order model and a small fit sample, we could lack some observation in the past; however, in this case, the real trouble comes from poor parameter estimates.

Example 11.14 Consider the model

$$Y_t = Y_{t-1} + \epsilon_t - \theta_1 \epsilon_{t-1}$$

In order to forecast Y_{t+1}, we write

$$\hat{Y}_{t+1} = Y_t + \hat{\epsilon}_{t+1} - \theta_1 \hat{\epsilon}_t$$

where the prediction of the future shock is $\hat{\epsilon}_{t+1} = 0$, and the past shock is estimated by the observed forecast error

$$\hat{\epsilon}_t = Y_t - \hat{Y}_t(t-1)$$

$\square$

If we want to forecast with horizon $h > 1$, we should rewrite Eq. (11.42) for the appropriate time subscript, and run a multistep forecasting procedure, whereby successive forecasts are generated and used. As the reader can imagine, when the moving-average order q is large, things are not as simple as in the example above. Indeed, forecasting based on ARIMA models is by no way a trivial business, and alternative approaches have been proposed. One possibility is to rewrite the model as an infinite-order moving average

$$Y_{t+h} = \delta + \sum_{k=1}^{\infty} \psi_k \epsilon_{t+h-k}$$

and plug estimates of random shocks. The procedures to accomplish all of this are beyond the scope of this book;[17] luckily, statistical software packages are available to carry out model estimation and forecasting. These software tools also provide the user with a prediction interval, using procedures which are not unlike those we used to find confidence intervals. The idea is surrounding the point forecast with an interval related to some standard prediction error, using quantiles from t or standard normal distributions. When using these procedures, we should keep in mind the following:

- Typically, it is assumed that random shocks are normal and uncorrelated.

- Since the mathematics of time series is quite involved, the only source of uncertainty accounted for is the realization of the future random shock. However, we have seen in Section 10.4 that uncertainty in parameter estimates also plays a role. Unfortunately, simple linear regression is a simple problem with an analytical solution, which lends itself to an accurate analysis; often, to estimate parameters of time series models,

---

[17]See, e.g., chapter 5 of Brockwell and Davis [3], where the recursive nature of time series forecasting is aptly illustrated.

numerical optimization is required, which makes a formal analysis quite difficult.

As a result, the forecast uncertainty could be underestimated. Hence, it is good practice to split available data into a fit and a test sample, and assess forecast errors out-of-sample before applying a model in a business setting.

Problems

11.1   Consider the demand data in the table below:

t	1	2	3	4	5	6
Y_t	35	50	60	72	83	90

We want to apply exponential smoothing with trend:

- Using a fit sample of size 3, initialize the smoother using linear regression.

- Choose smoothing coefficients and evaluate MAPE and RMSE on the test sample.

- After observing demand in the last time bucket, calculate forecasts with horizons $h = 2$ and $h = 3$.

11.2   The following table shows quarterly demand data for 3 consecutive years:

	Quarter			
Year	I	II	III	IV
2008	21	27	41	13
2009	19	32	42	12
2010	22	33	38	10

Choose smoothing coefficients and apply exponential smoothing with seasonality:

- Initial parameters are estimated using a fit sample consisting of two whole cycles.

- Evaluate MAD and MPE on the test sample, with $h = 1$.

- What is $F_{5,3}$?

11.3   In the table below, "–" indicates missing information and "??" is a placeholder for a future and unknown demand:

	Quarter			
Year	I	II	III	IV
2008	–	–	40	28
2009	21	37	46	30
2010	29	43	??	??

Initialize a smoother with multiplicative seasonality by using a fit sample of size 7.

11.4 We want to apply the Holt–Winter method, assuming a cycle of one year and a quarterly time bucket, corresponding to ordinary seasons. We are at the beginning of summer and the current parameter estimates are

- Level 80

- Trend 10

- Seasonality factors: winter 0.8, spring 0.9, summer 1.1, autumn 1.2

On the basis of these estimates, what is your forecast for next summer? If the demand scenario (summer 88, autumn 121, winter 110) is realized, what are MAD and MAD%?

11.5 Prove that the weights in Eq. (11.18) add up to one. (*Hint*: Use the geometric series.)

11.6 Prove Eqs. (11.31) and (11.32).

11.7 Consider a moving-average algorithm with time window n. Assume that the observed values are i.i.d. variables. Show that the autocorrelation function for two forecasts that are k time buckets apart is

$$\rho_k = \begin{cases} 1 - \frac{k}{n} & \text{if } k < n \\ 0 & \text{otherwise} \end{cases}$$

For further reading

- A classical reference on forecasting is the book by Makridakis et al. [6], which covers both regression and time series models, illustrating them with a vast array of applications. Among other things, the interested reader may find details on static time series decomposition approaches.

- Another general reference worth looking at is the text by Bowerman et al. [1], which offers a succinct treatment, without dwelling too much on technicalities. The text by Heij et al. [5], from which Example 11.13 was taken, offers a good compromise between theory and applications.

- There are quite advanced books on time series modeling, like the monumental text by Hamilton [4]. For a gentler treatment, a possible reference is that by Montgomery et al. [7], whose Chapter 5 has influenced the presentation in Section 11.6 above.

- We have dealt mostly with demand forecasting, but we did not cover many important issues, such as forecasting for brand-new products; interested readers may consult Chapter 3 of Ref. [2], which has influenced much of the presentation here.

- For further listening, the title of Section 11.5.1 is a homage to Jaco Pastorius' *Three Views of a Secret*.

REFERENCES

1. B.L. Bowerman, R.T. O'Connell, and A.B. Koehler, *Forecasting, Time Series, and Regression*, Brooks/Cole, Pacific Grove, CA, 2005.

2. P. Brandimarte and G. Zotteri, *Introduction to Distribution Logistics*, Wiley, New York, 2007.

3. P.J. Brockwell and R.A. Davis, *Time Series: Theory and Methods*, 2nd ed., Springer, New York, 1991.

4. J.D. Hamilton, *Time Series Analysis*, Princeton University Press, Princeton, NJ, 1994.

5. C. Heij, P. de Boer, P.H. Franses, T. Kloek, and H.K. van Dijk, *Econometric Methods with Applications in Business and Economics*, Oxford University Press, New York, 2004.

6. S. Makridakis, S.C. Wheelwright, and R.J. Hyndman, *Forecasting: Methods and Applications*, 3rd ed., Wiley, New York, 1998.

7. D.C. Montgomery, C.L. Jennings, and M. Kulahci, *Introduction to Time Series Analysis and Forecasting*, Wiley, New York, 2008.

Part III

Models for Decision Making

12

Deterministic Decision Models

In the last few chapters we have covered tools to represent some standard forms of uncertainty. Our main aims were to understand the relationship between variables of interest and possibly to forecast their future values. Understanding how a system works is clearly essential in all scientific disciplines, including the social ones. However, in management there is a further step: moving from knowledge discovery to decision making. So far, we have just hinted at decision models every now and then. In this chapter, we move on to a systematic treatment of quantitative models and methods for decision making. In this first step, we disregard uncertainty and deal with deterministic problems. Later, in Chapter 13, we merge decision models with probability and statistics to address the case of decision making under uncertainty. This will open up a world of challenging and rewarding models. Yet, we should always keep in mind that even the best decision model is always based on an approximate description of reality, and it should be regarded as a support tool, not a magical oracle. We will further insist on this in Chapter 14, where we outline a few complications arising in the practical world.

From a technical point of view, this chapter relies on concepts that were introduced in Chapters 2 and 3, such as:[1]

- Convex sets and convex/concave functions

- Local and global optimizers

- Quadratic forms and multivariable calculus

---

[1]See Sections 2.11 and 3.9.

There, we covered unconstrained optimization of functions of one variable; here we deal with problems involving possibly many decision variables and constraints. Apart from a very few lucky cases, there is no hope of solving such problems analytically, and we must rely on numerical solution methods. However, here we place much more emphasis on model building than model solving. Extremely efficient and reliable software packages are commercially available to solve rather large models; so, it can be argued that only model building is relevant. Nevertheless, a modicum of familiarity with the underlying solution strategies is needed to choose the right solution method and to understand when and why a model is easy or difficult to solve. Furthermore, model building and solving are not always disconnected; sometimes, the proper formulation of a model may greatly improve the computational performance in solving it.

We begin with a classification of decision models in Section 12.1; we draw the line between linear and nonlinear programming models, as well as between convex and nonconvex optimization problems. This classification has a quite practical purpose, as it is related to solution methods available to solve each class of problems. As we shall see, some models can be tackled by surprisingly fast methods that can solve large-scale problems with a reasonable computational effort; other models are a much harder nut to crack, and we should understand why. After this, we turn to model building. A few prototypical linear programming models are described in Section 12.2. Then, in Section 12.3, we illustrate a few tricks of the trade that are helpful in coping with less standard cases. The full power of quantitative modeling is unleashed in Section 12.4, where we see how to represent quite intricate problems mathematically by using integer programming techniques, involving logical decision variables. Section 12.5 is a bit more theoretical and deals with nonlinear programming; still, in this section we introduce quite relevant concepts, such as shadow prices, which have an important economic and managerial significance. Finally, in Section 12.6, we get a glimpse of standard solution algorithms like the simplex method for continuous linear programming and the branch and bound method for integer linear programming. That section may be safely skipped, if the reader so wishes, since its content is not used anywhere in the remainder of the book; yet, we should also stress the fact that these approaches are widely available in commercial software packages, and having at least a rough idea of how they work may help in using them properly.

12.1   A TAXONOMY OF OPTIMIZATION MODELS

In Part I we got acquainted with two elementary and prototypical optimization models, which we recall here for readers' convenience:

1. The *production mix optimization* model:[2]

$$\max \quad 45x_1 + 60x_2 \qquad (12.1)$$
$$\text{s.t.} \quad 15x_1 + 10x_2 \leq 2400 \qquad (12.2)$$
$$15x_1 + 35x_2 \leq 2400 \qquad (12.3)$$
$$15x_1 + 5x_2 \leq 2400 \qquad (12.4)$$
$$25x_1 + 15x_2 \leq 2400 \qquad (12.5)$$
$$0 \leq x_1 \leq 100 \qquad (12.6)$$
$$0 \leq x_2 \leq 50 \qquad (12.7)$$

Here the *decision variables* x_1 and x_2 represent production quantities of two items; in a real-life problem we have many more items, and their production could be restricted to integer amounts: $x_1, x_2 \in \{0, 1, 2, 3, \ldots\} = \mathbb{Z}_+$. In Eq. (12.1) we see a function that must be maximized, as it relates to profit. This *objective function* is not exactly our profit, as it does not consider fixed costs; nevertheless, it differs from true profit by a term which is constant with respect to the decisions that we are making at this level. Profit is maximized with respect to decision variables, *subject to* (s.t.) a set of *constraints*, which define a *feasible region*, also called *feasible set*. Inequalities (12.2)–(12.5) represent limitations due to finite capacity of the four resources that are used to manufacture the two item types. Finally, constraints (12.6) and (12.7) ensure that nonnegative amounts are produced, and that they do not exceed market demand bounds.

2. The *economic order quantity* model:[3]

$$\min_{Q \geq 0} \quad h\frac{Q}{2} + \frac{Ad}{Q} \qquad (12.8)$$

In this case, we have only one nonnegative decision variable Q, but the form of the objective function is definitely more complicated than Eq. (12.1). To begin with, it is nonlinear; then, it is not defined for $Q = 0$, and we should only consider the open interval $Q > 0$. Nevertheless, if we extend the objective function so that it takes the value $+\infty$ when $Q = 0$, we immediately see that, since we are minimizing a cost, we may consider the interval $Q \geq 0$ as the feasible region.[4]

---

[2]See Section 1.1.2.
[3]See Section 2.1.
[4]From a theoretical point of view, feasible regions with closed boundaries are always preferred, as otherwise we cannot even be sure that a minimum or a maximum exists. To see this, consider the problem $\min x$, subject to the bound $x > 2$. It is tempting to say that its solution is $x^* = 2$; however, this point is not feasible, as the inequality is strict. In fact, the problem has no solution. We may find the solution $x^* = 2$ if we consider the problem $\inf x$,

Looking at these two examples, we notice similarities and differences:

- In an optimization model we have an objective function that can be minimized or maximized. In the following, we will mostly refer to minimization problems, with no loss of generality. In fact, any maximization problem can be converted into an equivalent minimization by changing the sign of the objective function or, if you prefer, by flipping it upside down:

$$\max f(\mathbf{x}) \quad \Rightarrow \quad -\min[-f(\mathbf{x})]$$

- The objective function can be linear or nonlinear. In the production mix problem, we see an example of a linear objective function. A linear objective function is characterized by a sum like

$$f(\mathbf{x}) = \sum_{j=1}^{n} c_j x_j = \mathbf{c}^T \mathbf{x}$$

Each decision variable x_j is multiplied by a coefficient c_j, which in a minimization problem could be the cost of carrying out activity j at level x_j, such as producing an amount x_j of item j. The precise meaning of the coefficients depends on the problem. In the production mix problem, where the objective is maximized, they represent a contribution to profit. Recall that adding a constant term would make the function linear *affine*, but this would not change the optimal solution. Any other objective form, such as functions involving products of decision variables or powers with an exponent different from one, is nonlinear. In the EOQ model, the term $1/Q$ renders the problem nonlinear. Generally speaking, a nonlinear objective function makes the problem more complicated. As we shall see, we may sometimes approximate a nonlinear objective by a piecewise linear one.

- In the EOQ model, we have a very simple restriction on the decision variable, which is just restricted to nonnegative values. In this case, the feasible set is $\mathbb{R}_+$. In the production mix model, we have a list of more complicated constraints. Generally speaking, the feasible set is described by a set of constraints of the following forms:

    - *Inequality constraints* are represented by inequalities like $g(\mathbf{x}) \leq 0$ or $g(\mathbf{x}) \geq 0$. When dealing with a generic optimization model, we will typically use inequalities of the form

$$g(\mathbf{x}) \leq 0 \qquad\qquad (12.9)$$

which has a slightly different meaning that we leave to theoretically inclined treatments. By the same token, you may find sup rather than max. A theorem due to Weierstrass ensures that if a function is continuous, and the feasible set is closed and bounded, the function attains its maximum and minimum within that set. We will steer away from such technical complications.

There is no loss of generality in this choice, since we may always transform an inequality of the larger-than type as follows:

$$g(\mathbf{x}) \geq 0 \quad \Rightarrow \quad -g(\mathbf{x}) \leq 0$$

We have four such inequalities in the production mix problem; to cast constraint (12.2) into form (12.9), we just shift the constant term to the left:

$$15x_1 + 10x_2 - 2400 \leq 0$$

If the function g is linear (affine), i.e.

$$g(\mathbf{x}) = \sum_{j=1}^{n} a_j x_j - b = \mathbf{a}^T \mathbf{x} - b$$

we have a linear constraint. If function $g(\cdot)$ involves products of variables, general powers of them, or exponential functions, etc., the constraint is nonlinear.

- *Equality constraints* are represented by equations of the form $h(\mathbf{x}) = 0$. We do not have equality constraints in the two examples above, but we will see examples of them later. A generic linear equality constraint can be written as

$$\sum_{j=1}^{n} a_j x_j - b = \mathbf{a}^T \mathbf{x} - b = 0$$

Note that changing the sign of all the coefficients involved in this constraint has no effect on the feasible set. Equality constraints may be nonlinear as well.

- We may also have simple *lower bounds* on decision variables, like $x_j \geq l_j$, and *upper bounds*, like $x_j \leq u_j$. In the production mix model we see the most common example of lower bound, i.e., a non-negativity restriction on decision variables, as well as upper bounds enforced by market limitations. In principle, simple bounds could be treated as inequality constraints. However, most optimization algorithms treat them separately for efficiency reasons.

- Finally, in the following we will find examples of *integrality restrictions*, like $x_j \in \{0, 1, 2, 3, \ldots\} = \mathbb{Z}_+$. In both examples above, we could enforce such restrictions to make sure that we plan production of an integer amount of items; this may be not needed when we consider items that can be produced in amounts represented by continuous (real) variables, such as tons of a chemical element. We will see that, in general, restricting variables to integer values makes the solution of an optimization problem much harder. Whenever possible, it is advisable to get rid of such restrictions, even if it entails some degree of approximation. This is acceptable, for instance, in the case of items produced in large amounts, since rounding 1020.37 up or down implies a negligible error; this is not acceptable if production occurs on a very small scale, i.e., just a few items. The most common use of integer decision variables stems from the need to model logical decisions, such as "open a new manufacturing plant" or "do not open a new manufacturing plant." In such a case, we may represent our decision as a *binary* (or logical) decision variable that may take only one of two values: $x \in \{0, 1\}$. Clearly, binary variables are discrete in nature, and relaxing them to continuous values in the interval $[0, 1]$ makes no sense: Either you build a plant or not, as building 47% of it has no meaning.

In general and abstract terms, we may refer to an optimization problem in the following form

$$\min_{\mathbf{x} \in S} f(\mathbf{x}) \tag{12.10}$$

where S, the feasible set, is a subset of $\mathbb{R}^n$. If $S \equiv \mathbb{R}^n$, we have an *unconstrained* optimization problem; otherwise, we have a *constrained* optimization problem. An optimization model like (12.10) is also referred to as a *mathematical programming* model. The key feature is that there is a finite set of n decision variables $x_j, j = 1, \ldots, n$, collected into vector $\mathbf{x}$. However strange it may sound, there are problems featuring an *infinite* set of decision variables. One such example occurs when we are looking for a function $u(t)$ of time, where t ranges on the continuous interval $[0, T]$. The function $u(t)$ should satisfy some constraints and be optimal according to some well-defined criterion. These problems are quite common in optimal control theory, where $u(t)$ is a control input that should be applied to a system in order to force an optimal behavior in terms of desirable output. Optimal control problems are widely considered in economics as well, when time is assumed continuous in the model. In this book, we will consider only a discrete representation of time, which is partitioned into time periods (time buckets) indexed by $t = 1, 2, 3, \ldots, T$. Hence, we will only deal with mathematical programs.

When we consider the actual ways in which we may describe the feasible set S, we see that there is a rich set of possible structures of optimization models. Yet, this variety can be boiled down to a smaller set of prototypes by using simple manipulations. For instance, an inequality constraint may be transformed into an equality constraint by introducing a nonnegative slack variable s as follows:

$$g(\mathbf{x}) \leq 0 \Rightarrow \begin{cases} g(\mathbf{x}) + s = 0 \\ s \geq 0 \end{cases} \quad ; \quad g(\mathbf{x}) \geq 0 \Rightarrow \begin{cases} g(\mathbf{x}) - s = 0 \\ s \geq 0 \end{cases}$$

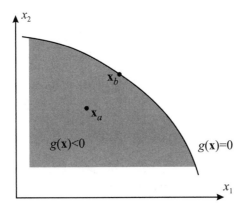

Fig. 12.1 Inequality constraints may be active/binding or not.

We may also go the other way around and transform an equality constraint into two inequality constraints:

$$h(\mathbf{x}) = 0 \quad \Rightarrow \quad \left\{ \begin{array}{l} h(\mathbf{x}) \leq 0 \\ -h(\mathbf{x}) \leq 0 \end{array} \right.$$

These tricks of the trade may be useful not only when stating general properties, but also when considering concrete solution methods.

A generic inequality constraint $g(\mathbf{x}) \leq 0$ defines a portion of n-dimensional space. In the plane, it may define a region like the shaded one in Fig. 12.1; to be precise, in the figure we observe a curve corresponding to equation $g(\mathbf{x}) = 0$ and a shaded region corresponding to points such that the strict inequality $g(\mathbf{x}) < 0$ applies. With reference to a specific point, an inequality constraint may be *active* or *inactive*. The constraint is active at point $\mathbf{x}_b$ in the figure, since there it is satisfied as an equality, $g(\mathbf{x}_b) = 0$; on the contrary, the constraint is inactive at $\mathbf{x}_a$, since $g(\mathbf{x}_a) < 0$. It is important to notice that we may freely move in a neighborhood $\mathbf{x}_a$ without going out of the feasible set, provided that our step is small enough; on the contrary, we cannot move at will around $\mathbf{x}_b$, since only some directions preserve feasibility. Furthermore, a little perturbation of the constraint could make x_b infeasible. If we consider the optimal solution $\mathbf{x}^*$, we also say that an inequality constraint is *binding* if $g(\mathbf{x}^*) = 0$, i.e., if it is active at the optimal solution; if $g(\mathbf{x}^*) < 0$, the constraint is *nonbinding*. A noteworthy specific case occurs when considering the nonnegativity requirement $x_j \geq 0$. When *all* the decision variables are strictly positive at the optimal point, i.e., when $x_j^* > 0$ for all j, we speak of an *interior* solution.

Example 12.1 For the optimal mix problem (12.1), we have observed that any mix feasible with respect to the second resource is also feasible with respect to the first and third ones. In fact, any point satisfying inequality

(12.3) will also satisfy inequalities (12.2) and (12.4), since item type 1 has the same resource requirement on all of the related resources, whereas item type 2 has the largest requirement on the second one, which is a bottleneck. Hence, (12.2) and (12.4) are redundant constraints and will never be binding for the optimal mix. The practical implication is that there is no point in increasing the availability of resources that are not saturated, as this would not improve profit. $\qquad\square$

It can be expected that some optimization problems are relatively easy to solve, whereas others can be quite challenging. Therefore, it is important to understand which features draw the line between easy and difficult models. As we have seen, important attributes are "linear" vs. "nonlinear," as well as "continuous" vs. "discrete." When dealing with calculus, we have also considered other important features, which relate to convexity.[5] We already know that even an unconstrained minimization problem involving a function of just one decision variable may be relatively difficult, as it may feature many local optima. In this chapter we generalize to problems involving multiple variables and constraints; in the following subsections, we illustrate all of these concepts by very simple examples.

12.1.1 Linear programming problems

A mathematical programming problem is called a *linear programming* (LP) problem when all the constraints and the objective function are expressed by linear affine functions, as in the following case:

$$\min \quad 2x_1 + 3x_2 + 3x_3$$
$$\text{s.t.} \quad x_1 + 2x_2 = 3$$
$$x_1 + x_3 \geq 3$$
$$x_1, x_2, x_3 \geq 0$$

An LP model can involve inequality and equality constraints, as well as simple bounds. The general form of a linear programming problem is

$$\min \quad \sum_{j=1}^{n} c_j x_j$$

$$\text{s.t.} \quad \sum_{j=1}^{n} a_{ij} x_j = b_i, \qquad i \in \mathcal{E}$$

$$\sum_{j=1}^{n} d_{ij} x_j \leq e_i, \qquad i \in \mathcal{I}$$

$$l_j \leq x_j \leq u_j, \qquad j = 1, \ldots, n$$

[5]See Section 2.11.

where we denote the set of equality constraints by $\mathcal{E}$ and the set of inequality constraints by $\mathcal{I}$; any of these sets may be empty. If variable x_j is unbounded from below, we may set $l_j = -\infty$; by the same token, we may set $u_j = +\infty$ if the corresponding variable is unbounded from above. An LP model can be conveniently written in matrix form as follows:

$$\min \quad \mathbf{c}^T \mathbf{x} \tag{12.11}$$
$$\text{s.t.} \quad \mathbf{A}\mathbf{x} = \mathbf{b}$$
$$\mathbf{D}\mathbf{x} \leq \mathbf{e}$$
$$\mathbf{l} \leq \mathbf{x} \leq \mathbf{u}$$

Of course, an inequality involving vectors is interpreted componentwise, i.e., we should think of it as the collection of several inequalities, one per component of the vector.

Example 12.2 Let us consider the LP problem:

$$\min \quad x_1 + 3x_3 + 5x_4$$
$$\text{s.t.} \quad x_1 + x_2 \leq 3$$
$$x_1 + x_3 + 4x_4 \geq 5$$
$$3x_1 - x_2 + x_3 = 6$$
$$x_1 + x_4 = 8$$
$$x_1, x_2, x_3, x_4 \geq 0$$

It may be cast in the matrix form (12.11) by defining the following matrices and vectors:

$$\mathbf{c} = \begin{bmatrix} 1 \\ 0 \\ 3 \\ 5 \end{bmatrix}, \quad \mathbf{b} = \begin{bmatrix} 6 \\ 8 \end{bmatrix}, \quad \mathbf{e} = \begin{bmatrix} 3 \\ -5 \end{bmatrix}, \quad \mathbf{l} = \begin{bmatrix} 0 \\ 0 \\ 0 \\ 0 \end{bmatrix}$$

$$\mathbf{A} = \begin{bmatrix} 3 & -1 & 1 & 0 \\ 1 & 0 & 0 & 1 \end{bmatrix}, \quad \mathbf{D} = \begin{bmatrix} 1 & 1 & 0 & 0 \\ -1 & 0 & -1 & -4 \end{bmatrix}$$

Please note the changes in sign needed to recast the second inequality in less-than form. The components of $\mathbf{u}$ can all be set to $+\infty$ or, more simply, the vector of upper bounds can just be disregarded. ▯

Using the transformations illustrated before, a generic LP model can also be cast into either of the following forms, involving one type of constraint:

- The *standard form*

$$\min \quad \mathbf{c}^T \mathbf{x}$$
$$\text{s.t.} \mathbf{A}\mathbf{x} = \mathbf{b} \tag{12.12}$$
$$\mathbf{x} \geq \mathbf{0}$$

- The *canonical form*

$$\max \quad \mathbf{c}^T \mathbf{x}$$
$$\text{s.t.} \quad \mathbf{Ax} \le \mathbf{b} \tag{12.13}$$
$$\mathbf{x} \ge \mathbf{0}$$

Note that all decision variables are restricted in sign. If a variable is free to take whatever sign, we may transform it into the difference of two nonnegative variables:

$$x_j = x_j^+ - x_j^-, \qquad x_j^+, x_j^- \ge 0.$$

The two variables should correspond to the positive and the negative part of x_j. However, this transformation introduces an ambiguity, at least in principle. For instance, if $x_j = 5$ we may write $5 = 5 - 0$, but also $5 = 8 - 3$. As a result, we have an infinite set of equivalent solutions, characterized by the same value of the objective function. Conceptually, this is not a relevant issue, even though it may be from a computational point of view. In practice, the careful implementation of solution methods handles free variables efficiently, but we need not be concerned with such technicalities.

The standard form (12.12) is more convenient computationally, as we shall see in Section 12.6.1. The canonical form (12.13), however, comes in handy to illustrate the *geometry* of an LP model. In fact, the feasible set for an LP problem in canonical form is expressed by a set of linear inequalities. Since each linear inequality is associated with a half-space, the feasible set is just the intersection of half-spaces, i.e., a polyhedron.[6] This is easy to visualize for a simple problem in two dimensions, where half-spaces are just half-planes. This case is illustrated in Fig. 12.2, together with the level curves of two possible objective functions, f_a and f_b. The feasible set is the shaded polyhedron, which is the intersection of six half-planes. Since the objective function is linear, its level curves are parallel lines. Depending on the coefficients in the objective function, these lines feature different slopes. If we assume a maximization problem, the figure illustrates that, depending on the slope, different optimal solutions are obtained, $\mathbf{z}_a$ or $\mathbf{z}_b$. If the level curves are parallel to the edge connecting $\mathbf{z}_a$ and $\mathbf{z}_b$, any solution on that edge is optimal. This shows that an LP problem need not have a unique solution. In fact, the following cases may apply:

1. There is a unique solution.

2. There are multiple equivalent solutions (an infinite number of them, actually).

3. There is no feasible solution; this may occur, e.g., when the intersection of the half-spaces is an empty set.

---

[6]See Fig. 1.3 for an illustration in the product mix case.

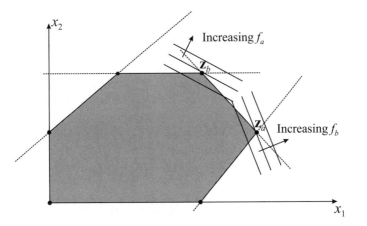

Fig. 12.2 Feasible set and optimal solutions for an LP problem.

4. The solution is unbounded, i.e., the optimal value of the objective goes to plus or minus infinity, for a maximization and a minimization problem, respectively.

It is easy to see that these cases are not limited to linear problems, as the same situations may arise in nonlinear problems as well.

Example 12.3 It is easy to build optimization problems which are, respectively:

1. Unbounded:

$$\begin{aligned} \max \quad & x_1^2 + x_2^2 \\ \text{s.t.} \quad & x_1 + x_2 \geq 4 \\ & x_1, x_2 \geq 0 \end{aligned}$$

2. Infeasible:

$$\begin{aligned} \max \quad & 2x_1 + 3x_2 \\ \text{s.t.} \quad & x_1 + x_2 \geq 4 \\ & 0 \leq x_1, x_2 \leq 1 \end{aligned}$$

3. Characterized by an infinite set of optima:

$$\begin{aligned} \max \quad & x_1 + x_2 \\ \text{s.t.} \quad & x_1 + x_2 \leq 4 \\ & x_1, x_2 \geq 0 \end{aligned}$$

The reader is urged to check all of this by drawing the feasible set and the level curves of the objective function, for the three problems above. □

It is also useful to notice that derivatives of a linear objective function $f(\mathbf{x}) = \mathbf{c}^T \mathbf{x}$ never vanish; its gradient is the vector of coefficients, $\nabla f(\mathbf{x}) = \mathbf{c}$, and it cannot be set to zero (for a meaningful problem). Hence, we need something more than stationarity conditions to solve a constrained optimization problem, and a linear one in particular. However, Fig. 12.2 suggests that, when looking for the solution of an LP problem, we may just consider the extreme points of the feasible set, i.e., the vertices of the polyhedron. Even though the feasible set in the figure includes an infinite number of points, only six of them are candidates for optimality. Even if there is an edge of equivalent optimal solutions, there is no loss in restricting our attention to one of the two extreme points of the edge. Indeed, this is a fundamental feature of LP problems that is exploited in the standard solution method: the simplex method, which is described later and is based on the idea of exploring a typically small subset of vertices to find the optimal solution. You may not really need to understand how this method works in depth. What is important is that this algorithm is widely available in commercial software tools, and it is normally able to solve rather large-scale problems, possibly involving several thousands of decision variables, in a few seconds.

In the LP models above, decision variables are real numbers, subject to joint restrictions represented by equality and inequality constraints, as well as lower and upper bounds. There are situations in which we must enforce some more restrictions on a subset of variables:

- We might require variables to take only integer values. For instance, we could specify that the number of items to be produced in the optimal mix problem be integer. In virtually all practically relevant cases, we need nonnegative integer variables, i.e., we require $x_j \in \mathbb{Z}_+$, for some j.

- Quite often, we need to express logical decisions, like "we place an order" or "we do not place any order," or "we start an activity" or "we don't." Such logical decisions are of all-or-nothing type and are typically represented by binary decision variables like $x_j \in \{0, 1\}$.

If the integrality condition is enforced on all of the decision variables, we have a *pure-integer linear programming* problem (ILP), such as

$$\min \quad \mathbf{c}^T \mathbf{x}$$
$$\text{s.t.} \quad \mathbf{A}\mathbf{x} \leq \mathbf{b}$$
$$\mathbf{x} \in \mathbb{Z}_+^n$$

Normally, in a pure ILP, we have only inequality constraints and bounds; equality constraints, i.e., equations involving integer numbers, are mathematically interesting and challenging, but they are uncommon in management. If the restriction is $\mathbf{x} \in \{0, 1\}^n$, i.e., we consider only binary variables, we have a *pure binary* LP problem. The most general case is a *mixed-integer linear programming* (MILP) problem, whereby integrality restrictions are enforced

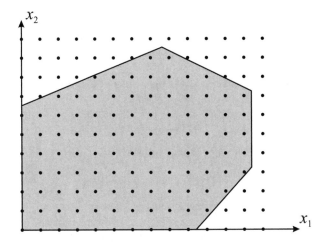

Fig. 12.3 The set of feasible solutions of a pure integer LP problem.

only on a subset $\mathcal{L}$ of variables:

$$\min \quad \sum_{j=1}^{n} c_j x_j$$

$$\text{s.t.} \quad \sum_{j=1}^{n} a_{ij} x_j = b_i, \qquad i \in \mathcal{E}$$

$$\sum_{j=1}^{n} d_{ij} x_j \le e_i, \qquad i \in \mathcal{I}$$

$$l_j \le x_j \le u_j, \qquad j = 1, \dots, n$$

$$x_j \in \mathbb{Z}, \qquad j \in \mathcal{L} \subseteq \{1, 2, \dots, n\}$$

In order to specify a binary decision variable, we can enforce the bounds $l_j = 0$ and $u_j = 1$ on an integer variable in $\mathcal{L}$.

It turns out that, as a general rule, integer programming problems are a much harder nut to crack than their continuous counterparts. To get a feeling for the reason behind this unpleasing state of affairs, let us have a look at Fig. 12.3. There, we see the feasible region of a pure ILP problem in two dimensions. If we drop the integrality requirements, we obtain the *continuous relaxation* of the ILP, whose feasible region is the shaded polyhedron depicted in the figure. The feasible region for the ILP problem consists of those points within the polyhedron that have integer coordinates. This immediately shows that we cannot apply familiar concepts from calculus: The derivative makes no sense, as we have a discrete set of points and we cannot take ordinary limits. However, we know that there is a way to solve a continuous LP. If we drop integrality requirements, we obtain an optimal solution corresponding to one of the extreme points (vertices) of the polyhedron in Fig. 12.3. In this case, we might expect that, although this vertex has noninteger coordinates in general, we could find the optimal solution by rounding the fractional solution of the continuous relaxation. For instance, in the optimal mix problem (12.1), the continuous LP has the optimal solution

$$x_1^* = 73.85, \qquad x_2^* = 36.92$$

whereas the corresponding ILP has optimal solution

$$x_1^* = 73, \qquad x_2^* = 37$$

It seems that by trying a few combinations of roundings, up and down, we should be able to find the optimal solution; some roundings might produce infeasible solutions, but when we get a feasible one, we just evaluate its objective function; by comparing all of the integer solutions found in this way, we spot the best rounding. Maybe, it is not trivial to *prove* that what we get is an optimal solution, but at least we should be able to find a pretty good one. Unfortunately, this approach does yield a good *heuristic* approach in some cases, but it does not work in general. If we have a large number n of integer decision variables, a rough calculation shows that we should try 2^n roundings. To see this, observe that we should round up or down each of the n variables, and this results in an exponential number of combinations. Even worse, it is not even true that we may find an optimal solution, or even a feasible one, by just rounding up and down. The following nice example shows that things may be difficult even in two dimensions.[7]

Example 12.4 Consider the following pure ILP problem:

$$
\begin{aligned}
\max \quad & x_1 + x_2 \\
\text{s.t.} \quad & 10x_1 - 8x_2 \leq 13 \\
& 2x_1 - 2x_2 \geq 1 \\
& x_1, x_2 \in \mathbb{Z}_+
\end{aligned}
$$

If we relax the integrality requirement, i.e., we just require $x_1, x_2 \geq 0$, we can apply the simplex method and find

$$x_1^* = 4.5, \qquad x_2^* = 4$$

with an optimal objective value 8.5. The reader is invited to try rounding the above solution up and down; unfortunately, the trivially rounded solutions are not feasible. In fact, the integer optimal solution is

$$x_1^* = 2, \qquad x_2^* = 1$$

---

[7]Example 12.4 is taken from Williams [13].

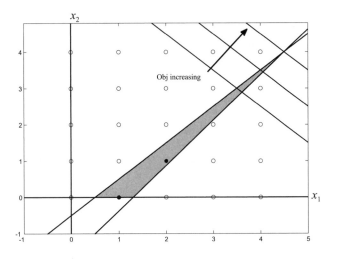

Fig. 12.4   Feasible solutions in Example 12.4.

with optimal value 3. We see that the continuous solution is quite far from the true integer minimizer; in this case it is even difficult to find a *feasible* solution by rounding, let alone the optimal one. The reason behind the trouble is illustrated in Fig. 12.4. The integer feasible set consists of only two points, $(1, 0)$ and $(2, 1)$. The feasible set of the continuous relaxation is the very narrow shaded triangle in Fig. 12.4, and it includes a many noninteger points quite far from the two integer solutions. □

Another interesting observation that we can draw from the example is that the continuous relaxation yields a bound on the true optimum value of the integer solution. In the example, we find an upper bound, 8.5., since we are maximizing. If we consider a minimization problem, the optimal value of the continuous relaxation yields a lower bound. To understand why, consider Fig. 12.3 again. The feasible set of the ILP is a subset of the feasible set of the continuous relaxation. Hence, when we drop the integrality requirements, we enlarge the feasible set. By doing so, we cannot increase cost, in a minimization problem, or decrease profit, in a maximization problem, since all of the integer points are still feasible and we are adding opportunities for optimization. Incidentally, in the last example, the upper bound we get, 8.5, is rather weak with respect to the true optimal value, which is 3. In other cases, the bounds yield better estimates of the optimal value. Indeed, this bounding approach is the starting point of the branch and bound method, which is the standard, commercially available approach to solving ILPs and MILPs. For the sake of the interested reader, we outline this algorithm in Section 12.6.2, which may be safely skipped at first reading. The branch and bound method is based on a search tree, on which each node corresponds to a continuous relaxation of the original MILP. The important message is that care must be taken when formulating MILPs, as tackling them may call for the solution of thousands of continuous relaxations to find the optimal integer solution.

12.1.2 Nonlinear programming problems

In an LP model, all the functions defining the objective and the constraints must be linear (affine). If even one function fails to meet this requirement, we face a *nonlinear programming* problem. For instance, the problem

$$\begin{aligned}
\min \quad & 2x_1 + 3x_2 + 3x_3 \\
\text{s.t.} \quad & x_1 + x_2^2 = 3 \\
& x_1 + x_3 \geq 3 \\
& x_1, x_2, x_3 \geq 0
\end{aligned}$$

is nonlinear because the first constraint involves a squared decision variable. The problem

$$\begin{aligned}
\min \quad & 2x_1 + 3x_2 x_3 \\
\text{s.t.} \quad & x_1 + 2x_2 = 3 \\
& x_1 x_3 \geq 3 \\
& x_1, x_2, x_3 \geq 0
\end{aligned}$$

is nonlinear because the objective function involves the product of two decision variables; the second constraint is nonlinear, too. Of course, we may have arbitrary nonlinearities in both the objective and the constraints. Now consider the nonlinear problem

$$\begin{aligned}
\min \quad & 2x_1^2 + 3x_2^2 + 3x_1 x_3 \\
\text{s.t.} \quad & x_1 + 2x_2 = 3 \\
& x_1 + x_3 \geq 3 \\
& x_1, x_2, x_3 \geq 0
\end{aligned}$$

We see that the feasible set is defined by linear functions, whereas the objective function is nonlinear. A more careful look at it, though, should ring a bell. The objective function involves quadratic terms and is an example of a quadratic form.[8] We know that, by introducing a suitable symmetric matrix, a quadratic form can be expressed in matrix form. A generic *quadratic*

---

[8]See Section 3.8.

programming problem can be represented as follows:

$$\min \quad \frac{1}{2}\mathbf{x}^T\mathbf{H}\mathbf{x} + \mathbf{c}^T\mathbf{x}$$
$$\text{s.t.} \quad \mathbf{A}\mathbf{x} = \mathbf{b}$$
$$\mathbf{D}\mathbf{x} \leq \mathbf{e}$$
$$\mathbf{l} \leq \mathbf{x} \leq \mathbf{u}$$

We see that, in general, the objective function may also involve a linear term (a constant term is irrelevant, as usual). The leading term includes a factor $\frac{1}{2}$ that is typically just included for convenience, as in so doing we see that the matrix $\mathbf{H}$ is actually the Hessian matrix of the objective function. We have already investigated convexity and concavity of quadratic forms. As it turns out, a quadratic programming problem may be very easy to solve or not, depending on these features. We recall convexity in the next section, but before that it is useful to consider two practical examples of nonlinear programming problems.

Example 12.5 In Example 8.5 we considered a simple portfolio optimization model. We must allocate our wealth among n assets, whose return is a random variable R_i, $i = 1, \ldots, n$. The decision variables are the fractions w_i of wealth that are allocated to each asset; these are also called *portfolio weights*. If we rule out short sales, natural constraints on decision variables are as follows:

$$\sum_{i=1}^{n} w_i = 1$$
$$w_i \geq 0, \qquad i = 1, \ldots, n$$

The first constraint ensures that we invest exactly our wealth, no more, no less. Nonnegativity constraints on portfolio weights ensure that no asset is sold short; we could also require that weights not exceed 100%, but this is redundant, given the two constraints above.

We also know from Example 8.5 that the portfolio return is the random variable

$$R_p = \sum_{i=1}^{n} w_i R_i$$

We cannot really maximize (nor minimize) a random variable, and a thorough treatment of this decision problem under uncertainty calls for some tools that we introduce in the next chapter. Nevertheless, we might pursue a rather simple and intuitive strategy. Most of us are risk-averse decision makers and do not want an overly uncertain return, such as the one that would result from choosing a very risky portfolio. What we certainly would like is a portfolio with a large expected return, but this desire must be tempered by a consideration of the risk involved. Hence, a possible way of framing the problem consists of

1. Choosing a minimum target expected return μ_t

2. Finding the portfolio with minimum variance, such that its expected return is not below the target μ_t

Taking advantage of what we learned about linear combinations of random variables in Chapter 8,[9] we recall that the variance of portfolio return is

$$\sigma_p^2 = \text{Var}\left(\sum_{i=1}^n w_i R_i\right) = \sum_{i=1}^n \sum_{j=1}^n w_i w_j \sigma_{ij} = \mathbf{w}^T \Sigma \mathbf{w}$$

where $\sigma_{ij} \equiv \text{Cov}(R_i, R_j)$ is the covariance of returns R_i and R_j, the symmetric covariance matrix Σ collects all such covariances, and $\mathbf{w}$ is the vector collecting portfolio weights. Let μ_i be the expected return of asset i. Then, we may formulate the following optimization problem:

$$\min \quad \mathbf{w}^T \Sigma \mathbf{w}$$

$$\text{s.t.} \quad \sum_{i=1}^n w_i \mu_i \geq \mu_t$$

$$\sum_{i=1}^n w_i = 1$$

$$w_i \geq 0, \qquad i = 1, \ldots, n$$

It is easy to see that this is a quadratic programming problem, since the objective function is a quadratic form and the constraints are linear.   □

Example 12.6 The EOQ model has plenty of limiting assumptions. One of them is that the model considers only one item, neglecting possible interactions with other ones. Furthermore, some assumptions concerning the involved costs are questionable as well. For instance, consider the fixed ordering charge. Some logistic channels are so efficiently organized that the need for including this charge should not be taken for granted. Imagine a retail store that receives a shipment each and every day from a large supplier. All of the item types are included in that shipment, so there is really no need to consider a fixed ordering charge per item; in fact such a charge is shared among all of the item types and is incurred anyway because shipments are arranged daily. Apparently, this would encourage ordering small lots quite often. However, there may be a limit on the *total* number of orders that are issued on the average. One possible reason is that orders must be inspected and tracked; if the number of persons in charge of purchasing activities is limited, an upper bound on the number of orders will result.

---

[9] See Section 8.3.2.

In the age of information systems and certified-quality suppliers, limiting the number of replenishment orders that we can issue may not seem warranted. However, building a model in this vein will turn out to be quite instructive. Then, on the basis of these considerations, let us extend the EOQ model to jointly order n items, minimizing the total average inventory holding cost, subject to an upper bound on the average number of orders issued over time. The information we need consists of

- The inventory holding cost h_i and the demand rate d_i for each item, $i = 1, \ldots, n$; as in the EOQ model, we assume constant demand rates.

- The upper bound U on the average number of orders issued per unit time.

The decision variables are the order quantities Q_i. The dynamics of inventory levels is the same as in the EOQ model. Hence, the average inventory cost for item i is $h_i Q_i / 2$, and the average number of orders issued per unit time is d_i / Q_i. Summing over items, we obtain the following model:

$$\min \quad \frac{1}{2} \sum_{i=1}^{n} h_i Q_i \tag{12.14}$$

$$\text{s.t.} \quad \sum_{i=1}^{n} \frac{d_i}{Q_i} \leq U \tag{12.15}$$

$$Q_i \geq 0, \qquad i = 1, \ldots, n$$

In this model, the objective function is linear, but the inequality constraint is not; hence, this is a nonlinear programming model. Later, in Example 12.22, we will show how such a model can be tackled, along with a quite instructive economic interpretation. □

12.1.3 Convex programming: difficult vs. easy problems

Let us consider an abstract mathematical programming problem:

$$\min \quad f(\mathbf{x}) \tag{12.16}$$

$$\text{s.t.} \quad \mathbf{x} \in S \subseteq \mathbb{R}^n$$

Intuition would suggest that an unconstrained problem, where $S \equiv \mathbb{R}^n$, is much easier to solve than a constrained one. Moreover, the same intuition would suggest that the larger the problem, in terms of the number of decision variables and constraints, the more difficult is solving it. In fact, this intuition is wrong. Some unconstrained problems with a few tens of variables are much more difficult to solve than some constrained problems involving thousands of variables. The key issue is convexity, which we introduced earlier.[10] In this

---

[10]See Section 2.11.

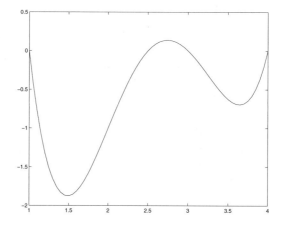

Fig. 12.5 Global and local optima for a polynomial.

section we extend those basic concepts a bit, and we summarize their impact on the difficulty of solving an optimization problem.

Solving problem (12.16) means finding a feasible global optimizer of function f. For readers' convenience, let us review the concepts of local and global optimality.

DEFINITION 12.1 *Given the optimization problem (12.16), a point* $\mathbf{x}^* \in S$ *is said to be a* **global minimizer** *if* $f(\mathbf{x}^*) \leq f(\mathbf{x})$, *for all* $\mathbf{x} \in S$. *We have a* **local minimizer** *if the condition holds only in the intersection between* S *and a neighborhood of* $\mathbf{x}^*$.

When dealing with maximization problems, we speak of maximizers; in general, we look for a global optimizer of the objective function. When one speaks of an "optimum," a little ambiguity arises, because it is not quite clear whether we mean the optimizer $\mathbf{x}^*$ or the optimal value $f(\mathbf{x}^*)$; usually, the context clarifies what we really mean. We may also use the notation $\mathbf{x}^* = \arg\min_{\mathbf{x} \in S} f(\mathbf{x})$ to indicate the optimizer.

Example 12.7 Consider the following unconstrained problem:

$$\min f(x) = x^4 - 10.5x^3 + 39x^2 - 59.5x + 30$$

The objective function is a polynomial, whose graph is depicted in Fig. 12.5. We observe that there are two local minimizers, one of which is also the global one, and one local maximizer. The function goes to infinity when x is large in absolute value, so there is no global maximizer. □

With a high-degree polynomial, there are potentially many local maxima and minima. Yet, polynomials in one variable are a relatively easy case; their derivative is very easy to find and is just a polynomial of lower degree. Therefore, application of the stationarity condition requires the solution of a nonlinear polynomial equation. While finding *all* of the roots of a generic nonlinear equation is not easy, efficient numerical methods are available to find all of the roots of a polynomial. Hence, we are able to find the whole set of stationary points of a polynomial, and it is an easy task to enumerate them to find the global optimizer, if any exists. However, this does not hold if we consider a function of several variables, with the possible complication of constraints. We already know that if a function of a single variable is convex and differentiable, then we do find a global minimizer by looking for a stationary point. Indeed, convexity is the basic feature that makes a mathematical problem relatively easy to solve.

DEFINITION 12.2 (Convex and concave problems) *Problem (12.16) is said to be a* **convex problem** *if f is a convex function and S is a convex set. Problem (12.16) is said to be a* **concave problem** *if f is a concave function and S is a convex set.*

Note that we are referring to a minimization problem. If we maximize a concave function over a convex set, we are actually dealing with a convex problem. Assuming that the optimization problem has a finite solution, the following properties can be proved.

PROPERTY 12.3 *In a convex problem a local minimizer is also a global minimizer.*

PROPERTY 12.4 *In a concave problem the global minimizer lies on the boundary of the feasible set.*

The first property essentially says that if we are able to find a local optimizer, we are done. The second property may look a bit harder to grasp, but an example will illustrate it.

Example 12.8 Consider the following one-dimensional problem:

$$\min \quad -(x-2)^2 + 3$$
$$\text{s.t.} \quad 1 \le x \le 4$$

This is a concave problem, since the leading term in the quadratic polynomial defining the objective is negative, and the second-order derivative is negative everywhere. In Fig. 12.6 we show the objective function and the feasible set. The stationarity point $x = 2$ is of no use to us, since it is a maximizer. We see that local minimizers are located at the boundary of the feasible set. A local minimizer lies at the left boundary, $x = 1$, and the global minimizer is located at the right boundary, $x = 4$.   ☐

Concave problems are not easy to solve, even though Property 12.4 helps a lot. What we wish for is convexity. In Chapter 2 we gave a general definition

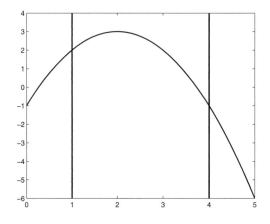

Fig. 12.6   A concave problem may have local optima, but they lie on the boundary of the feasible set.

of convex sets and functions, but how can we check convexity operationally? In the case of a convex and differentiable function of one variable, we know that convexity is linked to nonnegativity of the second-order derivative. If the second-order derivative is strictly positive, then the function is strictly convex. The following theorem generalizes the result to a function of multiple variables.

THEOREM 12.5 *Consider a function $f(\mathbf{x})$ defined on $\mathbb{R}^n$, and assume that its Hessian matrix exists at any point $\mathbf{x}$.[11] The function is convex if and only if its Hessian matrix is positive semidefinite for all $\mathbf{x}$.*

Referring back to Sections 3.8 and 3.9, we do understand what is the basic message of the theorem. A quadratic form is convex if its matrix is positive semidefinite. A generic function, provided that it is suitably differentiable, can be approximated by a second-order Taylor's expansion. If the Hessian matrix for all Taylor's expansions, taken at different points $\mathbf{x}$, is always positive semidefinite, then both the approximation and the original function are convex. From Section 3.8 we also know that this condition could be checked in terms of eigenvalues; more practical checks have been developed, which are a bit too technical for our purposes. There are practically relevant cases, in which we know for sure that an objective function is convex.

Example 12.9 In Example 12.5 we considered the constrained minimization of variance of portfolio return, which is expressed as $\mathbf{w}^T \mathbf{\Sigma} \mathbf{w}$. We know that

---

[11] The theorem can be stated in a less restrictive way, considering the function on an open convex subset of $\mathbb{R}^n$, but we prefer a simplified formulation to grasp the basic message.

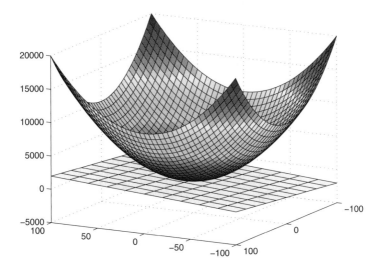

Fig. 12.7 A tangent plane underestimates a convex function.

the covariance matrix Σ is positive semi-definite for sure, otherwise variance could be negative for some setting of weights $\mathbf{w}$. Hence, the minimization of variance, subject to linear constraints, is a convex problem. ☐

We also know that a differentiable and convex function of a single variable has the property that, if we draw the tangent line to its graph at any point, this line lies below the function graph everywhere.[12] Using Taylor's expansion for functions of multiple variables, this property can be generalized as follows.

THEOREM 12.6 *Consider a convex function $f(\mathbf{x})$ defined on $\mathbb{R}^n$, and assume that its gradient vector $\nabla f(\mathbf{x})$ exists at any point $\mathbf{x}$. Then the following inequality holds for all $\mathbf{x}$ and $\mathbf{x}_0$:*

$$f(\mathbf{x}) \geq f(\mathbf{x}_0) + [\nabla f(\mathbf{x}_0)]^T (\mathbf{x} - \mathbf{x}_0)$$

The theorem can be easily visualized for $n = 2$, and it states that the function graph is always above the tangent plane at any point $\mathbf{x}_0$. Figure 12.7 was shown in Chapter 3 and is repeated here for readers' convenience; we observe that the tangent plane lies below the function graph.

Visualization in higher dimensions is impossible, but this theorem allows to prove very easily that stationarity is a necessary and sufficient condition

---

[12]See Fig. 2.28.

for global optimality in an unconstrained problem involving a differentiable convex function.

THEOREM 12.7 *If function f is convex and differentiable on $\mathbb{R}^n$, the point $\mathbf{x}^*$ is a global minimizer if and only if it satisfies the stationarity condition*

$$\nabla f(\mathbf{x}^*) = \mathbf{0}$$

PROOF If f is convex and differentiable, the application of Theorem 12.6 at point $\mathbf{x}^*$ yields

$$f(\mathbf{x}) \geq f(\mathbf{x}^*) + [\nabla f(\mathbf{x}^*)]^T (\mathbf{x} - \mathbf{x}^*)$$

for any $\mathbf{x}$. Since $\mathbf{x}^*$ is a stationarity point, for any $\mathbf{x}$ we have

$$f(\mathbf{x}) \geq f(\mathbf{x}^*) + \mathbf{0}^T (\mathbf{x} - \mathbf{x}^*) f(\mathbf{x}^*) = f(\mathbf{x}^*)$$

i.e., $\mathbf{x}^*$ is a global minimizer. ∎

Necessary and sufficient conditions like the one above are, regrettably, a scarce commodity in mathematical programming, unless we require convexity. To check convexity of the feasible region S, the following properties are useful:

- The set of points described by the inequality $g(\mathbf{x}) \leq 0$ is convex if function g is convex.[13]

- The intersection of convex sets is a convex set.[14]

We see immediately that if we have a set of inequality constraints

$$g_i(\mathbf{x}) \leq 0, \quad i \in \mathcal{I}$$

the resulting feasible set is convex if all functions g_i are convex. The case of equality constraints

$$h_i(\mathbf{x}) = 0, \quad i \in \mathcal{E}$$

is not as easy. To see this, let us rewrite the equality constraint as two inequalities:

$$h_i(\mathbf{x}) \leq 0, \qquad -h_i(\mathbf{x}) \leq 0$$

We have a convex set only if the function h_i is both convex and concave. This will be the case only if function h_i is affine, i.e., the constraint is of the form

$$h_i(\mathbf{x}) = \mathbf{a}_i^T \mathbf{x} - b_i = 0$$

---

[13] See Example 2.34.
[14] See problem 2.9.

Furthermore, any integrality requirement will make the feasible set nonconvex as well.[15]

The best way to wrap up the contents of this section is to list possible structures of mathematical programming problems and the corresponding solution difficulty.

- A *continuous linear programming problem* is both a convex and a concave problem, since the objective function is linear. This implies that a global optimizer can be found by exploring local optimizers on the boundary of the feasible set. The simplex algorithm, described later in Section 12.6.1, takes advantage of these features. More recently, an alternative class of methods, called *interior-point* methods, has been developed and offered in commercial packages. The bottom line is that we are able to solve rather large continuous LPs quite efficiently.

- *Mixed-integer linear programming* problems are nonconvex. Still, the relaxation of integrality requirements yields a continuous LP that can be used to find a bound (optimistic estimate) on the value of the objective function. This is exploited in branch-and-bound methods, described in Section 12.6.2, which are also widely available. Unfortunately, they are based on a tree search approach that calls for the solution of a possibly large number of LPs. Hence, solving a large-scale MILP to optimality may be rather difficult. Nevertheless, impressive improvement in state-of-the-art packages allow one to find a pretty good, if not optimal, solution in many practical cases.

- *Convex, continuous nonlinear programming* problems are relatively easy to solve, if all of the functions involved are differentiable. We will not consider methods for nonlinear programming problems in any detail; however, we describe some theoretical concepts in Section 12.5, since they have a very nice economic interpretation. There is no standard method to deal with nonlinear programming problems, and performance may depend on the exact kind of nonlinearity. Commercial tools offer a set of algorithms, from which the user should select the most appropriate one. Unlike linear programming codes, nonlinear ones require an initial solution from which a search process is started; however, if the problem is convex, the solution does not depend on this starting point. As a general rule, we are not able to solve very large-scale nonlinear problems as efficiently as the linear case. An exception is quadratic programming.

---

[15]Integrality requirements make the feasible set discrete; more precisely, we get disconnected subsets, possibly consisting of isolated points. Clearly, lines joining such points do not belong to the feasible set, which makes it nonconvex. Furthermore, it is not obvious how to define local optimality here, since there is no neighborhood in the sense of Definition 2.13. The bottom line is that, as a general rule with a very few exceptions, integrality requirements make an optimization problem much harder than a continuous one.

If the functions involved are nondifferentiable, additional complications arise. Luckily, most practical management problems may be formulated as LPs or MILPs, at least approximately.

- *Nonconvex, continuous nonlinear programming* problems are difficult, as a general rule. We may apply standard nonlinear programming methods, but the solution can be just a local optimizer, and it may depend on the initial point that we provide to start the search process. There is a set of global optimization methods, which may be rather time-consuming and are not quite general, as they may apply only to specific classes of problems. Still, for moderate-scale problems, we are able to find good solutions, even though commercial tools are not as standard and robust as in the linear programming case.

- *Nonlinear mixed-integer programming* problems are the hardest ones. Until recently, no commercial software tool was available. Now, some advanced packages are available to solve this class of problems; however, the problem size we are able to tackle efficiently is limited, and robustness may be an issue.

From this discussion, it is clear that there are pretty good reasons to represent a decision problem as a linear, possibly mixed-integer programming problem. Sometimes, we have to approximate the problem a bit to obtain a more tractable form. Luckily, there are many modeling tricks that we may use to this aim, and in the next sections many of them are illustrated.

12.2   BUILDING LINEAR PROGRAMMING MODELS

Continuous linear programming (LP) problems are convex mathematical programs, for which extremely efficient solution methods are widely available. Therefore, real-life and large-scale problems can actually be tackled, provided that we are able to cast the decision problem in LP form. To squeeze a problem into the LP paradigm, we need the ability of formalizing decisions, objectives, and constraints, which is more of an art than a science. The first step in learning the art of modeling is to get acquainted with a few standard problems, which can be regarded as paradigms, as well as building blocks for more realistic and interesting problems. In this section we introduce some of these stylized paradigms. Most real-life problems contain part or variations of these prototypes as submodels.

It is also important to draw the line between the abstract form of a model, capturing its structure, and the numerical problem to solve when we associate numerical information to all required data. It is customary to refer to the latter as a *problem instance*, which should not be confused with the problem itself. To illustrate how a problem is stated in its most general form, let us consider the familiar product mix problem, which we explored with a small-scale numerical example in Section 1.1. To generalize that specific problem instance to its general form, we need the following elements:

1. One or more families of *indices* to refer to real-life objects, like items to be produced and resources to be used. These indices correspond to *sets* and are typically used as subscripts when denoting data and decision variables. For instance, in the product mix we will refer to item types by subscript $i = 1, \ldots, N$, where N is the number of item types; by the same token, we consider M resource classes, indexed by $m = 1, \ldots, M$.

2. Then, we must list the data we need to gather so as to specify and solve a problem instance. In the product mix problem, we need:

   - Maximum demand, i.e., an upper bound for production and sales of each item: d_i, $i = 1, \ldots, N$
   - Contribution to profit p_i for each item type i
   - Resource availability R_m (per period) for each resource class $m = 1, \ldots, M$
   - Resource consumption r_{im}, for each item type i on each resource class m

3. Last but not least, we need to formalize the decisions we have to make. In this case, this task is quite easy, since we have to decide only how many items to produce for each item type; hence, the decision variables can be expressed as x_i, $i = 1, \ldots, N$. If production volume is high enough, these decision variables can be assumed as nonnegative real numbers in $\mathbb{R}_+$; otherwise, they have to be restricted to nonnegative integers in $\mathbb{Z}_+$.

(*Remark*: The importance of finding the right decision variables cannot be stressed enough. All too often, students struggle hopelessly to build an LP model because they fail to clarify which kind of decisions should be made. When this is clear, additional variables are typically needed to express constraints and objectives. Hence, model building is an iterative process in which a first set of "natural" decision variables is found; then, auxiliary variables are added when considering each constraint in turn and the objective function.)

Armed with the notation given above, it is now quite easy to state a general product mix problem:

$$\max \quad \sum_{i=1}^{N} p_i x_i \tag{12.17}$$

$$\text{s.t.} \quad \sum_{i=1}^{N} r_{im} x_i \leq R_m, \qquad m = 1, \ldots, M \tag{12.18}$$

$$0 \leq x_i \leq d_i, \qquad i = 1, \ldots, N \tag{12.19}$$

Let us consider each element of the model in more detail:

- Equation (12.17) is the objective function, i.e., the contribution to profit that we want to maximize.

- Equation (12.18) is a *set* of constraints, one for each resource class; for each m, we cannot exceed resource availability. Here we use a simple notation where all of the possible values of index m are listed. Alternatively, we could define a set $\mathcal{M} = \{1, 2, \ldots, m\}$ and write $m \in \mathcal{M}$. You may also find a more mathematically inclined notation, like $\forall m$ or $\forall m \in \mathcal{M}$. Here the *universal quantifier* $\forall$ is used, which simply means "for all." Then $\forall m \in \mathcal{M}$ means "for all elements m in the set $\mathcal{M}$."

- Finally, Eq. (12.19) states that decision variables are nonnegative real numbers not exceeding market limitations d_i for each item i. We recall that simple lower and upper bounds are typically considered separately from more complicated inequality constraints, for the sake of algorithmic efficiency. Insisting on the nonnegativity of x_i may seem redundant, as they enter an objective function that is maximized. However, if we forget that, it may be the case that an item with very low profit margin is produced in a negative quantity, in order to create a fictitious capacity in (12.18), i.e., to make room for the production of more profitable items. If necessary, we may also require integrality of decision variables. Enforcing $x_i \in \mathbb{Z}_+$ transforms the continuous LP above into a pure integer LP.

This is a rather simplistic model, but it is useful to state a couple of remarks in order to avoid common errors when modeling more complex problems.

- In an optimization model, there must be a well-defined objective function. Note that in (12.17) we have a sum over index i. This is the only subscript occurring in both p_i and x_i; hence, we have a well-defined expression. Always make sure that all of the subscripts are "covered" by a sum; for instance, an expression like

$$\sum_{i=1}^{N} c_{ij} x_{ij} \tag{12.20}$$

is *not* a well-defined expression since subscript j is not assigned. By a similar token, it does not make any sense to write something like

$$\max \sum_{i=1}^{N} c_{ij} x_{ij}, \qquad \forall j \tag{12.21}$$

Which sum for which j are we maximizing? We will see later how to define multiobjective problems in a meaningful way, but usually an expression like Eq. (12.21) is the result of a gross mistake.

- Also in constraints we must make sure that all of the involved subscripts are "covered." However, in the case of a constraint, subscripts may be covered either by a sum or a quantifier (but not both). For instance, in (12.18) we have a coefficient r_{im}, where subscript i is covered by the sum and m is covered by the quantifier. Clearly, you must make sure that your use of subscripts makes sense. In this case, we must write a capacity constraint for each resource class m; in each constraint, the sum must run over the whole set of item types using resource m. Also note how the sum couples different products; when capacity constraints are involved, we cannot plan production of each item separately, as they compete for the use of shared and limited resources.

Now we are ready to tackle some variations on production planning and some different classes of models.

12.2.1   Production planning with assembly of components

The naive production mix model is just a starting point in modeling production planning, as many issues that make real-life models interesting and challenging are blatantly disregarded. We will proceed step by step, showing how more realistic features may be represented. In this section we consider one such issue, related to purchasing raw materials or manufacturing intermediate components. In fact, the product mix model directly relates demand to manufacturing of items; this means that we are disregarding the possible presence of components and subassemblies. As an example, let us consider how to generalize the product mix model to a two-level *assembly-to-order* (ATO) model. In an ATO system, we do not produce end items directly. First we make (or buy) modules; then, modules and components are assembled into end items. If manufacturing such modules requires a relatively long lead time, production decisions should be based on forecasted demand. Only when demand for end items is realized (i.e., customer orders are received) is assembly started. An ATO system is two-level in the sense that we have production to stock for modules, before demand is known, and assembly to order for end items. The strategy makes sense when a huge variety of end items results from combining a relatively limited number of basic modules. The sheer number of possible combinations makes stocking end items impossible or way too costly, whereas the time needed to make modules prevents pure production to order. The idea is to exploit a risk pooling effect, by which demand for modules is much less uncertain than demand for end items. A quite familiar example of such a system may be observed when you order a pizza. The wide number of end items prevents preparing a pizza before receiving a customer's order. However, there is a limited number of basic ingredients, which can be prepared in advance and kept ready for use. A more industrial example is car or PC (personal computer) assembly, at least for those producers that offer a large degree of customization as a competitive weapon.

A true appreciation of the ATO system can be obtained only when accounting for uncertainty, something that we defer to chapter 13. Yet, as a first step, let us consider an unrealistic case characterized by deterministic and known demand for end items. Following the drill we have described before, let us define the basic sets we need, and the respective indices. We have

- A set $\mathcal{I}$ of modules, indexed by i

- A set $\mathcal{J}$ of end items, indexed by j

- A set $\mathcal{M}$ of resources, indexed by m

The data we need to tackle the problem extend what we have seen in the product mix problem:

- We need the demand d_j for each end item in $\mathcal{J}$; note that demand is given for end items, not modules.

- We need to know how end items are assembled starting from modules; this means that, in our two-level system, we must specify what in manufacturing parlance is the *gozinto* factor, i.e., the number g_{ij} of modules of type $i \in \mathcal{I}$ needed to assemble one end item of type $j \in \mathcal{J}$. This is a very simple example of a *bill of material* (BOM); true BOMs may entail many more levels and hundreds, if not thousands, of components.

- On the economic side, we need to know at which price p_j we sell end items, and the cost c_i of each module. In the naive mix problem, these two pieces of information are somehow merged into a contribution to profit; but here we make separate production and assembly decisions, and their economic features should be kept apart. For the sake of simplicity, we disregard assembly cost.

- On the technological side, we need the availability of each resource class $m \in \mathcal{M}$, as well as the amount r_{im} of resource m needed to produce one module i. In principle, we should also consider similar data for the assembly process. However, for an ATO system to be effective, assembly must be rather fast. Hence, we assume that assembly is never a bottleneck, or, in other words, that the capacity constraint for assembly will never be binding. Whenever we know that a constraint is redundant, there are at least a couple of reasons for not including it in the model. The first reason is that if a constraint is redundant, then we should eliminate it in order to speed up computation. Actually, state of the art solvers are able to preprocess a model formulation and to eliminate redundant constraints before starting the solution algorithm, but there is a second, and maybe more important reason to avoid useless constraint: Writing them requires collecting data, which is a costly and wasted effort.

Finally, we need to specify our decision variables:

- The amount of modules of type i that we produce is denoted by x_i.

- The amount of end items of type j that we assemble *and* sell is denoted by y_j.

Arguably, both decision variables should be restricted to integer values; for the sake of simplicity, we will assume that a continuous approximation is acceptable. It is important to stress that we are dealing with an unrealistic model, whereby demand is assumed certain. In real-life ATO systems, these two decisions are made at different time instants: We must plan production of module *now*, before we know actual demand; only after receiving customer orders do we start assembly. This will be essential later,[16] when we show how to represent uncertainty in demand. Our first step toward the right direction is the following deterministic LP model:

$$\max \quad -\sum_{i \in \mathcal{I}} c_i x_i + \sum_{j \in \mathcal{J}} p_j y_j \tag{12.22}$$

$$\text{s.t.} \quad \sum_{i \in \mathcal{I}} r_{im} x_i \le R_m, \qquad \forall m \in \mathcal{M} \tag{12.23}$$

$$\sum_{j \in \mathcal{J}} g_{ij} y_j \le x_i, \qquad \forall i \in \mathcal{I} \tag{12.24}$$

$$0 \le y_j \le d_j, \qquad \forall j \in \mathcal{J} \tag{12.25}$$

$$x_i \ge 0, \qquad \forall i \in \mathcal{I}$$

The objective function (12.22) is the difference between revenue from selling end items and cost from manufacturing modules. Constraints (12.23) and (12.25) are essentially the same capacity and market limitations as in the naive optimal mix model. The only constraint worthy of comment is (12.24), which links the two sets of decision variables. By multiplying the number of assembled end items by gozinto factors, and summing over end item types, we obtain the number of modules of type i needed to support assembly; this number is constrained by the number x_i of available modules. The careful reader will notice that this constraint could be rewritten as an equality constraint, as we will certainly not produce more modules than necessary. This is true in the idealized deterministic model, but later we will see that this is not the case when demand uncertainty enters the stage. For now, it may be useful to consider a toy example and analyze its solution.

Example 12.10 To set up a small problem instance, say that we own a (very) small firm, producing just 3 end item types (A_1, A_2, A_3), which are obtained by assembling 5 component types (c_1, c_2, c_3, c_4, c_5). The components we use for each end item are described by a bill of materials, which is flat (just

----

[16]See Section 13.3.1.

Table 12.1 Bill of materials for the assemble-to-order example.

	c_1	c_2	c_3	c_4	c_5
A_1	1	1	1	0	0
A_2	1	1	0	1	0
A_3	1	1	0	0	1

Table 12.2 Bill of resources, cost of components, and available capacity (Cap.).

	M_1	M_2	M_3	Cost
c_1	1	2	1	20
c_2	1	2	2	30
c_3	2	2	0	10
c_4	1	2	0	10
c_5	3	2	0	10
Cap.	800	700	600	

Table 12.3 Demand and selling price of end items.

	Demand	Selling Price
A_1	90	80
A_2	45	70
A_3	90	90

two levels: end items and components). The bill of materials is given in Table 12.1. Each entry in the table is a gozinto factor; for instance, to assemble an end item of type A_2, we need 1 component of type c_1, 1 of type c_2, and 1 of type c_4. From the bill of materials, we see that there are two common components, c_1 and c_2, while the remaining three are specific and characterize each end item. We assume that three resource types (M_1, M_2, M_3) are used for the production of components. Table 12.2 lists:

- The bill of resources, i.e., the time required on each resource to manufacture one component.

- The available capacity for each resource class.

- The cost of each component; this cost might include both direct variable production costs and material costs. Since we assume that assembly is not a bottleneck, we do not list any resource consumption or capacity information concerning assembly.

We note that the cost of every end item type is $20 + 30 + 10 = 60$; we do not make assembly cost explicit, but that would be easy to include. Other relevant data concern demand for end items and the price at which they are sold. They are given in Table 12.3. The last column displays the price at which end items are sold.[17] For all of the three end items, the selling price is larger than 60, the total component cost; however, A_3 looks more profitable, because its contribution to profit is $90 - 60 = 30$, whereas A_2 is the least profitable. We recall that this reasoning may be misleading, in that it does not take into account resource consumption. Solving the model, we obtain the following solution (rounded to 2 decimal digits):

$$x_1^* = 116.67, \qquad x_2^* = 116.67$$
$$x_3^* = 26.67, \qquad x_4^* = 0.00, \qquad x_5^* = 90.00$$
$$y_1^* = 26.67, \qquad y_2^* = 0.00, \qquad y_3^* = 90.00$$

The value of the objective function is 3233.33. In this very small example, we may easily interpret what this solution tries to accomplish. We assemble the maximum number of end items of type A_3, subject to its demand limitation $d_3 = 90$, since this is the most profitable one; this requires in turn the production of a corresponding number of common components c_1 and c_2, and of specific component c_5. Since the market limitation is binding for A_3, there is some capacity left, which is used to produce a limited amount of the specific component c_3, which is needed to assemble end item A_1, plus the corresponding number of common components. End item A_2 has the lowest selling price and is disregarded, as is its specific component c_4. It should be noted that, in general, one should not take for granted that the production of the highest profit item should be maximized; the consumption of available resources should be taken into account as well (as we learned by solving the production mix problem of Section 1.2). ▯

The numerical solution of this toy example is very easy to interpret, but it should also be taken with utmost care. Any experienced production planner would be very critical about its "extreme" nature: It is essentially a bet on demand of the most profitable item. This might make sense if we are quite sure about demand, but what if actual demand turns out to be rather different? We might end up with a very large number of unused components, with a possible loss. The specific component c_5 of A_3 is risky, since demand for this item could be much lower than expected, but any remaining component of type c_5 cannot be used for anything else. To add insult to injury, if demand for A_3 is low but demand for other end items is high, we might lack specific components for A_1 and A_2. If so, we will be unable to use available common components as well. We need a way to obtain robust solutions when facing uncertain scenarios. In Section 13.3.1 we show one possible approach to accomplish this aim.

---

[17] If we do not want to disregard assembly cost, we may substitute selling price by contribution to profit from assembling and selling an item.

12.2.2   A dynamic model for production planning

In the previous two models for production planning there is a major omission: They do not involve any inventory buildup and depletion. From the familiar EOQ model, we know that there is one possible reason for building inventory, i.e., the presence of fixed ordering cost. A similar reason, which may be more relevant when producing rather than just purchasing items, is the need for a setup time before starting production. Imagine that whenever you want to manufacture red pens you have to carry out a setup of machines, taking three hours. Clearly, your production run cannot just last 20 minutes, as you would waste a lot of production capacity. Modeling fixed charges and setup times calls for binary decision variables, and we will cover such models later in Section 12.4.4.

The EOQ model, by nature, disregards two more possible reasons for building inventories:

- Demand uncertainty, which can be hedged by setting a safety stock level. We have seen a very simple approach in Section 7.8.2. More sophisticated approaches may call for the modeling techniques that we describe in the next chapter.

- Demand variability, which is not necessarily due to uncertainty. In Chapter 11 we have considered seasonality in time series, which is a kind of *predictable* variability. Now imagine that demand is perfectly known, but features remarkable swings over time. How should we set our capacity level? One possibility is to set a production capacity level corresponding to peak demand, but when demand is low we incur the burden of a lot of expensive resources left unused. Another possibility is setting an intermediate production capacity level, building inventory when demand is low, and depleting inventory when demand is high.

In this section we extend the production mix model in order to account for predictable demand variability; this is our first example of a *dynamic decision model*. When building a dynamic model, the first issue is whether to represent a continuous or a discrete time. In production planning, discrete time is the standard choice. One good reason is that forecasting models we are familiar with are all based on discrete time buckets. Another reason is that when planning production with, say, weekly time buckets, we build a plan leaving room for lower and more detailed decision levels in charge of executing it. Given this discussion, the first ingredient that we need is a dynamic representation of time and demand. We assume that we have a collection of equal length time buckets, indexed by $t = 1, 2, \ldots, T$. Let us denote by d_{it} the demand for item $i = 1, 2, \ldots, N$,[18] *during* time bucket t. In this case, we may

---

[18]Note that, to be uniform with the way we denote the set of time buckets, we explicitly enumerate product types, rather than defining a set $\mathcal{I}$ and writing $i \in \mathcal{I}$, like we did in the

allow for a time-varying resource availability R_{mt} of each resource, indexed by $m = 1, \ldots, M$. Doing so, we may take into account several practical issues such as time-varying working calendars, holidays, or planned maintenance activities affecting production capacity.

Now what decision variables do we need? Clearly, the production amount is time-varying too, and we may denote by x_{it} the amount of item i produced *during* time bucket t. Now, we may immediately generalize the capacity constraints we are familiar with:

$$\sum_{i=1}^{N} r_{im} x_{it} \le R_{mt}, \qquad m = 1, \ldots, M, \ t = 1, \ldots, T$$

where r_{im} is the amount of resource m required for the production of one unit of item i. Do we need some more decision variables? From the EOQ model, we know that we must account for inventory holding costs. The most natural choice is to introduce a set of decision variables I_{it}, representing the inventory level of item i, *at the end* of time bucket t. Please note the difference between variables x_{it} and I_{it}: We observe inventory levels only at discrete time instants, without checking what happens within each time bucket; hence we assume that I_{it} refers to the level at the end of each period. By a similar token, we do not know exactly when the amount x_{it} is produced within each time bucket, nor when demand d_{it} is met. We assume that the time bucket is small enough that this is no concern and check inventory balance only at the end of time buckets. Indeed, the inventory balance is a fundamental equation in the model, linking the two sets of decision variables.[19] In order to formalize inventory balance, we must carefully specify which objective we are going to pursue – maximization of profit or minimization of cost – as they need not be equivalent.

The case of profit maximization   In the static optimal mix model, what we produce and what we sell are the same, since there is no inventory decoupling production and sales. In a dynamic model, we need to specify how much we sell of each item type i during each time bucket t. Let z_{it} be such a variable. Clearly, we cannot sell more than demand, so we should enforce the bound $z_{it} \le d_{it}$. In order to link all of the variables, we need to formalize a rather natural principle: The inventory at the end of time bucket t is the inventory at the beginning of time bucket t, plus what we produced during the time bucket, minus what we sold during the time bucket. Since inventory at the beginning of time bucket t is just inventory at the end of time bucket $t - 1$, we write

$$I_{it} = I_{i,t-1} + x_{it} - z_{it}, \qquad \forall i, t$$

ATO case. In practice, the two notations are equivalent and we may choose what we feel more comfortable with, case by case.

[19] See also Section 2.13.4.

Strictly speaking, we made a little mistake as far as time bucket $t = 1$ is concerned; for the first time bucket, the constraint involves the initial inventory level I_{i0}, which is *not* a decision variable and is part of the problem data. All the involved decision variables are restricted to nonnegative values.

To write the objective function, we need some fundamental economic information concerning each item:

- The inventory holding cost h_i

- The unit production cost c_i

- The selling price p_i, where we assume $p_i > c_i$

Then, we may write the following LP model:

$$\max \quad \sum_{i=1}^{N} \sum_{t=1}^{T} p_i z_{it} - \sum_{i=1}^{N} \sum_{t=1}^{T} c_i x_{it} - \sum_{i=1}^{N} \sum_{t=1}^{T} h_i I_{it} \qquad (12.26)$$

$$\text{s.t.} \quad I_{it} = I_{i,t-1} + x_{it} - z_{it}, \qquad i = 1, \ldots, N, \ \ t = 1, \ldots, T$$

$$\sum_{i=1}^{N} r_{im} x_{it} \le R_{mt}, \qquad m = 1, \ldots, M, \ \ t = 1, \ldots, T$$

$$z_{it} \le d_{it}, \qquad i = 1, \ldots, N, \ \ t = 1, \ldots, T$$

$$x_{it}, \ z_{it}, \ I_{it} \ge 0$$

We may make a few observations:

- If the capacity constraints are nonbinding, i.e., if coefficients R_{mt} are large enough, the optimal solution is obviously $x_{it}^* = d_{it}$. We satisfy the whole demand producing just in time. It is the capacity constraints that may call for decoupling between production and sales, as well as cause some lost demand.

- Even without solving the model, we may be sure that in the optimal solution $I_{iT}^* = 0$, i.e., inventories are depleted during the last time bucket. This happens because, as far as the model is concerned, the world stops at $t = T$. This kind of "end-of-horizon" effect may induce some distortions, unless T is large enough, and the solution we get is applied according to a "rolling horizon" strategy. The planning horizon rolls forward, in the sense that it initially involves time buckets $(1, \ldots, T)$, then $(2, \ldots, T+1)$, $(3, \ldots, T+2)$, and so on. Only the decision in the first time bucket is actually implemented, as the plan is revised after more information about demand is collected. By doing so, it is reasonable to expect that the end-of-horizon effect will be less critical; however, if the planning horizon is too long, we might lack reliable demand data.

The case of cost minimization   If we compare the model above with the EOQ model, we notice a few fundamental differences. Of course, demand is dynamic here, and we might introduce fixed charges or setup costs to make the model more complete and perhaps realistic. But a more fundamental difference is that when we maximize profit we do not take for granted that demand is fully satisfied. On the contrary, in the EOQ model we seek a solution allowing us to meet the whole demand, at minimal cost. In order to build a model more akin to EOQ, we should enforce full satisfaction of demand:

$$z_{it} = d_{it}, \qquad \forall i, t$$

Doing so has a significant impact on the objective function.

1. The first term becomes a constant term,

$$\sum_{i=1}^{N}\sum_{t=1}^{T} p_i z_{it} = \sum_{i=1}^{N}\sum_{t=1}^{T} p_i d_{it}$$

   and it may be dropped from the model. Now we are just maximizing a sum of terms multiplied by -1; hence we may transform the maximization into a minimization.

2. By a similar token, the second term, which is related to production cost, boils down to a constant as well, since what we produce of each item is just the sum of demand over the planning horizon, minus an initial inventory I_{i0}, plus a possible value I_{iT}, if this is explicitly enforced to avoid the end-of-horizon effect.

So, model (12.26) can be simplified as follows:

$$\min \quad \sum_{i=1}^{N}\sum_{t=1}^{T} h_i I_{it} \tag{12.27}$$

$$\text{s.t.} \quad I_{it} = I_{i,t-1} + x_{it} - d_{it}, \qquad i = 1, \ldots, N, \ \ t = 1, \ldots, T$$

$$\sum_{i=1}^{N} r_{im} x_{it} \leq R_{mt}, \qquad m = 1, \ldots, M, \ \ t = 1, \ldots, T$$

$$x_{it}, I_{it} \geq 0$$

In this model, it is important to emphasize the role of the nonnegativity condition for inventory levels. This, together with inventory balance, implies

$$I_{i,t-1} + x_{it} \geq d_{it}$$

i.e., production during time bucket t plus inventory carried over from the previous time bucket must be enough to cover demand, which is never back-logged. This opens a thorny practical issue: This model, unlike the previous ones, may fail to have a feasible solution. If demand is too large with respect to available capacity, no algorithm will be able to find a solution. This may be rather inconvenient; imagine a production planner sitting in front of a computer informing her that, alas, there is no feasible solution. It would be nice to get some diagnostic clue about which item or which resource seems to be critical and when. Later, in Section 12.3.4, we discuss elastic model formulations that are a possible way to overcome this difficulty.

As a final remark, we may note that, strictly speaking, the introduction of inventory decision variables is not necessary. Assuming full satisfaction of demand, we may substitute inventory holding with the difference between cumulative production and cumulative demand:

$$I_{it} = \sum_{k=1}^{t} x_{ik} - \sum_{k=1}^{t} d_{ik} + I_{i0}, \qquad \forall i, t$$

As a general rule, however, model readability should have a higher priority than the reduction of decision variables. Many years ago, when problems had to be solved by hand, such economies were worth pursuing. With lightning-fast hardware doing the work for us now, the amount of decision variables is much less of a concern. Furthermore, state-of-the-art solvers are able to carry out automatic preprocessing of a model, possibly eliminating redundant variables and constraints to improve its solvability, before starting the solution algorithm itself.

12.2.3 Blending models

In the production planning models that we have considered so far, there is a very precise way of producing each item type. When producing a car, you typically need an engine and four wheels. Factors cannot be substituted; there is no way to convince a customer to buy a car with 20 wheels and no engine. This is typical of discrete manufacturing, but there are situations allowing several ways to produce an item by blending ingredients, provided we satisfy the constraints that define an acceptable mix. This happens in the oil industry, where different chemical components are blended to obtain gasoline meeting some economical and regulatory constraints, as well as in certain food industries. In fact, the prototypical blending problem is the diet problem.

Example 12.11 (The diet problem) When specifying a diet, we want to make sure that we take a minimum amount of certain nutrients, like proteins and vitamins. On the other hand, we want to place an upper bound on other elements, like fat and sodium. We obtain such elements by eating different foods, that have different prices. Now suppose that we want to find the minimum cost diet, such that all of the elements stay within their prescribed bounds. This is a very crude approach, of course, as a diet has many more features that we should pay attention to, and we are disregarding the need for some day-to-day variation. Yet, it is a good starting point.

To formalize the problem, we define first the basic sets we are dealing with: the set $\mathcal{I}$ of elements or nutrients, indexed by i, and the set $\mathcal{J}$ of food types, indexed by j. The information we need is

- The cost c_j of each food type $j \in \mathcal{J}$.

- The content a_{ij} of each nutrient $i \in \mathcal{I}$ contained in a unit amount (measured in weight or volume) of food type $j \in \mathcal{J}$.

- The lower and upper bounds, l_i and u_i, on the daily assumption of each element; if one of the two bounds is missing, we may just set it to 0 or $+\infty$, respectively.

To specify a diet, we use as decision variable the amount x_j of each food $j \in \mathcal{J}$ that is consumed per day.[20] Then, the model is rather easy to write down:

$$\min \quad \sum_{j \in \mathcal{J}} c_j x_j$$

$$\text{s.t.} \quad l_i \le \sum_{j \in \mathcal{J}} a_{ij} x_j \le u_i, \qquad\qquad i \in \mathcal{I} \qquad\qquad (12.28)$$

$$x_j \ge 0 \qquad\qquad\qquad\qquad j \in \mathcal{J}$$

The model is pretty self-explanatory, but once again we invite the reader to observe the use of indices in constraint (12.28). The coefficients a_{ij} have a double subscript, and one is covered by the sum, while the other one is covered by the enumeration of constraints. $\square$

In the diet problem, the lower and upper bounds are given in *absolute* terms, since we specify the minimum and maximum amounts of proteins that one should consume per day. In other settings, blending constraints are more naturally expressed in percentage terms. To see a trivial example, imagine that you have to prepare a cocktail for your next party, blending three liquors, L_a, L_b, and L_c. The alcoholic content of each of the three ingredients is $d_a = 20°$, $d_b = 32°$, and $d_c = 55°$. You want to keep the damage that your guests will inflict on your home to an acceptable level, so the alcoholic content of your brew should be in the range between lower bound $L = 25°$ and upper bound $U = 30°$. If we assume that the alcoholic content of the mixture depends linearly on the percentage of each ingredient, we have the following constraint:

$$L \le d_a z_a + d_b z_b + d_c z_c \le U$$

---

[20]These decisions may be also subject to bounds on certain food amounts, in order to avoid a very unpleasing diet.

where variables z_a, z_b, and z_c are the *percentage* of each element in the cocktail. However, this is not very operational: If you have a given number of half-empty bottles of each ingredient, you would like to get a recipe giving you exactly *how much* of each liquor you should use. If we denote by x_a, x_b, and x_c the actual volume used of each liquor, respectively, the constraint above can be written as

$$L \leq \frac{d_a x_a + d_b x_b + d_c x_c}{x_a + x_b + x_c} \leq U$$

This form is not very pleasing however, as it is nonlinear, but it is very easy to get an equivalent linear expression. The amount of cocktail we mix, assuming there is no loss in volume, is

$$y = x_a + x_b + x_c$$

and the constraint above can be written as

$$Ly \leq d_a x_a + d_b x_b + d_c x_c \leq Uy$$

The introduction of decision variable y is not strictly necessary, but it helps to improve model readability.

Example 12.12 (Optimal mix in a process industry) Let us write an optimal mix model suited for process industries, whereby the blend of raw materials to produce end items features some recipe flexibility. We disregard capacity constraints, but we assume that raw materials are available in limited amounts, and we must make the best of them. Limited availability of raw materials may be due to budget limitations in purchasing, limited availability on the market, or limited stocking capacity. We have a set $\mathcal{I}$ of raw materials, indexed by i, and a set $\mathcal{J}$ of end products, indexed by j. Let us denote by $\mathcal{R}_j$ the subset of raw materials that we may use in blending end product j. In this simple example we go to the opposite extreme with respect to discrete manufacturing with fixed and very rigid bills of materials, as we assume that we may use any raw material in the proper subset to blend an end product. Real-life problems fall somewhere between these extremes. The only constraint we enforce on the mix of raw materials relates to a quality measure. Each raw material has some feature, which is measured by a score q_i (the larger, the better). When we mix ingredients, the quality score of the resulting blend is a linear combination, an average, if you prefer, of the scores of the inputs, with weights proportional to the percentage of each raw material used. The quality score of each end product must fall between lower and upper bounds L_j and U_j. The economic side of the coin is represented by the unit cost c_i of raw materials and the selling price p_j of end items. Demand for end items is bounded by d_j, and raw-material availability is represented by inventory level I_i.

Assuming that we want to maximize profit, which decision variables should we introduce? On the commercial side, we must decide the amount of each end product that we blend and sell. Let us denote this amount by y_j. On the blending side, we must decide how much of each raw material i we use in blending end product j. Let us denote this variable by x_{ij}, which is defined only for $i \in \mathcal{R}_j$. We may accomplish our aim by solving the following LP:

$$\max \quad \sum_{j \in \mathcal{J}} p_j y_j - \sum_{j \in \mathcal{J}} \sum_{i \in \mathcal{R}_j} c_i x_{ij} \qquad (12.29)$$

$$\text{s.t.} \quad y_j = \sum_{i \in \mathcal{R}_j} x_{ij}, \qquad\qquad \forall j \in \mathcal{J} \qquad (12.30)$$

$$L_j y_j \leq \sum_{i \in \mathcal{R}_j} q_i x_{ij} \leq U_j y_j, \qquad \forall j \in \mathcal{J} \qquad (12.31)$$

$$\sum_{j : i \in \mathcal{R}_j} x_{ij} \leq I_i, \qquad\qquad \forall i \in \mathcal{I} \qquad (12.32)$$

$$0 \leq y_j \leq d_j \qquad\qquad \forall j, \in \mathcal{J}$$

$$x_{ij} \geq 0, \qquad\qquad \forall j \in \mathcal{J}, i \in \mathcal{R}_j$$

Equation (12.29) is profit, expressed as revenue from sales, minus cost of raw materials; note that in the second term we have a double sum, where we first sum with respect to end products, and then with respect to each raw material that may be used to blend that end product. Equation (12.30) is essentially a balance constraint related to conservation of mass; when complex chemical reactions are involved, a more careful modeling may be needed to account for nonlinearities. Equation (12.31) ensures that quality stipulations are met. Finally, Eq. (12.32) ensures that we do not use more raw materials than available. The sum in Eq. (12.32) might look odd at first sight; the notation $j : i \in \mathcal{R}_j$ specifies the indices of end products for which raw material i is part of the recipe $\mathcal{R}_j$; formally, it denotes the subset of $j \in \mathcal{J}$ such that i is in $\mathcal{R}_j$. Note that index i is fixed by constraint enumeration, since we have one such constraint for each ingredient, and that by doing so we sum only over defined variables x_{ij}. In practice, this is essentially an inversion of the mapping represented by $\mathcal{R}_j$, which maps end item j into the set $\mathcal{R}_j$ of raw materials; given a raw material i, we want to sum over the set of end items that use it.[21]

Clearly, this is an oversimplified model. One questionable point concerns the cost of raw materials. From the perspective of a static model like this one, one could argue that this cost is sunk, and we should not consider it, as there is no value in keeping unused raw materials. In a realistic model, we would

---

[21]Some readers might find this way of working with sets a bit abstract. In practice, optimization models are expressed using special languages, like AMPL, that describe the structure of an optimization model. This structure is matched against data to produce a specific problem instance that is passed to a solver. These languages have quite powerful set manipulation tools, allowing for the expression of quite complicated models in a remarkably transparent way.

probably consider time and purchasing decisions; in many settings, the timing of purchase is important as some raw materials feature quite significant price swings.   ⬜

12.2.4 Network optimization

Many real-life optimization problems relate with transportation of items on a network. This is clearly a relevant class of problems in supply chain management, but also many telecommunications problems involve networks on which data flow, rather than physical commodities. More surprisingly, some dynamic problems may be represented as network models on which items flow in time, rather than space. In this section we describe two stylized models that are building blocks for dealing with more realistic problems. To begin with, we should formally define what we mean by a "network."

DEFINITION 12.8 (Network) *A network is a collection of* **nodes** *and* **arcs**. *Let $\mathcal{N}$ be the set of nodes, which are typically physical locations. Arcs represent links between pairs of locations and are actually pairs (i, j) of nodes. If the pair is ordered, we have a directed network; otherwise, we have an undirected network.*[22] *In a directed network, arcs have a specific direction along which items flow. Let us denote the set of arcs by $\mathcal{A}$, which is a subset of the Cartesian product of nodes, $\mathcal{A} \subseteq \mathcal{N} \times \mathcal{N}$. Numerical information may be attached to arcs, such as transportation capacity, and nodes, such as the maximum throughput of a logistic facility.*

A very simple two-level network is depicted in Fig. 12.8. This is a kind of network considered in our first model, which is known as the *transportation problem*, even though it is actually a very simplified view of a real-life transportation problem. The network is two-level since we have two disjoint sets of nodes: the set $\mathcal{S}$ of source nodes and the set $\mathcal{D}$ of destination nodes. Referring to the figure, we have $\mathcal{S} = \{A, B, C\}$ and $\mathcal{D} = \{1, 2, 3, 4, 5, 6\}$. Examples of (directed) arcs are $(A, 1)$ and $(C, 4)$; there is no $(B, 6)$ arc. Also, when there is no arc connecting two sources or two destinations; we say that the network is *bipartite*. Destination nodes may represent retail stores, which are characterized by a demand d_j, $j \in \mathcal{D}$, measured on a timespan of interest. In the prototype transportation model, only one type of commodity is considered. Source nodes might represent production plants, with a given limited capacity R_i, $i \in \mathcal{S}$, measured over the same timespan as demand. For each source–destination pair, i.e., for each arc (i, j), $i \in \mathcal{S}$, $j \in \mathcal{D}$, we have a unit transportation cost c_{ij}. This is considered as a variable linear cost; clearly, this is just a very rough approximation of a real-life transportation cost. The transportation problem consists in finding the minimum-cost set of flows, over

---

[22]To be precise, when dealing with an undirected network, we should speak of *vertices* and *edges*, rather than nodes and arcs.

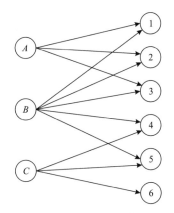

Fig. 12.8 A two-level, bipartite network corresponding to a transportation problem.

all of the links (i, j), such that demand is met and plant capacities are not exceeded.

To represent the transportation problem as a mathematical programming model, we have to find suitable decision variables first. In this case, it is fairly evident that we need one decision variable for each link (i, j); let x_{ij} be the flow on each arc. The resulting linear programming model is

$$\min \quad \sum_{i \in \mathcal{S}} \sum_{j \in \mathcal{D}} c_{ij} x_{ij} \tag{12.33}$$

$$\text{s.t.} \quad \sum_{i \in \mathcal{S}} x_{ij} = d_j, \qquad \forall j \in \mathcal{D} \tag{12.34}$$

$$\sum_{j \in \mathcal{D}} x_{ij} \leq R_i, \qquad \forall i \in \mathcal{S} \tag{12.35}$$

$$x_{ij} \geq 0$$

The objective function (12.33) is a sum over all the pairs of nodes, and it amounts to the total transportation cost. The expression above assumes that there is an arc for any source–destination pair, which need not be the case. We could think of associating a suitably high cost c_{ij} with nonexistent arcs, so that they are never used. A possibly more elegant solution is to sum directly over the set $\mathcal{A}$ of arcs: $\sum_{(i,j) \in \mathcal{A}} c_{ij} x_{ij}$. The constraint (12.34) ensures that demand is met at each destination node, by summing inflows from plants. The capacity constraint, limiting outflows from any source, is represented by Eq. (12.35). Notice the reversal of roles between subscripts i and j in constraints (12.34) and (12.35). This model is extremely simplistic and provides us with only a starting point for further modeling. To begin with, it is a static model ignoring time patterns in demand (demand variability). In principle, it is easy to extend the model to a dynamic one by introducing time-varying demand d_{jt} and inventory variables at nodes. By the same token, we could consider

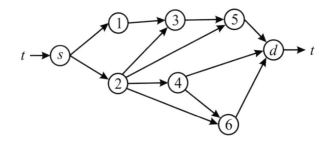

Fig. 12.9   A network corresponding to a minimum-cost flow problem.

diversified production costs across the plants, different items or families, and a more realistic transportation cost structure, possibly including fixed charges and economies of scale.

Another classical model is the *minimum-cost flow problem*, which features a multilevel network like the one depicted in Fig. 12.9. Sometimes, networks are arranged in well-defined layers, which may correspond to factories, warehouses, large distribution centers, and retail stores. To be as general as possible, let us deal with an unstructured network, in which one specific node denoted by s is the source of a given flow t, which must be routed to the destination node denoted by d. Each arc (i,j) in the network is directed and associated with a cost c_{ij}, and we want to route the flow from s to d at minimum cost. In a generic problem, we might have several commodities, more than one source and one destination, as well as capacities on both nodes and arcs. Let us consider a basic model, whereby flow on arc (i,j) cannot exceed an upper bound u_{ij}. Like in the transportation problem, we have one set of decision variables, the flow x_{ij} on arc (i,j). The only issue with this basic minimum-cost flow model is representing conservation of flows at each node: The flow going out of a node must equal the flow going into the node. When considering in- and outflows, we must account for arcs going into and out of a node, plus possible exogenous input and output flows. Let F_j be the net outflow of node $j \in \mathcal{N}$. The conservation of flows at node j reads as follows:

$$\sum_{(i,j)\in\mathcal{A}} x_{ij} = \sum_{(j,i)\in\mathcal{A}} x_{ji} + F_j$$

Note that if the net outflow F_j is positive, it represents the flow absorbed at node j; if it is negative, it represents flow generated at node j. The left-hand side of the equality represents the node inflow, i.e., the sum of flows on all arcs (i,j) entering node j. The right-hand side includes the node outflow, i.e., the sum of flows on all arcs (j,i) going out of node j, plus the net outflow F_j. In the simple network of Fig. 12.9, F_j is zero for all nodes except source and destination. With our convention, $F_s = -t$ and $F_d = t$, and we may write the model as follows:

$$\min \quad \sum_{(i,j)\in\mathcal{A}} c_{ij}x_{ij} \tag{12.36}$$

$$\text{s.t.} \quad \sum_{(i,j)\in\mathcal{A}} x_{ij} = \sum_{(j,i)\in\mathcal{A}} x_{ji} + F_j, \qquad \forall j \in \mathcal{N} \tag{12.37}$$

$$0 \leq x_{ij} \leq u_{ij}, \qquad\qquad\qquad \forall (i,j) \in \mathcal{A}$$

We express the objective function (12.36) as a sum over the arc set $\mathcal{A}$. We might also mention that not all of the flow balance constraints (12.37) need be written. Summing all of them, we get an identity $0 = 0$, implying that they are not all linearly independent. We may get rid of any one of them, but this is not relevant with respect to model formulation and is accommodated by solution procedures.

The transportation and the minimum-cost flow problems may be solved by any linear programming solver. Actually, they have very specific structures that may be exploited by special purpose solution procedures. State-of-the-art commercial solvers offer such procedures, and are even able to detect network structure automatically. Such algorithmic tricks are beyond the scope of this book, but there is one more feature that is worth mentioning. The decision variables x_{ij} are not restricted to integer values. However, if all of the problem data are integers, it can be shown that the solution of the problem will be integer even if we apply a solver for continuous LPs. This depends on the peculiar structure of the constraint matrix, which defines a polyhedron whose vertices have integer coordinates. Unfortunately this nice property is lost when we introduce different commodities in the model, competing for limited transportation capacity.

12.3   A REPERTOIRE OF MODEL FORMULATION TRICKS

The models we have described in the last section rely on two quite relevant limiting assumptions:

1. They are linear, whereas in real life we deal with economies or disec-onomies of scale that introduce nonlinearities in the model. Yet, the efficiency of LP solvers is so remarkable that, whenever we can, we should try to squeeze our model into a linear framework. This is rather natural when we have a piecewise linear objective function, as we discuss in full generality later, in Section 12.4.7.

2. They optimize a single objective, which amounts to say that whatever criteria we use in evaluating a solution can be aggregated into one func-tion, usually representing cost or profit. However, we might deal with conflicting criteria that cannot be expressed in common monetary terms.

Luckily, there is an array of modeling tricks that can be used to partially overcome these difficulties. In the next sections we illustrate a few of them, in order to show that the LP modeling framework is less restrictive than it might appear. We also consider a couple of approaches to deal with multiple objectives. Then, we discuss the use of elastic model formulations, as well as a general and powerful approach, modeling by columns, which may be used to model seemingly tough problems in a very simple way.

12.3.1   Alternative regression models

When dealing with simple linear regression, we typically use least squares to fit the coefficients of a simple linear model $y = a + bx$. Given a set of joint observations (x_i, y_i), $i = 1, \ldots, N$, we define residuals

$$e_i \equiv y_i - (a + bx_i), \qquad i = 1, \ldots, N$$

and minimize the sum of squared residuals:

$$\min_{a,b} \sum_{i=1}^{N} e_i^2$$

This is actually a quadratic program, but because of the simplicity of constraints, we know from Chapter 10 that a straightforward analytical solution is available, lending itself to statistical analysis. However, we also pointed out that alternative fitting procedures could be considered:[23]

- We may take the absolute value of residuals, rather than squaring them:

$$\min_{a,b} \sum_{i=1}^{N} \left| y_i - (a + bx_i) \right| \tag{12.38}$$

- We may also consider the minimization of the worst-case residual:

$$\min_{a,b} \left\{ \max_{i=1,\ldots,N} \left| y_i - (a + bx_i) \right| \right\} \tag{12.39}$$

Neither problem looks linear at all. In the first case, we are dealing with an absolute value, which is nonlinear and nondifferentiable. In the second case, we have a min–max problem which does look a bit awkward. In fact, both are easily converted into LP models by quite useful modeling tricks.

As a first step, consider the representation of absolute value $| x |$. The number x could be either positive or negative. If we denote its positive and

---

[23] See Section 10.1.1.

negative parts by $x^+ \equiv \max\{0, x\}$ and $x^- \equiv \max\{0, -x\}$, respectively, we see that x can be written as

$$x = x^+ - x^- \tag{12.40}$$

Then, the absolute value can be written as

$$|x| = x^+ + x^- \tag{12.41}$$

As we already remarked, there is a potential ambiguity in Eq. (12.40), since there is an infinite number of ways to express a number x as the difference of two nonnegative numbers x^+ and x^-; for instance $5 = 5 + 0 = 8 - 3$. So, it seems that we should write a nonlinear constraint like $x^+ \cdot x^- = 0$, to make sure that at most one of the two auxiliary variables is nonzero. However, since the absolute value enters the objective function (12.38), this condition will hold in the optimal solution; if we write $5 = 8 - 3$ rather than $5 = 5 - 0$, we increase the objective function by 6 without changing anything substantial. Hence, to cast (12.38) in linear form, we introduce auxiliary variables $w_i^+ \geq 0$ and $w_i^- \geq 0$ for each observation, and transform (12.38) as follows:

$$\min \quad \sum_{i=1}^{N} \left(w_i^+ + w_i^- \right)$$
$$\text{s.t.} \quad w_i^+ - w_i^- = y_i - a - bx_i, \qquad i = 1, \dots, N$$
$$w_i^+, w_i^- \geq 0, \qquad i = 1, \dots, N$$

It is important to realize that the decision variables in this model are the auxiliary variables w_i^+ and w_i^-, and the coefficients a and b; x_i and y_i are observed *data*.

By a similar token, problem (12.39) may be easily transformed into LP form by a standard modeling trick as well. To generalize a bit, let us consider the following min–max problem:

$$\min_{\mathbf{x} \in S} \left\{ \max_{i=1,\dots,N} g_i(\mathbf{x}) \right\}$$

The idea is that, for a given $\mathbf{x}$ in the feasible set S, we should evaluate each function $g_i(\mathbf{x})$ and pick the largest value; this defines the objective function, which should be minimized over the feasible set. To transform the model into a more familiar form, let us introduce the auxiliary variable z, and rewrite it as:

$$\min \quad z$$
$$\text{s.t.} \quad z \geq g_i(\mathbf{x}), \qquad i = 1, \dots, N$$
$$\mathbf{x} \in S$$

Depending on the nature of set S and functions $g_i(\mathbf{x})$, this may be a very tough or easy problem. Since the functions involved in (12.39) can be transformed in linear form, we obtain an easy LP:

$$\min \quad z$$
$$\text{s.t.} \quad z \geq w_i^+ + w_i^-, \qquad\qquad i = 1, \ldots, N$$
$$w_i^+ - w_i^- = y_i - a - bx_i, \qquad i = 1, \ldots, N$$
$$w_i^+, w_i^- \geq 0, \qquad\qquad i = 1, \ldots, N$$

12.3.2 Goal programming

The deviation variables that we have utilized in order to formulate alternative regression models as LPs have other uses as well. Let us consider a generic optimization problem over a feasible set S. A standard complication of real-life decision problems is that there is not just one criterion to evaluate the quality of a solution, but many. Say that such criteria are represented by functions $f_i(\mathbf{x})$, $i = 1, \ldots, M$. We would like to maximize some of these functions, whereas others should be minimized. Needless to say, alternative criteria are usually in conflict with each other, and there is no easy way to assess a satisfactory tradeoff. In the next section we consider one possible approach to multiobjective optimization, and in this section we consider a possible alternative, based on setting desirable targets, or goals, for each objective.

Even though it is impossible to find a solution optimizing all the criteria at the same time, we might be able to find values such that we would be satisfied with them. Let us denote these target values, or *goals*, by G_i^*. This means that if we could find a solution $\mathbf{x} \in S$, such that

$$f_i(\mathbf{x}) = G_i^*, \qquad i = 1, \ldots, M$$

we would be satisfied. Most likely, even this is impossible to accomplish, but we might settle for a solution with minimal deviations from prescribed goals. So, let us introduce deviations w_i^+ and w_i^- from goals, associated with penalties P_i^+ and P_i^-, respectively. A *goal programming* model may be stated as follows:

$$\min \quad \sum_{i=1}^{M} \left(P_i^+ w_i^+ + P_i^- w_i^- \right) \qquad\qquad (12.42)$$
$$\text{s.t.} \quad f_i(\mathbf{x}) = G_i^* + w_i^+ - w_i^-, \qquad i = 1, \ldots, M \qquad (12.43)$$
$$\mathbf{x} \in S \qquad\qquad (12.44)$$

If the functions f_i are linear, then this will be an LP problem. The difficulty, of course, is setting goals and penalties. For a given function f_i, we need not set two positive penalties. If the function is linked to a profit, we need not penalize a profit larger than a threshold goal, but we should penalize only underachievements.

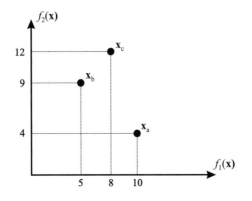

Fig. 12.10 Illustrating the concept of dominated solution.

The flexibility in setting goals and penalties might look quite confusing, but goal programming might be a useful framework for building an *interactive* decision support tool for an experienced user.

12.3.3 Multiobjective optimization

Goal programming is one way of dealing with conflicting objectives, but it requires the assessment of weights and targets. Unfortunately, it may be very difficult, or even unethical, to figure out weights. As an example, consider the tradeoff between the cost of a production process and its pollution level. Sometimes, we would like to visualize the tradeoff between conflicting requirements. Moreover, it would be nice to see which solutions can be safely ruled out, in a sense that we can illustrate by a small example.

Example 12.13 Consider a problem involving two objectives, $f_1(\mathbf{x})$ and $f_2(\mathbf{x})$, that we wish to minimize. Solutions are best visualized on a plane illustrating how solutions score with respect to both objectives. Consider the three alternative solutions $\mathbf{x}_a$, $\mathbf{x}_b$, and $\mathbf{x}_c$ depicted in Fig. 12.10. If we compare $\mathbf{x}_a$ and $\mathbf{x}_b$, which solution is the better one? Actually, it depends on our point of view: $\mathbf{x}_a$ is better in terms of the second objective, since $4 < 9$, but $\mathbf{x}_b$ is preferable in terms of the first objective, since $5 < 10$. Unless we assign some priority or weight to objectives, there is no way to rank the two alternatives. However, if we compare $\mathbf{x}_b$ against $\mathbf{x}_c$, the answer should be obvious: $\mathbf{x}_b$ is better from both points of view, so no decision maker interested in objectives f_1 and f_2 should choose $\mathbf{x}_c$, which is a "dominated" solution.   ⬜

From a mathematical perspective, each feasible solution is characterized by a vector of objective values; hence, we could consider a "vector" optimization problem:

$$\text{``min''} \quad \begin{bmatrix} f_1(\mathbf{x}) \\ f_2(\mathbf{x}) \end{bmatrix} \tag{12.45}$$

$$\text{s.t.} \quad \mathbf{x} \in S$$

However, stated as such, the problem has no meaning, and this is why we use "min." As the example above illustrates, vectors are not a well-ordered set. Given any pair of distinct points on the real line, we may say which one is larger, but we cannot rank vectors on a plane in the same way: The number 5 is larger than the number 2, but we cannot compare vectors $[10 \ 4]^T$ and $[5 \ 9]^T$ that easily. True, we can say that a vector is longer than another one by referring, e.g., to the Euclidean concept of vector length. However, this means that we are aggregating different components of the vector into a single number, using the Euclidean norm. In multiobjective optimization we could consider, e.g., weighted sums to transform a vector of objectives into a single one; this operation is called *scalarization*, as it transforms a multidimensional vector into a scalar, i.e., a single number. Choosing a scalarization is easy, if we can sensibly aggregate several objectives into a single one. However, this may be very difficult to do because:

- Objectives may be associated with different stakeholders, and assigning weights may be a tough "political" decision.

- Some objectives may be difficult to express and compare on a common ground, e.g., money.

Of course, in the end, we must choose *one* solution, and this choice may involve qualitative and political considerations that do not lend themselves to a quantitative assessment; nevertheless, we might try at least to spot the reasonable solutions to evaluate the tradeoffs between them. If anything, this should ease conflicts, or make them more transparent and objective. The intuition from Fig. 12.10 is that we may not be able to spot one "optimal" solution, but at least we may eliminate unreasonable alternatives from further consideration. In other words, we should just concentrate on *efficient* solutions.

DEFINITION 12.9 *Given the vector optimization problem (12.45), a feasible solution* $\mathbf{x}^*$ *is said to be an* **efficient**[24] *or* **nondominated** *solution, if there is no other solution* $\tilde{\mathbf{x}} \in S$ *such that*

$$f_1(\tilde{\mathbf{x}}) \leq f_1(\mathbf{x}^*) \qquad and \qquad f_2(\tilde{\mathbf{x}}) \leq f_2(\mathbf{x}^*)$$

---

[24]Often we speak of Pareto efficiency, in honor of Italian economist Vilfredo Pareto (1848–1923), who studied the allocation of goods among economic agents in these terms. It is worth noting that although he is best remembered as an economist, he had a degree in engineering. Later, in the 1950s, many scholars who eventually made a big name in economics also worked on inventory management and workforce planning. Maybe, this lesson in interdisciplinarity has been forgotten in times of over-specialization and "publish or perish" (or is it publish *and* perish?) syndrome plaguing academia.

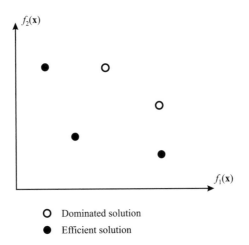

Fig. 12.11   Schematic illustration of the concept of efficient solution.

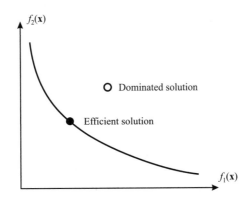

Fig. 12.12   The efficient frontier in the continuous case.

with a strict inequality for at least one of the two objectives. The set of nondominated solutions is called the **efficient frontier**.

The idea may be easily grasped by having a look at Fig. 12.11. We see that there is not necessarily *one* optimal solution, but rather a set of "reasonable" solutions to which we may restrict the choice, ruling out dominated alternatives. In a continuous mathematical program, the efficient frontier might be a continuous curve, as illustrated in Fig. 12.12. What we can do to help the decision maker is to generate the set of efficient solutions, which is a subset of the whole set of feasible solutions. To this aim, we can scalarize the problem according to some strategy, boiling the vector problem down to a family of single-objective optimization problems depending on one or more parameters. By changing the parameters, we may trace the efficient frontier.

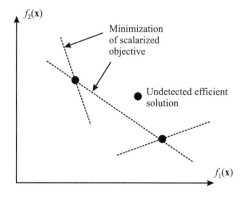

Fig. 12.13   A trivial scalarization may not be able to detect all of the efficient solutions.

The first and perhaps more intuitive approach is to devise a weighted linear combination of the two objectives. We introduce a parameter λ, bounded by 0 and 1, which expresses the relative importance of the objectives; letting λ span its range, we solve a sequence of problems

$$\text{min} \quad \lambda f_1(\mathbf{x}) + (1 - \lambda) f_2(\mathbf{x})$$
$$\text{s.t.} \quad \mathbf{x} \in S$$

Note that the parameter λ has no precise economic meaning, as it is just a tool for spanning the efficient frontier. This approach is clearly intuitive and related to the idea of varying a set of weights. We have the guarantee that all of the solutions we generate this way are efficient; however, there is no guarantee that *all* of the efficient solutions will be generated. The issue is illustrated in Fig. 12.13, where the dotted lines correspond to level curves of the scalarized objective, for different values of weights. It is easy to see that two of the three efficient solutions can be detected by changing weights, but one cannot. This does not occur in Fig. 12.12, where the plot of the efficient frontier looks essentially like a convex function. In a discrete optimization case, we cannot be sure to generate the whole efficient frontier by a trivial scalarization based on a linear combination of objectives. More complicated scalarizations have been proposed to overcome this difficulty.

An alternative approach is based on the idea of transforming one objective into a constraint. In other words, we can optimize f_1, subject to the constraint that f_2 cannot exceed some limit (or vice versa):

$$\text{min} \quad f_1(\mathbf{x})$$
$$\text{s.t.} \quad \mathbf{x} \in S,$$
$$f_2(\mathbf{x}) \leq \bar{f}_2$$

Solving a family of scalar problems for varying values of $\bar{f}_2$, we may trace the efficient frontier. It is worth noting that this second approach does not suffer from the aforementioned difficulty with the weighted combination approach, but the most important feature, arguably, is that it is more "readable" for a decision maker.[25] While the parameter λ has no obvious managerial meaning, the value of $\bar{f}_2$ is much more readable; it is a threshold level, which might be chosen by having a look at what competitors do. For instance, if we have to trade off service level against inventory holding cost, having an idea of what service level is offered by competitors helps a lot in choosing a sensible threshold.

12.3.4   Elastic model formulations

An optimization model need not have a unique optimal solution. As we have pointed out in Section 12.1.1, the following can occur:

- There are multiple optimal solutions. In such a case we may afford the luxury of using a secondary objective, which is optimized over the set of optimal solutions with respect to the original objective.

- The solution goes to infinity. If so, it is quite likely that there is something wrong with the model itself, probably a missing constraint.

- A common and much less pleasing occurrence, however, is when no feasible solution is found, because the constraints are too demanding and the feasible set is empty. To see a practically relevant example, consider the production planning model (12.27), whereby we want to satisfy demand at minimum cost. It is easy to see that if demand is too large with respect to available capacity, we may fail to satisfy it completely.

Commercial solvers are able to spot infeasible mathematical programs, but, from a practical perspective, we cannot just report that, leaving the decision maker without a clue. It would be nice to provide her with some usable information about what is causing infeasibility. This can be accomplished by an *elastic model formulation*. For the sake of concreteness, we illustrate the approach in the specific case of production planning, but the idea is quite general: We relax critical constraints and introduce a *penalty term* into the objective function. In our case, the infeasibility may be regarded as the result of excessive high demand or insufficient capacity. We may either relax the requirement that demand be completely satisfied, or we may relax capacity constraints.

The first route can be pursued by introducing a decision variable z_{it} representing *unmet* (i.e., unsatisfied) demand of item i during time bucket t. Now, rather than issuing d_{it} from inventory during each time bucket, we just issue

---

[25]On the other hand, we should also mention that sometimes the model resulting from a convex combination of objectives may be easier to solve from a computational perspective.

$d_{it} - z_{it}$. We also need to adjust the objective function by penalizing unmet demand; if we fail to do so, the obvious optimal solution would be to just shut down production. The resulting model is

$$\min \quad \sum_{i=1}^{N}\sum_{t=1}^{T} h_i I_{it} + \omega \sum_{i=1}^{N}\sum_{t=1}^{T} z_{it} \tag{12.46}$$

$$\text{s.t.} \quad I_{it} = I_{i,t-1} + x_{it} - d_{it} + z_{it}, \qquad \forall i,t$$

$$\sum_{i=1}^{N} r_{im} x_{it} \leq R_{mt}, \qquad \forall m,t$$

$$0 \leq z_{it} \leq d_{it}, \qquad \forall i,t \tag{12.47}$$

$$x_{it}, I_{it} \geq 0, \qquad \forall i,t$$

The coefficient ω in the objective (12.46) is a *penalty coefficient*, representing the cost of unmet demand. It is important to stress that ω is not just missed profit; if we want to maximize profit, we should use model (12.26). The penalty coefficient must be suitably large, so that in the optimal solution we have some $z_{it}^* > 0$, only if there is really no way to satisfy demand. Information about which items are critical, as well as when they are, allows us to spot which customer orders are creating the problem; this is helpful in supporting a negotiation process with customers, who may be willing to wait a little more for delivery or may accept a substitute product. All of these adjustments can actually be modeled and explicitly represented in an optimization model, but some decision makers may find themselves in trouble when required to quantify the costs of these actions. Of course, if some items are strategically more important than others, we may use different penalty coefficients ω_i.

As an alternative, we may relax capacity constraints by introducing auxiliary variables O_{mt}. The resulting model is

$$\min \quad \sum_{i=1}^{N}\sum_{t=1}^{T} h_i I_{it} + \omega \sum_{m=1}^{M}\sum_{t=1}^{T} O_{it} \tag{12.48}$$

$$\tag{12.49}$$

$$\text{s.t.} \quad I_{it} = I_{i,t-1} + x_{it} - d_{it}, \qquad \forall i,t$$

$$\sum_{i=1}^{N} r_{im} x_{it} \leq R_{mt} + O_{mt}, \qquad \forall m,t \tag{12.50}$$

$$O_{it} \geq 0, \qquad \forall m,t$$

$$x_{it}, I_{it}, \geq 0, \qquad \forall i,t$$

The nonnegative variables O_{mt} play the role of overtime capacity added to available capacity R_{mt} in constraint (12.50) and suitably penalized in the objective function (12.49). Observing which resources are critical, as well as when they are, may help in figuring out a way out of the dilemma. It is also possible to formulate a planning model allowing for true overtime work; in such a case, the variables O_{mt} should be multiplied by the actual cost of overtime, rather than by a large penalty coefficient. The model should also express limits on overtime, reflecting constraints on the amount and type of allowable overtime work.

12.3.5 Column-based model formulations

Sometimes, we face management problems with quite complicated constraints, which seem to defy the best modeling efforts. Column-based model formulations are a formidable tool, which is again best illustrated by a simple example, namely, a stylized staffing problem.

Imagine that we are running a post office, or something like that, with a lot of counters; alternatively, if you prefer, imagine a retail store with cashier's desks. During the day, we need not keep a fixed number of desks open; there are peak hours at which we need more people, and other time periods at which many desks may be left closed without running the risk of customers experiencing large waiting times at queues. Suppose that each day consists of eight opening hours, $t = 1, 2, \ldots, 8$, and that we have time-varying requirements represented by R_t, $t = 1, \ldots, 8$, i.e., the minimum number of desks/counters that should be open during time bucket t. We would like to find the minimal number of people to hire, taking into account the constraints on the way shifts must be organized. For instance, let us assume that each shift consists of four consecutive hours, with one hour of rest that may be either the second or the third one. To state it more clearly, the work pattern may be one of the following:

1. Work, rest, work, work

2. Work, work, rest, work

The pattern may start at $t = 1, 2, 3, 4, 5$; of course, no pattern can start during the last hours of each day. How many workers should we hire for each possible pattern, in order to find a staffing schedule with minimal cost?

This is a drastically simplified case, but expressing regulatory requirements on work patterns is a real-life issue in crew scheduling problems, for both airlines and train transportation companies. However, there is a standard way to represent the problem in terms of combining columns of a matrix. Recall that solving a system of linear equations $\mathbf{Ax} = \mathbf{b}$ is essentially the problem of expressing vector $\mathbf{b}$ as a linear combination of the columns of matrix $\mathbf{A}$.[26] Here we may define a set of columns representing the possible work patterns; each column is a vector of 8 elements, which may be 1 or 0, where a 1 in position t means that the worker is active during hour t. The

---

[26] See Section 3.5.

possible patterns are

$$
\begin{bmatrix} 1 \\ 1 \\ 0 \\ 1 \\ 0 \\ 0 \\ 0 \\ 0 \end{bmatrix}
\begin{bmatrix} 1 \\ 0 \\ 1 \\ 1 \\ 0 \\ 0 \\ 0 \\ 0 \end{bmatrix}
\begin{bmatrix} 0 \\ 1 \\ 1 \\ 0 \\ 1 \\ 0 \\ 0 \\ 0 \end{bmatrix}
\begin{bmatrix} 0 \\ 1 \\ 0 \\ 1 \\ 1 \\ 0 \\ 0 \\ 0 \end{bmatrix}
\begin{bmatrix} 0 \\ 0 \\ 1 \\ 1 \\ 0 \\ 1 \\ 0 \\ 0 \end{bmatrix}
\begin{bmatrix} 0 \\ 0 \\ 1 \\ 1 \\ 0 \\ 1 \\ 0 \\ 0 \end{bmatrix}
\begin{bmatrix} 0 \\ 0 \\ 0 \\ 1 \\ 1 \\ 0 \\ 1 \\ 0 \end{bmatrix}
\begin{bmatrix} 0 \\ 0 \\ 0 \\ 1 \\ 0 \\ 1 \\ 1 \\ 0 \end{bmatrix}
\begin{bmatrix} 0 \\ 0 \\ 0 \\ 0 \\ 1 \\ 1 \\ 0 \\ 1 \end{bmatrix}
\begin{bmatrix} 0 \\ 0 \\ 0 \\ 0 \\ 1 \\ 0 \\ 1 \\ 1 \end{bmatrix}
$$

There are 10 columns $\mathbf{a}_j$, $j = 1, \ldots, 10$, that can be grouped into a matrix $\mathbf{A} \in \mathbb{R}^{8,10}$, where element a_{tj} is set to 1 if the worker following pattern j is active at hour t, 0 otherwise. Let us denote by x_j the number of workers hired and following pattern j. The total number of workers active at hour t can be written as

$$
\sum_{j=1}^{10} a_{tj} x_j
$$

and this should not be less than the requirement R_t. We may minimize the total cost of staffing by solving the ILP:

$$
\min \sum_{j=1}^{10} x_j
$$

$$
\text{s.t.} \sum_{j=1}^{10} a_{tj} x_j \geq R_t, \qquad\qquad t = 1, \ldots, 8
$$

$$
x_j \in \mathbb{Z}_+, \qquad\qquad j = 1, \ldots, 10
$$

In this stylized case, we had just 10 columns. In more practical settings, we may have to do with a huge number of columns. Then, it may be a wiser strategy to generate only useful columns, which provide us with interesting building blocks that we may assemble to solve the overall problem. Such columns can be generated dynamically as needed. Quite sophisticated (and effective) *column generation* strategies have been devised to solve seemingly intractable problems. As they are typically quite problem-dependent, they are beyond the scope of the book, but it is important to understand that quite involved constraints may be taken into account in the process of generating columns; this level of complexity is separated from the issue of putting building blocks together, resulting in a nice decomposition of the overall problem.

12.4 BUILDING INTEGER PROGRAMMING MODELS

As we have already pointed out, integer programming models may pop up when there is a need to restrict purchase or production decisions to integer quantities, maybe multiples of a standard batch. However, the most common reason for using such models is by far the inclusion of logical decisions. In this section we use a set of illustrative examples to illustrate the basic techniques to represent such decisions by binary variables.

12.4.1 Knapsack problem

Let us consider a trivial model for capital budgeting decisions. We must allocate a given budget B of money to a set of N potential investments. For each investment opportunity, we know

- The initial capital outlay c_i, $i = 1, \ldots, n$

- The profit π_i that we will get from the investment (which we assume to be certain)

We would like to select the subset of investments that yields the highest total profit, subject to a limited budget B. This looks like a portfolio optimization model; the key difference is that our decision must be "all or nothing." For each investment opportunity we may decide whether we take it or leave it, but we cannot "buy" a fractional share of it. In typical portfolio optimization models, assets are assumed to be infinitely divisible, which may often be a reasonable approximation, e.g., for stocks, but not in this case. It may be helpful to think of our investments as projects that can be started or not.

The decision variables must reflect the logical nature of our decision. This is obtained by specifying the following decision variables:

$$x_i = \begin{cases} 1 & \text{if we invest in project } i \\ 0 & \text{otherwise} \end{cases}$$

Now it is easy to build an optimization model:

$$\max \quad \sum_{i=1}^{n} \pi_i x_i$$

$$\text{s.t.} \quad \sum_{i=1}^{n} c_i x_i \leq B$$

$$x_i \in \{0, 1\}$$

This model is grossly simplified, but it is a first example of an integer programming model; more precisely, it is a pure binary programming problem. It is also well known as the *knapsack problem*, as each investment may be interpreted as an object of given value π_i and volume c_i, and we want to determine the maximum value subset of objects that fit the knapsack capacity B. A model like this looks deceptively simple. However, it cannot be solved by ordinary optimization methods for continuous models. One could think of simply enumerating all of the feasible solutions, which are a finite set, in order to spot the best one. Unfortunately, this is not feasible in general, as the number of feasible solutions may be very large, even though finite. To see this, notice that there are n variables that can take two values; hence, there are 2^n possible variable assignments. Many of them would be ruled out by the budget constraint, but we see that the computational effort of complete enumeration grows exponentially with the size of the problem. A possible solution approach would be to rank the items in decreasing order of their return π_i/c_i and selecting them until the budget allows. This would work with divisible assets, but it does not guarantee the optimal solution in the discrete case. As a counterexample, let us consider the following problem:

$$\begin{aligned} \max \quad & 10x_1 + 7x_2 + 25x_3 + 24x_4 \\ \text{s.t.} \quad & 2x_1 + 1x_2 + 6x_3 + 5x_4 \le 7 \\ & x_i \in \{0, 1\} \end{aligned}$$

The returns are, respectively, $5.00, 7.00, 4.17, 4.80$. Hence, according to this logic we would select investment 2 first, then investment 1, and we would stop there, with profit 17, because no other investment fits the residual budget. This is a really bad solution, leaving much budget unused (4 units out of 7). There are two solutions that exploit the whole budget: $[1, 0, 0, 1]$, with total profit 34, and $[0, 1, 1, 0]$, with total profit 32. In this trivial case it is easy to see that the first one is optimal.

An important property of the knapsack problem is that if we solve its continuous relaxation, i.e., we relax the integrality requirement $x_i \in \{0, 1\}$ to $x_i \in [0, 1]$ and solve the problem as a continuous LP, we obtain a solution with at most one fractional variable. In the case above, $[1, 1, 0, 0.8]$. It is easy to see that this is the solution we obtain by applying the return-based heuristic above, allowing for a fractional investment in the fourth project, which is the third one in the ranking, in order to saturate the budget. This property can be exploited in ad hoc solution methods for the knapsack problem but, for the sake of generality, we illustrate later how the problem can be solved by general purpose branch and bound methods (see Examples 12.24 and 12.26).

12.4.2   Modeling logical constraints

In the knapsack model, each choice is independent from the other ones, as there is no link whatsoever among different projects. It may be the case that there are additional constraints on subsets of activities, taking into account their mutual relationships and overall requirements. Here are a few examples, where binary decision variables x_j are again related to the decision of starting activity j:

- Exactly one activity within a subset S must start (exclusive OR):

$$\sum_{j \in S} x_j = 1$$

- At least one activity within a subset S must start (inclusive OR):

$$\sum_{j \in S} x_j \geq 1$$

- At most one activity within a subset S may start:

$$\sum_{j \in S} x_j \leq 1$$

- If activity j is started, then activity k must start, too:

$$x_j \leq x_k$$

All the constraints above may be generalized to more complex situations that are relevant, for instance, if you want to enforce qualitative constraints on a portfolio of investments or activities.

12.4.3 Fixed-charge problem and semicontinuous decision variables

The knapsack model and its variants are pure binary programming models. In this section we get acquainted with a quite common mixed-integer model, arising when the cost structure related to an activity cannot be represented in simple linear terms. The *fixed-charge problem* is one such case. Let decision variable $x \geq 0$ represent the level of an activity. The total cost $\mathrm{TC}(x)$ related to the activity may consist of two terms:

- A variable cost, which is represented by a term like cx, where c is the unit variable cost.

- A fixed charge f, which is only paid if $x > 0$, i.e., if the activity is carried out at a positive level.

In principle, we could introduce a step function such as

$$\gamma(x) = \begin{cases} 1 & \text{if } x > 0 \\ 0 & \text{if } x = 0 \end{cases}$$

and express total cost as

$$\mathrm{TC}(x) = cx + f\gamma(x)$$

Unfortunately, the step function is nonlinear and discontinuous at the origin, as it jumps from 0 to 1. An alternative representation is obtained by introducing the following binary variable:

$$\delta = \begin{cases} 1 & \text{if } x > 0 \\ 0 & \text{otherwise} \end{cases}$$

Then we link x and δ by the so-called *big-M constraint*:

$$x \leq M\delta \tag{12.51}$$

where M is a suitably large constant. To see how Eq. (12.51) works, imagine that $\delta = 0$; then, the constraint reads $x \leq 0$, whatever the value of M is. This constraint, together with the standard nonnegativity restriction $x \geq 0$, enforces $x = 0$. If $\delta = 1$, the constraint is $x \leq M$, which is nonbinding if M is large enough; in practice, we should take a sensible upper bound on the level of activity x. In principle, the exact choice of the big-M is not relevant; however, the tighter bound we use, the faster a branch and bound solver will be in tackling the problem, as we shall see in Section 12.6.3. Apparently, the constraint allows for an absurd solution in which $\delta = 1$ and we pay the fixed charge f, but we do not carry out the related activity, i.e., we leave $x = 0$. While it is true that this is feasible for constraint (12.51), a solution like this will never be optimal for the objective function

$$\text{TC}(x) = cx + f\delta$$

Example 12.14 Suppose that we are given a set of activities, indexed by $i = 1, \ldots, n$. The level of activity i is measured by a nonnegative continuous variable x_i; the activity levels are subject to a set of constraints, formally expressed as $\mathbf{x} \in S$. Each activity has a cost proportional to the level x_i and a fixed charge f_i, which is paid whenever $x_i > 0$. Assume that we know an upper bound M_i on the level of activity i, and introduce a set of binary variables δ_i to represent fixed charges. Then, we should solve the following model:

$$\min \quad \sum_{i=1}^{n} (c_i x_i + f_i \delta_i)$$
$$\text{s.t.} \quad x_i \leq M_i \delta_i, \qquad \forall i$$
$$\mathbf{x} \in S$$
$$x_i \geq 0, \ \delta_i \in \{0, 1\}, \quad \forall i$$

□

In the next sections we illustrate a few classical examples of problems involving the big-M trick of the trade. Another common requirement on the level of an activity is that, if it is undertaken, its level should be in the interval $[m_i, M_i]$.

Note that this is *not* equivalent to requiring that $m_i \leq x_i \leq M_i$. Rather, we want something like

$$x_i \in \{0\} \cup [m_i, M_i]$$

which is a nonconvex set (recall that the union of convex sets need not be convex). Using the same trick as above, we may just write

$$x_i \geq m_i \delta_i, \qquad x_i \leq M_i \delta_i$$

These constraints define a *semicontinuous decision variable*. Semicontinuous variables may be used, e.g., in the following cases:

- We are blending an end product by using chemicals, and we have a choice between several ingredients; however, there is no point in using a very small amount of a raw material, as this implies that some piece of blending equipment will get dirty for nothing.

- We want to build a financial portfolio, but we do not want to include assets with a very small weight, as this will just make the portfolio hard to manage because of transaction costs that are incurred whenever an asset is bought or sold.

12.4.4   Lot-sizing with setup times and costs

A classical example involving fixed charges is the lot-sizing model, which is essentially a generalization of the basic EOQ model to take into account multiple items, limited production capacity, and time-varying demand. To see why such a model arises, note that in the multiperiod planning models (12.26) and (12.27) we did not consider at all the need for machine setup before starting production. Suppose that, in order to produce a lot of item i, we need to spend a setup time r'_{im} for each resource m. This setup time does not depend on the lot size, and it gives us an incentive to stock an item, rather than producing it in each time bucket. By the same token, we may have a fixed cost f_i associated with each setup for item i; this may depend, e.g., on material which is scrapped at the beginning of a lot because of the need of adjusting machines. In purchasing, setup times play no role, but we may need to tackle similar issues, e.g., when there is a fixed component in the transportation cost. The decision of producing a lot of item i during time bucket t is a logical decision; either we do it or we do not. Hence, we introduce a binary decision variable

$$\delta_{it} = \begin{cases} 1 & \text{if we carry out a setup for item } i \text{ during time bucket } t \\ 0 & \text{otherwise} \end{cases}$$

and link x_{it} and δ_{it} using the big-M constraint

$$x_{it} \leq M_{it} \delta_{it}$$

In practice, one way to quantify the big-M is to consider that there is no economic reason to produce more than what we can sell in the remaining time to the end of the planning horizon; therefore, we may choose

$$M_{it} = \sum_{\tau=t}^{T} d_{i\tau} \tag{12.52}$$

The resulting model, in the case of cost minimization, is a fairly straightforward extension of (12.27):

$$\min \quad \sum_{i=1}^{N} \sum_{t=1}^{T} (h_i I_{it} + f_i \delta_{it}) \tag{12.53}$$

$$\text{s.t.} \quad I_{it} = I_{i,t-1} + x_{it} - d_{it}, \qquad\qquad i = 1, \ldots, N, \ \ t = 1, \ldots, T,$$

$$\sum_{i=1}^{N} (r_{im} x_{it} + r'_{im} x_{it}) \le R_{mt}, \qquad m = 1, \ldots, M, \ \ t = 1, \ldots, T$$

$$x_{it} \le M_{it} \delta_{it}, \qquad\qquad\qquad i = 1, \ldots, N, \ \ t = 1, \ldots, T$$

$$x_{it}, \ z_{it}, \ I_{it} \ge 0; \quad \delta_{it} \in \{0, 1\}$$

This is a rather innocent-looking MILP problem, which can be solved by commercial branch and bound code. In practice, it is very hard to solve to optimality; in Section 12.6.3 we consider a suitable reformulation that improves model solvability considerably.

12.4.5   Plant location

In the network optimization models of Section 12.2.4, we have taken the network structure as given. Hence, the decisions we had to make were tactical or operational, and just linked to flow routing. However, at a more strategic level, we have to make decisions concerning:

- The location (or relocation) of production plants

- The sizing (or the expansion) of production capacities

- The capacity and location planning for distribution centers

- The allocation of retail stores to distribution centers

As far as the last point is concerned, we have considered a purely exogenous demand, which should be met at minimum cost. However, there are problems, such as the location of retail stores, in which the demand is a *result* of our decisions.

What we describe here is a straightforward extension of the transportation problem, whereby source nodes are just *potential* locations of plants. We should decide where a plant must be opened, within a set of predefined options, taking into account the related costs. Such decisions, as well as the related variables, are logical in nature: Either we open a plant, or we do not. This is a typical setting in which binary decision variables are used:

$$y_i = \begin{cases} 1 & \text{if source node } i \text{ is opened} \\ 0 & \text{otherwise} \end{cases}$$

When opening a plant, the related costs include a fixed component, linked to the binary decision variables y_i. Finding a good solution calls for trading off the cost of opening a plant against transportation costs. In fact, to minimize the transportation cost, we should open plants close to destinations, but this increases the cost of the network structure. We should also note that there is a timing difference between the two decision levels. We open plants now, and then we transport items over a possibly long time period. Therefore, when considering a fixed charge for opening a plant, we must be careful and make it comparable to transportation costs; we should transform a one-time-only cost into a kind of per-unit-of-time fee, say, a monthly or yearly fee. If we measure demand and transportation costs on some time unit, we must somehow amortize opening costs to make all of them comparable. Doing so, we end up with a fixed charge for operating plant i, denoted by f_i. The classical *plant location model*, where one item type is considered, has the following form:

$$\min \quad \sum_{i \in \mathcal{S}} f_i y_i + \sum_{i \in \mathcal{S}} \sum_{j \in \mathcal{D}} c_{ij} x_{ij} \qquad (12.54)$$

$$\text{s.t.} \quad \sum_{i \in \mathcal{S}} x_{ij} = d_j, \qquad\qquad \forall j \in \mathcal{D}$$

$$\sum_{j \in \mathcal{D}} x_{ij} \le R_i y_i, \qquad\qquad \forall i \in \mathcal{S} \qquad (12.55)$$

$$x_{ij} \ge 0, \ y_i \in \{0, 1\}$$

Comparing this model against the transportation problem, we see two basic differences:

1. There is an additional term in the objective function (12.54), related to fixed charges.

2. The capacity constraint (12.55) does not include a given capacity, but a capacity depending on our strategic decisions. If a plant is not opened ($y_i = 0$), there can be no flow going out of the corresponding node.

Since the model includes binary decision variables, it must be solved by mixed-integer programming methods such as branch and bound. Leaving solution issues aside, it is important to realize that the main difference between the two sets of decision variables is not really due to integrality requirements. One set of variables is related to strategic decisions, which are not easy to change on a short timescale; these should be considered as *design variables*, and are the true output of the model. Another set of variables is related to tactical decisions: Transportation decisions, should the demand pattern change, can be adapted on a short notice, subject to plant capacity constraints; these should be considered as *control variables*. The flow variables x_{ij} are not meant for immediate implementation. Rather, their role in a strategic model is to "anticipate" the effects of tactical decisions that will be made later. Therefore, the second term in the objective function (12.54) is actually an *anticipation function*. Such functions are common in hierarchical optimization models, where we need a link between different decision layers. Such a link will be approximate in nature, but we should mention that in the basic plant location model we are making two gross mistakes:

- On the one hand, we assume deterministic demand, but since the time horizon is relatively long, we should consider an uncertain demand.

- On the other hand, a linear cost structure cannot account for economies of scale which may characterize true transportation costs. Hence, the anticipation function should be nonlinear.

Later, we will see how such limitations can be overcome.

12.4.6 An optimization model for portfolio tracking and compression

The portfolio optimization model that we have considered in Example 12.5 does not place any restriction on the composition of the portfolio. In practice, bounds are enforced, e.g., to limit exposure to certain risk factors; for instance, we might wish to limit exposure to emerging markets or to the energy sector. Another practical issue that is worth mentioning concerns the cardinality of a portfolio, i.e., the number of different stock names that we include. If we have a universe of n assets, indexed by $i = 1, \ldots, n$, enforcing a cardinality constraint means limiting the number of assets with a positive weight in the portfolio. This aims at simplifying portfolio management and the analysis of the related risk, but arguably the best reason for considering cardinality-constrained portfolios is index tracking. In fact, equilibrium models like CAPM, if taken literally, imply that there is no way to systematically outperform the market. Hence, a sensible strategy is to invest in a way that mimics a broad market index. However, doing this might require the inclusion of too many assets in the portfolio, with a possible increase in management and transaction costs, which is really unpleasing in a portfolio just tracking an index, since its main virtue should be a low cost; indeed, since this is a passive management style, there is no reason why a customer should pay high fees, which can only be justified by an active strategy based on thorough financial analysis.

Based on these considerations, we want to build a portfolio tracking a target portfolio with a limited number of assets. This is also called portfolio compression, and the target portfolio can be an index or whatever you like. As usual, we denote the random return of asset i by R_i and its weight by $w_i \geq 0$, ruling out short selling. We are also given the composition of a benchmark portfolio, expressed by weights w_i^b. The return of our portfolio is

$$R_p = \sum_{i=1}^{n} w_i R_i$$

and the return of the benchmark portfolio is

$$R_b = \sum_{i=1}^{n} w_i^b R_i$$

We want to track the benchmark with a *cardinality-constrained tracking portfolio*, i.e., a portfolio including at most $C_{\max}$ assets.

The first thing we need is a way to measure the distance between our portfolio and the benchmark. A trivial distance metric can be defined, based on absolute values of portfolio weight discrepancy:[27]

$$\sum_{i=1}^{n} \left| w_i - w_i^b \right|$$

By proper model formulation, this metric yields a MILP model. However, this distance metric does not take into account many important facets of the problem. To see this, consider two assets, i_1 and i_2, with a strong positive correlation, and assume that $w_{i_1}^b = 0.07$, $w_{i_2}^b = 0.04$. The trivial distance measure above would include the following two terms:

$$\left| w_{i_1} - 0.07 \right| + \left| w_{i_2} - 0.04 \right|$$

If $w_{i_1} = 0.11$ and $w_{i_2} = 0$, how much is the distance from the target? In the limit, if the correlation coefficient is $\rho_{i_1,i_2} = 1$, the actual distance is zero, but the above distance misses the point completely. Hence, we might consider correlations and covariances as well, and focus on actual portfolio return. An alternative distance metric is *tracking error variance* (TEV), defined as

$$\text{TEV} = \text{Var}\left(R_p - R_b\right)$$

---

[27] In fact, this is an alternative norm to the Euclidean one, which squares differences. Sometimes the norm based on absolute values is referred to as L_1 norm, whereas the Euclidean norm is the L_2 norm.

To relate TEV to portfolio weights, let us rewrite its definition and use what we know about linear combinations of random variables:

$$
\begin{aligned}
\text{TEV} &= \text{Var}\left[\sum_{i=1}^{n} w_i R_i - \sum_{i=1}^{n} w_i^b R_i\right] \\
&= \text{Var}\left[\sum_{i=1}^{n} \left(w_i - w_i^b\right) R_i\right] \\
&= \sum_{i=1}^{n}\sum_{j=1}^{n}(w_i - w_i^b)\sigma_{ij}(w_j - w_j^b)
\end{aligned}
$$

where σ_{ij} is the usual covariance between the returns of assets i and j.

In order to express the cardinality constraint, we need to introduce a set of binary variables δ_i, one for each asset, modeling the inclusion of asset i in the tracking portfolio:

$$
\delta_i = \begin{cases} 1 & \text{if asset } i \text{ is included in the tracking portfolio} \\ 0 & \text{otherwise} \end{cases}
$$

Binary variables δ_i and continuous variables w_i should be linked by a big-M constraint such as

$$
w_i \leq M\delta_i
$$

where M is the by now familiar suitably large constant. A natural choice is to set $M = 1$, as no portfolio weight should exceed 100%. Since we know that performance of branch and bound methods depends on the strength of the continuous relaxation, the model can be improved by tightening the constraint; if we know that, because of policy restrictions, there is an upper bound $\overline{w}_i < 1$ on the weight of asset i, we may write the constraint as

$$
w_i \leq \overline{w}_i \delta_i
$$

Straightforward minimization of TEV, subject to a cardinality constraint, is accomplished by the following model:

$$
\min \quad \sum_{i=1}^{n}\sum_{j=1}^{n}(w_i - w_i^b)\sigma_{ij}(w_j - w_j^b) \tag{12.56}
$$

$$
\begin{aligned}
\text{s.t.} \quad & \sum_{i=1}^{n} w_i = 1 \\
& \sum_{i=1}^{n} \delta_i \leq C_{\max} \\
& w_i \leq \overline{w}_i \delta_i, && \forall i \\
& w_i \geq 0, \quad \delta_i \in \{0,1\}, && \forall i
\end{aligned}
$$

This model is a mixed-integer quadratic programming problem. Since the covariance matrix is positive semidefinite, the continuous relaxation of binary variables ($\delta_i \in [0, 1]$) yields a convex quadratic program which is solved very efficiently. In fact, recent commercial software packages can tackle the problem; for large-scale problem instances, specific approximate algorithms have also been devised.

12.4.7   Piecewise linear functions

Sometimes, there are nonlinear relationships between variables, which cannot be disregarded, as forcing the problem into a linear framework would result in a blatantly inadequate model. For instance, we may have to do with economies or diseconomies of scale:

- The total transportation cost may depend in a nonlinear way on the volume of shipped goods on a timespan of interest.

- The cost of purchasing items may be affected by volume discount opportunities.

- The transaction cost of a trade on financial markets may depend on the amount that we buy or sell.

One possibility would be to resort to nonlinear programming solvers to cope with a nonlinear formulation. Unfortunately, there are two possible complications:

- The model may involve integer variables, and solvers for nonlinear integer programming are nowhere as fast and robust as the linear ones.

- We may need to minimize a concave cost function, which makes the overall problem nonconvex; the net result is that commercial solvers may get trapped in a locally optimal solution.

An alternative strategy is to approximate the nonlinear cost function, or the nonlinear link between decision variables, by a piecewise linear function. In the following, we will consider the case in which a nonlinear function $y = f(x)$ describes the cost of an activity carried out at level x. We may fix a few selected points $x^{(i)}$, evaluate the corresponding values $y^{(i)} = f\left(x^{(i)}\right)$, and interpolate linearly between points $\left(x^{(i)}, y^{(i)}\right)$. Such points are usually referred to as *knots*. A few examples are shown in Fig. 12.14. We see that the points $x^{(i)}$ are the breakpoints separating subintervals on which the function is linear. We also see that we may get a convex function, or a concave one, or a function that is neither convex nor concave. There are several strategies to choose the number of knots and placing them in order to find a satisfactory approximation of the function. In other cases, a piecewise linear function arises naturally in the application, as in the case of a supplier offering discount opportunities

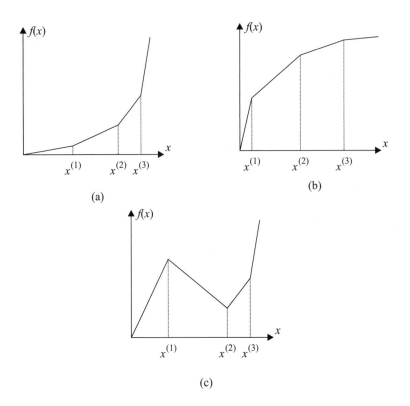

Fig. 12.14 Piecewise linear functions: (a) convex, (b) concave, (c) neither convex nor concave.

based on volume; then, the breakpoints specify a price schedule depending on the number of purchased items. As far as we are concerned, we assume that we are given a list of knots describing the function or, alternatively, a list of breakpoints and slopes.

Consider the function in Fig. 12.14(a). Rather than giving the list of knots, we may list breakpoints $\left(x^{(1)}, x^{(2)}, x^{(3)}\right)$ and slopes (c_1, c_2, c_3, c_4). Note that, without loss of generality, we assume that the function graph starts at the origin. The function could be described as follows:

$$
f(x) = \begin{cases}
c_1 x, & 0 \le x \le x^{(1)} \\
c_2 \left(x - x^{(1)}\right) + c_1 x^{(1)}, & x^{(1)} \le x \le x^{(2)} \\
c_3 \left(x - x^{(2)}\right) + c_1 x^{(1)} + c_2 \left(x^{(2)} - x^{(1)}\right), & x^{(2)} \le x \le x^{(3)} \\
c_4 \left(x - x^{(3)}\right) + c_1 x^{(1)} + c_2 \left(x^{(2)} - x^{(1)}\right) \\
\qquad + c_3 \left(x^{(3)} - x^{(2)}\right), & x \ge x^{(3)}
\end{cases}
$$

Of course, this is correct if the function is continuous. If $c_1 < c_2 < c_3 < c_4$ (increasing marginal costs), then $f(x)$ is convex [Fig. 12.14(a)]; if $c_1 > c_2 >$

$c_3 > c_4$ (decreasing marginal costs), the function is concave [Fig. 12.14(b)]; for arbitrary slopes c_i, the function is neither convex nor concave [Fig. 12.14(c)].

The convex case is easy and it can be coped with by continuous LP models. The function $f(x)$ can be converted to a linear form by introducing four auxiliary variables z_1, z_2, z_3, z_4, one for each subinterval. Then, we can express x as the sum of each single piece:

$$x = z_1 + z_2 + z_3 + z_4 \tag{12.57}$$
$$0 \leq z_1 \leq x^{(1)}$$
$$0 \leq z_2 \leq \left(x^{(2)} - x^{(1)} \right)$$
$$0 \leq z_3 \leq \left(x^{(3)} - x^{(2)} \right)$$
$$0 \leq z_4 \tag{12.58}$$

Finally, the function $f(x)$ is rewritten as

$$f(x) = c_1 z_1 + c_2 z_2 + c_3 z_3 + c_4 z_4$$

The approach is correct if the auxiliary variables in Eq. (12.57) are activated in the right sequence. Clearly, z_2 should be positive only if z_1 is at its upper bound; z_3 should be positive only if both z_1 and z_2 are at their upper bounds, etc. Since the slopes are increasing, we are sure that this is the case at the optimal solution. There is no incentive to use the more expensive variable z_2 rather than z_1, unless the latter is saturated. Hence, if we have a piecewise convex function, the whole model can be reformulated as an LP problem.

Unfortunately, if the function is not convex this is not guaranteed at all. If $c_2 < c_1$, then the solver will use the cheaper variable z_2 before z_1. This is no surprise, after all; if the problem is nonconvex, there is no way to express it as a convex LP problem. However, we may trade one nonconvexity for another one, i.e., we may devise a modeling trick based on binary decision variables. To get a clue on how a general piecewise linear function may be modeled, it is more convenient to describe the function by the list of its knots $\left(x^{(i)}, y^{(i)} \right)$, where $y^{(i)} = f \left(x^{(i)} \right)$. An example is shown in Fig. 12.15. There, we have four knots for $i = 0, 1, 2, 3$. Now consider the line segment from knot $(x^{(i)}, y^{(i)})$ to knot $(x^{(i+1)}, y^{(i+1)})$. From our knowledge of convex sets, we know that this line segment can be expressed by taking a convex combination of its extreme points:

$$x = \lambda x^{(i)} + (1 - \lambda)x^{(i+1)}$$
$$y = \lambda y^{(i)} + (1 - \lambda)y^{(i+1)}$$

where $0 \leq \lambda \leq 1$. Now, what about forming a convex combination of the four knots? If we take a linear combination of four points, with nonnegative

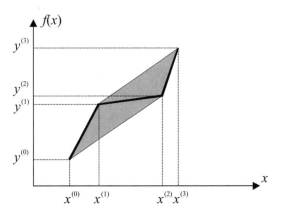

Fig. 12.15 Modeling a piecewise linear function.

weights adding up to 1, we have

$$x = \sum_{i=0}^{3} \lambda_i x^{(i)}$$

$$y = \sum_{i=0}^{3} \lambda_i y^{(i)}$$

$$\sum_{i=0}^{3} \lambda_i = 1, \qquad \lambda_i \geq 0$$

However, this is not really what we want, since by doing so we obtain the convex hull of the four knots, i.e., the shaded area in Fig. 12.15. However, we are close to accomplish our aim. What is wrong with the convex hull? The issue is that we take a convex combination of four points, but we should only take a combination involving two *consecutive* knots, in order to describe each linear piece. So, we should allow only pairs of "adjacent" coefficients λ_i to be strictly positive. This is accomplished by introducing a binary decision variable s_i, $i = 1, 2, 3$, for each line segment $(i - 1, i)$. Then, we may describe the nonlinear link between x and $y = f(x)$ by the following set of constraints:

$$x = \sum_{i=0}^{3} \lambda_i x^{(i)}$$

$$y = \sum_{i=0}^{3} \lambda_i y^{(i)}$$

$$0 \leq \lambda_0 \leq s_1$$
$$0 \leq \lambda_1 \leq s_1 + s_2$$
$$0 \leq \lambda_2 \leq s_2 + s_3$$
$$0 \leq \lambda_3 \leq s_3$$
$$\sum_{i=1}^{3} s_i = 1, \qquad s_i \in \{0, 1\}$$

We see that if $s_1 = 1$, we can only combine knots for $i = 0$ and $i = 1$, i.e., we describe the first linear piece. If $s_2 = 1$, we can only combine knots for $i = 1$ and $i = 2$, i.e., we are on the second linear piece, and so on.

The overall model is a MILP problem and we see that the nonconvexity of the cost function has been replaced by a nonconvexity in the feasible set, as we have introduced binary decision variables. Branch and bound codes are able to cope with the model above, and there is a clear tradeoff between the accuracy of our approximation of the nonlinear function $f(x)$, which would require as many knots as possible, and the need of keeping the number of decision variables as low as possible. The transformations above may look a bit tricky, but luckily many commercial tools allow the user to describe a piecewise linear function in readable terms, involving breakpoints and slopes; then the software is able to build the corresponding model, invoking the LP solver or the MILP one depending on the convex or nonconvex nature of the function involved.

12.5 NONLINEAR PROGRAMMING CONCEPTS

In this section and the next one, we consider the solution of a mathematical programming problem. We will do so essentially for linear programs, continuous and mixed-integer ones, but it is also important to get a feeling for more general, theoretical concepts in nonlinear programming. We will not cover nonlinear programming methods, but we will stress the economic interpretation of some fundamental concepts.

The constrained optimization problem

$$\begin{aligned} \min \quad & f(\mathbf{x}) & (12.59) \\ \text{s.t.} \quad & h_j(\mathbf{x}) = 0, & j = 1, \ldots, m \\ & g_k(\mathbf{x}) \leq 0, & k = 1, \ldots, l \end{aligned}$$

is a nonlinear programming problem if even one function among f, h_j, or g_k is nonlinear. A stationarity condition for the objective function does not help in finding an optimizer; to see why, a look at the following counterexample suffices:

$$\min_{2 \leq x \leq 3} x^2$$

The function is stationary at the origin, but this point is outside the feasible region. The obvious optimizer $x^* = 2$ is not a stationarity point, and it is the lower bound on x that determines the optimal solution. However, assuming that all of the involved functions are well-behaved enough, in terms of differentiability, we can try to generalize stationarity concepts to find candidate optimal points.

12.5.1 The case of equality constraints: Lagrange multipliers

For the sake of simplicity, we start by considering the equality constrained case:

$$\min \quad f(\mathbf{x}) \tag{12.60}$$
$$\text{s.t.} \quad h_j(\mathbf{x}) = 0, \qquad j = 1, \dots, m$$

which can be dealt with by the classical *Lagrange multiplier* method.

THEOREM 12.10 *Assume that the functions f and h_j in problem (12.60) meet some differentiability requirements, that the point $\mathbf{x}^*$ is feasible, and that the constraints satisfy a suitable regularity property in $\mathbf{x}^*$. Then, a necessary condition for local optimality of $\mathbf{x}^*$ is that there exist numbers λ_j^*, $j = 1, \dots, m$, called* **Lagrange multipliers**, *such that*

$$\nabla f(\mathbf{x}^*) + \sum_{j=1}^{m} \lambda_j^* \nabla h_j(\mathbf{x}^*) = \mathbf{0}$$

The reader has undoubtedly noticed that we have been very loose in stating the conditions of the theorem. In fact, our aim is just to appreciate the concept of Lagrange multiplier and its economical interpretation, and we can do it without getting bogged down in too many technicalities. However, it is important to realize that the theorem is somewhat weak. It holds under technical conditions,[28] which we do not describe in detail; furthermore, it is only a *necessary* (hence, not sufficient) condition for *local* (hence, not global) optimality. The good news is that it can be shown that the condition of the theorem is necessary and sufficient for a convex optimization problem, assuming differentiability of the involved functions.

---

[28] Differentiability is an obvious requirement, since if it does not hold, we cannot take the derivatives involved in the condition. The "regularity" conditions are known in the literature as constraint qualification conditions and assume many forms. One such condition is that the gradients of functions h_j are linearly independent at $\mathbf{x}^*$. That this condition makes some sense is not too difficult to understand. The stationarity condition in Theorem 12.10 states that the gradient of the objective can be expressed as a linear combination of the gradient of the constraints. There are cases in which this is impossible; one such case occurs if the gradients of the constraints are parallel, as shown later in Example 12.17.

To interpret the condition above, we may observe that it generalizes the stationarity condition; the trick is requiring stationarity not for the objective function, but for the following *Lagrangian function*:

$$\mathcal{L}(\mathbf{x}, \boldsymbol{\lambda}) = f(\mathbf{x}) + \sum_{j=1}^{m} \lambda_j h_j(\mathbf{x}) = f(\mathbf{x}) + \boldsymbol{\lambda}^T \mathbf{h}(\mathbf{x}) \qquad (12.61)$$

In practice, the "recipe" requires us to augment the objective function by the constraints, which are multiplied by the Lagrange multipliers, and to enforce stationarity both with respect to the decision variables $\mathbf{x}$:

$$\nabla_{\mathbf{x}} \mathcal{L}(\mathbf{x}, \boldsymbol{\lambda}) = \nabla f(\mathbf{x}) + \sum_{j=1}^{m} \lambda_j \nabla h_j(\mathbf{x}) = \mathbf{0} \qquad (12.62)$$

and with respect to the multipliers, which actually boils down to the original equality constraints:

$$\nabla_{\boldsymbol{\lambda}} \mathcal{L}(\mathbf{x}, \boldsymbol{\lambda}) = \begin{bmatrix} h_1(\mathbf{x}) \\ h_2(\mathbf{x}) \\ \vdots \end{bmatrix} = \mathbf{0} \qquad (12.63)$$

The mechanism can be best clarified by an example, but it is important to note that the conditions above yield a sensible system of equations. We have n decision variables and m equality constraints ($m < n$); Eqs. (12.62) and (12.63) yield a system of $n + m$ (possibly) nonlinear equations to find the n values x_i^* and the m multipliers λ_j^*.

Example 12.15 Consider the quadratic programming problem:

$$\min \quad x_1^2 + x_2^2 \qquad (12.64)$$
$$\text{s.t.} \quad x_1 + x_2 = 4 \qquad (12.65)$$

Since this quadratic form is convex, we may use Theorem 12.10 to find the global optimum. We associate the constraint with a multiplier λ, and form the Lagrangian function:

$$\mathcal{L}(x_1, x_2, \lambda) = x_1^2 + x_2^2 + \lambda(x_1 + x_2 - 4)$$

The stationarity conditions,

$$\frac{\partial \mathcal{L}}{\partial x_1} = 2x_1 + \lambda = 0$$

$$\frac{\partial \mathcal{L}}{\partial x_2} = 2x_2 + \lambda = 0$$

$$\frac{\partial \mathcal{L}}{\partial \lambda} = x_1 + x_2 - 4 = 0$$

are just a system of linear equations, whose solution yields $x_1^* = x_2^* = 2$ and $\lambda^* = -4$. We may notice that the equality constraint can also be written as $4 - x_1 - x_2 = 0$; if we do so, we have only a change in the sign of the multiplier, which is inconsequential. □

Proving Theorem 12.10 rigorously would call for some additional concepts beyond the scope of the book, but we may at least get a better feeling for it by trying a justification based on Taylor's expansions. How can we characterize the optimal solution $\mathbf{x}^*$? It should be a point such that there is no way of improving it, without violating constraints. If we want to improve on $\mathbf{x}^*$, we should look for a displacement $\boldsymbol{\epsilon}$, such that

$$f(\mathbf{x}^* + \boldsymbol{\epsilon}) - f(\mathbf{x}^*) < 0 \qquad (12.66)$$

This condition characterizes $\boldsymbol{\epsilon}$ as a *descent direction*. Of course, the new solution must be feasible, i.e., for each equality constraint we must have $h_j(\mathbf{x}^* + \boldsymbol{\epsilon}) = 0$. If we apply Taylor's expansion to constraints, we obtain

$$h_j(\mathbf{x}^* + \boldsymbol{\epsilon}) \approx h_j(\mathbf{x}^*) + \left[\nabla h_j^*\right]^T \boldsymbol{\epsilon}$$

where ∇h_j^* is a shorthand for the gradient at the optimal point, $\nabla h_j^* = \nabla h_j(\mathbf{x}^*)$. But since $\mathbf{x}^*$ is feasible, this boils down to

$$\left[\nabla h_j^*\right]^T \boldsymbol{\epsilon} = \mathbf{0} \qquad (12.67)$$

This condition characterizes feasible directions and states that the displacement must be orthogonal to the gradient of the constraints. If we also apply Taylor's expansion to the objective function, we may rewrite (12.66) as

$$f(\mathbf{x}^* + \boldsymbol{\epsilon}) - f(\mathbf{x}^*) \approx [\nabla f^*]^T \boldsymbol{\epsilon} < \mathbf{0} \qquad (12.68)$$

The two conditions (12.67) and (12.68) are not compatible if the gradient of the objective at $\mathbf{x}^*$, ∇f^*, is a linear combination of gradients of the constraints:

$$\nabla f^* + \sum_{j=1}^{m} \lambda_j^* \nabla h_j^* = \mathbf{0} \qquad (12.69)$$

In fact, if we take the inner product of both sides of Eq. (12.69) with $\boldsymbol{\epsilon}$, we find:

$$[\nabla f^*]^T \boldsymbol{\epsilon} + \sum_{j=1}^{m} \lambda_j^* [\nabla h_j^*]^T \boldsymbol{\epsilon} = \mathbf{0}^T \boldsymbol{\epsilon} = \mathbf{0}$$

If Eq. (12.68) holds, then $\boldsymbol{\epsilon}$ cannot be a descent direction. But Eq. (12.69) is exactly the requirement of Theorem 12.10. It is important to stress that this is no rigorous proof at all, but only a heuristic justification. In fact, it may give the impression that the stationarity of the Lagangian is a sufficient

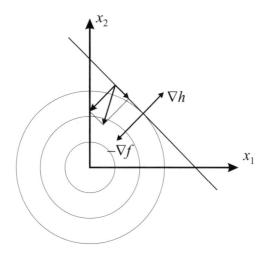

Fig. 12.16 A quadratic programming example: geometric interpretation of Lagrange conditions.

condition for local optimality, whereas it is actually a necessary condition, assuming regularity of constraints. A correct proof requires the implicit function theorem, which is beyond the scope of this book. However, this justification does provide us with some value, as shown in the next example.

Example 12.16 We may get an intuitive feeling for the Lagrange conditions by taking a look at Fig. 12.16, where we see the level curves of the objective function (12.64) of Example 12.15, a set of concentric circles, and the feasible region corresponding to Eq. (12.65), a line. From a geometric perspective, the problem calls for finding the closest point to the origin on the line $x_1 + x_2 = 4$. We note that the optimizer is where this line is tangent to the level curve associated with the lowest value of the objective. From an analytical viewpoint, the gradient of the objective function $f(\mathbf{x}) = x_1^2 + x_2^2$ is

$$\nabla f(x_1, x_2) = \begin{bmatrix} \dfrac{\partial f}{\partial x_1} \\[2mm] \dfrac{\partial f}{\partial x_2} \end{bmatrix} = \begin{bmatrix} 2x_1 \\ 2x_2 \end{bmatrix}$$

This gradient, changed in sign, is a vector pointing toward the origin, which is the steepest-descent direction for the objective. At point $\mathbf{x}^* = (2, 2)$ the gradient is $[4,\ 4]^T$. The gradient of the constraint $h(\mathbf{x}) = x_1 + x_2 - 4$ is

$$\nabla h(x_1, x_2) = \begin{bmatrix} \dfrac{\partial h}{\partial x_1} \\[2mm] \dfrac{\partial h}{\partial x_2} \end{bmatrix} = \begin{bmatrix} 1 \\ 1 \end{bmatrix}$$

Note that this vector is orthogonal to the feasible region and is parallel to the gradient of the objective at the optimizer. If we multiply the gradient of the constraint by $\lambda^* = -4$ and we add the result to the gradient of the objective, we get the null vector, as required. Actually, all of this boils down to requiring that the negative of the gradient, $-\nabla f^*$, which is the steepest descent direction, be orthogonal to the constraints at the optimizer; this means that further improvements could be obtained only by going out of the feasible region, which is forbidden. The last condition is what characterizes the optimizer.

If we consider point $(1, 3)$ on the line, the negative of the gradient there is $[-2, \ -6]^T$ which points towards the origin but is not a feasible direction. However, we may decompose this vector as

$$\begin{bmatrix} -2 \\ -6 \end{bmatrix} = \begin{bmatrix} -4 \\ -4 \end{bmatrix} + \begin{bmatrix} 2 \\ -2 \end{bmatrix}$$

The first vector is parallel to the gradient of the constraint, and it is an infeasible direction; the second vector is orthogonal to the gradient of the constraint, and it is the feasible component of the objective gradient, representing a step along the line. This decomposition in a feasible and infeasible component of the desired displacement are illustrated in Fig. 12.16 as well. At the optimal solution, there is no feasible component of the (negative) objective gradient.

□

As we pointed out, the justification we offered yields a very useful interpretation, but it is not quite a proof. For instance, can we always express the gradient of the objective as a linear combination of the gradient of the constraints? The following example shows a pathological case where we are in trouble.

Example 12.17 (The role of constraint qualification) To understand the issue behind the constraint qualification condition, consider the problem

$$\begin{aligned} \min \quad & x_1 + x_2 \\ \text{s.t.} \quad & h_1(\mathbf{x}) = x_2 - x_1^3 = 0 \\ & h_2(\mathbf{x}) = x_2 = 0 \end{aligned}$$

It is easy to see that the feasible set is the single point $(0, 0)$, which is the (trivial) optimal solution. Let us ignore this fact and build the Lagrangian function

$$\mathcal{L}(x_1, x_2, \lambda_1, \lambda_2) = x_1 + x_2 + \lambda_1(x_2 - x_1^3) + \lambda_2 x_2$$

The stationarity conditions yield the system

$$\frac{\partial \mathcal{L}}{\partial x_1} = 1 - 3\lambda_1 x_1^2 = 0$$

$$\frac{\partial \mathcal{L}}{\partial x_2} = 1 + \lambda_1 + \lambda_2 = 0$$

$$\frac{\partial \mathcal{L}}{\partial \lambda_1} = x_2 - x_1^3 = 0$$

$$\frac{\partial \mathcal{L}}{\partial \lambda_2} = x_2 = 0$$

Unfortunately, this system of equations has no solution; the first equation requires $x_1 \neq 0$, which is not compatible with the last two equations. This is due to the fact that the gradients of the two constraints are parallel at the origin:

$$\nabla h_1(0,0) = \begin{bmatrix} -3x_1^2 \\ 1 \end{bmatrix}_{\mathbf{x}=\mathbf{0}} = \begin{bmatrix} 0 \\ 1 \end{bmatrix}$$

$$\nabla h_2(0,0) = \begin{bmatrix} 0 \\ 1 \end{bmatrix}_{\mathbf{x}=\mathbf{0}} = \begin{bmatrix} 0 \\ 1 \end{bmatrix}$$

and they are not a basis able to express the gradient of f:

$$\nabla f(0,0) = \begin{bmatrix} 1 \\ 1 \end{bmatrix}_{\mathbf{x}=\mathbf{0}} = \begin{bmatrix} 1 \\ 1 \end{bmatrix}$$

We say that the origin is not a regular point, as the constraint qualification conditions do not hold there. □

The reader is urged to draw a diagram for the last example, in order to visualize the issues involved; constraint qualification conditions ensure that such difficulties do not arise.

12.5.2 Dealing with inequality constraints: Karush–Kuhn–Tucker conditions

Consider the following problem, featuring only inequality constraints:

$$\min \quad f(\mathbf{x})$$
$$g_k(\mathbf{x}) \leq 0, \qquad k = 1, \ldots, l$$

In order to characterize a (locally) optimal solution, we may follow the same reasoning as in the equality-constrained case; if $\mathbf{x}^*$ is a locally optimal solution, then we cannot find a feasible descent direction at $\mathbf{x}^*$. A fundamental observation is that an inequality constraint can be either active at $\mathbf{x}^*$, $g_k(\mathbf{x}^*) = 0$, or inactive, $g_k(\mathbf{x}^*) < 0$. Let $\mathcal{A}$ denote the set of active constraints at $\mathbf{x}^*$.

- An active inequality constraint is similar to an equality constraint, with a significant difference. Unlike the case of equality constraints, here we have more degrees of freedom to move around $\mathbf{x}^*$ by a displacement $\boldsymbol{\epsilon}$, as we should just make sure that $g_k(\mathbf{x}^* + \boldsymbol{\epsilon}) \leq 0$; we are free to move along the boundary of the feasible set, but we may also move toward its *interior*. Formally, under regularity conditions on the constraints, we may characterize a descent direction as follows. On the basis of a first-order Taylor's expansions, a direction $\boldsymbol{\epsilon}$ is feasible at $\mathbf{x}^*$ if

$$g_k(\mathbf{x}^* + \boldsymbol{\epsilon}) \approx g_k(\mathbf{x}^*) + [\nabla g_k^*]^T \boldsymbol{\epsilon} \leq 0$$

Since the constraint is active at $\mathbf{x}^*$ and $g_k(\mathbf{x}^*) = 0$, a feasible direction is characterized by the condition

$$[\nabla g_k^*]^T \boldsymbol{\epsilon} \leq 0$$

This condition should be compared with Eq. (12.67).

- If a constraint is inactive, a point $\mathbf{x}^* + \boldsymbol{\epsilon}$ will stay feasible for any displacement, provided that this is small enough. An inactive constraint does not contribute to defining the locally optimal solution $\mathbf{x}^*$ and can be disregarded.

The following example illustrates these considerations.

Example 12.18 Consider the nonlinear programming problem

$$\begin{aligned} \min \quad & x_1 + x_2 \\ \text{s.t.} \quad & x_1 x_2 \geq 4 \\ & x_1, x_2 \geq 0 \end{aligned}$$

The feasible region is shaded in Fig. 12.17. Taking advantage of this picture, and noting the obvious symmetry of the problem, it is easy to see that the optimal solution is $x_1^* = x_2^* = 2$, the nonnegativity constraints are inactive, and the constraint

$$g(\mathbf{x}) = 4 - x_1 x_2 \leq 0$$

is active. The gradient of the objective function is

$$\nabla f(\mathbf{x}) = \begin{bmatrix} 1 \\ 1 \end{bmatrix}$$

In order to improve the objective, we should move along the direction $-\nabla f = [-1, -1]^T$ (southwest), as illustrated in Fig. 12.17. The gradient of the constraint at $\mathbf{x}^*$ is

$$\nabla g(\mathbf{x}^*) = \begin{bmatrix} -x_2 \\ -x_1 \end{bmatrix}_{\mathbf{x}=\mathbf{x}^*} = \begin{bmatrix} -2 \\ -2 \end{bmatrix}$$

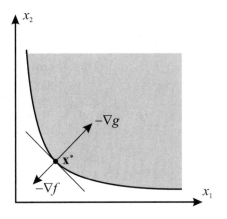

Fig. 12.17 Optimality conditions with an inequality constraint.

Looking again at Fig. 12.17, we observe that at point $\mathbf{x}^*$ we may not only move along the tangent line to the constraint, but also toward the interior of the feasible set, along the direction $-\nabla g(\mathbf{x}^*) = [2, 2]^T$. We notice once again that the gradients $\nabla f(\mathbf{x}^*)$ and $\nabla g(\mathbf{x}^*)$ are parallel, like in Example 12.16, and there exists a number μ such that

$$\nabla f(\mathbf{x}^*) + \mu \nabla g(\mathbf{x}^*) = \mathbf{0}$$

However, this is not sufficient. We observe that $\mu = 0.5 > 0$. The Lagrange multiplier, for an inequality constraint, should be positive. In our example, this means that the two gradients point toward opposite directions; otherwise, we could improve the objective by moving toward the interior of the feasible region. To see this, imagine that the gradient $-\nabla f$ in Fig. 12.17 points northeast, rather than southwest.   ▯

The example seems to suggest that there is no feasible descent direction at $\mathbf{x}^*$, if we can express the gradient of the objective as a linear combination of the gradients of the active constraints:

$$\nabla f^* + \sum_{k \in \mathcal{A}} \mu_k^* \nabla g_k^* = \mathbf{0} \tag{12.70}$$

This is not unlike the case of equality constraints, but now we have a nonnegativity restriction on the multipliers associated with inequality constraints: $\mu_k^* \geq 0$. This condition may be extended to include the inactive constraints, for which $g_k(\mathbf{x}^*) < 0$; however, we must enforce a further condition: $\mu_k^* = 0$ for inactive constraints. We may express this condition by requiring $\mu_k^* g_k(\mathbf{x}^*) = 0$ for all inequality constraints.[29] Everything is wrapped up into the following

---

[29] Once again, this line of reasoning is not quite correct, and it should be taken only as an intuitive justification. A correct proof requires a bit of hard work.

theorem, which allows us to cope with the general problem, including both equality and inequality constraints.

THEOREM 12.11 (Karush–Kuhn–Tucker conditions) *Assume that the functions* f, h_j, g_k *in problem (12.59) are continuously differentiable, and that* $\mathbf{x}^*$ *is feasible and satisfies a constraint qualification condition. Then a necessary condition for the local optimality of* $\mathbf{x}^*$ *is that there exist numbers* λ_j^* $(j = 1, \ldots, m)$ *and* $\mu_k^* \geq 0$ $(k = 1, \ldots, l)$ *such that*

$$\nabla f(\mathbf{x}^*) + \sum_{j=1}^m \lambda_j^* \nabla h_j(\mathbf{x}^*) + \sum_{k=1}^l \mu_k^* \nabla g_k(\mathbf{x}^*) = \mathbf{0} \tag{12.71}$$

$$\mu_k^* g_k(\mathbf{x}^*) = 0, \qquad k = 1, \ldots, l \tag{12.72}$$

These conditions are called **Karush–Kuhn–Tucker (KKT) conditions**.

The condition (12.71) is the familiar stationarity requirement for the Lagrangian function:

$$\mathcal{L}(\mathbf{x}, \boldsymbol{\lambda}, \boldsymbol{\mu}) = f(\mathbf{x}) + \sum_{j=1}^m \lambda_j h_j(\mathbf{x}) + \sum_{k=1}^l \mu_k g_k(\mathbf{x}) \tag{12.73}$$

Note that all of the constraints, equalities and inequalities, are included in the Lagrangian function, multiplied by the respective multipliers. The conditions (12.72) are known as *complementary slackness* conditions. They may be interpreted by noting that if a constraint is inactive at $\mathbf{x}^*$, i.e., if $g_k(\mathbf{x}^*) < 0$, the corresponding multiplier must be zero; by the same token, if the multiplier μ_k^* is strictly positive, the corresponding constraint must be active (which roughly means that it could be substituted by an equality constraint without changing the optimal solution).

Note again that the multipliers for inequality constraints are restricted in sign. Moreover, the KKT conditions are, like Theorem 12.10, rather weak, as they are only necessary conditions for local optimality, and they further require differentiability properties and the additional constraint qualification condition (see Example 12.17). They are, however, necessary and sufficient for global optimality in the convex and differentiable case. Finding a solution, at least in principle, requires the solution of a system of equations and inequalities, as illustrated by the following example.

Example 12.19 Consider the convex problem

$$\begin{aligned}
\min \quad & x_1^2 + x_2^2 \\
\text{s.t.} \quad & x_1 \geq 0 \\
& x_2 \geq 3 \\
& x_1 + x_2 = 4
\end{aligned}$$

First write the Lagrangian function

$$\mathcal{L}(\mathbf{x}, \boldsymbol{\mu}, \lambda) = x_1^2 + x_2^2 - \mu_1 x_1 - \mu_2(x_2 - 3) + \lambda(x_1 + x_2 - 4)$$

A set of numbers satisfying the KKT conditions can be found by solving the following system of equations and inequalities:

$$2x_1 - \mu_1 + \lambda = 0$$
$$2x_2 - \mu_2 + \lambda = 0$$
$$x_1 \geq 0, \qquad\qquad x_2 \geq 3$$
$$x_1 + x_2 = 4$$
$$\mu_1 x_1 = 0 \qquad\qquad \mu_1 \geq 0$$
$$\mu_2(x_2 - 3) = 0 \qquad \mu_2 \geq 0$$

We may proceed with a case-by-case analysis exploiting the complementary slackness conditions. If a multiplier is strictly positive, the corresponding inequality is active, which helps us in finding the value of a decision variable.

Case 1 ($\mu_1 = \mu_2 = 0$) In this case, the inequality constraints are dropped from the Lagrangian function. From the stationarity conditions we obtain the system

$$2x_1 + \lambda = 0$$
$$2x_2 + \lambda = 0$$
$$x_1 + x_2 - 4 = 0$$

This yields a solution $x_1 = x_2 = 2$, which violates the second inequality constraint.

Case 2 ($\mu_1, \mu_2 \neq 0$) The complementary slackness conditions immediately yield $x_1 = 0, x_2 = 3$, violating the equality constraint.

Case 3 ($\mu_1 \neq 0, \mu_2 = 0$) We obtain

$$x_1 = 0$$
$$x_2 = 4$$
$$\lambda = -2x_2 = -8$$
$$\mu_1 = \lambda = -8$$

The KKT conditions are not satisfied since the value of μ_1 is negative.

Case 4 ($\mu_1 = 0, \mu_2 \neq 0$) We obtain

$$x_2 = 3$$
$$x_1 = 1$$
$$\lambda = -2$$
$$\mu_2 = 4$$

This solution satisfies all the necessary KKT conditions.

Since this is a convex problem, we have obtained the global optimum. Note how nonzero multipliers correspond to the active constraints, whereas the inactive constraint $x_1 \geq 0$ is associated with a multiplier $\mu_1 = 0$.   □

The procedure above is obviously cumbersome and cannot be applied to a large scale problem. In practice, the KKT conditions are the theoretical background of many numerical methods for nonlinear programming. Furthermore, the analysis can be extended to include the Hessian matrix of the second-order derivatives; doing so, it is also possible to find sufficient conditions for local optimality in the nonconvex case.

12.5.3   An economic interpretation of Lagrange multipliers: shadow prices

Lagrange multipliers play a major role in optimization theory, as well as in economics. Indeed, within the economic community, they are rather known as *shadow prices*,[30] due to their important economical interpretation, which we illustrate in this section. Consider an equality-constrained problem and apply a small perturbation to the constraints:

$$h_j(\mathbf{x}) = \epsilon_j, \qquad j = 1, \ldots, m$$

Let $\boldsymbol{\epsilon}$ be a vector collecting these perturbations. Solving the perturbed problem by the Lagrange multiplier method, we get a new solution $\mathbf{x}^*(\boldsymbol{\epsilon})$ and a new multiplier vector $\boldsymbol{\lambda}^*(\boldsymbol{\epsilon})$, both depending on $\boldsymbol{\epsilon}$. An interesting question is how these perturbations affect the optimal solution and its corresponding value; in other words, we should be interested in the sensitivity

$$\frac{f(\mathbf{x}^*(\boldsymbol{\epsilon}))}{d\epsilon_j}, \qquad j = 1, \ldots, m$$

To be precise, we should notice that this derivative need not exist in general, as the differentiability of $f(\mathbf{x}^*(\boldsymbol{\epsilon}))$ cannot be guaranteed; nevertheless, when the derivative exists, it is related to the Lagrange multiplier for constraint j.

To see why, let us consider the Lagrangian function for the perturbed problem

$$\mathcal{L}(\mathbf{x}, \boldsymbol{\lambda}, \boldsymbol{\epsilon}) = f(\mathbf{x}) + \sum_{j=1}^{m} \lambda_j (h_j(\mathbf{x}) - \epsilon_j) \tag{12.74}$$

---

[30] Another common name for Lagrange multipliers is "dual variables," as opposed to the original variables $\mathbf{x}$ of the problem which are referred to as "primal." This alternative name is related to the theory of duality in mathematical programming.

Equality constraints must be satisfied by the optimal solution of the perturbed problem, too. Hence

$$f^* = f(\mathbf{x}^*(\boldsymbol{\epsilon})) = \mathcal{L}(\mathbf{x}^*(\boldsymbol{\epsilon}), \boldsymbol{\lambda}^*(\boldsymbol{\epsilon}), \boldsymbol{\epsilon}) \qquad (12.75)$$

Now we can find the derivative of the optimal value with respect to each component of $\boldsymbol{\epsilon}$:

$$\frac{df^*}{d\epsilon_j} = \frac{d\mathcal{L}}{d\epsilon_j} = \underbrace{[\nabla_{\mathbf{x}}\mathcal{L}]^T \frac{\partial \mathbf{x}}{\partial \epsilon_j} + [\nabla_{\boldsymbol{\lambda}}\mathcal{L}]^T \frac{\partial \boldsymbol{\lambda}}{\partial \epsilon_j}}_{=0} + \frac{\partial \mathcal{L}}{\partial \epsilon_j} = \frac{\partial \mathcal{L}}{\partial \epsilon_j} = -\lambda_j \qquad (12.76)$$

where we have used the stationarity condition of $\mathcal{L}$. Thus, we conclude that Lagrange multipliers are, apart from a change in sign, sensitivity measures of the optimal value with respect to perturbations in the right-hand side of constraints.

Example 12.20 Consider the quadratic programming problem:

$$\min \quad (x_1 - 2)^2 + (x_2 - 2)^2$$
$$\text{s.t.} \quad x_1 + x_2 = b$$

where b is a parameter, and let us investigate how the optimal value changes as a function of b. In fact, the optimal value of the objective is a function $q(b) = f(x_1^*, x_2^*; b)$, and in this very simple case we may find this function explicitly. To this aim, we may eliminate the constraint in order to obtain an equivalent unconstrained problem. From the constraint we get $x_2 = b - x_1$, and plugging this into the objective function yields the unconstrained problem

$$\min \quad (x_1 - 2)^2 + (b - 2 - x_1)^2$$

Then, setting the first-order derivative with respect to x_1 to zero, we find $x_1^* = b/2$. This also implies $x_2^* = b/2$. This solution may also be easily checked geometrically, since the problem ask us to find a point on the line $x_1 + x_2 = b$, such that the distance from point $(2, 2)$ is minimal; see Fig. 12.18. The optimal value as a function of b is then

$$q(b) = 2\left(\frac{b}{2} - 2\right)^2$$

If we take its derivative with respect to b, we obtain

$$\frac{dq}{db} = b - 4$$

This shows that the optimal value will decrease, if we increase b when the line is below the point $(2, 2)$ (the line gets closer to that point); if the line is above that point, increasing b will increase the distance.

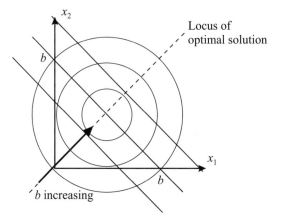

Fig. 12.18 Illustrating Lagrange multipliers as sensitivity measures.

Now let us set all of this geometric intuition aside and apply the Lagrange multiplier approach. First, we build the Lagrangian function

$$\mathcal{L}(x_1, x_2, \lambda) = (x_1 - 2)^2 + (x_2 - 2)^2 + \lambda(x_1 + x_2 - b)$$

The stationarity conditions are

$$\frac{\partial \mathcal{L}}{\partial x_1} = 2(x_1 - 2) + \lambda = 0$$

$$\frac{\partial \mathcal{L}}{\partial x_2} = 2(x_2 - 2) + \lambda = 0$$

$$\frac{\partial \mathcal{L}}{\partial \lambda} = x_1 + x_2 - b = 0$$

and we find

$$x_1^* = x_2^* = \frac{b}{2}, \qquad \lambda^* = 4 - b$$

This confirms that, apart from a change in sign, the multiplier is actually the derivative of the optimal value with respect to b.[31]   □

The theorem above applies to the case of an equality constraint, but what about inequality constraints? Thanks to the complementary slackness condition, we can extend the result to this case as well. There are two possibilities:

1. If an inequality constraint is inactive, the sensitivity with perturbations of its right-hand side is zero, because small changes of the constraint

---

[31] The change in sign is not really relevant, as it depends on how we build the Lagrangian function, but differentiability of the value function $q(b)$ is not guaranteed in general.

have no effect. In this case, complementary slackness ensures that the corresponding multiplier is zero.

2. If an inequality constraint is active, it basically behaves as an equality constraint and the above theorem applies, with the additional caveat concerning the sign of the multiplier. In this case, how we write the perturbed constraint is essential:

$$g_i(\mathbf{x}) \le \epsilon_k$$

We know that $\mu_i^* \ge 0$, which implies that $df^*/d\epsilon_k \le 0$. But this makes sense, as enlarging the feasible region can only decrease cost.

Example 12.21 Let us illustrate the meaning of Lagrange multipliers in the case of the optimal mix problem (12.1). Any software tool for linear programming yields the value of the optimal contribution to profit, $\pi_0 = 5538.4615$, as well as the value of the multipliers for the four capacity constraints (12.2 – 12.5):

$$0.0000, \ 1.2692, \ 0.0000, \ 1.0385$$

Not surprisingly, the multipliers for redundant capacity constraints are zero. Now, let us tackle two questions:

1. If we were to increase capacity, which resource should be our top priority?

2. How much money should we be willing to pay for one extra hour?

The first answer is that the second resource is more important, as it is associated with the largest multiplier. Note that here we are maximizing profit, rather than minimizing cost, so we expect that the optimal profit will increase by 1.2692 monetary units for each additional hour on the second resource type. In fact, if we solve the problem by changing the right-hand side of constraint (12.3) we obtain the following optimal objective values:

$$\begin{cases} \pi_1 = 5539.7307 & \text{if capacity is 2401 hours} \\ \pi_2 = 5665.3846 & \text{if capacity is 2500 hours} \end{cases}$$

It is easy to check that $\pi_1 - \pi_0 = 1.2692$ and $\pi_2 - \pi_0 = 126.92$. Therefore, we should not be willing to pay more than 126.92 monetary units for 100 additional hours for the second resource type. Of course, multipliers are only "local" sensitivities: If we increase a capacity beyond a limit, then the corresponding resource constraints will not be binding anymore. □

In the optimal mix above, how should we measure multipliers associated with capacity constraints? They are sensitivities of profit (monetary units) with respect to resource availability (hours); hence they are measured in money

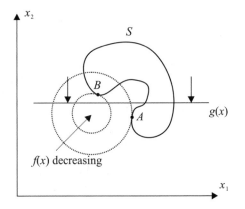

Fig. 12.19 Counterexample showing that a constraint may be relevant even if it has a null multiplier.

per unit amount of resource and are interpreted as prices. This is where the term *shadow price* comes from.

One word of caution, before we leave this section: It may be tempting to conclude that if a constraint is associated with a zero multiplier, then it can be dropped from the optimization problem without changing the optimal solution. The counterexample of Fig. 12.19 shows that this is not the case. Here we have a convex quadratic objective, whose level curves are concentric circles; the feasible region is the portion of the "bean" S below the constraint $g(\mathbf{x}) \le 0$, which is actually an upper bound on x_2. The optimal solution is the point A, and the constraint $g(\mathbf{x}) \le 0$ is inactive at that point; however, if we eliminate the constraint, the optimal solution is B (A is still a locally optimal solution). It is true that the solution does not move at all for small perturbations of the constraint; but dropping the constraint is a different matter. It is also worth noting that the source of the trouble here is that the overall problem is nonconvex.

Example 12.22 Let us solve the nonlinear programming model of Example 12.6. There, we built the nonlinear programming model

$$\min \quad \frac{1}{2}\sum_{i=1}^{n} h_i Q_i \tag{12.77}$$

$$\text{s.t.} \quad \sum_{i=1}^{n} \frac{d_i}{Q_i} \le U \tag{12.78}$$

$$Q_i \ge 0, \qquad i = 1, \ldots, n$$

to minimize inventory cost in an EOQ-like setting, subject to an upper bound on the average number of replenishment orders. Before applying the theory we know, it is advisable to streamline the model a little bit. In fact, we observe that the nonnegativity requirements may be dropped, if we assume a strictly positive, interior optimal solution $Q_i^* > 0$. We will check later that this is indeed the case, but it is easy to see that no order size can be zero, since the function involved in the inequality constraint (12.15) is not defined there and goes to infinity for small order sizes, thus exceeding the bound U. Furthermore, the inequality (12.78) may be replaced by an equality constraint. To see this, observe that the objective function calls for small order quantities, which has the effect of increasing the average number of issued orders. However, since the objective is linear, we should reduce order sizes until the ordering capacity constraint is saturated, i.e., the ordering capacity constraint is binding. Since there is no reason to leave any ordering capacity underutilized, we may rewrite the model as follows:

$$\min \quad \frac{1}{2} \sum_{i=1}^{n} h_i Q_i \tag{12.79}$$

$$\text{s.t.} \quad \sum_{i=1}^{n} \frac{d_i}{Q_i} = U \tag{12.80}$$

Now we see that there is no need for applying the KKT conditions; we just introduce a Lagrange multiplier λ and build the Lagrangian function

$$
\begin{aligned}
\mathcal{L}(Q_1, \ldots, Q_n, \lambda) &= \frac{1}{2} \sum_{i=1}^{n} h_i Q_i + \lambda \left(\sum_{i=1}^{n} \frac{d_i}{Q_i} - U \right) \\
&= \sum_{i=1}^{n} \left(\frac{1}{2} h_i Q_i + \frac{\lambda d_i}{Q_i} \right) - \lambda U
\end{aligned}
$$

A key observation is that the Lagrangian function, for a given value of the multiplier λ, can be decomposed with respect to items. Moreover, the cost for each item should be quite familiar; it is just the cost function (12.8) for the EOQ model, in which the fixed ordering charge A is replaced by the multiplier λ. Hence, the application of the stationarity conditions with respect to the order size Q_i yields

$$Q_i^* = \sqrt{\frac{2\lambda d_i}{h_i}}, \qquad i = 1, \ldots, n$$

This is just the traditional EOQ formula, but which value of λ should we use? Since the equality constraint (12.80) must be satisfied (stationarity condition of the Lagrangian with respect to λ), we should find a multiplier such that the average number of orders is exactly U. Such a value is easily found numerically by trial and error. The important point is that we get an EOQ-like solution, in which the Lagrange multiplier plays the role of a fixed ordering charge, making too frequent orders unattractive. This example contributes to explain the popularity of the EOQ model, even when the introduction of a fixed ordering charge is questionable; the problem is that this charge should depend on data, rather than being given a priori. In old practice, the fixed ordering charge was often overstated, resulting in an unnecessary inventory buildup. ⬜

12.6   A GLANCE AT SOLUTION METHODS

In this section we outline two standard solution methods:

1. The simplex method for continuous LP problems (Section 12.6.1)

2. The LP-based branch and bound method for MILP problems (Section 12.6.2)

Both are widely available in commercial software packages, but even a cursory knowledge of their internal working may be a useful asset; nevertheless, readers may safely skip this section. We should also stress the fact that we are going to describe basic solution strategies, leaving aside a lot of issues concerning robustness and efficiency. They should be regarded as useful starting points to understand what commercial tool achieve, as well as their potential pitfalls.

The simplex method is a remarkably fast method, able to solve rather large-scale problem instances, and it was invented in 1947 by George Dantzig. More recently, alternative algorithms have been proposed, generally labeled *interior-point* methods, which may be more efficient for some problems. The state of the art for branch and bound methods is a bit less happy, because they are generally slower and there are some kind of problems that are very difficult to solve to optimality. One possible alternative is the development of heuristic approaches to find a satisfactory solution, possibly a near-optimal one. Heuristics can be quite powerful but are typically problem-specific and are beyond the scope of this book. Another issue is that, more often than not, they require *ad hoc* software development. Since branch and bound code is available on the shelves, it may be more advisable to make the best of it, when possible. In fact, branch and bound methods may be applied as heuristics, if we give up the guarantee of finding an optimal solution. Furthermore, we may sometimes reformulate the model in order to improve its solvability; this is where model building and model solving meet each other. We illustrate the idea in Section 12.6.3.

12.6.1   Simplex method

The simplex method is the standard algorithm to solve LP problems and is based on the following observations:

- An LP problem is a convex problem; hence, if we find a locally optimal solution, we have also found a global optimizer.

- An LP problem is a concave problem, too; hence, we know that we may restrict the search for the optimal solution to the boundary of the feasible set, which is a polyhedron. Actually, it can be shown that there is an optimal solution corresponding to a *vertex*, or extreme point, of the polyhedron.

- So, we need a clever way to explore extreme points of the feasible set. Geometrically, we may imagine moving from a vertex to a neighboring one, trying to improve the objective function. When there is no neighboring vertex improving the objective, we have found a local optimizer, which is also global.

It is useful to visualize this process, by referring to Fig. 1.3. Imagine starting from a trivially feasible solution, the origin M_0. We may improve this production plan by moving along the edges of the polyhedron; one possible path is (M_0, M_1, M_2, M_3); another path is (M_0, M_4, M_3). Both paths lead to the optimizer, even though one seems preferable, as it visits less vertices. In large-scale problems, only a small subset of vertices is actually visited to find an optimizer.

Now, in order to obtain a working algorithm, we should translate this geometric intuition into algebraic terms. The first step is transforming the LP problem in standard form,

$$\min \quad \mathbf{c}^T \mathbf{x}$$
$$\text{s.t.} \quad \mathbf{A}\mathbf{x} = \mathbf{b}$$
$$\mathbf{x} \geq \mathbf{0}$$

where $\mathbf{x} \in \mathbb{R}^n$, $\mathbf{c} \in \mathbb{R}^n$, $\mathbf{A} \in \mathbb{R}^{m,n}$, and $\mathbf{b} \in \mathbb{R}^m$. Clearly, the problem makes sense only if the matrix $\mathbf{A}$ has less rows than columns, i.e., if $m < n$. If so, the set of equality constraints, regarded as a system of linear equations, has an infinite set of solutions, which may be considered as ways of expressing the right-hand side vector $\mathbf{b}$ as linear combinations of columns of $\mathbf{A}$. Let us express $\mathbf{A}$ in terms of its column vectors $\mathbf{a}_j$, $j = 1, \ldots, n$:

$$\mathbf{A} = \begin{bmatrix} \vdots & \vdots & & \vdots \\ \mathbf{a}_1 & \mathbf{a}_2 & \cdots & \mathbf{a}_n \\ \vdots & \vdots & & \vdots \end{bmatrix}$$

where $\mathbf{a}_j \in \mathbb{R}^m$. There are n columns, but a subset $\mathcal{B}$ of m columns suffices to express $\mathbf{b}$ as follows:

$$\sum_{j \in \mathcal{B}} \mathbf{a}_j x_j = \mathbf{b}$$

To be precise, we should make sure that this subset of columns is a basis, i.e., that they are linearly independent; in what follows, we will cut a few corners and assume that this is the case. A solution of this system, in which $n - m$

variables are set to zero, and only m are allowed to assume a nonzero value is a *basic solution*. Hence, we can partition the vector $\mathbf{x}$ into two subvectors: the subvector $\mathbf{x}_B \in \mathbb{R}^m$ of the *basic variables* and the subvector $\mathbf{x}_N \in \mathbb{R}^{n-m}$ of the *nonbasic variables*. Using a suitable permutation of the variable indices, we may rewrite the system of linear equations

$$\mathbf{A}\mathbf{x} = \mathbf{b}$$

as

$$[\mathbf{A}_B \mathbf{A}_N] \begin{bmatrix} \mathbf{x}_B \\ \mathbf{x}_N \end{bmatrix} = \mathbf{A}_B \mathbf{x}_B + \mathbf{A}_N \mathbf{x}_N = \mathbf{b} \qquad (12.81)$$

where $\mathbf{A}_B \in \mathbb{R}^{m,m}$ is nonsingular and $\mathbf{A}_N \in \mathbb{R}^{m,n-m}$. However, in the LP model we have a nonnegativity restriction on variables. A basic solution where $\mathbf{x}_B \geq \mathbf{0}$ is called a *basic feasible solution*. The simplex algorithm relies on a fundamental result, that we state loosely and without proof: *Basic feasible solutions correspond to extreme points of the polyhedral feasible set of the LP problem.*

Solving an LP amounts to finding a way to express $\mathbf{b}$ as a least-cost linear combination of at most m columns of $\mathbf{A}$, with nonnegative coefficients. Assume that we have a basic feasible solution $\mathbf{x}$; we will consider later how to obtain an initial basic feasible solution. If $\mathbf{x}$ is basic feasible, it may be written as

$$\mathbf{x} = \begin{bmatrix} \mathbf{x}_B \\ \mathbf{x}_N \end{bmatrix} = \begin{bmatrix} \hat{\mathbf{b}} \\ \mathbf{0} \end{bmatrix}$$

where

$$\hat{\mathbf{b}} = \mathbf{A}_B^{-1}\mathbf{b} \geq \mathbf{0}$$

The objective function value corresponding to $\mathbf{x}$ is

$$\hat{f} = [\mathbf{c}_B^T \ \mathbf{c}_N^T] \begin{bmatrix} \hat{\mathbf{b}} \\ \mathbf{0} \end{bmatrix} = \mathbf{c}_B^T \hat{\mathbf{b}} \qquad (12.82)$$

Now we should look for neighboring vertices improving this value. Neighboring vertices may be obtained by swapping a column in the basis with a column outside the basis. This means that one nonbasic variable is brought into the basis, and one basic variable leaves the basis.

To assess the potential benefit of introducing a nonbasic variable into the basis, we should express the objective function in terms of nonbasic variables. To this aim, we rewrite the objective function in (12.82), making its dependence on nonbasic variables explicit. Using Eq. (12.81), we may express the basic variables as

$$\mathbf{x}_B = \mathbf{A}_B^{-1}(\mathbf{b} - \mathbf{A}_N \mathbf{x}_N) = \hat{\mathbf{b}} - \mathbf{A}_B^{-1}\mathbf{A}_N \mathbf{x}_N \qquad (12.83)$$

Then we rewrite the objective function in terms of nonbasic variables only

$$
\begin{aligned}
\mathbf{c}^T\mathbf{x} &= \mathbf{c}_B^T\mathbf{x}_B + \mathbf{c}_N^T\mathbf{x}_N \\
&= \mathbf{c}_B^T\left(\hat{\mathbf{b}} - \mathbf{A}_B^{-1}\mathbf{A}_N\mathbf{x}_N\right) + \mathbf{c}_N^T\mathbf{x}_N \\
&= \mathbf{c}_B^T\hat{\mathbf{b}} + \left(\mathbf{c}_N^T - \mathbf{c}_B^T\mathbf{A}_B^{-1}\mathbf{A}_N\right)\mathbf{x}_N \\
&= \hat{f} + \hat{\mathbf{c}}_N^T\mathbf{x}_N
\end{aligned}
$$

where

$$
\hat{\mathbf{c}}_N^T = \mathbf{c}_N^T - \mathbf{c}_B^T\mathbf{A}_B^{-1}\mathbf{A}_N \tag{12.84}
$$

The quantities $\hat{\mathbf{c}}_N$ are called *reduced costs*, as they measure the marginal variation of the objective function with respect to the nonbasic variables. If $\hat{\mathbf{c}}_N \geq \mathbf{0}$, it is not possible to improve the objective function; in such a case, bringing any nonbasic variable into the basis at some positive value cannot reduce the overall cost. Therefore, the current basis is optimal if $\hat{\mathbf{c}}_N \geq \mathbf{0}$. If, on the contrary, there exists a $q \in N$ such that $\hat{c}_q < 0$, it is possible to improve the objective function by bringing x_q into the basis. A simple strategy is to choose q such that

$$
\hat{c}_q = \min_{j \in N} \hat{c}_j \tag{12.85}
$$

This selection does not necessarily result in the best performance of the algorithm; we should consider not only the rate of change in the objective function, but also the value attained by the new basic variable. Furthermore, it may happen that the entering variable is stuck to zero and does not change the value of the objective. In such a case, there is danger of cycling on a set of bases; ways to overcome this difficulty are well explained in the literature.

When x_q is brought into the basis, a basic variable must "leave" the basis in order to maintain $\mathbf{A}\mathbf{x} = \mathbf{b}$. To spot the leaving variable, we can reason as follows. Given the current basis, we can use it to express both $\mathbf{b}$ and the column $\mathbf{a}_q$ corresponding to the entering variable:

$$
\mathbf{b} = \sum_{i=1}^{m} x_{B(i)}\mathbf{a}_{B(i)} \tag{12.86}
$$

$$
\mathbf{a}_q = \sum_{i=1}^{m} d_i\mathbf{a}_{B(i)} \tag{12.87}
$$

where $B(i)$ is the index of the ith basic variable $(i = 1, \ldots, m)$, $\mathbf{a}_{B(i)}$ is the corresponding column, and

$$
\mathbf{d} = \mathbf{A}_B^{-1}\mathbf{a}_q \tag{12.88}
$$

If we multiply Eq. (12.87) by a number θ and subtract it from Eq. (12.86), we obtain

$$
\mathbf{b} = \sum_{i=1}^{m} \left(x_{B(i)} - \theta d_i\right)\mathbf{a}_{B(i)} + \theta\mathbf{a}_q \tag{12.89}
$$

From Eq. (12.89) we see that θ is the value of the entering variable in the new solution and the values of the current basic variables are affected in a way depending on the sign of d_i. If $d_i \leq 0$, $x_{B(i)}$ remains nonnegative when x_q increases. But if there is an index i such that $d_i > 0$, then we cannot increase x_q at will, since there is a limit value for which a currently basic variable becomes zero. This limit value is attained by the entering variable x_q, and the first current basic variable that gets zero leaves the basis:

$$x_q = \min_{\substack{i=1,\ldots,m \\ d_i > 0}} \frac{\hat{b}_i}{d_i} \qquad (12.90)$$

If $\mathbf{d} \leq \mathbf{0}$, there is no limit on the increase of x_q, and the optimal solution is unbounded.

In order to start the iterations, an initial basis is needed. One possibility is to introduce a set of auxiliary *artificial variables* $\mathbf{z}$ in the constraints:

$$\mathbf{Ax} + \mathbf{z} = \mathbf{b} \qquad (12.91)$$
$$\mathbf{x}, \mathbf{z} \geq \mathbf{0}$$

The artificial variables can be regarded as residuals, in the same vein as residuals in linear regression. Assume also that the equations have been rearranged in such a way that $\mathbf{b} \geq \mathbf{0}$. Clearly, a basic feasible solution of the system (12.91) where $\mathbf{z} = \mathbf{0}$ is also a basic feasible solution for the original system $\mathbf{Ax} = \mathbf{b}$. In order to find such a solution, we can introduce and minimize an auxiliary function ϕ as follows

$$\min \phi = \sum_{i=1}^{m} z_i \qquad (12.92)$$

using the simplex method itself. Finding an initial basic feasible solution for this artificial problem is trivial: $\mathbf{z} = \mathbf{b}$. If the optimal value of (12.92) is $\phi^* = 0$, we have found a starting point for the original problem; otherwise, the original problem is infeasible.

The reader is urged to keep in mind that what we have stated here is only the *principle* of the simplex method. Many things may go wrong with such a naive idea, and considerable work is needed to make it robust and efficient. Indeed, even though the simplex method dates back to 1947, it is still being improved today. We leave these refinement of the simplex method to the specialized literature and illustrate its application to the familiar optimal mix problem.[32]

---

[32]In LP textbooks, calculations are efficiently organized in a *tableau*. The tableau is an excellent tool, if the purpose is torturing students by forcing them to solve toy LP problems using pencil, paper, and pocket calculators. I refrain from doing so, as I prefer to emphasize the link between the simplex method and the linear algebraic concepts we have used in describing it.

Example 12.23 A first step, which is actually carried out by good solvers, is to preprocess the formulation eliminating redundant constraints. This results in the streamlined problem

$$\max \quad 45x_1 + 60x_2$$
$$\text{s.t.} \quad 15x_1 + 35x_2 \leq 2400$$
$$25x_1 + 15x_2 \leq 2400$$
$$0 \leq x_2 \leq 50$$

We should rewrite it in standard form, by introducing three slack variables $s_1, s_2, s_3 \geq 0$:[33]

$$\min \quad -45x_1 - 60x_2$$
$$\text{s.t.} \quad 15x_1 + 35x_2 + s_1 = 2400$$
$$25x_1 + 15x_2 + s_2 = 2400$$
$$x_2 + s_3 = 50$$
$$x_1, x_2, s_1, s_2, s_3 \geq 0$$

In matrix terms, we have:

$$\mathbf{A} = \begin{bmatrix} 15 & 35 & 1 & 0 & 0 \\ 25 & 15 & 0 & 1 & 0 \\ 0 & 1 & 0 & 0 & 1 \end{bmatrix}, \quad \mathbf{c} = \begin{bmatrix} -45 \\ -60 \\ 0 \\ 0 \\ 0 \end{bmatrix}, \quad \mathbf{b} = \begin{bmatrix} 2400 \\ 2400 \\ 50 \end{bmatrix}$$

Finding an initial feasible solution is easy; as a starting basis, we consider $\{s_1, s_2, s_3\}$, which corresponds to a production plan where $x_1 = x_2 = 0$, i.e., the origin M_0 in Fig. 1.3. The corresponding basis matrix is $\mathbf{A}_B = \mathbf{I}$, i.e., a 3×3 identity matrix. Using the notation we introduced, we have

$$\mathbf{A}_N = \begin{bmatrix} 15 & 35 \\ 25 & 15 \\ 0 & 1 \end{bmatrix}, \quad \mathbf{c}_N = \begin{bmatrix} -45 \\ -60 \end{bmatrix}, \quad \mathbf{c}_B = \begin{bmatrix} 0 \\ 0 \\ 0 \end{bmatrix}, \quad \hat{\mathbf{b}} = \begin{bmatrix} 2400 \\ 2400 \\ 50 \end{bmatrix}$$

Since the cost coefficients of the basic variables are zero, application of Eq. (12.84) yields the reduced costs $[-45, -60]$. To improve the mix, according to Eq. (12.85) we should bring x_2 into the basis. Geometrically, we move vertically along an edge of the polyhedron, from M_0 to M_1. By the way, this is sensible as item 2 has the largest contribution to profit, but we observe that this is not necessarily the best choice, as we start moving along the longer of the two paths that lead to the optimal solution M_3; in fact, selecting the

---

[33] In practice, the simplex method works with simple bounds in a different way, but this is beyond the scope of the book.

nonbasic variable with the most negative reduced cost is not the strategy of choice of modern solvers. Now we should apply Eq. (12.90) to find the variable leaving the basis. Since the matrix of basic columns is the identity matrix, applying (12.88) is easy:

$$\mathbf{d} = \mathbf{I}^{-1}\mathbf{a}_2 = \begin{bmatrix} 35 \\ 15 \\ 1 \end{bmatrix}$$

and

$$x_2 = \min\left\{\frac{2400}{35}, \frac{2400}{15}, \frac{50}{1}\right\} = \min\{68.5714, 160, 50\} = 50$$

We see that the slack variable s_3 leaves the basis, as x_2 reaches its upper bound.

Now, let us avoid terribly boring calculations and cut a long story short. Repeating these steps, we would bring x_1 into the basis, making one capacity constraint binding by driving to zero the corresponding slack variable. We know that both capacity constraints are binding at the optimal solution M_3. Let us just check that this is indeed the optimal solution. The extreme point M_3 corresponds to the basis $\{x_1, x_2, s_3\}$; therefore, we have

$$\mathbf{A}_B = \begin{bmatrix} 15 & 35 & 0 \\ 25 & 15 & 0 \\ 0 & 1 & 1 \end{bmatrix}, \ \mathbf{A}_N = \begin{bmatrix} 1 & 0 \\ 0 & 1 \\ 0 & 0 \end{bmatrix}, \ \mathbf{c}_B = \begin{bmatrix} -45 \\ -60 \\ 0 \end{bmatrix}, \ \mathbf{c}_N = \begin{bmatrix} 0 \\ 0 \end{bmatrix}$$

Finding the corresponding basic feasible solution requires solving a system of linear equations. Formally:

$$\mathbf{x}_B = \hat{\mathbf{b}} = \mathbf{A}_B^{-1}\mathbf{b} = \begin{bmatrix} 73.8462 \\ 36.9231 \\ 13.0769 \end{bmatrix}$$

Now we need the reduced costs of s_1 and s_2

$$\hat{\mathbf{c}}_N^T = \begin{bmatrix} 0, & 0 \end{bmatrix} - \begin{bmatrix} -45, & -60, & 0 \end{bmatrix} \begin{bmatrix} 15 & 35 & 0 \\ 25 & 15 & 0 \\ 0 & 1 & 1 \end{bmatrix}^{-1} \begin{bmatrix} 1 & 0 \\ 0 & 1 \\ 0 & 0 \end{bmatrix} = \begin{bmatrix} 1.2692, & 1.0385 \end{bmatrix}$$

As both reduced costs are positive, we have indeed found the optimal basis. We note that in practice no matrix inversion is required; we may just solve systems of linear equations using by the Gaussian elimination process of Section 3.2.2.

In closing this example, the very careful reader might notice that the reduced costs above are exactly the shadow prices of capacity constraints that we have seen in Example 12.21. Indeed, there is a connection, which we cannot pursue here; we just notice that the simplex method yields shadow prices as a very useful by-product.   ☐

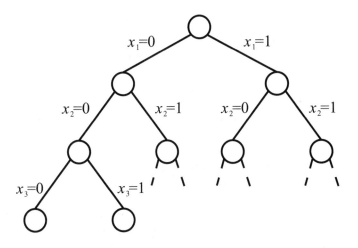

Fig. 12.20   Search tree for a pure binary programming problem.

12.6.2   LP-based branch and bound method

When dealing with a MILP, the simplex method cannot guarantee an integer solution; more often than not, we obtain a fractional solution. There is an even more troubling fact, which is essentially due to the lack of convexity. In fact, if some magical oracle hands over the optimal solution of a continuous LP, we know how to verify its optimality by checking reduced costs. Unfortunately there is no such condition for integer LPs; to find the optimal solution, we should explore the whole set of feasible solutions, at least in principle. Enumerating the whole set of solutions is conceptually easy for a pure binary program, such as the knapsack problem of Section 12.4.1. It is natural to consider a search tree like the one depicted in Fig. 12.20. At each node, we branch on a variable generating two subproblems. At the root node in the figure, we branch on x_1, defining a left subproblem in which $x_1 = 0$ (i.e., item 1 is not included in the knapsack), and a right subproblem in which $x_1 = 1$ (i.e., item 1 is not included in the knapsack). At each level of the tree we branch on a different variable, and the leaves of the tree correspond to a candidate solution. Not all of the leaves correspond to feasible solutions, because of the budget constraint. Hence, many solutions will be excluded from the search process. Yet, it is easy to see that this is not a very practical solution approach. If there are n binary variables, there are potentially 2^n solutions to check, resulting in a combinatorial explosion. We say that this enumeration process has *exponential complexity* in the worst case.

Things would definitely look brighter, if we could find a way to "prune" the tree by eliminating some of its branches. We may avoid branching at a node of the tree if we are sure that the nodes beneath that one cannot yield an optimal solution; but how can we conclude this without considering all of these nodes? We have already observed that if we relax the integrality requirements in a

MILP and solve the corresponding continuous LP, we obtain a *bound* on the optimal value of the MILP.[34] In the familiar production mix example, solving the continuous LP we get an optimal contribution to profit 5538.46. This is an *optimistic estimate* on the true optimal contribution to profit, 5505, which we obtain when requiring integrality of the decision variables. In the case of a maximization problem, the optimistic estimate is an *upper bound UB* on the optimal objective value (in a minimization problem, the optimistic estimate is a *lower bound*). Such bounds may be calculated at any node of the search tree for a MILP by relaxing the free integer decision variables to continuous values; free variables are those that have not been fixed to any value by previous branching decisions. This is called *continuous relaxation* or *LP-based relaxation*.

Now suppose that in the process of exploring the search tree we find a feasible (integer), though not necessarily optimal solution. Then, its value is a lower bound LB on the optimal objective value for a maximization problem, since the optimal profit cannot be lower than the value of any feasible solution (a feasible solution yields an upper bound for a minimization problem). It is easy to see that by comparing lower and upper bounds, we may eliminate certain nodes from the search tree. For instance, imagine that we know a feasible solution of a knapsack problem, such that its value is 100; so, we have a lower bound $LB = 100$ on the value of the optimal knapsack. Say that the LP relaxation at a node in the search tree yields an upper bound $UB = 95$. Then, we observe that $LB > UB$ and immediately conclude that this node can be safely eliminated: We do not know what is the value of the best solution in the branches below that node, but we know that it cannot be better than the feasible solution we already know. The roles of lower and upper bounds are reversed for a minimization problem. The following example shows in detail LP-based branch and bound for a knapsack problem.[35]

Example 12.24 Consider the knapsack problem

$$\max \quad 10x_1 + 7x_2 + 25x_3 + 24x_4$$
$$\text{s.t.} \quad 2x_1 + 1x_2 + 6x_3 + 5x_4 \leq 7$$
$$x_i \in \{0, 1\}$$

We first solve the root problem of the tree (P_0 in Fig. 12.21), which is the continuous relaxation of the binary problem, obtained by relaxing the integrality of decision variables and requiring $0 \leq x_i \leq 1$. Solving the problem, we get the solution

$$x_1 = 1, \quad x_2 = 1, \quad x_3 = 0, \quad x_4 = 0.8$$

---

[34]See also Example 12.4.
[35]We should mention that there are specific branch and bound approaches for the knapsack problem, exploiting its structure, but our aim is illustrating a more general idea in the simplest case.

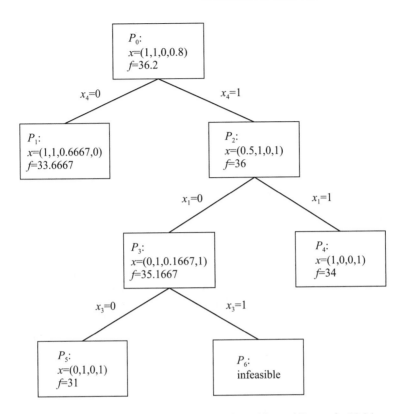

Fig. 12.21   Search tree for the knapsack problem of Example 12.24.

with objective value 36.2. This is an upper bound and, since all of the problem data are integers, we can immediately conclude that the optimal value cannot exceed 36. We observe that the last variable is fractional, and we branch on it, generating two more subproblems: In P_1 we set $x_4 = 0$, and in P_2 we set $x_4 = 1$. Note that we are free to branch on other variables as well, but this choice seems to make more sense; there is the possibility of obtaining an integer solution before reaching the bottom level of the tree. Solving P_1 yields

$$\mathbf{x} = (1, 1, 0.6667, 0), \quad f = 33.6667$$

whereas P_2 yields

$$\mathbf{x} = (0.5, 1, 0, 1), \quad f = 36$$

It is important to notice that both upper bounds are smaller than the bound obtained at the root node. This makes sense, because we are adding constraints, and the value of the optimal continuous solution cannot increase by doing so. Now we should choose which branch of the tree (i.e., which subproblem) to explore first. One possible rule is to look down the most promising branch. Hence, we generate two more subproblems from P_2, branching on the fractional variable x_1. Setting $x_1 = 0$ results in subproblem P_3, whose solution yields

$$\mathbf{x} = (0, 1, 0.1667, 1), \quad f = 35.1667$$

Setting $x_1 = 1$ results in subproblem P_4, whose solution yields our first integer solution

$$\mathbf{x} = (1, 0, 0, 1), \quad f = 34$$

Note that we have found an integer solution without reaching a leaf node; when this happens, i.e., when the continuous relaxation yields an integer solution, there is no point in further branching at that node. Having found our first integer solution, we also have a lower bound (34) on the optimal value. Comparing this with the root bound, we could conclude that the optimal value must either be 34, 35, or 36. Actually, checking the bound on the lower-level, still-active nodes, we may immediately conclude that the optimal value must be either 34 or 35, the bound we obtain from node P_3. Subproblem P_1 may be immediately eliminated, as its upper bound is lower than the value of our feasible solution. We say that the subproblem has been "fathomed," or that the tree "has been pruned." Clearly, the earlier we prune the tree, the more subproblems we eliminate.

We are not done, though, because node P_3 looks promising. There, we branch on the fractional variable x_3. Setting $x_3 = 0$ results in subproblem P_4, whose solution yields a new integer solution

$$\mathbf{x} = (0, 1, 0, 1), \quad f = 31$$

This solution is worse than the first one; yet, it is the best we can do down that branch, so we may prune the node. Setting $x_3 = 1$ results in the infeasible subproblem P_4; indeed, if we try putting both items 1 and 3 into the knapsack, we exceed its capacity $(2 + 6 > 7)$. Now we may report solution $\mathbf{x} = (1, 0, 0, 1)$ as optimal. We have explored a fair amount of nodes, but many of them have been eliminated in the process.   ☐

Let us generalize the above scheme a bit. The idea of branching consists of taking a problem $P(S)$, with feasible set S, and generating a set of suproblems by partitioning S into a collection of subsets $S_1, \ldots, S_q$ such that

$$S = S_1 \cup S_2 \cup \cdots \cup S_q$$

Then, for a minimization problem, we have

$$\min_{\mathbf{x} \in S} f(\mathbf{x}) = \min_{i=1,\ldots,q} \left\{ \min_{\mathbf{x} \in S_i} f(\mathbf{x}) \right\}$$

The rationale behind this decomposition of the feasible set is that solving the problems over smaller sets should be easier; or, at least, the bounds obtained by solving the relaxed problems should be tighter. For efficiency reasons it is advisable, but not strictly necessary, to partition the set S in such a way that subsets are disjoint:

$$S_i \cap S_j = \emptyset, \qquad i \neq j$$

Of course, in a MILP model involving binary and continuous variables, we branch only on binary variables. But what about general integer variables? It is clearly unreasonable to create a branch for each possible integer value. The standard strategy is again to generate only two subproblems, branching on a fractional variable as follows. Assume that an integer variable x_j takes a noninteger value $\bar{x}_j$ in the optimal solution of the relaxed subproblem. Then two subproblems are generated; in the *downchild* we add the constraint[36]

$$x_j \leq \lfloor \bar{x}_j \rfloor$$

to the formulation; in the *upchild* we add

$$x_j \geq \lfloor \bar{x}_j \rfloor + 1$$

For instance, if $\bar{x}_j = 4.2$, we generate two subproblems with the addition of constraints $x_j \leq 4$ (for the downchild) and $x_j \geq 5$ (for the upchild).

Now we may state a branch and bound algorithm in fairly general terms. To fix ideas, we refer to a minimization problem; it is easy to adapt the idea to a maximization problem. Given subproblem $P(S_k)$, let $\nu[P(S_k)]$ denote the value of its optimal solution, and let $\beta[P(S_k)]$ be a lower bound:

$$\beta[P(S_k)] \leq \nu[P(S_k)]$$

Note that $P(S_k)$ can be fathomed only by comparing the lower bound $\beta[P(S_k)]$ with an upper bound on $\nu[P(S)]$. It is *not* correct to fathom $P(S_k)$ on the basis of a comparison with a subproblem $P(S_i)$ such that

$$\beta[P(S_i)] < \beta[P(S_k)]$$

Fundamental branch and bound algorithm

1. *Initialization.* The list of open subproblems is initialized to $P(S)$; the value of the incumbent solution ν^* is set to $+\infty$. At each step, the incumbent solution is the best integer solution found so far.

2. *Selecting a candidate subproblem.* If the list of open subproblems is empty, stop: The incumbent solution $\mathbf{x}^*$, if any has been found, is optimal; if $\nu^* = +\infty$, the original problem was infeasible. Otherwise, select a subproblem $P(S_k)$ from the list.

---

[36]The notation $\lfloor x \rfloor$ corresponds to the "floor" operator, which rounds a fractional number down: $\lfloor 4.7 \rfloor = 4$.

3. *Bounding.* Compute a lower bound $\beta(S_k)$ on $\nu[P(S_k)]$ by solving a relaxed problem $P(\overline{S}_k)$. Let $\overline{\mathbf{x}}_k$ be the optimal solution of the relaxed subproblem.

4. *Prune by optimality.* If $\overline{\mathbf{x}}_k$ is feasible, prune subproblem $P(S_k)$. Furthermore, if $f(\overline{\mathbf{x}}_k) < \nu^*$, update the incumbent solution $\mathbf{x}^*$ and its value ν^*. Go to step 2.

5. *Prune by infeasibility.* If the relaxed subproblem $P(\overline{S}_k)$ is infeasible, eliminate $P(S_k)$ from further consideration. Go to step 2.

6. *Prune by bound.* If $\beta(S_k) \geq \nu^*$, eliminate subproblem $P(S_k)$ and go to step 2.

7. *Branching.* Replace $P(S_k)$ in the list of open subproblems with a list of child subproblems $P(S_{k1})$, $P(S_{k2}),\ldots, P(S_{kq})$, obtained by partitioning S_k; go to step 2.

A thorny issue is which variable we should branch on. Similarly, we should decide which subproblem we select from the list at step 2 of the branch and bound algorithm. As it is often the case, there is no general answer; software packages offer different options to the user, and some experimentation may be required to come up with the best strategy.

Many years ago, the ability of branch and bound methods to solve realistically sized models was quite limited. Quite impressive improvements have been made in commercial branch and bound packages and these, together with the availability of extremely fast and cheap computers, have made branch and bound a practical tool for business management. Nevertheless, many practical problems remain, for which finding the optimal solution requires a prohibitive computational effort. Domain-dependent heuristics have been developed, but we should note that the above branch and bound scheme can be twisted to yield high-quality heuristics by a simple trick. We should just relax the condition $\beta(S_k) \geq \nu^*$ as follows:

$$\beta(S_k) \geq \nu^* - \epsilon$$

where ϵ is a given threshold representing the minimal absolute improvement over the incumbent that we require to explore a branch. Doing so, we might miss the true optimal solution, but we might prune many additional branches, with the guarantee that the difference between the best integer solution found and the optimal solution is bounded by ϵ. If we prefer to state the threshold in relative terms, we can prune a node when

$$\beta(S_k) \geq \nu^*(1 - \epsilon)$$

In this case, ϵ is related to a guarantee on the maximum percentage suboptimality. If $\epsilon = 0.01$, we know that maybe we missed the optimal solution, but this would improve the best integer solution we found by at most 1%. It is up to the user to find the best tradeoff between computational effort and solution quality. Another interesting way to reduce computational effort is by reformulating the model in order to improve its solvability, as we illustrate next.

12.6.3 The impact of model formulation

We have seen that commercial branch and bound procedures compute bounds by LP-based (continuous) relaxations. Given a MILP problem

$$P(S) \qquad \min \quad \mathbf{c}^T \mathbf{x} + \mathbf{d}^T \mathbf{y}$$
$$\text{s.t.} \quad \mathbf{Ax} + \mathbf{Ey} \leq \mathbf{b}$$
$$\mathbf{x} \in \mathbb{R}_+^{n_1}, \qquad \mathbf{y} \in \mathbb{Z}_+^{n_2}$$

where S denotes its feasible set, the continuous relaxation is obtained by relaxing the integrality constraints, which yields the relaxed feasible set $\overline{S}$ and the relaxed problem:

$$P(\overline{S}) \qquad \min \quad \mathbf{c}^T \mathbf{x} + \mathbf{d}^T \mathbf{y}$$
$$\text{s.t.} \quad \mathbf{Ax} + \mathbf{Ey} \leq \mathbf{b}$$
$$\begin{bmatrix} \mathbf{x} \\ \mathbf{y} \end{bmatrix} \in \mathbb{R}_+^{n_1 + n_2}$$

If we could find the convex hull of S, which is a polyhedron, the application of LP methods on that set would automatically yield an integer solution. Unfortunately, apart from a few lucky cases, finding the convex hull is as hard as solving the MILP problem. A less ambitious task is to formulate a model in such a way that its relaxed region $\overline{S}$ is as close as possible to the convex hull of S; in fact, the smaller $\overline{S}$, the larger the lower bound (for a minimization problem) and we know that tighter bounds make pruning more effective. In the following example, we show how careful model formulation may help.

Example 12.25 (Plant location reformulation of lot-sizing problems)
When modeling fixed-charge problems, we link a continuous variable x and a binary variable δ by the big-M constraint

$$x \leq M\delta \tag{12.93}$$

where M is any upper bound on x. In principle, M may be a very large number, but to get a tight relaxation, we should make it as small as possible. To see why, consider Fig. 12.22, where we illustrate the feasible region associated with constraint (12.93). The feasible set consists of the origin and the vertical line corresponding to $\delta = 1$ and $x \geq 0$. When we solve the continuous relaxation, we drop the integrality constraint on δ and replace it by $\delta \in [0, 1]$. This

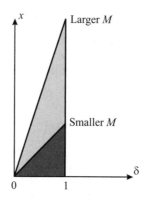

Fig. 12.22   The impact of big-M on a continuous relaxation.

results in the shaded triangles in the figure, whose area depends on a line with slope M. It is easy to see that if M is large, the resulting feasible set for the relaxed problem is large as well. In fact, models with big-M constraints are notoriously hard to solve, and the lot-sizing problem is a well-known example. However, sometimes we may improve the tightness of bounds by resorting to clever reformulations. We illustrate the idea for the minimum-cost version (12.53) of the model.

The naive model formulation is based on production variables x_{it}, representing how much we produce of item i during time bucket t. This continuous variable is related to a binary setup variable δ_{it} by a fixed-charge constraint such as

$$x_{it} \leq \left(\sum_{\tau=t}^{T} d_{i\tau} \right) s_{it}, \qquad \forall i, t \tag{12.94}$$

Here, the big-M is given by the total demand of item i, over the time buckets from t to T. One way to reduce this big-M is to disaggregate the production variable x_{it}, introducing a set of decision variables y_{itp}, which represent the amount of item i produced during time bucket t *to satisfy the demand during time bucket* $p \geq t$. This new variable represents a disaggregation of the original variable x_{it}, since

$$x_{it} = \sum_{p=t}^{T} y_{itp}$$

This reformulation is related to a sort of plant location problem, whereby locations are "in time," rather than "in space." This is illustrated in Fig. 12.23. If we "open the plant" in time bucket 1, we pay the setup cost; then material can flow outside that supply period in order to meet demand at destination nodes. Clearly, if we open a plant in time bucket t, we may only use its outflow to meet demand at later time buckets $p \geq t$.

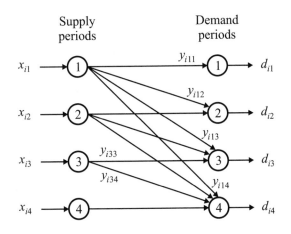

Fig. 12.23 Interpreting the plant location reformulation of lot sizing problems.

Doing so, we introduce more decision variables, but now the link between continuous and binary setup variables is

$$y_{itp} \le d_{ip}s_{it}, \qquad \forall i,t,p \ge t \qquad (12.95)$$

This constraint involves a much smaller big-M; indeed, if we sum constraints (12.95) over p, we find the aggregate constraint (12.94). Now we may also get rid of inventory variables; the amount corresponding to y_{itp} is held in inventory for $(p-t)$ time buckets; hence the corresponding holding cost is

$$h_i(p-t)y_{itp}$$

Finally, we obtain the following model:

$$\min \quad \sum_{t=1}^{T}\sum_{i=1}^{N}\left(f_i\delta_{it} + \sum_{p=t+1}^{T}(p-t)h_iy_{itp} \right)$$

$$\text{s.t.} \quad \sum_{i=1}^{N}\sum_{p=t}^{T} r_{im}y_{itp} + \sum_{i=1}^{N} r'_{im}\delta_{it} \le R_{mt}, \qquad \forall m,t$$

$$y_{itp} \le d_{ip}s_{it}, \qquad \forall i,t,p \ge t$$

$$\sum_{t=1}^{p} y_{itp} = d_{ip}, \qquad \forall i,p$$

$$y_{itp} \ge 0, \qquad \forall i,t,p \ge t$$

$$\delta_{it} \in \{0,1\}, \qquad \forall i,t$$

It is also worth noting that this model formulation allows us to consider perishable items in a quite natural way. If the shelf life of an item is, say, 3

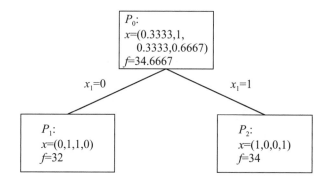

Fig. 12.24   Search tree for the knapsack problem with cover inequalities.

time buckets, we will not define variables y_{itp} for $p - t > 3$; you may visualize the idea by just dropping some arcs in Fig. 12.23.   □

The reformulation we have just considered may look a bit counter-intuitive, and it is not always easy to find a suitable model reformulation like this one. Luckily, many tricks to strengthen a model formulation may be automated.

Example 12.26 (Cover inequalities) Let us consider the knapsack problem of Example 12.24 again:

$$\max \quad 10x_1 + 7x_2 + 25x_3 + 24x_4$$
$$\text{s.t.} \quad 2x_1 + 1x_2 + 6x_3 + 5x_4 \le 7$$
$$x_i \in \{0, 1\}$$

If we observe the budget constraint, it is easy to see that items 1 and 3 cannot be both selected, as their total weight is 8, it exceeds the available budget. Hence we might add the constraint

$$x_1 + x_3 \le 1$$

which is obviously redundant in the discrete domain, but is *not* redundant in the continuous relaxation. By the same token, we could add the following constraints

$$x_3 + x_4 \le 1$$
$$x_1 + x_2 + x_4 \le 2$$

Such additional constraints are called *cover inequalities* and may contribute to strengthen the bound from the LP relaxation, cutting the CPU time considerably. Solving the strengthened model formulation results in the search tree depicted in Fig. 12.24. We may notice that the root subproblem P_0 yields a stronger bound than the plain knapsack model ($34.66667 < 36.2$). What is more striking is that, if we branch on x_1, we immediately get two integer solutions and the search process can be stopped. ⬜

Cover inequalities may be automatically generated and are only one of the types of *cuts* that have been introduced in state-of-the-art software packages implementing branch and bound. The term "cuts" stems from the fact that these additional constraints cut portions of the polyhedron of the continuous relaxation, strengthening bounds. Moreover, clever and effective heuristics have also been devised to generate good integer solutions as soon as possible in the search tree. This generates good upper bounds that help in further pruning the tree. Indeed, the improvement in commercial branch and bound packages over the last 10 years or so has been dramatic, allowing the solution of problems that were intractable a while ago.

Problems

12.1   Assume that functions $f_i(\mathbf{x})$, $i = 1, \ldots, m$, are convex. Prove that the function

$$g(\mathbf{x}) = \sum_{i=1}^{m} \alpha_i f_i(\mathbf{x})$$

where $\alpha_i > 0$, is convex.

12.2   Is the function $f(x) = xe^{2x}$ convex? Does the function feature local minima? What can you conclude?

12.3   Consider the domain defined by the intersection of planes:

$$3x + y + z = 5$$
$$x + y + z = 1$$

Find the point on this domain which is closest to the origin.

12.4   Solve the optimization problem

$$\begin{aligned} \max \quad & xyz \\ \text{s.t.} \quad & x + y + z \leq 1 \\ & x, y, z \geq 0 \end{aligned}$$

How can you justify intuitively the solution you find?

12.5   Consider the constrained problem:

$$\begin{aligned} \min \quad & x^3 - 3xy \\ \text{s.t.} \quad & 2x - y = -5 \\ & 5x + 2y \geq 37 \\ & x, y \geq 0 \end{aligned}$$

- Is the objective function convex?

- Apply the KKT conditions; do we find the true minimizer?

12.6   In Example 12.12 we considered a single-period blending problem with limited availability of raw materials. In practice, we should account for the possibility of purchasing raw materials at a time-varying cost and storing them.

- Extend the model to a multiperiod decision model with purchase decisions, assuming that you know the future prices of raw materials and that storage capacity is unlimited. (*Note*: of course, assuming that future prices are known may be unrealistic; however, commodity derivatives could be used to eliminate uncertainty.)

- Assume that raw materials must be stored in separate tanks, which are available in a limited number. Hence, you may only store up to a given number of raw material types. How can you model this additional constraint?

12.7   Extend the production planning model (12.27) in order to take maintenance activities into account. More precisely, we have M resource centers, and each one must be shut down for exactly one time bucket within the planning horizon. Furthermore, since the maintenance department has quite limited personnel, we can maintain at most two resource centers per time bucket.

12.8   Extend the knapsack problem to cope with logical precedence between activities. For instance, say that activity 1 can be selected only if activities 2, 3, and 4 are selected. Consider alternative model formulations in terms of branch and bound efficiency.

12.9   In Section 12.4.2 we have illustrated a few ways to represent logical constraints. Suppose that activity i must be started *if and only if* both activities j and k are started. By introducing customary binary variables, it is tempting to write a constraint like $x_i = x_j x_k$; unfortunately, this is a bad nonlinear constraint. How can we express this logical constraint linearly? Generalize the idea and find a way to linearize the product $\prod_{i=1}^{n} x_i$ of n binary variables.

12.10   In the minimum cost lot-sizing problem, we assumed that demand must be satisfied immediately; by a similar token, in the maximum profit lot-sizing model, we assumed that any demand which is not satisfied immediately is lost. In other words, in both cases we assumed that customers are impatient.

- Write a model for cost minimization, assuming that customers are willing to wait, but there is a penalty. More precisely, backlog is allowed, which can be represented as "negative inventory holding." Clearly, the backlog cost b_i must be larger than the holding cost h_i. Build a model to minimize cost.

- Now assume that customers are indeed patient, but they are willing to wait only for two time buckets; after two time buckets, any unsatisfied demand is lost. Build a model to maximize profit.

- In the classical lot-sizing model, we implicitly assume that each customer order may be satisfied by items that were produced in different batches. In some cases, this is not acceptable; one possible reason is due to lot tracing; another possible reason is that there are little differences among batches (e.g., in color), that customers are not willing to accept. Then, we should explicitly account for individual order sizes and due dates. Build a model to maximize profit.

- As a final generalization, assume that customers are impatient and that they order different items together (each order consists of several lines, specifying item type and quantity). If you cannot satisfy the whole order immediately, it is lost. Build a model to maximize profit.

12.11 In the portfolio optimization models that we considered in this chapter, risk is represented by variance or standard deviation of portfolio return. An alternative is using MAD (mean absolute deviation):

$$\mathrm{E}\left\{\left|\sum_{i=1}^{n} R_i w_i - \mathrm{E}\left[\sum_{k=1}^{n} R_k w_k\right]\right|\right\}$$

where R_i is the random return of asset i and w_i is its portfolio weight. Suppose that we do not trust any probability distribution for return, but we have a time series of historical data. Let r_{it} be the observed return of asset i in time bucket t, $t = 1, \ldots, T$.

- Build a MILP model to find the minimum MAD portfolio subject to the following constraints:

  - Short selling is not allowed.
  - Expected return should not be below a given target.
  - To avoid a fragmented portfolio, no more than $k < n$ assets can be included in the portfolio, and if an asset is included, there is a lower bound on its weight.
  - Assets are partitioned according to industrial sectors (e.g., banks, energy, chemicals, etc), as well as according to geographic criteria (Asia, Europe, etc.). For each set of assets, overall lower and upper bounds are to be satisfied.

- What is the danger of this modeling approach, based on observed time series?

12.12   A telecommunication network is a set of nodes and directed arcs on which data packets flow. We assume that the flow between each pair of nodes is known and constant over time; please note that the matrix of such flows need not be symmetric, and that packets labeled with a source/destination pair (s, d) are a commodity on their own. Nodes are both source and destination of data packets to and from other nodes, respectively; they can be also used as intermediate nodes for routing, as some pairs of nodes may not be connected directly. Both arcs and nodes are subject to a capacity constraint in terms of packets that they can transport and route over a time frame.

From an operational point of view, we would like to route all of the traffic, in such a way that no network element (node or arc) is congested. For the sake of simplicity, let us assume that a network element is congested when its traffic load exceeds 90% of its nominal capacity (in practice, congestion is a nonlinear phenomenon). We measure network congestion by the number of network elements whose traffic load exceeds this limit.

- Build a model to minimize network congestion, which has an impact on quality of service.

- Extend the model to include capacity expansion opportunities. For each network element, we may expand capacity either by 25% or by 70%; each expansion level is associated with a fixed cost. Build a MILP model to find a tradeoff between quality of service and network cost.

For further reading

- We have been far from rigorous in stating and proving optimality conditions. Readers looking for a more complete treatment, including second-order conditions and handy criteria to check convexity, may refer to Ref. [10] or [11].

- A good reference on linear programming, including alternatives to the classical simplex method, is Ref. [12]. An in-depth treatment of nonlinear programming can be found in Ref. [2], which also illustrates many solution algorithms; see also Ref. [7].

- Integer programming is thoroughly dealt with in Ref. [15]. We just mentioned powerful methods based dynamic column generation; see Ref. [6] for an extensive treatment and many examples.

- We did not cover at all heuristic methods for integer programming models and discrete optimization, but there is a huge literature on ad hoc methods. Each variant of problem may be tackled by a specific approach. To avoid getting lost, it is useful to have a grasp of general principles that can be tailored to specific problems. For instance, tabu search is described in Ref. [8], and Ref. [9] is devoted to genetic algorithms.

- The bibliography on optimization methods is quite rich, but unfortunately the same cannot be said with respect to model *building*. A welcome exception is the text by Williams [14], which shows well-crafted examples taken from a wide range of application domains. The same author has also written a book on model solving, Ref. [13], from which Example 12.4 has been taken.

- It is also useful to have a look at books illustrating the application of optimization modeling to specific domains. Readers interested in optimization models for manufacturing management may consult Ref. [3]; models for distribution logistics are described in Ref. [4]. Optimization models in finance are treated at an introductory level in Ref. [1], which also describes nonlinear programming algorithms. At a more advanced level, Refs. [5] and [16] are useful readings.

- From a practical perspective, optimization modeling is of no use if it is not complemented by a working knowledge of commercial optimization software. There is a wide array of both solvers and modeling tools. Solvers are the libraries implementing solution methods, which may be hard to use without a suitable interface enabling the user to express the model in a natural way. I definitely suggest having a look at

  ◇ http://www.gurobi.com
  ◇ http://www.ampl.com

Other useful links are

  ◇ http://www.informs.org
  ◇ http://www.gams.com
  ◇ http://www.lindo.com
  ◇ http://www.solver.com
  ◇ http://www.maximal-usa.com

REFERENCES

1. M. Bartholomew-Biggs, *Nonlinear Optimization with Financial Applications*, Kluwer Academic Publishers, New York, 2005.

2. M.S. Bazaraa, H.D. Sherali, and C.M. Shetty, *Nonlinear Programming. Theory and Algorithms*, 2nd ed., Wiley, Chichester, West Sussex, UK, 1993.

3. P. Brandimarte and A. Villa, *Advanced Models for Manufacturing Systems Management*, CRC Press, Boca Raton, FL, 1995.

4. P. Brandimarte and G. Zotteri, *Introduction to Distribution Logistics*, Wiley, New York, 2007.

5. G. Cornuejols and R. Tütüncü, *Optimization Methods in Finance*, Cambridge University Press, New York, 2007.

6. G. Desaulniers, J. Desrosiers, and M.M. Solomon, eds., *Column Generation*, Springer, New York, 2005.

7. R. Fletcher, *Practical Methods of Optimization*, 2nd ed., Wiley, Chichester, West Sussex, UK, 1987.

8. F.W. Glover and M. Laguna, *Tabu Search*, Kluwer Academic, Dordrecht, The Netherlands, 1998.

9. Z. Michalewicz, *Genetic Algorithms + Data Structures = Evolution Programs*, Springer-Verlag, Berlin, 1996.

10. C.P. Simon and L. Blume, *Mathematics for Economists*, W.W. Norton, New York, 1994.

11. R.K. Sundaram, *A First Course in Optimization Theory*, Cambridge University Press, Cambridge, UK, 1996.

12. R.J. Vanderbei, *Linear Programming: Foundations and Extensions*, Kluwer Academic, Dordrecht, The Netherlands, 1996.

13. H.P. Williams, *Model Solving in Mathematical Programming*, Wiley, Chichester, 1993.

14. H.P. Williams, *Model Building in Mathematical Programming*, 4th ed., Wiley, Chichester, 1999.

15. L.A. Wolsey, *Integer Programming*, Wiley, New York, 1998.

16. S. Zenios, *Practical Financial Optimization*, Wiley-Blackwell, Oxford, 2008.

13

Decision Making Under Risk

This chapter represents a synthesis of what we have become acquainted with so far. Decision making under uncertainty is a quite challenging topic, merging probability theory and statistics with optimization modeling. This mix may result in quite demanding mathematics, which we will avoid by focusing on fundamental concepts and a few illustrative toy examples to clarify them.

One preliminary question that we should address is: Which kind of uncertainty should we consider? In this chapter we take a rather standard view, i.e., that uncertainty may be represented by the classical tools of probability and statistics. In fact, this is not to be taken for granted, as there are quite different kinds of uncertainty. Compare the roll of a die against the production decision for a brand-new and truly innovative product. In the first case we do not know which number will be drawn, and betting on it means making a risky decision. However, we have no doubt about the rules of the game. In other words, we have a well-defined probability distribution of a random variable, and we just do not know in advance its realization. In the second case, we do not even know the probability distribution, which will be more subjective than fact-based. In extreme cases, even the very use of probabilities is questionable. It has been proposed to distinguish between *decision making under risk* and *decision making under uncertainty*. Strictly speaking, what we deal here with is decision making under risk, as we assume a known probability distribution. True uncertainty is a more elusive concept, possibly involving beliefs, rather than frequentist concepts. We will consider issues related to subjective probability in the next chapter. Here we introduce the fundamental concepts of risk aversion and the way we may account for it when making decisions. We also outline alternative frameworks, addressing additional issues like robustness, disappointment, and regret.

Our first step in this chapter is the formalization of decision trees, which is the subject of Section 13.1. Then, we consider the attitude toward risk. Much theory concerning random variables revolves around expected values; in previous chapters we have considered, for instance, the maximization of expected profit in newsvendor problems. However, such an approach may lead to unreasonable solutions for some decision problems, and this motivates the need to represent risk aversion. In Section 13.2 we introduce concepts related to utility theory and risk measures. Then, we start considering the extension of optimization models, namely, linear programming models, to decision making under risk. In Section 13.3 we consider two-stage stochastic linear programming, which is extended to the multistage case in Section 13.4. Stochastic linear programs can be quite challenging to solve, but we will not consider specific solution methods that have been proposed; what is really important is to consider a couple of basic examples, in order to understand the value of this modeling framework and the qualitative difference between solutions obtained when uncertainty is disregarded and those obtained by considering a set of alternative scenarios. Finally, in Section 13.5, we close the chapter by outlining some further developments related to robustness and regret.

13.1 DECISION TREES

Decision trees are a natural way to describe decision problems under risk, involving a sequence of decisions among a finite set of alternative options and a set of discrete scenarios, modeling uncertain outcomes that follow our decisions. Actually, we have already dealt with decision trees informally in earlier examples.[1] Now we should treat this formalism more systematically, by distinguishing two kinds of node:

- *Decision nodes*, represented by squares, correspond to discrete choices between mutually exclusive alternatives, as depicted in Fig. 13.1(a). At these nodes, the decision maker must choose one among multiple available options.

- *Chance nodes*, represented by circles, correspond to the realization of random outcomes, as depicted in Fig. 13.1(b). Each outcome i is associated with a probability π_i; clearly, the probabilities for the random outcomes at each chance node add up to 1.

A decision tree consists of a set of decision and chance nodes, as shown in Fig. 13.2. Typically, decision and chance nodes are interspersed, but we may

---

[1]See Section 1.2.3 and Example 6.10.

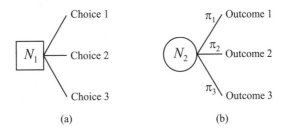

Fig. 13.1 Node types in a decision tree: (a) decision nodes, where choices are made; (b) chance nodes, where random outcomes are selected by "nature" according to probabilities.

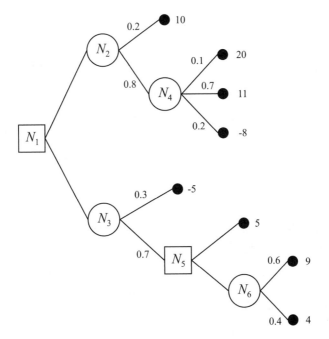

Fig. 13.2 A sample decision tree.

have two chance nodes or two decision nodes in sequence. We also have *terminal nodes*, represented by bullets. Typically, terminal nodes are labeled with a payoff, which is essentially the final monetary value of a sequence of decisions and random outcomes. It may be helpful to associate cash flows with intermediate nodes to clarify the economic impact of each decision. When time plays a significant role, cash flows may have to be discounted.

Solving the problem means choosing a *strategy*, i.e., selecting, for each decision node we *might* visit, one among multiple alternative decisions. In the case of a complex decision tree, when the strategy is implemented most decision nodes will not be reached, as the sequence of random outcomes will generate a path that does not visit those nodes. However, a strategy must plan for every possible contingency. Assuming that payoffs have a monetary nature, the most natural criterion to follow in building the strategy is the maximization of the *expected monetary value* (EMV). When a decision node is followed by a set of chance nodes, we should label each chance node with an EMV, which allows us to choose the best action at the decision node. The labeling process should go backward in time, starting from terminal nodes, and it is best illustrated by a simple example.

Example 13.1 Let us solve the decision problem of Fig. 13.2. At chance node N_4, we calculate the EMV E_4 resulting from the three successive terminal nodes:

$$E_4 = 0.1 \times 20 + 0.7 \times 11 + 0.2 \times (-8) = 8.1$$

Then, to calculate the EMV E_2 for chance node N_2 we consider the expected value E_4 and the value of the sibling terminal node, which yields

$$E_2 = 0.2 \times 10 + 0.8 \times 8.1 = 8.48$$

These calculations are reflected in Fig. 13.3, where nodes are successively labeled. If, at decision node N_1 we choose the upbranch, the EMV of the decision is 8.48. The downbranch is a bit more complicated, as we must consider another decision node. We start labeling chance node N_6:

$$E_6 = 0.6 \times 9 + 0.4 \times 4 = 7$$

At decision node N_5 we should compare the upbranch, with EMV 5, against the downbranch with EMV 7. If we accept the idea of just considering expected values, disregarding risk, we should choose the downbranch. In the figure, this decision is represented by barring (cutting) the suboptimal path corresponding to the upbranch, and labeling node N_5 with the value of the optimal decision, $\max\{5, 7\} = 7$. Now we label node N_3 with the EMV

$$0.3 \times (-5) + 0.7 \times 7 = 4.9$$

Taking the maximum between 4.9, and the previously computed EMV 8.48, we conclude that the best initial decision is to follow the upbranch. After that, there are no more decisions to make, and everything is in the hands of Mother Nature.   ⬜

In this trivial example, along the path following the first decision, we do not have to make any other choice. In realistic cases, decisions are made sequentially, after gathering further information. Moreover, such decisions may represent successive investments, calling for a sequence of cash flows. In this example we did not consider intermediate cash flows, which should be properly weighted probabilistically and possibly discounted, in order to take the time value of money into due account.

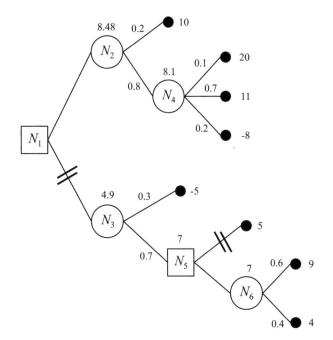

Fig. 13.3 Solution of the decision tree of Fig. 13.2.

Example 13.2 As a more meaningful example, let us represent the problem of Example 6.10 as a decision tree. Figure 13.4 depicts the corresponding decision tree, including its solution. The decision tree allows us to visualize the strategy:

- At decision node N_1, we choose to run the customers' survey, which costs €4000 and provides us with a less uncertain forecast of new product success, one way or another. Note once again that the survey does *not* change the unconditional probability of success of the new product.

- If the result of the customers' survey is promising, we prefer producing; otherwise, we sell the license.

The strategy does not consist of a deterministic sequence of actions; rather, for each random outcome, we have a course of action. Furthermore, in drawing the tree we subtracted the cost of the survey from the payoff of terminal nodes. This does not reflect the true logical sequence of cash flows, but if there is no discounting, there is no real difference. However, when time is of the essence and discounting is needed, it may be much simpler and more informative to associate cash flows with intermediate nodes. ⧠

The decision trees we have considered are rather trivial, and software packages for decision analysis provide the user for more options in order to represent cash flow timing and discounting. Of course, the true difficulty is in

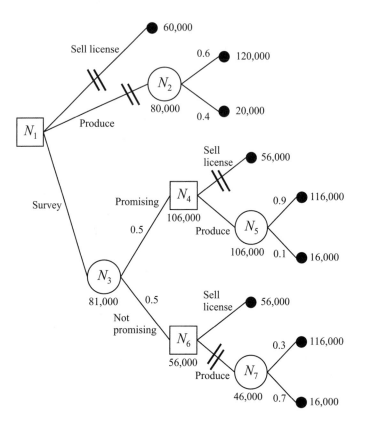

Fig. 13.4   Decision tree for new product launch.

estimating cash flows and the probabilities at chance nodes; typically, a thorough sensitivity analysis is needed in order to check the robustness of the recommended strategy. It should be mentioned, however, that the very process of structuring a decision tree has value in itself, as it forces the decision maker(s) to lay down the structure of the problem, the decisions to be made, their logical sequence, and the related risks and opportunities. This thinking process in itself may be more valuable than a recommendation relying on questionable estimates of probabilities. Furthermore, decision trees also allow estimation of the value of information. For the tree of Fig. 13.4 we know that the value of the *partial* information provided by the customers' survey is not larger than €5000. In the next section we show how to value *perfect* information.

13.1.1   Expected value of perfect information

Decision trees are a very simple tool for framing decision problems with a discrete set of alternatives and a discrete representation of uncertainty. We may start moving to more complicated cases by expressing a decision problem under risk in a more general way:

$$f^* = \min_{\mathbf{x} \in S} \mathrm{E}_\omega [f(\mathbf{x}, \omega)] \tag{13.1}$$

It is important to understand what problem (13.1) represents:

- First, we pick up a feasible solution $\mathbf{x}$ in the feasible set S.

- Then, a random event ω occurs, random variables are realized, and we have a cost $f(\mathbf{x}, \omega)$ depending on both our decision (variables under our control) and random risk factors (variables not under our control). If the function $f(\mathbf{x}, \omega)$ represent a profit, we should change the problem to a maximization.

This is a *here-and-now decision*, as we must make a decision before observing the realization of random risk factors; all we can do is look for a solution that is the best one "on average," which is what is obtained by minimizing the expected value of the cost; the notation $\mathrm{E}_\omega[\cdot]$ points out that expectation depends on random event ω. The optimal solution yields an expected value f^*.

It would be very, very nice to postpone decisions and make them *after* we observe the realization of risk factors. This *wait-and-see solution* would, no doubt, be better than the here-and-now decision, as we could adapt our choice to the specific realized contingency. Unfortunately, in most real-life situations, we cannot wait and see and decide under *perfect information*. Nevertheless, we can estimate the theoretical value of perfect information. This is obtained by swapping minimization and expectation in (13.1):

$$f_{\mathrm{PI}}^* = \mathrm{E}_\omega \left[\min_{\mathbf{x} \in S} f(\mathbf{x}, \omega) \right] \tag{13.2}$$

The subscript in f_{PI}^* tells us that this is the expected value of cost if we could optimize with perfect information, after observing event ω. It stands to reason, and it can be shown formally, that for a minimization problem

$$f_{\mathrm{PI}}^* \le f^*$$

The difference in cost is the *expected value of perfect information* (EVPI):

$$\mathrm{EVPI} \equiv f^* - f_{\mathrm{PI}}^* \tag{13.3}$$

When dealing with a maximization problem, the terms in the difference represent profits, rather than costs, and should be swapped. The EVPI tells us something about the impact of uncertainty and is best illustrated by a toy example.

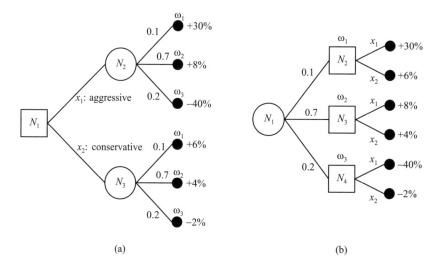

(a)   (b)

Fig. 13.5 Schematic illustration of EVPI – the value of (waiting for) perfect information: (a) here-and-now and (b) wait-and-see decisions.

Example 13.3 Consider a stylized investment problem. We must select an investment strategy between two possibilities, aggressive and conservative. These two choices are associated with decisions x_1 and x_2, respectively; so, the feasible set is $S = \{x_1, x_2\}$. Uncertainty is represented by three possible states of the economy, which we interpret as follows:

- ω_1, with probability 0.1, is a "very bullish" economy, in which the aggressive strategy yields 30% and the conservative one yields a much less exciting +6%.

- ω_2, with probability 0.7, is a "moderately bullish" economy, in which the aggressive strategy yields 8% and the conservative one yields +4%.

- ω_3, with probability 0.2, is a "bearish" economy, in which the aggressive strategy loses an epic 40% and the conservative one limits the loss to a moderate 2%.

The here-and-now decision consists of selecting a strategy before observing returns. The corresponding decision tree is depicted in Fig. 13.5(a). If we take expected return as our objective function, to be maximized, we should take the maximum between

$$f_1 = 0.1 \times 30\% + 0.7 \times 8\% + 0.2 \times (-40\%) = 6\%$$

and

$$f_2 = 0.1 \times 6\% + 0.7 \times 4\% + 0.2 \times (-2\%) = 3\%$$

So, we would choose the aggressive strategy, and $f^* = 6\%$. Now, how much would be the value of clairvoyance? We should restructure the decision tree as in Fig. 13.5(b). If we could invest after observing return, the expected return before the realization of the state of the economy would be

$$f^*_{\text{PI}} = 0.1 \times \max\{30\%, 6\%\} + 0.7 \times \max\{8\%, 4\%\}$$
$$+0.2 \times \max\{-40\%, -2\%\} = 8.2\%$$

The EVPI, in terms of percentage returns, is

$$\text{EVPI} = f^*_{\text{PI}} - f^* = 8.2\% - 6\% = 2.2\%.$$

Of course, we swap terms with respect to Definition (13.3), as we seek to maximize return. The return could be translated into monetary terms by assuming an invested wealth. □

The careful reader, of course, will notice that there is a missing piece in the above example: If an investor really had clairvoyance, in the case of the very bullish scenario, she would actually borrow money to pursue an aggressive strategy; even a small difference in return can yield a huge amount of money, if a correspondingly huge wealth is invested. This strategy is based on *leveraging*, and has been pursued in recent times of low interest rates. As the 2008 financial crisis has clearly shown, such leveraged strategies are quite risky. In fact, another fundamental missing piece in the example is the role of risk: Wise investors are risk-averse. In the next section we take a look at modeling risk aversion and measuring risk.

13.2 RISK AVERSION AND RISK MEASURES

So far, when dealing with a decision problem under risk, we have used expected profit or expected cost as the criterion of choice. We did so, e.g., for the newsvendor problem,[2] as well as for the decision trees of the previous section. But does this actually make sense? The following examples show that this need not be the case.

Example 13.4 (A single bet vs. many repeated bets) Consider the following offer by a professor. A fair coin is flipped: If it lands tail, you win €10, otherwise you lose €5. When offered this lottery, most students would be willing to play. The reasoning is that the expected win (€2.5) is positive. But then, the same should apply if we make things more interesting by scaling the lottery up: You may win €10 *million* or lose €5 *million*. No sensible person would play this game, unless he could afford losing €5 million without changing his lifestyle. However, if one could play the game many times, settling the score only at the end, one should probably accept. Of

---

[2]See Section 7.4.4.

course, playing the game a lot of times reduces variance; in the limit, the law of large numbers applies and risk disappears. However, if you can be thrown out of business in the short term, after a couple of losses, bright long-term prospects will be little solace, if any. □

Example 13.5 (Putting all of your eggs in one basket) Consider an investor who must allocate her wealth to n assets. The return of each asset, indexed by $i = 1, \ldots, n$, is a random variable R_i with expected value $\mu_i = \mathrm{E}[R_i]$. We have introduced this kind of choice in Example 12.5, and we know that asset allocation decisions may be expressed by decision variables w_i representing the fraction of wealth invested in asset i. If we rule out short-selling, these decision variables are naturally bounded by $0 \le w_i \le 1$. If we assume that the investor should just maximize expected return, she should solve the problem

$$\max \quad \sum_{i=1}^{n} \mu_i w_i$$

$$\text{s.t.} \quad \sum_{i=1}^{n} w_i = 1$$

$$w_i \ge 0$$

However, its solution is quite trivial; she should simply pick the asset with maximum expected return, $i^* = \arg\max_{i=1,\ldots,n} \mu_i$, and set $w_{i^*} = 1$. It is easy to see that this concentrated portfolio is a very dangerous bet. In practice, portfolios are diversified, which means that decisions depend on something beyond expected values. Furthermore, one would also add some additional constraints on portfolio composition, bounding exposure to certain geographic areas or types of industry, and they would render the trivial solution above infeasible. However, it may be necessary to add many such additional constraints to find a sensible solution; this means that the solution is basically shaped by the user who enforces these bounds. Incidentally, if shortselling is allowed, the decision variables are unrestricted, and the expected value of future wealth goes to infinity. In fact, one would short-sell assets with low expected return, to make money to be invested in the most promising asset. This is clearly unreasonable. □

Example 13.6 (St. Petersburg paradox) Consider the following proposition. You are offered a lottery, whose outcome is determined by flipping a fair and memoryless coin. The coin is flipped until it lands tail. Let k be the number of times the coin lands head; then, the payoff you get is $\$2^k$. Now, how much should you be willing to pay for this lottery? We may consider this as an asset pricing problem, and set the expected value of the payoff as the fair price for this rather peculiar asset. The probability of winning $\$2^k$ is the probability of having k consecutive heads followed by one tail, which stops the game, after $k+1$ flips of the coin. Given the independence of events, the probability of this sequence is $1/2^{k+1}$, i.e., the product of individual event probabilities. Then, the expected value of the payoff is

$$\sum_{k=0}^{\infty} \frac{1}{2^{k+1}} 2^k = \frac{1}{2} \times 1 + \frac{1}{4} \times 2 + \frac{1}{8} \times 4 + \cdots = \frac{1}{2} + \frac{1}{2} + \frac{1}{2} + \cdots = +\infty$$

This game looks so beautiful that we should be willing to pay any amount of money to play it! No one would probably do so. Again, we see that expected values do not tell the whole story. □

These examples should suffice to convince us that considering expected monetary values, whether costs or profits, is not enough to fully address decision making under risk. We should find a way to account for the natural tendency to avoid unnecessary or excessive risk. Sometimes, ad hoc tools are used. For instance, when evaluating an investment, cash flows are discounted by a rate accounting for risk.[3] In this section we try to find a more general framework, which is provided by utility theory. This is a standard approach in classical economics, fraught with many difficulties and shortcomings; yet, it is a useful conceptual tool. Later, we consider more practical ways to account for and measure risk.

13.2.1   A conceptual tool: the utility function

The idea that most decision makers are risk-averse is intuitively clear, but what does *risk aversion* really mean? A theoretical answer, commonly put forward in economic theory, can be found by assuming that decision makers order uncertain outcomes by a utility function rather than by straightforward expected monetary values. To introduce the concept, let us consider simple lotteries. A lottery is represented by a random variable X that assumes values x_i with probabilities p_i; the decision maker should select among alternative lotteries and may also combine them, forming new random variables. For instance, consider an agent who has to choose between the following two lotteries:

1. Lottery a_1, which is actually deterministic and has a sure payoff μ.

2. Lottery a_2, which has two equally likely payoffs $\mu + \delta$ and $\mu - \delta$.

The two lotteries are clearly equivalent in terms of expected payoff, but a risk-averse agent will arguably select lottery a_1. More generally, if we have a random variable X and we add a *mean-preserving spread*, i.e., a random variable ϵ with $E[\epsilon] = 0$, this addition is not welcome by a risk-averse decision maker.

---

[3]We did so for the growth option example of Section 1.2.2.

Given a set of lotteries, the agent should be able to pick the preferred one; more precisely, given any pair of lotteries, the agent should be able to tell which one she prefers or decide that she is indifferent to a choice between them. If so, we have a preference relationship among lotteries. Since preference relationships are a bit cumbersome and are not easy to deal with, we could map each lottery to a number, measuring the attractiveness of that lottery to the agent, and use the standard ordering of numbers to rank lotteries. For arbitrary preference relationships, a function representing them may not exist but, under a set of more or less reasonable assumptions,[4] such a mapping does exist and can be represented by a *utility function*. A particularly simple form of utility function, which looks reasonable but is justified by specific hypotheses on the preference relationship that it models, is the *Von Neumann–Morgenstern utility*, defined as

$$U(a) = \sum_{i=1}^{n} p_i u(x_i)$$

The definition involves a function $u(\cdot)$, and a is a lottery with n outcomes x_i and probabilities p_i. The function $u(\cdot)$ is the utility of a certain payoff, and $U(\cdot)$ is the *expected utility*. If $u(x) \equiv x$, then the utility function boils down to the expected value of the payoff. Alternative choices of the utility function u model different attitudes toward risk. For business problems, it is reasonable to assume that utility $u(\cdot)$ is an increasing function, since we prefer more wealth to less.

In the case of the two lotteries above, preference for a_1 is expressed by

$$U(a_1) = u(\mu) \geq \tfrac{1}{2}u(\mu - \delta) + \tfrac{1}{2}u(\mu + \delta) = U(a_2)$$

Since the inequality is not strict, we should say that lottery a_1 is at least as preferred as a_2, and the decision maker could be indifferent between the two. More generally, if we have two possible outcomes, x_1 and x_2, with probabilities $p_1 = p$ and $p_2 = 1 - p$, respectively, a risk-averse decision maker would prefer not taking chances:

$$u(\mathrm{E}[X]) = u(px_1 + (1 - p)x_2) \geq pu(x_1) + (1 - p)u(x_2) = \mathrm{E}[u(X)] \quad (13.4)$$

This condition basically states that the function $u(\cdot)$ is *concave*. Figure 13.6 illustrates the role of concavity. In Theorem 6.8 we have introduced Jensen's inequality for convex functions. Indeed, Eq. (13.4) is just a specific case of Jensen's inequality for concave functions of a random variable X:

$$u(\mathrm{E}[X]) \geq \mathrm{E}[u(X)] \quad (13.5)$$

---

[4]The discussion of these assumptions is best left to books on microeconomics or decision theory; we should mention that most of them seem rather innocent and reasonable under most circumstances, but they may lead to surprising effects in paradoxical examples.

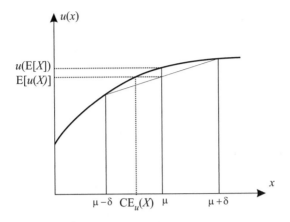

Fig. 13.6 How concave utility functions imply risk aversion; the certainty equivalent is also shown.

It is fundamental to observe that the specific numerical value that the utility function assigns to a lottery is irrelevant per se; only the *relative* ordering of alternatives is essential. In fact, we speak of *ordinal* rather than cardinal utility. Given the linearity of expectation, we also see that an affine transformation of the utility function $u(\cdot)$ has no effect, provided it is increasing: If we use $au(x)+b$ instead of $u(x)$, where $a > 0$, the relative ranking of alternatives is preserved.

How can we say something about the properties of a specific utility function? In particular, we would like to come up with some way to measure risk aversion. We have said that a risk-averse decision maker would prefer a certain payoff rather than an uncertain one, when the expected values are the same. She would take the gamble only if the expected value of the risky lottery were suitably larger than the certain payoff. In other words, she requires a *risk premium*. The risk premium depends partly on the risk attitude of the decision maker, and partly on the uncertainty of the gamble itself. We will denote the risk premium by $\rho_u(X)$; note that this is a number that a decision maker with utility $u(\cdot)$ associates with a random variable X. The risk premium is defined by the condition

$$u(\mathrm{E}[X] - \rho_u(X)) = U(X) \tag{13.6}$$

The risk premium implicitly defines a *certainty equivalent*, i.e., a sure and guaranteed payoff $\mathrm{CE}_u(X)$ such that the agent would be indifferent between this certain amount and the uncertain lottery:

$$\mathrm{CE}_u(X) \equiv \mathrm{E}[X] - \rho_u(X)$$

Note that the certainty equivalent is smaller than the expected value, and the difference is larger when the risk premium is larger. These concepts may be better grasped by looking again at Fig. 13.6.

A difficulty with the risk premium concept is that it mixes the intrinsic risk of a lottery with the risk attitude of the decision maker. We might wish to separate the two sides of the coin. Consider a lottery $X = x + \tilde{\epsilon}$, where x is a given number, and $\tilde{\epsilon}$ is a random variable with $\mathrm{E}[\tilde{\epsilon}] = 0$ and $\mathrm{Var}(\tilde{\epsilon}) = \sigma^2$. Assume that the random variable $\tilde{\epsilon}$ is a "small" perturbation, in the sense that any possible realization ϵ is a relatively small number.[5] Hence, we may approximate both sides of Eq. (13.6) by Taylor's expansions. Consider, for instance, the expression $u(x + \epsilon)$. Since only numbers are involved here, we may write

$$u(x + \epsilon) \approx u(x) + \epsilon u'(x) + \tfrac{1}{2}\epsilon^2 u''(x)$$

Using this approximation for the random variable $\tilde{\epsilon}$, under the assumption that its realization are small enough, and taking expected values, we may approximate the right-hand side of (13.6) as follows:

$$\begin{aligned} U(X) \equiv \mathrm{E}[u(x)] &\approx \mathrm{E}\left[u(x) + \tilde{\epsilon}u'(x) + \tfrac{1}{2}\tilde{\epsilon}^2 u''(x)\right] \\ &= u(x) + \mathrm{E}[\tilde{\epsilon}]u'(x) + \tfrac{1}{2}\mathrm{E}[\tilde{\epsilon}^2]u''(x) \\ &= u(x) + 0 \cdot u'(x) + \tfrac{1}{2}\mathrm{Var}(\tilde{\epsilon})u''(x) \\ &= u(x) + \tfrac{1}{2}\sigma^2 u''(x) \end{aligned}$$

In the second-to-last (penultimate) line we have used $\mathrm{Var}(\tilde{\epsilon}) = \mathrm{E}[\tilde{\epsilon}^2] - \mathrm{E}^2[\tilde{\epsilon}] = \mathrm{E}[\tilde{\epsilon}^2] - 0$. We may also approximate the left-hand side of (13.6), which involves only numbers, by a first-order expansion around $\mathrm{E}[X] = x$:

$$u(\mathrm{E}[X] - \rho_u(X)) \approx u(x) - \rho_u(X)u'(x)$$

Equating the two approximations and rearranging yields

$$\rho_u(X) = -\frac{1}{2}\frac{u''(x)}{u'(x)}\sigma^2 \tag{13.7}$$

Since we assume the utility function to be concave and increasing, the right-hand side of Eq. (13.7) is positive.[6] We observe that the risk premium is factored as the product of one term depending on the subjective agent's risk aversion and another one depending on the objective uncertainty of the lottery. This justifies the following definition of the *coefficient of absolute risk aversion*:

$$R_u^a(x) \equiv -\frac{u''(x)}{u'(x)} \tag{13.8}$$

---

[5]For the sake of convenience, in this section we denote by $\tilde{\epsilon}$ a random variable and by ϵ a realization of that variable. This notation is common in economics; in statistics, one typically uses X and x with the corresponding pair of meanings.

[6]We recall that, for a differentiable concave function of one variable, we have $u''(x) \leq 0$; see Section 2.11.

We have said that, given the linearity of the expectation operator, transforming the utility function $u(x)$ by an increasing affine transformation is inconsequential. Indeed, the definition of the risk aversion coefficient is consistent with this observation, as it is easy to see that the coefficients for $u(x)$ and $au(x) + b$ are the same. Note that the coefficient $R_u^a(x)$ does not depend on uncertainty, but it does depend on the expected value of the lottery. From an investor's perspective, this implies that risk aversion depends on the current level of wealth. The more concave the utility function, the larger the risk aversion.

By a similar token, we may define the *coefficient of relative risk aversion*. This is motivated by considering a multiplicative, rather than additive, shock on an expected value x: $X = x(1 + \tilde{\epsilon})$. Using a similar reasoning, we get

$$\rho_u(X) = -\frac{1}{2}\frac{u''(x)}{u'(x)}x\sigma^2$$

which suggests the definition

$$R_u^r(x) \equiv -\frac{u''(x)x}{u'(x)} \tag{13.9}$$

Example 13.7 (A few standard utility functions) A typical utility function is logarithmic utility:[7]

$$u(x) = \log(x) \tag{13.10}$$

Clearly this makes sense only for positive values of wealth. It is easy to check that, for the logarithmic utility, we have

$$R_u^a(x) = \frac{1}{x}, \qquad R_u^r(x) = 1$$

Hence, logarithmic utility has decreasing absolute risk aversion, but constant relative risk aversion. We say that logarithmic utility belongs to the families of *decreasing absolute risk aversion* (DARA) and *constant relative risk aversion* (CRRA) utility functions. We may also consider the exponential utility function

$$u(x) = -e^{-\alpha x} \tag{13.11}$$

for $\alpha > 0$. Note that this is an increasing function, and it is easy to interpret the parameter α:

$$R_u^a(x) = -\frac{-\alpha^2 e^{-\alpha x}}{\alpha e^{-\alpha x}} = \alpha$$

Hence, we see that the exponential utility is *constant absolute risk aversion* (CARA). It is important to remark that some utility functions have been used

---

[7]In the following text we will use the notation log, rather than ln, to denote the *natural* logarithm.

because they are easy to manipulate, but this does not imply that they always model realistic investors' behavior.[8]

Another common utility function is the quadratic utility:

$$u(x) = x - \frac{\lambda}{2}x^2 \tag{13.12}$$

Note that this function is not monotonically increasing and makes sense only for $x \in [0, 1/\lambda]$. Another odd property of quadratic utility is that it is *increasing absolute risk aversion* (IARA):

$$R_u^a(x) = \frac{\lambda}{1 - \lambda x} \Rightarrow \frac{dR_u^a(x)}{dx} = \frac{\lambda^2}{(1 - \lambda x)^2} > 0$$

This implies, for instance, that an investor becomes more risk averse if her wealth increases, which is usually considered at odds with standard investors' behavior. Nevertheless, we may also see that quadratic utility emphasizes the role of variance, since for this utility

$$U(X) = \mathrm{E}\left[X - \frac{\lambda}{2}X^2\right] = \mathrm{E}[X] - \frac{\lambda}{2}\left(\mathrm{Var}(X) + \mathrm{E}^2[X]\right)$$

A decision maker with quadratic utility is basically concerned only with the expected value and the variance of an uncertain outcome.   ⬚

Specifying a utility function may be a difficult task, since assessing the tradeoff between expected payoff and risk is far from trivial. This may be of no concern in economics, if the aim is to build a model explaining some observed behavior and qualitative insights that are of interest; however, in business decision making, it is an issue. In the following sections, we consider more operational approaches.

13.2.2   Mean–risk optimization

If asked about our utility function, we would hardly be able to give a sensible answer. However, in real life, we do trade off expectations against risk; to do so, we need a way to measure risk.

DEFINITION 13.1 *A risk measure is a function $\rho(X)$, mapping a random variable X into the set of nonnegative real numbers $\mathbb{R}_+$.*

In practice, the random variable X could be a random profit or a random return on an investment; generally speaking, it represents the consequence of our decision. The idea behind a risk measure is that the larger its value, the

---

[8]See Problem 13.6.

riskier our choice are. We should not confuse such a risk measure with a coefficient of risk aversion. The latter is a measure of the subjective attitude of a decision maker toward risk; the former is an objective measure of the riskiness of a lottery or a prospect, and not a measure of *perceived* risk. Apparently, the most natural risk measure we may come up with is related to variance or standard deviation. Standard deviation is more natural in the sense that it is measured in the same units as the corresponding expectation; however, variance may be more convenient when building optimization models. In fact, minimization of variance is equivalent to minimization of standard deviation, but the former may result in a convex optimization problem, whereas the latter does not. Later, we will see that variance and standard deviation might not make very good risk measures, but they are a suitable starting point.

Now, given a random variable X representing profit or return, we should find a tradeoff between its expected value $E[X]$ and a risk measure $\rho(X)$. In Section 12.3.3 we have considered multiobjective optimization, and we have defined the concept of efficient solution and efficient frontier. Even if we are not able to spot a single optimal solution trading off expectation and risk, we might trace the efficient frontier with respect to these conflicting objectives.

Example 13.8 (Mean–variance portfolio efficiency) We are already familiar with elementary portfolio choice problems.[9] Given a universe of assets with random return R_i, $i = 1, \ldots, n$, expected return μ_i, and covariance matrix $\Sigma = [\sigma_{ij}]$, we should select portfolio weights w_i adding up to one. Given a vector $\mathbf{w}$ representing the portfolio, its expected return and variance are given as follows:

$$\mu_p = \sum_{i=1}^{n} \mu_i w_i = \boldsymbol{\mu}^T \mathbf{w}, \quad \sigma_p^2 = \sum_{i=1}^{n} \sum_{j=1}^{n} w_i \sigma_{ij} w_j = \mathbf{w}^T \Sigma \mathbf{w}$$

The standard deviation of return, σ_p, might be considered as a risk measure, as it provides us with a measure of dispersion. Then, we may trade off expected return μ_p against σ_p. Of course the tradeoff may be unclear, but it can be visualized by tracing the frontier of mean–variance efficient portfolios, depicted in Fig. 13.7.[10] The efficient frontier is bent toward the northwest, since we want to maximize μ_p and minimize σ_p. Portfolios on this frontier are called *mean–variance efficient*, since variance and standard deviation may be switched from one to another depending on convenience. A portfolio is efficient if it is not possible to obtain a higher expected return without increasing risk or, seeing things the other way around, if it is not possible to decrease risk without decreasing expected return.

It is important to emphasize that we are talking about *portfolios*. If we compare two assets, it is tempting to say that the decision problem is trivial

---

[9]See Example 12.5.
[10]Efficient frontiers for multiobjective optimization problems, as well as models to trace them, are discussed in Section 12.3.3.

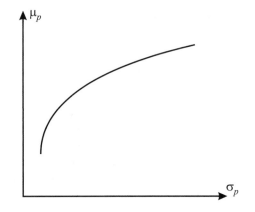

Fig. 13.7   Mean–variance efficient portfolio frontier.

when $\mu_1 > \mu_2$ and $\sigma_1^2 = \sigma_{11} < \sigma_{22} = \sigma_2^2$. In this case, asset 1 has a larger expected return than asset 2, and it is also less risky; hence, a naive argument would lead to the conclusion that asset 2 should not be considered at all. Actually, this may not be the case, since we have neglected the possible correlation between the two assets. The inclusion of asset 2 may, in fact, be beneficial in reducing risk, particularly if its return is negatively correlated ($\sigma_{12} < 0$) with the return of asset 1.

To trace the efficient frontier, there are two main possibilities. One is to scalarize the vector of two objectives and solve

$$\max \quad \boldsymbol{\mu}^T \mathbf{w} - \frac{\lambda}{2} \mathbf{w}^T \boldsymbol{\Sigma} \mathbf{w}$$

$$\text{s.t.} \quad \sum_{i=1}^{n} w_i = 1$$

$$w_i \geq 0$$

for various values of coefficient λ. This coefficient penalizes variance, and can be interpreted as a risk aversion coefficient. By changing the value of λ, we can trace the efficient set. From a mathematical perspective, this is a quadratic programming model; since the covariance matrix is positive semidefinite, it is a convex problem. We see the reason for using variance, rather than standard deviation, in this kind of optimization models: Variance is a convex quadratic form with respect to portfolio weights; standard deviation involves a square root that makes things a bit more complicated. Incidentally, we see that the problem is somewhat related to a quadratic utility function.[11] Unfortunately, it is a bit difficult to get a feeling for the parameter λ. Alternatively, we may

---

[11] It can be shown that the mean–variance framework is compatible with the utility framework if the utility function is quadratic, which may be debatable since we have seen that

use the constraint approach, leading to the solution of the problem

$$\min \quad \mathbf{w}^T \Sigma \mathbf{w}$$
$$\text{s.t.} \quad \mathbf{w}^T \boldsymbol{\mu} = \mu_{\mathrm{t}} \qquad\qquad (13.13)$$
$$\sum_{i=1}^{n} w_i = 1$$
$$w_i \geq 0$$

for various values of the target return μ_{t}. Again, this is a convex quadratic programming problem. For an investor, specifying a target return may be more intuitive than struggling with risk aversion coefficient λ.[12] Anyway, common sense and experience suggest risk aversion coefficients in the range between 2 and 4.   ⧠

The mean–variance efficiency framework has played a pivotal role in the development of modern financial theory, even though it has quite significant practical limitations:

- The estimation of the relevant data may not be easy.

- We are neglecting transaction costs.

- We are assuming that correlations do not change over time, even under severe market conditions.

More generally, the use of a risk measure based on variance or standard deviation is questionable. In fact, variance measures both positive and negative deviations with respect to the expected return. However, this symmetry does not make economic sense: We are certainly not annoyed by extra profits. When dealing with a symmetric distribution like the normal one, standard deviation makes sense, since the good and the bad tails of the distributions have the same shape. However, this does not necessarily apply to skewed distributions. It may also be argued that not only skewness, but also kurtosis can play a role in measuring risk: Kurtosis might do a better job at measuring extreme risks, i.e., those associated with fat tails. More recently, alternative risk measures have been proposed, based on quantiles.

---

quadratic utility features increasing absolute risk aversion, or if random returns are normally distributed, which is rather at odds with observed data.

[12]Some optimization models include discrete decision variables; they can be used to constrain portfolio cardinality, i.e., the number of different assets included in the portfolio. This makes the problem nonconvex, and it can be shown that the first scalarization approach does not guarantee that the whole efficient frontier is traced. The constraint approach does not suffer from this difficulty; see, e.g., Chapter 12 of Ref. [10].

13.2.3   Quantile-based risk measures: value at risk

Given the limitations of standard deviation and variance as risk measures, alternative ones have been proposed. To be specific, we will refer once more to a financial investment problem, where risk is related to portfolio loss. The most widely known such measure is *value at risk* [VaR; not to be confused with variance (Var)]. The VaR concept was introduced as an easy-to-understand measure of portfolio risk. In fact, measuring, monitoring, and managing risk are fundamental activities for any portfolio manager, but another fundamental side of the coin is the possibility of communicating about risk with top management, through a single number making sense for a broad class of assets. Bonds and stocks involve different forms of risk and derivatives, if used for speculation, may be even riskier. Basically, VaR aims at measuring the maximum portfolio loss one could suffer, over a given time horizon, within a given confidence level. Technically speaking, it is a quantile of the probability distribution of loss. Let L_T be a random variable representing the loss, in absolute value, of a portfolio over a holding period of length T. Then, VaR at confidence level $1 - \alpha$ can be defined as the smallest number $\text{VaR}_{1-\alpha}$ such that

$$P(L_T \leq \text{VaR}_{1-\alpha}) \geq 1 - \alpha \qquad (13.14)$$

This definition is consistent with Definition 7.2 of a quantile of a discrete probability distribution. If L_T is a continuous random variable and its CDF is invertible, we may rewrite Eq. (13.14) as

$$P(L_T \leq \text{VaR}_{1-\alpha}) = 1 - \alpha \qquad (13.15)$$

In the following discussion, unless otherwise noted, we will assume for simplicity that loss is a continuous random variable. For instance, if we set $\alpha = 0.05$, we obtain VaR at 95%. The probability that the loss exceeds VaR is α.

Actually, there are different definitions of VaR, which are best clarified by relating the distribution of loss to the distribution of return. Let W_0 be the initial portfolio wealth. If R_T is the random return over the holding period, the future wealth is

$$W_T = W_0(1 + R_T)$$

A loss occurs when the wealth increment

$$\delta W = W_T - W_0 = W_0 R_T$$

turns out to be negative. So, the *absolute loss* over the holding period is

$$L_T = -W_0 R_T$$

Let us assume that the holding period return has a continuous distribution with density $f_{R_T}(r)$. Equation (13.15) implies that loss will exceed VaR with a low probability:

$$P(L_T \geq \text{VaR}_{1-\alpha}) = \alpha$$

where we may indifferently write "$\geq$" or "$>$," since the distribution involved is continuous. This can be rewritten as follows:

$$P(L_T \geq \text{VaR}_{1-\alpha}) = P(-W_0 R_T \geq \text{VaR}_{1-\alpha})$$

$$= P\left(R_T \leq -\frac{\text{VaR}_{1-\alpha}}{W_0}\right)$$

$$= P(R_T \leq r_\alpha) = \alpha \qquad (13.16)$$

where we have defined

$$r_\alpha \equiv -\frac{\text{VaR}_{1-\alpha}}{W_0}$$

The return r_α will be negative in most practical cases and is just the quantile at level α of the distribution of portfolio return:

$$P(R_T \leq r_\alpha) = \int_{-\infty}^{r_\alpha} f_{R_T}(r)\, dr = \alpha$$

The quantile r_α is obviously associated with a critical wealth w_α, which is the wealth we end up with if our loss is exactly the VaR:

$$w_\alpha = W_0 - \text{VaR}$$

We may interpret r_α as the worst-case return with confidence level α; if $\alpha = 0.05$, return will be worse than r_α only in 5% of the cases. By the same token, we will end up with a wealth lower than w_α only with probability α.

Now we may define an *absolute* VaR as

$$\text{VaR} = W_0 - w_\alpha = -W_0 r_\alpha \qquad (13.17)$$

Note once again that the critical return r_α is usually negative and VaR (absolute value of loss) is positive. We may also define a *relative* VaR, where the reference value to define loss is the expected value of future wealth. Let us denote the expected holding period return by $\mu = E[R_T]$. Then

$$E[W_T] = (1 + \mu)W_0$$

Relative VaR is defined as

$$\text{VaR} = E[W_T] - w_\alpha = -W_0(r_\alpha - \mu) \qquad (13.18)$$

The definitions in Eqs. (13.17) and (13.18) may yield approximately the same VaR over a short time horizon, say, a few days. In this case volatility dominates drift[13] and $E[W_T] \approx W_0$. This assumption is not unreasonable, as bank

---

[13]Intuitively, *drift* is related to expected return, and *volatility* is related to standard deviation. On a short time interval of length δt, drift scales linearly with δt, whereas volatility is proportional to $\sqrt{\delta t}$, which means that when the time interval tends to zero, drift goes to zero more rapidly than does volatility. This square-root rule was discussed in Section 7.7.1.

regulations require the use of a risk measure in order to set aside enough cash to be able to cover *short-term* losses.

Computing VaR is easy if we assume that return or, equivalently, loss are normally distributed. Let us assume that the holding period return R_T is normally distributed and that

$$E[R_T] \approx 0, \qquad \text{Var}(R_T) = \sigma^2$$

Then, there is no difference between absolute and relative VaR. Since loss is $L_T = -W_0 R_T$, we obtain $L_T \sim \mathcal{N}(0, W_0^2 \sigma^2)$. We should not overlook the fact that we are taking advantage of the symmetry of the normal distribution of return with respect to its expected value, which is 0. Hence, the critical return r_α is, in absolute value, equal to the quantile $r_{1-\alpha}$. Then, to compute $\text{VaR}_{1-\alpha}$, we may use the familiar standardization/destandardization drill for normal variables:

$$P(L_T \leq \text{VaR}_{1-\alpha}) = P\left(\frac{L_T - 0}{W_0 \sigma} \leq \frac{\text{VaR}_{1-\alpha} - 0}{W_0 \sigma}\right) = P\left(Z \leq \frac{\text{VaR}_{1-\alpha}}{W_0 \sigma}\right)$$

where Z is a standard normal variable. We have just to find the standard quantile $z_{1-\alpha}$ and set

$$\text{VaR}_{1-\alpha} = z_{1-\alpha} \sigma W_0$$

Example 13.9 You have invested \$100,000 in Quacko Corporation stock shares. Daily volatility is 2%. VaR at 95% level is

$$\$100,000 \times 0.02 \times z_{0.95} = \$100,000 \times 0.02 \times 1.6449 = \$3,289.71$$

We are "95% sure" that we will not lose more than \$3,289.71 in one day. VaR at 99% level is

$$\$100,000 \times 0.02 \times z_{0.99} = \$100,000 \times 0.02 \times 2.3263 = \$4652.70$$

Clearly, increasing the confidence level by 4% has a significant effect, since we are working on the tail of the distribution.   □

A commonly proposed scaling procedure allows us to compute VaR on multiple time periods. If we assume that daily returns are a stream of i.i.d. normal variables with standard deviation σ, over a time period spanning T days, the application of the square-root rule yields

$$\text{VaR}_{1-\alpha}(T \text{ days}) = z_{1-\alpha} \sqrt{T} \sigma W_0 = \sqrt{T} \text{VaR}_{1-\alpha}(1 \text{ day})$$

Needless to say, this simple-minded approach hides a few dangers. To begin with, when we consider longer time periods, expected return does play a role, and we should clarify whether we are interested in the relative or absolute VaR. Furthermore, we are assuming that returns on consecutive days can be just summed, disregarding compounding effects; this allows us to consider return over T days as the sum of T normal variables, which is normal again. Last but not least, we are assuming that there is no correlation between the returns on consecutive days.[14] The very assumption of normality of returns can be dangerous, as the normal distribution has a relatively low kurtosis; alternative distributions have been proposed, featuring fatter tails, in order to better account for tail-risk, which is what we are concerned about in risk management. Nevertheless, the calculation based on the normal distribution is so simple and appealing that it is tempting to use it even when we should rely on more realistic models. In practice, we are not interested in VaR for a single asset, but in VaR for the whole portfolio. Again, a normality assumption streamlines our task considerably.

Example 13.10 Suppose that we hold a portfolio of two assets. The portfolio weights are $w_1 = \frac{2}{3}$ and $w_2 = \frac{1}{3}$, respectively. We also assume that the returns of the two assets have a jointly normal distribution; the two daily volatilities are $\sigma_1 = 2\%$ and $\sigma_2 = 1\%$, respectively, and the correlation is $\rho = 0.7$. Let the time horizon be $T = 10$ days; despite this, we assume again that expected holding period return is zero. To obtain portfolio risk, we first compute the variance of the holding period return:

$$\sigma_p^2 = \begin{bmatrix} w_1 & w_2 \end{bmatrix} \begin{bmatrix} \sigma_1^2 T & \rho\sigma_1\sigma_2 T \\ \rho\sigma_1\sigma_2 T & \sigma_2^2 T \end{bmatrix} \begin{bmatrix} w_1 \\ w_2 \end{bmatrix} = 0.0025111$$

Hence, $\sigma_p = 0.05011$. If the overall portfolio value is \$10 million, and the required confidence level is 99%, we obtain

$$\text{VaR}_{0.99} = z_{0.99} \cdot \sigma_p \cdot W_0 = 2.3263 \times 0.05011 \times 10^7 = \$1,165,709$$

◻

Once again, we stress that the calculations in Example 13.10 are quite simple (maybe *too* simple) since they rely on a few rather critical assumptions. Alternative ways of estimating VaR have been proposed:

- *Monte Carlo simulation.* In the previous examples, we have taken advantage of the analytical tractability of the normal distribution, and the fact the return of stock shares was the only risk factor involved. Other risk factors may be involved, such as inflation and interest rates, and the portfolio can include derivatives, whose value is a complicated function of underlying asset values. Even if we assume that the underlying risk factors are normally distributed, the portfolio value may be a nonlinear function of them, and the analytical tractability of the normal distribution is lost. In this case, we may resort to Monte Carlo simulation,[15]

---

[14] We may also say that returns are not autocorrelated. Lack of correlation is equivalent to independence if we assume normality; otherwise, we should require that daily returns are independent of each other.

[15] See Section 9.7.

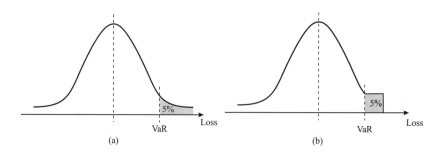

Fig. 13.8 Value at risk can be the same in quite different situations.

which is a remarkably flexible tool, even though not necessarily the most efficient one.

- *Historical VaR.* What we have illustrated so far is a *parametric* approach, since it relies on a theoretical probability distribution, not necessarily normal. One advantage of the normal distribution is that it simplifies the task of characterizing the joint distribution of returns, since we need only a correlation matrix. However, we know that correlations need not capture dependence between random variables. Alternative distributions can be used, possibly requiring sophisticated numerical methods or Monte Carlo simulation, but it is generally difficult to capture dependence. Rather than assuming a specific joint distribution, we may rely on a *nonparametric* approach based on historical data. The advantage of historical data is that they should naturally capture dependence. Hence, we may combine them, according to *bootstrapping* procedures, to generate future scenarios and estimate VaR by historical simulation.

Whatever approach we use to compute VaR, it is not free from some fundamental flaws, which depend on its definition as a quantile, and we should be well aware of them. For instance, a quantile cannot distinguish between different tail shapes. Consider the two densities in Fig. 13.8. In Fig. 13.8(a) we see a normal loss distribution and its 95% VaR, which is just its quantile at probability level 95%; the area of the right tail is 5%. In Fig. 13.8(b) we see a sort of truncated distribution obtained by appending a uniform tail to a normal density, which accounts for 5% of the total probability. By construction, VaR is the same in both cases, since the areas of the right tails are identical. However, we should not associate the same measure of risk with the two distributions. In the case of the normal distribution there is no upper bound to loss; in the second case, there is a clearly defined worst-case loss. Hence, the risk for density (a) should be larger than with density (b), but VaR does not indicate any difference between them. In order to discriminate the two cases, we may consider the expected value of loss *conditional* on being on the right (unfortunate) tail of the loss distribution. This conditional expectation yields the midpoint of the uniform tail in the truncated density; conditional expected value is larger in the normal case, because of its unbounded support. This observation has led to the definition of alternative risk measures, such as *conditional value at risk* (CVaR), which is the expected value of loss, conditional on being to the right of VaR.

Risk measures like VaR or CVaR could also be used in portfolio optimization, by solving mathematical programs with the same structure as problem (13.13), where variance is replaced by such measures. The resulting optimization problem can be rather difficult. In particular, it may lack the convexity properties that are so important in optimization. It turns out that minimizing VaR, when uncertainty is modeled by a finite set of scenarios (which may be useful to capture complex distributions and dependencies among asset prices), is a nasty nonconvex problem, whereas minimizing CVaR is (numerically) easier as it yields a convex optimization problem.[16]

There is one last issue with VaR that deserves mention. Intuitively, risk is reduced by diversification. This should be reflected by any risk measure $\rho(\cdot)$ we consider. A little more formally, we should require a *subadditivity* condition like

$$\rho(A + B) \le \rho(A) + \rho(B)$$

where A and B are two portfolio positions. The following counterexample is often used to show that VaR lacks this property.

Example 13.11 (VaR is not subadditive) Let us consider two corporate bonds, A and B, whose issuers may default with probability 4%. Say that, in the case of default, we lose the full face value, $100 (in practice, we might partially recover the face value of the bond). Let us compute the VaR of each bond with confidence level 95%. Since loss has a discrete distribution in this example, we should use the more general definition of VaR provided by (13.14). The probability of default probability is 4%, and $1 - 0.04 = 0.96 > 0.95$; therefore, we find

$$\text{VaR}(A) = \text{VaR}(B) = \text{VaR}(A) + \text{VaR}(B) = \$0$$

Now what happens if we hold both bonds and assume independent defaults? We will suffer:

- A loss of $0, with probability $0.96^2 = 0.9216$

- A loss of $100, with probability $2 \times 0.96 \times 0.04 = 0.0768$

- A loss of $200, with probability $0.04^2 = 0.0016$

Note that the probability of losing $0 is smaller than 95%, but

$$\text{P}(L_T \le 100) = 0.9216 + 0.0768 > 0.95$$

---

[16]We illustrate this later in Section 13.3.3.

Hence, with that confidence level, $\text{VaR}(A + B) = 100 > \text{VaR}(A) + \text{VaR}(B)$, which means that diversification increases risk, if we measure it by VaR.   □

Subadditivity is one of the properties that any sensible risk measure should enjoy. The term *coherent risk measure* has been introduced to specify a risk measure that meets a set of sensible requirements. VaR is not a coherent risk measure, whereas it can be shown that CVaR is.

13.3   TWO-STAGE STOCHASTIC PROGRAMMING MODELS

So far, in terms of concrete procedures, we have considered only decision trees, which are well suited to cope with discrete decisions, when uncertainty can be represented by a finite set of scenarios. More generally, we would like to solve a problem like

$$\min_{\mathbf{x} \in S} \text{E}_{\omega}[f(\mathbf{x}, \omega)]$$

where S is a subset of $\mathbb{R}^n$, and the expectation can be taken with respect to a multidimensional continuous distribution. The objective function is not necessarily an expected cost or an expected profit (to be maximized), and it could be related to a utility function. Unfortunately, this optimization problem involves an expectation, which in turn involves a multidimensional integral, if the underlying distribution is continuous. It is a safe bet that a problem like this is almost intractable in all but very simple cases. One ingredient to build a tractable model is the approximation of continuous distributions by a discrete set of scenarios, i.e., a discrete probability distribution, which boils the nasty multidimensional integral down to a more manageable sum. Indeed, this is what is typically done to build decision trees, where the description of uncertainty is discrete by nature.

We should also note that, in practice, it is very difficult to specify a high-dimensional distribution describing the uncertainty we face. The multivariate normal distribution is an exception, as in order to describe it we just need expected values, variances, and the correlation matrix. In other cases, a set of well-crafted scenarios may be the raw material we may start from. They may be obtained by Monte-Carlo simulation of a possibly complex dynamic model, or by taking advantage of historical data; scenarios can also be the result of a discussion with a group of domain experts, if past data are not relevant or not available altogether. However, unlike the case with decision trees, here we want to deal with complex decisions, rather than the choice among a finite and small number of alternatives. As a concrete illustration of these concepts, we will treat the stochastic extension of linear programming (LP) models.

Consider the following deterministic LP model (in canonical form):

$$\min \quad \mathbf{c}^T \mathbf{x}$$

$$\text{s.t.} \quad \mathbf{Ax} \geq \mathbf{b}$$
$$\mathbf{x} \geq \mathbf{0}$$

We may try to deal with uncertainty by making randomness in the data explicit in the model. In the most general case, we may have randomness in all of our data, which could be represented by random variables $\mathbf{c}(\omega)$, $\mathbf{A}(\omega)$, and $\mathbf{b}(\omega)$, depending on an underlying event ω. However, we cannot simply translate the model above to something like

$$\min \quad \mathbf{c}(\omega)^T \mathbf{x} \qquad\qquad (13.19)$$
$$\text{s.t.} \quad \mathbf{A}(\omega)\mathbf{x} \geq \mathbf{b}(\omega) \qquad\qquad (13.20)$$
$$\mathbf{x} \geq \mathbf{0}$$

To begin with, the objective function (13.19) does not make sense, since minimizing a function of a random variable has no clear meaning. Still, we could solve this issue simply by considering its expected value. The real issue is that we should not require that the constraints (13.20) are satisfied for every event ω. In some cases, doing so would yield a "fat" solution, which is expected to be quite costly. In other cases, it would be simply impossible to do so. To see this, consider a simple inventory control system operating under a reorder point policy. If demand is assumed normal, 100% service level would imply setting the reorder point to infinity. By the same token, imagine that the model involves equality constraints; equality constraints for different scenarios are most likely to be inconsistent. Hence, we must relax constraints in some sensible way.

One possible approach to relax the requirements is to allow for a violation in a small subset of scenarios. In other words, we settle for a suitably high probability $\beta < 1$ to satisfy constraints. The *chance-constrained programming* approach deals with models of the form

$$\min \quad \mathbf{c}^T \mathbf{x}$$
$$\text{s.t.} \quad \mathbf{Ax} \geq \mathbf{b}$$
$$P\{\mathbf{G}(\omega)\mathbf{x} \geq \mathbf{h}(\omega)\} \geq \beta$$
$$\mathbf{x} \geq \mathbf{0}$$

Note that here we require a high probability of satisfying the *joint* set of constraints; alternatively, we might enforce the condition for each constraint separately. This modeling framework has a clear interpretation in terms of reliability of the solution. Nevertheless, there are a few difficulties:

- In technical terms, there is no guarantee that the resulting optimization problem is convex in general; nonconvexity may arise with discrete probability distributions.[17]

---

[17]From a technical point of view, the reason for this difficulty is that the union of convex sets is nonconvex in general.

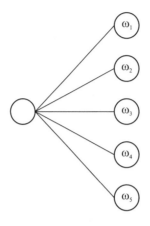

Fig. 13.9 Scenario tree for a two-stage problem.

- From a more practical perspective, we are not saying anything about what will happen if constraints are violated. Probably, corrective actions will be taken, but they are left outside the model. Furthermore, our previous discussion about VaR shows that disregarding a small set of scenarios is dangerous, if they are unlikely, but associated with catastrophic consequences.

- Finally, the above framework is *static*: We do not account for a dynamic decision process, whereby decisions may be adapted when uncertainty is progressively resolved.

An alternative framework to deal with stochastic optimization is *stochastic programming with recourse*. To get a feeling for this modeling framework, consider Fig. 13.9.

- We have a scenario tree. The root of the tree (node on the left) represents the current state, here and now; the nodes on the right represent different future states of nature, or scenarios. Each scenario has some probability of occurrence, which can be an objective measure derived by statistical information or a subjective measure of likelihood. As we have noted before, scenarios can result from a suitable discretization of a continuous probability distribution, or a set of plausible forecasts by a pool of experts.

- We should make a set of decisions now, but in the future, when uncertainty is at least partially resolved, we might take some action in order to "adjust" our previous decisions on the basis of additional information that we have gathered. These adjustments are called *recourse actions*.

- We want to find a set of decisions *now* so as to optimize the sum of two terms:

1. The cost of immediate actions, which are certain
2. The expected cost of the future recourse actions, which are uncertain

In order to get acquainted with this modeling framework, we consider once again the assembly-to-order production planning model of Section 12.2.1, allowing for demand uncertainty.

13.3.1   A two-stage model: assembly-to-order production planning

In Section 12.2.1 we dealt with a production planning problem within an assembly-to-order (ATO) framework. There, we disregarded demand uncertainty and built a deterministic LP model. Now, in order to make the model a bit more realistic, we represent demand uncertainty by a scenario tree and adopt a two-stage stochastic linear programming framework:

1. First, we decide how many units of each component we should build, subject to manufacturing capacity constraints. This first-stage decision sets the total production cost.

2. After receiving customer orders, we use available components to assemble finished goods. The assembly plan should maximize revenue; the cost term in the profit function is fixed by the previous decision (if we neglect assembly cost); if components are insufficient to meet customer orders, we lose profit opportunities; if too many components are available, they are scrapped, with a possibly considerable loss of money. We see that this problem is a generalization of the basic newsvendor model. The optimal use of available components is a second-stage decision, contingent on the realization of a specific demand scenario.

Of course, we cannot maximize profit, because it is a function of random variables (demand for end items), as well as our decisions; nevertheless, we may maximize its expected value.[18] To be concrete, let us solve the same problem instance as in Example 12.10. Economic and technical data are given in Tables 12.1, 12.2, and 12.3. The only difference is in demand uncertainty. There, we assumed perfect knowledge of end-item demand. Now we assume the scenarios represented in Table 13.1. Let us denote the set of scenarios by $\mathcal{S}$. For each scenario $s \in \mathcal{S}$, we have demand d_j^s, for all end items $j \in \mathcal{J}$, and the probability π^s. In the table, we consider only three scenarios, S_1, S_2, and S_3. We also show the expected value of demand, under the assumption that the three scenarios are equally likely. Please note that expected demand is just the deterministic demand that we used in Table 12.3, when we solved the deterministic version of the model. Hence, the solution we obtained there

---

[18]In this first example we disregard risk aversion; in Section 13.3.3 we extend the model by considering CVaR as a risk measure.

Table 13.1   Demand scenarios and expected value of demand.

	S_1	S_2	S_3	Expected demand
A_1	100	50	120	90
A_2	50	25	60	45
A_3	100	110	60	90

is what we would get, if we ignored demand uncertainty by considering only expected demand. We recall that the deterministic solution was

$$x_1^* = 116.67, \quad x_2^* = 116.67, \quad x_3^* = 26.67, \quad x_4^* = 0.00, \quad x_5^* = 90.00$$
$$y_1^* = 26.67, \quad y_2^* = 0.00, \quad y_3^* = 90.00$$

with "optimal profit" 3233.33. We should immediately understand that this is the value of the optimal profit in the deterministic model, but is not the *true* optimal profit, as this does not make any sense in decision-making under risk. To build a two-stage stochastic programming model, we need to introduce suitable decision variables at each stage of the decision process:

1. The first-stage decision is the amount of components that we produce: x_i, $i \in \mathcal{I}$; this decision is the same as in the deterministic model, as here-and-now decisions are not scenario-contingent.

2. The second-stage decision is the amount of end items that we assemble, contingent on the demand scenario: y_j^s, $j \in \mathcal{J}$, $s \in \mathcal{S}$. Note that, with respect to the deterministic model, we add a scenario superscript, since second-stage decisions are scenario-contingent.

Now we may extend the deterministic model as follows:

$$\max \quad - \sum_{i \in \mathcal{I}} c_i x_i + \sum_{s \in \mathcal{S}} \pi^s \left(\sum_{j \in \mathcal{J}} p_j y_j^s \right) \tag{13.21}$$

$$\text{s.t.} \quad \sum_{i \in \mathcal{I}} r_{im} x_i \leq R_m, \qquad\qquad \forall m \in \mathcal{M} \tag{13.22}$$

$$y_j^s \leq d_j^s, \qquad\qquad \forall j \in \mathcal{J}, s \in \mathcal{S} \tag{13.23}$$

$$\sum_{j \in \mathcal{J}} g_{ij} y_j^s \leq x_i, \qquad\qquad \forall i \in \mathcal{I}, s \in \mathcal{S} \tag{13.24}$$

$$y_j^s, x_i \geq 0$$

The big change in this model, with respect to the expected demand model [(12.22)–(12.24)], is that demand uncertainty is taken into account explicitly.

In practice, we just implement the production plan (first-stage decisions x_i) and develop a contingency plan for the assembly operations (second-stage decisions y_j^s). Only when demand is realized do we choose one among the contingency plans. With three scenarios, we have three contingency plans. Clearly, there is little hope to fully capture demand uncertainty with a hand-ful of scenarios, and actual realized demand will probably differ from that assumed in any scenario. In practice, once demand is realized, we simply have to write a second model for assembly decisions, where we need to meet the realized demand with a limited availability of components, so as to max-imize revenue. The aim of scenarios is to make here-and-now decisions not myopic and as robust as we can; second-stage decisions need not be immedi-ately implemented.

Going into the details of the model above, the objective function (13.21) consists of a first-stage (deterministic) term, accounting for the cost of com-ponents, along with a second-stage term, which is the expected revenue from selling end items (not including component cost); the expected value is com-puted by summing the revenues for each scenario s, weighted by the scenario probabilities π^s. The capacity constraint (13.22) is unchanged, because it per-tains to the first-stage decisions only. The market demand constraint (13.23) is now scenario-dependent, as it considers the stochastic demand d_j^s. Finally, constraint (13.24) links the two stages, stating that assembly is constrained by component availability, for each end item j and each scenario s. By solv-ing this model with the numerical data of our toy example, we obtain the following solution:

$$x_1^* = 115.71, \qquad x_2^* = 115.71, \qquad x_3^* = 52.86$$
$$x_4^* = 2.86, \qquad x_5^* = 62.86$$
$$y_1^{1*} = 52.86, \qquad y_2^{1*} = 0.00, \qquad y_3^{1*} = 62.86$$
$$y_1^{2*} = 50.00, \qquad y_2^{2*} = 2.86, \qquad y_3^{2*} = 62.86$$
$$y_1^{3*} = 52.86, \qquad y_2^{3*} = 2.86, \qquad y_3^{3*} = 60.00$$

The expected profit for this solution is 2885.71. As we pointed out, the real outcome of the model is the set of the first-stage decision variables x_i^*. Ob-serving the component production plan, we immediately see a qualitative difference with respect to the model disregarding uncertainty: It is less ex-treme. We do not produce a large amount of component c_5, because we do not place a risky bet on high sales of A_3. In fact, scenario S_3 would prove a disaster for the deterministic solution; in that scenario, sales are lower for A_3, but we could not react because we do not have enough specific components for the other end items. So, 30 specific components c_5 would be scrapped; furthermore common components would be scrapped as well, since they can-not be used to assemble other end items for the lack of the related specific components.[19]   The stochastic model, on the contrary, reduces production of c_5 and increases production of specific component c_3, which is needed to support assembly and sales of A_1; even a small amount of component c_4 is produced, in order to support the least profitable end item A_2, which helps in using common components when sales are low for other end items. While there is a big difference in terms of specific components, we see that, as far as common components are concerned, the solutions of the deterministic and the stochastic solutions are essentially the same. There is a good reason for this, as common components are a flexible resource, which can be exploited to assemble different end items. Moreover, the demand for common components is the sum of the individual demands for the end items, and by aggregating demand we often reduce uncertainty. Indeed, this *risk-pooling* effect is what we try to exploit in assemble-to-order systems. However, it is also important to note that when end-item demands are strongly correlated, the risk-pooling effect is considerably reduced. In such a case, we should expect that even the produced quantities of common components differ in the deterministic and the stochastic models. Another relevant factor is capacity: If this is so tight that we may sell whatever we are able to produce, a simple deterministic model could be a viable option.

But how do the two solutions compare in terms of profit? The objective function from the solution of the second model is 2885.71; apparently, the stochastic solution is worse than the deterministic solution, whose optimal profit was 3233.33. But this comparison makes no sense, as doing so we are actually comparing two different situations, rather than two different solutions. This simply proves that we would rather face a certain demand rather than an uncertain one. The objective function of the first model is neither the true profit, which is uncertain, nor its expected value. It would be the optimal profit, if we knew that the average demand scenario will be realized for sure. In the first model [(12.22)–(12.24)] we *pretend* to know the end-item demand, and we get the illusion of higher profits. In order to compare the two solutions, we should fix the production plans for components that the two models propose, and then we should solve a set of second-stage problems, where we optimize assembly of end items subject to component availability, for different demand scenarios. More formally, given the vector $\mathbf{x}^*$ of first-stage optimal decisions of whatever model, we should solve the following second-stage (recourse) problem for each scenario s in $\mathcal{S}$:

$$R^s(\mathbf{x}^*) \equiv \max \quad \sum_{j \in \mathcal{J}} p_j y_j^s$$

---

[19]This applies under our assumption of limited time window for sales and lack of any salvage value of unused components.

$$\text{s.t.} \quad y_j^s \le d_j^s, \qquad\qquad\qquad \forall j \in \mathcal{J}$$

$$\sum_{j \in \mathcal{J}} g_{ij} y_j^s \le x_i^*, \qquad\qquad \forall i \in \mathcal{I}$$

$$y_j^s \ge 0$$

where $R^s(\mathbf{x}^*)$ is the optimal revenue that we collect under scenario s, given the first-stage solution $\mathbf{x}^*$, when making optimal use of the available components to meet demand. Note that in this model the component availability x_i^* is given, either by the stochastic or by the deterministic, expected-value model. Whatever the case, the resulting expected revenue is

$$\sum_s \pi^s R^s(\mathbf{x}^*)$$

Expected profit for an arbitrary solution is obtained by subtracting its first-stage cost from this second-stage expected revenue.[20] To evaluate the deterministic solution, we should plug it into this model. In the case of scenario S_1, the optimal assembly and sales plan is

$$y_1^* = 26.67, \qquad y_2^* = 0.00, \qquad y_3^* = 90.00$$

and the same holds for S_2. The bad news is that if scenario S_3 occurs, we are in trouble, because the high-risk solution does not fit demand very well. The optimal assembly and sales plan would be

$$y_1^* = 26.67, \qquad y_2^* = 0.00, \qquad y_3^* = 60.00$$

This is a pretty bad scenario with low sales and corresponding low profit. As we said, we must compute revenue for each scenario, multiply it by its probability, sum everything to get the expected value, and subtract the component cost from the first stage. The expected profit from the deterministic solution turns out to be 2333.33, and is much lower than what the objective function of the deterministic model [(12.22)–(12.24)] predicts (3233.33), based on one average-case scenario. The percentage improvement of the stochastic solution with respect to the deterministic one is, in terms of expected profit for the three scenarios,

$$\frac{2885.71 - 2333.33}{2333.33} \approx 23.67\%$$

Clearly, we cannot extrapolate general results from a small toy example. Indeed, the advantage of using a stochastic model is striking here, because

---

[20] We are evaluating expected profit *in-sample*, i.e., by using the same set of scenarios that are used in the stochastic model; we could use a much larger set of out-of-sample scenarios to get a more reliable estimate. This is feasible in practice, since solving a large number of small LP problems usually takes much less CPU time than solving one large-scale stochastic LP model.

specific components have a large impact. In a case featuring much more component commonality, the result would be less impressive. Furthermore, we have assumed that unused components are scrapped, which need not be the case. They could have some salvage value, and we could have a multistage problem where remaining components can be used at later stages.

13.3.2   The value of the stochastic solution

In Section 13.1.1 we defined EVPI, which is not only a way to price perfect information, but also a measure of the impact of uncertainty. If EVPI is low, uncertainty is not that relevant in the decision. However, EVPI is in most cases a theoretical construct, as we cannot trade the unpleasing here-and-now decision problem for the reassuring wait-and-see one. In the example above, what we have done is more practical: We assessed the value of solving a stochastic model against the solution of a much simpler deterministic model, based on expected values of uncertain parameters. Here we formalize the concept, using the framework of Section 13.1.1. There, we defined the here-and-now problem

$$\min_{\mathbf{x} \in S} \mathrm{E}_\omega[f(\mathbf{x}, \omega)] \qquad (13.25)$$

which yields an objective value f^* and a decision vector $\mathbf{x}^*$.

As we have seen, we could disregard uncertainty and solve a deterministic problem based on expected values. Using a somewhat sloppy notation, let us denote by $\bar{\omega} = \mathrm{E}[\omega]$ the expected value of the problem data and define the deterministic "expected value" problem

$$f^*_{\mathrm{EV}} = \min_{\mathbf{x} \in S} f(\mathbf{x}, \bar{\omega})$$

which yields the "expected value solution" $\bar{\mathbf{x}}(\bar{\omega})$. This model also yields a value of the objective function, but, as we have seen, the solution $\bar{\mathbf{x}}(\bar{\omega})$ must be evaluated within the actual uncertain setting. Doing so yields the expected value of the expected value solution:

$$f_{\mathrm{EEV}} = \mathrm{E}_\omega[f(\bar{\mathbf{x}}(\bar{\omega}), \omega)]$$

What we should compare is f_{EEV} against f^*. The *value of the stochastic solution* (VSS) is defined as

$$\mathrm{VSS} = f_{\mathrm{EEV}} - f^*$$

for a minimization problem.[21] When VSS is large, the additional effort in generating scenarios and solving the much more complicated stochastic programming effort does pay off. As a final remark, we should note that in this

---

[21] It can be shown that VSS is nonnegative; see the text by Birge and Louveaux [7].

discussion we have taken for granted that the scenario tree describes uncertainty adequately. In other words, solutions are compared *in sample*. If we do not feel too comfortable with this assumption, we may compare solutions out-of-sample on a much larger set of scenarios. This is also a good way to check the validity of the selected scenario generation approach and the robustness of the solutions that we obtain.

13.3.3   A mean–risk formulation of the assembly-to-order problem

Mean–risk formulations are based on the idea of trading off expected profit (or return) against a risk measure. Classical mean–variance portfolio optimization relies on an analytical representation of variance, which leads to an easy convex quadratic programming problem. This need not be the case if we choose another risk measure. Value at risk is easy to evaluate and optimize under a normality assumption, but it may turn awkward in general, as its minimization may result in a nonconvex optimization problem. Conditional VaR is better behaved from this point of view, and it may lead to a (stochastic) linear programming model formulation.[22] Hence, we may consider minimizing CVaR, at some confidence level $1 - \alpha$, subject to a constraint on expected profit or loss; by changing this target expectation, we may trace an efficient frontier of solutions.

Let $f(\mathbf{x}, \mathbf{Y})$ be a loss or cost function, depending on a vector of decision variables $\mathbf{x}$ and a vector of random variables $\mathbf{Y}$ with joint density $g_{\mathbf{Y}}(\mathbf{y})$, and consider function $F_{1-\alpha}(\mathbf{x}, \zeta)$ defined as

$$F_{1-\alpha}(\mathbf{x}, \zeta) = \zeta + \frac{1}{\alpha} \int [f(\mathbf{x}, \mathbf{y}) - \zeta]^+ g_{\mathbf{Y}}(\mathbf{y}) d\mathbf{y}$$

where $[z]^+ \equiv \max\{z, 0\}$, and $\zeta \in \mathbb{R}$ is an auxiliary variable. It can be shown that minimization of CVaR, at confidence level $1 - \alpha$, is accomplished by the minimization of $F_{1-\alpha}(\mathbf{x}, \zeta)$ with respect to its arguments. In a stochastic linear programming model based on discrete scenarios, if we denote by $f(\mathbf{x}, \mathbf{y}^s)$ the loss in scenario s, $s \in \mathcal{S}$, the minimization of CVaR is equivalent to the solution of the LP model

$$\min \quad \zeta + \frac{1}{\alpha} \sum_{s \in \mathcal{S}} \pi^s z^s$$

$$\text{s.t.} \quad z^s \geq f(\mathbf{x}, \mathbf{y}^s) - \zeta, \qquad s \in \mathcal{S}$$

$$z^s \geq 0, \qquad s \in \mathcal{S}$$

where π^s is the probability of scenario $s \in \mathcal{S}$, subject to the additional constraints depending on the specific model.

---

[22] In this section, we rely on results from Rockafellar and Uryasev [26, 27], which we take for granted, thereby cutting a few corners.

The application of this result to the ATO problem is rather straightforward:

$$\min \quad \zeta + \frac{1}{\alpha} \sum_{s \in \mathcal{S}} \pi^s z^s$$

$$\text{s.t.} \quad \sum_{i \in \mathcal{I}} r_{im} x_i \le R_m, \qquad\qquad \forall m \in \mathcal{M}$$

$$y_j^s \le d_j^s, \qquad\qquad \forall j \in \mathcal{J}, s \in \mathcal{S}$$

$$\sum_{j \in \mathcal{J}} g_{ij} y_j^s \le x_i, \qquad\qquad \forall i \in \mathcal{I}, s \in \mathcal{S}$$

$$\sum_{s \in \mathcal{S}} \pi^s \left(\sum_{j \in \mathcal{J}} p_j y_j^s \right) - \sum_{i \in \mathcal{I}} c_i x_i \ge \beta \qquad\qquad (13.26)$$

$$z^s \ge \sum_{i \in \mathcal{I}} c_i x_i - \sum_{j \in \mathcal{J}} p_j y_j^s - \zeta, \qquad \forall s \in \mathcal{S} \qquad (13.27)$$

$$\zeta \in \mathbb{R}, \quad x_i, y_j^s, z^s \ge 0$$

Decision variables, parameters, and constraints have the same meaning as in the previous ATO models. The only constraint worth mentioning is Eq. (13.26), which sets a lower bound β on expected profit. Also note that we change the sign of profit in Eq. (13.27), which should refer to a loss function.

13.4   MULTISTAGE STOCHASTIC LINEAR PROGRAMMING WITH RECOURSE

Multistage stochastic programming formulations arise naturally as a generalization of two-stage models. At each stage, we gather new information and we make decisions accordingly, taking into account immediate costs and expected future recourse cost. The resulting decision process may be summarized as follows:[23]

- At the beginning of the first time period (at time $t = 0$) we select the decision vector $\mathbf{x}_0$; this decision has a deterministic immediate cost $\mathbf{c}_0^T \mathbf{x}_0$ and must satisfy constraints

$$\mathbf{A}_{00} \mathbf{x}_0 = \mathbf{b}_0$$

- At the beginning of the second time period we observe random data $(\mathbf{A}_{10}, \mathbf{A}_{11}, \mathbf{c}_1, \mathbf{b}_1)$ depending on event ω_1; then, on the basis of this information, we make decision $\mathbf{x}_1$; this second decision has an immediate

---

[23]See, e.g., Ref. [28] for a more detailed discussion.

cost $\mathbf{c}_1^T \mathbf{x}_1$ and must satisfy the constraint

$$\mathbf{A}_{10}\mathbf{x}_0 + \mathbf{A}_{11}\mathbf{x}_1 = \mathbf{b}_1$$

Note that these data are not known at time $t = 0$, only at time $t = 1$; the new decision depends on the realization of these random variables and is also affected by the previous decision.

- We repeat the same scheme as above for time periods up to $H - 1$, where H is our planning horizon.

- At the beginning of the last time period H, we observe random data $(\mathbf{A}_{H,H-1}, \mathbf{A}_{HH}, \mathbf{c}_H, \mathbf{b}_H)$ depending on event ω_H; then, on the basis this information we make decision $\mathbf{x}_H$, which has an immediate cost $\mathbf{c}_T^H \mathbf{x}_H$ and must satisfy the constraint

$$\mathbf{A}_{H,H-1}\mathbf{x}_{H-1} + \mathbf{A}_{HH}\mathbf{x}_H = \mathbf{b}_H$$

From the point of view of time period $t = 0$, the decisions $\mathbf{x}_1, \ldots, \mathbf{x}_H$ are random variables, as they will be adapted to the realization of the stochastic process. However, the only information we may use in making each decision consists on the history so far. The resulting dynamic decision process can be appreciated by the following recursive formulation of the multistage problem:

$$
\begin{array}{c}
\min \\
\mathbf{A}_{00}\mathbf{x}_0 = \mathbf{b}_0 \\
\mathbf{x}_0 \geq \mathbf{0}
\end{array}
\mathbf{c}_0^T \mathbf{x}_0 + \mathrm{E}
\left[
\begin{array}{c}
\min \\
\mathbf{A}_{10}\mathbf{x}_0 + \mathbf{A}_{11}\mathbf{x}_1 = \mathbf{b}_1 \\
\mathbf{x}_1 \geq \mathbf{0}
\end{array}
\mathbf{c}_1^T \mathbf{x}_1
\right.
$$

$$
\left.
+ \mathrm{E}
\left[
\cdots + \mathrm{E}
\left[
\begin{array}{c}
\min \\
\mathbf{A}_{H,H-1}\mathbf{x}_{H-1} + \mathbf{A}_{HH}\mathbf{x}_H = \mathbf{b}_H \\
\mathbf{x}_t \geq \mathbf{0}
\end{array}
\mathbf{c}_H^T \mathbf{x}_H
\right]
\right]
\right]
$$

In this formulation, we see that decision $\mathbf{x}_t$ depends directly only on the previous decisions $\mathbf{x}_{t-1}$. In general, decisions may depend on all of the past history, leading to a slightly more complicated model. However, we may often introduce additional state variables, such that the above formulation applies. For instance, in a production planning model we may "forget" the past produced quantities if we know the current inventory levels. It should be noted that, in practice, the real output of the above model is the set of immediate decisions $\mathbf{x}_0$. The remaining decision variables could be regarded as contingent plans, which are implemented in time, much in the vein of a feedback control policy; however, in a practical setting, it is more likely that the model will be solved again and again according to a rolling-horizon logic. While this formulation points out the dynamic optimization nature of multistage problems, we usually resort to deterministic equivalents based on discrete scenario trees. We illustrate the idea with a toy financial planning problem.

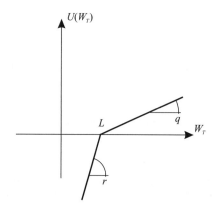

Fig. 13.10   Piecewise linear concave utility function.

13.4.1   A multistage model: asset–liability management

The best way to introduce multistage stochastic models is a simple asset–liability management (ALM) model.[24] We have an initial wealth W_0, that should be properly invested in such a way to meet a liability L at the end of the planning horizon H. If possible, we would like to own a terminal wealth W_H larger than L; however, we should account properly for risk aversion, since there could be some chance to end up with a terminal wealth that is not sufficient to pay for the liability, in which case we will have to borrow some money. A nonlinear, strictly concave utility function of the difference between the terminal wealth W_H, which is a random variable, and the liability L would do the job, but this would lead to a nonlinear programming model. As an alternative, we may build a piecewise linear utility function like the one illustrated in Fig. 13.10. The utility is zero when the terminal wealth W_H matches the liability exactly. If the slope r penalizing the shortfall is larger than q, this function is concave (but not strictly).

The portfolio consists of a set of I assets. For simplicity, we assume that we may rebalance it only at a discrete set of time instants $t = 1, \ldots, H - 1$, with no transaction cost; the initial portfolio is chosen at time $t = 0$, and the liability must be paid at time H. Time period t is the period between time instants $t - 1$ and t. In order to represent uncertainty, we may build a tree like that in Fig. 13.11, which is a generalization of the two-stage tree of Fig. 13.9. Each node n_k in the tree corresponds to an event, where we should make some decision. We have an initial node n_0 corresponding to time $t = 0$. Then, for each event node, we have two branches; each branch is labeled by a conditional probability of occurrence, $P(n_k \mid n_i)$, where $n_i = a(n_k)$ is the immediate predecessor of node n_k. Here, we have two nodes at time $t = 1$ and

---

[24]The numerical example is taken from the book by Birge and Louveaux [7, pp. 20–28].

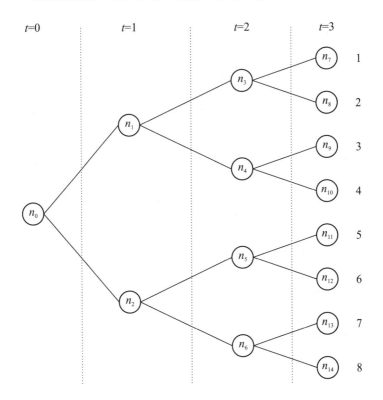

Fig. 13.11 Scenario tree for a simple asset–liability management problem.

four at time $t = 2$, where we may rebalance our portfolio on the basis of the previous asset returns. Finally, in the eight nodes corresponding to $t = 3$, the leaves of the tree, we just compare the terminal wealth with the liability and evaluate the utility function. Each node of the tree is associated with the set of asset returns during the corresponding time period. A scenario consists of an event sequence, i.e., a sequence of nodes in the tree, along with the associated asset returns. We have 8 scenarios in Fig. 13.11. For instance, scenario 2 consists of the node sequence (n_0, n_1, n_3, n_8). The probability of each scenario depends on the conditional probability of each node on its path. If each branch at each node is equiprobable, i.e., the conditional probabilities are always $\frac{1}{2}$, each scenario in the figure has probability $\pi^s = \frac{1}{8}$, for $s = 1, \ldots, 8$. The branching factor may be arbitrary in principle; the more branches we use, the better our ability to model uncertainty; unfortunately, the number of nodes grows exponentially with the number of stages, as well as the computational effort.

At each node in the tree, we must make a set of decisions. In practice, we are interested in the decisions that must be implemented here and now, i.e., those corresponding to the first node of the tree; the other (recourse) decision variables are instrumental to the aim of devising a robust plan, but they are not implemented in practice, as the multistage model is solved on a rolling-horizon basis. This suggests that, in order to model the uncertainty as accurately as possible with a limited computational effort, a possible idea is to branch many paths from the initial node, and less from the subsequent nodes. Each decision at each stage may depend on the information gathered so far, but not on the future; this requirement is called a *nonanticipativity condition*. Essentially, this means that decisions made at time t must be the same for scenarios that cannot be distinguished at time t.[25] To build a model ensuring that the decision process makes sense, there are two choices:

- We can introduce a set of decision variables x_{it}^s, representing wealth allocated to asset i at time t on scenario s; we should force decision variables to take the same value when appropriate, by writing explicit nonanticipativity constraints for scenarios that cannot be distinguished at time t.

- We can associate decision variables with nodes in the scenario trees and write the model in a way that relates each node to its predecessors.

We will illustrate the second alternative in detail, using the following numerical data:

- The initial wealth is 55.

- The target liability is 80.

- There are two assets, say, stocks and bonds; hence, $I = 2$.

- In the scenario tree of Fig. 13.11 we have up- and downbranches; in the (lucky) upbranches, total return is 1.25 for stocks and 1.14 for bonds; in the (bad) downbranches, total return is 1.06 for stocks and 1.12 for bonds. We see that bonds play the role of safer assets here. We also see that returns are a sequence of i.i.d. random variables, but more realistic scenarios can be defined.

- The reward rate q for excess wealth above the target liability is 1.

- The penalty rate r for the shortfall below the target liability is 4.

Let us introduce the following notation:

- $\mathcal{N}$ is the set of event nodes; in our case

$$\mathcal{N} = \{n_0, n_1, n_2, \ldots, n_{14}\}$$

---

[25]You may refer back to Section 7.10 to see an abstract discussion of measurability of random variables and the role of information. Technically, decisions are a stochastic process, which must be adapted to the filtration generated by the data.

- Each node $n \in \mathcal{N}$, apart from the root node n_0, has a unique direct predecessor node, denoted by $a(n)$: for instance, $a(n_3) = n_1$.

- There is a set $\mathcal{S} \subset \mathcal{N}$ of leaf (terminal) nodes; in our case

$$\mathcal{S} = \{n_7, \ldots, n_{14}\};$$

for each node $s \in \mathcal{S}$ we have surplus and shortfall variables w_+^s and w_-^s, related to the difference between terminal wealth and liability.

- There is a set $\mathcal{T} \subset \mathcal{N}$ of intermediate nodes, where portfolio rebalancing may occur after the initial allocation in node n_0; in our case

$$\mathcal{T} = \{n_1, \ldots, n_6\}$$

for each node $n \in \{n_0\} \cup \mathcal{T}$ there is a decision variable x_{in}, expressing the money invested in asset i at node n.

With this notation, the model may be written as follows:

$$\max \quad \sum_{s \in \mathcal{S}} \pi^s (q w_+^s - r w_-^s)$$

$$\text{s.t.} \quad \sum_{i=1}^{I} x_{i,n_0} = W_0$$

$$\sum_{i=1}^{I} R_{i,n} x_{i,a(n)} = \sum_{i=1}^{I} x_{in}, \qquad \forall n \in \mathcal{T}$$

$$\sum_{i=1}^{I} R_{is} x_{i,a(s)} = L + w_+^s - w_-^s, \qquad \forall s \in \mathcal{S}$$

$$x_{in}, w_+^s, w_-^s \geq 0$$

where $R_{i,n}$ is the total return for asset i during the period that *leads to* node n, and π^s is the probability of reaching the terminal node $s \in \mathcal{S}$; this probability is the product of all the conditional probabilities on the path that leads from root node n_0 to leaf node s. This is an LP model that may be easily solved by the simplex algorithm, resulting in the solution of Table 13.2. We may notice that in the last period the portfolio is not diversified, since the whole wealth is allocated to one asset, and we should wonder if this makes sense. Actually, it is a consequence of two features of this toy model:

- We are approximating a nonlinear utility function by a piecewise linear function, and this may imply "local" risk neutrality, so that we only care about expected return; we should use either a nonlinear programming model or a more accurate representation of utility with more linear pieces.

Table 13.2 Investment strategy for a simple ALM problem.

Node	Stocks	Bonds
n_0	41.4793	13.5207
n_1	65.0946	2.16814
n_2	36.7432	22.368
n_3	83.8399	0
n_4	0	71.4286
n_5	0	71.4286
n_6	64	0

- The scenario tree has a very low branching factor, and this does not represent uncertainty accurately.

However, the portfolio allocation in the last time period is not necessarily a critical output of the model: the real stuff is the *initial* portfolio allocation. As we pointed out, the decision variables for future stages have the purpose of avoiding a myopic policy, but they are not meant to be implemented.

13.4.2   Asset–liability management with transaction costs

To give the reader an idea of how to build nontrivial financial planning models, we generalize a bit the model formulation of the previous section, in order to account for proportional transaction costs. The assumptions and the limitations behind this extended model are the following:

- We are given a set of initial holdings for each asset; this is a more realistic assumption, since we should use the model to rebalance the portfolio periodically, according to a rolling-horizon strategy.

- We take proportional (linear) transaction costs into account; the transaction cost is a percentage c of the traded value, for both buying and selling an asset.

- We want to maximize the expected utility of the terminal wealth.

- There is a stream of uncertain liabilities that we have to meet.

- We do not consider the possibility of borrowing money; we assume that all of the available wealth at each rebalancing period is invested in the available assets; actually, the possibility of investing in a risk-free asset is implicit in the model.

- We do not consider the possibility of investing new cash at each rebalancing date (as would be the case, e.g., for a pension fund).

Some of the limitations of the model may easily be relaxed. The important point we make is that when transaction costs are involved, we have to introduce new decision variables to express the amount of assets (number of shares, not the monetary value) held, sold, and bought at each rebalancing date. We use a notation which is similar to that used in the previous ALM formulation:

- $\mathcal{N}$ is the set of nodes in the tree; n_0 is the root node.

- The (unique) predecessor of node $n \in \mathcal{N} \backslash \{n_0\}$ is denoted by $a(n)$; the set of terminal nodes is denoted by $\mathcal{S}$; as in the previous formulation, each of these nodes corresponds to a scenario, which is the sequence of event nodes along the unique path leading from n_0 to $s \in \mathcal{S}$, with probability π^s.

- $\mathcal{T} = \mathcal{N} \backslash (\{n_0\} \cup \mathcal{S})$ is the set of intermediate trading nodes.

- L^n is the liability we have to meet in node $n \in \mathcal{N}$; liabilities are node dependent and stochastic.

- c is the percentage transaction cost.

- $\overline{h}_i^{n_0}$ is the initial holding for asset $i = 1, \ldots, I$ at the root node.

- P_i^n is the price for asset i at node n.

- z_i^n is the amount of asset i purchased at node n.

- y_i^n is the amount of asset i sold at node n.

- x_i^n is the amount of asset i we hold at node n, after rebalancing.

- W^s is the wealth at terminal node $s \in \mathcal{S}$.

- $u(w)$ is the utility for wealth w; this function is used to express utility of terminal wealth.

On the basis this notation, we may write the following model:

$$\max \quad \sum_{s \in S} \pi^s u(W^s) \tag{13.28}$$

$$\text{s.t.} \quad x_i^{n_0} = \overline{h}_i^{n_0} + z_i^{n_0} - y_i^{n_0}, \qquad \forall i \tag{13.29}$$

$$x_i^n = x_i^{a(n)} + z_i^n - y_i^n, \qquad \forall i, \forall n \in \mathcal{T} \tag{13.30}$$

$$(1-c) \sum_{i=1}^{I} P_i^n y_i^n - (1+c) \sum_{i=1}^{I} P_i^n z_i^n = L^n, \qquad \forall n \in \mathcal{T} \cup \{n_0\}$$

$$\tag{13.31}$$

$$W^s = \sum_{i=1}^{I} P_i^s x_i^{a(s)} - L^s, \qquad \forall s \in \mathcal{S} \tag{13.32}$$

$$x_i^n, z_i^n, y_i^n, W^s \geq 0 \tag{13.33}$$

The objective (13.28) is the expected utility of the terminal wealth; if we approximate this nonlinear concave function by a piecewise linear concave function, we get an LP problem (as we did in Section 12.4.7). Equation (13.29) expresses the initial asset balance, taking the current holdings into account; the asset balance at intermediate trading dates is taken into account by Eq. (13.30). Equation (13.31) ensures that enough cash is generated by selling assets in order to meet the liabilities; we may also reinvest the proceeds of what we sell in new asset holdings; note how the transaction costs are expressed for selling and purchasing. Equation (13.32) is used to evaluate terminal wealth at leaf nodes; note here that we have not taken into account the need to sell assets in order to generate the cash required by the last liability; but this would make only sense if the whole fund is liquidated at the end of the planning horizon. If so, we could rewrite Eq. (13.32) as

$$W^s = (1 - c) \sum_{i=1}^{I} P_i^s x_i^{a(s)} - L^s$$

In practice, we would repeatedly solve the model on a rolling-horizon basis, so the exact expression of the objective function is a bit debatable. The role of terminal utility is just to ensure that we are left in a good position at the end of the planning horizon.

This model can be generalized in a number of ways, which are left as an exercise to the reader. The most important point is that we have assumed that the liabilities must be met. This may be a very hard constraint; if extreme scenarios are included in the formulation, as they should be, it may well be the case that the model above is infeasible. Therefore, the formulation should be relaxed in a sensible way; we could consider the possibility of borrowing cash; we could also introduce suitable penalties for not meeting the liabilities. In principle, we could also require that the probability of not meeting the liabilities is small enough; this leads to chance-constrained formulations, for which we refer the reader to the literature.

13.4.3   Scenario generation for stochastic programming

Multistage stochastic programming is a very powerful modeling framework, and it can be extended to cope with risk measures like CVaR, as we have seen in Section 13.3.3. However, the approach can be only as good as the scenario tree on which it is based. Given a multivariate probability distribution characterizing uncertainty, the most obvious way to generate a tree is to draw a random sample using the Monte Carlo principles that we have outlined in Section 9.7. Unfortunately, one clear issue is the exponential growth of the number of tree branches and nodes needed to adequately represent uncertainty. This may be somewhat countered by sophisticated solution strategies. However, it is also important to generate scenarios in a way that even with a limited set of them, uncertainty is captured in a suitable way. By "suitable" we mean in such a way that the first-stage solution is good enough; we are not really interested in representing uncertainty per se on a long time horizon. We list here a few ideas that can be used in practice:

- *Variance reduction* strategies are a Monte Carlo sampling technique to improve the quality of estimates for a given sample size. From inferential statistics we know that the width of a confidence interval is related to the standard deviation of the sample mean, $\sigma/\sqrt{n}$. This expression shows the impact of standard deviation σ of observations and sample size n. A brute force approach entails increasing n, but since the square root is a concave function, this is less and less effective. Apparently, there is little we can do about σ; in fact, clever sampling can be used to reduce this factor.

- *Low-discrepancy sequences* were introduced as a method used to evaluate multidimensional integrals numerically with a limited number of function evaluations. Unlike Monte Carlo methods, low-discrepancy sequences do not resort to a statistical framework in any way, as they are a deterministic approach to spread observations in the most regular way on a region.

- *Moment and property matching* is a rather natural idea. Imagine taking a sample from a multivariate distribution with expected value vector $\boldsymbol{\mu}$ and covariance matrix $\boldsymbol{\Sigma}$. A good sample should have sample means and sample covariances as close as possible to these values. We may extend the idea to skewness, kurtosis, and other properties. A perfect matching of the properties of a sample against the required values is impossible to obtain, but we may get the best approximation by solving a least-squares problem.

- *Optimal approximation of probability measures* is a formal approach to scenario reduction, based on the idea of approximating a continuous distribution with a discrete one. In order to pursue this approach, there is a need for a precise definition of "distance" between probability measures, as well as for suitable computational procedures.

For more details on these advanced approaches, we refer to the references at the end of this chapter.

13.5 ROBUSTNESS, REGRET, AND DISAPPOINTMENT

Proper scenario generation is needed to reduce sampling errors and successfully apply stochastic programming models. However, there may be more fundamental flaws in the approach:

- How is our information reliable when we assume a probability distribution? The best scenario generation will not help if we assume a probability distribution that has little to do with the true one. There are cases in which we have so little information, that building a full-fledged multivariate probability distribution amounts to pulling a heroic and futile stunt. The dependence between variables can be difficult to capture by a correlation matrix, as this only picks up linear dependencies. Furthermore, autocorrelation over time may also be an issue. In other words, there may be considerable uncertainty about the uncertainty model itself. In some case, all we are able to specify is an interval of sensible values for uncertain parameters, without attaching any probabilistic information.

- Even if the representation of uncertainty is adequate, we may look for a conservative solution. A solution which is good in an "average" sense may prove unacceptable in some extreme scenarios.

- Psychological research on the behavior of decision makers facing uncertainty has revealed patterns that are not fully compatible with the maximization of expected utility. Mechanisms such as regret and disappointment may lead to different decisions.

Chapter 14 illustrates a few issues with standard decision making procedures in a world of multiple stakeholders and subjective probabilities. In this section we just mention a couple of approaches that have been proposed to improve model-based decisions under risk.

13.5.1   Robust optimization

Robust optimization is a label that has been attached to a fairly wide variety of optimization modeling frameworks. A rather confusing feature is that "robust" may refer to our inability to represent uncertainty reliably within a probabilistic framework; alternatively, "robust" may refer to decision-makers' attitude towards risk taking. In this section we mainly refer to the first meaning, which essentially questions the reliability of probability distributions used in making decisions. Sometimes, all we know about uncertain parameters is that they are bounded to lie in a certain region. For instance, we could say that a certain quantity β is not smaller than 5 and not larger than 15, but we are not able to say if the distribution on that support is uniform, beta, triangular, or whatever. Some values of these parameters could be particularly nasty, but even if we know that this is an unlikely occurrence, we do not want to rely on this; on the one hand, we may feel uncomfortable in estimating small probabilities; on the other hand, ethical reasons may prevent us from accepting a solution that could prove a disaster, even with a negligible probability.

All of this is problematic in terms of optimality, but even more so in terms of feasibility. Quite often, one would want to find a solution that is feasible even in the worst possible setting of the parameters, as well as good enough.

One way of formulating the problem is the following. Consider a function $f(\mathbf{x}, \boldsymbol{\beta})$, where $\mathbf{x} \in S$ is the usual vector of decision variables, which should stay within the feasible set S. The vector $\boldsymbol{\beta}$ represents a vector of uncertain parameters, which are bounded by a set Ω. Note that we do not characterize this uncertainty in probabilistic terms. Then we can formulate the worst-case optimization problem as

$$\min_{\mathbf{x} \in S} \left\{ \max_{\boldsymbol{\beta} \in \Omega} f(\mathbf{x}, \boldsymbol{\beta}) \right\} \tag{13.34}$$

You may interpret the problem as a sort of game between you and Nature. You choose $\mathbf{x}$ in order to minimize cost and then, given your choice, she will choose the worst $\boldsymbol{\beta}$ for you. What you should do is anticipating her nasty behavior by finding a robust solution. Solving a problem like (13.34) requires methods which are quite outside the scope of this book,[26] and we refer the reader to the listed references. This may lead to an overly expensive solution, which is usually referred to as *fat* solution. A chance-constrained approach can be used to enforce a reasonable degree of reliability, but we have already discussed the hidden dangers in this choice. Whether this is in fact an issue depends on the problem at hand and on what is at stake: A few bucks or human lives?

A possible intermediate approach is based on the idea of assuming a probabilistic framework, but relying on a *set* of probability measures, rather than just one. Denoting by $\mathcal{P}$ the set of plausible probability measures, we may tackle the following problem:

$$\min_{\mathbf{x} \in S} \left\{ \max_{\mathbb{Q} \in \mathcal{P}} \mathrm{E}^{\mathbb{Q}} \left[f(\mathbf{x}, \omega) \right] \right\}$$

where $\mathrm{E}^{\mathbb{Q}}[\cdot]$ denotes the expectation under a probability measure $\mathbb{Q}$ within the set $\mathcal{P}$. Again, this typically results in an intractable problem that can be suitably approximated by numerical techniques.

13.5.2 Disappointment and regret in decision making

When making decisions under risk and uncertainty in our lives, we rarely set up a utility function to formalize the problem we are facing. We come up with a solution but, unfortunately, sometimes we must admit that we were wrong. Indeed, disappointment and regret are emotions that we have all experienced. A discussion of disappointment and regret may seem more akin to psychology than business decision making. However, they can provide an explanation for some typical paradoxes of decision-making models based on the assumption of full rationality.

---

[26] Note that we are also disregarding feasibility issues: Is $\mathbf{x}$ acceptable for any setting of $\boldsymbol{\beta}$?

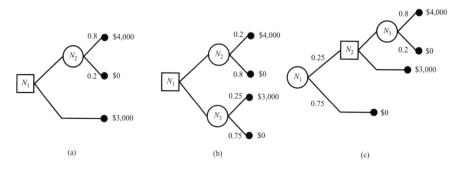

Fig. 13.12 An example illustrating the effect of disappointment.

Example 13.12 Consider the decision trees depicted in Fig. 13.12.[27] In Fig. 13.12(a), the decision maker must choose between a lottery, with expected payoff

$$0.8 \times \$4,000 + 0.2 \times \$0 = \$3,200$$

and $3,000 for sure. Most people prefer the certain payoff, which is consistent with the idea of risk aversion. In Fig. 13.12(b) we are comparing two lotteries, whose expected payoffs are

$$0.2 \times \$4,000 + 0.8 \times \$0 = \$800$$
$$0.25 \times \$3,000 + 0.75 \times \$0 = \$750$$

respectively. In this case, empirical research shows that most people prefer the first lottery.[28] Arguably the increase in win probability is not enough to compensate for the reduction in the payoff. Now consider Fig. 13.12(c). Here, the root of the decision tree is a chance node, rather than a decision node. This basically means that a biased coin is flipped first; with probability 0.75 the game stops immediately; with probability 0.25, the decision maker is faced with the same situation depicted in decision tree (a). If we see things this way, common sense suggests that the decision should be the same as in case (a): Go for the sure payoff.[29] Furthermore, the decision maker should not object if requested to make her choice before the flip of the coin, thus defining a strategy. However, if we carry out a few calculations, it is easy to see that tree (a) is nothing but tree (b) in disguise:

- If the strategy selects the lower branch for the decision node in (c), there is an *a priori* probability $0.25 \times 1 = 0.25$ of winning $3,000.

- If the strategy selects the upper branch for the decision node in (c), there is an *a priori* probability $0.25 \times 0.8 = 0.2$ of winning $4,000.

---

[27] This example is based on the paper by Bell [3], which refers back to Ref. [18].
[28] See Ref. [18].
[29] This is an example of a seemingly harmless *substitution* principle.

Now we see a contradiction with the standard choice in case (b). A bit of reflection suggests that a rational decision maker should either select the upper or the lower branches in all of the three cases. But this is at odds with what is empirically observed. Hence, we must be missing something. ⬜

A possible explanation of the weird finding of Example 13.12 is based on *disappointment* and *regret*. Imagine taking your chances in the case of Fig. 13.12(a). If you go for the lottery and you lose, you will certainly regret your decision, since you have wasted $3,000 for sure. Then, rather than feeling so sorry, it might be better to choose the safer option. In Fig. 13.12(b), the probability of winning is low for both lotteries. You are likely lose in both cases: Whatever you choose, you will not feel too sorry. Hence, you may just go for the lottery with higher payoff in case of win, which is also the lottery with the higher expected payoff. Now consider the situation of Fig. 13.12(c). If, after the first flip, you are out of the game, so be it. You will be sorry, but there is nothing to regret, as this outcome did not depend on your choice. But in the lucky case you enter the lottery at the second stage, you will certainly be disappointed if you end up losing $4,000 rather than grabbing $3,000 for sure. Hence, chances are that in this case you will go for the safer option, even though this contradicts the choice in case (b). A formal framework for modeling disappointment has been proposed in Ref. [3], where expected economic utility is integrated with a psychological view of decision maker's behavior. In fact, the paradigm of expected utility is at odds with quite some share of observed behavior, and this has suggested alternative frameworks.

So far, we have been a bit ambiguous when speaking about disappointment and regret. They are certainly related, but different emotions.[30] To see the difference, consider the following experiments:[31]

- You spin a simple fortune wheel, where you may just win or loose. There is a high probability that you will win the prize, but unfortunately the outcome is unsatisfactory. You will be disappointed, but no decision was involved: You are comparing the actual outcome with a better one that might have resulted.

- You must spin one of two fortune wheels, say, A and B. Suppose that you select wheel A. Then both wheels are spun, and you lose. To add insult to injury, you are shown that the result from wheel B is "win." Then, you will experience regret, as a better outcome would have resulted, had you made a different choice.

It may seem that the role of emotions in decision making is disconnected from business management. However, there are indeed cases in which regret is what you should keep under control. Common features of such cases are:

---

[30]See, e.g., Ref. [33] for a thorough discussion. Evidence suggests that these emotions activate sections of our brain in different ways; see Ref. [11].
[31]See Ref. [21].

- You are facing a nonrepetitive situation, where the law of large numbers does not apply; there is no point in making a decision that would prove the best one in the long run.

- You are evaluated ex post. Expectation is by definition ex ante, but imagine a situation in which your performance is assessed in comparison to what would have been the optimal decision, had you known the true value of uncertain parameters.

Min-max regret models have been proposed to address such situations. Consider the familiar cost function $f(\mathbf{x}, \omega)$, where ω is a random event. To be specific, say that uncertainty may be represented by a set of scenarios ω_s, $s \in \mathcal{S}$. If you knew that scenario ω_s is going to occur, you would solve the problem

$$\min_{\mathbf{x} \in \mathcal{S}} f(\mathbf{x}, \omega_s)$$

resulting in an optimal solution $\mathbf{x}_s^*$ and cost $f_s^* = f(\mathbf{x}_s^*, \omega_s)$. Note that this is a rather simple problem that can be solved even for a rather large number of scenarios.[32] Unfortunately, you have to solve the here-and-now problem, resulting in solution x_h^*; whatever model formulation you choose, some value of the objective function will result, but this is *not* what will affect your future career; the true *ex post* cost is what matters to your boss. The cost of the here-and-now decision, in scenario s, is $\hat{f}_s = (x_h^*, \omega_s)$. Unless you are *very* lucky, this will not be the optimal cost in scenario s, and you will experience a *regret*, formally defined as

$$R_s = \hat{f}_s - f_s^*$$

The idea is that, if scenario s occurs, you will discover that you had an extra cost R_s; the larger R_s, the larger the regret. However unfair this may sound, you could be just evaluated this way in real life: You might make a good and sensible solution, which turns out to be awkward if a bad scenario occurs. Then, a safer strategy could be to minimize maximum regret. By introducing an auxiliary variable ζ, this min–max model can be expressed as follows:

$$\begin{aligned}
\min \quad & \zeta \\
\text{s.t.} \quad & \zeta \geq f(\mathbf{x}, \omega_s), \qquad \forall s \in \mathcal{S} \\
& \zeta \geq 0, \ \mathbf{x} \in \mathcal{S}
\end{aligned}$$

Several variations of the model, possibly including expected regret, may be formulated. If the original model was a linear programming problem, so will be the formulation based on regret.

---

[32] We recall that it is usually faster to solve a large number of small scenario-dependent problems, than a single large-scale stochastic or robust problem.

Problems

13.1 Consider again the "fancy coin flipping" example of Section 1.2.3, i.e., the decision of producing a movie or not. Formalize the problem with a proper decision tree.

13.2 The Research and Development (R&D) division of your firm has developed a new product that could be immediately launched on the market. If so, the probability of success is 60%, in which case profit has a present value of €10 million (discounting all future cash flows). In case of flop, you will lose €2 million. You might also delay product launch by 6 months, in order to improve its design. This delay has advantages and disadvantages:

- Product improvement would cost €500,000, but it increases success probability to 90% (without changing hypothetical profits and losses); however, this applies only if no competitor takes advantage of your delay and enters the market, eroding your share (see the last bullet below).

- Delaying product launch has a financial impact, as cash flows are delayed; let us assume that the effect of delay is accounted for by a 5% discount rate for the 6 months. (Note: This is the rate applying to the 6 months, not to 1 year.)

- Delaying product launch has the effect of leaving room for the entry of competitors; we assume that the probability of this entry is 50% and its effect is not on success probability (which is still 90%), but on cash flows: Profit is halved, and loss is doubled, depending on product success or flop.

1. Assuming that you are risk-neutral, what would you do?

2. In practice all of the above probabilities are the result of educated guesswork. In particular, the entry probability is quite uncertain. Hence, you want to carry out a sensitivity analysis. What is the value of the entry probability (if it exists) that would make you change your mind with respect to the decision above?

13.3 You are the manager of a pension fund, and your fee depends on the return attained. You can play it safe and allocate wealth to a risk-free portfolio earning 4% per year. Alternatively, you can pursue an active portfolio management strategy, whose a return is normally distributed, with expected value 8% and standard deviation 10%. Your fee depends on the earned return according to the following tabulation:

Return (R)	Fee
$R < 0\%$	\$0
$0\% \leq R < 3\%$	\$50,000
$3\% \leq R < 9\%$	\$100,000
$9\% \leq R$	\$200,000

- Which is your best strategy, if you are risk-neutral?

- What is the standard deviation of your fee, if you take the active strategy?

13.4   A decision maker with a quadratic utility function of the form (13.12) is offered the following lottery:

Probability	Payoff
0.20	\$10,000
0.50	\$50,000
0.30	\$100,000

If the risk aversion coefficient is $\lambda = 1/150,000$:

- What is the certainty equivalent of the lottery?

- What is the risk premium?

13.5   You own a plant whose value is \$100,000. In case of a fire, the value of your property might be significantly reduced or even destroyed, depending on how severe the accident is. Let us represent risk by the following scenarios:

Scenario	Value (\$)	Probability
1	100,000	0.95
2	50,000	0.04
3	1	0.01

For each scenario, we have the value of your property and a probability. Clearly, in scenario 1 there is no fire and no loss. Assuming that your risk aversion is represented by a logarithmic utility function, what is the maximum insurance premium that you would be willing to pay? (*Hint*: The insurance would pay \$0, \$50,000, \$99,999, respectively, in the three scenarios.)

13.6   An investor has an initial wealth W_0 that must be allocated between a risk-free asset, with certain return r_f, and a risky asset. We assume a simple binomial model of uncertainty, like we did in Section 3.1. The return of the risky asset will either be R_u or R_d, with probabilities p_u and p_d, respectively.

- Find the optimal allocation of wealth if the investor has an exponential utility function, like Eq. (13.11), with absolute risk aversion α.

- How does the value of the initial wealth influence the result?

13.7 You have invested \$150,000 in stock shares of Doom and \$200,000 in stock shares of Mishap. Assume that daily returns follow a multivariate normal distribution; daily volatilities for the two stock shares are 2% and 3%, respectively, and correlation is 0.8. Find the one-day VaR at 99% confidence level.

13.8 We know that VaR, in general, is not a subadditive risk measure. Consider a portfolio of two assets, with jointly normal returns.

- Show that, in this specific case, VaR is a subadditive risk measure.

- Is standard deviation a subadditive measure in the case above? What can we say in general?

13.9 Consider the plant location model of Section 12.4.5 [see Eqs. (12.54–12.55)]. Adapt the model to cope with uncertain demand scenarios, building a two-stage stochastic linear programming model with recourse.

13.10 Consider a point-to-point transportation network consisting of M nodes. By "point-to-point" we mean that, given transportation requirements between all pair of nodes, there is a direct transportation link from each origin to each destination, without transshipment through intermediate nodes (this should be understood as a simplification of the real-life problem). Each node, in general, is both a source and a destination for many such requirements. Transportation requires containers of given volume capacity, measured in the same units as transportation requirements. At present, we know how many empty containers are present at each node in the network. Before transportation needs arise, we should consider repositioning containers, so that we will be in a better position to meet all requirements. Repositioning is carried out now, and transportation occurs in the next period. We know the repositioning cost for each container, for each pair of nodes in the network (say that they essentially depend on traveled distance). What we do not know are the future transportation requirements, but we are able to generate a set of plausible scenarios. Each scenario is essentially a matrix describing the transportation requirement for each pair of nodes (the diagonal elements are zero), associated with a probability. If, when requirements are revealed, we do not have enough containers to satisfy them at a node, we have to rent containers, which implies a rather high fixed cost per container. For the sake of simplicity, let us assume that such cost does not depend on nodes, nor traveled distances. On the one hand, we would not like to move too many containers; on the other one, we do not want to spend too much money to rent containers if spikes in transportation demand occur at some nodes. Build a two-stage stochastic programming model to minimize the total expected cost.

For further reading

- Extensive examples of modeling by decision trees can be found, e.g., in the book by Bertsimas and Freund [5].

- We have taken for granted that expected value of an additive utility can be used to model preferences. However, the validity of the approach depends on critical assumptions concerning the preference structure we are representing. A textbook treatment can be found in Chapter 6 of Ref. [22].

- An extensive discussion of value at risk can be found in Ref. [17]. We have also seen that VaR lacks some fundamental properties that make a coherent measure of risk; the seminal paper on coherent risk measures is Ref. [2]. CVaR optimization is dealt with in the papers by Rockafellar and Uryasev [26, 27]. It is also important to be aware of the literature questioning some basic assumptions behind risk measurement and management; see, e.g., Refs. [8] and [25].

- When faced with uncertain problem data, sensitivity analysis is often applied; however its utility is questioned in Ref. [31], providing support for stochastic programming.

- The first historical source on stochastic programming models is the paper by Dantzig [12]. Short tutorials on stochastic programming can be found in Refs. [6], [28], and [30]. An extensive treatment is offered by Refs. [7] and [19].

- Stochastic programming has been proposed for a wide array of applications:

  - For an illustration of stochastic programming models in finance, see Refs. [34] and [35].
  - The application to capacity planning in the automotive industry is described in Ref. [14]; lot-sizing under demand uncertainty is discussed in Ref. [9]; for a survey on applications to manufacturing systems, see Ref. [1].
  - Energy applications are discussed in Ref. [32].
  - An early application to telecommunication network planning is described in Ref. [29].
  - On the marketing side, an application to airline revenue management is described in Ref. [24].

- Scenario generation is a challenging facet of stochastic programming. In this context, analyzing sensitivity and stability of solution with respect to selected scenarios is fundamental, as discussed in Ref. [13].

Furthermore, a good scenario tree should capture the relevant uncertainty, without making the overall model intractable; the generation of scenarios that match moments and other properties of the underlying probability distribution is discussed in Ref. [16]; another idea is to sample a large tree, which is then reduced minimizing some measure of error, as discussed in Ref. [15]. For an elementary introduction to variance reduction and low discrepancy sequences in Monte Carlo simulation, see Ref. [10].

- A demanding but comprehensive treatment of robust optimization can be found in Ref. [4]. See also Ref. [20], where the case for regret-based models is made. Sometimes, uncertainty is described by giving *intervals* for uncertain parameters; see, e.g., Ref. [23] for such an example.

REFERENCES

1. A. Alfieri and P. Brandimarte, Stochastic programming models for manufacturing applications, in A. Matta and Q. Semeraro, eds., *Design of Advanced Manufacturing Systems*, Springer, Dordrecht, 2005.

2. P. Artzner, F. Delbaen, J.-M. Eber, and D. Heath, Coherent measures of risk, *Mathematical Finance*, **9**:203–228, 1999.

3. D.E. Bell, Disappointment in decision making under uncertainty, *Operations Research*, **33**:1–27, 1985.

4. A. Ben-Tal, L. El Ghaoui, and A. Nemirovski, *Robust Optimization*, Princeton University Press, Princeton, NJ, 2009.

5. D. Bertsimas and R.M. Freund, *Data, Models and Decisions: The Fundamentals of Management Science*, Dynamic Ideas, Belmont, MA, 2004.

6. J.R. Birge, Stochastic programming computation and applications, *INFORMS Journal of Computing*, **9**:111–133, 1997.

7. J.R. Birge and F. Louveaux, *Introduction to Stochastic Programming*, Springer-Verlag, New York, 1997.

8. R. Bookstaber, *A Demon of Our Own Design: Markets, Hedge Funds, and the Perils of Financial Innovation*, Wiley, New York, 2008.

9. P. Brandimarte, Multi-item capacitated lot-sizing with demand uncertainty, *International Journal of Production Research*, **44**:2997–3022, 2006.

10. P. Brandimarte, *Numerical Methods in Finance and Economics: A Matlab-Based Introduction*, 2nd ed., Wiley, New York, 2006.

11. H.F. Chua, R. Gonzalez, S.F. Taylor, R.C. Welsh, and I. Liberzon, Decision-related loss: regret and disappointment, *NeuroImage*, **47**:2031–2040, 2009.

12. G.B. Dantzig, Linear programming under uncertainty, *Management Science*, **1**:197–206, 1955.

13. J. Dupačová, Stability and sensitivity analysis for stochastic programming, *Annals of Operations Research*, **27**:115–142, 1990.

14. G.D. Eppen, R.K. Martin, and L. Schrage, A scenario approach to capacity planning, *Operations Research*, **37**:517–527, 1989.

15. H. Heitsch and W. Römisch, Scenario reduction algorithms in stochastic programming, *Computational Optimization and Applications*, **24**:187–206, 2003.

16. K. Hoyland and S.W. Wallace, Generating scenario trees for multistage decision problems, *Management Science*, **47**:296–307, 2001.

17. P. Jorion, *Value at Risk: The New Benchmark for Controlling Financial Risk*, 3rd ed., McGraw-Hill, New York, 2007.

18. D. Kahneman and A. Tversky, Prospect theory: an analysis of decision under risk, *Econometrica*, **47**:263–291, 1979.

19. P. Kall and S.W. Wallace, *Stochastic Programming*, Wiley, Chichester, 1994.

20. P. Kouvelis and G. Yu, *Robust Discrete Optimization and Its Applications*, Kluwer Academic Publishers, Dordrecht, 1997.

21. F. Marcatto and D. Ferrante, The regret and disappointment scale: an instrument for assessing regret and disappointment in decision making, *Judgment and Decision Making*, **3**:87–99, 2008.

22. A. Mas-Colell, M.D. Whinston, and J.R. Green, *Microeconomic Theory*, Oxford University Press, New York, 1995.

23. H.E. Mausser and M. Laguna, A heuristic to minimax absolute regret for linear programs with interval objective function coefficients, *European Journal of Operational Research*, **117**:157–174, 1999.

24. Andris Möller, W. Römisch, and K. Weber, Airline network revenue management by multistage stochastic programming, *Computational Management Science*, **4**:355–377, 2008.

25. R. Rebonato, *Plight of the Fortune Tellers: Why We Need to Manage Financial Risk Differently*, Princeton University Press, Princeton, NJ, 2007.

26. R.T. Rockafellar and S. Uryasev, Optimization of conditional value-at-risk, *The Journal of Risk*, **2**:21–41, 2000.

27. R.T. Rockafellar and S. Uryasev, Conditional value-at-risk for general loss distributions, *Journal of Banking and Finance*, **26**:1443–1471, 2002.

28. A. Ruszczyński and A. Shapiro, Stochastic programming models, in A. Ruszczyński and A. Shapiro, eds., *Stochastic Programming*, Elsevier, Amsterdam, 2003.

29. S. Sen, R.D. Doverspike, and S. Cosares, Network planning with random demand, *Telecommunications Systems*, **3**:11–30, 1994.

30. S. Sen and J.L. Higle, An introductory tutorial on stochastic programming models, *Interfaces*, **29**:33–61, 1999.

31. S.W. Wallace, Decision making under uncertainty: is sensitivity analysis of any use? *Operations Research*, **48**:20–25, 2000.

32. S.W. Wallace and S.-E. Fleten, Stochastic programming models in energy, in A. Ruszczynski and A. Shapiro, eds., *Stochastic Programming*, Elsevier, Amsterdam, 2003.

33. M. Zeelenberg, W.W. van Dijk, A.S.R. Manstead, and J. van der Pligt, The experience of regret and disappointment, *Cognition and Emotion*, **12**:221–230, 1998.

34. S. Zenios, *Practical Financial Optimization*, Wiley-Blackwell, Oxford, 2008.

35. W.T. Ziemba and J.M. Mulvey, eds., *Worldwide Asset and Liability Modeling*, Cambridge University Press, New York, 1998.

14

Multiple Decision Makers, Subjective Probability, and Other Wild Beasts

The previous chapters have presented a rather standard view of quantitative modeling. When dealing with probabilities, we have often taken for granted a frequentist perspective; our approach to statistics, especially in terms of parameter estimation, has been an orthodox one. Actually, these are not the only possible viewpoints. In fact, probability and statistics are a branch of mathematics at the boundary with philosophy of science, and as such they are not free from heated controversy. This might sound like a matter of academic debate, but it is not. The "death of probability" was invoked in the wake of the 2008 financial turmoil, when the quantitative modeling approach in finance has been blamed as one of the root causes of the disaster. Of course, truth always lies somewhere between extremes, but this is reason enough to see the need for an eye opening chapter, illustrating alternative views that have been put forward, like subjective probabilities and Bayesian statistics. A similar consideration applies to the chapters on decision models. There, we have also followed a standard route, implicitly assuming that decisions are made by one person keeping all problem dimensions under direct control. We have hinted at some difficulties in trading off multiple and conflicting *objectives*, when dealing with multiobjective optimization in Section 12.3.3. However, we did not fully address the thorny issues that are raised when multiple *decision makers* are involved. On the one hand, they can be interested in different objectives; on the other one, they may behave without coordinating their decisions with other actors. Decisions involving not necessarily cooperative actors are the subject of game theory. Rather surprising results are obtained when multiple players interact, possibly leading to suboptimal decisions; here, we do not mean suboptimal only for a single decision maker, but, for the whole set of them. Finally, standard models assume that uncertainty is exogenous, whereas there are many practical situations in which decisions do influence uncertainty, such as big trades on thin and illiquid financial markets or inventory management decisions affecting demand. When all of the above difficulties compound, multiple decision makers can influence one another through decisions, behavior, and information flows, possibly leading to instability due to vicious feedbacks. Such mechanisms have been put forward as an explanation of some major financial crashes.

We are certainly in no position to deal with such demanding topics extensively. They all would require their own (voluminous) book and the technicalities involved are far from trivial. However, I strongly believe that there must be room for a chapter fostering critical thinking about quantitative models. This is not to say that what we have dealt with so far is useless. On the contrary, it must be taken with a grain of salt and properly applied, keeping in mind that we could be missing quite important points, rendering our analysis irrelevant or even counterproductive. Unlike the rest of the book, the aim of this chapter is not to provide readers with working knowledge and ready-to-use methods; rather, a sequence of simple and stylized examples will hopefully stimulate curiosity and further study.

In Section 14.1 we discuss general issues concerning the difference between decision making under risk, the topic of the previous chapter, and decision making under uncertainty, which is related to a more radical view. In Section 14.2 we begin formalizing decision problems characterized by the presence of multiple noncooperative decision makers, setting the stage for the next sections, where we discuss the effects of conflicting viewpoints and introduce game theory. In Section 14.3 we illustrate the effect of misaligned incentives in a stylized example involving two decision makers in a supply chain. The two stakeholders aim at maximizing their own profit, and this results in a solution that does *not* maximize the overall profit of the supply chain. Such noncooperative behavior is the subject of game theory, which is the topic of Section 14.4. We broaden our view by outlining fundamental concepts about equilibrium, as well as games with sequential or simultaneous moves. Very stylized examples will illustrate the ideas, but this section is a bit more abstract than usual. Therefore, in Section 14.5, we show a more practical example related to equilibrium in traffic networks; this example, known as Braess' paradox, shows that quite counterintuitive outcomes may result from noncooperative decision making. In Section 14.6 we discuss how the dynamic interaction among multiple actors may lead to instability and, ultimately, to disaster, by analyzing a couple of real-life financial market crashes. Finally, we close the chapter by providing the reader with a scent of Bayesian statistics in Section 14.7. Bayesian learning is related to parameter estimation issues in orthodox statistics; there, the basic concept is that parameters are unknown *numbers*; within this alternative framework, we may cope with probabilistic knowledge about parameters, possibly subjective in nature. As a concrete example, we outline the Black–Litterman portfolio optimization model, which can be interpreted as a Bayesian approach.

14.1 WHAT IS UNCERTAINTY?

When we flip a fair coin, we are uncertain about the outcome. However, we are pretty sure about the rules of the game: The coin will either land head or tail, and to all practical purposes we assume that the two outcomes are equally likely. However, what about an alien who has never seen a coin and does not take our probabilities for granted? Probably, it would face a more radical form of uncertainty, where the probabilities themselves are uncertain, and not only the outcome of the flip.[1] By a similar token, sometimes we have plenty of reliable and relevant data about a random phenomenon, possibly featuring significant variability, which suggests the application of a frequentist concept of probability. Again, coming up with a good decision in such a setting may be far from trivial, but at least we might feel confident about our representation of uncertainty. Unfortunately, we are not always so lucky. Sometimes, we do not have relevant data, as we are facing a brand-new situation, as is the case when launching a truly innovative product. In other cases, the situation is so risky that we cannot have blind faith in statistical data analysis. How about an event with a very low probability, but potentially catastrophic?

Example 14.1 Suppose that we are interested in investigating the safety of an airport, in terms of its ability to manage the takeoff–landing traffic. Apparently, we should consider the statistics of accidents that may be blamed on the airport. However, hopefully, data on such accidents are so scarce that they are hardly relevant. In such a case, we should also consider the *near misses*, i.e., events that did not actually result in a disaster, but indicate that something is not working as it should. By the same token, car insurance companies are also interested in the driving habits of a potential customer, and not only in his accident track record.   ⬚

This example shows that sometimes past statistics are not quite relevant because of lack of data. In other cases, even if we have plenty of data, a structural change in the phenomenon may make them irrelevant. In Chapter 5 we have pointed out that there are different interpretations of probabilities. Indeed, the cases above illustrate both the classical concept of probability, based on

---

[1]We illustrate a Bayesian framework to learn such probabilities in Example 14.14.

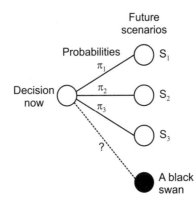

Fig. 14.1   Schematic illustration of different kinds of uncertainty.

the idea that there are underlying equally likely outcomes, and the frequentist concept. However, probability may also be a measure of belief in a scientific theory, possibly including subjective elements. Some have even questioned the use of probability as a model of uncertainty. Alternative frameworks have been proposed, like fuzzy sets. We will stick to a probabilistic framework, but it is important to understand a few basic issues, with reference to Fig. 14.1. There, we use a familiar scenario tree (or fan) to represent uncertainty. We have to make a decision here and now, but its ultimate consequence depends on which future scenario will occur. According to a simple probabilistic view, each scenario S_i is associated with a well-known probability π_i, $i = 1, 2, 3$. Unfortunately, things are not always that easy.

14.1.1   The standard case: decision making under risk

Let us compare two random experiments: fair coin flipping and the draw of a multidimensional random variable, with a possibly complicated joint probability density. The two cases may look quite different. The first one can be represented by a quite simple Bernoulli random variable, and calculating expected values of whatever function of the outcome is pretty straightforward. On the contrary, calculating an expectation in the second case may involve a possibly awkward multidimensional integral. However, the difference is more technical and computational than conceptual, as in both cases we assume that we have a full picture of uncertainty. The important point is that the risk we are facing is linked to the realization of a random variable, which is perfectly known from a probabilistic perspective. With reference to Fig. 14.1, we are pretty sure about:

1. The possible future scenarios, since we know exactly what might possibly happen, and there is no possibility unaccounted for.

2. The probabilities of these scenarios, whatever meaning we attach to
them.

Hence, we have a full picture of the scenario tree. This standard case has
been labeled as *decision under risk*, to draw the line between this and more
radical views about uncertainty. Risk may be substantial; to see this, consider
a one-shot decision when there is a very dangerous, but very unlikely scenario.
Which decision model is appropriate to cope with such a case? There is no
easy answer, but, at least, we have no uncertainty about our description of
uncertainty itself.

14.1.2 Uncertainty about uncertainty

If we are about to launch a brand-new product, uncertainty about future sales
is rather different from that in the previous case. Maybe, we know pretty well
what may happen, so that the scenarios in Fig. 14.1 are known. However, it
is quite hard to assess their probabilities. The following definitions, although
not generally accepted, have been proposed:

- *Risk* is related to uncertainty about the realization of a random variable
  whose distribution is known.

- *Uncertainty*, in the strict sense, is related to:

  - The parameters of a probability distribution whose qualitative form
    is known

  - A probability distribution whose shape itself is unknown

We see that there are increasing levels of uncertainty. We may be pretty
sure that the probability distribution of demand is normal, but we are not
quite sure about its parameters; this is where we started feeling the need
for inferential statistics and parameter estimation. Nonparametric statistics
comes into play when we even question the type of probability distribution
we should use. But even if we are armed with a formidable array of statistical
techniques, we may lack the necessary data to apply them. Probability in this
setting tends to be subjective and based on expert opinion.

Whatever the level of uncertainty, if the probabilities π_i are not reliable, a
more robust decision making model is needed, as we pointed out in Section
13.5.1; however, there may be something more at play, such as prior opinions
and their revision by a learning mechanism. In Chapters 9 and 10 we have
taken an orthodox view of statistics, based on the fundamental assumption
that parameters are *unknown numbers*. Then, we try to estimate unknown
parameters using estimators; estimators are random variables, and their realized value, the estimate, depends on data collected by random sampling.
Given this framework, there are two consequences:

1. There is no probabilistic knowledge associated with parameters, and we never speak about the probability distribution of a parameter.[2]

2. There is no room for subjective opinions, and collected data are the only sensible information that we should use.

However, when facing new decision making problems, relying on subjective views may be not only appropriate, but also necessary. This is feasible within a Bayesian framework, whereby parameters are regarded as random variables themselves. Within this framework, the distribution that we associate with a parameter depicts our limited state of knowledge, possibly subjective in nature. As we outline in Section 14.7, in the Bayesian approach we start with a prior distribution, which reflects background information and subjective knowledge, or lack thereof; as new information is collected by random sampling, this is reflected by an update of the prior distribution, leading to a posterior distribution by the application of Bayes' theorem.

Issues surrounding orthodox and Bayesian statistics are quite controversial, but one thing is sure: The probabilities π_i in Fig. 14.1 may not be reliable, and we might move move from decision making under risk to decision-making under uncertainty.

14.1.3   Do black swans exist?

The most troublesome case is when some scenarios are particularly dangerous, yet quite unlikely. How can we trust estimates of *very low* probabilities? To get the message, consider financial risk management. Here we need to work with extreme events (stock market crashes, defaults on sovereign debt, etc.), whose probabilities can be very low and very difficult to assess, because of limited occurrence of such events in the past. Can we trust our ability to estimate the probability of a rare event? And what if we are missing some scenarios completely?

Example 14.2 (Barings Bank and Kobe earthquake) Barings bank was founded in 1762 and it had a long and remarkable history, which came to an abrupt end in 1995, when it was purchased by another bank for the nominal price of £1. The bank went bankrupt as a consequence of highly leveraged[3] and risky positions in derivatives, taken by a rogue trader who managed to hide his trading activity behind some glitches in the internal risk management

---

[2]Sometimes, it is stated that a confidence interval contains the uncertain parameter with a given probability; this is correct if we refer to an interval *estimator*, i.e., a pair of random variables. On the contrary, it is plain wrong if we refer to an interval *estimate*, i.e., to a pair of numbers realized after random sampling. See the discussion in Section 9.2.2.

[3]A leveraged position is a trade in which you borrow money to invest in risky positions. If things turn out well, this has the effect to enhance your gain; in case of loss, this is magnified, possibly leading to bankruptcy.

system of the bank. These strategies lead to disaster when Nikkei, the Tokyo stock market index, went south. When does a stock market crash? This can be the result of real industrial or economical problems, or maybe the financial distress of the banking system. Risk management models should account for the uncertainty in such underlying factors, and even unlikely extreme scenarios should enter the picture, when taking very risky positions. However, many have attributed that drop in the Nikkei index to swinging market mood after an earthquake stroke Kobe. Luckily, the earthquake was not hard enough to cause a real economic crisis; yet, its effect was pretty concrete on Barings, which had to face huge losses, ultimately leading to its demise.

Very sophisticated models have been built for financial applications, accounting for a lot of micro- and macro-economic factors, but it must take much imagination to build one considering the potential impact of an earthquake. □

Indeed, the most radical form of uncertainty is when we cannot even trust our view about the possible scenarios. To reinforce the concept, imagine asking someone about the probability of a subprime mortgage crisis twenty years *before* 2007. In Fig. 14.1 we have depicted a black scenario with which it is impossible to associate a probability, for the simple reason that we do not even know beforehand that such a scenario may occur. In common parlance, these scenarios are referred to as "black swans."[4] When black swans are involved, measuring risk is quite difficult if not impossible. The term *Knightian uncertainty* was proposed much earlier to refer to unmeasurable uncertainty, after the economist Frank Knight drew the line between risk and uncertainty.[5] It is quite difficult to assess the impact of unmeasurable uncertainty on a decision model, of course; maybe, there are cases in which we should just refrain from making decisions that can lead to disaster, however small its probability might look like (if nothing else, because of ethical reasons).

14.1.4 Is uncertainty purely exogenous?

The scenario tree of Fig. 14.1 may apply, e.g., to a two-stage stochastic programming problem. In a multistage stochastic programming model we have to make a sequence of decisions; a multistage scenario tree, like the one shown in Fig. 13.11, may be used to depict uncertainty. Even if we take for granted that sensible probabilities can be assigned and that no black swan is lurking somewhere, how can we be sure that our sequence of decisions will not affect uncertainty?

---

[4] A few centuries ago, in Europe, it was common wisdom that all swans were white, for the simple reason that no one ever observed a black one. The picture changed after 1697, when one was found in Australia. See N.N. Taleb, *The Black Swan: The Impact of the Highly Improbable*, 2nd ed., Random House Trade Paperbacks, 2010.

[5] See F.H. Knight, *Risk, Uncertainty, and Profit*, Houghton Mifflin, Boston, 1921.

Example 14.3 Setting inventory levels at a retail store is a rather standard decision making problem under risk. Typically, the task requires choosing a model of demand uncertainty, which is an input to the decision procedure. However, which comes first: Our decision or demand uncertainty? Indeed, our very decision may affect uncertainty. Marketing studies show that the amount of items available on the shelves may affect demand. To see why, imagine buying the very last box of a product on a shelf, when there is plenty of a similar item just below. In this case, consumers' psychology plays an important role, but even in a strict business-to-business problem, which need not involve such issues, an array of stockouts may be fatal to your customer demand. A naive newsvendor model may suggest a low service level because of low profit margins; since the order quantity should be the corresponding quantile of the probability distribution of demand, it will be very low as well, resulting in frequent stockouts. What such a model disregards is that the distribution itself will change as a consequence of our decision, if we offer a consistently low service level and keep disappointing customers. Even worse, this is likely to have an impact on the demand for other items as well. In practice, if a firm offers a catalogue of 1,000 items, it may well be the case that only 10% of them are profitable; the remaining ones are needed nonetheless, to support sales of profitable items.   ▯

The line we are drawing here is between *endogenous* and *exogenous* uncertainty. Standard decision models may fail to consider how decisions affect uncertainty, which is clearly relevant for sequential decision making. These issues may be exacerbated by the presence of multiple actors, possibly influencing each other by means of actions and information flows, giving rise to feedback effects. A quite relevant example of such nasty mechanisms is represented by financial markets instability and liquidity crises. We consider a couple of such stories in Section 14.6. But even if we disregard risk and uncertainty, the presence of multiple actors may have a relevant impact, as we illustrate in the next sections.

14.2 DECISION PROBLEMS WITH MULTIPLE DECISION MAKERS

Consider the decision problem

$$\max \quad \pi_1(x_1, x_2) + \pi_2(x_1, x_2) \qquad (14.1)$$
$$\text{s.t.} \quad x_1 \in S_1, \ x_2 \in S_2$$

The objective function (14.1) can be interpreted in terms of a profit depending on two decision variables, x_1 and x_2, which must stay within feasible sets S_1 and S_2, respectively. Note that, even though the constraints on x_1 and x_2 are separable, we cannot decompose the overall problem, since the two decisions interact through the two profit functions $\pi_1(x_1, x_2)$ and $\pi_2(x_1, x_2)$. Nevertheless, using the array of optimization methods of Chapter 12, we should be able to find optimal decisions, x_1^* and x_2^*, yielding the optimal total profit

$$\pi_{1+2}^* = \pi_1(x_1^*, x_2^*) + \pi_2(x_1^*, x_2^*)$$

In doing so, we assume that there is either a single stakeholder in charge of making both decisions, or a pair of cooperative decision makers, in charge of choosing x_1 and x_2, respectively, but sharing a common desire to maximize the overall sum of profits. But how about the quite realistic case of two *noncooperative* decision makers, associated with profit functions $\pi_1(x_1, x_2)$ and $\pi_2(x_1, x_2)$, respectively?

Decision maker 1 wishes to solve the problem

$$\max \quad \pi_1(x_1, x_2) \tag{14.2}$$
$$\text{s.t.} \quad x_1 \in S_1$$

whereas decision maker 2 wishes to solve the problem

$$\max \quad \pi_2(x_1, x_2) \tag{14.3}$$
$$\text{s.t.} \quad x_2 \in S_2$$

Unfortunately, these two problems, stated as such, make no sense. Which value of x_2 should we consider in problem (14.2)? Which value of x_1 should we consider in problem (14.3)? We must clarify how the two decision makers make their moves.

1. One possibility is that the two decision makers act sequentially. For instance, decision maker 1 might select $x_1 \in S_1$ before decision maker 2 selects $x_2 \in S_2$. In this case, we may say that decision maker 1 is the *leader*, and decision maker 2 is the *follower*. In making her choice, decision maker 1 could try to anticipate the reaction of decision maker 2 to each possible value of x_1.

2. Another possibility is that the two decisions are made simultaneously. Unfortunately, the conceptual tools that we have developed so far do not help us in making any sensible prediction about the overall outcome of such a simultaneous decision.

In Section 14.3 we illustrate the first case with a concrete, although stylized, example. Then, in section 14.4 we consider the second case as well, introducing a general theory of noncooperative games. Game theory aims at finding a sensible prediction of an *equilibrium solution* (x_1^e, x_2^e), which depends on the precise assumptions that we make about the structure of the game. Whatever equilibrium solution we obtain, it cannot yield an overall profit larger than π_{1+2}^*, as the following inequality necessarily holds:

$$\pi_{1+2}^e = \pi_1(x_1^e, x_2^e) + \pi_2(x_1^e, x_2^e) \leq \pi_1(x_1^*, x_2^*) + \pi_2(x_1^*, x_2^*) = \pi_{1+2}^*$$

If this inequality were violated, (x_1^*, x_2^*) would not be the optimal solution of problem (14.1). This means that if decentralize decisions, the overall system is likely to lose something.

14.3   INCENTIVE MISALIGNMENT IN SUPPLY CHAIN MANAGEMENT

The last point that we stressed in the previous section is the potential difficulty due to the interaction of multiple noncooperative, if not competitive, decision makers. The example we consider is a generalization of the newsvendor model:[6]

1. Unlike the basic model, there are two decisions to be made. The ordering decision follows the same logic as the standard case, but there is another one, related to product quality, which influences the probability distribution of demand.

2. Since there are two decisions involved, we should distinguish two cases:

   • In the *integrated* supply chain, there is only one decision maker in charge of both decisions.

   • In the *disintegrated* supply chain, there are two decision makers; a producer, who is in charge of setting the level of her product quality, and a distributor, who is in charge of deciding his order quantity.

In order to be able to find analytical solutions, we depart from the usual assumption of normal demand, and we suppose a uniform distribution, say, between 500 and 1000. We will rely on the following notation; the lower bound of the distribution support is denoted by a and its width by w; hence, the expected value is $a + w/2$. In the numerical example, $a = 500$ and $w = 1000 - 500 = 500$. The item has a production cost of $c = €0.20$ and sells at a price $p = €1.00$. To keep things as simple as possible, we suppose that the unsold items are just scrapped, and there is no salvage value. With the numbers above, we see that service level should be[7]

$$\beta = \frac{m}{m + c_u} = \frac{p - c}{(p - c) + (c - 0)} = \frac{p - c}{p} = \frac{1 - 0.2}{1} = 80\%$$

where $m = p - c$ is the profit margin and $c_u = c$ is the cost of unsold items, when there is no salvage value. The optimal order quantity is

$$Q^* = a + w\beta = 500 + 500 \times 0.8 = 900 \qquad (14.4)$$

Since the probability distribution is uniform, we may easily compute the expected profit. This requires calculating fairly simple integrals involving the

---

[6]This example has been inspired by the Harvard Business School (HBS) case [19], to which we refer for additional questions and issues.

[7]See Section 7.4.4.

constant demand density

$$f_D(x) = \begin{cases} \dfrac{1}{w} = \dfrac{1}{500}, & \text{for } a \leq x \leq a + w \\[2mm] 0, & \text{otherwise} \end{cases}$$

The expected profit depends on Q and it amounts to the expected revenue minus cost:

$$\begin{aligned} \Pi(Q) &= \int_{-\infty}^{+\infty} p \min\{x, Q\} f_D(x) dx - cQ \\ &= \int_a^Q \frac{p}{w} x \, dx + \int_Q^{a+w} \frac{p}{w} Q \, dx - cQ \\ &= \frac{p}{w} \frac{Q^2 - a^2}{2} + \frac{p}{w} Q(a + w - Q) - cQ \\ &= (p - c)Q + \frac{p}{w} \left(\frac{Q^2 - a^2 + 2Qa - 2Q^2}{2} \right) \\ &= mQ - \frac{p}{2w} (Q - a)^2 \end{aligned} \qquad (14.5)$$

This expression of expected profit may be interpreted as profit related to the purchased quantity, minus the expected lost revenue *due to unsold items*.[8] To get an intuitive feeling for Eq. (14.5), we may refer to Fig. 14.2, where lost revenue is plotted against demand, for a given order quantity Q. When demand is at its lower bound, $D = a$, lost revenue is $p(Q - a)$; when demand is $D = Q$, lost revenue is 0. For intermediate values of demand, lost revenue is a decreasing function of demand and results in the triangle illustrated in the figure. To evaluate the expected value of lost revenue, we should integrate this function, multiplied by the probability density $1/w$. But this is simply the area of the triangle in Fig. 14.2, $P(Q - a)^2/2$, times $1/w$, which in fact yields the second term in Eq. (14.5). With our numerical data, the optimal expected profit is

$$\Pi(Q^*) = 0.8 \times 900 - \frac{1}{2 \times 500} (900 - 500)^2 = 560$$

We may also express the expected profit as a function of the service level β, by plugging Eq. (14.4) into Eq. (14.5):

$$\Pi(\beta) = m(a + \beta w) - \frac{pw\beta^2}{2} \qquad (14.6)$$

---

[8]There is a potential source of confusion here, as we may lose potential revenue because of a stockout, i.e., when demand is larger than the available inventory. We refer here to revenue that we lose because of unsold items, i.e., when demand is smaller than the available inventory.

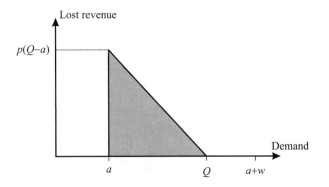

Fig. 14.2 A geometric illustration of Eq. (14.5).

So far, we have always assumed that a demand distribution cannot be influenced by the producer. However, she could improve the product or adopt marketing strategies to change the distribution a bit. The result depends on the effort she spends, which in turn has a cost. Let us measure the amount of effort by h, with unit cost 10. To model the effect of h on demand, we make the following assumptions:

1. For the sake of simplicity, we assume a pure shift in the uniform distribution of demand. Its lower bound a is shifted, but its support width w does not change.[9]

2. The shift in a should be a concave function of h, to represent the fact that effort is decreasingly effective for increasing levels (diminishing marginal return). One possible building block that we can use is the square-root function.

Hence, we represent the lower extreme of the distribution support by the following function of h:

$$a + 100\sqrt{h}$$

For instance, with our numerical parameters, if $h = 1$ the lower bound a shifts from 500 to 600. Using Eq. (14.6), we see that, if we include the effort h, the expected profit becomes

$$\Pi(\beta, h) = m\left(a + 100\sqrt{h} + \beta w\right) - \frac{pw\beta^2}{2} - 10h \qquad (14.7)$$

Note that the optimal service level does not change, according to our model, since we do not change either the unit cost of the item or its selling price. To

---

[9]Earlier in this chapter we insisted on the potential effect of our decisions on uncertainty. In this case, we assume that the effort decision affects only the location of the demand distribution, but not its dispersion. Nevertheless, uncertainty is indeed partially endogenous in our stylized example.

find the optimal effort, we should take the first-order derivative of expected profit (14.7) with respect to h and set it to 0:

$$\frac{100m}{2\sqrt{h}} - 10 = 0 \quad \Rightarrow \quad h^* = \left(\frac{100m}{2 \times 10}\right)^2 = 16$$

We notice that this condition equates the marginal increase of expected profit contribution from production and sales with the marginal cost of the effort. With this effort, the new probability distribution is uniform between 900 and 1400, the optimal produced quantity increases to $Q^* = 1300$, and by applying Eq. (14.7) the new total expected profit is

$$0.8 \times (500 + 100\sqrt{16} + 0.8 \times 500) - \frac{1 \times 0.8^2 \times 500}{2} - 10 \times 16 = 720$$

This is the optimal profit resulting from the *joint* optimal decisions concerning effort level h and purchased amount Q.

The above calculations are formally correct, but potentially flawed: They rely on the assumption that there is *one* decision maker maximizing overall expected profit and in control of both decisions. What if we have a supply chain with different stakeholders in charge of each decision, with possibly conflicting objectives? To see the impact of misaligned incentives, let us assume that there are two stakeholders:

1. The *producer*, who is in charge of determining the quality of the product and its potential for sales, through the effort h

2. The *distributor*, who is in charge of determining the order quantity Q, depending on the probability distribution of demand and the prices at which he can buy and sell

We cannot analyze such a system if we do not clarify not only who is in charge of deciding what, but also when and on the basis of what information. So, to be specific, let as assume the following:

- The producer is the *leader* and the distributor is the *follower*, in the sense that the producer decides her effort level h first, and then the distributor will select his order quantity Q. It must be stressed that, in deciding the effort level h, the producer should somehow anticipate how her choice will influence the choice of Q by the distributor.

- The producer is also the leader in the sense that she is the one in charge of pricing decisions, which we assume given. She keeps the selling price fixed at €1.00 and sells each item to the distributor for a price of €0.80, which is the purchase cost from the viewpoint of the distributor. Please note that the profit margin for the distributor is just €0.20; furthermore, he is the one facing all of the risk, as items have no salvage value and there is no buyback agreement in case of unsold items.

- Last but not least, everything is common knowledge, in the sense that the producer knows how the distributor is going to make his decision and both agree on the probability distribution of demand and how this is affected by the effort level h.

The last point implies that when the producer makes her decision, she can anticipate what the optimal decision of the distributor is going to be; hence, she can build a *reaction function*, also known as *best response function*, describing how the order quantity Q is influenced by her effort level h. From the point of view of the follower, i.e., the distributor, once the effort h is decided by the leader, the only thing he can do is to determine the order quantity as a function of h. Under our simple assumption of a pure shift in the demand distribution, we have

$$Q^*(h) = a + 100\sqrt{h} + \beta w$$

This expression look just the same as before but there is a fundamental difference. Under our assumptions, the production cost and the selling price do not change, but now, since the profit margin is shared, the optimal service level for the distributor is only

$$\beta = \frac{(1.0 - 0.8)}{(1.0 - 0.8) + (0.8 - 0.0)} = 0.2$$

There is a remarkable reduction in the service level, from 80% to 20%, because of the high price at which the distributor buys from the producer. Indeed, the split of the profit is €0.60 to the producer and €0.20 to the distributor. Plugging numbers, the optimal order quantity is

$$Q^*(h) = 500 + 100\sqrt{h} + 0.2 \times 500 = 600 + 100\sqrt{h}$$

This is how the producer can anticipate the effect of her choice of h on the decision of the distributor, which in turn influences her profit. The profit from the point of view of the producer is

$$0.6 \times \left[600 + 100\sqrt{h}\right] - 10h$$

Applying the first-order optimality condition, we obtain the optimal effort, and then the optimal order quantity:

$$\frac{0.6 \times 100}{2\sqrt{h^*}} = 10 \quad \Rightarrow \quad h^* = 9 \quad \Rightarrow \quad Q^* = 900$$

The new optimal order quantity depends on the fact that the shifted probability distribution of demand, with that level of effort, is uniform between 800 and 1300, and optimal service level is just 20%. A first observation that we can make is that both effort and order quantity are reduced when supply chain management is decentralized. This implies that consumers will receive a worse product and a reduced service level. What is also relevant, though, is the change in profit for the two players. The profit of the producer is

$$0.6 \times 900 - 10 \times 9 = €450$$

This profit is considerably reduced with respect to expected profit for the producer when she also manages distribution, which is €720. Of course, this is not quite surprising, since the producer has given up a fraction of her profit margin, leaving it to the distributor. Unlike the producer, the distributor faces an uncertain profit. Its expected value can be obtained from Eq. (14.5), after adjusting the parameters to reflect the reduced profit margin

$$mQ - \frac{p}{2w}(Q-a)^2 = 0.2 \times 900 - \frac{1}{2 \times 500}(900 - 800)^2 = €170$$

The expected profit for the distributor is less than the (certain) profit of the producer, but this is no surprise after all, considering how margins are split between the two parties. Last but not least, the total profit for the disintegrated supply chain is

$$€450 + €170 = €620$$

which is €100 less than the total profit for the fully integrated supply chain. Again, this is reasonable, in light of the concepts of Section 14.2.

These observations raise a couple of points:

1. One might well wonder why the producer should delegate distribution to someone else. Actually, there is a twofold answer:

   - Her profit now is a *certain* amount and not an expected value, since the whole risk is transferred to the distributor.

   - We did not consider fixed costs related to distribution; they may not impact optimal decisions at the tactical or operational level, but they do have an impact on the bottom line and on strategic decisions.

   We may also add further considerations concerning the fact that a distributor is probably better equipped at making demand forecasts and has a better incentive to offer additional services to the customer.

2. In splitting profit margins and decentralizing decisions, something has been lost for everyone: the producer, the distributor, and, last but not least, the customers, who receive less quality and less service. In principle, we could try to find the "optimal split" of profit margins, i.e., the allocation of margins that maximizes the total profit. Of course, this is hardly feasible in practice, as it would require cooperation between the two stakeholders; furthermore, in general, the overall profit will remain

suboptimal, anyway. There are practical ways to realign the incentives by shifting fixed amounts of money between the parties, as a lump payment does not affect the above decisions, since it is a constant amount and does not affect the calculation of the derivatives that are involved in the optimality conditions. An alternative is to redistribute risks by introducing buyback contracts;[10] if risk is shared between the two stakeholders, the optimal service level for the distributor is increased. The best strategy depends on the specific problem setting, as well as on the relative strengths of the parties involved.

In closing this section, we should note that, although this supply chain problem involves uncertainty in demand, this is not the key point. Coordination issues arise in purely deterministic problems, like the ones we show in Sections 14.4 and 14.5.

14.4 GAME THEORY

In the previous section we have considered a case in which two stakeholders, a producer and a distributor, make their decisions in a specific order. The producer (leader) determines product quality, as well as the probability distribution of demand as a consequence; the distributor (follower) chooses the order quantity. In other cases, however, decisions are taken simultaneously, at least in principle. In fact, the term "simultaneously" need not be taken literally; the point is that no player has any information about what others have previously chosen and cannot use this as an input for her decision. Predicting the outcome of the joint decisions is no easy task in general, and it is the aim of game theory. There is no hope to treat this challenging subject adequately in a few pages, but for our purposes it is quite enough to grasp a few fundamental concepts; these will be illustrated by very stylized examples in this section; in the next one, we consider a paradoxical result in traffic networks, which is a result of noncooperation between decision makers, namely, drivers in that case.

For the sake of simplicity, we consider a very stylized setting:

- There are only two decision makers (players); each player has an objective (payoff) that she wants to maximize and there is no form of cooperation.

- Only one decision has to be made; hence, we do not consider sequential games in which multiple decisions are made over time.

---

[10] See Problem 7.9.

- We assume complete information and common knowledge.[11] Formalizing these concepts precisely is not that trivial, but (very) loosely speaking they mean that there is no uncertainty about the data of the problem nor about the mechanisms that map decisions into payoffs. The two players agree on their view of the world, the rules of the game, and know the incentives of the other party; furthermore, each player knows that the other one has all of the relevant information.

There are different ways to represent a game. The best way to understand the basic concepts is by considering the situation in which players must choose within a very small and discrete set of available actions, and the game is represented in normal form.

14.4.1 Games in normal form

The standard example to illustrate the normal form representation of a simple game is the prisoner's dilemma, which is arguably the prototypical example of a two-player game. The prisoner's dilemma has been phrased in many different ways;[12] in the next example we use what is closest to a business management setting.

Example 14.4 (Prisoner's dilemma) Consider two firms, say, A and B, which have to set the prices at which their products are sold. The products are equivalent, so price is a major determinant of sales. On the one hand, firm A would like to keep its price high, to keep revenues high as well; on the other one, if the competitor firm B lowers its price, it will erode the market share of firm A. Indeed, sometimes a price war erupts, reducing profits for both firms. To represent the problem, let us assume that there are only two possible prices, low and high. Hence, we consider only a discrete set of possible *actions*; it is also possible to formalize a game with a continuum of actions represented by real numbers. The following outcomes could result, depending on firms' actions:

- If firm A sets a high price and firm B sets a low price, firm A will be wiped off the market and firm B will get a huge reward; we obtain a symmetric outcome if we swap firms' actions.

- If both firms set a low price, the result of this price war will be a fairly low profit for both firms.

---

[11] Since we do not consider multistage games, we do not deal with *perfect* information. Perfect information means that each player knows which moves have been played by the other players in the previous stages of the game. Complete information refers to the structure of the game and the payoffs of the other players.

[12] In the literal statement of the dilemma, two prisoners are kept in separate cells and are invited to confess their crime, in exchange for a reduction in punishment; if both refuse to confess, they will be released, but if only one prisoner confesses, the other one will be severely punished.

Table 14.1   Representation of prisoner's dilemma in normal form.

	Firm B	
Firm A	Low	High
Low	(1, 1)	(3, 0)
High	(0, 3)	(2, 2)

- If firms collude and both of them set a high price, the result will be a fairly high profit for both of them.

Depending on the action selected by all of the players, they will receive a *payoff*. Unlike the optimization models that we have described in Chapters 12 and 13, the payoff for each player is a function of the decisions of *all* of the players. Representing the game in normal form requires to specify the payoff to each player, for any combination of actions. The normal form of prisoner's dilemma is illustrated in Table 14.1. Firm A is the *row player* and firm B is the *column player*. Each cell in the table shows the payoff to firms A and B, respectively; the first number is the payoff for the row player, and the second number is the payoff to the column player:

- If both firms play *high*, the payoff is 2 for both of them.

- If firm A plays *high*, but firm B plays *low*, the payoff is 0 for the former and 3 for the latter; these payoffs are swapped if firm A plays *low*, but firm B plays *high*.

- If both firms play *low*, the ensuing price war results in a payoff 1 for each firm.

Actions are selected simultaneously, and we need a sensible way of predicting the result of the game.   ⫿

It is easy to see that the normal form of a game is appropriate for a small and nonsequential game, and different representations might be more suitable in other cases. In particular, sequential games involve the selection of multiple actions by each player; each action, in general, may depend on the previous choices by the other players. When there is a discrete set of actions available at each stage of the game, this can be represented in *extensive form* by a *tree*, much like the decision trees of Section 13.1. Indeed, decision trees may be regarded as a multistage game between the decision maker and nature, which randomly selects an outcome at each chance node. Solving a decision tree requires the specification of a *strategy*, i.e., a selection of a choice for each decision node. By a similar token, a multistage game requires the specification of a strategy for each player, i.e., a mapping from each state/node in the tree to the set of available actions of the player that must make a choice at that stage of the game. We will not consider the extensive form of a game in this book. Furthermore, we consider only *pure strategies*, whereby one action is selected by a player. In *mixed strategies*, each action is selected with a certain probability. Even though we neglect these more advanced concepts, we see that actions are only the building blocks of strategies. Therefore, in the following text we will use the latter term, even though in our very simple examples actions and strategies coincide.

Now we are able to formalize a simple game involving n players; to specify such a game we need:

- The set of *available strategies* S_i for each player, $i = 1, \ldots, n$; in other words, we need the set of available strategies for each player.

- The set of *payoff functions* $\pi_j(s_1, s_2, \ldots, s_n)$ for each player $j = 1, \ldots, n$; the payoff depends on the set of strategies $s_i \in S_i$ selected by all of the players, within the respective feasible set S_i.

Note that, given a set of strategies, the payoff is known, as there is no uncertainty involved; furthermore, each player knows the set of available strategies of the other players, as well as their payoff functions. Hence, all of the players have a clear picture of the incentives of the other players and there is no hidden agenda. Clearly, this is only the simplest kind of game one can consider; partial information and uncertainty are involved in more realistic models. Now the problem is to figure out which outcome should be expected as a result of the game, where by *outcome* we mean a vector of strategies $(\tilde{s}_1, \tilde{s}_2, \ldots, \tilde{s}_n)$, one per player.

14.4.2 Equilibrium in dominant strategies

Sometimes, it is fairly easy to argue which outcome is to be expected. If we consider the strategies for firm A in Table 14.1, we see that:

- If firm B plays *high*, firm A is better off by playing *low*, since the payoff 3 is larger than 2.

- If firm B plays *low*, firm A is better off by playing *low*, since the payoff 1 is larger than 0.

So, whatever firm B plays, firm A is better off by playing *low*. The symmetry of the game implies that the same consideration applies to firm B, which will also play *low*. If there is a single strategy that is preferred by a player, whatever the other players do, it is fairly easy to predict her move. A formalization of this observation leads to the concept of dominant strategy.[13]

---

[13]In our simple context of nonsequential games, we could speak of dominant *actions*.

DEFINITION 14.1 (Strictly dominant strategy) *A strategy s_i^a strictly dominates strategy s_i^b for player i if*

$$\pi_i(s_1, s_2, \ldots, s_i^a, \ldots, s_n) > \pi_i(s_1, s_2, \ldots, s_i^b, \ldots, s_n)$$

for any possible combination of strategies selected by the other players. There is a strictly dominant strategy s_i^* for player i if it strictly dominates all of the alternatives:

$$\pi_i(s_1, s_2, \ldots, s_i^*, \ldots, s_n) > \pi_i(s_1, s_2, \ldots, s_i, \ldots, s_n)$$

for all alternative strategies $s_i \neq s_i^*$ of player i and for all possible strategies s_j of players $j \neq i$.

In the above definition we only consider *strict* domination, which involves strict inequalities. For the sake of brevity, in the following we will often just speak of dominant or dominated strategies, leaving the "strict" qualifier aside; actually, some results in game theory do require strict dominance, but we will not be too precise. Clearly, dominant strategies need not exist, but if one exist for a player, it is an easy matter to predict her behavior. It is also important to notice that, in such a case, the prediction assumes only player's rationality, and there is no overly stringent requirement concerning what she knows or assumes to know about the other players. Furthermore, if there is a dominant strategy for each player, it is also easy to predict the overall outcome of the game.

DEFINITION 14.2 (Equilibrium in dominant strategies) *An outcome $(\tilde{s}_1, \tilde{s}_2, \ldots, \tilde{s}_n)$ is an equilibrium in dominant strategies if $\tilde{s}_i \in S_i$ is a dominant strategy for each player i, $i = 1, \ldots, n$.*

If an outcome is an equilibrium in dominant strategies, given that no rational player will play a dominated strategy, it is sensible to predict that this outcome will be the result of the game. In the prisoner's dilemma, the strategy *low* is dominant for both players, and we may argue that the outcome will indeed be (*low, low*). An important observation is that the resulting payoff for the firms is $(1, 1)$; if the two firms selected (*high, high*), the payoff would be $(2, 2)$, which is higher for both players. Using the terminology of multiobjective optimization,[14] the second outcome would be preferred by both players and is Pareto-dominant. However, the lack of coordination between players results in a lower payoff to both of them. The problem with the outcome (*low, low*) is that it is an *unstable* equilibrium. Incidentally, we observe that certain firms are often prone to collude on prices, and this is exactly why anti-trust authorities have been set up all around the globe; hence, one could wonder if the above prediction makes empirical sense. The key issue here is

---

[14]See Section 12.3.3.

Table 14.2 Battle of the sexes.

| Juliet | Romeo | |
	Horror	Shopping
Horror	(1, 3)	(0, 0)
Shopping	(0, 0)	(3, 1)

that we are considering only a one-shot game. Collusion may be the outcome of more complicated, as well as more realistic, multistage models that capture strategic interaction between firms.

It is reasonable to expect that an equilibrium in dominant strategies may only be found in trivial games. The following is a classical example in which no equilibrium in dominant strategies exists.

Example 14.5 (Battle of the sexes) After a long week of hard work, Romeo and Juliet have to decide how to spend their Saturday afternoon. There are two choices available:

1. Attending a horror movie festival, which we denote as strategy *horror* and is much preferred by Romeo.

2. Going on a shoe shopping spree, which we denote as strategy *shopping* and is much preferred by Juliet.

Despite their differences in taste, Romeo and Juliet are pretty romantic lovers, so they would prefer to spend their time together anyway. Their preferences may be represented by the payoffs in Table 14.2. We note that if the two players select different strategies, the payoff is zero to both of them, as they will be alone. If the outcome is (*shopping, shopping*), the resulting payoff will be 3 to Juliet, who would be very happy, and 1 to Romeo, who at least will spend his time and more with his sweetheart. Payoffs are reversed for outcome (*horror, horror*). The game is simple enough, and in practice it can be seen as a stylized model of two firms that should agree on a standard to make their respective products compatible; these games are called *coordination games*. However, it is easy to see that there is no dominant strategy for either player. For instance, Romeo would play *horror*, if he knew that Juliet is about to play *horror*; however, if he knew that Juliet is going to play *shopping*, he would change his mind.   ▢

Equilibrium in dominant strategies is based on a very restrictive requirement, but we may try to guess the outcome in a slightly more elaborated way, by assuming that all players are rational and that everything is common knowledge. The idea is *iterated elimination of dominated strategies* and is best illustrated by an example.

Table 14.3 Predicting an outcome by iterated elimination of dominated strategies.

(a)

Player 1	Player 2		
	Left	Center	Right
Top	$(1, 0)$	$(1, 2)$	$(0, 1)$
Bottom	$(0, 3)$	$(0, 1)$	$(2, 0)$

(b)

Player 1	Player 2	
	Left	Center
Top	$(1, 0)$	$(1, 2)$
Bottom	$(0, 3)$	$(0, 1)$

(c)

Player 1	Player 2	
	Left	Center
Top	$(1, 0)$	$(1, 2)$

Example 14.6 Consider the game in Table 14.3. Table 14.3(a) describes the game in normal form; note that the row player 1 has two possible strategies, *top* and *bottom*, whereas the column player 2 has three available strategies, *left*, *center*, and *right*. Let us compare strategies *center* and *right* from the viewpoint of player 2. Checking her payoffs, we see that, whatever the choice of player 1, it is better for player 2 to choose *center* rather than *right*, as $2 > 1$ and $1 > 0$; indeed, *center* dominates *right*. Eliminating the dominated strategy from further consideration, we obtain Table 14.3(b). By a similar token, from the viewpoint of player 1, we see that strategy *bottom* is dominated by *top*, since $1 > 0$. Eliminating the dominated strategy, we get Table 14.3(c). Now, we see that the payoff to player 1 is 1 in any case, whereas player 2 will definitely play *center*, since $2 > 0$. Then, we predict the outcome (*top*, *center*). □

The procedure above sounds quite reasonable. However, a few considerations are in order.

- We should wonder whether the resulting outcome, if any, depends on the sequence that we follow in eliminating dominated strategies. However, it

Table 14.4   Can you trust your opponent's rationality?

| Player 1 | Player 2 | |
	Left	Right
Top	(1, 0)	(1, 1)
Bottom	(-1000, 0)	(2, 1)

can be shown that *strict* dominance makes sure that whatever sequence we take, we will find the same result.

- We may fail to find a prediction, when strict dominance does not apply.

- We have to assume that all players are rational and that everything is common knowledge. This, for instance requires that each of the two players knows that the other one is rational, that she knows that the other one knows that she is rational, and so on. As we pointed out, formalizing the intuitive idea of common knowledge is not quite trivial, but the following example shows that this assumption may be critical.

Example 14.7 Trusting the other player's rationality can be dangerous indeed. Consider the game in Table 14.4. If player 2 is rational, she should choose the strictly dominant strategy *right*, because it ensures a payoff $1 > 0$, whatever the choice of player 1 is. If we eliminate *left*, it is easy to see that the outcome should be (*bottom*, *right*), since by playing *bottom* player 1 gets a payoff $2 > 1$. However, playing *bottom* is quite risky for player 1 since, if player 2 makes the wrong choice, his payoff will be -1000. ▯

14.4.3   Nash equilibrium

The concepts that we have used so far make sense, but they are a bit too restrictive and limit the set of games for which we may make reasonable predictions. A better approach, in a sense that we should clarify, is Nash equilibrium. Before formalizing the concept, imagine a game in which there is one sensible prediction of the outcome. If the prediction makes sense, all players should find it acceptable, in the sense that they would not be willing to deviate from the prescribed strategy, if no other player deviates as well: The equilibrium should be *stable*.

DEFINITION 14.3 (Nash equilibrium) *An outcome* $(s_1^*, s_2^*, \ldots, s_n^*)$ *is a Nash equilibrium if no player would gain anything by choosing another strategy, provided that the other players do not change their strategy; in other words, no player i has an incentive to deviate from strategy* s_i^*. *Let us denote*

Table 14.5 Finding Nash equilibrium for the game in Table 14.3.

| | Player 2 | | |
Player 1	Left	Center	Right
Top	($\underline{1}$, 0)	($\underline{1}$, $\underline{2}$)	(0, 1)
Bottom	(0, $\underline{3}$)	(0, 1)	($\underline{2}$, 0)

the set of strategies played by all players but i as

$$s^*_{-i} \equiv \left(s^*_1, \ldots, s^*_{i-1}, s^*_{i+1}, \ldots, s^*_n\right)$$

This can be used to streamline notation as follows:

$$\left(s^*_i, s^*_{-i}\right) = \left(s^*_1, s^*_2, \ldots, s^*_i, \ldots, s^*_n\right)$$

Then, for any player i, s^*_i is the best response to s^*_{-i}:

$$\pi_i\left(s^*_i, s^*_{-i}\right) \geq \pi_i\left(s_i, s^*_{-i}\right), \qquad i = 1, \ldots, n, \ \forall s_i \in S_i \tag{14.8}$$

The best way to grasp Nash equilibria is by a simple example, in which we also illustrate a possible way to find them for two-players games with discrete sets of actions.

Example 14.8 To find a Nash equilibrium, we can:

1. Consider each player in turn, and find her best strategy for each possible strategy of the opponent; for the row player, this means considering each column (strategy of the column player) and find the best row (best response of row player) by marking the corresponding payoff; rows and columns are swapped for the column player.

2. Check if there is any cell in which both payoffs are marked; all such cells are Nash equilibria.

Let us apply the idea to the game in Table 14.3; the results are shown in Table 14.5. If we start with row player 1, we see that her best response to *left* is *top*, as $1 > 0$; then we underline the payoff 1 in the top-left cell; by the same token, player 1 should respond with *top* to *center*, and with *bottom* to *right*. Then we proceed with column player 2; we notice that her best response to *top* is *center*, since $2 = \max\{0, 2, 1\}$; if player 1 plays *bottom*, player 2 should choose *left*. In this game, there is no need to break ties, as for each row there is exactly one payoff preferred by the column player, and for each column there is exactly one payoff preferred by the row player. The outcome (*top, center*) is the only cell with both payoffs marked, and indeed it is a Nash equilibrium. To better grasp what Nash equilibrium is about, note that no player has an incentive to deviate from this outcome, if the other player does not.   ⬜

We see that the Nash equilibrium in the example is the same outcome that we predicted by iterated elimination of (strongly) dominated strategies. Indeed, it can be proved that:

- If iterated elimination of (strictly) dominated strategies results in a unique outcome, then this is a Nash equilibrium.

- If a Nash equilibrium exists, it is not eliminated by the iterated elimination of (strictly) dominated strategies.

We should note that Nash equilibria may not exist,[15] and they need not be unique (see problem 14.1). However, they are a more powerful tool than elimination of dominated strategies, as they may provide us with an answer when the use of dominated strategies fail, without contradicting the prediction of iterated elimination, when this works. In other words, since Nash equilibria are based on less restrictive assumptions than iterated elimination, it fails less often.[16] Last but not least, the prediction suggested by Nash equilibria makes sense and sounds plausible enough.

Now it is natural to wonder about more complicated settings, like the case in which each player has a continuum of strategies at her disposal. To this aim, it is useful to interpret Nash equilibria in terms of best response functions.

DEFINITION 14.4 (Best response function) *The best response function for each player i is a function*

$$s_i^* = R_i(s_{-i})$$

mapping the strategies of the other players into the strategy s_i^* maximizing the payoff of i in response to s_{-i}.

We immediately see that a Nash equilibrium is a solution of a system of equations defined by the best response functions of all players. For a two-player game, we should essentially solve the following system of two equations

$$\begin{cases} s_1^* = R_1(s_2^*) \\ s_2^* = R_2(s_1^*) \end{cases}$$

---

[15] To be precise, for the simple games that we consider here, featuring finite sets of possible actions, they do exist if we allow for mixed strategies, but this is outside the scope of the book.

[16] Actually, they never fail in relatively simple games, if we allow for mixed strategies.

14.4.4   Simultaneous vs. sequential games

In this section we consider Nash equilibrium for the case in which a continuum of infinite actions is available to each player. To be specific, we analyze the behavior of two firms competing with each other in terms of quantities produced. Both firms would like to maximize their profit, but they influence each other since their choices of quantities have an impact on the price at which the product is sold on the market. This price is common to both firms, as we assume that they produce a perfectly identical product and there is no possibility of differentiating prices. This kind of competition is called *Cournot competition*; the case in which firms compete on prices is called *Bertrand competition*, but it will not be analyzed here. We start with the case of simultaneous moves, which leads to the *Cournot–Nash equilibrium*.[17] Then we will consider sequential moves.

To clarify these concepts it is useful to tackle a simple model, in which we assume that each firm has a cost structure involving only a variable cost:

$$\mathrm{TC}_i(q_i) = c_i q_i, \qquad i = 1, 2$$

Here, TC_i denotes total cost for firm i, c_i is the variable cost, and q_i is the amount produced by firm $i = 1, 2$. The total amount available on the market is $Q = q_1 + q_2$, and it is going to influence price. In Section 2.11.5 we introduced the concepts of demand and inverse demand functions, relating price and demand. There, we assumed the simplest relationship, i.e., a linear one. By a similar token, we assume here that there is a linear relationship between total quantity produced and price:

$$P(Q) = a - bQ, \qquad a, b > 0; \; a \geq c_i$$

Incidentally, this stylized model assumes implicitly that all produced items are sold on the market. Then, the profit for firm i is

$$\pi_i(q_1, q_2) = P(q_1 + q_2)q_i - \mathrm{TC}_i(q_i) = [a - b(q_1 + q_2)]q_i - c_i q_i$$

Note that the profit of each firm is influenced by the decisions of the competitor. Assuming that the two firms make their decisions simultaneously, it is natural to wonder what the Nash equilibrium will be. Note that we assume complete information and common knowledge, in the sense that each player has all of the above information and knows that the other player has such information. We can find the equilibrium by finding the best response function $R_i(q_j)$ for each firm, i.e., a function giving the profit-maximizing quantity q_i^* for firm i, for each possible value of q_j set by firm j. Enforcing the stationarity condition for the profit of firm 1, we find

$$\frac{\partial \pi_1(q_1, q_2)}{\partial q_1} = a - 2bq_1 - bq_2 - c_1 = 0 \quad \Rightarrow \quad R_1(q_2) = \frac{a - c_1}{2b} - \frac{1}{2}q_2 \quad (14.9)$$

---

[17]It is interesting to note that the Cournot duopoly model anticipated Nash equilibrium by more than one century, although in a limited setting.

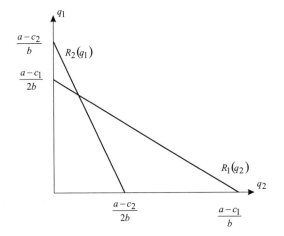

Fig. 14.3   Finding a Nash equilibrium in Cournot competition.

By the same token, for firm 2 we have

$$R_2(q_1) = \frac{a - c_2}{2b} - \frac{1}{2}q_1 \qquad (14.10)$$

To solve the problem, we should find where the two response functions intersect; in other words, we should solve the system of equations

$$\begin{cases} q_1^c = R_1(q_2^c) \\ q_2^c = R_2(q_1^c) \end{cases}$$

where we use the superscript "c" to denote Cournot equilibrium. In our case, we are lucky, since response functions are linear. In particular, both response functions are downward-sloping lines, as illustrated in Fig. 14.3. Hence, to find the Nash equilibrium we simply solve the system of linear equations

$$\begin{cases} q_1^c = \dfrac{a - c_1}{2b} - \dfrac{1}{2}q_2^c \\ q_2^c = \dfrac{a - c_2}{2b} - \dfrac{1}{2}q_1^c \end{cases}$$

which yields

$$q_1^c = \frac{a - 2c_1 + c_2}{3b}, \quad q_2^c = \frac{a - 2c_2 + c_1}{3b} \qquad (14.11)$$

The resulting equilibrium price turns out to be

$$p^c = \frac{a + c_1 + c_2}{3} \qquad (14.12)$$

and the profit of each firm is

$$
\begin{aligned}
\pi_i^c &= (p^c - c_i)q_i = \left(\frac{a + c_i + c_j}{3} - c_i\right)\left(\frac{a - 2c_i + c_j}{3b}\right) \\
&= \frac{(a - 2c_i + c_j)^2}{9b} = b(q_i^c)^2
\end{aligned}
\tag{14.13}
$$

It is interesting to note that if a firm manages to reduce its cost, it will increase its produced quantity and profit as well. We leave this check as an exercise. It is also worth noting that if the firms have the same production technology (i.e., $c_1 = c_2$), then we have a symmetric solution $q_1^c = q_2^c$, as expected.

So far, we have assumed that the two competing firms play simultaneously. In the supply chain management problem of Section 14.3, since there are two different types of actions, it is more natural to assume that one of the two players moves first. Hence, we may also wonder what happens in the quantity game of this section if we assume that firm 1, the leader, sets its quantity q_1 before firm 2, the follower. Unlike the simultaneous game, firm 2 knows the decision of firm 1 before making its decision; thus, firm 2 has perfect information. The analysis of the resulting sequential game leads to *von Stackelberg equilibrium*. Firm 1 makes its decision knowing the best response function for firm 2, as given in Eq. (14.10). Hence, the leader's problem is

$$
\max_{q_1} \pi_1^s = P\left(q_1 + R_2(q_1)\right)q_1 - c_1 q_1 = \left[a - b\left(q_1 + \frac{a - c_2}{2b} - \frac{q_1}{2}\right)\right]q_1 - c_1 q_1
$$

where the superscript "s" refers to von Stackelberg competition. Applying the stationarity condition yields

$$
q_1^s = \frac{a - 2c_1 + c_2}{2b} = \frac{3}{2}q_1^c
\tag{14.14}
$$

We see that firm 1 produces more in this sequential game than in the Cournot game. If we plug this value into the best response function $R_2(q_1)$, we obtain

$$
\begin{aligned}
q_2^s &= \frac{a - c_2}{2b} - \frac{a - 2c_2 + c_1}{4b} \\
&= \frac{a - 3c_2 + 2c_1}{4b} = \frac{a - 2c_2 + c_1 + (c_1 - c_2)}{4b} \\
&= \frac{3}{4}q_2^c + \frac{c_1 - c_2}{4b}.
\end{aligned}
\tag{14.15}
$$

We see that the output of firm 2 is a fraction of that of the Cournot game, plus a term that is positive if firm 1 is less efficient than firm 2. Now it would be interesting to compare the profits for the two firms under this kind of game. This is easy to do when marginal production costs are the same; we illustrate the idea with a toy numerical example.

Example 14.9 Two firms have the same marginal production cost, $c_1 = c_2 = 5$, and the market is characterized by the price/quantity function

$$
P(Q) = 120 - Q
$$

In this example we compare three cases:

1. The two firms collude and work together as a cartel. We may also consider the two firms as two branches of a monopolist firm. Note that if the two marginal costs were different, one of the two branches would be just shut down (assuming infinite production capacity, as we did so far).

2. The firms do not cooperate and move simultaneously (Cournot game).

3. The firms do not cooperate and move sequentially (von Stackelberg game).

In the first case, we just need to work with the aggregate output Q. The monopolist solves the problem

$$\max \pi^m = (120 - Q)Q - 5Q$$

where superscript "m" indicates that we are referring to the monopolist case. We solve the problem by applying the stationarity condition

$$120 - 2Q - 5 = 0 \qquad \Rightarrow \qquad Q^m = 57.50$$

which yields the following market price and profit:

$$p^m = 120 - 57.5 = 62.50, \qquad \pi^m_{1+2} = (62.50 - 5) \times 57.50 = 3306.25$$

In the second case, the solution given by (14.11) is symmetric:

$$q^c_1 = q^c_2 = \frac{120 - 10 + 5}{3} = 38.33$$

The overall output and price are

$$Q^c = 2 \times 38.33 = 76.77, \qquad p^c = 120 - 76.77 = 43.33$$

respectively. The profit for each firm is

$$\pi^c_1 = \pi^c_2 = (q^c_1)^2 = 1469.19$$

Note that the total overall profit is

$$\pi^c_{1+2} = 2 \times 1469.19 = 2938.89 < 3306.25 = \pi^m_{1+2}$$

In fact, the monopolist would restrict output to increase price, resulting in a larger overall profit than with the Cournot competition. So, collusion results in a larger profit than competition, which is no surprise.

Let us consider now the von Stackelberg sequential game. Using (14.14) and (14.15), we see that

$$q^s_1 = \frac{120 - 10 + 5}{2} = 57.5, \qquad q^s_1 = \frac{120 - 10 + 5}{4} = 28.75$$

Table 14.6   Battle of the sexes, alternative version.

Morticia	Romeo	
	Cinema	Restaurant
Cinema	(5, -100)	(0, 1)
Restaurant	(0, 1)	(5, -100)

from which we see that, with respect to the simultaneous game, the output of firm 1 is increased whereas the output of firm 2 is decreased. The total output and price are

$$Q^s = 57.5 + 28.75 = 86.25, \qquad p^s = 120 - 86.25 = 33.75$$

respectively. The price is lower than in both previous cases, and the distribution of profit is now quite asymmetric:

$$\pi_1^s = (33.75 - 5) \times 57.5 = 1653.13$$
$$\pi_2^s = (33.75 - 5) \times 28.75 = 826.56$$
$$\pi_{1+2}^s = 1653.13 + 826.56 = 2479.69$$

The overall profit for the sequential game is lower than for the simultaneous one; however, the leader has a definite advantage and its profit is larger in the sequential game. □

The toy example above shows that the privilege of moving first may yield an advantage to the leader. Given the structure of the game, it is easy to see that the leader of the sequential game cannot do worse than in the simultaneous game; in fact, she could produce the same amount as in the Cournot game, anyway. However, this need not apply in general. In particular, when there are asymmetries in information or things are random, the choice of the leader, or its outcome when there is uncertainty, could provide the follower with useful information. The following example shows that being the first to move is not always desirable.

Example 14.10 (Battle of the sexes, alternative versions) Let us consider again the battle of the sexes of Example 14.5, where now we assume that Juliet has the privilege of moving first. Given the payoffs in Table 14.2, she knows that, whatever her choice, Romeo will play the move that allows him to enjoy her company. Hence, she will play *shopping* for sure. In this case, Juliet does not face the uncertainty due to the presence of two Nash equilibria in Table 14.2 and is certainly happy to move first. The situation is quite different for the payoffs in Table 14.6. In this case, Romeo is indifferent between going to cinema or restaurant. What he really dreads is an evening

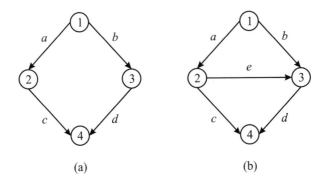

Fig. 14.4 An example of the Braess' paradox.

with Morticia. It is easy to see that this game has no Nash equilibrium, as one of the two players has always an incentive to deviate. An equilibrium can be found if we admit mixed strategies, in which players select a strategy according to a probability distribution, related to the uncertainty about the move of the competitor. We do not consider mixed strategies here, but the important point in this case is that no player would like to move first. □

We noted that the first version of the battle of the sexes is a stylized coordination game for two firms that should adopt a common standard; in this second version, one firm wants to adopt the same standard as the competitor, whereas the other firm would like to select a different one.

14.5 BRAESS' PARADOX FOR TRAFFIC NETWORKS

The result of the collective interaction of noncooperative players may be occasionally quite surprising. We illustrate here a little example of the Braess' paradox for traffic networks.[18] Imagine a traffic network consisting of links such as road segments, bridges, and whatnot. Most of us had some pretty bad experiences with traffic jams. Intuition would suggest that adding a link to the network should improve the situation or, to the very least, should not make things worse.

Consider the network in Fig. 14.4.[19] In Fig. 14.4(a) we have a four-node

---

[18]For a full treatment of the Braess' paradox see Ref. [3]. The numerical example in this section has been taken from a presentation by A. Nagurney, which can be downloaded from http://supernet.som.umass.edu/ (CMS 2010 plenary presentation).
[19]See Section 12.2.4 for an introduction to network flow optimization.

directed network, where the following costs are associated with each arc:

$$C_a(f_a) = 10f_a$$
$$C_b(f_b) = f_b + 50$$
$$C_c(f_c) = f_c + 50$$
$$C_d(f_d) = 10f_d$$

The meaning of these costs should be clarified: They are the traveling costs for an individual driving along each arc, where the flows f_a, f_b, f_c, and f_d are the *total* number of drivers using the corresponding arc. Keep in mind that in game theory the payoff to each player depends on the decisions of the other players as well, and players do not cooperate. In fact, the paradox applies to a problem where drivers have to choose a path independently of each other, but they interact through the level of congestion. This is *not* a classical network flow problem, where a centralized planner optimizes the overall flow minimizing total cost. Each driver selects a route in a noncooperative way, and we should look for an equilibrium. Traffic equilibrium is beyond the scope of the book, but we may take advantage of the symmetry of the example above. Imagine that the total travel demand is 6, i.e., there are six drivers that must go from node 1 to node 4. There are two possible paths, i.e., sequences of arcs: $P_1 = (a, c)$ and $P_2 = (b, d)$. Let us denote by $f_{P_1}^*$ and $f_{P_2}^*$ the traffic flows along each path at equilibrium. The cost for path P_1 is

$$C_a(f_{P_1}) + C_c(f_{P_1}) = 10f_{P_1} + (f_{P_1} + 50) = 11f_{P_1} + 50$$

But this happens to be the same as the cost for path P_2:

$$C_b(f_{P_2}) + C_d(f_{P_2}) = (f_{P_2} + 50) + 10f_{P_2} = 11f_{P_2} + 50$$

Since the problem is symmetric, intuition suggests that, at equilibrium, the total travel demand is split in two: $f_{P_1}^* = f_{P_2}^* = 3$. Then, the cost of each path for each driver is

$$C_{P_1}^* = C_{P_2}^* = 11 \times 3 + 50 = 83$$

No single driver has an incentive to deviate, because if a driver switches to the other path, her cost would rise to $11 \times 4 + 50 = 94$. Hence, this is a Nash equilibrium.

Now consider the network 14.4(b), where arc e has been added, whose cost is

$$C_e(f_e) = f_e + 10$$

Now, we have an additional path $P_3 = (a, e, d)$. Having opened a new route, some drivers have an incentive to deviate. For instance, if a driver moves from path P_1 to path P_3, her cost would be

$$(10 \times 3) + (1 + 10) + (10 \times 4) = 81$$

What new equilibrium will emerge? Let us try a symmetric solution again: $x_{P_1}^* = x_{P_2}^* = x_{P_3}^* = 2$. At this level of flow, the costs of the three paths are

$$C_{P_1}^* = (10 \times 4) + (2 + 50) = 92$$
$$C_{P_2}^* = (2 + 50) + (10 \times 4) = 92$$
$$C_{P_3}^* = (10 \times 4) + (2 + 10) + (10 \times 4) = 92$$

We observe that the three costs are the same, and no one has an incentive to deviate. However, with this equilibrium, every driver is worse off!

The peculiarity of the example above might suggest that this kind of problem is too pathological to actually arise. However, real-life cases have been reported in the literature, where closing a link in a network has improved the situation or adding a new one has created a congestion.[20] The crux of the problem, once again, is the collective behavior of several noncooperative decision makers. In business management, this is a common occurrence that should not be dismissed.

14.6 DYNAMIC FEEDBACK EFFECTS AND HERDING BEHAVIOR

Game theory in its simplest form does not consider dynamics, as it revolves around a static equilibrium concept: It posits a situation such that no player has an incentive to deviate. But how is that equilibrium reached dynamically? And what about the disorderly interaction of many stakeholders, maybe *stockholders* in financial markets? Addressing such issues is beyond the scope of this book, but we illustrate their relevance with two real-life examples; they show that the effects of such interactions may be quite nasty and that uncertainty may not always be adequately represented by an exogenous probability distribution.

Example 14.11 (Long-Term Capital Management) Many financial equilibrium models, like CAPM, assume that no player is big enough to influence markets. However, in specific conditions, players may turn out to be so big that their actions on thin or illiquid markets are significant, with perverse effects. Long-Term Capital Management (LTCM) was a successful hedge fund, which has become famous for its demise.[21] The fund took hugely leveraged positions, betting on spreads between securities, such as bonds. A simplified explanation of the strategy is the following:

- The required yield on bonds depends on many factors, including the credit standing of the issuer. If you do not trust the ability of the issuer

[20]See http://supernet.som.umass.edu/facts/braess.html
[21]Two Nobel Laureates, Robert Merton and Myron Scholes, were involved in the fund, and this has contributed to the fame of this case. In fact, it is often written that they *founded* LTCM, which is false in many respects. For a full account of the story, see Ref. [18].

to repay its debt, you require a higher interest rate to buy its bond, whose price is reduced. If the balance sheet of the issuer is rock solid, you settle for a lower yield, implying a higher price. The difference in required yields is the spread.

- If a bond issuer is in trouble, but you believe that its difficulties are overstated by the market and that it will recover, you could buy its bonds (which are cheaper than they should) and take a short position in high-quality bonds (selling them short), because sooner or later, according to your view, the two bond prices will converge.[22]

In 1998, the strategy backfired because of a default on Russian bonds; markets got nervous and everyone rushed to sell risky bonds to buy safe ones. This "flight to quality" widened the spreads, resulting in huge losses for LTCM. In such a case, if your positions are leveraged, your creditors get nervous as well, and start asking you to repay your debt; in financial parlance, you get a *margin call*. This implies that you should liquidate some of the securities in your portfolio to raise some cash. Unfortunately, this exacerbates your trouble, since prices are further depressed by your sales. In normal times, a small trade on a very liquid market should not move prices too much. But if troubled markets get illiquid and you try unloading a huge position in an asset, you get a nasty vicious circle: The more you sell, the more money you lose, the more margin calls you get, the more you should sell. In the end, a committee of bankers had to rescue and bail out the fund in order to avoid a dangerous market crash.   ☐

Example 14.12 (The Black Monday crash of 1987) Portfolio insurance is a portfolio management strategy that aims at keeping the value of a portfolio of assets from falling below a given target. The idea is to create a synthetic put option by proper dynamic trading. To cut a long story short, the idea is that when asset value falls, one should sell a fraction of the assets in well-determined proportions. On Monday, 19 October, 1987, stock markets around the world crashed. In what has been aptly named the Black Monday, the Dow Jones Industrial Average index dropped by 22.61%. An explanation of this crash was put forward, which blames portfolio insurance. The idea is rather simple; the market goes south and you start selling to implement dynamic portfolio insurance. Unfortunately, you are not alone, as many other players do the same; hence, there is a further drop in prices that in turn triggers further sales. The result is a liquidity and feedback disaster, exacerbated by the use of automated, computer-based trading systems.[23]   ☐

The above stories, and all of the similar ones, are controversial; there is no general agreement that portfolio insurance has caused the Black Monday

---

[22]The practical implementation of this strategy is not that simple, and it may require the use of financial derivatives.

[23]This explanation is very simplified. For a much better account see, e.g., Ref. [2].

crash. Whatever your opinion is, feedback effects and the partial endogenous character of uncertainty cannot be disregarded.

14.7 SUBJECTIVE PROBABILITY: THE BAYESIAN VIEW

In all the preceding chapters concerning probability and statistics we took a rather standard view. On the one hand, we have introduced events and probabilities according to an axiomatic approach. On the other hand, when dealing with inferential statistics, we have followed the orthodox approach: Parameters are unknown numbers, that we try to estimate by squeezing information out of a random sample, in the form of point estimators and confidence intervals. Since parameters are numbers, when given a specific confidence interval, we cannot say that the true parameter is contained there with a given probability; this statement would make no sense, since we are comparing only known and unknown numbers, but no random variable is involved. So, there is no such a thing as "probabilistic knowledge" about parameters, and data are the only source of information; any other knowledge, objective or subjective, is disregarded. The following example illustrates the potential difficulties induced by this view.[24]

Example 14.13 Let X be a uniformly distributed random variable, and let us assume that we do not know where the support of this distribution is located, but we know that its width is 1. Then, $X \sim U[\mu - 0.5, \mu + 0.5]$, where μ is the unknown expected value of X, as well as the midpoint of the support. To estimate μ we take a sample of $n = 2$ independent realizations X_1 and X_2 of the random variable. Now consider the order statistics

$$X_{(1)} = \min\{X_1, X_2\}, \qquad X_{(2)} = \max\{X_1, X_2\}$$

and the confidence interval

$$\mathcal{I} = [X_{(1)}, X_{(2)}] \tag{14.16}$$

What is the confidence level of $\mathcal{I}$, i.e., the probability $P\{\mu \in \mathcal{I}\}$? Both observations have a probability 0.5 of falling to the left or to the right of μ. The confidence interval will not contain μ if both fall on the same half of the support. Then, since X_1 and X_2 are independent, we have

$$\begin{aligned} P\{\mu \notin \mathcal{I}\} &= P\{X_1 < \mu, X_2 < \mu\} + P\{X_1 > \mu, X_2 > \mu\} \\ &= 0.5 \times 0.5 + 0.5 \times 0.5 = 0.5 \end{aligned}$$

So, the confidence level for $\mathcal{I}$ is the complement of this probability, i.e., 50%. Now suppose that we observe $X_1 = 0$ and $X_2 = 0.6$. What is the probability

---

[24]This example is taken from Ref. [12], page 45, which in turn refers back to Ref. [6].

that μ is included in the confidence interval $\mathcal{I}$ resulting from Eq. (14.16), i.e., $\mathrm{P}\{0 \leq \mu \leq 0.6\}$? In general, this question does not make any sense, since μ is a number. But in this specific case, we have some additional knowledge leading to the conclusion that the expected value is included in the interval $[0, 0.6]$ with probability 1. In fact, if $X_{(1)} = 0$, we may conclude $\mu \leq 0.5$; by the same token, if $X_{(2)} = 0.6$, we may conclude $\mu \geq 0.1$. Since the confidence interval $\mathcal{I}$ includes the interval $[0.1, 0.5]$, we would have good reasons to claim that $\mathrm{P}\{0 \leq \mu \leq 0.6\} = 1$. But again, this makes no sense in the orthodox framework. By a similar token, if we get $X_1 = 0$ and $X_2 = 0.001$, we would be tempted to say that such a small interval is quite unlikely to include μ, but there is no way in which we can express this properly, within the framework of orthodox statistics.   □

On one hand, the example illustrates the need to make our background knowledge explicit. In the Bayesian framework, it can be argued that unconditional probabilities do not exist, in the sense that probabilities are always conditional on background knowledge and assumptions. On the other hand, we see the need of a way to express subjective views, which may be revised after collecting empirical data. Bayesian estimation has been proposed to cope with such issues.

14.7.1   Bayesian estimation

Consider the problem of estimating a parameter θ, characterizing the probability distribution of a random avaliable X. We have some prior information about θ, that we would like to express in a sensible way. We might assume that the unknown parameter lies anywhere in the unit interval $[0, 1]$, or we might assume that it is close to some number μ, but we are somewhat uncertain about it. Such a knowledge or subjective view may be expressed by a probability density $p(\theta)$, which is called the *prior* distribution of θ. In the first case, we might associate a uniform distribution with θ; in the second case the prior could be a normal distribution with expected value μ and variance σ^2. Note that this is the variance that we associate with the parameter, which is a random variable rather than a number, and not the variance of the random variable X itself.

In Bayesian estimation, the prior is merged with experimental evidence by using Bayes' theorem. Experimental evidence consists of independent observations $X_1, \ldots, X_n$ from the unknown distribution. Here and in the following we mostly assume that random variable X is continuous, and we speak of densities; the case of discrete random variables and probability mass functions is similar. We also assume that the values of the parameter θ are not restricted to a discrete set, so that the prior is a density as well. Hence, let us denote the density of X by $f(x \mid \theta)$, to emphasize its dependence on parameter θ. Since a random sample consists of independent random variables, their joint distribution, *conditional* on θ, is

$$f_n(x_1, \ldots, x_n \,|\, \theta) = f(x_1 \,|\, \theta) \cdot f(x_2 \,|\, \theta) \cdots f(x_n \,|\, \theta)$$

The conditional density $f_n(x_1, \ldots, x_n \,|\, \theta)$ is also called the *likelihood function*, as it is related to the likelihood of observing the data values $x_1, \ldots, x_n$, given the value of the parameter θ; also notice the similarity with the likelihood function in maximum likelihood estimation.[25] Note that what really matters here is that the observed random variables $X_1, \ldots, X_n$ are independent *conditionally* on θ. Since we are speaking about $n+1$ random variables, we could also consider the joint density

$$g(x_1, \ldots, x_n, \theta)$$

but this will not be really necessary for what follows. Given the joint conditional distribution $f_n(x_1, \ldots, x_n \,|\, \theta)$ and the prior $p(\theta)$, we can find the marginal density of $X_1, \ldots, X_n$ by applying the total probability theorem:[26]

$$g_n(x_1, \ldots, x_n) = \int_\Omega f_n(x_1, \ldots, x_n \,|\, \theta) p(\theta) d\theta$$

where we integrate over the domain Ω on which θ is defined, i.e., the support of the prior distribution. Now what we need is to invert conditioning, i.e., we would like the distribution of θ conditional on the observed values $X_i = x_i$, $i = 1, \ldots, n$, i.e.

$$p_n(\theta \,|\, x_1, \ldots, x_n)$$

This *posterior* density should merge the prior and the density of observed data conditional on the parameter. This is obtained by applying Bayes' theorem to densities, which yields

$$p_n(\theta \,|\, x_1, \ldots, x_n) = \frac{g(x_1, \ldots, x_n, \theta)}{g_n(x_1, \ldots, x_n)} = \frac{f_n(x_1, \ldots, x_n \,|\, \theta) p(\theta)}{g_n(x_1, \ldots, x_n)} \qquad (14.17)$$

Note that the posterior density involves a term $g_n(x_1, \ldots, x_n)$ which does not really depend on θ. Its role is to normalize the posterior distribution, so that its integral is 1. Sometimes, it might be convenient to rewrite Eq. (14.17) as

$$p_n(\theta \,|\, x_1, \ldots, x_n) \propto f_n(x_1, \ldots, x_n \,|\, \theta) p(\theta) \qquad (14.18)$$

where the symbol $\propto$ means "proportional to." In plain English, Eq. (14.17) states that the posterior is *proportional* to the product of the likelihood function $f_n(x_1, \ldots, x_n \,|\, \theta)$ and the prior distribution $p(\theta)$:

$$\text{posterior} \propto \text{prior} \times \text{likelihood}$$

---

[25] See Section 9.9.3.
[26] Here we strain a bit some concepts related to conditional densities, which we just outlined in Chapter 8, but the intuition should be clear.

What we are saying is that, given some prior knowledge about the parameter and the distribution of observations conditional on the parameter, we obtain an updated distribution of the parameter, conditional on the actually observed data.

Example 14.14 (Bayesian learning and coin flipping) We tend to take for granted that coins are fair, and that the probability of getting head is $1/2$. Let us consider flipping a possibly unfair coin, with an unknown probability θ of getting head. In order to learn this unknown value, we flip the coin repeatedly, i.e., we run a sequence of independent Bernoulli trials with unknown parameter θ.[27] If we do not know anything about the coin, we might just assume a uniform prior

$$p(\theta) = 1, \qquad 0 \le \theta \le 1$$

If we flip the coin N times, we know that the probability of getting H heads is related to the binomial probability distribution

$$f_N(H \,|\, \theta) \propto \theta^H (1 - \theta)^{N-H} \tag{14.19}$$

This is our likelihood function. If we regard this expression as the probability of observing H heads, given θ, this should actually be the probability mass function of a binomial variable with parameters θ and N, but we are disregarding the binomial coefficient [see Eq. (6.16)], which does not depend on θ and just normalizes the distribution. If we multiply this likelihood function by the prior, which is just 1, we obtain the posterior density for θ, given the number of observed heads:

$$p_N(\theta \,|\, H) \propto \theta^H (1 - \theta)^{N-H}, \qquad 0 \le \theta \le 1 \tag{14.20}$$

Equations (14.19) and (14.20) look like the same thing, because we use a uniform prior, but they very different in nature. Equation (14.20) gives the posterior density of θ, conditional on the fact that we observed H heads and $N - H$ tails. If we look at it this way, we recognize the shape of a beta distribution, which is a density, rather than a mass function. To normalize the posterior, we should multiply it by the appropriate value of the beta function.[28] Again, this normalization factor does not depend on θ and can be disregarded.

In Fig. 14.5 we display posterior densities, normalized in such a way that their maximum is 1, after flipping the coin N times and having observed H heads. The plot in Fig. 14.5(a) is just the uniform prior. Now imagine that the first flip lands head. After observing the first head, we know for sure that

---

[27] This example is based on Chapter 2 of the text by Sivia and Skilling [23].
[28] See Section 7.6.2.

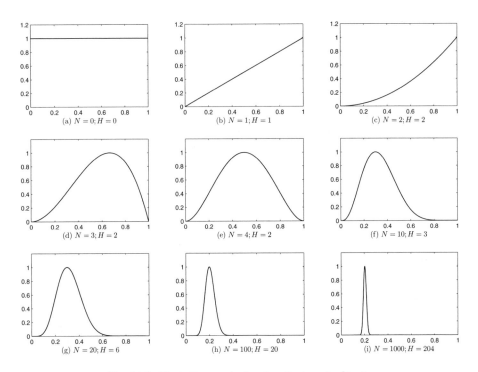

Fig. 14.5 Updating posterior density in coin flipping.

$\theta \neq 0$; indeed, if θ were zero, we could not observe any head. The posterior is now proportional to a triangle:

$$p_1(\theta \mid 1) \propto \theta^1 (1 - \theta)^{1-1} = \theta, \qquad 0 \leq \theta \leq 1$$

This triangle is shown in Fig. 14.5(b). If we observe another head in the second flip, the updated posterior density is a portion of a parabola, as shown in Fig. 14.5(c):

$$p_2(\theta \mid 2) \propto \theta^2 (1 - \theta)^{2-2} = \theta^2, \qquad 0 \leq \theta \leq 1$$

If we get tail at the third flip, we rule out $\theta = 1$ as well. Proceeding this way, we get beta distributions concentrating around the true (unknown) value of θ. Incidentally, the figure has been obtained by Monte Carlo simulation of coin flipping with $\theta = 0.2$. □

Armed with the posterior density, there are different ways of obtaining a Bayes' estimator. Figure 14.5 would suggest taking the mode of the posterior, which would spare us the work of normalizing it. However, this need not be the most sensible choice. If we consider the expected value for the posterior distribution, we obtain

$$E\left[\theta \,|\, X_1 = x_1, \ldots, X_n = x_n\right] = \int_\Omega \theta f(\theta \,|\, x_1, \ldots, x_n) d\theta \qquad (14.21)$$

There are different ways of framing the problem, that are a bit beyond the scope of the book,[29] but one thing that we can immediately appreciate is the challenge we face. The estimator above involves what looks like an intimidating integral, but our task is even more difficult in practice, because finding the posterior density may be a challenging computational exercise as well. In fact, given a prior, there is no general way of finding a closed-form posterior; things can really get awkward when *multiple* parameters are involved. Moreover, there is no guarantee that the posterior distribution $p_n(\theta \,|\, x_1, \ldots, x_n)$ will belong to the same family as does the prior $p(\theta)$. However, there are some exceptions. A family of distributions is called a *conjugate family of priors* if, whenever the prior is in the family, the posterior is too. The following example illustrates the idea.

Example 14.15 (A normal prior) Consider a sample $(X_1, \ldots, X_n)$ from a normal distribution with unknown expected value θ and known variance σ_0^2. Then, given our knowledge about the multivariate normal, and taking advantage of independence among observations, we have the following likelihood function:

$$f_n(x_1, \ldots, x_n \,|\, \theta) = \frac{1}{(2\pi)^{n/2}\sigma_0^n} \exp\left\{-\sum_{i=1}^n \frac{(x_i - \theta)^2}{2\sigma_0^2}\right\}$$

Let us assume that the prior distribution of θ is normal, too, with expected value μ and σ:

$$p(\theta) = \frac{1}{\sqrt{2\pi}\sigma} \exp\left\{-\frac{(\theta - \mu)^2}{2\sigma^2}\right\}$$

To get the posterior, we may simplify our work by considering in each function only the part that involves θ, wrapping the rest within a proportionality constant. In more detail, the likelihood function can be written as

$$f_n(x_1, \ldots, x_n \,|\, \theta) \propto \exp\left\{-\frac{1}{2\sigma_0^2}\sum_{i=1}^n (x_i - \theta)^2\right\} \qquad (14.22)$$

---

[29] One way to frame the problem is by introducing a *loss function* that accounts for the cost of a wrong estimate. It can be shown that Eq. (14.21) yields the optimal estimator when the loss function has a certain quadratic form.

We may simplify the expression further, by observing that

$$\sum_{i=1}^{n} (x_i - \theta)^2 = \sum_{i=1}^{n} (x_i - \bar{x} + \bar{x} - \theta)^2$$

$$= \sum_{i=1}^{n} (x_i - \bar{x})^2 + \sum_{i=1}^{n} (\bar{x} - \theta)^2$$

$$= \sum_{i=1}^{n} (x_i - \bar{x})^2 + n (\theta - \bar{x})^2$$

where $\bar{x}$ is the average of x_i, $i = 1, \ldots, n$. Then, we may include terms not depending on θ into the proportionality constant and rewrite (14.22) as

$$f_n(x_1, \ldots, x_n \,|\, \theta) \propto \exp \left\{ -\frac{n}{2\sigma_0^2} (\theta - \bar{x})^2 \right\} \qquad (14.23)$$

By a similar token, we may rewrite the prior as

$$p(\theta) \propto \exp \left\{ -\frac{(\theta - \mu)^2}{2\sigma^2} \right\} \qquad (14.24)$$

Multiplying Eqs. (14.23) and (14.24), we obtain the posterior

$$p_n(\theta \,|\, x_1, \ldots, x_n) \propto \exp \left\{ -\frac{1}{2} \left[\frac{n}{\sigma_0^2} (\theta - \bar{x})^2 + \frac{1}{\sigma^2} (\theta - \mu)^2 \right] \right\} \qquad (14.25)$$

Again, we should try to include θ within one term; to this aim, we use a bit of tedious algebra[30] and rewrite the argument of the exponential as follows:

$$\frac{n}{\sigma_0^2} (\theta - \bar{x})^2 + \frac{1}{\sigma^2} (\theta - \mu)^2 = \frac{1}{\xi^2} (\theta - \nu)^2 + \frac{n}{\sigma_0^2 + n\sigma^2} (\bar{x} - \mu)^2 \qquad (14.26)$$

where

$$\nu = \frac{n\sigma^2 \bar{x} + \sigma_0^2 \mu}{n\sigma^2 + \sigma_0^2} \qquad (14.27)$$

$$\xi^2 = \frac{\sigma^2 \sigma_0^2}{n\sigma^2 + \sigma_0^2} \qquad (14.28)$$

Finally, this leads to

$$p_n(\theta \,|\, x_1, \ldots, x_n) \propto \exp \left\{ -\frac{1}{2\xi^2} (\theta - \nu)^2 \right\} \qquad (14.29)$$

---

[30] See Problem 14.8.

Disregarding the normalization constant, we immediately recognize the familiar shape of a normal density, with expected value ν. Then, given an observed sample mean $\overline{X}$ and a prior μ, Eq. (14.27) tells us that the Bayes' estimator of θ can be written as

$$\begin{aligned}
E\left[\theta \mid X_1, \ldots, X_n\right] &= \frac{n\sigma^2}{n\sigma^2 + \sigma_0^2}\overline{X} + \frac{\sigma_0^2}{n\sigma^2 + \sigma_0^2}\mu \\[2mm]
&= \frac{\dfrac{n}{\sigma_0^2}}{\dfrac{n}{\sigma_0^2} + \dfrac{1}{\sigma^2}}\overline{X} + \frac{\dfrac{1}{\sigma^2}}{\dfrac{n}{\sigma_0^2} + \dfrac{1}{\sigma^2}}\mu \qquad (14.30)
\end{aligned}$$

Eq. (14.30) has a particularly nice and intuitive interpretation: The posterior estimator is a weighted average of the sample mean $\overline{X}$ (the new evidence) and the prior μ, with weights that are inversely proportional to σ_0^2/n, the variance of sample mean, and σ^2, the variance of the prior. The more reliable a term, the larger its weight in the average.   □

In the example, it may sound a little weird to assume that the variance σ_0^2 of X is known, but not its expected value. Of course, one can extend the framework to cope with estimation of multiple parameters, but there are cases in which we are more uncertain about expected value than variance, as we see in the following section.

14.7.2   A financial application: The Black–Litterman model

We considered portfolio optimization in Example 12.5 and in Section 13.2.2. For the sake of convenience, let us reconsider the problem here. We must allocate our wealth among n risky assets and a risk-free one. The returns of the risky assets are a vector of random variables with expected value μ and covariance matrix Σ; let r_f be the return of the risk-free asset. Let w_0 be the weight of the risk free asset in the portfolio, whereas the weights of the risky assets are denoted by w_i, $i = 1, \ldots, n$, collected into vector $\mathbf{w} \in \mathbb{R}^n$. We assume that short sales and cash borrowing are possible; hence, we do not enforce any nonnegativity restriction on portfolio weights. Then, the expected value of portfolio return is

$$w_0 r_f + \sum_{i=1}^{n} w_i \mu_i = w_0 + \mathbf{w}^T \mu$$

and its variance is $\mathbf{w}^T \Sigma \mathbf{w}$. Note that w_0 does not affect variance. If we assume a quadratic utility function, we are essentially lead to consider a mean-variance objective, with risk aversion coefficient λ, resulting in the following quadratic programming problem:

$$\max \quad w_0 r_{\mathrm{f}} + \sum_{i=1}^{n} w_i \mu_i - \frac{1}{2} \lambda \mathbf{w}^T \boldsymbol{\Sigma} \mathbf{w}$$

$$\text{s.t.} \quad w_0 + \sum_{i=1}^{n} w_i = 1$$

The factor $\frac{1}{2}$ is not really essential, as it may be included in the risk aversion coefficient, but it simplifies the derivatives we take below. This constrained optimization problem may be transformed into an unconstrained one by eliminating w_0. Plugging

$$w_0 = 1 - \sum_{i=1}^{n} w_i \tag{14.31}$$

into the objective, we obtain

$$\left(1 - \sum_{i=1}^{n} w_i\right) r_{\mathrm{f}} + \sum_{i=1}^{n} w_i \mu_i - \frac{1}{2} \lambda \mathbf{w}^T \boldsymbol{\Sigma} \mathbf{w}$$

$$= r_{\mathrm{f}} + \sum_{i=1}^{n} w_i (\mu_i - r_{\mathrm{f}}) - \frac{1}{2} \lambda \mathbf{w}^T \boldsymbol{\Sigma} \mathbf{w}$$

$$= r_{\mathrm{f}} + \sum_{i=1}^{n} w_i \mu_{e,i} - \frac{1}{2} \lambda \mathbf{w}^T \boldsymbol{\Sigma} \mathbf{w}$$

where $\mu_{e,i} \equiv \mu_i - r_{\mathrm{f}}$ is the *expected excess return* of asset i. Since the leading r_{f} term is inconsequential, the portfolio choice problem can be restated as

$$\max \quad \boldsymbol{\mu}_e^T \mathbf{w} - \frac{1}{2} \lambda \mathbf{w}^T \boldsymbol{\Sigma} \mathbf{w}$$

which is a convex optimization problem. Then, to find optimal portfolio weights, we just enforce stationarity conditions:

$$\boldsymbol{\mu}_e - \lambda \boldsymbol{\Sigma} \mathbf{w} = 0 \quad \Rightarrow \quad \mathbf{w}^* = \frac{1}{\lambda} \boldsymbol{\Sigma}^{-1} \boldsymbol{\mu}_e \tag{14.32}$$

Solving a system of linear equations is easy enough, and we can find the optimal portfolio weights w_i, $i = 1, \ldots, n$, for the risky assets; then, using Eq. (14.31) we also get the weight of the risk-free asset. One of the difficulties of this simple framework is the estimation of problem inputs,[31] such as the vector of expected returns $\boldsymbol{\mu}$ and the covariance matrix $\boldsymbol{\Sigma}$. Practical experience shows that if we take a simplistic approach and use straightforward sample estimates of these parameters, quite unreasonable portfolios are obtained.

---

[31] See Example 9.21.

Indeed, they may include very large positions in a few assets, possibly with large short positions as well.

Example 14.16 Consider a universe consisting of seven assets. The correlations among their excess returns, over an investment horizon, are given by the following symmetric table:[32]

$$
\begin{array}{ccccccc}
1.000 & 0.488 & 0.478 & 0.515 & 0.439 & 0.512 & 0.491 \\
\cdot & 1.000 & 0.664 & 0.655 & 0.310 & 0.608 & 0.779 \\
\cdot & \cdot & 1.000 & 0.861 & 0.355 & 0.783 & 0.668 \\
\cdot & \cdot & \cdot & 1.000 & 0.354 & 0.777 & 0.653 \\
\cdot & \cdot & \cdot & \cdot & 1.000 & 0.405 & 0.306 \\
\cdot & \cdot & \cdot & \cdot & \cdot & 1.000 & 0.652 \\
\cdot & \cdot & \cdot & \cdot & \cdot & \cdot & 1.000
\end{array}
$$

Let us assume that the vector of volatilities is

$$16\%,\ 20.3\%,\ 24.8\%,\ 27.1\%,\ 21\%,\ 20\%,\ 18.7\%$$

Based on these data, we may easily build the covariance matrix Σ. If the risk aversion coefficient λ is set to 2.5 and the expected excess return is 7% for each asset, the resulting portfolio weights are

$$
\mathbf{w} = \begin{bmatrix}
0.7136 \\
0.1528 \\
-0.0172 \\
-0.3348 \\
0.2792 \\
0.3362 \\
0.3733
\end{bmatrix}
$$

This solution might look unreasonable because weights do not add up to 1, but we should remember that there is a weight w_0 for the risk free asset, and therefore

$$w_0 = 1 - \sum_{i=1}^{7} w_i = -0.5032$$

The portfolio looks a bit extreme, as there are quite large weights. This is partially a consequence of a relatively small risk aversion; experience suggests that λ can be in the range from 2 to 4. In practice, by forbidding short sales and adding policy constraints bounding portfolio weights, we might get a more sensible portfolio, but doing so essentially means that we are shaping our strategy using constraints.

---

[32] The numerical data have been taken from [13].

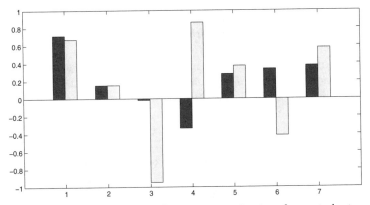

Fig. 14.6 Portfolio instability with respect to estimates of expected return.

A very surprising finding is the dramatic effect of changing estimates of expected excess returns on portfolio weights. Let us assume that the vector of expected (excess) returns is changed to

$$7\%,\ 7\%,\ 4.5\%,\ 9.5\%,\ 7\%,\ 4.5\%,\ 7\%$$

This could be a way to translate an investor's view, feeling that some assets will outperform other assets. The new portfolio is

$$\mathbf{w} = \begin{bmatrix} 0.6707 \\ 0.1537 \\ -0.9476 \\ 0.8621 \\ 0.3717 \\ -0.4156 \\ 0.5747 \end{bmatrix}$$

The change in portfolio composition is illustrated in Fig. 14.6. For each asset we display a pair of bars corresponding to portfolio weights; the second portfolio is associated with the right (white) bar within each pair. The dramatic change in the portfolio weights is rather evident and can be explained by the change in expected return of asset 3, 4, and 6, with respect to the first portfolio choice problem. But is this change justified, or is just due to estimation errors? The problem that we are highlighting is that an error in the estimate of expected value may have a significant impact on portfolio choice. ☐

It is a common opinion that estimating expected return is even more troublesome than estimating covariances, in the sense that it has a stronger effect on optimized portfolios. A cynical view states that portfolio optimization is the best way to maximize the effect of estimation errors. Even if we refrain from being so drastic, it is certainly true that most financial analysts would feel comfortable only with the estimation of a few expected returns, for those market segments on which they have the most experience. In practice, analysts would not just estimate a parameter based on past observations; what about *predicting* an expected return for the future? There is no doubt that portfolio management should look forward, rather than backward, but again, analysts may do this for limited market segments. For the rest, they might just share the market consensus and go for a passive management strategy. We also recall that a passive management strategy is a consequence of the capital asset pricing model (CAPM).[33] According to a passive strategy, one should just hold the market portfolio, i.e., a portfolio with stock share weights shaped after market capitalization of the corresponding firms. If we accept the validity of CAPM, we may even find expected excess returns implied by the market portfolio. According to CAPM, for each asset $i = 1, \ldots, n$, we have

$$\mu_i - r_f = \beta_i (\mu_m - r_f)$$

where μ_m is the expected return of the market portfolio, which can be approximated by a broad index. This relationship may be restated in terms of expected excess returns:

$$\mu_{e,i} = \beta_i \mu_{e,m}$$

We also note that expected excess returns can be interpreted as *risk premia*, as they state by how much the expected return of a risky asset exceeds the risk-free return r_f. The coefficient β_i looks much like the slope of a linear regression, as it is given by

$$\beta_i = \frac{\mathrm{Cov}(R_i, R_m)}{\sigma_m^2}$$

where the numerator is the covariance between the return of asset i and the return of the market portfolio, and σ_m^2 is the variance of market portfolio return. If we denote by $\mathbf{w}_m$ the weights of the market portfolio, its return is

$$R_m = \sum_{j=1}^{n} w_{m,j} R_j$$

Hence, we have

$$\beta_i = \frac{\mathrm{Cov}\left(R_i, \sum_{j=1}^{n} w_{m,j} R_j\right)}{\sigma_m^2} = \frac{\sum_{j=1}^{n} w_{m,j} \mathrm{Cov}\left(R_i, R_j\right)}{\sigma_m^2}$$

Thus, we may collect the "asset betas" β_i into vector $\boldsymbol{\beta}$ and rewrite CAPM as

$$\boldsymbol{\mu}_e = \delta \boldsymbol{\Sigma} \mathbf{w}_m \tag{14.33}$$

---

[33] See Example 9.21.

where the constant

$$\delta = \frac{\mu_m - r_{\mathrm{f}}}{\sigma_m^2} \tag{14.34}$$

trades off the risk premium $(\mu_m - r_{\mathrm{f}})$ against squared market volatility. Incidentally, a comparison between Eqs. (14.32) and (14.33) suggests some link between δ and an average risk aversion coefficient. The important contribution of Eq. (14.33) is that it yields a consensus market view of the expected excess returns, implied by equilibrium. This might be considered as a starting point of an estimation process, which is not just backward-looking and based on historical data, but also forward-looking.

If we look forward, rather than backward, we are immediately lead to a question: How can we include subjective views that an investor might have? For instance, an investor could think that an asset will outperform another one by, say, 5%. Black and Litterman proposed an approach whereby expected returns are estimated based on two primary inputs:

- A forecast implied by market equilibrium

- A set of subjective views on expected returns

The way the two ingredients are blended depends on the uncertainty that is associated with each of them, and can be interpreted as a Bayesian estimate blending subjective expectations with an outside input.[34]

One way of stating the model is the following:

1. The vector of expected excess returns is considered as a vector of random variables $\boldsymbol{\theta}$ with multivariate normal distribution:

$$\boldsymbol{\theta} \sim \mathcal{N}(\boldsymbol{\mu}_e, \tau\boldsymbol{\Sigma})$$

   where $\boldsymbol{\mu}_e$ is implied by market equilibrium and $\boldsymbol{\Sigma}$ is the estimated covariance matrix. Note that we are trusting the estimate of covariance matrix in providing us the covariance matrix $\tau\boldsymbol{\Sigma}$ of the prior distribution; in fact, one of the main difficulties of the Bayesian framework is specifying sensible multivariate priors. The rationale here is that if assets are correlated, the error in estimating their risk premia will be as well. The parameter τ can be used to fine tune the degree of confidence in the market view.

2. Subjective views about risk premia and expected returns can be expressed as linear relationships. For instance, say that we believe that asset 2 will outperform asset 5 by 2%. Then, we could write a condition such as

$$\mu_2 - \mu_5 = 0.02$$

---

[34] See Ref. [21] for a detailed proof.

In this case, using expected return or risk premia is inconsequential. Several similar views could be expressed in terms of excess returns and collected in the following matrix form:

$$\mathbf{P}\boldsymbol{\mu}_e = \mathbf{q}$$

The view above would correspond to a row in matrix $\mathbf{P}$, with elements set to 1 in columns 2 and 5, zero otherwise; the corresponding element in vector $\mathbf{q}$ would be 0.02. Of course, subjective views are uncertain as well, and we might express this as

$$\mathbf{P}\boldsymbol{\theta} \sim \mathcal{N}(\mathbf{q}, \boldsymbol{\Omega})$$

where $\boldsymbol{\Omega}$ is typically a diagonal matrix whose elements are related to the confidence in subjective views.

The Black-Litterman model, relying on Bayesian estimation, yields the following estimate of risk premia:

$$\mu_{\mathrm{BL}} = \left((\tau\boldsymbol{\Sigma})^{-1} + \mathbf{P}^T\boldsymbol{\Omega}^{-1}\mathbf{P} \right)^{-1} \left((\tau\boldsymbol{\Sigma})^{-1}\boldsymbol{\mu}_e + \mathbf{P}^T\boldsymbol{\Omega}^{-1}\mathbf{q} \right)$$

The proof of this relationship is definitely beyond the scope of the book, but it is quite instructive to note its similarity with Eq. (14.30). The interpretation is again that the estimate blends subjective views and objective data, in a way that reflects their perceived reliability.[35] Several variations of the Black–Litterman model have been proposed, but the essential message is that subjective and more or less objective views may be blended within a Bayesian framework. In fact, Bayesian approaches have also been proposed in marketing, where they are most relevant to cope with brand-new products and markets, where past data are quite scarce or irrelevant.

Problems

14.1 Find the Nash equilibria in the games in Tables 14.2 and 14.4. Are they unique?

14.2 Consider the Cournot competition outcome of Eqs. (14.12) and (14.13). Analyze the sensitivity of the solution with respect to innovation in production technology, i.e., how a reduction in production cost c_i for firm i affects equilibrium quantities, price, and profit.

14.3 Two firms have the same production technology, represented by the cost function:

$$\mathrm{TC}_i(q_i) = \begin{cases} F + (q_i)^2 & \text{if } q_i > 0 \\ 0 & \text{if } q_i = 0 \end{cases}, \qquad i = 1, 2$$

---

[35] There is also an alternative way of interpreting Black–Litterman model in terms of robust "shrinkage" estimators; see, e.g., Chapter 9 of Ref. [8].

The cost function involves a fixed cost F and a squared term implying a diseconomy of scale; a firm will produce only when its profit is positive. The two firms compete on quantity, and price is related to total quantity $Q = q_1 + q_2$ by the linear function $P(Q) = 100 - Q$. Find the maximum F such that both firms engage in production, assuming a Cournot competition.

14.4   Two firms have a production technology involving a fixed cost and constant marginal cost, as represented by the cost function:

$$\mathrm{TC}_i(q_i) = \begin{cases} F_i + c_i q_i & \text{if } q_i > 0 \\ 0 & \text{if } q_i = 0 \end{cases}, \qquad i = 1, 2$$

Analyze the von Stackelberg equilibrium when firm 1 is the leader and the two firms compete on quantities.

14.5   Consider the data of the Braess' paradox example in Section 14.5, but imagine that a central planner can assign routes to drivers, in order to minimize total travel cost. Check that adding the new link e, as in Fig. 14.4(b), cannot make the total cost worse.

14.6   Consider again the Bayesian coin flipping experiment of Example 14.14, where the prior is uniform. If we use Eq. (14.21) to find the Bayesian estimator, what is the estimate of θ after the first head? And after the second head?

14.7   Consider the fraction θ of defective items in a batch of manufactured parts. Say that the prior distribution of θ is a beta distribution with parameters $\alpha_1 = 5$ and $\alpha_2 = 10$ (see Section 7.6.2 for a description of the beta distribution). A new batch of 30 items is manufactured and two of them are found defective. What is the posterior estimate of θ, if we use Eq. (14.21)?

14.8   Prove the identity in Eq. (14.26).

For further reading

- Game theory is dealt with at an introductory level in books on industrial organization, such as Refs. [4] and [22]. See Ref. [11] for a thorough introduction and Ref. [10] for an advanced treatment.

- For Bayesian statistics, you may see Ref. [6] or the more recent version [7]. These books, like the text by Casella and Berger [5], deal with Bayesian statistics within a framework that is compatible with the orthodox view. A more radical approach is advocated by Jaynes in Ref. [16]. An intermediate treatment can be found in Ref. [17].

- For a scientific perspective on Bayesian data analysis, see the treatise by Sivia and Skilling [23].

- The Black–Litterman portfolio management approach was introduced in Ref. [1]; see also Chapter 9 of Ref. [8], or Refs. [13] and [21].

- Interesting and instructive accounts of some financial mishaps can be found in Refs. [2] and [15]. For discussions on black swans, as well as the meaning and the role of probabilities and quantitative modeling in finance, you may refer, e.g., to Refs. [9], [14], and [20].

REFERENCES

1. F. Black and R. Litterman, Global portfolio optimization, *Financial Analysts Journal*, **48**:28–43, 1992.

2. R. Bookstaber, *A Demon of Our Own Design: Markets, Hedge Funds, and the Perils of Financial Innovation*, Wiley, New York, 2008.

3. D. Braess, A. Nagurney, and T. Wakolbinger, On a paradox of traffic planning, *Transportation Planning*, **39**:446–450, 2005.

4. L.M.B. Cabral, *Introduction to Industrial Organization*, MIT Press, Cambridge, MA, 2000.

5. G. Casella and R.L. Berger, *Statistical Inference,* 2nd ed., Duxbury, Pacific Grove, CA, 2002.

6. M.H. DeGroot, *Probability and Statistics*, Addison-Wesley, Reading, MA, 1975.

7. M.H. DeGroot and M.J. Schervish, *Probability and Statistics,* 3rd ed., Addison-Wesley, Boston, 2002.

8. F.J. Fabozzi, S.M. Focardi, and P.N. Kolm, *Quantitative Equity Investing: Techniques and Strategies*, Wiley, New York, 2010.

9. S.M. Focardi and F.J. Fabozzi, Black swans and white eagles: on mathematics and finance, *Mathematical Methods of Operations Research*, **69**:379–394, 2009.

10. D. Fudenberg and J. Tirole, *Game Theory*, MIT Press, Cambridge, MA, 1991.

11. R. Gibbons, *Game Theory for Applied Economists*, Princeton University Press, Princeton, NJ, 1992.

12. I. Gilboa, *Theory of Decision Under Uncertainty*, Cambridge University Press, New York, 2009.

13. G. He and R. Litterman, *The Intuition Behind Black–Litterman Model Portfolios*, Technical report, Investment Management Research, Goldman & Sachs Company, 1999. A more recent version can be downloaded from http://www.ssrn.org.

14. Glyn A. Holton, Defining risk, *Financial Analysts Journal*, **60**:19–25, 2004.

15. L.L. Jacque, *Global Derivative Debacles: From Theory to Malpractice*, World Scientific, Singapore, 2010.

16. E.T. Jaynes, *Probability Theory: The Logic of Science*, Cambridge University Press, Cambridge, UK, 2003.

17. P.M. Lee, *Bayesian Statistics: An Introduction,* 3rd ed., Oxford University Press, New York, 2004.

18. R. Lowenstein, *When Genius Failed: The Rise and Fall of Long-Term Capital Management*, Random House Trade Paperbacks, New York, 2001.

19. V.G. Narayanan and A. Raman, *Hamptonshire Express*, Harvard Business School, 2002, Business Case no. 9-698-053.

20. R. Rebonato, *Plight of the Fortune Tellers: Why We Need to Manage Financial Risk Differently*, Princeton University Press, Princeton, NJ, 2007.

21. S. Satchell and A. Scowcroft, A demystification of the Black–Litterman model: managing quantitative and traditional portfolio construction, *Journal of Asset Management*, **1**:138–150, 2000.

22. O. Shy, *Industrial Organization: Theory and Applications*, MIT Press, Cambridge, MA, 1995.

23. D.S. Sivia and J. Skilling, *Data Analysis: A Bayesian Tutorial,* 2nd ed., Oxford University Press, Oxford, 2006.

Part IV

Advanced Statistical Modeling

15

Introduction to Multivariate Analysis

Multivariate analysis is the more-or-less natural extension of elementary inferential statistics to the case of multidimensional data. The first difficulty we encounter is the representation of data. How can we visualize data in multiple dimensions, on the basis of our limited ability to plot bidimensional and tridimensional diagrams? In Section 15.1 we show that this is just one of the many issues that we may have to face. The richness of problems and applications of multivariate analysis has given rise to a correspondingly rich array of methods. In the next two chapters we will outline a few of them, but in Section 15.2 we offer a more general classification. Finally, the mathematics involved in multivariate analysis is certainly not easier than that involved in univariate inferential statistics. Also probability theory in the multidimensional case is more challenging than what we have seen in the first part of the book, and the limited tools of correlation analysis should be expanded. However, for the limited purposes of the following treatment, we just need a few additional concepts; in Section 15.3 we illustrate the important role of linear algebra and matrix theory in multivariate methods.

15.1 ISSUES IN MULTIVARIATE ANALYSIS

In the next sections we briefly outline the main complication factors that arise when dealing with multidimensional data. Some of them are to be expected, but some are a bit surprising. Getting aware of these difficulties provides the

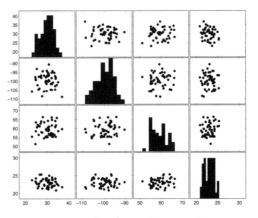

Fig. 15.1 A matrix of scatterplots.

motivation for studying the wide array of sometimes quite complex methods that have been developed.

15.1.1   Visualization

The first and most obvious difficulty we face with multivariate data is visualization. If we want to explore the association between variables, one possibility is to draw scatterplots for each pair of them; for instance, if we have 4 variables, we may draw a matrix of scatterplots, like the one illustrated in Fig. 15.1. The matrix of plots is symmetric, and the histograms of each single variable are displayed on the diagonal. Clearly, this is a rather partial view, even though it can help in spotting pairwise relationships. Many fancy methods have been proposed to obtain a more complete picture of multivariate data, such as drawing human faces, whose features are related to data characteristics; however, they may be rather hard to interpret. A less trivial approach is based on data reduction. Quite often, even though there are many variables, we may take linear combinations of them in such a way that a limited number of such transformations includes most of the really interesting information. Such an approach is *principal component analysis*, which is illustrated in Chapter 17.

15.1.2   Complexity and redundancy

Visualization is not the only reason why we need data reduction methods. Quite often, multivariate data stem from the administration of a questionnaire to a sample of respondents; each question corresponds to a single variable, and a set of answers by a single respondent is a multivariate observation. It is customary to ask respondents many related questions, possibly in order to check the coherence in their answers. However, an unpleasing consequence is that some variables may be strictly related, if not redundant. On the one hand, this motivates the use of data reduction methods further. On the other hand, this may complicate the application of rather standard approaches, such as multiple linear regression. In Chapter 16, we will see that a strong correlation between variables may result in unreliable regressed models; this issue is known as *collinearity*. By reducing the number of explanatory variables in the regression model, we may ease collinearity issues. Another common issue is that when a problem has multiple dimensions, it is difficult to group similar observations together. For instance, a common task in marketing is customer segmentation. In Chapter 17 we also outline *clustering methods* that may be used to this aim.

15.1.3 Different types of variables

In standard inferential statistics one typically assumes that data consist of real or integer numbers. However, data may be qualitative as well, and the more dimensions we have, the more likely the joint presence of quantitative and qualitative variables will be. In some cases, dealing with qualitative variables is not that difficult. For instance, if we are building a multiple regression model that includes one or more qualitative explanatory variables, we may represent them as binary (dummy) variables, where 0 and 1 correspond to "false" and "true," respectively. However, things are not that easy if it is the regressed variable that is binary. For instance, we might wish to estimate the probability of an event on the basis of explanatory variables; this occurs when we are evaluating the creditworthiness of a potential borrower, on the basis of a set of personal characteristics, and the output is the probability of default. Another standard example is a marketing model to predict purchasing decisions on the basis of product features. Adapting linear regression to this case calls for less obvious transformations, leading to *logistic regression*, which is also considered in Chapter 16. In other cases, the qualitative nature of data calls for methods that are quite different from those used for quantitative variables.

15.1.4 Adapting statistical inference procedures

The core topics in statistical inference are point and interval parameter estimation, hypothesis testing, and analysis of variance. Some of the related procedures are conceptually easy to adapt to a multivariate case. For instance, maximum likelihood estimation is not quite different, even though it is going to prove computationally more challenging, thus requiring numerical optimization methods for the maximization of the likelihood function. In other cases, things are not that easy and may call for the introduction of new classes of multivariable probability distributions to characterize data. The

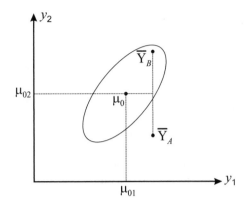

Fig. 15.2   A bidimensional hypothesis test.

following example shows that a straightforward extension of single-variable (univariate) methods may not be appropriate.

Example 15.1 Let us consider a hypothesis test concerning the mean of a multivariate probability distribution. We want to check the hypothesis that the expected value of a jointly normal random variable $\mathbf{Y} \sim \mathcal{N}(\boldsymbol{\mu}, \boldsymbol{\Sigma})$ is $\boldsymbol{\mu}_0$. Therefore, we test the null hypothesis

$$H_0 \colon \boldsymbol{\mu} = \boldsymbol{\mu}_0$$

against the alternative one

$$H_\mathrm{a} \colon \boldsymbol{\mu} \neq \boldsymbol{\mu}_0$$

The random vector $\mathbf{Y}$ has components Y_1 and Y_2; let us denote the two components of vector $\boldsymbol{\mu}_0$ by μ_{01} and μ_{02}, respectively. One possible approach would be to calculate the sample mean of each component, $\overline{Y}_1$ and $\overline{Y}_2$, and run two univariate tests, one for μ_{01} and one for μ_{02}. More precisely, we could test the null hypothesis $H_0 \colon \mu_1 = \mu_{01}$ on the basis of the sample mean $\overline{Y}_1$, and $H_0 \colon \mu_2 = \mu_{02}$ on the basis of the sample mean $\overline{Y}_2$. Then, if we reject even one of the two null univariate hypotheses, we reject the multivariate hypothesis as well. Unfortunately, this approach may fail, as illustrated in Fig. 15.2. In the figure, we show an ellipse, which is a level curve of the joint PDF, and two possible sample means, corresponding to vectors $\overline{\mathbf{Y}}_A$ and $\overline{\mathbf{Y}}_B$. The rotation of the ellipse corresponds to a positive correlation between Y_1 and Y_2. If we account for the nature of the PDF of jointly normal variables (see Section 8.4), it turns out that the acceptance region for the test above should be an ellipse. Let us assume that the ellipse in Fig. 15.2 is the acceptance region for the test; then, in the case of $\overline{\mathbf{Y}}_A$, the null hypothesis should be rejected; in the case of $\overline{\mathbf{Y}}_B$, we do not reject H_0. On the contrary, testing the two hypotheses separately results in a rectangular acceptance region around $\boldsymbol{\mu}_0$ (the rectangle is a square if the two standard deviations are the same). We immediately observe that, along each dimension, the distance between $\boldsymbol{\mu}_0$ and $\overline{\mathbf{Y}}_A$ and the distance between $\boldsymbol{\mu}_0$ and $\overline{\mathbf{Y}}_B$ are exactly the same, in absolute value. Hence, if we run separate tests, we either reject or accept the null hypothesis for both sample means, which is not correct. The point is that a rectangular acceptance region, unlike the elliptical one in Fig. 15.2, does not account for correlation. □

Incidentally, a proper analysis of the testing problem in Example 15.1 leads to consider the test statistic

$$T^2 = n(\overline{\mathbf{Y}} - \boldsymbol{\mu}_0)^T \mathbf{S}^{-1}(\overline{\mathbf{Y}} - \boldsymbol{\mu}_0)$$

where $\mathbf{S}^{-1}$ is the inverse of the sample covariance matrix. Hence, distributional results for univariate statistics, which involve a t distribution, must be extended by the introduction of a new class of probability distributions, namely, Hotelling's T^2 distribution.

15.1.5 Missing data and outliers

Outliers and wrong data are quite common in data analysis. If data are collected automatically, and they are engineering measurements, this may not be a tough issue; however, when people are involved, either because we are collecting data using questionnaires, or because we are investigating a social system, things may turn out to be a nightmare. Having to cope with multidimensional data will naturally exacerbate the issue. While an outlier may be easy to "see" in one-dimensional data, possibly using a boxplot,[1] it is not obvious at all how an outlier is to be spotted in 10 or more dimensions. Again, data reduction techniques may help. Missing data may also be the consequence of real or perceived redundancy. The following example illustrates a counterintuitive effect.

Example 15.2 Imagine collecting data in a city logistics problem. One of the most important measures is the percentage saturation of vehicles. No one would like the idea of half-empty vehicles polluting air more than necessary in congested urban traffic. So, one would naturally want to investigate the real level of vehicle saturation to check whether some improvement can be attained by proper reorganization. However, capacity is multidimensional. Probably, the most natural capacity measure that comes to mind is volume capacity.[2] However, weight may be the main issue with certain types of items; before dismissing this issue, think of the impact of weight on the space needed to brake a fully loaded truck. Moreover, if small parcels are delivered, the binding

---

[1]See Section 4.4.1.
[2]This is likely to stem from our struggle with many suitcases that do not like the idea of fitting the trunk of our cheap car.

Table 15.1 Difficulties with missing multidimensional data.

time	volume	weight	max
100%	–	–	100%
40%	20%	45%	45%
–	50%	60%	60%
70%	35%	52.5%	68.33%

constraint on a vehicle tour will be neither volume nor weight, but time, since driving shifts are constrained.

Now, imagine administering a questionnaire to truck drivers, asking them an estimate of their average saturation level, in percentage, with respect to the three dimensions of capacity. The result might look like the fictional data in Table 15.1. There, we show the hypothetical result of three interviews. The first truck driver answered that he is 100% saturated in terms of time; the binding factor here is the number of deliveries, and he did not provide any answer in terms of the other two capacity dimensions, as they are not relevant to him and he had no clue. The second driver was quite thorough, whereas the third one did not consider time. In the table, we also give the maximum saturation percentage in the last column, over the three capacity dimensions, for each driver. The last row gives the average over drivers, for each dimension. Finally, we also consider the average of the maximum saturation for each driver. Do you see something wrong here?

The average of the maxima is

$$\frac{100 + 45 + 60}{3} = 68.33\%$$

but if we take the average saturation with respect to time, we get a larger number:

$$\frac{100 + 40}{2} = 70\%$$

This should not be the case, however: How can the average of maxima across dimensions (68.33%) be less than the average with respect to one dimension (70%)? This surprising fact is, of course, the result of missing data. ▯

The example above may look somewhat pathological, since very few data are displayed; on the contrary, this is what happened in a real-life case, and we display fictional data in Table 15.1 just to illustrate the issue more clearly.[3]

---

[3]The case was an investigation of city logistics in Turin and Piedmont Region, carried out by a colleague of mine. And if you think that a more thorough interview technique

The higher the number of dimensions, the more severe the issues with missing data will be. The hard way to solve this issue is to discard incomplete data, but this may considerably reduce the sample. Another strategy is to fill the holes by using regression models. We may fit regression models with available data, and we compute the missing pieces of information as a function of what is available. Clearly, this does sound a bit arbitrary, but it may be better than ending up with a very small and useless set of complete data.

15.2 AN OVERVIEW OF MULTIVARIATE METHODS

Multivariate methods can be classified along different features:

- *Confirmatory vs. exploratory.* This feature refers to the general aim of a method, as some are aimed at confirming a theory or a hypothesis, and others are aimed at analyzing data and discovering hidden patterns.

- *Metric vs. nonmetric.* This feature refers to the kind of variables that the method is able to deal with, i.e., quantitative or qualitative. We recall that sometimes numerical codes are associated with qualitative variables, which have no real content. In such a case, we speak of *nominal scales.*[4] When variables are quantitative, we distinguish the following types of scale:

  - *Ordinal scales*, where variables have numerical values that can sensibly ordered, but their differences have no meaning. As an example, imagine a set of customers ranking different brands by assigning a numerical evaluation.

  - *Interval scales*, where differences between numerical values have a meaning, but there is no "natural" origin of the scale. For instance, consider temperatures, which can be measured with different scales.

  - *Ratio scales*, where there is an objective reference point acting as the origin of the scale.

- *Interdependence vs. dependence.* When we are focusing on dependence, there is a clear separation between the set of independent variables (e.g., factors) and the set of dependent variables (e.g., effects). One such case is simple linear regression. In interdependence analysis there is no such clear-cut distinction.

---

would solve the issue, try stopping a voluminous truck driver, at 8 a.m., in the middle of a congested road, for a long and amicable conversation. Needless to say, "voluminous " refers to the driver.

[4]See Section 4.1.1.

In the following sections we outline some multivariate methods, suggesting a classification along the above dimensions. We do not aim at being comprehensive; the idea is getting to appreciate the richness of this field of statistics, as well as the classification above in concrete terms.

15.2.1   Multiple regression models

In regression models there is a clear separation between the regressed variable and the regressors (explanatory variables):

$$Y = \beta_0 + \beta_1 X_1 + \beta_2 X_2 + \cdots + \beta_m X_m + \epsilon$$

This does not necessarily mean that there is a causal relationship, but it is enough to classify regression models as dependence models. Regression models arise naturally for dealing with metric variables, but we may use binary variables to model qualitative features in a limited way. We may use regression models for confirming a theory, by testing the significance of individual coefficients or the overall significance of the whole model, as we show in Chapter 16. However, we may also use the technology as an exploratory tool, by running a sequence of regression models involving different sets of regressors. Furthermore, logistic regression models allow for a qualitative regressed variable taking only two values. Alternative methods, such as *discrete discriminant analysis*, may be used for the case of a dependent categorical variable assuming more than two values.

15.2.2   Principal component analysis

Principal component analysis (PCA) is a data reduction method. Technically, we take a vector of random variables $\mathbf{X} \in \mathbb{R}^m$, and we transform it to another vector $\mathbf{Z} \in \mathbb{R}^m$, by a linear transformation represented by a square matrix $\mathbf{A} \in \mathbb{R}^{m,m}$. In more detail we have

$$
\begin{aligned}
Z_1 &= a_{11}X_1 + a_{12}X_2 + \cdots + a_{1m}X_m \\
Z_2 &= a_{21}X_1 + a_{22}X_2 + \cdots + a_{2m}X_m \\
&\vdots \\
Z_m &= a_{m1}X_1 + a_{m2}X_2 + \cdots + a_{nn}X_m
\end{aligned}
$$

These equations should not be confused with regression equations. The transformed Z_i variables are not observed and used in a fitting procedure; indeed, there is no error term. They are just transformations of the original variables, which are not classified as dependent or independent. Hence, PCA is an interdependence technique, aimed at metric data, and used for exploratory purposes. In Section 17.2 we show that, by taking suitable combinations, we may find a small subset of Z_i variables, the principal components, that explain most of the variability in the original variables X_i. By disregarding the less relevant components, we reduce data dimensionality without losing a significant portion of information.

15.2.3 Factor analysis

Factor analysis is another interdependence technique, which shares some theoretical background with PCA, as we show in Section 17.3. Factor analysis can be used for data reduction, too, but it should not be confused with PCA, as in factor analysis we are looking for hidden factors that may explain common sources of variance between variables. Formally, we aim at finding a model such as

$$
\begin{aligned}
Y_1 &= \mu_1 + \lambda_{11} F_1 + \cdots + \lambda_{1m} F_m + \epsilon_1 \\
Y_2 &= \mu_2 + \lambda_{21} F_1 + \cdots + \lambda_{2m} F_m + \epsilon_2 \\
&\vdots \qquad \vdots \\
Y_p &= \mu_p + \lambda_{p1} F_1 + \cdots + \lambda_{pm} F_m + \epsilon_p
\end{aligned}
$$

where the variables Y_i are what we observe, F_j are common underlying factors, ϵ_i are individual sources of variability, and m is significantly smaller than p. This may look like a set of regression models, but the main difference is that factors are *not* directly observable. We are trying to uncover hidden factors, which have to be interpreted. Even though the above equations suggest a dependence structure between the observations and the underlying factors, there is no dependence structure among the observations themselves; hence, factor analysis is considered an interdependence method for dealing with metric data. It is natural to use factor analysis for exploratory purposes, but it can also be used for confirmatory purposes.

15.2.4 Cluster analysis

The aim of cluster analysis is categorization, i.e., the creation of groups of objects according to their similarities. The idea is hinted at in Fig. 15.3. There are other methods, such as discriminant analysis, essentially aimed at separating groups of observations. However, they differ in the underlying approach, and some can only deal with metric data. Cluster analysis relies on a distance measure; hence, provided we are able to define a distance with respect to qualitative attributes, it can cope with nonmetric variables. There is an array of cluster analysis methods, which are exploratory and aimed at studying interdependence; they are outlined in Section 17.4.

15.2.5 Canonical correlation

Consider two sets of variables that are collected in vectors $\mathbf{X}$ and $\mathbf{Y}$, respectively, and imagine that we would like to study the relationship between

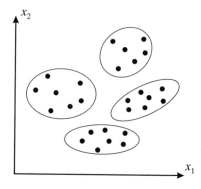

Fig. 15.3 Four bidimensional clusters.

the two sets. One way for doing so is by forming two linear combinations, $Z = \mathbf{a}^T \mathbf{X}$ and $W = \mathbf{b}^T \mathbf{Y}$, in such a way that the correlation $\rho_{Z,W}$ is maximized. This is what is accomplished by canonical correlation, or canonical analysis. Essentially, the idea is to study the relationship between groups of metric variables by relating low-dimensional projections. When $\mathbf{Y}$ reduces to a scalar random variable Y, canonical correlation boils down to multiple linear regression. To illustrate one potential application of canonical correlation, consider a set of marketing activities (advertising effort, packaging quality and appeal, pricing, bundling of products and services, etc.) and a set of corresponding customer behaviors (willingness to purchase, brand loyalty, willingness to pay, etc.); clearly, to apply the approach, we must introduce a metric scale to measure each variable. It is natural to consider the variables in the second set as dependent variables; therefore, this is a case in which we wish to explore dependence, but canonical correlation can also be used to investigate interdependence. In fact, canonical correlation forms the basis for other multivariate analysis techniques.

15.2.6 Discriminant analysis

Consider a firm that, on the basis of a set of variables measuring customer attributes, wishes to discriminate between purchasers and nonpurchasers of a product of service. In concrete, the firm has collected a sample of consumers and, given their attributes and observed behavior, wants to find a way to classify them. Two-group discriminant analysis aims at finding a function of variables, called a *discriminant function*, which best separates the two groups. This sounds much like cluster analysis, but the mechanism is quite different:

- Cluster analysis relies on a measure of distance and tries to find groups such that the distance within groups is small, and distance between groups is large.

- Discriminant analysis relies on a discriminant function $f(\mathbf{x})$, possibly a linear combination of variables, and a threshold value γ such that if $f(\mathbf{x}_a) \leq \gamma$, object a with attributes $\mathbf{x}_a$ is classified in one group; if $f(\mathbf{x}_a) > \gamma$, object a is assigned to the other group.

Another fundamental difference is that, in discriminant analysis, clusters are known a priori and are used for learning. Discriminant analysis can be generalized to multiple groups, for both exploratory and confirmatory purposes.

15.2.7   Structural equation models with latent variables

Consider the relationship between the following variables:

- Self-esteem and job satisfaction

- Customer satisfaction and repurchase intention

The assumption that these variables are somehow related makes sense, but unfortunately they are not directly observable; they are *latent variables*. Nevertheless, imagine that we wish to build a model expressing the dependence between latent variables. For instance, we may consider the *structural equation*

$$\zeta = \gamma \xi + \nu$$

where ζ and ξ are latent variables, ν is an error term, and γ is an unknown parameter. If we want to estimate the parameter, we need to relate the latent variables to observable variables, which play the role of *measurements*. Imagine that the latent variable ξ can be related to observable variables X_1 and X_2 by the following *measurement model*:

$$X_1 = \alpha_1 \xi + \epsilon_1$$
$$X_2 = \alpha_2 \xi + \epsilon_2$$

where α_1 and α_2 are unknown parameters, and ϵ_1 and ϵ_2 are errors. This is an interdependence model, quite similar to factor analysis. By the same token, the measurement model for ζ can be something like

$$Z_1 = \beta_1 \zeta + \eta_1$$
$$Z_2 = \beta_2 \zeta + \eta_2$$
$$Z_3 = \beta_3 \zeta + \eta_3$$

The overall model can be depicted as in Fig. 15.4. We stress again that a structural model with latent variables includes both dependence and interdependence components. Methods have been proposed to estimate the unknown parameters, combining ideas from regression and factor analysis.

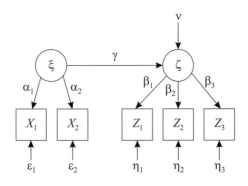

Fig. 15.4   Schematic illustration of a structural model with latent variables.

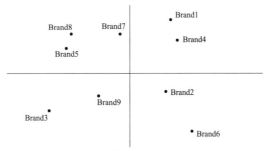

Fig. 15.5   Schematic illustration of multidimensional scaling.

15.2.8   Multidimensional scaling

Multidimensional scaling is a family of procedures that aim at producing a low-dimensional representation of object similarity/dissimilarity. Consider n brands and a similarity matrix, whose entry d_{ij} measures the distance between brands i and j, as perceived by consumers. This matrix is a direct input of multidimensional scaling, whereas other methods aim at computing distances. Then, we want to find a representation of brands as points on a plane, in such a way that the geometric (Euclidean) distance δ_{ij} between points is approximately proportional to the perceived distance between brands:

$$\delta_{ij} \approx \alpha d_{ij}, \qquad i, j = 1, \ldots, n;\ i \neq j$$

for some irrelevant constant α. The idea is illustrated in Fig. 15.5. In a marketing context, for instance, multidimensional scaling can help researchers to understand how consumers perceive brands and how product features relate to each other. We observe that multidimensional scaling procedures accomplish a form of dimensionality reduction, are exploratory in nature, and can be classified as interdependence methods.

Table 15.2 Correspondence analysis works on a two-way contingency table.

| Attribute | Snacks | | | |
	S_1	S_2	$\cdots$	S_n
A_1	N_{11}	N_{12}	$\cdots$	N_{1n}
A_2	N_{21}	N_{22}	$\cdots$	N_{2n}
$\vdots$	$\vdots$	$\vdots$		$\vdots$
A_m	N_{m1}	N_{m2}	$\cdots$	N_{mn}

15.2.9 Correspondence analysis

Correspondence analysis is a graphical technique for representing the information included in a two-way contingency table containing frequency counts. For example, Table 15.2 lists the number of times an attribute (crispy, sugar-free, good with coffee, etc.) is used by consumers to describe a snack (cookies, candies, muffins, etc.).[5] The method deals with two categorical or discrete quantitative variables and aims at visualizing how row and column profiles relate to each other, by developing indexes that are used as coordinates for depicting row and column categories on a plane. Again, this can be used for assessing product positioning, among other things. Correspondence analysis is another exploratory–interdependence method, which can be considered as a factorial decomposition of a contingency table.

This cursory and superficial overview should illustrate well the richness of multivariate analysis and its potential for applications. We should also mention that:

- The boundaries between multivariate analysis tools are not quite sharp, as some methods can be considered as specific cases of other methods.

- Methods can be combined in practice. For instance, in order to ease the task of a cluster analysis algorithm, we may first reduce problem dimensionality by principal component analysis.

15.3 MATRIX ALGEBRA AND MULTIVARIATE ANALYSIS

The methods that we describe in the next two chapters rely heavily on matrices and linear algebra, which we have covered in Chapter 3. In this section we

---

[5]For a complete example, see p. 311 of the book by Myers and Mullen [4].

discuss a few more concepts that are useful in multivariate analysis. Unfortunately, when moving to multivariate statistics, we run out of notation. As usual, capital letters will refer to random quantities, with boldface reserved for random vectors such as $\mathbf{X}$ and $\mathbf{Z}$; elements of these vectors will be denoted by X_i and Z_i, and scalar random variables will be denoted by Y as usual. Lowercase letters, such as $\mathbf{x}$ and x, refer to numbers or specific realizations of random quantities $\mathbf{X}$ and x, respectively. We will also use matrices such as $\boldsymbol{\Sigma}$, $\mathbf{S}$, and $\mathbf{A}$; usually, there is no ambiguity between matrices and random vectors. However, we also need to represent the whole set of observations in matrix form. Observation k is a vector $\mathbf{X}^{(k)} \in \mathbb{R}^p$, with elements $X_j^{(k)}$, $j = 1, \ldots, p$, corresponding to single variables or dimensions. Observations are typically collected into matrices, where columns correspond to single variables and rows to their joint realizations (observations). The whole dataset will be denoted by $\mathcal{X}$, to avoid confusion with vector $\mathbf{X}$. The element $[\mathcal{X}]_{kj}$ in row k and column j of the data matrix is the element j of observation k, i.e., $X_j^{(k)}$:

$$
\mathcal{X} = \begin{bmatrix}
X_1^{(1)} & X_2^{(1)} & \cdots & X_p^{(1)} \\
X_1^{(2)} & X_2^{(2)} & \cdots & X_p^{(2)} \\
\vdots & \vdots & \ddots & \vdots \\
X_1^{(n)} & X_2^{(n)} & \cdots & X_p^{(1)}
\end{bmatrix}
$$

For instance, by using the data matrix $\mathcal{X}$, we may express the column vector of sample means in the compact form

$$
\overline{\mathbf{X}} = \frac{1}{n} \mathcal{X}^T \mathbf{1}_n \tag{15.1}
$$

Here, $\mathbf{1}_n \in \mathbb{R}^n$ is a column vector with n elements set to 1, not to be confused with the identity matrix $\mathbf{I}_n \in \mathbb{R}^{n,n}$. A useful matrix is

$$
\mathbf{J}_n = \mathbf{1}_n \mathbf{1}_n^T = \begin{bmatrix}
1 & 1 & \cdots & 1 \\
1 & 1 & \cdots & 1 \\
\vdots & \vdots & \ddots & \vdots \\
1 & 1 & \cdots & 1
\end{bmatrix}
$$

Example 15.3 (The centering matrix) When we premultiply a vector $\mathbf{X} \in \mathbb{R}^n$, consisting of univariate observations $X_1, \ldots, X_n$, by the matrix

$$
\mathbf{H}_n = \mathbf{I}_n - \frac{1}{n} \mathbf{1}_n \mathbf{1}_n^T = \mathbf{I}_n - \frac{1}{n} \mathbf{J}_n
$$

we are subtracting the sample mean $\overline{X}$ from all elements of $\mathbf{X}$:

$$
\begin{aligned}
\mathbf{H}_n\mathbf{X} &= \left(\mathbf{I}_n - \frac{1}{n}\mathbf{1}_n\mathbf{1}_n^T\right)\mathbf{X} \\
&= \mathbf{X} - \frac{1}{n}\mathbf{1}_n\sum_{i=1}^{n}X_i \\
&= \left[X_1 - \overline{X}, X_2 - \overline{X}, \ldots, X_n - \overline{X}\right]^T
\end{aligned}
$$

Not surprisingly, the matrix $\mathbf{H}_n$ is called *centering matrix* and may be used with a data matrix $\mathcal{X}$ in order to obtain the matrix of centered data

$$
\mathcal{X}_c = \mathbf{H}_n\mathcal{X}
$$

To understand how this last formula works, you should think of the data matrix as a bundle of column vectors, each one corresponding to a single variable. □

15.3.1 Covariance matrices

Given a random vector $\mathbf{X} \in \mathbb{R}^p$ with expected value $\boldsymbol{\mu}$, the covariance matrix can be expressed as

$$
\boldsymbol{\Sigma} = \mathrm{E}[(\mathbf{X} - \boldsymbol{\mu})(\mathbf{X} - \boldsymbol{\mu})^T]
$$

Note that inside the expectation we are multiplying a column vector $p \times 1$ and a row vector $1 \times p$, which does result in a square matrix $p \times p$. It may also be worth noting that there is a slight inconsistency of notation, since we denote variance in the scalar case by σ^2, but we do not use $\boldsymbol{\Sigma}^2$ here, as this would be somewhat confusing. The element in row i and column j of matrix $\boldsymbol{\Sigma}$, $[\boldsymbol{\Sigma}]_{ij}$, is the covariance σ_{ij} between X_i and X_j. Consistently, we should regard the variance of X_i as $\mathrm{Cov}(X_i, X_i) = \sigma_{ii}$. We may also express the covariance matrix as

$$
\boldsymbol{\Sigma} = \mathrm{E}\left[\mathbf{X}\mathbf{X}^T\right] - \boldsymbol{\mu}\boldsymbol{\mu}^T
$$

This is just a vector generalization of Eq. (8.5). If we consider a linear combination Z of variables $\mathbf{X}$, i.e.,

$$
Z = \mathbf{a}^T\mathbf{X}
$$

then the variance of Z is

$$
\sigma_Z^2 = \mathbf{a}^T\boldsymbol{\Sigma}\mathbf{a} \tag{15.2}
$$

where $\boldsymbol{\Sigma}$ is the covariance matrix of $\mathbf{X}$. By a similar token, let us consider a linear transformation from random vector $\mathbf{X}$ to random vector $\mathbf{Z}$, represented by the matrix $\mathbf{A}$, i.e.

$$
\mathbf{Z} = \mathbf{A}\mathbf{X}
$$

It turns out[6] that the covariance matrix of $\mathbf{Z}$ is

$$\Sigma_{\mathbf{Z}} = \mathbf{A}\Sigma\mathbf{A}^T \tag{15.3}$$

By recalling that a linear combination of jointly normal variables is normal, the following theorem can be immediately understood.

THEOREM 15.1 *Let $\mathbf{X}$ be a vector of n jointly normal variables with expected value $\boldsymbol{\mu}$ and covariance matrix Σ. Given a matrix $\mathbf{A} \in \mathbb{R}^{m,n}$, the transformed vector $\mathbf{A}\mathbf{X}$, taking values in $\mathbb{R}^m$, has a jointly normal distribution with expected value $\mathbf{A}\boldsymbol{\mu}$ and covariance matrix $\mathbf{A}\Sigma\mathbf{A}^T$.*

The above properties refer to covariance matrices, i.e., to probabilistic concepts. The same results carry over to the *sample covariance matrix*, which we denote by $\mathbf{S}$. Again, there is a bit of notational inconsistency with respect to sample variance S^2 in the scalar case, but we will think of sample variance in terms of sample covariance, $S_j^2 = S_{jj}$, and adopt this notation, which is consistent with the use of Σ for a covariance matrix. The sample covariance matrix may be expressed in terms of random observation vectors $\mathbf{X}^{(k)}$:

$$\begin{aligned}
\mathbf{S} &= \frac{1}{n-1}\sum_{k=1}^{n}\left(\mathbf{X}^{(k)} - \overline{\mathbf{X}}\right)\left(\mathbf{X}^{(k)} - \overline{\mathbf{X}}\right)^T \\
&= \frac{1}{n-1}\left(\sum_{k=1}^{n}\mathbf{X}^{(k)}\left[\mathbf{X}^{(k)}\right]^T - n\overline{\mathbf{X}}\,\overline{\mathbf{X}}^T\right) \tag{15.4}
\end{aligned}$$

The expression in Eq. (15.4) is a multivariable generalization of the familiar way of rewriting sample variance; see Eq. (9.7). It is also fairly easy to show that we may write the sample covariance matrix in a very compact form using the data matrix $\mathcal{X}$. The sum in Eq. (15.4) can be expressed as $\mathcal{X}^T\mathcal{X}$, and by rewriting the vector of sample means as in Eq. (15.1) we obtain

$$\begin{aligned}
\mathbf{S} &= \frac{1}{n-1}\left[\mathcal{X}^T\mathcal{X} - n\left(\frac{1}{n}\mathcal{X}^T\mathbf{1}_n\right)\left(\frac{1}{n}\mathbf{1}_n^T\mathcal{X}\right)\right] \\
&= \frac{1}{n-1}\left[\mathcal{X}^T\mathcal{X} - \frac{1}{n}\mathcal{X}^T\mathbf{J}_n\mathcal{X}\right] \\
&= \frac{1}{n-1}\mathcal{X}^T\left[\mathbf{I}_n - \frac{1}{n}\mathbf{J}_n\right]\mathcal{X} \tag{15.5}
\end{aligned}$$

From a computational point of view, Eq. (15.5) may not be quite convenient; however, these ways of rewriting the sample covariance matrix may come in handy when proving theorems and analyzing data manipulations. If the

---

[6]See Chapter 3 of the textbook by Rencher [5] for a detailed treatment.

data are already centered, then expressing the sample covariance matrix is immediate:

$$\mathbf{S} = \frac{1}{n-1}\mathcal{X}_c^T \mathcal{X}_c$$

We should also note the following properties, that generalize what we are familiar with in the scalar case:

$$E[\mathbf{AX} + \mathbf{b}] = \mathbf{A}E[\mathbf{X}] + \mathbf{b} = \mathbf{A}\boldsymbol{\mu} + \mathbf{b} \qquad (15.6)$$

$$\text{Cov}(\mathbf{AX} + \mathbf{b}) = \text{Cov}(\mathbf{AX}) = \mathbf{A}\text{Cov}(\mathbf{X})\mathbf{A}^T \qquad (15.7)$$

where $\mathbf{b}$ is an arbitrary vector of real numbers.

If we need the sample correlation matrix $\mathbf{R}$, consisting of sample correlation coefficients R_{ij} between X_i and X_j, we may introduce the diagonal matrix of sample standard deviations

$$\mathbf{D} = \begin{bmatrix} \sqrt{S_{11}} & 0 & \cdots & 0 \\ 0 & \sqrt{S_{22}} & \cdots & 0 \\ \vdots & \vdots & \ddots & \vdots \\ 0 & 0 & \cdots & \sqrt{S_{pp}} \end{bmatrix}$$

and then let

$$\mathbf{R} = \mathbf{D}^{-1}\mathbf{S}\mathbf{D}$$

15.3.2   Measuring distance and the Mahalanobis transformation

In Section 3.3.1 we defined the concept of vector norm, which can be used to measure the distance between points in $\mathbb{R}^p$. We might also define the distance between observed vectors in the same way, but in statistics we typically want to account for the covariance structure as well. As an introduction, consider the distance between the realization of a random variable X and its expected value μ, or between two realizations X_1 and X_2. Does a distance measure based on a plain difference, such as $|X - \mu|$ or $|X_1 - X_2|$, make sense? Such a measure is highly debatable, from a statistical perspective, as it disregards dispersion altogether. A suitable measure should be expressed in terms of number of standard deviations, which leads to the *standardized distances*

$$\frac{|X - \mu|}{\sigma}, \qquad \frac{|X_1 - X_2|}{\sigma}$$

Alternatively, we may consider the squared distances

$$D^2(X, \mu) = \frac{(X - \mu)^2}{\sigma^2}, \qquad D^2(X_1, X_2) = \frac{(X_1 - X_2)^2}{\sigma^2}$$

To generalize the idea to the distance between observation vectors $\mathbf{X}^{(1)}$ and $\mathbf{X}^{(2)}$, we may rely on the covariance matrix and define the squared distance

$$D^2\left(\mathbf{X}^{(1)}, \mathbf{X}^{(2)}\right) = \left(\mathbf{X}^{(1)} - \mathbf{X}^{(2)}\right)^T \boldsymbol{\Sigma}^{-1} \left(\mathbf{X}^{(1)} - \mathbf{X}^{(2)}\right)$$

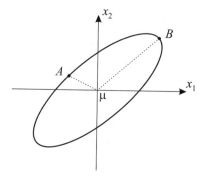

Fig. 15.6   Illustration of Mahalanobis distance.

where $\boldsymbol{\Sigma}^{-1}$ is the inverse of the covariance matrix. More often than not, we do not know the underlying covariance matrix, and we have to replace it with the sample covariance matrix $\mathbf{S}$. We may also express the distance with respect to the expected value in the same way:

$$D^2(\mathbf{X}, \boldsymbol{\mu}) = (\mathbf{X} - \boldsymbol{\mu})^T \boldsymbol{\Sigma}^{-1} (\mathbf{X} - \boldsymbol{\mu})$$

The last expression should be familiar, since it is related to the argument of the exponential function that defines the joint PDF of a multivariable normal distribution.[7] We also recall that the level curves of this PDF are ellipses, whose shape depends on the correlation between variables. This is very helpful in understanding the rationale behind the definition of the distances described above, which are known as *Mahalanobis distances*. Consider the two points A and B in Fig. 15.6. The figure is best interpreted in terms of a bivariate normal distribution with expected value $\boldsymbol{\mu}$; the ellipse is a level curve of its PDF. Geometrically, if we consider standard Euclidean distance, the points A and B do not have the same distance from $\boldsymbol{\mu}$. However, if we account for covariance by Mahalanobis distance, we see that the two points have the same distance from $\boldsymbol{\mu}$. Strictly speaking, we cannot compare the probabilities of outcomes A and B, as they are both zero; nevertheless, the two points are on the same "isodensity" curve and are, in a loose sense, equally likely.

Mahalanobis distance can also be interpreted as a Euclidean distance modified by a suitably chosen weight matrix, which changes the relative importance of dimensions. Measuring distances is essential in clustering algorithms, as we will see in Section 17.4.1. Finally, Mahalanobis distance may also be interpreted in terms of a transformation, called *Mahalanobis transformation*. Consider the square root of the covariance matrix, i.e., a symmetric matrix $\boldsymbol{\Sigma}^{1/2}$ such that

$$\boldsymbol{\Sigma}^{1/2}\boldsymbol{\Sigma}^{1/2} = \boldsymbol{\Sigma}$$

[7]See Section 8.4.

and the transformation

$$\mathbf{Z} = \left(\boldsymbol{\Sigma}^{1/2}\right)^{-1} (\mathbf{X} - \boldsymbol{\mu})$$

where $\mathbf{X}$ is a random variable with expected value $\boldsymbol{\mu}$ and covariance matrix $\boldsymbol{\Sigma}$. Clearly, this transformation is just an extension of familiar standardization of a scalar random variable. The distance between $\mathbf{X}$ and $\boldsymbol{\mu}$ can be expressed in terms of standardized variables as follows:

$$
\begin{aligned}
D^2\left(\mathbf{X}, \boldsymbol{\mu}\right) &= (\mathbf{X} - \boldsymbol{\mu})^T \boldsymbol{\Sigma}^{-1} (\mathbf{X} - \boldsymbol{\mu}) \\
&= (\mathbf{X} - \boldsymbol{\mu})^T \left(\boldsymbol{\Sigma}^{1/2}\boldsymbol{\Sigma}^{1/2}\right)^{-1} (\mathbf{X} - \boldsymbol{\mu}) \\
&= \left[\left(\boldsymbol{\Sigma}^{1/2}\right)^{-1}(\mathbf{X} - \boldsymbol{\mu})\right]^T \left[\left(\boldsymbol{\Sigma}^{1/2}\right)^{-1}(\mathbf{X} - \boldsymbol{\mu})\right] \\
&= \mathbf{Z}^T\mathbf{Z}.
\end{aligned}
$$

Now, using Eqs. (15.3) and (15.7), we find that the covariance matrix of the standardized variables is

$$
\begin{aligned}
\boldsymbol{\Sigma}_{\mathbf{Z}} &= \mathrm{Cov}\left[\left(\boldsymbol{\Sigma}^{1/2}\right)^{-1}(\mathbf{X} - \boldsymbol{\mu})\right] \\
&= \left(\boldsymbol{\Sigma}^{1/2}\right)^{-1}\boldsymbol{\Sigma}\left(\boldsymbol{\Sigma}^{1/2}\right)^{-1} \\
&= \left(\boldsymbol{\Sigma}^{1/2}\right)^{-1}\boldsymbol{\Sigma}^{1/2}\boldsymbol{\Sigma}^{1/2}\left(\boldsymbol{\Sigma}^{1/2}\right)^{-1} \\
&= \mathbf{I}_n
\end{aligned}
$$

Thus, we see that Mahalanobis transformation yields a set of uncorrelated and standardized variables.

For further reading

- The theory behind multivariate analysis may be rather challenging, but a readable treatment can be found in Refs. [1] and [5].

- See Chapter 1 of Ref. [6] for a discussion of dependence vs. interdependence techniques, as well as the different types of scale.

- One of the most fertile application domains of multivariate methods is marketing; see Ref. [4] for an illustration of this kind of applications.

- A very nice compromise between the need of explaining the theory behind the methods, which is the only way of truly understanding their pitfalls, and the urge of presenting concrete and real-life applications has been struck in the book by Lattin et al. [3].

- The link between matrix algebra and multivariate statistics is well documented in the treatise by Harville [2].

REFERENCES

1. W. Härdle and L. Simar, *Applied Multivariate Statistical Analysis,* 2nd ed., Springer, Berlin, 2007.

2. D.H. Harville, *Matrix Algebra from a Statistician's Perspective*, Springer, New York, 1997.

3. J.M. Lattin, J.D. Carroll, and P.E. Green, *Analyzing Multivariate Data,* Brooks/Cole, Belmont, CA, 2003.

4. J.H. Myers and G.M. Mullet, *Managerial Applications of Multivariate Analysis in Marketing,* South-Western, Cincinnati, OH, 2003.

5. A.C. Rencher, *Methods of Multivariate Analysis,* 2nd ed., Wiley, New York, 2002.

6. S. Sharma, *Applied Multivariate Techniques,* Wiley, Hoboken, NJ, 1996.

16

Advanced Regression Models

In this chapter we extend the simple linear regression concepts that were introduced in Chapter 10. The first quite natural idea is building a linear regression model involving more than one regressor. Finding the parameters by ordinary least squares (OLS) is a rather straightforward exercise, as we see in Section 16.1. What is much less straightforward is the statistical side of the coin, since the presence of multiple variables introduces some new issues. In Section 16.2 we consider the problems of testing a multiple regression model, selecting regressor variables, and assessing forecasting uncertainty. We do so for the simpler case of nonstochastic regressors and under restrictive assumptions about the errors, i.e., independence, homoskedasticity, and normality. Even within this limited framework we may appreciate issues like bias from omitted variables and multicollinearity. An understanding of their impact is essential from the users' point of view. In fact, given the computational power of statistical software, it is tempting to build a huge model encompassing a rich set of explanatory variables. In practice, this may be a dangerous route, and a sound *parsimony* principle should be always kept in mind.

Multiple linear regression models often include categorical regressors, which are typically accounted for by dummy, or binary, variables. When it is the regressed variable that is categorical, a crude application of regression modeling may lead to nonsensical results. As an example, we could try to model purchasing decisions in discrete terms (yes/no); a regression model can be adopted, estimating the purchase probability of a consumer as a function of explanatory variables related to product features and consumer's profile. However, a standard linear regression model could well predict probabilities that are smaller than zero or larger than one. In Section 16.3 we consider logistic regression, a possible approach to cope with a categorical regressed variable, based on a nonlinear transformation of the output of a linear regression model. There are many settings in which nonlinearity in data must be explictly recognized, leading to nonlinear regression. This is a quite difficult topic, but in Section 16.4 we introduce some modeling tricks to transform a nonlinear regression model to a linear one.

16.1 MULTIPLE LINEAR REGRESSION BY LEAST SQUARES

Running a linear regression with multiple explanatory variables is a rather straightforward extension of what we have seen in Chapter 10, especially if we assume fixed, deterministic values of the regressors. The underlying statistical model is

$$Y = \beta_0 + \beta_1 x_1 + \beta_2 x_2 + \cdots + \beta_q x_q + \epsilon$$

We avoid using α to denote a constant term, so that we may group parameters into a vector $\boldsymbol{\beta} \in \mathbb{R}^{q+1}$. The model is estimated on the basis of a sample of n observations,

$$Y_i, x_{i1}, x_{i2}, \ldots, x_{iq}, \quad i = 1, \ldots, n$$

which are collected in vector $\mathbf{Y}$ and matrix $\mathcal{X}$:

$$\mathbf{Y} = \begin{bmatrix} Y_1 \\ Y_2 \\ Y_3 \\ \vdots \\ Y_n \end{bmatrix}, \quad \mathcal{X} = \begin{bmatrix} 1 & x_{11} & x_{12} & \cdots & x_{1q} \\ 1 & x_{21} & x_{22} & \cdots & x_{2q} \\ 1 & x_{31} & x_{32} & \cdots & x_{3q} \\ \vdots & \vdots & \vdots & & \vdots \\ 1 & x_{n1} & x_{n2} & \cdots & x_{nq} \end{bmatrix}$$

The data matrix $\mathcal{X} \in \mathbb{R}^{n,q+1}$ collects observed values of the regressor variables, and it includes a leading column of ones. This makes notation more uniform; we may think that coefficient β_0 is associated with a stream of constant observations $x_{i0} = 1$. To estimate $\boldsymbol{\beta}$, we apply ordinary least squares, on the basis of the following regression equations:

$$Y_i = b_0 + b_1 x_{i1} + b_2 x_{i2} + \cdots + b_1 x_{iq} + e_i, \quad i = 1, \ldots, n$$

where e_i is the residual for observation i, and b_j is the estimate of parameter β_j, $j = 0, \ldots, q$. If we collect residuals in vector $\mathbf{e} \in \mathbb{R}^n$ and coefficients in vector $\mathbf{b}$, the regression equations may be rewritten in the following convenient matrix form:

$$\mathbf{Y} = \mathcal{X}\mathbf{b} + \mathbf{e} \tag{16.1}$$

Using least squares, we aim at minimizing the sum of squared residuals, which is just the squared norm of vector $\mathbf{e}$:

$$\min_{\mathbf{b}} \sum_{i=1}^{n} e_i^2 = \|\mathbf{e}\|^2 = \mathbf{e}^T \mathbf{e} \tag{16.2}$$

Now we only have to follow the familiar least-squares drift, but in matrix terms. The concepts of Section 3.9.1, concerning derivatives of quadratic forms, come in quite handy here. To see why, let us rewrite Eq. (16.2):

$$
\begin{aligned}
\mathbf{e}^T\mathbf{e} &= (\mathbf{Y} - \mathcal{X}\mathbf{b})^T (\mathbf{Y} - \mathcal{X}\mathbf{b}) \\
&= \mathbf{Y}^T\mathbf{Y} - \mathbf{Y}^T\mathcal{X}\mathbf{b} - \mathbf{b}^T\mathcal{X}^T\mathbf{Y} + \mathbf{b}^T\mathcal{X}^T\mathcal{X}\mathbf{b} \\
&= \mathbf{Y}^T\mathbf{Y} - 2\mathbf{Y}^T\mathcal{X}\mathbf{b} + \mathbf{b}^T\mathcal{X}^T\mathcal{X}\mathbf{b}
\end{aligned}
\tag{16.3}
$$

This is a function of the vector of coefficients $\mathbf{b}$ and includes a constant term, a linear term, and a quadratic form. Furthermore the matrix $\mathcal{X}^T\mathcal{X}$ is square, symmetric, and positive semidefinite, implying that the associated quadratic form is convex. Hence, we are minimizing a convex function, and stationarity conditions are sufficient for optimality; we must just take the gradient, i.e., the vector of partial derivatives of the quadratic form with respect to each coefficient b_j, and set it to zero. From Section 3.9.1, we recall the following rules to obtain the gradient of linear and quadratic functions of multiple variables:

$$
\begin{aligned}
h(\mathbf{z}) = \mathbf{a}^T\mathbf{x} &\Rightarrow \nabla h(\mathbf{z}) = \mathbf{a} \\
g(\mathbf{z}) = \mathbf{z}^T\mathbf{A}\mathbf{z} &\Rightarrow \nabla g(\mathbf{z}) = 2\mathbf{A}\mathbf{z}
\end{aligned}
$$

for a column vector $\mathbf{a}$ and a square matrix $\mathbf{A}$. By applying these rules to (16.3), we immediately get the optimality conditions:

$$
-2\mathcal{X}^T\mathbf{Y} + 2\mathcal{X}^T\mathcal{X}\mathbf{b} = 0 \quad \Rightarrow \quad \mathcal{X}^T\mathcal{X}\mathbf{b} = \mathcal{X}^T\mathbf{Y}
$$

This is just a system of linear equations; the reader is urged to check the size of each matrix involved and to verify that all of the sizes match; in particular, the square matrix $\mathcal{X}^T\mathcal{X}$ belongs to the space $\mathbb{R}^{q+1,q+1}$. To solve this system, formally, we have just to invert a matrix:

$$
\mathbf{b} = \left(\mathcal{X}^T\mathcal{X}\right)^{-1}\mathcal{X}^T\mathbf{Y}
\tag{16.4}
$$

Can something go wrong with this matrix inversion? The answer is "definitely yes," and it is fairly easy to see why, by a proper interpretation of the regression equation (16.1). What we are doing is trying to express a vector $\mathbf{Y} \in \mathbb{R}^n$ as a linear combination of $q + 1$ vectors:

$$
\mathbf{Y} = \beta_0 \begin{bmatrix} 1 \\ 1 \\ 1 \\ \vdots \\ 1 \end{bmatrix} + \beta_1 \begin{bmatrix} x_{11} \\ x_{21} \\ x_{31} \\ \vdots \\ x_{n1} \end{bmatrix} + \beta_2 \begin{bmatrix} x_{12} \\ x_{22} \\ x_{32} \\ \vdots \\ x_{n2} \end{bmatrix} + \cdots + \beta_q \begin{bmatrix} x_{1q} \\ x_{2q} \\ x_{3q} \\ \vdots \\ x_{nq} \end{bmatrix}
$$

$$
= \beta_0 \mathbf{1}_n + \beta_1 \mathbf{x}_1 + \beta_2 \mathbf{x}_2 + \cdots + \beta_q \mathbf{x}_q
$$

where $\mathbf{x}_j$, $j = 1,\ldots,q$, is a column vector collecting the n observations x_{ij} of variable j, and $\mathbf{1}_n \in \mathbb{R}^n$ is a vector consisting of elements equal to 1.

Since $n > q + 1$, there is little hope of succeeding, and we must settle for an optimal approximation, whereby we project the vector $\mathbf{Y} \in \mathbb{R}^n$ onto a subspace of $q+1$ vectors, in such a way as to minimize the norm of the residual vector $\mathbf{e} = \mathbf{Y} - \mathcal{X}\mathbf{b}$. In general, we cannot take for granted that these $q + 1$ vectors are linearly independent; if they are not, even the coefficients in this approximation will not be well defined, since one of the basis columns can be expressed as a linear combination of some other columns. So, in order to avoid trouble, the vectors $\mathbf{1}_n$ and $\mathbf{x}_j$ should be linearly independent, which amounts to saying that the data matrix $\mathcal{X}$ is full-rank.[1] If so, it turns out that the matrix $\mathcal{X}^T \mathcal{X}$ is nonsingular and Eq. (16.4) makes sense. Actually, there is an even subtler issue: Even if the columns of $\mathcal{X}$ are linearly independent, some regressor variables could be strongly correlated. Even in such a case, it is unlikely that random sampling will result in truly linearly dependent columns; however, the $\mathcal{X}^T \mathcal{X}$ could be close to singular, resulting in unstable estimates of the regression parameters. This issue is called *multicollinearity* and is outlined in Section 16.2.1.

16.2 BUILDING, TESTING, AND USING MULTIPLE LINEAR REGRESSION MODELS

The least-squares approach to estimating parameters of a multiple regression model is a fairly straightforward extension of simple linear regression. What is not so easy is the extension of the statistical testing procedures, which present more variants when multiple variables are involved. Nevertheless, the necessary intuition for understanding what commercially available statistical software tools offer is not difficult to grasp, if one is armed with a solid understanding of simple regression. However, when multiple variables are involved, new issues arise, which we outline in the following subsections.

16.2.1 Selecting explanatory variables: collinearity

When selecting variables, there are a few issues and tradeoffs involved.

- We might wish to include variables that are not directly observable, and we have to settle for an observable *proxy*. If so, we must ensure that the variable we include is an adequate substitute.

- Our choices are also affected by the availability of data. If data collection has been carried out and there is no possibility for adding variables and observations, we have to live with it. If data have still to be gathered, cost issues may arise.

---

[1]The data matrix $\mathcal{X} \in \mathbb{R}^{n,q+1}$ has more rows than columns, in nonpathological cases. Hence, the maximum rank it may have is $q + 1$. We also say that it has full *column* rank.

Table 16.1 Sample data illustrating issues in regressor selection.

Y_i	x_{i1}	x_{i2}	Y_i	x_{i1}	x_{i2}
-15.394	2.7023	-5.3178	13.8214	4.3418	-1.0346
-37.077	-0.9968	-4.7894	42.9428	7.2003	0.2621
46.2448	4.376	-0.0892	-4.6615	4.1778	-5.8078
-10.354	4.863	-0.0315	-10.619	3.7131	-10.176
4.6302	0.5606	-1.6227	-27.627	1.503	-5.187
70.7628	7.5727	1.9977	6.7162	4.8832	-5.5274
77.4447	7.5675	3.6647	-6.5051	-0.0085	-4.2406
-7.6364	3.8871	-6.9382	54.5676	6.143	0.3391
15.0487	4.9819	-2.2778	65.3855	8.8707	6.3123
20.6672	4.5239	-3.0824	0.2551	1.9247	-2.7682
-19.91	3.4399	-8.5812	22.6257	6.574	-3.0003
26.1656	6.1774	-0.4347	29.0148	7.762	1.2267
-40.13	2.2351	-7.7927	-76.072	-0.7812	-10.054
90.4386	10.5496	6.7681	-16.093	-0.3229	-6.5207
-43.971	3.5908	-5.9036	18.2359	5.7134	-1.7836

- If observations of many variables are available, it may not be clear which subset of regressors offers the best explanatory or predictive power. Model building procedures have been proposed to come up with the best subset of variables, such as forward selection (a stepwise procedure in which one regressor is added at a time) and backward elimination (a stepwise procedure in which one variable is omitted at a time).

Apparently, we should aim at finding the model with the largest R^2 coefficient, and, arguably, the more variables we include, the better model we obtain. However, the following examples show that subtle difficulties may be encountered.

Example 16.1 (Omitted variables and bias) Let us consider the sample data in Table 16.1. If we regress a model specified as

$$Y = \beta_0 + \beta_1 x_1 + \beta x_2 + \epsilon$$

we obtain the following estimates, including a 95% confidence interval for each parameter:

$$b_0 = 0.8507 \qquad (-18.2734, \ 19.9749)$$
$$b_1 = 5.3863 \qquad (2.2421, \ 8.5306)$$
$$b_2 = 5.1532 \qquad (3.0311, \ 7.2753)$$

The R^2 coefficient is 0.8424. Now, let us repeat the application of least squares, omitting the variable x_2:

$$b_0 = -37.6442 \qquad (-52.1744, \, -23.1140)$$
$$b_1 = 11.1039 \qquad (8.2734, \, 13.9343)$$

The R^2 coefficient now is 0.6975. We see that by omitting a variable, we reduce R^2, which may not be so surprising. We also see that the estimate of β_1 changes considerably and is increased. Actually, data have been generated by Monte Carlo sampling on the basis of the model

$$Y = 3 + 5x_1 + 6x_2 + \epsilon$$

The standard deviation of error was $\sigma_\epsilon = 20$, and the values of the regressors x_1 and x_2 have been obtained by sampling a multivariate normal distribution with:

$$\mu_1 = 4, \quad \mu_2 = -3, \quad \sigma_1 = 3, \quad \sigma_2 = 4, \quad \rho = 0.6$$

When the two regressors are used, R^2 is less than ideal because of the limited sample size and the large variability of errors. When we omit the second explanatory variable, we loose explanatory power, but there is a subtler effect: The estimate of β_1 is biased. In fact, since the two regressors are positively correlated, the coefficient of x_1 in the second regression increases, because part of the effect of x_2 is attributed to x_1. This is an example of *distortion by omitted variables*.

If correlation is negative, we get a lower estimate. For instance, repeating the experiment with $\rho = -0.6$, we obtain

$$b_0 = 6.2703 \qquad (-4.6660, \, 17.2066)$$
$$b_1 = 4.0314 \qquad (1.7020, \, 6.3609)$$
$$b_2 = 5.1532 \qquad (3.0311, \, 7.2753)$$

with $R^2 = 0.5002$, when regressing against both variables. If we omit x_2, we find

$$b_0 = 0.7558 \qquad (-13.7744, \, 15.2860)$$
$$b_1 = 1.5039 \qquad (-1.3266, \, 4.3343)$$

with $R^2 = 0.0406$. In the last case we see a dramatic reduction of R^2, and a negative bias in the estimate of β_1.   ▯

The numbers involved in the example above should be considered with due care, as we are reporting results obtained with one Monte Carlo sample, and no general conclusion can be drawn. What is really important is that when a variable is omitted, this may affect the estimate of parameters associated with other variables. Then, one might argue that it is better to stay on the

Table 16.2 Sample data illustrating the effect of collinearity.

Sample	b_0	b_1	b_2	R^2
1	35.3849	0.1007	10.9388	0.8518
	(-39.6703, 110.4402)	(-11.9250, 12.1265)	(2.0416, 19.8360)	
2	24.8326	1.2393	9.1750	0.8503
	(-53.6050, 103.2701)	(-11.3710, 13.8495)	(-0.0404, 18.3905)	
3	-5.0036	5.7263	5.0299	0.7819
	(-78.9659, 68.9588)	(-6.0237, 17.4762)	(-3.9615, 14.0213)	
4	69.2543	-6.9745	13.9512	0.7769
	(-22.5259, 161.0345)	(-21.8495, 7.9004)	(2.6788, 25.2237)	
5	-30.6348	11.7167	1.6350	0.8404
	(-98.8312, 37.5617)	(0.9933, 22.4401)	(-6.3319, 9.6019)	

safe side, and include as many variables as we can. The following example shows that this is not the case.

Example 16.2 (Multicollinearity and instability) Let us repeat the experiment of Example 16.1, but this time let $\rho = 0.98$. This means that we are regressing against two strongly correlated regressors. Rather than running one regression, we use Monte Carlo sampling to generate $n = 30$ observations, on which we apply OLS, and the procedure is repeated 5 times. In Table 16.2 we show the estimates of the three parameters, along with their 95% confidence intervals, for each sample. The results are quite striking. If we compare the estimates for all samples, we notice a lot of variability: for instance, coefficient b_1 is negative in sample 4 and positive in sample 5. Furthermore, even looking at one sample at a time, we are often unsure of the sign of an effect. It is tempting to attribute all of this to the limited number of observations (30) and to large variability in the underlying errors; however, we see that R^2 is not too bad in the five samples. In fact, the problem is the high correlation between the regressors, which leads to difficulties because of *multicollinearity*.

$\square$

To better understand the difficulty revealed by Example 16.2, we should reflect on the meaning of parameter β_1. This is supposed to be the increment in the expected value of Y, when x_1 is incremented by 1, and x_2 is kept fixed. However, this makes no real sense here, because x_1 and x_2 are strongly correlated. We cannot explore the effect of a variation in one variable, while keeping the other one fixed. Furthermore, because of this strong link, the data matrix $\mathcal{X}$ has two columns that are linearly related to each other, although not really linearly dependent; as a result, the matrix $(\mathcal{X}^T \mathcal{X})$ is not really singular, but solving the system of linear equations is critical since a little change in the inputs results in a large change in the solution. This is a problem called *numerical instability* in numerical analysis.

So, we see that we should include neither too few, nor too many explanatory variables. Finding the right model requires for a bit of experimentation, as well as some understanding of the underlying phenomenon. As the following example illustrates, we cannot trust a linear regression model as a black box.

Example 16.3 (Lurking variables) Consider a regression model used to study the impact of car features on mileage, i.e., miles per gallon or kilometers per liter of petrol. There are many characteristics of a car that can explain mileage; let us choose the turn circle, measured in feet or meters. If you regress mileage on turn circle, you will find a positive coefficient, and a significant model. But is this a *sensible* model? No one would really think that turn circle determines mileage, but then why is the model statistically significant? Actually, the turn circle is positively correlated with vehicle length, which in turn is positively correlated with vehicle weight. Arguably, it is weight the main determinant of mileage, which is somehow surrogated by turn circle in the model. If you include weight and the other characteristics, the coefficient of turn circle will be not statistically significant.

Weight, in this case, is the **lurking variable**.[2] As another example, imagine regressing sales against expenditure in advertisement. Any marketeer will be delighted by a regression model showing the positive impact of such expenditure on sales. But imagine that advertisements are part of a more general campaign in which prices are reduced. How can you be sure that the real driver of sales is not just price reduction?   ☐

These examples show that the interpretation of a regression model cannot rely on pure statistical concepts. Regression models may be able to detect association, not causation. A statistically significant model need not be significant from a business perspective, and domain-specific knowledge is needed to correctly analyze model results.

16.2.2 Testing a multiple regression model

To investigate the statistical validity of a multiple regression model, the first step is to check the variance of the estimators. In this case, we have multiple estimators, so we should check their covariance matrix:

$$\text{Var}(\mathbf{b}) = \text{E}\left[(\mathbf{b} - \boldsymbol{\beta})(\mathbf{b} - \boldsymbol{\beta})^T\right] \tag{16.5}$$

---

[2]We encountered lurking variables when dealing with correlation analysis in Section 9.4.5.

Using Eq. (16.4), we see that

$$
\begin{aligned}
\mathbf{b} - \boldsymbol{\beta} &= \left(\mathcal{X}^T\mathcal{X}\right)^{-1}\mathcal{X}^T\mathbf{Y} - \boldsymbol{\beta} \\
&= \left(\mathcal{X}^T\mathcal{X}\right)^{-1}\mathcal{X}^T(\mathcal{X}\boldsymbol{\beta} + \boldsymbol{\epsilon}) - \boldsymbol{\beta} \\
&= \boldsymbol{\beta} + \left(\mathcal{X}^T\mathcal{X}\right)^{-1}\mathcal{X}^T\boldsymbol{\epsilon} - \boldsymbol{\beta} \\
&= \left(\mathcal{X}^T\mathcal{X}\right)^{-1}\mathcal{X}^T\boldsymbol{\epsilon} \qquad (16.6)
\end{aligned}
$$

The familiar assumptions about errors, in the multivariate case, can be expressed as

$$
E[\boldsymbol{\epsilon}] = \mathbf{0}, \quad \text{Var}(\boldsymbol{\epsilon}) = \sigma^2\,\mathbf{I}
$$

i.e., the expected value is zero and the covariance matrix is the identity matrix times a common variance σ^2, since errors are mutually independent and identically distributed. Incidentally, in what follows we also assume normality of errors. The first implication of Eq. (16.6) is that ordinary least-squares estimators are, under the standard assumptions, unbiased:

$$
E[\mathbf{b} - \boldsymbol{\beta}] = \mathbf{0}
$$

Note that this is easy to obtain if we consider the data matrix $\mathcal{X}$ as a given matrix of fixed numbers; if we consider stochastic regressors, some more work is needed. To find the covariance matrix of estimators, we substitute (16.6) into (16.5):

$$
\begin{aligned}
\text{Cov}(\mathbf{b}) &= E\left\{\left[\left(\mathcal{X}^T\mathcal{X}\right)^{-1}\mathcal{X}^T\boldsymbol{\epsilon}\right]\left[\left(\mathcal{X}^T\mathcal{X}\right)^{-1}\mathcal{X}^T\boldsymbol{\epsilon}\right]^T\right\} \\
&= E\left\{\left(\mathcal{X}^T\mathcal{X}\right)^{-1}\mathcal{X}^T\boldsymbol{\epsilon}\boldsymbol{\epsilon}^T\mathcal{X}\left(\mathcal{X}^T\mathcal{X}\right)^{-1}\right\} \qquad (16.7) \\
&= \left(\mathcal{X}^T\mathcal{X}\right)^{-1}\mathcal{X}^T E\left[\boldsymbol{\epsilon}\boldsymbol{\epsilon}^T\right]\mathcal{X}\left(\mathcal{X}^T\mathcal{X}\right)^{-1} \\
&= \left(\mathcal{X}^T\mathcal{X}\right)^{-1}\mathcal{X}^T\sigma^2\mathbf{I}\mathcal{X}\left(\mathcal{X}^T\mathcal{X}\right)^{-1} \\
&= \sigma^2\left(\mathcal{X}^T\mathcal{X}\right)^{-1}\mathcal{X}^T\mathcal{X}\left(\mathcal{X}^T\mathcal{X}\right)^{-1} \\
&= \sigma^2\left(\mathcal{X}^T\mathcal{X}\right)^{-1} \qquad (16.8)
\end{aligned}
$$

In (16.7), we have used the fact that, since $\mathcal{X}^T\mathcal{X}$ is a symmetric matrix, the transposition of its inverse just yields its inverse

$$
\left[\left(\mathcal{X}^T\mathcal{X}\right)^{-1}\right]^T = \left(\mathcal{X}^T\mathcal{X}\right)^{-1}
$$

What we are still missing is a way to estimate error variance σ^2. Not surprisingly, the problem is solved by an extension of what we know from the theory of simple regression. Let us define the sum of squared residuals

$$
\text{SSR} \equiv \sum_{i=1}^{n}(Y_i - \hat{Y}_i)^2 = (\mathbf{Y} - \mathcal{X}\mathbf{b})^T(\mathbf{Y} - \mathcal{X}\mathbf{b})
$$

It can be shown that

$$E[\text{SSR}] = \sigma^2(n - q - 1)$$

where q is the number of regressors. Then, the following unbiased estimator of σ^2 is obtained:

$$S^2 = \frac{\text{SSR}}{n - q - 1}$$

The first thing to notice is that if we plug $q = 1$, we get the familiar result from simple regression.[3] We wee that the more regressors we use, the more degrees of freedom we lose. Indeed, confidence intervals and hypothesis testing for *single* parameters, under a normality assumption, is not different from the case of a single regressor; the only caution is that we must account for the degrees of freedom we lose for each additional regressor.

If we test the significance of a *single* parameter, essentially we run a familiar t test. In the case of multiple regressors, however, it may be more informative to test the *whole* model, or maybe a subset of parameters. This leads to an F test, based on the analysis of variance concepts that we outlined in Section 10.3.4. The F statistic is

$$F = \frac{\sum_{i=1}^{n}(\hat{Y}_i - \overline{Y})^2/q}{\sum_{i=1}^{n}(Y_i - \hat{Y}_i)^2/(n - q - 1)}$$

This is an F variable with q and $n - q - 1$ degrees of freedom, whose quantiles are tabulated and can be used to check overall significance. To understand this ratio, we should regard it as the ratio of two terms:

1. Explained variability, in the form of a sum of squares divided by q degrees of freedom (note that q is the total number of regression coefficients minus 1)

2. Unexplained variability, the sum of squared residuals divided by $n - q - 1$ degrees of freedom

Finally, the R^2 coefficient can also be calculated along familiar lines. However, when using the coefficient of determination with multiple variables, there is a subtle issue we should consider. What happens if we add another regressor? Clearly, R^2 cannot decrease if we add one more opportunity to fit empirical data, but this does not necessarily mean that the model has improved. The more variables we consider, the more uncertainty we have in estimating their parameters. As Example 16.2 shows, we may have trouble with uncertain estimates because of collinearity. Hence, adding a possibly correlated regressor

---

[3]See Eq. (10.19).

may be far from beneficial, and we need a measure to capture the tradeoff between adding one variable and the trouble this may cause. The *adjusted R^2* coefficient has been proposed, which accounts for the degrees of freedom that are lost because of additional regressors:

$$R^2_{\text{adj}} \equiv 1 - (1 - R^2)\frac{(\text{total df})}{(\text{error df})} = 1 - (1 - R^2)\frac{n-1}{n-q-1} \qquad (16.9)$$

To see the rationale behind the adjustment, recall that R^2 is the ratio of explained variability and total variability:[4]

$$R^2 \equiv \frac{\text{ESS}}{\text{TSS}} = 1 - \frac{\text{SSR}}{\text{TSS}}$$

In the adjusted R^2, we divide the two terms of the ratio by their degrees of freedom:

$$R^2_{\text{adj}} = 1 - \frac{\text{SSR}/(n-q-1)}{\text{TSS}/(n-1)}$$

which leads to (16.9). The sum of squared residuals SSR cannot increase, if we add regressors. However, the term $(n-q-1)$ may decrease faster than SSR, if the added regressors do not contribute much explanatory power. The net result is that an additional regressor may actually result in a decrease of R^2_{adj}.

16.2.3 Using regression for forecasting and explanation purposes

We have seen that omitting variables may result in biased estimates, or even in debatable models where significant coefficients are associated with regressor variables that may even have no real impact on the response variable. However, a rather cynical point of view could be that, as long as the model does a good job at forecasting, no one should care. This is an opinion that should not be dismissed harshly. We should wonder what our real aim is when building a regression model. If we are interested only in a forecasting model, then maybe proper variable selection is not an issue, as long as bad choices do not result in a model with very little predictive power. This could be the point of view of an engineer or someone who is just interested in the decisions that are based on model output; if the decisions are satisfactory, so be it. Arguably, the viewpoint of a sociologist would be rather different, as she would probably really like to understand what drives a phenomenon. There is nothing wrong in either reasoning; they are just two different ways of using modeling tools. Indeed, many forecasting modeling frameworks that we do not consider here, like neural networks, have been criticized because they do not offer any explanation for their output, yet they may be practically useful.

---

[4]See Eq. (10.21).

Leaving such philosophical considerations aside, when we have estimated a multiple linear regression model, we might be interested in forecasting. Given the model[5]

$$\hat{Y} = b_0 + b_1 x_1 + b_2 x_2 + \cdots + b_q x_q = \mathbf{b}^T \mathbf{x}$$

we should just plug $\mathbf{x}_0$ to obtain a point forecast $\hat{Y}_0 = \mathbf{b}^T \mathbf{x}_0$. However, we have repeatedly observed that a single, point forecast may be of little use, and a suitable prediction interval should be devised. Given an estimate $\hat{\sigma}_\epsilon$ and a confidence level α_2, if we assume normality of errors, it is tempting to build a prediction interval like

$$\hat{Y}_0 \pm t_{1-\alpha/2, n-q-1} \hat{\sigma}_\epsilon$$

where in selecting the t distribution we account for estimated coefficients in setting the degrees of freedom. However, we know from Section 10.4 that doing so is not quite correct, as we would consider the uncertainty only in the realization of the error term, and not the uncertainty in the estimate of coefficients. To properly cast the problem, we should evaluate the variance of the forecast error:

$$\text{Var}\left(Y_0 - \mathbf{b}^T \mathbf{x}_0\right)$$

where the random response is given by

$$Y_0 = \boldsymbol{\beta}^T \mathbf{x}_0 + \epsilon$$

Since the estimate of $\mathbf{b}$ depends only on previous realizations of the errors that affect Y, Y_0 is independent from them, and we may write

$$\text{Var}\left(Y_0 - \mathbf{b}^T \mathbf{x}_0\right) = \text{Var}(Y_0) + \text{Var}\left(\mathbf{b}^T \mathbf{x}_0\right) = \sigma_\epsilon^2 + \text{Var}\left(\mathbf{b}^T \mathbf{x}_0\right)$$

Now, to evaluate the second term, we may take advantage of Eq. (16.8):

$$\text{Var}\left(\mathbf{b}^T \mathbf{x}_0\right) = \mathbf{x}_0^T \text{Cov}(\mathbf{b})\mathbf{x}_0 = \sigma_\epsilon^2 \mathbf{x}_0^T \left(\mathcal{X}^T \mathcal{X}\right)^{-1} \mathbf{x}_0.$$

Then, if we knew the "true" standard deviation of errors σ_ϵ, we could conclude

$$\frac{Y_0 - \mathbf{b}^T \mathbf{x}_0}{\sigma_\epsilon \sqrt{1 + \mathbf{x}_0^T \left(\mathcal{X}^T \mathcal{X}\right)^{-1} \mathbf{x}_0}} \sim \mathcal{N}(0, 1)$$

Plugging the estimate $\hat{\sigma}_\epsilon$, we obtain a t distribution with $n - q - 1$ degrees of freedom, and the following prediction interval with confidence level $1 - \alpha$:

$$\mathbf{b}^T \mathbf{x}_0 \pm \hat{\sigma}_\epsilon t_{1-\alpha/2, n-q-1} \sqrt{1 + \mathbf{x}_0^T \left(\mathcal{X}^T \mathcal{X}\right)^{-1} \mathbf{x}_0}$$

---

[5]Consistently with Eq. (16.1), we include a constant 1 in the vector $\mathbf{x}$ of explanatory variables.

16.3   LOGISTIC REGRESSION

Consider the following questions:

- Given a set of characteristics, such as age, sex, cholesterol level, and smoking habits, what is the probability that a person will die because of a heart disease in the next 10 years?

- Given household income, number of children, education level, etc., what is the probability that a household will subscribe to a package of telephone/Web services?

- Given occupation, age, education, income, loan amount, etc., what is the probability that a homeowner will default on his mortgage payments?

All of these questions could be addressed by building a statistical model, possibly a regression model, but they have a troubling feature in common: The response variable is either 0 or 1, where we interpret 0 as "it did not happen" and 1 as "it happened." So far, we have considered the possibility of regression models with qualitative variables as regressors, but here it is the regressed variable that is qualitative, and represented by a Bernoulli random variable taking values 1 or 0 with probabilities π and $1 - \pi$, respectively. We could generalize the problem to a multinomial variable, taking values within a discrete and finite set, but for the sake of simplicity we will stick to the simple binary case.

One possibility for building a statistical model relying on linear regression would be a relationship such as

$$p = \boldsymbol{\beta}^T \mathbf{x} + \epsilon \tag{16.10}$$

where p is the probability of the event of interest, and ϵ is an error whose probability distribution must be chosen. Note that we are not using a binary variable as the response variable, since we are not predicting the occurrence of the event, but its probability. This is possible, e.g., using *linear discriminant analysis*, whereby we identify a linear combination $\boldsymbol{\beta}^T \mathbf{x}$, a threshold level γ, and we predict

$$Y = \begin{cases} 1 & \text{if } \boldsymbol{\beta}^T \mathbf{x} \geq \gamma \\ 0 & \text{otherwise} \end{cases}$$

However, for the purpose of forecasting, estimating a probability may have some advantages, so we will pursue this alternative approach. To fit a model such as (16.10), we could use a dataset in which $Y_i \in \{0, 1\}$ and apply familiar OLS. However, this idea suffers from a significant drawback: When the event is very likely or very unlikely, the response may well be a probability larger than 1 or smaller than 0, respectively. So, we cannot take such a simplistic linear regression approach. To overcome the difficulty, we may adopt a non-linear transformation of the output, which can be sensibly interpreted as a

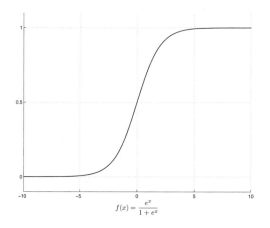

$$f(x) = \frac{e^x}{1 + e^x}$$

Fig. 16.1 The logistic function.

probability. This may be accomplished by using the logistic function

$$f(z) = \frac{\exp(z)}{1 + \exp(z)} \tag{16.11}$$

which is plotted in Fig. 16.1. This is where the term *logistic* regression comes from. The nonlinear transformation allows to express the probability p as

$$p = \frac{\exp(\boldsymbol{\beta}^T \mathbf{x})}{1 + \exp(\boldsymbol{\beta}^T \mathbf{x})} = \frac{1}{1 + \exp(-\boldsymbol{\beta}^T \mathbf{x})} \tag{16.12}$$

Now the probability p is correctly bounded within the interval $[0, 1]$. To gain a better insight into the meaning of parameters β_j, we need a more convenient form that is obtained by solving Eq. (16.12) for the exponential:

$$\exp(\boldsymbol{\beta}^T \mathbf{x}) = \frac{p}{1 - p}$$

Taking logarithms of both sides and adding an error term, we obtain the statistical model

$$\ln\left(\frac{p}{1 - p}\right) = \boldsymbol{\beta}^T \mathbf{x} + \epsilon \tag{16.13}$$

The ratio $p/(1 - p)$ provides us with equivalent information in terms of *odds*. Odds are well known to people engaged in betting. When the odds are "1 to 1," this means that the ratio is $1/1$, which in turn implies

$$\frac{p}{1 - p} = 1 \quad \Rightarrow \quad p = 0.5$$

When we say "it's 10 to 1 that....," we mean that the event is quite likely; indeed:

$$\frac{p}{1 - p} = 10 \quad \Rightarrow \quad p = 0.909$$

The logarithm of the odds ratio is known as a *logit* function. From Eq. (16.13) we can interpret the parameter β_j as the increment in the logit, i.e., the logarithm of the odds ratio, for a unit increment in x_j.

Now we need to find a way to fit the regression coefficients against a set of observations $\mathbf{x}_i$ and $Y_i \in \{0, 1\}$, $i = 1, \ldots, n$. The nonlinear transformation operated by the logistic function precludes application of straightforward least squares; the common approach used to estimate a logistic regression model exploits parameter estimation by the maximum-likelihood approach.[6] To build the maximum-likelihood function, we observe again that Y_i is the realization of a Bernoulli variable, which may be regarded as a binomial variable when only one experiment is carried out.[7] Hence

$$\mathrm{P}(Y = y_i) = p_i^{y_i}(1 - p_i)^{1-y_i}$$

where we write y_i as a number, since we are referring to the observed value of the random variable Y_i. The probability p_i depends on observation $\mathbf{x}_i$ and parameters $\boldsymbol{\beta}$. Of course, we assume independence of the errors, so observations are independent and the likelihood function is just the product of individual probabilities:

$$L = \prod_{i=1}^{n} p_i^{y_i}(1 - p_i)^{1-y_i}$$

$$= \prod_{i=1}^{n} \left(\frac{\exp(\boldsymbol{\beta}^T \mathbf{x}_i)}{1 + \exp(\boldsymbol{\beta}^T \mathbf{x}_i)} \right)^{y_i} \left(\frac{1}{1 + \exp(\boldsymbol{\beta}^T \mathbf{x}_i)} \right)^{1-y_i}$$

The task of maximizing L can be somewhat simplified by taking its logarithm:

$$\ln L = \sum_{i=1}^{n} y_i \left(\frac{\exp(\boldsymbol{\beta}^T \mathbf{x}_i)}{1 + \exp(\boldsymbol{\beta}^T \mathbf{x}_i)} \right) + \sum_{i=1}^{n} (1 - y_i) \left(\frac{1}{1 + \exp(\boldsymbol{\beta}^T \mathbf{x}_i)} \right)$$

Efficient nonlinear programming algorithms are available to maximize log-likelihood numerically and estimate a logistic regression model. Commercial statistical packages are widely available to carry out the task.

16.3.1 A digression: logit and probit choice models

The concepts behind logistic regression and the logit function have also been proposed as a tool to model brand choice in marketing applications. Since choice models are a good way to see integrated use of decision and statistical models, we outline the approach in this section. Consider an individual who chooses between two brands. Ideally, we could model her choice in terms of a

---

[6]See Section 9.9.3.
[7]See Section 6.5.5.

multiattribute utility function $u(\mathbf{x})$ depending on the features of each brand. If such features are collected into vectors $\mathbf{x}_1$ and $\mathbf{x}_2$, the individual would choose brand 1 if

$$u(\mathbf{x}_1) > u(\mathbf{x}_2)$$

Of course, we do not know the utility function of the individual; furthermore, there may be factors at play that are not included in the vector $\mathbf{x}$; indeed, the choice may also depend on factors that are not really related to the product itself, but to price, type of display at a supermarket, etc. Last but not least, we are actually interested in modeling the choice of a "typical" individual, so that we observe different individuals, resulting in some randomness in the choice. A simple approach is to split utility into a term that can be related to product attributes and a term that is completely random:

$$U_{it} = v_{it} + \epsilon_{it}$$

where $i = 1, 2$ refers to brand and t to time or observation number. The simplest model for attributes is $v_{it} = \boldsymbol{\beta}^T \mathbf{x}_{it}$, where $\mathbf{x}_{it}$ is the vector of attributes of brand i when offered at time t. The individual chooses brand 1 at time t if

$$\boldsymbol{\beta}^T \mathbf{x}_{1t} + \epsilon_{1t} > \boldsymbol{\beta}^T \mathbf{x}_{2t} + \epsilon_{2t}$$

or, in other words, when the random effect is not larger than the difference in utilities linked to attributes:

$$\epsilon_{2t} - \epsilon_{1t} < \boldsymbol{\beta}^T (\mathbf{x}_{1t} - \mathbf{x}_{2t})$$

If we introduce the random variable $\eta_t = \epsilon_{2t} - \epsilon_{1t}$, the probability of choosing brand 1 is given by its CDF

$$F_\eta \left[\boldsymbol{\beta}^T (\mathbf{x}_{1t} - \mathbf{x}_{2t}) \right] = \mathrm{P} \left\{ \eta \le \boldsymbol{\beta}^T (\mathbf{x}_{1t} - \mathbf{x}_{2t}) \right\}$$

where we also assume that the distribution of η does not depend on time. Choice models differ in the assumed probability distribution for η.

- If a normal distribution is assumed, a *probit* model is obtained. A possible disadvantage of this choice is that, since the CDF of a normal variable is not available in closed form, we do not find an explicit functional form for the probability of choice.

- An alternative choice, which defines a *logit* model, is the logistic distribution, which is characterized by a CDF of the form (16.11). Taking the derivative of the logistic function, we see that its PDF is given by

$$f_\eta(z) = \frac{\exp(-z)}{[1 + \exp(-z)]^2}$$

This PDF, as shown in Fig. 16.2, has a shape similar to a normal distribution, but its tails are a bit different (and fatter).

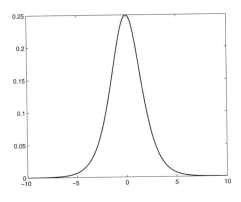

Fig. 16.2 The PDF of the logistic distribution.

If we assume a logit choice model, the probability of choosing brand 1 over brand 2 at time t is given by

$$p_{1t} = \frac{1}{1 - \exp\left\{-\boldsymbol{\beta}^T(\mathbf{x}_{1t} - \mathbf{x}_{2t})\right\}}$$

$$= \frac{\exp\left(\boldsymbol{\beta}^T\mathbf{x}_{1t}\right)}{\exp\left(\boldsymbol{\beta}^T\mathbf{x}_{1t}\right) + \exp\left(\boldsymbol{\beta}^T\mathbf{x}_{2t}\right)}$$

Probit and logit models can be extended to cope with the more complex setting of a choice between $m > 2$ alternatives. In such a case, the probability of choosing brand k is just the probability that the random variable U_{kt} is the largest in the set of variables U_{jt}, $j = 1, \ldots, m$. Here we must specify a joint distribution of a vector $\boldsymbol{\epsilon}_t$ collecting the random terms associated with each possible choice. Different assumptions lead to different choice models, like multinomial probit and multinomial logit models. Each model has advantages and disadvantages in terms of readability and ease of estimation, whose discussion is beyond the scope of the book. One point worth emphasizing, from a managerial perspective, is the sensibility of model predictions. It turns out that the multinomial logit model implies a property called *independence of irrelevant alternatives* (IIA). In fact, using simplifying assumptions about the errors, the predicted probability of choosing brand k is given by

$$p_k = \frac{\exp(v_k)}{\displaystyle\sum_{j=1}^{m} \exp(v_j)}$$

where v_j is a linear function of attributes. The denominator in this ratio is a normalizing factor, which does not influence the ratio of choice probabilities

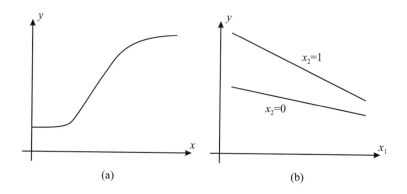

Fig. 16.3   Two cases requiring a nonlinear regression.

of two brands, say, k and l:

$$\frac{p_k}{p_l} = \frac{\exp(v_k)}{\exp(v_l)}$$

If we introduce another brand, this ratio will not change. In other words, the model implies that the introduction of a new alternative will diminish the probability of choosing the preexisting alternative proportionally. However, an often cited counterexample involves two soft-drinks, say a cola and a lemonade. Assume that the choice probabilities are the same, 50% and 50%. If we add a sugar-free cola, it is not quite sensible to assume that the two probabilities will change by the same amount; sales of cola are likely to be much more influenced by the introduction of a similar product. Alternative choice models have been proposed in order to overcome this limitation.

16.4   A GLANCE AT NONLINEAR REGRESSION

Logistic regression introduces a nonlinear transformation to account for the qualitative nature of the response variable. But even when considering a quantitative response, we may be forced to consider nonlinearity. Figure 16.3 shows two examples.

- In Fig. 16.3(a) we observe two effects: a threshold and a saturation effect. It may be helpful to interpret this case in concrete terms: Let us assume that we are regressing sales against advertisement expenditure. A linear model would be unable to account for the fact that, unless some threshold is exceeded, no effect will be discernible. Consider the effect of a couple of spots at 3 a.m. on some unknown local TV station. This is why the picture shows an initial portion where the sales level is constant. On the other hand, if you swamp all of the major networks with spots, sales cannot grow to infinity; the impact will be less and less affective, possibly even counterproductive. In such a case, we could imagine a more general regression model based on a nonlinear functional form such as

$$Y = h(\mathbf{x}, \boldsymbol{\beta}) + \epsilon \qquad (16.14)$$

or

$$Y = h(\mathbf{x}, \boldsymbol{\beta})g(\epsilon) \qquad (16.15)$$

for suitable functions $h(\cdot, \cdot)$ and $g(\cdot)$. Note that we should not take for granted that the error component must be additive.

- The case of Fig. 16.3(b) is a bit different. Here we observe the interaction between the quantitative variable x_1 and the qualitative (categorical) variable x_2. Within a linear regression framework, we may consider a specification such as

$$Y = \beta_0 + \beta_1 x_1 + \beta_2 x_2 + \epsilon$$

where $x_2 \in \{0, 1\}$. However, in this case the effect of x_2 would be a plain *shift* of a linear function. In Fig. 16.3(b) we observe that when $x_2 = 1$, we have a change in slope as well. Hence, we have a more complex interaction between the two variables, which cannot be captured by a linear form. We have to consider a model such as

$$Y = \beta_0 + \beta_1 x_1 + \beta_2 x_2 + \beta_3 x_1 x_2 + \epsilon \qquad (16.16)$$

Given these introductory examples, we should be convinced that sometimes a nonlinear regression model is warranted. Unfortunately, this may be a sort of quantum leap in terms of complexity:

1. We must choose a sensible functional form $h(\mathbf{x}, \boldsymbol{\beta})$, depending on a limited set of parameters.

2. We must find a suitable parameter estimation procedure. In principle, this should not be that difficult. On the basis of Eq. (16.14) and armed with n observations $(Y_i, \mathbf{x}_i)$, $i = 1, \ldots, n$, we may think of solving a *nonlinear* least-squares optimization problem:

$$\min_{\mathbf{b}} \sum_{i=1}^{n} \left(Y_i - h(\mathbf{x}_i, \mathbf{b}) \right)^2$$

Indeed, this is a quite common practice, often referred to as *model calibration*.[8] Powerful software is available to tackle such a problem numerically. Unfortunately, there is a major problem here. First, the resulting

---

[8] As a concrete example of model calibration, imagine that Y_i are observed prices of financial assets, and $\mathbf{x}_i$ is a vector of observable financial variables. The function $h(\cdot, \cdot)$ is a pricing model depending on parameters $\boldsymbol{\beta}$ that are calibrated from quoted prices. These parameters may be used in a variety of ways, both for risk management purposes and for pricing other financial assets that are not traded on regulated markets, such as over-the-counter (OTC) derivatives.

optimization model, unlike ordinary least squares, need not be convex; and second, ensuring good properties of estimators is nontrivial.

3. We must choose a proper statistical framework to tackle issues such as bias and uncertainty in estimates, hypothesis testing, diagnostics, etc. This last point is definitely beyond the scope of this book, and the reader is referred to the listed references.

Fortunately, there are cases in which all of the above is not as awful as it may look. In the remainder of the section we digress a bit on different forms of nonlinear regression, and their implications. In fact, from a users' perspective, the ability to see advantages and disadvantages of different model formulations is by far the most important one; other tasks may be left to suitable statistical software.

16.4.1 Polynomial regression

A good starting point is *polynomial regression*. When facing a clearly nonlinear data pattern, like the one in Fig. 16.3(a), we may try to come up with a suitable approximation of the nonlinear function relating data. In principle, polynomials provide us with an arbitrary degree of flexibility.[9] Let us take a closer look at a model of polynomial regression, in the case of one regressor:

$$Y = \beta_0 + \beta_1 x + \beta_2 x^2 + \beta_3 x^3 + \cdots + \beta_m x^m + \epsilon \qquad (16.17)$$

In this case, there is one regressor variable that is raised to powers in the range from 0 to m. The function is clearly nonlinear in x. However, there is a very important point: The model is nonlinear in the variables, but it is linear in the parameters. To see this, imagine plugging observations x_i, $i = 1, \ldots, n$. We obtain a function that is linear in β_j, $j = 0, 1, \ldots, m$. We could introduce variables $z_j = x^j$, and a multiple linear regression model would result. The same applies to the model with interactions, represented by Eq. (16.16). So, we see that sometimes nonlinear regression can be tackled by the machinery of classical linear regression. There are, however, a few traps:

- Polynomial regression may be quite dangerous, as it is tempting to introduce a high-order polynomial to improve fit. The result can be an overfitted model that offers very poor performance out of the sample. Furthermore, given the oscillatory nature of polynomials,[10] extrapolation outside the dataset may result in meaningless predictions.

- While polynomial regression does not result in additional difficulties if we treat the regressors as given numbers, establishing results for stochastic regressors is not trivial.

---

[9]One of the many Weierstrass theorems states that a continuous function can be approximated to arbitrary accuracy using a polynomial with suitably high degree.
[10]See Section 2.3.2.

16.4.2 Data transformations

If we plug values for the regressor variables into Eq. (16.17), we obtain a linear function of parameters β_j. Now, let us consider an exponential functional form such as

$$Y = e^{\beta_0 + \beta_1 x_1 + \beta_2 x_2}$$

If we plug values for x_1 and x_2, we do not obtain a linear function of parameters β_j, $j = 0, 1, 2$. However, it is easy to see that a proper transformation can lead us back to the linear case. Taking the logarithm of both sides of the equality yields

$$\log Y = \beta_0 + \beta_1 x_1 + \beta_2 x_2$$

This looks a bit like logistic regression, but the underlying model is of a completely different nature. The logarithm is just one of the possible transformations that can be used to linearize a relationship. Another common example is a regression model of the form

$$Y = \beta_0 x_1^{\beta_1} x_2^{\beta_2}$$

which can be transformed to

$$\log Y = \beta_0 + \beta_1 \log x_1 + \beta_2 \log x_2$$

In this case, we take the logarithm of both the regressed and the regressor variables. In other cases, we may also adopt a model such as

$$Y = \beta_0 + \beta_1 \log x_1 + \beta_2 \log x_2$$

We see that with a little ingenuity, many tricks can be used to our advantage. However, we stress again that we must be careful and not forget statistical issues. If we want to introduce errors, should we use a form such as

$$Y = \beta_0 x_1^{\beta_1} x_2^{\beta_2} + \epsilon$$

or the following equation?

$$Y = \beta_0 x_1^{\beta_1} x_2^{\beta_2} e^{\epsilon}$$

An answer can be given only if we choose an appropriate distribution for ϵ and, more importantly, if we know the underlying phenomena well enough to appreciate the sensibility of each assumption. To illustrate, imagine that ϵ is assumed normal. The first model may make no sense if Y is restricted to nonnegative values, as, in principle, a negative realization of the normal error might result in a negative value of Y. The second model would look more sensible from this perspective, since by taking the exponential of a normal variable we obtain a lognormal variable.[11] Clearly, each choice must be carefully pondered and has an impact on the statistical side of regression, in terms

---

[11] See Section 7.7.2.

of testing the model and assessing uncertainty in both parameter estimates and model predictions.

Problems

16.1   Apply the formulas of multiple regression to the case of a single regressor, and verify that the familiar formulas for simple regression are obtained.

16.2   Use the concepts that we have introduced in Chapter 2 to check the qualitative properties of the logistic function of Eq. (16.11).

For further reading

- Readers interested in an introductory but rigorous treatment of regression models can consult Ref. [8]; see also Ref. [7].

- A more thorough and challenging treatment can be found in books specifically aimed at regression models (e.g., Refs. [1] and [3]).

- A treatment geared to applications can be found in Refs. [4], [5], and [6]. In particular, Chapter 13 of Ref. [4] is recommended for further reading on logit choice models, whereas Ref. [5] is a standard reference on forecasting models, including but not limited to regression models.

- Finally, we should also mention references dealing with econometric models, which often illustrate regression models with a good balance between mathematical rigor and real-life applications. An introductory source is Ref. [9], while Ref. [2] is a more advanced reading.

REFERENCES

1. N.R. Draper and H. Smith, *Applied Regression Analysis,* 3rd ed., Wiley, New York, 1998.

2. W.H. Greene, *Econometric Analysis,* 6th ed., Prentice-Hall, Upper Saddle River, NJ, 2008.

3. R.R. Hocking, *Methods and Applications of Linear Models: Regression and the Analysis of Variance,* 2nd ed., Wiley, Hoboken, NJ, 2003.

4. J.M. Lattin, J.D. Carroll, and P.E. Green, *Analyzing Multivariate Data,* Brooks/Cole, Belmont, CA, 2003.

5. S. Makridakis, S.C. Wheelwright, and R.J. Hyndman, *Forecasting: Methods and Applications,* 3rd ed., Wiley, New York, 1998.

6. D.C. Montgomery, C.L. Jennings, and M. Kulahci, *Introduction to Time Series Analysis and Forecasting,* Wiley, New York, 2008.

7. A.C. Rencher, *Methods of Multivariate Analysis,* 2nd ed., Wiley, New York, 2002.

8. S.M. Ross, *Introduction to Probability and Statistics for Engineers and Scientists,* 4th ed., Elsevier Academic Press, Burlington, MA, 2009.

9. J.H. Stock and M.W. Watson, *Introduction to Econometrics,* 2nd ed., Addison-Wesley, Boston, 2006.

17

Dealing with Complexity: Data Reduction and Clustering

This is certainly an age in which we do not suffer from scarcity of data. Using information infrastructures and the Web, we may collect plenty of observations of many variables, resulting in rich datasets waiting for analysis, maybe *too* rich. We sometimes need to simplify data in order to visualize them, to discover patterns, and to make decisions based on them. In this chapter we outline some of the most relevant techniques, which have several applications in supply chain management, marketing, finance, and related fields. First, we motivate the need for data reduction in Section 17.1; this is often a preliminary step to make the application of other quantitative methods possible. Principal component analysis (PCA), the subject of Section 17.2, is a nice illustration of the role played by linear algebra in multivariate statistics; be sure to master the material on eigenvalues and eigenvectors from Chapter 3 before getting here. Section 17.3 illustrates factor analysis, which shares some of the technical machinery of PCA, but takes a different view. Factor analysis is an example of the statistical techniques trying to find latent, i.e., not directly observable, variables that may help in understanding an otherwise too complicated phenomenon. The chapter closes with Section 17.4, which outlines a range of techniques collectively known as cluster analysis. This set of methods aims at grouping observations into similar clusters, implicitly discovering some common features. Again, this has plenty of applications as tariff design and market segmentation, just to name a couple of them.

17.1   THE NEED FOR DATA REDUCTION

Consider a sample of observations $\mathbf{X}^{(k)} \in \mathbb{R}^p$, $k = 1, \ldots, n$. Each observation $\mathbf{X}^{(k)}$ consists of a vector of p elements $X_i^{(k)}$. If $p = 2$, visualizing observations is easy, but this is certainly no piece of cake for large values of p. Hence, we need some way to reduce data dimensionality, by mapping observations in $\mathbb{R}^p$ to observations in a lower-dimensional space $\mathbb{R}^q$, where q is possibly much smaller than p. Reducing data should be helpful in

- Simplifying our analysis

- Improving regression (when too many variables are considered, numerical results may not be stable because of collinearity)

- Speeding up Monte Carlo simulation by concentrating sampling on the most relevant dimensions

- Classifying patterns

- Spotting outliers

One possible strategy for reducing dimensionality of data is to discard some components of each observed vector. For instance, we could consider only the first component X_1 of each observation. However, by adopting such a crude strategy, we will miss much information.

Example 17.1 As a concrete example, imagine analyzing grades obtained by a sample of students on different subjects. Comparing students on the basis of their performance obtained in only one subject, whatever it is, does not seem quite appropriate. Taking the average of the grades is a better choice, but it does not make much sense, either, even though this is what is typically done, since it aggregates different kinds of ability. We might have different subjects that are in fact strongly related to one another; the consequence is that some grades might be strongly correlated, whereas other subjects shed some light on unrelated forms of intelligence. Generalizing a bit, we might take linear combinations of grades. In other words, we could consider a small set of linear combinations of variables for weight vectors $\mathbf{u}_j$:

$$Z_j = \mathbf{u}_j^T \mathbf{X} = \sum_{i=1}^{p} u_{ij} X_i$$

For each weighting scheme j, we find a new variable Z_j. But how can we find a sensible way to assign weights? We should find a limited set of variables Z_j, which really contribute the most information. By "most information," we mean combinations that maximize variance and are uncorrelated. The rationale is that we should keep just a few variables Z_j that account for most observed variability among students and are not redundant in terms of information they provide. This is what principal component analysis does.

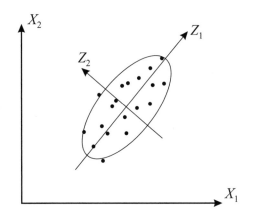

Fig. 17.1 Introducing PCA.

Alternatively, we could try to find underlying factors, which are latent as they are not directly observed, which should explain students' performance. This is the approach taken by factor analysis. Each observation $\mathbf{X}^{(k)}$ should be rewritten as a linear function of just a very few factors f_j. Such latent factors might correspond to basic abilities that may have different degrees of influence on results obtained in different subjects.

Alternatively, we might just try to find meaningful groups of similar students, which may be ideal candidates for different types of jobs or further study, or maybe for supplemental lecture hours in order to help them. This is what cluster analysis is all about.   ⬚

17.2 PRINCIPAL COMPONENT ANALYSIS (PCA)

To introduce principal component analysis (PCA) in the most intuitive way, let us have a look at Fig. 17.1, which shows a scatterplot of observations in two dimensions, X_1 and X_2. The observations are clearly correlated, and correlation is positive. Now, consider the two axes referring to variables Z_1 and Z_2. Of course, we may represent the very same set of observations in terms of Z_1 and Z_2. This is just a change of coordinates, accomplishing two objectives:

- The data have been centered, which is obtained by subtracting the sample mean $\mathbf{X} - \overline{\mathbf{X}}$; as we observed, this shift in the origin of data is often advisable to improve numerical stability in data analysis algorithms.[1]

---

[1]See Example 4.13 for a numerical illustration. We have defined the centering matrix in Example 15.3.

- The variables Z_1 and Z_2 are uncorrelated; this is accomplished by a rotation of the axes.

We recall from Section 3.4.3 that to rotate a vector we multiply it by an orthogonal matrix. Since orthogonal matrices enjoy plenty of nice properties, we might be able to find a linear transformation of variables with some interesting structure. We notice that variables Z_1 and Z_2, regarded as coordinates, are parallel to the axes of an ellipsoid associated with the scatterplot, and the first variable is associated with the largest variance. In fact, there are two different ways to introduce PCA:

1. PCA is a means of taking linear combinations of variables, in such a way that they are uncorrelated.

2. PCA is a way to combine variables so that the first component has maximal variance, the second one has the next maximal variance, and so forth.

Let us pursue both points of view.

17.2.1   A geometric view of PCA

The linear data transformation, including centering, can be written as

$$\mathbf{Z} = \mathbf{A}(\mathbf{X} - \overline{\mathbf{X}})$$

where $\mathbf{A} \in \mathbb{R}^{p,p}$. We assume that data have already been centered, in order to ease notation. Hence

$$
\begin{aligned}
Z_1 &= a_{11}X_1 + a_{12}X_2 + \cdots + a_{1p}X_p \\
Z_2 &= a_{21}X_1 + a_{22}X_2 + \cdots + a_{2p}X_p \\
&\ \ \vdots \qquad \vdots \\
Z_p &= a_{p1}X_1 + a_{p2}X_2 + \cdots + a_{pp}X_p
\end{aligned}
$$

The Z_i variables are called *principal components*: Z_1 is the first principal component. We recall that the matrix $\mathbf{A}$ rotating axes is orthogonal:

$$\mathbf{A}^T\mathbf{A} = \mathbf{I}$$

Now, let us consider the sample covariance matrix of $\mathbf{X}$, i.e., $\mathbf{S_X}$. Since we assume centered data, we recall from Section 15.3.1 that this matrix is given as follows:

$$\mathbf{S_X} = \frac{1}{n-1}\mathcal{X}^T\mathcal{X}$$

Now, we may also find the corresponding sample covariance matrix for Z, $\mathbf{S_Z}$, taking advantage of the results of Section 15.3.1. However, we would like to find a matrix $\mathbf{A}$ such that the resulting principal components are uncorrelated; in other words, $\mathbf{S_Z}$ should be diagonal:

$$\mathbf{S_Z} = \mathbf{A S_X A}^T = \begin{bmatrix} S^2_{Z_1} & & & \\ & S^2_{Z_2} & & \\ & & \ddots & \\ & & & S^2_{Z_p} \end{bmatrix}$$

where $S^2_{Z_i}$ is the sample variance of each principal component. The matrix $\mathbf{A}$ should diagonalize the sample covariance matrix $\mathbf{S_X}$, and we have already seen such a diagonalization in Eq. (3.16). To diagonalize $\mathbf{S_X}$, we should consider the product

$$\mathbf{P}^T \mathbf{S_X P} = \begin{bmatrix} \lambda_1 & & & \\ & \lambda_2 & & \\ & & \ddots & \\ & & & \lambda_p \end{bmatrix}$$

where matrix $\mathbf{P}$ is orthogonal and its columns consist of the normalized eigenvectors of the sample covariance matrix; since this is symmetric, its eigenvectors are indeed orthogonal.[2] The diagonalized matrix consists of the eigenvalues λ_i, $i = 1, \ldots, p$, of the sample covariance matrix $\mathbf{S_X}$. Putting everything together, we see that the rows $\mathbf{a}_i^T$, $i = 1, \ldots, p$, of matrix $\mathbf{A}$ should be the normalized eigenvectors of the sample covariance matrix:

$$\mathbf{A} = \begin{bmatrix} \mathbf{a}_1^T \\ \mathbf{a}_2^T \\ \ldots \\ \mathbf{a}_p^T \end{bmatrix}$$

We also see that the sample variances of the principal components Z_i are the eigenvalues of $\mathbf{S_X}$:

$$S^2_{Z_i} = \lambda_i, \qquad i = 1, \ldots, p$$

If we sort eigenvalues in decreasing order, we see that indeed Z_1 is the first principal component, accounting for most variability. Then, the second principal component Z_2 is orthogonal to Z_1 and is the second in rank. The fraction of variance explained by the first q components is

$$\frac{\lambda_1 + \lambda_2 + \cdots \lambda_q}{\lambda_1 + \lambda_2 + \cdots + \lambda_p} = \frac{\lambda_1 + \lambda_2 + \cdots + \lambda_q}{\sum_{j=1}^{p} [\mathbf{S_Z}]_{jj}}$$

Taking the first few components, we can account for most variability and reduce the problem dimension by replacing the original variables by the principal components.

---

[2]See Theorem 3.10.

17.2.2   Another view of PCA

Another view is obtained by interpreting the first principal component in terms of orthogonal projection. Consider a unit vector $\mathbf{u} \in \mathbb{R}^p$, and imagine projecting the observed vector $\mathbf{X}$ on $\mathbf{u}$. This yields a vector parallel to $\mathbf{u}$, of length $\mathbf{u}^T \mathbf{X}$. Since $\mathbf{u}$ has unit length, the projection of observation $\mathbf{X}^{(k)}$ on $\mathbf{u}$ is

$$\mathbf{P}_{\mathbf{X}^{(k)}} = \left(\mathbf{u}^T \mathbf{X}^{(k)} \right) \mathbf{u}$$

We are projecting p-dimensional observations on just one axis, and of course we would like to have an approximation that is as good as possible. More precisely, we should find $\mathbf{u}$ in such a way that the distance between the originally observed vector $\mathbf{X}^{(k)}$ and its projection is as small as possible. If we have a sample of n observations, we should minimize the average distance

$$\sum_{k=1}^{n} \| \mathbf{X}^{(k)} - \mathbf{P}_{\mathbf{X}^{(k)}} \|^2$$

which looks much like a least-squares problem. This amounts to an orthogonal projection of the original vectors on $\mathbf{u}$, where we know that the original and the projected vectors are orthogonal.[3] Hence, we can apply the Pythagorean theorem to rewrite the problem:

$$\| \mathbf{X}^{(k)} - \mathbf{P}_{\mathbf{X}^{(k)}} \|^2 = \| \mathbf{X}^{(k)} \|^2 - \| \mathbf{P}_{\mathbf{X}^{(k)}} \|^2$$

Therefore, we essentially want to maximize

$$\sum_{i=1}^{n} \| \mathbf{P}_{\mathbf{X}_i} \|^2$$

subject to the condition $\| \mathbf{u} \| = 1$. The problem can be restated as

$$\max \quad \mathbf{u}^T \mathcal{X}^T \mathcal{X} \mathbf{u}$$
$$\text{s.t.} \quad \mathbf{u}^T \mathbf{u} = 1$$

But we know that, assuming data are centered, the sample covariance matrix is $\mathbf{S}_X = \mathcal{X}^T \mathcal{X} / (n - 1)$; hence, the problem is equivalent to

$$\max \quad \mathbf{u}^T \mathbf{S}_X \mathbf{u} \tag{17.1}$$
$$\text{s.t.} \quad \mathbf{u}^T \mathbf{u} = 1 \tag{17.2}$$

In plain English, what we want is finding one dimension on which multi-dimensional data should be projected, in such a way that the variance of

---

[3] See Example 3.7.

the projected data is maximized. This makes sense from a least-squares perspective, but it also have an intuitive appeal: The dimension along which we maximize variance is the one providing the most information.

To solve the problem above, we may associate the constraint (17.2) with a Lagrange multiplier λ and augment the objective function (17.1) to obtain the Lagrangian function:[4]

$$\mathcal{L}(\mathbf{u}, \lambda) = \mathbf{u}^T \mathbf{S_X} \mathbf{u} + \lambda(1 - \mathbf{u}^T \mathbf{u}) = \mathbf{u}^T(\mathbf{S_X} - \lambda \mathbf{I})\mathbf{u} + \lambda$$

The gradient of the Lagrangian function with respect to $\mathbf{u}$ is

$$2(\mathbf{S_X} - \lambda \mathbf{I})\mathbf{u}$$

and setting it to zero yields the first-order optimality condition

$$\mathbf{S_X} \mathbf{u} = \lambda \mathbf{u}$$

This amounts to saying that λ must be an eigenvalue of the sample covariance matrix, but which one? We can rewrite the objective function (17.1) as follows:

$$\mathbf{u}^T \mathbf{S_X} \mathbf{u} = \lambda \mathbf{u}^T \mathbf{u} = \lambda$$

Hence, we see that λ should be the largest eigenvalue of $\mathbf{S_X}$, $\mathbf{u}$ is the corresponding normalized eigenvector, and we obtain the same result as in the previous section. Furthermore, we should continue on the same route, by asking for another direction in which variance is maximized, subject to the constraint that it is orthogonal to the first direction we found. Since eigenvectors of a symmetric matrix are orthogonal, we see that indeed we will find all of them, in decreasing order of the corresponding eigenvalues.

17.2.3   A small numerical example

Principal component analysis in practice is carried out on sampled data, but it may be instructive to consider an example where both the probabilistic and the statistical sides are dealt with.[5] Consider first a random variable with bivariate normal distribution, $\mathbf{X} \sim \mathcal{N}(\mathbf{0}, \mathbf{\Sigma})$, where

$$\mathbf{\Sigma} = \begin{bmatrix} 1 & \rho \\ \rho & 1 \end{bmatrix}$$

[4]Strictly speaking, we are in trouble here, since we are maximizing a *convex* quadratic form. However, we may replace the equality constraint by an inequality constraint, which results in a concave problem, i.e., a problem in which the optimal solution is on the boundary of the feasible solution (see Section 12.1.3). It turns out that the solution we pinpoint using the Lagrange multiplier method is the right one.

[5]This example has been adapted from Chapter 9 of Ref. [1].

and $\rho > 0$. Essentially X_1 and X_2 are standard normal variables with positive correlation ρ. To find the eigenvalues of $\boldsymbol{\Sigma}$, we must find its characteristic polynomial and solve the corresponding equation

$$\left| \begin{array}{cc} 1 - \lambda & \rho \\ \rho & 1 - \lambda \end{array} \right| = (1 - \lambda)^2 - \rho^2 = 0$$

This yields the two eigenvalues $\lambda_1 = 1 + \rho$ and $\lambda_2 = 1 - \rho$. Note that the two eigenvalues are positive, since ρ is a correlation coefficient. To find the first eigenvalue, we consider the system of linear equations:

$$\left[\begin{array}{cc} -\rho & \rho \\ \rho & -\rho \end{array} \right] \left[\begin{array}{c} u_1 \\ u_2 \end{array} \right] = (1 + \rho) \left[\begin{array}{c} u_1 \\ u_2 \end{array} \right]$$

Clearly, the two equations are linearly dependent and any vector such that $u_1 = u_2$ is an eigenvector. By a similar token, any vector such that $u_1 = -u_2$ is an eigenvector corresponding to λ_2. Two normalized eigenvectors are

$$\gamma_1 = \frac{1}{\sqrt{2}} \left[\begin{array}{c} 1 \\ 1 \end{array} \right], \qquad \gamma_2 = \frac{1}{\sqrt{2}} \left[\begin{array}{c} 1 \\ -1 \end{array} \right]$$

These are the rows of the transformation matrix

$$\mathbf{Z} = \mathbf{A}(\mathbf{X} - \boldsymbol{\mu}) = \frac{1}{\sqrt{2}} \left[\begin{array}{cc} 1 & 1 \\ 1 & -1 \end{array} \right] \left[\begin{array}{c} X_1 \\ X_2 \end{array} \right]$$

Since we are dealing with standard normals, $\boldsymbol{\mu} = \mathbf{0}$ and the first principal component is

$$Z_1 = \frac{X_1 + X_2}{\sqrt{2}}$$

The second principal component is

$$Z_2 = \frac{X_1 - X_2}{\sqrt{2}}$$

As a further check, let us compute the variance of the first principal component:

$$\text{Var}(Z_1) = \tfrac{1}{2} \left[\text{Var}(X_1) + \text{Var}(X_2) + 2\text{Cov}(X_1, X_2) \right] = 1 + \rho = \lambda_1$$

Figure 17.2 shows the level curves of the joint density of $\mathbf{X}$ when $\rho = 0.85$:

$$f_{\mathbf{X}}(\mathbf{x}) = \frac{1}{2\pi(1 - \rho^2)} \exp \left\{ -\frac{1}{2} \mathbf{x}^T \boldsymbol{\Sigma}^{-1} \mathbf{x} \right\}$$

Since correlation is positive, the main axis of the ellipses has positive slope. It is easy to see that along that direction we have the largest variability.

As we noted, practical PCA is carried out on sampled data. Figure 17.3 shows a sample of size 200 from the above bivariate normal distribution, with $\rho = 0.85$. The sample statistics are

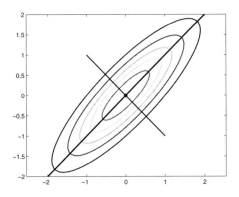

Fig. 17.2 Level curves of a multivariate normal with $\rho = 0.85$.

$$\overline{\mathbf{X}} = \begin{bmatrix} -0.0395 \\ -0.0568 \end{bmatrix}, \qquad \mathbf{S_X} = \begin{bmatrix} 0.8269 & 0.7118 \\ 0.7118 & 0.8768 \end{bmatrix}$$

Since the sample size is not very large, we see that estimated parameters are not too close to what is expected. Nevertheless, the observation cloud (scatterplot) displayed in Fig. 17.3(a) clearly shows positive correlation. The matrix of normalized eigenvectors is

$$\mathbf{C} = \begin{bmatrix} 0.6946 & -0.7194 \\ 0.7194 & 0.6946 \end{bmatrix}$$

Apart from a sign, the values in this matrix, if the estimates were perfect, should be $1/\sqrt{2} = 0.7071$. The eigenvalues of the sample covariance matrix are

$$\lambda_1 = 1.5641, \qquad \lambda_2 = 0.1396$$

and indeed:

$$\mathbf{C S_X C}^T = \begin{bmatrix} 1.5641 & 0 \\ 0 & 0.1396 \end{bmatrix}$$

Note that data need be centered, since the sample means are not zero. The two small plots in Figs. 17.3(b) and (c) show the two principal components, i.e., the projections of the original data. We clearly see that the first principal component accounts for most variability, precisely

$$\frac{1.5641}{1.5641 + 0.1396} = 91.81\%$$

17.2.4 Applications of PCA

Principal component analysis can be applied in a marketing setting when questionnaires are administered to potential customers asking for a quantitative evaluation along many dimensions. Many such questions are, or are

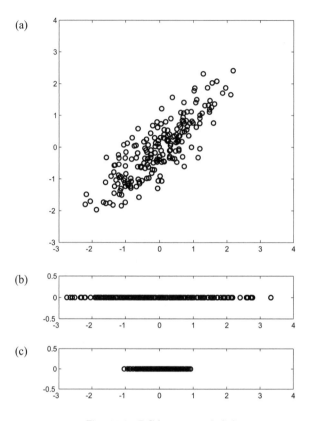

Fig. 17.3 PCA on sampled data.

perceived as, redundant. Spotting the few principal components may help in assessing which product features, or combination thereof, are most important. They can also tell groups of customers apart, helping in market segmentation.

PCA can be applied in a financial setting as well, to reduce the dimensionality of the *term structure of interest rates*, which specifies an interest rate for different time periods. Let $R_{t,t+\tau}$ be the annualized interest rate for a loan starting at time t and maturing at time $t + \tau$, where τ is the length of corresponding time period. Usually, such rates are not constant and are increasing with τ. There are different theories trying to explain the term structure; one of the possible explanations is based on liquidity risk, which is higher for longer maturities. If we observe these rates over time, they are going to change randomly. However, such changes must preserve some basic consistency between rates, in order to prevent arbitrage opportunities. Measuring and managing interest rate risk is complicated by the presence of so many interrelated risk factors; also Monte Carlo simulation can be challenging. One possible approach to make the problem more manageable is to replace the term structure by a few key risk factors, possibly obtained by PCA.[6]

17.3 FACTOR ANALYSIS

The rationale behind factor analysis may be best understood by a small numerical example.

Example 17.2 Consider observations in $\mathbb{R}^5$ and the correlation matrix of their component variables $X_1, X_2, \ldots, X_5$:

$$\begin{bmatrix} 1.00 & 0.90 & 0.05 & 0.05 & 0.05 \\ 0.90 & 1.00 & 0.05 & 0.05 & 0.05 \\ 0.05 & 0.05 & 1.00 & 0.90 & 0.90 \\ 0.05 & 0.05 & 0.90 & 1.00 & 0.90 \\ 0.05 & 0.05 & 0.90 & 0.90 & 1.00 \end{bmatrix}$$

Does this suggest some structure? We see that X_1 and X_2 seem to be strongly correlated with each other, whereas they seem weekly correlated with X_3, X_4, and X_5. The latter variables, on the contrary, seem to be strongly correlated with one another. This suggests that the components of the random vector $\mathbf{X}$ have some latent structure. In particular, we may imagine the existence of two factors, f_1 and f_2, that explain the two groups of variables. Some additional "noise" component must exist, otherwise we would have a perfectly block-structured matrix, but it seems that we might reduce dimensionality from 5 to 2, without losing much information. $\square$

The example displays an obvious structure, which may need some data transformation work to be discovered in more general cases. Factor analysis (FA) is somewhat related to principal component analysis, as both may be regarded as data reduction procedures; however

- In PCA we build linear combinations of observable variables; in FA we look for unobservable underlying factors.

- In PCA we want to explain most of observed variance; in FA we work with covariances and correlations.

To express FA in formulas, let us introduce a vector $\mathbf{f}$ of m factors, f_j, $j = 1, \ldots, m$, where $m < p$. These factors are common to all p components X_i of a random vector $\mathbf{X}$; these are also associated with specific factors ϵ_i, $i = 1, \ldots, p$,

---

[6]See, e.g., Chapters 3 and 6 of L. Martellini, P. Priaulet, S. Priaulet, *Fixed-Income Securities: Valuation, Risk Management, and Portfolio Strategies*, Wiley, New York, 2003.

resulting in the following set of relationships:

$$X_1 = \mu_1 + \lambda_{11} f_1 + \cdots + \lambda_{1m} f_m + \epsilon_1$$
$$X_2 = \mu_2 + \lambda_{21} f_1 + \cdots + \lambda_{2m} f_m + \epsilon_2$$
$$\vdots \quad \vdots$$
$$X_p = \mu_p + \lambda_{p1} f_1 + \cdots + \lambda_{pm} f_m + \epsilon_p$$

The coefficients λ_{ij} are called *factor loadings* and are related to the impact of common factor f_j on component X_i. The following conditions are typically assumed:

- $\mathrm{E}[f_j] = 0$, $\mathrm{Var}(f_j) = 1$
- $\mathrm{E}[\epsilon_i] = 0$, $\mathrm{Var}(\epsilon_i) = \psi_i$ (specific variance)
- $\mathrm{Cov}(f_j, f_k) = 0$ for $j \neq k$ (uncorrelated factors)
- $\mathrm{Cov}(\epsilon_i, \epsilon_k) = 0$ for $i \neq k$ (uncorrelated specific components)
- $\mathrm{Cov}(\epsilon_i, f_j) = 0$

These assumptions imply that $\mathrm{E}[X_i] = \mu_i$ and all of the involved factors, common and specific, are mutually uncorrelated; in other words, the model assumes that the factors f_j represent whatever is "common" between the components. Since common factors are uncorrelated, we have an *orthogonal factor model*. Since specific factors are uncorrelated as well, we may also speak of a *diagonal model*.[7] This way of writing FA bears some resemblance to multiple linear regression, but we should note some key differences:

- We are relating factors to many variables X_i simultaneously, not a single one.
- In linear regression we use observable explanatory variables; factors f_j are latent (unobservable).

We may also express FA in matrix form:

$$\mathbf{X} = \boldsymbol{\mu} + \boldsymbol{\Lambda}\mathbf{f} + \boldsymbol{\epsilon} \tag{17.3}$$

where the *loading matrix* $\boldsymbol{\Lambda} \in \mathbb{R}^{p,m}$ collects the factor loadings λ_{ij}. In order to make the idea more precise and operational, we need to express the covariance matrix of $\mathbf{X}$:

$$\boldsymbol{\Sigma} = \mathrm{Cov}(\boldsymbol{\Lambda}\mathbf{f}) + \mathrm{Cov}(\boldsymbol{\epsilon})$$
$$= \boldsymbol{\Lambda}\,\mathrm{Cov}(\mathbf{f})\boldsymbol{\Lambda}^T + \boldsymbol{\Psi}$$
$$= \boldsymbol{\Lambda}\mathbf{I}\boldsymbol{\Lambda}^T + \boldsymbol{\Psi}$$
$$= \boldsymbol{\Lambda}\boldsymbol{\Lambda}^T + \boldsymbol{\Psi}$$

---

[7]See Example 9.21.

An important consequence of the above assumptions is that the matrix $\boldsymbol{\Psi}$ is diagonal; in fact, its diagonal contains the specific variances ψ_i.

The main task of FA is to find the loading matrix $\boldsymbol{\Lambda}$. There is a host of approaches for doing so, including methods based on maximum-likelihood estimation, and commercial software packages for multivariate analysis offer the user plenty of choices. A most important point to notice is that, whatever method we use, *factors are not unique*. To see this, consider an orthogonal matrix $\mathbf{T}$, representing a vector rotation; since $\mathbf{T}\mathbf{T}^T = \mathbf{I}$, we may rewrite (17.3) as

$$\mathbf{y} = \boldsymbol{\mu} + \boldsymbol{\Lambda}\mathbf{T}\mathbf{T}^T\mathbf{f} + \boldsymbol{\epsilon}$$
$$= \boldsymbol{\mu} + \boldsymbol{\Lambda}^*\mathbf{f}^* + \boldsymbol{\epsilon}$$

where $\boldsymbol{\Lambda}^* = \boldsymbol{\Lambda}\mathbf{T}$ and $\mathbf{f}^* = \mathbf{T}^T\mathbf{f}$. This amounts to rotating the factors, and it is easy to see that the covariance matrix $\boldsymbol{\Sigma}$ can be expressed in terms of $\boldsymbol{\Lambda}^*$ as well:

$$\boldsymbol{\Sigma} = \boldsymbol{\Lambda}^*(\boldsymbol{\Lambda}^*)^T + \boldsymbol{\Psi}$$
$$= (\boldsymbol{\Lambda}\mathbf{T})(\boldsymbol{\Lambda}\mathbf{T})^T + \boldsymbol{\Psi}$$
$$= \boldsymbol{\Lambda}\mathbf{T}\mathbf{T}^T\boldsymbol{\Lambda}^T + \boldsymbol{\Psi}$$
$$= \boldsymbol{\Lambda}\boldsymbol{\Lambda}^T + \boldsymbol{\Psi}$$

This shows that the choice of factors is not unique. Software tools also offer many factor rotation strategies, which may help in finding a sensible interpretation of the factors. Generally, finding such an interpretation is difficult when many factor loadings are large for all of the variables; factor rotation may be used to find a meaningful structure. Doing so is not trivial at all, as it requires experience and domain-specific knowledge.

Example 17.3 Consider the data-generating model of Eq. (17.3), where:

$$\boldsymbol{\mu} = \begin{bmatrix} 40 \\ -30 \\ 60 \\ 100 \\ -40 \end{bmatrix}, \quad \boldsymbol{\Lambda} = \begin{bmatrix} 10 & 1 \\ 1 & 8 \\ 2 & -10 \\ 11 & 3 \\ -3 & 6 \end{bmatrix}$$

The two factors f_1 and f_2 are independent standard normal variables, and the five specific factors ϵ_i, $i = 1, \ldots, 5$, are independent normal variables with expected value 0 and standard deviation 5. Of course, in real life, we do not know the underlying data-generating process, but let us see what we can recover by sampling $n = 1000$ observations by a Monte Carlo method and applying factor analysis on the resulting data. By using a commercial software package,[8] we find the following estimates:

$$\hat{\boldsymbol{\Lambda}} = \begin{bmatrix} -0.0896 & 0.8687 \\ 0.8403 & 0.2362 \\ -0.8865 & 0.0194 \\ 0.0709 & 0.9122 \\ 0.7738 & -0.2421 \end{bmatrix}$$

$$\hat{\boldsymbol{\Psi}} = \begin{bmatrix} 0.2374 & & & & \\ & 0.2382 & & & \\ & & 0.2138 & & \\ & & & 0.1628 & \\ & & & & 0.3426 \end{bmatrix}$$

This looks disappointing at first sight, as $\hat{\boldsymbol{\Lambda}}$ does not look quite like $\boldsymbol{\Lambda}$. However, we should take two points into account:

1. Factors may need to be rotated to find a meaningful pattern.

2. For numerical convenience, factor analysis is applied to *standardized* observations.

The last consideration suggests that we may check the estimates by comparing the estimated correlation matrix from the factor model with the straightforward sample correlation matrix $\mathbf{R}$. The sample correlation matrix for the random sample was

$$\mathbf{R} = \begin{bmatrix} 1.0000 & 0.1275 & 0.0924 & 0.7861 & -0.2831 \\ 0.1275 & 1.0000 & -0.7407 & 0.2764 & 0.5917 \\ 0.0924 & -0.7407 & 1.0000 & -0.0426 & -0.6912 \\ 0.7861 & 0.2764 & -0.0426 & 1.0000 & -0.1635 \\ -0.2831 & 0.5917 & -0.6912 & -0.1635 & 1.0000 \end{bmatrix}$$

We urge the reader to compare this estimate with the true correlation matrix for the data-generating model. The correlation matrix from the factor model, due to standardization, is just the covariance matrix

$$\hat{\boldsymbol{\Lambda}}\hat{\boldsymbol{\Lambda}}^T + \hat{\boldsymbol{\Psi}} = \begin{bmatrix} 1.0000 & 0.1299 & 0.0963 & 0.7861 & -0.2797 \\ 0.1299 & 1.0000 & -0.7403 & 0.2750 & 0.5930 \\ 0.0963 & -0.7403 & 1.0000 & -0.0452 & -0.6907 \\ 0.7861 & 0.2750 & -0.0452 & 1.0000 & -0.1660 \\ -0.2797 & 0.5930 & -0.6907 & -0.1660 & 1.0000 \end{bmatrix}$$

We see that indeed factor analysis recreates the correlation matrix rather well. With a smaller sample and a stronger impact of specific factors (their

---

[8]This numerical example has been solved using the Statistics Toolbox of MATLAB; see http:// www.mathworks.com/. In particular, the function factoran has been used, with default settings.

standard deviation is only 5 in our little experiment), the results can be less reassuring. Furthermore, the task of finding the right rotation and a useful interpretation remains a challenge.   □

17.4 CLUSTER ANALYSIS

The aim of cluster analysis is to search for patterns in a dataset by grouping similar items. The number of groups (clusters) need not be fixed in advance, and, in fact, there is an array of different methods, which share a common need: the definition of a distance between observations, which is used to measure similarity or dissimilarity. Observations within a cluster should be similar one to another and dissimilar from items in other clusters. We first outline a few methods to measure distance, and then we describe the two main families of clustering methods, hierarchical and nonhierarchical.

17.4.1 Measuring distance

Given two observations $\mathbf{X}, \mathbf{Y} \in \mathbb{R}^p$, we may define various distance measures between them, such as

- The Euclidean distance

$$d(\mathbf{X}, \mathbf{Y}) = \sqrt{\sum_{j=1}^{p} (X_j - Y_j)^2} = \sqrt{(\mathbf{X} - \mathbf{Y})^T (\mathbf{X} - \mathbf{Y})}$$

- The Mahalanobis distance[9]

$$d(\mathbf{X}, \mathbf{Y}) = \sqrt{(\mathbf{X} - \mathbf{Y})^T \mathbf{S}^{-1} (\mathbf{X} - \mathbf{Y})}$$

  where $\mathbf{S}$ is the (sample) covariance matrix.

Other distances may be defined to account for categorical variables; we may also assign weights to variables to express their relative importance. These distances measure the dissimilarity between two single observations, but when we aggregate several observations into clusters, we need some way to measure distance between clusters, i.e., between *sets* of observations. Four ways of defining distance between clusters are illustrated in Fig. 17.4:

- Figure 17.4(a) illustrates the *single linkage* (*nearest-neighbor*) distance

$$D(A, B) = \min\{d(\mathbf{X}, \mathbf{Y}); \quad \mathbf{X} \in A, \ \mathbf{Y} \in B\}$$

[9]See Section 15.3.2.

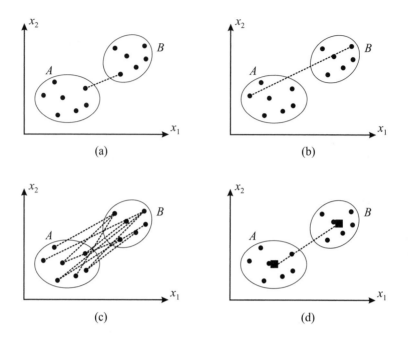

Fig. 17.4 Four different distances between clusters.

This distance between clusters A and B is given by the smallest distance between an element of A and an element of B.

- Fig. 17.4(b) illustrates the *complete linkage (farthest-neighbor)* distance

$$D(A, B) = \max\{d(\mathbf{X}, \mathbf{Y}); \quad \mathbf{X} \in A, \ \mathbf{Y} \in B\}$$

This distance between clusters A and B is given by the largest distance between an element of A and an element of B.

- Figure 17.4(c) illustrates the *average linkage distance*

$$D(A, B) = \frac{1}{N_A N_B} \sum_{i=1}^{N_A} \sum_{j=1}^{N_B} d(\mathbf{X}_i, \mathbf{Y}_j)$$

This distance between clusters A and B is given by the average distance between any pair consisting of an element of A and an element of B; N_A and N_B are the numbers of elements in A and B, respectively.

- Figure 17.4(d) illustrates the *centroid distance*

$$D(A, B) = d(\overline{\mathbf{X}}_A, \overline{\mathbf{Y}}_B)$$

In this case we take the sample mean along each dimension for observations in cluster A, and define the centroid $\overline{\mathbf{X}}_A$; the centroid $\overline{\mathbf{X}}_B$ for

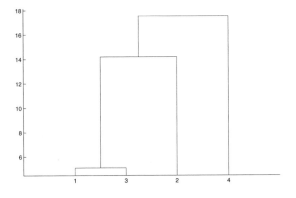

Fig. 17.5 A dendrogram from hierarchical clustering.

cluster B is defined in the same way, and the distance between clusters is given by the distance between these two "representative" observations, which are essentially the barycenter of each set [the black squares in Fig. 17.4(d)].

17.4.2 Hierarchical methods

A first class of approaches to cluster formation is based on the sequential construction of a tree (*dendrogram*), that leads us to form clusters; Fig. 17.5 shows a simple dendrogram. The leaves of the tree correspond to objects; branches of the tree correspond to sequential groupings of observations and clusters. Since a tree suggests a natural hierarchy, methods in this class are called *hierarchical*. Methods differ in the way clusters are built:

- In *divisive* methods we start from one big cluster and proceed to build disjoint smaller clusters.

- In *agglomerative* methods, increasingly large clusters are built by merging smaller ones.

The second approach is more common and is best illustrated by a simple example.

Example 17.4 Consider four observations collected in the following matrix:

$$\mathcal{X} = \begin{bmatrix} 3 & 6 & 9 & -1 \\ 10 & -2 & 3 & 5 \\ -1 & 3 & 9 & -2 \\ 20 & 7 & 3 & -3 \end{bmatrix}$$

Remember that rows correspond to observations and columns to variables; so, the first observation is $\mathbf{X}_1 = [3, 6, 9, -1]^T$. Using Euclidean distance, we find the following distance matrix:

$$\mathbf{D} = \begin{bmatrix} 0.0000 & 13.6015 & 5.0990 & 18.1659 \\ 13.6015 & 0.0000 & 15.1987 & 15.6525 \\ 5.0990 & 15.1987 & 0.0000 & 22.2261 \\ 18.1659 & 15.6525 & 22.2261 & 0.0000 \end{bmatrix}$$

Now, let us build clusters using an agglomerative (bottom up) strategy based on the centroid distance between clusters. The first iteration involves the distances between single observations, and we immediately see in matrix $\mathbf{D}$ that the two closest observations are $\mathbf{X}_1$ and $\mathbf{X}_3$, whose distance is 5.0990. The centroid of these two observations is

$$\overline{\mathbf{X}}_A = [1.0, 4.5, 9.0, -1.5]^T$$

Now we compute the following distances:

$$D(\overline{\mathbf{X}}_A, \mathbf{X}_2) = 14.1951, \quad D(\overline{\mathbf{X}}_A, \mathbf{X}_4) = 20.1370$$

By comparing these distances with $d(\mathbf{X}_2, \mathbf{X}_4) = 15.6525$, we conclude that we should merge cluster A with $\mathbf{X}_2$, obtaining a new cluster B consisting of three observations; this is finally merged with the remaining observation $\mathbf{X}_4$ and the process stops, resulting in the dendrogram of Fig. 17.5.[10] We observe that the centroid of cluster B is

$$\overline{\mathbf{X}}_B = [4.0000, 2.3333, 7.0000, 0.6667]^T$$

and its distance from $\mathbf{X}_4$ is 17.5278, which is the height of the dendrogram. In fact, the height of the vertical bars in the dendrogram is related to distances.

☐

In the example, we always merge a cluster with a single observation, but this need not be the case in general, as we may well merge clusters. We should note that in this class of methods we do not necessarily specify the number of clusters a priori. Given the tree, we may decide where to draw a horizontal line separating clusters, based on the distances. Also note that the approach is not iterative, but one-shot: In agglomerative methods we build the tree all the way up, without revising previous decisions.

17.4.3 Nonhierarchical clustering: k-means

The best-known method in the class of nonhierarchical clustering algorithms is the k-*means* approach. In the k-means method, unlike with the hierarchical ones, observations can be moved from one cluster to another in order to

---

[10]The dendrogram has been obtained by applying the Statistics Toolbox of MATLAB.

minimize a joint measure of the quality of clustering; hence, the method is iterative in nature. The starting point is the selection of k *seeds*, one per cluster, which are used to "attract" items into the cluster. The seeds should be natural attractors; i.e., they should be rather different from one another. There is no point in selecting two seeds that are close to each other, as they would be natural candidates for being placed in the same cluster. We also notice that in this approach, we have to specify in advance the number of clusters that we wish to form.

There are several variations on the theme, but a possible k-means algorithm is as follows:

- *Step 1*: Select k initial seeds.

- *Step 2*: Find an initial partitioning by assigning each observation to the closest cluster; in doing so, we may use the distance between the observation to be assigned and the centroid of each cluster; the centroid of a cluster is updated whenever a new element is added.

- *Step 3*: Reassign selected observations to another cluster if this improves an overall measure of the quality of clustering; update centroids accordingly, and stop when no further improvement is possible.

One possible measure of overall clustering quality in step 3 is the *error sum of squares of the partition* (ESS)

$$\text{ESS} \equiv \sum_{i=1}^{n} (\mathbf{X}_i - \overline{\mathbf{X}}_{C(i)})^T (\mathbf{X}_i - \overline{\mathbf{X}}_{C(i)})$$

where $C(i)$ is the index of the cluster to which observation $\mathbf{X}_i$ is currently assigned, and $\overline{\mathbf{X}}_{C(i)}$ is its centroid. Clearly, the procedure is heuristic, because we may get stuck in a local minimum, which may depend on the initial seeds.

For further reading

- For more details on PCA see, e.g., Chapters 4, 12, and 4 of Refs. [2], [3], and [4], respectively.

- There are actually two possible uses of factor analysis: exploratory and confirmatory. They are dealt with in Chapters 5 and 6 of Ref. [2], respectively. See also Chapter 13 of Ref. [3] and Chapters 5 and 6 of Ref. [4].

- The same references cited above may be consulted for cluster analysis: [2, Chapter 8]; [3, Chapter 14]; [4, Chapter 7].

REFERENCES

1. W. Härdle and L. Simar, *Applied Multivariate Statistical Analysis,* 2nd ed., Springer, Berlin, 2007.

2. J.M. Lattin, J.D. Carroll, and P.E. Green, *Analyzing Multivariate Data,* Brooks/Cole, Belmont, CA, 2003.

3. A.C. Rencher, *Methods of Multivariate Analysis,* 2nd ed., Wiley, New York, 2002.

4. S. Sharma, *Applied Multivariate Techniques*, Wiley, Hoboken, NJ, 1996.

Index